Bimonoidal Categories, E_n-Monoidal Categories, and Algebraic K-Theory

Volume III: From Categories to Structured Ring Spectra

Mathematical
Surveys
and
Monographs
Volume 285

Bimonoidal Categories, E_n-Monoidal Categories, and Algebraic K-Theory

Volume III: From Categories to Structured Ring Spectra

Niles Johnson
Donald Yau

AMERICAN
MATHEMATICAL
SOCIETY
Providence, Rhode Island

2020 *Mathematics Subject Classification.* Primary 18D20, 18F25, 18M05, 18M15, 18M50, 18M60, 18M65, 19D23, 55P43, 55P48.

For additional information and updates on this book, visit
www.ams.org/bookpages/surv-285

Library of Congress Cataloging-in-Publication Data

Names: Johnson, Niles, author. | Yau, Donald Y. (Donald Ying), 1977– author.
Title: Bimonoidal categories, E_n-monoidal categories, and algebraic K-theory : Volume III : from categories to structured ring spectra / Niles Johnson, Donald Yau.
Description: Providence, Rhode Island : American Mathematical Society, [2024] | Series: Mathematical surveys and monographs, 0076-5376 ; volume 285 | Includes bibliographical references and index.
Identifiers: LCCN 2024025161 | ISBN 9781470478117 (paperback) | ISBN 9781470478483 (ebook)
Subjects: LCSH: Categories (Mathematics) | K-theory. | AMS: Category theory; homological algebra – Categories with structure – Enriched categories (over closed or monoidal categories). | Category theory; homological algebra – Categories and geometry – Algebraic K-theory and L-theory. | K-theory – Higher algebraic K-theory – Symmetric monoidal categories. | Algebraic topology – Homotopy theory – Spectra with additional structure (E_∞, A_∞, ring spectra, etc.). | Algebraic topology – Homotopy theory – Loop space machines, operads.
Classification: LCC QA169 .J593 2024 | DDC 514/.23–dc23/eng/20240904
LC record available at https://lccn.loc.gov/2024025161

Mathematical Surveys and Monographs ISSN: 0076-5376 (print); 2331-7159 (online)
DOI: https://doi.org/10.1090/surv/285

∞ The paper used in this book is acid-free and falls within the guidelines established to ensure permanence and durability.
Visit the AMS home page at `https://www.ams.org/`

10 9 8 7 6 5 4 3 2 1   29 28 27 26 25 24

The first author dedicates this book to Nemili, Linus, and Kavya.
The second author dedicates this book to Jacqueline.

Contents

This is a three-volume work; each volume consists of two parts. The preface and backmatter are identical across all three volumes. The table of contents for each volume is followed by a copy of the tables of contents for the other two volumes.

Reference Convention. In each of Volumes II and III, the numbering of parts, chapters, and pages restarts at 1. Page numbers are prefixed with a volume number so that, for example, page n of Volume II is printed II.n. References (chapters, definitions, theorems, equations, etc.) to a different volume are preceded by I, II, or III according to the volume. For example, $a.b.c$ in Volume I is denoted by I.$a.b.c$ in Volumes II and III, and Chapter N in Volume III is denoted by Chapter III.N in Volumes I and II.

Main Contents of Volume I

Main Contents of Volume II

Preface

Bimonoidal categories are categorical analogues of rings without additive inverses. They have been actively studied in category theory, homotopy theory, and algebraic K-theory since around 1970. There is an abundance of new applications and questions of bimonoidal categories in mathematics and other sciences. This work provides the first unified treatment of bimonoidal and higher ring-like categories, their connection with algebraic K-theory and homotopy theory, and applications to quantum groups and topological quantum computation. With ample background material, extensive coverage, detailed presentation of both well-known and new theorems, and a list of open questions, this work is a user-friendly resource for beginners and experts alike.

Bimonoidal and E_n-Monoidal Categories

A *bimonoidal category* C is a categorical analogue of a rig, which is a ring without additive inverses. In this categorification, the addition, multiplication, 0, and 1 of a rig are replaced by functors and objects in a bimonoidal category. Rig axioms are replaced by natural structure morphisms, along with suitable coherence axioms of their own.

More specifically, in place of the rig addition and multiplication, C has two monoidal structures

$$\big(\mathsf{C}, \oplus, \mathbb{0}, \alpha^{\oplus}, \lambda^{\oplus}, \rho^{\oplus}, \xi^{\oplus}\big) \quad \text{and} \quad \big(\mathsf{C}, \otimes, \mathbb{1}, \alpha^{\otimes}, \lambda^{\otimes}, \rho^{\otimes}\big).$$

The first is symmetric monoidal, and called the *additive structure*. The second is plain monoidal, and called the *multiplicative structure*. As with plain monoidal categories, there are variants with braided or symmetric multiplicative structure, and a variety of intermediate multiplicative structures parametrized by E_n-operads.

In place of distributivity relations in a rig, a bimonoidal category has natural *distributivity monomorphisms* for objects A, B, and C:

$$A \otimes (B \oplus C) \xrightarrow{\delta^l_{A,B,C}} (A \otimes B) \oplus (A \otimes C)$$

$$(A \oplus B) \otimes C \xrightarrow{\delta^r_{A,B,C}} (A \otimes C) \oplus (B \otimes C).$$

These data are required to satisfy a finite list of axioms that (i) are checkable in practice and (ii) ensure that (symmetric/braided) bimonoidal categories have good coherence and other categorical properties. An important special case is a *tight* bimonoidal category, in which the distributivity monomorphisms δ^l and δ^r are *isomorphisms*.

A number of examples, arising in both algebraic and homotopical contexts, are discussed throughout the text. Here, we summarize three important ones. More examples are discussed in the next section about quantum science.

(1) The category of finite dimensional complex vector spaces, $\mathsf{Vect}^{\mathbb{C}}$, is a tight symmetric bimonoidal category with its additive and multiplicative structures given by the usual direct sum and tensor product of vector spaces. More generally, each distributive symmetric monoidal category is a tight symmetric bimonoidal category.
(2) The nonnegative integers and permutations form the objects and the morphisms of a tight symmetric bimonoidal category Σ, called the finite ordinal category.
(3) May's bipermutative categories, with the additional axiom $\xi^{\otimes}_{-,\mathbb{0}} = \mathrm{Id}$, are tight symmetric bimonoidal categories.

The definition and coherence theorems for symmetric bimonoidal categories are due to Laplaza [**Lap72a, Lap72b**]. These theorems and their plain bimonoidal analogues are discussed in detail in Part I.1. In addition to providing completely detailed proofs, we also correct some subtle and nontrivial inaccuracies in the original statements and proofs. See Sections I.3.11 and I.4.7 for related discussion. Just as applications of monoidal categories heavily depend on Mac Lane's coherence theorem, Laplaza's two coherence theorems for symmetric bimonoidal categories, as well as their plain and braided analogues, are crucial to their applications.

Part I.2 applies Laplaza's coherence theorems to prove a number of theorems about bimonoidal categories in the context of 2-dimensional categories. These include existence of a bi-initial object, confirming a conjecture of Baez [**Bae18**], and a symmetric monoidal bicategory of matrices, $\mathsf{Mat}^{\mathsf{C}}$, constructed from a tight symmetric bimonoidal category C. (Note the unfortunately subtle confluence of terminology that "symmetric monoidal *bicategory*" and "symmetric *bimonoidal* category" refer to wildly distinct algebraic structures.) In the case $\mathsf{C} = \mathsf{Vect}^{\mathbb{C}}$, $\mathsf{Mat}^{\mathsf{C}}$ is the symmetric monoidal bicategory of *coordinatized 2-vector spaces*, one version of the 2-vector spaces introduced by Kapranov and Voevodsky in [**KV94**].

Braided bimonoidal categories, along with their corresponding coherence and strictification theorems, are discussed in Part II.1. These structures are of interest for applications in quantum science, discussed below. The relevant coherence results are new, confirming a conjecture of Blass and Gurevich [**BG20a**].

Part II.2 introduces a similar but distinct categorification of rigs, called E_n-*monoidal categories*. These have *factorization morphisms* in place of distributivity monomorphisms, and are significant for the K-theoretic applications in Part 2. The E_n-monoidal structure is a generalization of n-fold monoidal structure due to Balteanu, Fiedorowicz, Schwänzel, and Vogt [**BFSV03**]. Special cases of E_n-monoidal categories include, or are closely related to, the *bipermutative categories* of May [**May77**], the *ring categories* of Elmendorf and Mandell [**EM06**], and the *braided bimonoidal categories* of Richter [**Ric10**].

Applications in Quantum Science

Due to the ubiquity of ring-like structures and categories, bimonoidal categories are increasingly applied in a variety of disciplines in mathematics and other formal sciences related to quantum algebra. To support readers with a variety of interests, this work includes summary and introduction to several such applications.

Quantum Groups. The first part of Chapter II.3 extends a well-known fact in quantum group theory. We observe that the category of left modules over a braided bialgebra, which is also known as a quasitriangular bialgebra in the literature, is a tight braided bimonoidal category.

Topological Quantum Computation. The second part of Chapter II.3 discusses applications of braided bimonoidal categories to topological quantum computation (TQC). We prove that the Fibonacci anyons and the Ising anyons, which are two of the most important models in TQC, are both tight braided bimonoidal categories.

Centers. Monoidal, braided monoidal, and symmetric monoidal categories are connected by the Drinfeld center and the symmetric center. Kassel [**Kas95**] and Majid [**Maj91**] explain the relationship between the Drinfeld center construction, due to Drinfeld in unpublished work, and modules over the Drinfeld double of a a finite dimensional Hopf algebra. (See Note II.1.7.2 for further explanation and context.) Bimonoidal and ring categorical analogues of these center constructions are discussed in Chapters II.4 and II.9.

Reversible Programming. Section I.2.6 is a brief illustration that symmetric bimonoidal categories naturally arise in reversible programming of finite types. We observe that there is a symmetric bimonoidal groupoid whose objects are syntax of finite types. Note I.2.7.5 directs the reader to further applications in the sheet diagrams of [**CDH∞**] and the work of [**Hin13**] on quantum circuits.

Applications in Algebraic K-Theory

Uses of additive and multiplicative categorical structure in homotopy theory are among the earliest and most well-developed applications of the material from Volumes I and II of this work. Volume III focuses on those applications, beginning with the work of Segal [**Seg74**] that assembles structured ring spectra from permutative categories. Under the Segal K-theory functor K^{Se}, the symmetric monoidal structure of a permutative category C results in the additive structure in the cohomology theory represented by the spectrum $\mathsf{K}^{\mathsf{Se}}(\mathsf{C})$.

The computational importance of multiplicative structure in cohomology motivates significant interest in multiplicative structure for the representing spectra, leading to the highly-structured spectra that are presaged in work of Adams [**Ada95**] and realized in the S-modules of Elmendorf-Kriz-Mandell-May [**EKMM97**], the symmetric spectra of Hovey-Shipley-Smith [**HSS00**], and the orthogonal spectra of Mandell-May-Schwede-Shipley [**MMSS01**], among other equivalent models. Thus, there is corresponding interest in bimonoidal structures for the input categories.

One difficulty, however, is that Segal K-theory does *not* preserve the multiplicative aspect of such structures. As a resolution, the work of Elmendorf and Mandell [**EM06, EM09**] introduces an alternative construction that (a) is suitably equivalent to Segal's construction and (b) preserves multiplicative structure. This is known as Elmendorf-Mandell K-theory, K^{EM}.

The formalism in which these statements can be made precise is that of multicategories, where multilinear functors between permutative categories encode the relevant multiplicative structure. In these terms, the essential difference between the Segal and Elmendorf-Mandell constructions is that the latter is *multifunctorial*, while the former is merely functorial.

Part 1 develops the necessary supporting theory of enriched monoidal categories and multicategories, but is also of independent interest. Part 2 contains the applications to algebraic K-theory, including a review of the relevant background in homotopy theory and detailed treatments of both the Segal and Elmendorf-Mandell constructions.

The second half of Part 2 applies the Elmendorf-Mandell K-theory multifunctor K^{EM} to the E_n-monoidal categories in Part II.2 to produce structured ring spectra. We prove in detail that K^{EM} sends

- small ring categories to strict ring symmetric spectra,
- small bipermutative categories to E_∞-symmetric spectra,
- small braided ring categories to E_2-symmetric spectra, and
- small E_n-monoidal categories to E_n-symmetric spectra for $2 \leq n < \infty$.

The strict ring and E_∞ cases are from [**EM06, EM09**]. The $1 < n < \infty$ cases are new results.

Audience and Features

This work is aimed at graduate students and advanced researchers with an interest in category theory, homotopy theory, algebraic K-theory, and their applications. Below are some features that make this work a unique and user-friendly resource.

Unified Presentation: The literature on bimonoidal categories, higher ring-like categories, enriched monoidal categories, multicategories, and their connection with algebraic K-theory, homotopy theory, and the sciences is scattered across many journal articles over several decades, with varying definitions, notations, and terminology. This work presents these topics in a unified manner, with both well-known and new theorems.

Background Material: To make this work self-contained and to bring the reader quickly up to speed, there is extensive background material on

- basic category theory (Chapter I.1),
- 2-dimensional categories (Chapter I.6),
- braided structures (Chapter II.1),
- abelian categories (Section II.2.3),
- braided, also known as quasitriangular, bialgebras (Section II.3.1),
- enriched monoidal categories (Chapters 1, 2, and 3),
- pointed objects and pointed diagram categories (Chapter 4),
- enriched multicategories (Chapters 5 and 6), and
- homotopy theory (Chapter 7).

These chapters and sections form a substantial portion of this work.

Open Questions: Appendix A discusses open questions related to the topics of this work. The reader is encouraged to take advantage of these open questions and use them as a springboard to read the main text.

Detailed Discussion: This work contains many highly detailed and carefully structured proofs for both known and new theorems. For each major result, our discussion has much more detail than one would normally find in the literature. Our detailed discussion has several purposes.

Exercises with Solutions. Our detailed presentation makes the material accessible to a diverse audience, including those who are new to bimonoidal and higher ring-like categories and algebraic K-theory. Students are encouraged to regard the numerous detailed proofs as exercises with full solutions. Each result whose proof has many different parts has been carefully structured to make it easy for the reader to jump forward and backward.

Axioms. Symmetric bimonoidal categories are defined by 24 axioms, and the list of axioms for (braided) bimonoidal categories is similarly substantial. Our detailed discussion helps the reader see exactly where these axioms are used and why they are needed.

Laplaza's Theorems. The Coherence Theorems I.3.9.1 and I.4.4.3 for symmetric bimonoidal categories are central results in this subject that have been cited and used numerous times in the literature. Their original proofs given by Laplaza in [**Lap72a, Lap72b**] were written in outline form, with much detail and some cases in the proofs completely omitted. Moreover, Laplaza's original proofs and statements of these theorems have several subtle and nontrivial inaccuracies that have never been made explicit before and are not easy to spot. For both archival and educational purposes, we present fully detailed proofs of these theorems and correct the inaccuracies. Sections I.3.11 and I.4.7 have more related discussion.

K-Theory Multifunctors. The K-theory multifunctors in Chapters 9 and 10, due to Elmendorf-Mandell [**EM06, EM09**], are fundamental constructions for multiplicative structure of algebraic K-theory spectra. They are essential for our development of E_n-monoidal symmetric spectra from corresponding structure on small permutative categories. We use the theory of enriched monoidal categories and enriched multicategories from Part 1 to give complete explanations of the constructions and their properties. This treatment corrects an inaccuracy in the statement of [**EM09**, Theorem 1.3] and some other statements about expanding the domain of the K-theory multifunctor. The basic issue has to do with monoidal units and, to the authors' knowledge, has not been previously explained. See Note 10.8.2 for further discussion.

Reading Guides: In addition to a detailed introduction, almost every chapter has a brief *Reading Guide* that provides an alternative to reading that chapter linearly. Our presentation in the main text follows a straightly logical order and has a lot of detail. By following the reading guide, it is possible to first obtain a bird's-eye view of that chapter before digesting all the detail. The end of this Preface also includes several thematic reading guides for salient topics that span multiple chapters.

Motivation and Explanation: Main definitions and results are often preceded by discussion that motivates the upcoming definitions and proofs. Whenever useful, definitions and results are followed by a detailed explanation that interprets and unpacks the various components. In the text, these are clearly marked as *Motivation* and *Explanation*, respectively. Examples include Motivation I.2.1.1, Explanation I.2.4.7, and Section I.4.1.

Organization: There are extensive cross-references throughout the text. In addition to a detailed index, there are lists of main facts and notations, each organized by chapters. While the text follows a strictly logical order, it is not necessary to read the chapters in a linear order. The reader can jump straight to a section and use the extensive cross-references to fill in the necessary definitions and facts.

Part and Chapter Summaries

Part I.1: Symmetric Bimonoidal Categories

This part studies symmetric bimonoidal categories and bimonoidal categories (Chapter I.2). It presents highly detailed proofs of Laplaza's Coherence Theorems for symmetric bimonoidal categories (Chapters I.3 and I.4), May's Strictification Theorem for tight symmetric bimonoidal categories (Chapter I.5), and their non-symmetric analogues for bimonoidal categories. The only prerequisite for this part is some basic knowledge of category theory, which is summarized in Chapter I.1.

Part I.2: Bicategorical Aspects of Symmetric Bimonoidal Categories

Applying Laplaza's Coherence Theorems, this part proves several new theorems on the connection between symmetric bimonoidal categories and bicategories. All the necessary definitions of 2-dimensional category theory are summarized in Chapter I.6. The first main result is a confirmation of Baez's Conjecture (Chapter I.7) that proves the existence of a bi-initial object in a 2-category of symmetric bimonoidal categories. Chapter I.8 proves that a matrix construction $\mathsf{Mat}^{\mathsf{c}}$ sends each tight symmetric bimonoidal category to a symmetric monoidal bicategory.

Part II.1: Braided Bimonoidal Categories

Starting with a preliminary chapter on the braid groups and braided monoidal categories, this part is a detailed study of braided bimonoidal categories (Chapter II.2), which are strictly more general than Richter's [**Ric10**] and the BD categories of Blass-Gurevich [**BG20a**]. This part discusses applications to quantum groups and topological quantum computation (Chapter II.3), bimonoidal centers (Chapter II.4), coherence and strictification of braided bimonoidal categories (Chapters II.5 and II.6), and the braided versions of Baez's Conjecture and the matrix construction (Chapters II.7 and II.8). Our coherence and strictification theorems confirm the Blass-Gurevich Conjecture. The main theorems in Parts I.1 and I.2 are used in this part.

Part II.2: E_n-Monoidal Categories

This part studies a closely related variant of bimonoidal categories, called ring categories, and their bipermutative, braided, and higher analogues, called E_n-monoidal categories. Ring and bipermutative categories are due to Elmendorf-Mandell [**EM06, EM09**]. An E_n-monoidal category combines n ring categories with a common additive structure and an n-fold monoidal category as in [**BFSV03**]. The categories in this part are applied in Part 2 to obtain E_n-symmetric spectra via algebraic K-theory. This part is independent of the earlier parts, except for some definitions and statements of theorems.

Part 1: Enriched Monoidal Categories and Multicategories

To prepare for Part 2, this part lays the groundwork on enriched monoidal categories (Chapters 1, 2, and 3), smash products (Chapter 4), and multicategories (Chapters 5 and 6). In addition to their roles in the Segal K-theory functor and the Elmendorf-Mandell K-theory multifunctor, the detailed discussion of enriched monoidal categories—including change of enrichment, coherence, self-enrichment, and the Enriched Yoneda Lemma—and multicategories is also of independent interest. These chapters assume only a basic knowledge of monoidal categories, as summarized in Section 1.1.

Part 2: Algebraic K-Theory

This part studies the interconnection between E_n-monoidal categories (Part II.2), homotopy theory (Chapter 7), and algebraic K-theory. The first half discusses in detail the Segal K-theory functor (Chapter 8) and the Elmendorf-Mandell K-theory multifunctor (Chapters 9 and 10) from small permutative categories to symmetric spectra. The second half (Chapters 11, 12, and 13) applies the K-theory multifunctor to small ring, bipermutative, braided ring, and E_n-monoidal categories to obtain, respectively, strict ring, E_∞-, E_2-, and E_n-symmetric spectra. These structured ring spectra are fundamental objects in homotopy theory. Our discussion shows how they arise from E_n-monoidal categories via algebraic K-theory.

In the main text, each chapter starts with a detailed introduction. A summary of each chapter follows.

Part I.1: Symmetric Bimonoidal Categories

Chapter I.1: Basic Category Theory

To make this book self-contained, this chapter reviews the basics of category theory, starting from the definitions of categories, functors, and natural transformations. Then it discusses adjunctions, equivalences of categories, (co)limits, (co)ends, and Kan extensions. The remaining sections review (symmetric) monoidal categories, (symmetric) monoidal functors, monoidal natural transformations, and their coherence theorems.

Chapter I.2: Symmetric Bimonoidal Categories

This chapter introduces symmetric bimonoidal categories and bimonoidal categories. Then we prove Laplaza's Theorem I.2.2.13 that says that half of the 24 symmetric bimonoidal category axioms are formal consequences of the other 12 axioms. The weaker bimonoidal analogue is Proposition I.2.2.14. The remaining sections discuss examples of symmetric bimonoidal categories, including distributive symmetric monoidal categories, the finite ordinal category Σ, a variant Σ', and left and right bipermutative categories. The finite ordinal category Σ is an important part of (i) the distortion category $\mathcal{D}$ (Chapter I.4) used in Laplaza's Second Coherence Theorem I.4.4.3, (ii) Baez's Conjecture (Chapter I.7), and (iii) the braided version of Baez's Conjecture (Chapter II.7). Section I.2.6 contains an application of symmetric bimonoidal categories to reversible programming of finite types.

Chapter I.3: Coherence of Symmetric Bimonoidal Categories

This chapter proves Laplaza's First Coherence Theorem I.3.9.1 for symmetric bimonoidal categories that satisfy a monomorphism assumption. This assumption is automatically satisfied if tightness—that is, the invertibility of the distributivity morphisms δ^l and δ^r is assumed, but the general form of this theorem only requires that the distributivity morphisms be natural monomorphisms. The analogue of this coherence theorem for bimonoidal categories is Theorem I.3.10.7. Section I.3.11 discusses the main differences between this chapter and Laplaza's original work in **[Lap72a]**.

Chapter I.4: Coherence of Symmetric Bimonoidal Categories II

This chapter proves Laplaza's Second Coherence Theorem I.4.4.3 for symmetric bimonoidal categories that satisfy the same monomorphism assumption as in Theorem I.3.9.1. The analogue of this coherence theorem for bimonoidal categories is Theorem I.4.5.8. Section I.4.7 discusses the main differences between this chapter and Laplaza's original work in [**Lap72b**]. Both Coherence Theorems I.3.9.1 and I.4.4.3 say that some formal diagrams in certain symmetric bimonoidal categories commute. The first theorem has an assumption called *regularity* on the common domain of the two paths involved, which is analogous to Mac Lane's Coherence Theorem I.1.3.3 for monoidal categories. The second theorem has an assumption about the two paths themselves, which is reminiscent of the Joyal-Street Coherence Theorem II.1.6.3 for braided monoidal categories. In Chapter II.5, we observe that the second, but not the first, theorem has a braided analogue.

Chapter I.5: Strictification of Tight Symmetric Bimonoidal Categories

This chapter proves May's Strictification Theorem I.5.4.6 of *tight* symmetric bimonoidal categories to right bipermutative categories. The latter are tight symmetric bimonoidal categories whose additive structures and multiplicative structures are both permutative categories, and whose structure morphisms $\lambda^\bullet$, $\rho^\bullet$, δ^r, and $\xi^\otimes_{-,\mathbb{0}}$ are identities. Unlike the Coherence Theorems I.3.9.1 and I.4.4.3, the strictification theorem requires the tightness assumption. Our detailed proofs show exactly where the invertibility of δ^l and δ^r is used. Theorem I.5.4.7 is another version of the strictification theorem involving *left* bipermutative categories, in which δ^l, instead of δ^r, is the identity. Theorems I.5.5.11 and I.5.5.12 are the corresponding strictification results for tight bimonoidal categories. Section I.5.6 briefly discusses the history of related strictification theorems and claims. The proofs in this chapter are repurposed in Chapter II.6 to prove the strictification form of the Blass-Gurevich conjecture for braided bimonoidal categories.

Part I.2: Bicategorical Aspects of Symmetric Bimonoidal Categories

Chapter I.6: Definitions from Bicategory Theory

Without assuming any knowledge of 2-dimensional categories, in this chapter we review the basics of 2-/bicategories, pasting diagrams, lax functors, lax transformations, modifications, and adjunctions in bicategories. Then it reviews multiplicative structures, including monoidal bicategories, their braided, sylleptic, and symmetric analogues, the Gray tensor product for 2-categories, (permutative) Gray monoids, and permutative 2-categories. Most of these topics are discussed in detail in the book [**JY21**].

Chapter I.7: Baez's Conjecture

This chapter proves Baez's Conjecture (Theorems I.7.8.1 and I.7.8.3). Section I.7.1 defines a 2-category $\mathsf{Bi}_\mathsf{r}^\mathsf{fsy}$ with *flat* small symmetric bimonoidal categories as objects. Flatness (Definition I.3.9.9) is much weaker than tightness, and it guarantees that the Coherence Theorems I.3.9.1 and I.4.4.3 are applicable. The first version of Baez's Conjecture (Theorem I.7.8.1) says that the finite ordinal category Σ is a lax bicolimit of the 2-functor $\varnothing \longrightarrow \mathsf{Bi}_\mathsf{r}^\mathsf{fsy}$. Another version is Theorem I.7.8.3, which says that the variant Σ' of Σ is also such a lax bicolimit. We emphasize that our proof of Baez's Conjecture does *not* use the Strictification Theorems I.5.4.6 and I.5.4.7. This allows us to use flat small symmetric bimonoidal categories in the 2-category $\mathsf{Bi}_\mathsf{r}^\mathsf{fsy}$, instead of the smaller class of tight ones. Section I.7.9 discusses the relationship between our version of Baez's Conjecture and the more restricted version in [**CDH∞, Elg21**] for rig categories, which are multiplicatively nonsymmetric and tight.

Chapter I.8: Symmetric Monoidal Bicategorification

This chapter proves Theorem I.8.15.4. It says that, for each tight symmetric bimonoidal category C, a matrix construction $\mathsf{Mat}^{\mathsf{C}}$ is a symmetric monoidal bicategory, with no strict structure morphisms in general. Therefore, the construction $\mathsf{Mat}^{\mathsf{C}}$ is a direct connection between tight symmetric bimonoidal categories and symmetric monoidal bicategories. The objects in $\mathsf{Mat}^{\mathsf{C}}$ are nonnegative integers. Its 1-/2-cells are matrices whose entries are objects/morphisms in C. The horizontal composition in the bicategory $\mathsf{Mat}^{\mathsf{C}}$ uses the usual matrix product. The monoidal composition in its monoidal bicategory structure uses the matrix tensor product, which is also known as the Kronecker product. The category of coordinatized 2-vector spaces, which is $\mathsf{Mat}^{\mathsf{C}}$ with $\mathsf{C} = \mathsf{Vect}^{\mathbb{C}}$, is such a symmetric monoidal bicategory. This chapter uses the Coherence Theorems I.3.9.1 and I.3.10.7 and the graph theoretic machinery in Chapter I.3, but neither the Coherence Theorem I.4.4.3 nor the Strictification Theorems I.5.4.6 and I.5.4.7.

Part II.1: Braided Bimonoidal Categories

Chapter II.1: Preliminaries on Braided Structures

To prepare for the rest of Part II.1, this chapter discusses the braid groups and braided monoidal categories. First it defines the braid groups and proves some useful properties for sum braids and block braids. Then it reviews braided monoidal categories and proves some basic properties, including two manifestations of the third Reidemeister move. Next it proves in detail that the Drinfeld center of a monoidal category is a braided monoidal category and that the symmetric center of a braided monoidal category is a symmetric monoidal category. Then it recalls the Joyal-Street Coherence Theorem II.1.6.3 for braided monoidal categories.

Chapter II.2: Braided Bimonoidal Categories

This chapter defines braided bimonoidal categories. They are defined using 12 of the 24 Laplaza axioms of a symmetric bimonoidal category, together with two additional axioms that are variants of the only two Laplaza axioms involving the braiding $\xi^{\otimes}$. In a symmetric bimonoidal category, each of these two variant axioms is equivalent to the original Laplaza axiom. This is reminiscent of the fact that a braided monoidal category has two hexagon axioms, which are equivalent to each other in a symmetric monoidal category. A *tight* braided bimonoidal category—that is, one with invertible distributivity morphisms δ^l and δ^r—is equivalent to a BD category in the sense of Blass and Gurevich [**BG20a**]. The first main observation in this chapter is Theorem II.2.2.1, which says that each braided bimonoidal category satisfies all 24 Laplaza axioms. Therefore, a symmetric bimonoidal category is precisely a braided bimonoidal category whose braiding satisfies the symmetry axiom. The second main result in this chapter says that an abelian category with a compatible (symmetric/braided) monoidal structure is a tight (symmetric/braided) bimonoidal category. The additive structure comes from the abelian structure, and the multiplicative structure comes from the monoidal structure. The braided case of this result is due to Blass and Gurevich [**BG20a**].

Chapter II.3: Applications to Quantum Groups and Topological Quantum Computation

This chapter shows that braided bimonoidal categories arise naturally in quantum groups and topological quantum computation (TQC). The first main observation is Theorem II.3.2.19. It says that for a (symmetric/braided) bialgebra A, the category $\mathsf{Mod}(A)$ of left A-modules, equipped with the usual direct sum and tensor product, is a tight (symmetric/braided) bimonoidal category. This is an extension of the important fact in quantum group theory that, for a braided bialgebra A, $\mathsf{Mod}(A)$ is a braided monoidal category. Next we prove in detail that Fibonacci anyons and Ising anyons, which are two central models in TQC, are both tight braided bimonoidal categories. In each case, the additive structure comes from an abelian category structure, and the multiplicative structure comes from the fusion rules of anyons.

Chapter II.4: Bimonoidal Centers

This chapter generalizes the Drinfeld center of a monoidal category and the symmetric center of a braided monoidal category (Sections II.1.4 and II.1.5) to the bimonoidal setting. Generalizing the Drinfeld center, Theorem II.4.4.3 says that, for each tight bimonoidal category C, the bimonoidal Drinfeld center $\overline{\mathsf{C}}^{\mathsf{bi}}$ is a tight braided bimonoidal category. Tightness is required for this theorem because the invertibility of δ^l and δ^r is used in the construction of $\overline{\mathsf{C}}^{\mathsf{bi}}$. The proof of this theorem is another good illustration of the axioms of a braided bimonoidal category, since we will use all 24 Laplaza axioms and the two variant axioms in the braided case. Generalizing the symmetric center, Theorem II.4.5.3 says that, for each braided bimonoidal category C, the bimonoidal symmetric center $\mathsf{C}^{\mathsf{sym}}$ is a symmetric bimonoidal category.

Chapter II.5: Coherence of Braided Bimonoidal Categories

This chapter proves the Coherence Theorem II.5.4.4 for braided bimonoidal categories that satisfy a monomorphism assumption. As in the symmetric case (Theorems I.3.9.1 and I.4.4.3), the monomorphism assumption in Theorem II.5.4.4 is automatically satisfied if tightness is assumed. This theorem is the braided analogue of Laplaza's Second Coherence Theorem I.4.4.3. It uses a braided version $\mathcal{D}^{\mathsf{br}}$ of the distortion category that involves the symmetric groups and the braid groups to keep track of, respectively, the additive symmetry $\xi^{\oplus}$ and the braiding $\xi^{\otimes}$. Reminiscent of the Joyal-Street Coherence Theorem II.1.6.3 for braided monoidal categories, Theorem II.5.4.4 says that, if two paths have the same image in the braided distortion category $\mathcal{D}^{\mathsf{br}}$, then they have the same value in the braided bimonoidal category in question. This condition of having the same image in $\mathcal{D}^{\mathsf{br}}$ is very much checkable in practice. In fact, the proofs of the main results in Chapters II.6, II.7, and II.8 all use Theorem II.5.4.4 and involve checking this condition many times. In [**BG20a**], Blass and Gurevich conjectured the existence of a coherence theorem for their BD categories, which are equivalent to our tight braided bimonoidal categories. Theorem II.5.4.4 confirms the Blass-Gurevich Conjecture in the form of commutative formal diagrams.

Chapter II.6: Strictification of Tight Braided Bimonoidal Categories

This chapter proves two Strictification Theorems II.6.3.6 and II.6.3.7 for tight braided bimonoidal categories. As in the symmetric case (Theorems I.5.4.6 and I.5.4.7), strictification requires the tightness assumption because the construction of the strictification uses the invertibility of the distributivity morphisms δ^l and δ^r. A *right permbraided category* is a tight braided bimonoidal category with both the additive and the multiplicative structures strict monoidal, and with identities for the structure morphisms $\lambda^\bullet$, $\rho^\bullet$, δ^r, $\xi^\otimes_{-,\mathbb{0}}$, and $\xi^\otimes_{\mathbb{0},-}$. Theorem II.6.3.6 says that each tight braided bimonoidal category is adjoint equivalent to a right permbraided category via strong braided bimonoidal functors. Theorem II.6.3.7 is the analogue that strictifies each tight braided bimonoidal category to a *left* permbraided category, in which δ^l, instead of δ^r, is the identity. Theorems II.6.3.6 and II.6.3.7 are two further positive answers to the Blass-Gurevich Conjecture [**BG20a**] in the form of strictification.

Chapter II.7: The Braided Baez Conjecture

This chapter proves the braided version of Baez's Conjecture. Section II.7.1 defines the 2-category $\mathsf{Bi}_\mathsf{r}^{\mathsf{fbr}}$ with *flat* small braided bimonoidal categories as objects. As in the symmetric case, flatness (Definition II.5.4.5) is much weaker than tightness, and it guarantees that the Braided Bimonoidal Coherence Theorem II.5.4.4 is applicable. The first version of the Braided Baez Conjecture (Theorem II.7.3.4) says that the finite ordinal category Σ is a lax bicolimit of the 2-functor $\varnothing \longrightarrow \mathsf{Bi}_\mathsf{r}^{\mathsf{fbr}}$. Another version is Theorem II.7.3.6, which says that the variant Σ' of Σ is also such a lax bicolimit. Also like the symmetric case, the proofs of the Braided Baez Conjecture do *not* use the Strictification Theorems II.6.3.6 and II.6.3.7. This allows us to use flat small braided bimonoidal categories in the 2-category $\mathsf{Bi}_\mathsf{r}^{\mathsf{fbr}}$, instead of the smaller class of tight ones. The reader may wonder why the finite ordinal category Σ and its variant Σ' are bi-initial in both the symmetric case (Theorems I.7.8.1 and I.7.8.3) and the braided case. This is analogous to the fact that the ring of integers is initial in both the category of rings and the category of commutative rings.

Chapter II.8: Monoidal Bicategorification

The main Theorem II.8.4.7 in this chapter says that, for each tight braided bimonoidal category C, the matrix construction Mat^C is a monoidal bicategory. While most of the definitions for Mat^C are the same as in the symmetric case in Chapter I.8, there are two subtleties. First, in the current braided case, the lax functoriality constraint $\boxtimes^2$ of the monoidal composition $\boxtimes$ in Mat^C has two additional conditions about the braided distortions of the two paths involved; see (II.8.2.15) and (II.8.2.16). These conditions about the braided distortions are necessary because a braid is not determined by its underlying permutation, and the braided distortion category $\mathcal{D}^{\mathsf{br}}$ involves the braid groups. The second subtle point is that, even if C is a tight braided bimonoidal category, the monoidal bicategory Mat^C does *not* seem to have any reasonable braided monoidal bicategory structure in general. We will explain this point in more detail near the end of Section II.8.4. The difficulty once again comes from the fact that the braided distortion category $\mathcal{D}^{\mathsf{br}}$ involves the braid groups, and a braid with an identity underlying permutation is usually not the identity braid.

Part II.2: E_n-Monoidal Categories

Chapter II.9: Ring, Bipermutative, and Braided Ring Categories

This chapter discusses ring and bipermutative categories in the sense of Elmendorf-Mandell and the braided version. The main difference between these categorical notions and their bimonoidal counterparts in Parts I.1 and II.1 is that ring categories have generally non-invertible *factorization morphisms*

$$(A \otimes C) \oplus (B \otimes C) \xrightarrow{\partial^l_{A,B,C}} (A \oplus B) \otimes C$$
$$(A \otimes B) \oplus (A \otimes C) \xrightarrow{\partial^r_{A,B,C}} A \otimes (B \oplus C)$$

that go in the opposite direction as the distributivity morphisms δ^r and δ^l. Ring categories with invertible factorization morphisms are special cases of tight bimonoidal categories, so the latter's strictification theorems in Chapter I.5 also apply to such ring categories. The bipermutative and braided analogues are also true. Similar to the endomorphism rig of a commutative monoid, each small permutative category C yields an endomorphism ring category $\mathsf{Perm}^{\mathsf{su}}(\mathsf{C};\mathsf{C})$. Similar to the reduction of Laplaza's axioms in symmetric bimonoidal categories in Section I.2.2 and the braided version in Theorem II.2.2.1, about half of the ring category axioms are redundant in a bipermutative category and a braided ring category. This is an extension of an observation in [**EM06**, Fig. 1]. Moreover, the Drinfeld center and the symmetric center have natural analogues for these ring-like categories. As we will discuss in Chapters 11 and 12, the Elmendorf-Mandell K-theory multifunctor sends small ring, braided ring, and bipermutative categories to, respectively, strict ring, E_2-, and E_∞-symmetric spectra. The strict ring and E_∞ cases are due to Elmendorf-Mandell [**EM06, EM09**], and the E_2 case is new.

Chapter II.10: Iterated and E_n-Monoidal Categories

Keeping in mind that the ring-like categories in Chapter II.9 correspond to E_n-symmetric spectra for $n \in \{1, 2, \infty\}$ via algebriac K-theory, this chapter discusses the categorical structure for the general E_n cases. An n-fold monoidal category in the sense of [**BFSV03**] has n monoidal structures $\otimes_1, \ldots, \otimes_n$ that are strictly associative and unital and interact via the exchange natural transformations

$$(A \otimes_j B) \otimes_i (C \otimes_j D) \xrightarrow{\eta^{i,j}_{A,B,C,D}} (A \otimes_i C) \otimes_j (B \otimes_i D)$$

for $1 \leq i < j \leq n$. Monoids in the monoidal category of small n-fold monoidal categories are precisely small $(n+1)$-fold monoidal categories. We introduce the notion of an *E_n-monoidal category* as a permutative category $(\mathsf{C}, \oplus)$ equipped with an n-fold monoidal structure $\{\otimes_i, \eta^{i,j}\}$ and factorization morphisms $\{\partial^{l,i}, \partial^{r,i}\}$ for each monoidal structure $\otimes_i$, such that (i) each $(\oplus, \otimes_i, \partial^{l,i}, \partial^{r,i})$ is a ring category and (ii) several axioms relating $\eta^{i,j}$, $\partial^{l,i}$, and $\partial^{r,i}$ hold. Ring categories are E_1-monoidal categories. Braided ring categories and bipermutative categories are special cases of, respectively, E_2- and E_n-monoidal categories for $n \geq 2$. Moreover, each small category generates a free E_n-monoidal category. In Chapter 13, we will show that the Elmendorf-Mandell K-theory of a small E_n-monoidal category is an E_n-symmetric spectrum for $n \geq 1$.

Part 1: Enriched Monoidal Categories and Multicategories

Chapter 1: Enriched Monoidal Categories

This chapter gives the basic definitions and properties for enriched monoidal categories, including plain, braided, and symmetric variants. Definition 1.4.25 describes 2-categories of each, with 1- and 2-cells given by appropriately monoidal enriched functors and natural transformations, respectively. For our applications to K-theory in Part 2, the enriching category V is symmetric monoidal closed.

However, our treatment in this chapter addresses the more general case that V is merely monoidal, with additional assumptions about braided or symmetric monoidal structure stated as necessary.

Section 1.5 discusses the important special case V = Cat, the category of small categories with its Cartesian product. Explanation 1.5.3 describes how the monoidal V-categories in this case are strict versions of monoidal bicategories. The braided and symmetric cases are similarly compared.

Chapter 2: Change of Enrichment

This chapter describes change of enriching category induced by a symmetric monoidal functor, showing that monoidal structures are preserved. Sections 2.1 through 2.4 give a thorough treatment of 2-functoriality results. As an application, Corollary 2.4.17 shows that taking underlying categories gives a 2-functor from small monoidal V-categories, V-functors, and V-natural transformations to ordinary monoidal categories, functors, and natural transformations. Similar statements hold for the braided and symmetric cases.

A partial reverse of Corollary 2.4.17 is given in Theorem 2.5.1. The theorem shows that, for given V-enriched data, various enriched monoidal axioms are satisfied if and only if the corresponding monoidal axioms for the underlying data are satisfied. This provides a mechanism to lift ordinary monoidal structures to enriched monoidal structures.

Sections 2.5 and 2.6 apply Theorem 2.5.1 to lift coherence and strictification results for ordinary monoidal, braided, and symmetric monoidal categories to their enriched counterparts. The Enriched Monoidal Coherence Theorem 2.5.6 and Enriched Epstein's Coherence Theorem 2.5.8 play a significant role in subsequent chapters.

Chapter 3: Self-Enrichment and Enriched Yoneda

This chapter restricts to the case that V is a symmetric monoidal closed category. Theorem 3.3.2 shows, via Theorem 2.5.1, that the canonical enrichment of V over itself is symmetric monoidal as a V-category. The next several sections develop the theory of V-enriched co/ends followed by the V-Yoneda Lemma (Theorem 3.6.9) and an equivalent form called the V-Yoneda Density Theorem 3.7.8. These are applied to develop the Day convolution and internal hom for enriched diagram categories (Theorem 3.7.22). The remainder of the chapter discusses additional theory of enriched diagram categories and tensor/cotensor structures that will be important for the development of enriched K-theory functors in Part 2.

Chapter 4: Pointed Objects, Smash Products, and Pointed Homs

This chapter gives the definitions and properties of smash products and pointed homs. These will be used throughout Part 2, and the smash product of pointed multicategories, developed in Chapter 5, will be particularly important.

Section 4.3 uses the Day convolution and internal hom to develop symmetric monoidal closed structure for pointed diagram categories. The results are summarized in Theorem 4.3.37. Applications of this material appear in Chapters 8, 9, and 10, where the Segal and Elmendorf-Mandell K-theory constructions are given via certain pointed diagram categories.

Chapter 5: Multicategories

This chapter gives relevant background on multicategories, multifunctors, and multinatural transformations. Theorem 5.5.14 shows that the category of small multicategories is complete and cocomplete. The Boardman-Vogt tensor product of multicategories, and the associated smash product for pointed multicategories, are developed in Section 5.6. The corresponding internal hom and its pointed variant are developed in Section 5.7.

Chapter 6: Enriched Multicategories

This chapter develops basic definitions and properties for enriched multicategories. One of our important applications, developed in Section 6.3, is the enriched multicategory associated to an enriched symmetric monoidal category. Our first use of this is in Section 6.4 where we describe the Cat-enriched multicategory structure on $\mathsf{Multicat}$, the category of small multicategories. It is induced by showing that the tensor product makes $\mathsf{Multicat}$ symmetric monoidal as a Cat-enriched category (Theorem 6.4.3). The pointed variant, with the smash product of small pointed multicategories, is given in Theorem 6.4.4 and will be essential for Part 2.

Sections 6.5 and 6.6 cover our second important application of enriched multicategories. The category $\mathsf{PermCat}^{\mathsf{su}}$, consisting of small permutative categories and strictly unital symmetric monoidal functors, has a Cat-enriched multicategory structure given by multilinear functors and multilinear transformations (Definitions 6.5.4 and 6.5.11). Propositions 6.5.10 and 6.5.13 show that this Cat-enriched multicategory structure is induced from that of small pointed multicategories and their smash product. Section 6.6 gives a second, direct proof of the Cat-enriched multicategory axioms.

Part 2: Algebraic K-Theory

Chapter 7: Homotopy Theory Background

This chapter gives relevant background from homotopy theory. Sections 7.1 and 7.2 introduce simplicial sets and simplicial homotopy, along with the nerve and geometric realization functors. The category of symmetric spectra, with its symmetric monoidal closed structure, is presented in Sections 7.3 through 7.6. Then, Sections 7.7 and 7.8 give a short review of Quillen model categories and a number of key examples.

Chapter 8: Segal K-Theory of Permutative Categories

This chapter presents the K-theory functor K^{Se} due to Segal [**Seg74**]. Its inputs are small permutative categories and its outputs are symmetric spectra. Section 8.3 describes the key construction as given by Segal. Sections 8.4 and 8.5 describe an equivalent construction that compares more readily with the K-theory multifunctor of Elmendorf-Mandell, K^{EM}.

Chapter 9: Categories of $\mathcal{G}_*$-Objects

This chapter is the first of two that replace the Segal K-theory functor with a simplicially-enriched multifunctor due to Elmendorf-Mandell [**EM06, EM09**]. This chapter focuses on the replacement of Γ-categories and Γ-simplicial sets with pointed diagrams out of a larger indexing category $\mathcal{G}$. The construction of symmetric spectra from such diagram categories is given in Section 9.3 and is denoted $\mathsf{K}^{\mathcal{G}}$. Sections 9.2 and 9.4 use the material from Part 1 to explain that the new diagram categories and the new functor $\mathsf{K}^{\mathcal{G}}$ are symmetric monoidal, in the enriched sense of Chapter 1, over the category of pointed simplicial sets.

Chapter 10: Elmendorf-Mandell K-Theory of Permutative Categories

This chapter is the second of two that replace the Segal K-theory functor with a simplicially enriched multifunctor due to Elmendorf-Mandell [**EM06, EM09**]. This chapter focuses on the construction of $\mathcal{G}_*$-categories from small permutative categories, replacing Segal's construction of Γ-categories from the same. Additional material from Part 1 is used throughout the chapter to explain that the multi/categories and multi/functors are enriched either in the symmetric monoidal sense of Chapter 1 or in the multicategorical and multifunctorial sense of Chapter 6. Section 10.6 contains the proof that the Segal and Elmendorf-Mandell K-theory symmetric spectra associated to a small permutative category C are level equivalent (Theorem 10.6.10). Because K^{EM} is an enriched multifunctor, it preserves operad actions. We state this result as Theorem 10.3.33 and apply it in Chapters 11, 12, and 13.

Chapter 11: K-Theory of Ring and Bipermutative Categories

This is the first of three chapters that contain algebraic K-theory applications of the ring-like categories in Part II.2. The main K-theory results in this chapter, Corollaries 11.3.16 and 11.6.12, are from [**EM06, EM09**], and they are the E_1 and the E_∞ cases. These results state that the Elmendorf-Mandell K-theory multifunctor K^{EM} sends (i) small ring categories to strict ring symmetric spectra and (ii) small bipermutative categories to E_∞-symmetric spectra. They are obtained by combining the multifunctor K^{EM} and the fact that the associative operad and the Barratt-Eccles operad parametrize, respectively, ring and bipermutative category structures on small permutative categories. Since the associative operad has monoids as algebras and the Barratt-Eccles operad is an E_∞-operad, the K-theory results follow.

Chapter 12: K-Theory of Braided Ring Categories

This chapter contains the E_2 analogues of the results in Chapter 11. The first part of this chapter discusses the braid operad Br, which generalizes the Barratt-Eccles operad. This is a categorical E_2-operad (Theorem 12.2.4) whose algebras in Cat are small braided strict monoidal categories (Proposition 12.3.22). The main categorical input is Theorem 12.4.5, which says that Br parametrizes braided ring category structures, as in Chapter II.9, on small permutative categories. Applying the K-theory multifunctor K^{EM}, it follows that K^{EM} sends small braided ring categories to E_2-symmetric spectra (Corollary 12.5.3). The K-theory result, Corollary 12.5.3, and the main categorical input, Theorem 12.4.5, are new results.

Chapter 13: K-Theory of E_n-Monoidal Categories

This chapter contains the general E_n analogues for $n \geq 1$ of the categorical and K-theory results in Chapters 11 and 12. The first part of this chapter discusses the n-fold monoidal category operad Mon^n. This is a categorical E_n-operad (Theorem 13.2.1) whose algebras in Cat are small n-fold monoidal categories (Proposition 13.3.18) as in Chapter II.10. The main categorical input is Theorem 13.4.12, which says that Mon^n parametrizes E_n-monoidal category structures on small permutative categories. Applying the K-theory multifunctor K^{EM}, it follows that K^{EM} sends small E_n-monoidal categories to E_n-symmetric spectra for $n \geq 1$ (Corollary 13.5.2). As in Chapter 12, the K-theory result, Corollary 13.5.2, and the main categorical input, Theorem 13.4.12, are new results.

Appendix A: Open Questions

This chapter discusses open questions related to the topics of this work. We encourage the reader to read these open questions at any time and use them as additional motivation for the main text.

Reading Guides

Supplementing the chapter introductions and individual reading guides therein, the following guides describe themes that span multiple chapters. For especially broad topics, we include a selection of general references for background or further reading. The Notes section at the end of each chapter provides additional references relevant to the content of that chapter.

Category Theory. For a refresher of basic category theory, including braided and symmetric monoidal categories, read Chapters I.1 and II.1. For bicategories and 2-categories, read Chapter I.6. For abelian categories, read Section II.2.3. For enriched category theory, read Section 1.2 and Chapter 3. Bimonoidal categories are built upon monoidal categories. Thus, a thorough understanding of the definitions and coherence of monoidal categories is necessary to understand bimonoidal categories and their coherence.

References for basic category theory include [**Awo10, BK00, Gra18, Lei14, Rie16, Rom17, Sim11**]. References for more advanced category theory include [**Bor94a, Bor94b, Bor94c, ML98, Mit65, Sch72**]. References for Abelian categories include [**EGNO15, Fre03, Mit65**]. References for enriched categories include [**Bor94b, Cru09, For04, Kel05**]. References for ends and coends include [**Day70, DK69, Lor21**]. For further reference on 2-dimensional categories, we highly recommend [**JY21**].

Symmetric Bimonoidal Categories. To review the axioms of a symmetric bimonoidal category, read Sections I.2.1 and I.2.2. For Laplaza's Coherence Theorems for symmetric bimonoidal categories, read Theorems I.3.9.1 and I.4.4.3. For strictification theorems, read Theorems I.5.4.6, I.5.4.7, I.5.5.11, and I.5.5.12. For their bimonoidal analogues, read Theorems I.3.10.7 and I.4.5.8. For Baez's Conjecture, read Definition I.7.1.8 and Theorem I.7.8.1. The introductions of Chapters I.3, I.4, I.5, and I.7 have more detailed description and reading suggestions. This material on (symmetric) bimonoidal categories is used extensively in Volume II in the discussion of braided bimonoidal categories.

Braided Bimonoidal Categories. To review the axioms of a braided bimonoidal category, read Definition II.2.1.29 and Theorem II.2.2.1. For their coherence and strictification (i.e., the Blass-Gurevich Conjecture), read Theorems II.5.4.4, II.6.3.6, and II.6.3.7. For the braided version of Baez's Conjecture, read Definition II.7.1.5 and Theorem II.7.3.4. The introductions of these chapters have more detailed discussion and reading guides. This material on braided bimonoidal categories is not heavily used in Volume III in our discussion of algebraic K-theory. Instead, these braided structures provide a great illustration of the coherence theory in Volume I and have many applications in the sciences, some of which are discussed in this work.

Applications to Quantum Groups, TQC, and Programming. For applications to quantum group theory, read Sections II.3.1 and II.3.2. For applications to topological quantum computation, read Theorem II.2.4.22 and Sections II.3.3 through II.3.6. For applications to reversible programming, read Section I.2.6. Our treatment is self-contained and assumes no prior knowledge of these topics. These applications are not used in Volume III. Instead, they illustrate the much larger scientific context where categorical structures discussed in this text are applied.

Enriched Monoidal Categories. To review enriched monoidal categories, read Sections 1.3 through 1.5. For their coherence theory, read Sections 2.5 and 2.6. For change of enrichment, read Proposition 2.1.2, Theorems 2.2.7, 2.3.7, and 2.4.10, and Corollary 2.4.17. For symmetric monoidal closed structures in the pointed context, read Theorems 4.1.8, 4.2.3, 4.3.19, and 4.3.37. Much of this material is known to experts, but it is not easily accessible in the literature in a unified format. Our discussion is self-contained and highly detailed, so the reader can thoroughly learn these topics just from our text. This material is necessary to understand the intricate multicategorical properties of Elmendorf-Mandell K-theory discussed in later chapters.

Multicategories and Operads. To review enriched multicategories, read Section 6.1. The Boardman-Vogt tensor product of multicategories and its pointed variant are discussed in Sections 5.6 and 6.4. For the passage from enriched symmetric monoidal categories to enriched multicategories, read Sections 5.3 and 6.3. For the categorically-enriched multicategory of small permutative categories, read Sections 6.5 and 6.6. For the associative operad, the Barratt-Eccles operad, the braid operad, and the n-fold monoidal category operad, read Definitions 11.1.1, 11.4.10, 12.1.2, and 13.1.12. Elmendorf-Mandell K-theory is an enriched multifunctor, which can transport operadic algebras in small permutative categories to the same type of operadic algebras in symmetric spectra. This material is necessary to describe the enriched multifunctoriality of Elmendorf-Mandell K-theory and its applications to highly structured ring spectra. References for multicategories and operads include [**Fre17, MSS02, May72, Yau16**].

Ring, Bipermutative, Braided Ring, and E_n-Monoidal Categories. For the definitions of ring, bipermutative, braided ring, and E_n-monoidal categories, read Definitions II.9.1.2, II.9.3.2, II.9.5.1, II.10.1.1, and II.10.7.2. For their coherence theory, read Corollaries II.9.1.19, II.9.1.20, II.9.3.12, and II.9.3.13, Theorem II.10.6.8, and Question A.2.1. For their description in terms of operads, read Theorems 11.2.16, 11.5.5, 12.4.5, and 13.4.12. The fact that these monoidal categories with extra structures are algebras over operads is the precise reason why they are sent by Elmendorf-Mandell K-theory to structured ring spectra of the corresponding types.

Basic Homotopy Theory. For a brief review of simplicial objects and the nerve construction, read Sections 7.1 and 7.2. To review symmetric spectra, read Sections 7.3 and 7.4. For the smash product of symmetric spectra, read Sections 7.5 and 7.6. Model category theory is reviewed in Sections 7.7 and 7.8, but later chapters do not use model categories in any way. While we assume some basic homotopy theory, we do not assume any prior knowledge of symmetric spectra. Our discussion of symmetric spectra is very gentle and contains a lot of details that are not explicitly available elsewhere in the literature. The main point of both Segal K-theory and Elmendorf-Mandell K-theory is the construction of symmetric spectra from purely categorical data. A detailed understanding of symmetric spectra and their smash product is necessary to fully appreciate the K-theory constructions of Segal and Elmendorf-Mandell.

References for homotopy theory include [**BR20, May99, MP12, Mil20, Ric20, Rie14**]. References for simplicial homotopy theory include [**Cur71, GZ67, GJ09, May92**]. References for further background and a broader perspective on algebraic K-theory include [**Mil71, Qui73, Ros95, Wal85, Wei13**].

Segal K-Theory. To review the passage from Γ-simplicial sets to symmetric spectra, read Definitions 7.4.5, 8.1.8, and 8.2.5. For the passage from small permutative categories to Γ-categories, read Definitions 8.1.17, 8.3.1, 8.3.6, 8.3.9, and 8.3.12 and Proposition 8.4.8. To review Segal's K-theory construction, from small permutative categories to symmetric spectra via Γ-categories, read Definitions 8.4.10 and 8.5.1 and Theorem 8.5.2. Although Segal K-theory does not generally preserve multiplicative structures, it is a fundamental tool in this subject and the main motivation for Elmendorf-Mandell K-theory.

Elmendorf-Mandell K-Theory. To review the passage from $\mathcal{G}_*$-simplicial sets to symmetric spectra, read Sections 9.1 through 9.3. Its symmetric monoidality is discussed in Theorem 9.4.9. For the passage from small permutative categories to $\mathcal{G}_*$-categories, read Section 10.4. For Elmendorf-Mandell K-theory, from small permutative categories to symmetric spectra via $\mathcal{G}_*$-categories, read Lemma 10.2.14 and Corollary 10.3.24. For the comparison between Segal K-theory and Elmendorf-Mandell K-theory, read Theorem 10.6.10. For the passage from categorical data to highly structured ring spectra via Elmendorf-Mandell K-theory, read Theorem 10.3.33 and Corollaries 11.3.16, 12.5.3, and 13.5.2. This material is the main point of Volume III; all preceding chapters of Volume III are preparation for it.

Part 1

Enriched Monoidal Categories and Multicategories

CHAPTER 1

Enriched Monoidal Categories

Part 1 is about enriched monoidal categories and enriched multicategories. We will use this material in the following two ways.

- Many of the K-theory constructions presented in Chapters 8, 9, and 10 are enriched, as monoidal functors or as multifunctors. We use the general theory presented here to explain the relevant enrichments of their co/domains and the (multi-)functors themselves.
- The operads that describe E_n-monoidal structure, presented in Chapters 11, 12, and 13, are 1-object enriched multicategories. That point of view is important for the preservation of E_n-monoidal structure by the (enriched) Elmendorf-Mandell K-theory multifunctor.

All of the necessary background on 2-categories and bicategories can be found in Chapter I.6 and [**JY21**].

This chapter gives introductory definitions and basic results about monoidal structure for enriched categories. We do not assume any prior knowledge of enriched category theory. Throughout this chapter $\mathsf{V} = (\mathsf{V}, \otimes)$ will denote a monidal category and will be assumed either braided or symmetric, as needed. We give a short review of monoidal categories, along with their braided and symmetric variants, in Section 1.1. We then review the basic definitions of V-category, V-functor, and V-natural transformation in Section 1.2.

In Section 1.3, under the assumption that V is braided monoidal, we define the tensor product of enriched categories and show that it defines a monoidal product for the category of V-categories (Theorem 1.3.35). In Section 1.4 we use the tensor product to define monoidal V-categories, V-functors, and V-natural transformations along with braided and symmetric variants.

In Section 1.5 we apply the definitions of Section 1.4 in the case $(\mathsf{V}, \otimes) = (\mathsf{Cat}, \times)$ to obtain definitions of plain, braided, and symmetric monoidal structure for a 2-category. We explain how these definitions are a special, more strict, case of the general definitions for monoidal bicategories from Section I.6.4. In Theorem 1.5.5 we show, for general braided monoidal V, that the 2-category of V-categories is monoidal in the more strict Cat-enriched sense.

1.1. Review of Monoidal Categories

For the reader's convenience we recall definitions and basic properties of monoidal categories, functors, and natural transformations along with the braided and symmetric variants. These definitions along with basic properties will be used throughout this and several subsequent chapters. At the end of the section we include statements of the coherence and strictification theorems, which we will use in a few key places. With the exception of Lemma 1.1.34, all of this material has been covered with more detail previously in Chapters I.1 and II.1.

Monoidal Categories.

Definition 1.1.1. A *monoidal category* is a tuple

$$(\mathsf{C}, \otimes, \mathbb{1}, \alpha, \lambda, \rho)$$

consisting of

- a category C;
- a functor $\otimes: \mathsf{C} \times \mathsf{C} \longrightarrow \mathsf{C}$, which is called the *monoidal product*;
- an object $\mathbb{1} \in \mathsf{C}$, which is called the *monoidal unit*;
- a natural isomorphism

$$(X \otimes Y) \otimes Z \xrightarrow[\cong]{\alpha_{X,Y,Z}} X \otimes (Y \otimes Z)$$

  for all objects $X, Y, Z \in \mathsf{C}$, which is called the *associativity isomorphism*; and
- natural isomorphisms

$$\mathbb{1} \otimes X \xrightarrow[\cong]{\lambda_X} X \quad \text{and} \quad X \otimes \mathbb{1} \xrightarrow[\cong]{\rho_X} X$$

  for all objects $X \in \mathsf{C}$, which are called the *left unit isomorphism* and the *right unit isomorphism*, respectively.

These data are subject to the following two axioms.

The Unity Axiom: The diagram

(1.1.2)
$$\begin{array}{ccc} (X \otimes \mathbb{1}) \otimes Y & \xrightarrow{\alpha_{X,\mathbb{1},Y}} & X \otimes (\mathbb{1} \otimes Y) \\ {\scriptstyle \rho_X \otimes 1_Y} \downarrow & & \downarrow {\scriptstyle 1_X \otimes \lambda_Y} \\ X \otimes Y & = & X \otimes Y \end{array}$$

is commutative for all objects $X, Y \in \mathsf{C}$.

The Pentagon Axiom: The pentagon

(1.1.3)
$$\begin{array}{ccccc} & & (W \otimes X) \otimes (Y \otimes Z) & & \\ & {\scriptstyle \alpha_{W \otimes X, Y, Z}} \nearrow & & \searrow {\scriptstyle \alpha_{W,X,Y \otimes Z}} & \\ ((W \otimes X) \otimes Y) \otimes Z & & & & W \otimes (X \otimes (Y \otimes Z)) \\ {\scriptstyle \alpha_{W,X,Y} \otimes 1_Z} \downarrow & & & & \uparrow {\scriptstyle 1_W \otimes \alpha_{X,Y,Z}} \\ (W \otimes (X \otimes Y)) \otimes Z & & \xrightarrow{\alpha_{W, X \otimes Y, Z}} & & W \otimes ((X \otimes Y) \otimes Z) \end{array}$$

is commutative for all objects $W, X, Y, Z \in \mathsf{C}$.

This finishes the definition of a monoidal category. A monoidal category is *strict* if α, λ, and ρ are identity natural transformations. ◇

In a monoidal category, the equality

(1.1.4)
$$\lambda_{\mathbb{1}} = \rho_{\mathbb{1}}: \mathbb{1} \otimes \mathbb{1} \longrightarrow \mathbb{1}$$

and the commutative diagrams

$$(1.1.5)\quad \begin{array}{ccc} (\mathbb{1}\otimes X)\otimes Y & \xrightarrow{\alpha_{\mathbb{1},X,Y}} & \mathbb{1}\otimes(X\otimes Y) \\ {\scriptstyle \lambda_X\otimes 1_Y}\downarrow & & \downarrow{\scriptstyle \lambda_{X\otimes Y}} \\ X\otimes Y & = & X\otimes Y \end{array} \qquad \begin{array}{ccc} (X\otimes Y)\otimes\mathbb{1} & \xrightarrow{\alpha_{X,Y,\mathbb{1}}} & X\otimes(Y\otimes\mathbb{1}) \\ {\scriptstyle \rho_{X\otimes Y}}\downarrow & & \downarrow{\scriptstyle 1_X\otimes\rho_Y} \\ X\otimes Y & = & X\otimes Y \end{array}$$

are formal consequences of the monoidal category axioms. These two diagrams are called the *left unity diagram* and the *right unity diagram*, respectively.

Definition 1.1.6. For monoidal categories C and D, a *monoidal functor*

$$(F, F^2, F^0) : \mathsf{C} \longrightarrow \mathsf{D}$$

consists of

- a functor $F : \mathsf{C} \longrightarrow \mathsf{D}$;
- a natural transformation, which is called the *monoidal constraint*,

$$(1.1.7)\qquad FX\otimes FY \xrightarrow{F^2} F(X\otimes Y) \in \mathsf{D}$$

  for objects $X, Y \in \mathsf{C}$; and
- a morphism, which is called the *unit constraint*,

$$(1.1.8)\qquad \mathbb{1}^{\mathsf{D}} \xrightarrow{F^0} F\mathbb{1}^{\mathsf{C}} \in \mathsf{D}.$$

These data are required to satisfy the following associativity and unity axioms.

Associativity: The diagram

$$(1.1.9)\quad \begin{array}{ccc} (FX\otimes FY)\otimes FZ & \xrightarrow{\alpha^{\mathsf{D}}} & FX\otimes(FY\otimes FZ) \\ {\scriptstyle F^2\otimes 1_{FZ}}\downarrow & & \downarrow{\scriptstyle 1_{FX}\otimes F^2} \\ F(X\otimes Y)\otimes FZ & & FX\otimes F(Y\otimes Z) \\ {\scriptstyle F^2}\downarrow & & \downarrow{\scriptstyle F^2} \\ F((X\otimes Y)\otimes Z) & \xrightarrow{F\alpha^{\mathsf{C}}} & F(X\otimes(Y\otimes Z)) \end{array}$$

is commutative for all objects $X, Y, Z \in \mathsf{C}$.

Unity: The diagrams

$$(1.1.10)\quad \begin{array}{ccc} \mathbb{1}^{\mathsf{D}}\otimes FX & \xrightarrow{\lambda^{\mathsf{D}}_{FX}} & FX \\ {\scriptstyle F^0\otimes 1_{FX}}\downarrow & & \uparrow{\scriptstyle F\lambda^{\mathsf{C}}_X} \\ F\mathbb{1}^{\mathsf{C}}\otimes FX & \xrightarrow{F^2} & F(\mathbb{1}^{\mathsf{C}}\otimes X) \end{array} \qquad \begin{array}{ccc} FX\otimes\mathbb{1}^{\mathsf{D}} & \xrightarrow{\rho^{\mathsf{D}}_{FX}} & FX \\ {\scriptstyle 1_{FX}\otimes F^0}\downarrow & & \uparrow{\scriptstyle F\rho^{\mathsf{C}}_X} \\ FX\otimes F\mathbb{1}^{\mathsf{C}} & \xrightarrow{F^2} & F(X\otimes\mathbb{1}^{\mathsf{C}}) \end{array}$$

are commutative for all objects $X \in \mathsf{C}$. They are called the *left unity diagram* and the *right unity diagram*, respectively.

This finishes the definition of a monoidal functor. A monoidal functor (F, F^2, F^0) is often abbreviated to F.

Moreover, a monoidal functor (F, F^2, F^0) is said to be

- *unital* if F^0 is an isomorphism;
- *strictly unital* if F^0 is the identity morphism;
- *strong* if F^0 and the components of F^2 are isomorphisms; and
- *strict* if F^0 and the components of F^2 are identity morphisms. ◇

Definition 1.1.11. Suppose

$$\mathsf{C} \xrightarrow{\ F\ } \mathsf{D} \xrightarrow{\ G\ } \mathsf{E}$$

are monoidal functors. Their *composite*

$$(GF, (GF)^2, (GF)^0) : \mathsf{C} \longrightarrow \mathsf{E}$$

is the monoidal functor with underlying functor GF and the structure morphisms

$$(GF)^0 : \mathbb{1} \xrightarrow{\ G^0\ } G\mathbb{1} \xrightarrow{\ G(F^0)\ } GF\mathbb{1}$$

$$(GF)^2 : GFA \otimes GFB \xrightarrow{\ G^2\ } G(FA \otimes FB) \xrightarrow{\ G(F^2)\ } GF(A \otimes B)$$

for objects $A, B \in \mathsf{C}$. ◇

Definition 1.1.12. For monoidal functors $F, G : \mathsf{C} \longrightarrow \mathsf{D}$, a *monoidal natural transformation* $\theta : F \longrightarrow G$ is a natural transformation between the underlying functors such that the diagrams

(1.1.13)
$$\begin{array}{ccc} FX \otimes FY & \xrightarrow{\theta_X \otimes \theta_Y} & GX \otimes GY \\ {\scriptstyle F^2}\downarrow & & \downarrow{\scriptstyle G^2} \\ F(X \otimes Y) & \xrightarrow{\theta_{X \otimes Y}} & G(X \otimes Y) \end{array} \qquad \begin{array}{ccc} \mathbb{1}^{\mathsf{D}} & \xrightarrow{F^0} & F\mathbb{1}^{\mathsf{C}} \\ \| & & \downarrow{\scriptstyle \theta_{\mathbb{1}^{\mathsf{C}}}} \\ \mathbb{1}^{\mathsf{D}} & \xrightarrow{G^0} & G\mathbb{1}^{\mathsf{C}} \end{array}$$

are commutative for all objects $X, Y \in \mathsf{C}$. ◇

Braided Monoidal Categories.

Definition 1.1.14. A *braided monoidal category* is a pair (C, ξ) consisting of the following data.

- $(\mathsf{C}, \otimes, \mathbb{1}, \alpha, \lambda, \rho)$ is a monoidal category.
- ξ is a natural isomorphism

(1.1.15)
$$X \otimes Y \xrightarrow[\cong]{\ \xi_{X,Y}\ } Y \otimes X$$

for objects $X, Y \in \mathsf{C}$, which is called the *braiding*.

These data are required to satisfy the *Hexagon Axioms*, stating the commutativity of the following two diagrams, called the *left hexagon diagram* and the *right hexagon diagram*, respectively, for objects $X, Y, Z \in \mathsf{C}$.

(1.1.16)
$$\begin{array}{ccc} (Y \otimes X) \otimes Z & \xrightarrow{\alpha} & Y \otimes (X \otimes Z) \\ {\scriptstyle \xi_{X,Y} \otimes 1_Z}\uparrow & & \downarrow{\scriptstyle 1_Y \otimes \xi_{X,Z}} \\ (X \otimes Y) \otimes Z & & Y \otimes (Z \otimes X) \\ {\scriptstyle \alpha}\downarrow & & \uparrow{\scriptstyle \alpha} \\ X \otimes (Y \otimes Z) & \xrightarrow{\xi_{X,YZ}} & (Y \otimes Z) \otimes X \end{array} \qquad \begin{array}{ccc} X \otimes (Z \otimes Y) & \xrightarrow{\alpha^{-1}} & (X \otimes Z) \otimes Y \\ {\scriptstyle 1_X \otimes \xi_{Y,Z}}\uparrow & & \downarrow{\scriptstyle \xi_{X,Z} \otimes 1_Y} \\ X \otimes (Y \otimes Z) & & (Z \otimes X) \otimes Y \\ {\scriptstyle \alpha^{-1}}\downarrow & & \uparrow{\scriptstyle \alpha^{-1}} \\ (X \otimes Y) \otimes Z & \xrightarrow{\xi_{XY,Z}} & Z \otimes (X \otimes Y) \end{array}$$

This finishes the definition of a braided monoidal category. A braided monoidal category is *strict* if the underlying monoidal category is strict. ◇

Definition 1.1.17. For braided monoidal categories C and D, a *braided monoidal functor* $(F, F^2, F^0)\colon \mathsf{C} \longrightarrow \mathsf{D}$ is a monoidal functor between the underlying monoidal categories such that the diagram

(1.1.18)
$$\begin{array}{ccc} FX \otimes FY & \xrightarrow[\cong]{\xi_{FX,FY}} & FY \otimes FX \\ {\scriptstyle F^2}\downarrow & & \downarrow{\scriptstyle F^2} \\ F(X \otimes Y) & \xrightarrow[\cong]{F\xi_{X,Y}} & F(Y \otimes X) \end{array}$$

is commutative for all objects $X, Y \in \mathsf{C}$. A braided monoidal functor is said to be *strong* (respectively, *strict*, *unital*, or *strictly unital*) if the underlying monoidal functor is so. ◇

Explanation 1.1.19. The two hexagon diagrams in (1.1.16) may be visualized as the braids, read bottom-to-top,

in the braid group B_3, with the braiding ξ interpreted as the generating braid $s_1 \in B_2$. On the left, the two strings labeled by Y and Z cross over the string labeled by X. The two composites along the boundary of the left hexagon diagram (1.1.16) correspond to passing Y and Z over X either one at a time, or both at once. On the right, the string labeled by Z crosses over the two strings labeled by Y and X. The two composites along the boundary of the right hexagon diagram (1.1.16) likewise correspond to the two ways of passing Z over X and Y. ◇

We note several useful consequences of the hexagon axioms (1.1.16); proofs of these are given in Section II.1.3.

In each braided monoidal category (C, ξ), the following two unit diagrams are commutative for all objects $X \in \mathsf{C}$.

(1.1.20)
$$\begin{array}{ccc} X \otimes \mathbb{1} & \xrightarrow{\xi_{X,\mathbb{1}}} & \mathbb{1} \otimes X \\ {\scriptstyle \rho_X}\downarrow & & \downarrow{\scriptstyle \lambda_X} \\ X & = & X \end{array} \qquad \begin{array}{ccc} \mathbb{1} \otimes X & \xrightarrow{\xi_{\mathbb{1},X}} & X \otimes \mathbb{1} \\ {\scriptstyle \lambda_X}\downarrow & & \downarrow{\scriptstyle \rho_X} \\ X & = & X \end{array}$$

In each braided monoidal category (C, ξ), the equality

(1.1.21)
$$\xi_{\mathbb{1},\mathbb{1}} = 1_{\mathbb{1}\otimes\mathbb{1}}\colon \mathbb{1} \otimes \mathbb{1} \longrightarrow \mathbb{1} \otimes \mathbb{1}$$

holds.

In each braided monoidal category (C, ξ), the following diagram is commutative for all objects $A, B, C \in \mathsf{C}$.

(1.1.22)
$$\begin{array}{ccc} (A \otimes B) \otimes C & \xrightarrow{\alpha_{A,B,C}} & A \otimes (B \otimes C) \\ {\scriptstyle \xi_{A,B}\otimes 1_C}\swarrow & & \searrow{\scriptstyle 1_A \otimes \xi_{B,C}} \\ (B \otimes A) \otimes C & & A \otimes (C \otimes B) \\ {\scriptstyle \xi_{B\otimes A,C}}\searrow & & \swarrow{\scriptstyle \xi_{A,C\otimes B}} \\ C \otimes (B \otimes A) & \xrightarrow{\alpha^{-1}_{C,B,A}} & (C \otimes B) \otimes A \end{array}$$

Symmetric Monoidal Categories.

Definition 1.1.23. A *symmetric monoidal category* is a monoidal category

$$(\mathsf{C}, \otimes, \mathbb{1}, \alpha, \lambda, \rho)$$

equipped with a natural isomorphism

$$X \otimes Y \xrightarrow[\cong]{\xi_{X,Y}} Y \otimes X$$

for objects $X, Y \in \mathsf{C}$, which is called the *braiding* or the *symmetry isomorphism*, that satisfies the following axioms.

The Symmetry Axiom: The diagram

$$\begin{array}{ccc} X \otimes Y & \xrightarrow{\xi_{X,Y}} & Y \otimes X \\ & \searrow^{=} & \downarrow \xi_{Y,X} \\ & & X \otimes Y \end{array} \tag{1.1.24}$$

is commutative for all objects $X, Y \in \mathsf{C}$.

The Unit Axiom: The diagram

$$\begin{array}{ccc} X \otimes \mathbb{1} & \xrightarrow{\xi_{X,\mathbb{1}}} & \mathbb{1} \otimes X \\ \rho_X \downarrow & & \downarrow \lambda_X \\ X & = & X \end{array} \tag{1.1.25}$$

is commutative for all objects $X \in \mathsf{C}$.

The Hexagon Axiom: The diagram

$$\begin{array}{ccc} X \otimes (Z \otimes Y) & \xrightarrow{1_X \otimes \xi_{Z,Y}} & X \otimes (Y \otimes Z) \\ \alpha \nearrow & & \searrow \alpha^{-1} \\ (X \otimes Z) \otimes Y & & (X \otimes Y) \otimes Z \\ \xi_{X \otimes Z, Y} \searrow & & \nearrow \xi_{Y,X} \otimes 1_Z \\ Y \otimes (X \otimes Z) & \xrightarrow{\alpha^{-1}} & (Y \otimes X) \otimes Z \end{array} \tag{1.1.26}$$

is commutative for all objects $X, Y, Z \in \mathsf{C}$.

This finishes the definition of a symmetric monoidal category. A *permutative category* is a symmetric monoidal category whose underlying monoidal category is strict.

A *symmetric monoidal functor* between symmetric monoidal categories is a monoidal functor that satisfies (1.1.18). A symmetric monoidal functor is said to be *strong* (respectively, *strict*, *unital*, or *strictly unital*) if the underlying monoidal functor is so.

Proposition II.1.3.36 shows that a symmetric monoidal category is precisely a braided monoidal category whose braiding satisfies the symmetry axiom (1.1.24). ◇

Definition 1.1.27. We let $\mathsf{PermCat}$ denote the 2-category of small permutative categories, symmetric monoidal functors, and monoidal natural transformations. We also use the following locally-full sub 2-categories consisting of the same objects but restricting the 1-cells.

- $\mathsf{PermCat}^{\mathsf{u}}$ has 1-cells given by unital symmetric monoidal functors.
- $\mathsf{PermCat}^{\mathsf{su}}$ has 1-cells given by strictly unital symmetric monoidal functors.
- $\mathsf{PermCat}^{\mathsf{sus}}$ has 1-cells given by strictly unital strong symmetric monoidal functors.

In each case the 2-cells are given by monoidal natural transformations. ◇

Definition 1.1.28. A symmetric monoidal category C is *closed* if, for each object X, the functor

$$- \otimes X : \mathsf{C} \longrightarrow \mathsf{C}$$

admits a right adjoint, which is denoted by $[X, -]$ and is called the *internal hom*. ◇

Coherence and Strictification Theorems. We recall the statements of monoidal coherence and strictification theorems together with the braided and symmetric variants. More detailed discussion has been given in Sections I.1.3 and II.1.6.

We recall from Definitions I.1.3.1 and I.1.3.2 the notions of (normalized) words and canonical maps. We then state the monoidal coherence and strictification theorems.

Definition 1.1.29. A *word* of length $n \geq 0$ is defined inductively as follows.

- The only word of length 0 is the symbol e.
- The only word of length 1 is the symbol $-$.
- If u and v are words of lengths m and n, respectively, then $u \,\square\, v$ is a word of length $m + n$.

Moreover, we define the following.

- A *left normalized word* is the word e, $-$, $u \,\square\, e$, or $u \,\square\, -$, with u a left normalized word.
- A *right normalized word* is the word e, $-$, $e \,\square\, u$, or $- \,\square\, u$, with u a right normalized word.
- For a monoidal category C, each word w of length n determines a functor $w : \mathsf{C}^n \longrightarrow \mathsf{C}$ by interpreting
  - the length 0 word e as the constant functor at $\mathbb{1}$;
  - the length 1 word $-$ as the identity functor 1_{C}; and
  - $\square$ as the monoidal product in C.

  We also call this functor a *word*. ◇

Definition 1.1.30. For a monoidal category $(\mathsf{C}, \otimes, \mathbb{1}, \alpha, \lambda, \rho)$, *canonical maps* are natural isomorphisms between words of the same length, defined inductively by the following four conditions.

- The identity morphism of $\mathbb{1}$ is a canonical map.
- The identity natural transformation of 1_{C} is a canonical map.
- α, λ, ρ, and their inverses are canonical maps.
- Canonical maps are closed under $\otimes$ and vertical composites. ◇

Theorem 1.1.31 (Mac Lane's Coherence). *Suppose u and v are words $\mathsf{C}^n \longrightarrow \mathsf{C}$ of the same length in a monoidal category C. Then there exists a unique canonical map $u \longrightarrow v$.*

Theorem 1.1.32 (Mac Lane's Strictification). *For each monoidal category* C*, there exist a strict monoidal category* C_{st} *and an adjoint equivalence*

$$\mathsf{C}_{\mathsf{st}} \underset{R}{\overset{L}{\rightleftarrows}} \mathsf{C}$$

with (i) both L *and* R *strong monoidal functors, and (ii)* $LR = 1_{\mathsf{C}}$*.*

Explanation 1.1.33. We note that the equality $LR = 1_{\mathsf{C}}$ in Theorem 1.1.32 is an equality of monoidal functors, and hence entails that the monoidal and unit constraints $(LR)^2$ and $(LR)^0$, respectively, are identities. The same holds for the Braided and Symmetric Strictification Theorems 1.1.39 and 1.1.42 below. ◇

The following lemma shows that the unit of the adjunction in Theorem 1.1.32 is a monoidal natural transformation. The same conclusion holds for the Braided and Symmetric Strictification Theorems 1.1.39 and 1.1.42 below.

Lemma 1.1.34. *Suppose given an adjoint equivalence of monoidal categories*

$$\mathsf{D} \underset{R}{\overset{L}{\rightleftarrows}} \mathsf{C}$$

with (i) both L *and* R *strong monoidal functors, and (ii)* $LR = 1_{\mathsf{C}}$*. Then the unit*

$$\eta \colon 1_{\mathsf{D}} \longrightarrow RL$$

is a monoidal natural transformation.

Proof. We must show that the following diagrams commute for each $X, Y \in \mathsf{D}$. In the diagram below we let $\varepsilon = 1_{1_{\mathsf{C}}}$ denote the counit of the adjunction.

$$\begin{array}{ccc} X \otimes Y & \xrightarrow{\eta_X \otimes \eta_Y} & RLX \otimes RLY \\ {\scriptstyle 1_{X\otimes Y}}\downarrow & & \downarrow{\scriptstyle (RL)^2} \\ X \otimes Y & \xrightarrow{\eta_{X\otimes Y}} & RL(X \otimes Y) \end{array} \qquad \begin{array}{ccc} \mathbb{1}^{\mathsf{D}} & \xrightarrow{1} & \mathbb{1}^{\mathsf{D}} \\ \| & & \downarrow{\scriptstyle \eta_{\mathbb{1}^{\mathsf{D}}}} \\ \mathbb{1}^{\mathsf{D}} & \xrightarrow{(RL)^0} & RL\mathbb{1}^{\mathsf{D}} \end{array} \tag{1.1.35}$$

Since (L, R) is an equivalence, it suffices to see that the diagrams commute after applying L. Applying L to the left hand diagram of (1.1.35) yields the outer diagram below, where we recall the monoidal constraint $(RL)^2$ is given by the composite of $R(L^2)$ and R^2 (Definition 1.1.11).

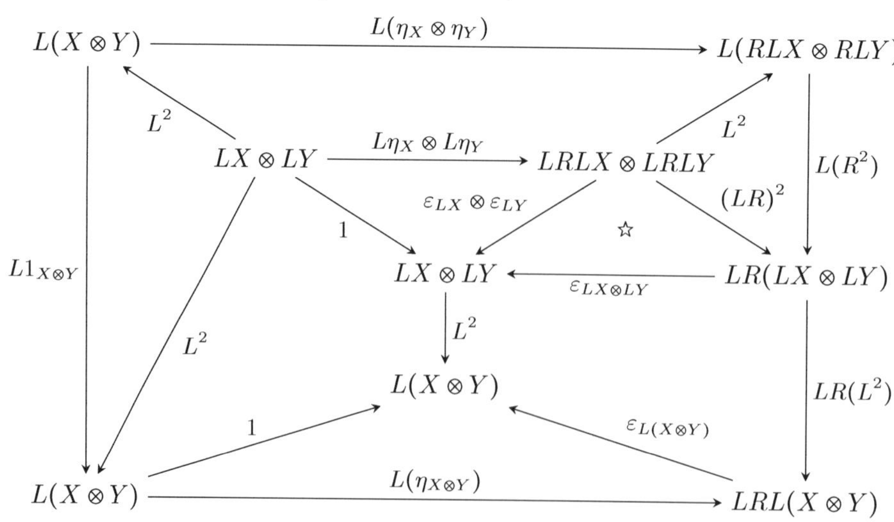

Each of the quadrilaterals in the diagram above commutes by naturality of either L^2 or ε. The triangle ☆ commutes because $LR = 1_{\mathsf{C}}$ and in particular the monoidal constraint $(LR)^2$ is the identity. The remaining two middle triangles commute by the triangle identities for the unit and counit of the adjunction. The outer triangle at left commutes by functoriality of L and the triangle at right commutes by definition of $(LR)^2$.

For the right hand diagram of (1.1.35), a similar analysis shows that the diagram commutes after applying L. This analysis uses the triangle diagrams for the adjunction and the condition $(LR)^0 = 1$ implied by $LR = 1_{\mathsf{C}}$. □

The braided coherence and strictification theorems involve the concepts of braided canonical map and underlying braid from Definitions II.1.6.1 and II.1.6.2. In the latter, B_n denotes the nth braid group (Definition II.1.1.1). See Section II.1.1 for further explanations of braids.

Definition 1.1.36. In a braided monoidal category (C, ξ), a *braided canonical map* is a natural isomorphism between permuted words of the same length that has the same definition as a canonical map by also allowing the braiding ξ and its inverse. For a symmetric monoidal category, a braided canonical map is also called a *permuted canonical map*. ◇

Definition 1.1.37. In a braided monoidal category $(\mathsf{C}, \otimes, \mathbb{1}, \alpha, \lambda, \rho, \xi)$, each braided canonical map ϕ between permuted words of the same length n has an *underlying braid* $\mathsf{br}(\phi) \in B_n$ defined as follows.

- $\mathsf{br}(1_{\mathbb{1}}) = \mathrm{id}_0 \in B_0$.
- The identity natural transformation of 1_{C} has underlying braid $\mathrm{id}_1 \in B_1$.
- The vertical composite $\phi'\phi$ of two braided canonical maps has underlying braid the product $\mathsf{br}(\phi')\mathsf{br}(\phi)$.
- For braided canonical maps ϕ_1 and ϕ_2, the underlying braid of $\phi_1 \otimes \phi_2$ is
$$\mathsf{br}(\phi_1 \otimes \phi_2) = \mathsf{br}(\phi_1) \oplus \mathsf{br}(\phi_2)$$
with $\oplus$ the sum braid (II.1.1.10).
- For permuted words u, v, and w, the associativity isomorphism
$$\alpha_{u,v,w} : (u \otimes v) \otimes w \longrightarrow u \otimes (v \otimes w)$$
has underlying braid
$$\mathsf{br}(\alpha_{u,v,w}) = \mathrm{id} \in B_{|u|+|v|+|w|}.$$
- The unit isomorphisms
$$\lambda : \mathbb{1} \otimes u \longrightarrow u \quad \text{and} \quad \rho : u \otimes \mathbb{1} \longrightarrow u$$
have underlying braids
$$\mathsf{br}(\lambda_u) = \mathsf{br}(\rho_u) = \mathrm{id} \in B_{|u|}.$$
- The braiding
$$\xi_{u,v} : u \otimes v \longrightarrow v \otimes u$$
has underlying braid
$$\mathsf{br}(\xi_{u,v}) = b^{\oplus}_{|u|,|v|} \in B_{|u|+|v|}$$
with the right-hand side the elementary block braid (II.1.2.4).

This finishes the definition of the underlying braid. ◇

Theorem 1.1.38 (Braided Coherence). *In a braided monoidal category* C*, two braided canonical maps with the same (co)domain are equal if their underlying braids are equal.*

Theorem 1.1.39 (Braided Strictification). *For each braided monoidal category* C*, there exist a braided strict monoidal category* C_{st} *and an adjoint equivalence*

$$\mathsf{C}_{\mathsf{st}} \underset{R}{\overset{L}{\rightleftarrows}} \mathsf{C}$$

with (i) both L *and* R *strong braided monoidal functors and (ii)* $LR = 1_{\mathsf{C}}$.

The symmetric coherence and strictification theorems involve the concepts of permuted word and permuted canonical map from Definition I.1.3.6.

Definition 1.1.40. The symmetric group on n letters is denoted by Σ_n. Suppose C is a monoidal category.

- For a word w of length n and a permutation $\sigma \in \Sigma_n$, the *permuted word* $w\sigma\colon \mathsf{C}^n \longrightarrow \mathsf{C}$ is the composite functor $w \circ \sigma$, where $\sigma\colon \mathsf{C}^n \longrightarrow \mathsf{C}^n$ is given by
$$\sigma(x_1, \ldots, x_n) = \big(x_{\sigma^{-1}(1)}, \ldots, x_{\sigma^{-1}(n)}\big)$$
with the x_i's all objects, or all morphisms, in C.
- For a symmetric monoidal category (C, ξ), a *permuted canonical map* is a natural isomorphism between permuted words of the same length, defined as in Definition 1.1.30 by also allowing the symmetry isomorphism ξ. ◇

Theorem 1.1.41 (Symmetric Coherence). *Suppose* $u\sigma$ *and* $v\tau$ *are two permuted words of the same length in a symmetric monoidal category* C*. Then there exists a unique permuted canonical map* $u\sigma \longrightarrow v\tau$.

Theorem 1.1.42 (Symmetric Strictification). *For each symmetric monoidal category* C*, there exist a permutative category* C_{st} *and an adjoint equivalence*

$$\mathsf{C}_{\mathsf{st}} \underset{R}{\overset{L}{\rightleftarrows}} \mathsf{C}$$

with (i) both L *and* R *strong symmetric monoidal functors, and (ii)* $LR = 1_{\mathsf{C}}$.

We also recall Epstein's Coherence Theorem I.1.3.12 for monoidal and symmetric monoidal functors. This involves the concept of F-iterate from Definition I.1.3.11.

Definition 1.1.43. Suppose $(F, F^2, F^0)\colon \mathsf{C} \longrightarrow \mathsf{D}$ is a (symmetric) monoidal functor between (symmetric) monoidal categories.

(1) The set of *F-iterates* is the set of functors $\mathsf{C}^n \longrightarrow \mathsf{D}$ for $n \geq 1$ defined inductively by the following two conditions.
  - $Fw\colon \mathsf{C}^n \longrightarrow \mathsf{D}$ is an F-iterate for each (permuted) word $w\colon \mathsf{C}^n \longrightarrow \mathsf{C}$ of length n.
  - If $G\colon \mathsf{C}^m \longrightarrow \mathsf{D}$ and $H\colon \mathsf{C}^n \longrightarrow \mathsf{D}$ are F-iterates, then so is the composite
$$\mathsf{C}^{m+n} = \mathsf{C}^m \times \mathsf{C}^n \xrightarrow{G \times H} \mathsf{D} \times \mathsf{D} \xrightarrow{\;\otimes\;} \mathsf{D}.$$

(2) The set of *F-coherent maps* is the set of natural transformations between F-iterates defined inductively as follows.
  - Suppose θ is a (permuted) canonical map in C that does not involve $\mathbb{1}^{\mathsf{C}}$, λ^{C}, or ρ^{C}. Then $1_F * \theta$ is an F-coherent map.

- The identity natural transformation, α^{D}, their inverses, and ξ^{D} in the symmetric case, applied to F-iterates are F-coherent maps.
- F^2 is an F-coherent map.
- F-coherent maps are closed under vertical composites and $\otimes^{\mathsf{D}}$. ◇

Theorem 1.1.44 (Epstein's Coherence). *Suppose $F: \mathsf{C} \longrightarrow \mathsf{D}$ is a (symmetric) monoidal functor between (symmetric) monoidal categories, and $G, H: \mathsf{C}^n \longrightarrow \mathsf{D}$ are F-iterates. Then there exists at most one F-coherent map from G to H.*

1.2. Enriched Categories, Functors, and Natural Transformations

In this section we recall some basic definitions regarding categories, functors, and natural transformations that are enriched over a monoidal category $\mathsf{V} = (\mathsf{V}, \otimes, \mathbb{1}, \alpha, \lambda, \rho)$. The material in this section does not assume that V has a braiding until the discussion of opposites in Definition 1.2.16 and following.

Definition 1.2.1. Suppose V is a monoidal category. A V-*category* C, also called a *category enriched in* V, consists of:

- a class $\mathsf{Ob}(\mathsf{C})$ of objects in C;
- for each pair of objects X, Y in C, an object $\mathsf{C}(X,Y)$ in V, called the *hom object* with domain X and codomain Y;
- for each triple of objects X, Y, Z in C, a morphism
$$\mathsf{C}(Y,Z) \otimes \mathsf{C}(X,Y) \xrightarrow{m_{X,Y,Z}} \mathsf{C}(X,Z)$$
in V, called the *composition*; and
- for each object X in C, a morphism
$$\mathbb{1} \xrightarrow{i_X} \mathsf{C}(X,X)$$
in V, called the *identity* of X.

These data are required to make the *associativity diagram*

(1.2.2)
$$\begin{array}{ccc}
\big(\mathsf{C}(Y,Z) \otimes \mathsf{C}(X,Y)\big) \otimes \mathsf{C}(W,X) & \xrightarrow{\alpha} & \mathsf{C}(Y,Z) \otimes \big(\mathsf{C}(X,Y) \otimes \mathsf{C}(W,X)\big) \\
 & & \downarrow{\scriptstyle 1\otimes m} \\
\downarrow{\scriptstyle m\otimes 1} & & \mathsf{C}(Y,Z) \otimes \mathsf{C}(W,Y) \\
 & & \downarrow{\scriptstyle m} \\
\mathsf{C}(X,Z) \otimes \mathsf{C}(W,X) & \xrightarrow{m} & \mathsf{C}(W,Z)
\end{array}$$

and the *unity diagram*

(1.2.3)
$$\begin{array}{ccccc}
\mathbb{1} \otimes \mathsf{C}(X,Y) & \xrightarrow{\lambda} & \mathsf{C}(X,Y) & \xleftarrow{\rho} & \mathsf{C}(X,Y) \otimes \mathbb{1} \\
\downarrow{\scriptstyle i_Y \otimes 1} & & \| & & \downarrow{\scriptstyle 1\otimes i_X} \\
\mathsf{C}(Y,Y) \otimes \mathsf{C}(X,Y) & \xrightarrow{m} & \mathsf{C}(X,Y) & \xleftarrow{m} & \mathsf{C}(X,Y) \otimes \mathsf{C}(X,X)
\end{array}$$

commute for objects W, X, Y, Z in C. This finishes the definition of a V-category. A V-category C is *small* if $\mathsf{Ob}(\mathsf{C})$ is a set. ◇

Next we recall functors, natural transformations, and adjunctions in the enriched setting. In the next few definitions, the reader will notice that we recover the plain categorical notions from Section I.1.1 when $\mathsf{V} = \mathsf{Set}$.

Definition 1.2.4. Suppose C and D are V-categories. A V-*functor* $F : \mathsf{C} \longrightarrow \mathsf{D}$ consists of:

- an assignment on objects

$$\mathsf{Ob}(\mathsf{C}) \longrightarrow \mathsf{Ob}(\mathsf{D}), \qquad X \longmapsto FX;$$

  and
- for each pair of objects X, Y in C, a morphism

$$\mathsf{C}(X,Y) \xrightarrow{F_{X,Y}} \mathsf{D}\big(FX, FY\big)$$

  in V.

These data are required to satisfy the following two axioms.

Composition: For each triple of objects X, Y, Z in C, the diagram

$$\begin{array}{ccc} \mathsf{C}(Y,Z) \otimes \mathsf{C}(X,Y) & \xrightarrow{\quad m \quad} & \mathsf{C}(X,Z) \\ {\scriptstyle F\otimes F}\downarrow & & \downarrow{\scriptstyle F} \\ \mathsf{D}(FY,FZ) \otimes \mathsf{D}(FX,FY) & \xrightarrow{\ m\ } & \mathsf{D}(FX,FZ) \end{array} \tag{1.2.5}$$

in V is commutative.

Identities: For each object $X \in \mathsf{C}$, the diagram

$$\begin{array}{ccc} \mathbb{1} & \xrightarrow{\ i_X\ } & \mathsf{C}(X,X) \\ \| & & \downarrow{\scriptstyle F} \\ \mathbb{1} & \xrightarrow{\ i_{FX}\ } & \mathsf{D}(FX,FX) \end{array} \tag{1.2.6}$$

in V is commutative.

Moreover:

- For V-functors $F : \mathsf{C} \longrightarrow \mathsf{D}$ and $G : \mathsf{D} \longrightarrow \mathsf{E}$, their composition

$$GF : \mathsf{C} \longrightarrow \mathsf{E}$$

  is the V-functor defined by composing the assignments on objects and forming the composite

$$(GF)_{X,Y} = G_{FX,FY}F_{X,Y} : \mathsf{C}(X,Y) \longrightarrow \mathsf{E}(GFX, GFY)$$

  in V on hom objects.
- The *identity* V-*functor* of C, denoted $1_{\mathsf{C}} : \mathsf{C} \longrightarrow \mathsf{C}$, is given by the identity morphism on $\mathsf{Ob}(\mathsf{C})$ and the identity morphism $1_{\mathsf{C}(X,Y)}$ for objects X, Y in C. ◇

Definition 1.2.7. Suppose $F, G : \mathsf{C} \longrightarrow \mathsf{D}$ are V-functors between V-categories C and D.

(1) A V-*natural transformation* $\theta : F \longrightarrow G$ consists of a morphism

$$\theta_X : \mathbb{1} \longrightarrow \mathsf{D}(FX, GX)$$

in V, called a *component* of θ, for each object X in C, such that the following diagram commutes for objects X, Y in C.

(1.2.8)
$$\begin{tikzcd} & \mathbb{1} \otimes \mathsf{C}(X,Y) \arrow[r, "\theta_Y \otimes F"] & \mathsf{D}(FY,GY) \otimes \mathsf{D}(FX,FY) \arrow[dr, "m"] & \\ \mathsf{C}(X,Y) \arrow[ur, "\lambda^{-1}", "\cong"'] \arrow[dr, "\rho^{-1}"', "\cong"] & & & \mathsf{D}(FX,GY) \\ & \mathsf{C}(X,Y) \otimes \mathbb{1} \arrow[r, "G \otimes \theta_X"] & \mathsf{D}(GX,GY) \otimes \mathsf{D}(FX,GX) \arrow[ur, "m"'] & \end{tikzcd}$$

(2) The *identity* V*-natural transformation of* F, denoted by $1_F : F \longrightarrow F$, is defined by the component

$$(1_F)_X = i_{FX} : \mathbb{1} \longrightarrow \mathsf{D}(FX, FX)$$

for each object X in C. ◇

We give an alternative characterization of V-naturality in Lemma 2.1.11 below. Just as for ordinary natural transformations, there are two types of compositions for V-natural transformations.

Definition 1.2.9. Suppose $\theta : F \longrightarrow G$ is a V-natural transformation for V-functors $F, G : \mathsf{C} \longrightarrow \mathsf{D}$.

(1) Suppose $\phi : G \longrightarrow H$ is another V-natural transformation for a V-functor $H : \mathsf{C} \longrightarrow \mathsf{D}$. The *vertical composition*

$$\phi\theta : F \longrightarrow H$$

is the V-natural transformation whose component $(\phi\theta)_X$ is the composite

$$\begin{tikzcd} \mathbb{1} \arrow[r, "(\phi\theta)_X"] \arrow[d, "\lambda^{-1}"', "\cong"] & \mathsf{D}(FX, HX) \\ \mathbb{1} \otimes \mathbb{1} \arrow[r, "\phi_X \otimes \theta_X"] & \mathsf{D}(GX, HX) \otimes \mathsf{D}(FX, GX) \arrow[u, "m"'] \end{tikzcd}$$

in V for each object X in C.

(2) Suppose $\theta' : F' \longrightarrow G'$ is a V-natural transformation for V-functors $F', G' : \mathsf{D} \longrightarrow \mathsf{E}$ with E a V-category. The *horizontal composition*

$$\theta' * \theta : F'F \longrightarrow G'G$$

is the V-natural transformation whose component $(\theta' * \theta)_X$, for an object X in C, is defined as the composite

(1.2.10)
$$\begin{tikzcd} \mathbb{1} \arrow[r, "(\theta' * \theta)_X"] \arrow[dd, "\lambda^{-1}"', "\cong"] & \mathsf{E}(F'FX, G'GX) \\ & \mathsf{E}(F'GX, G'GX) \otimes \mathsf{E}(F'FX, F'GX) \arrow[u, "m"'] \\ \mathbb{1} \otimes \mathbb{1} \arrow[r, "\theta'_{GX} \otimes \theta_X"] & \mathsf{E}(F'GX, G'GX) \otimes \mathsf{D}(FX, GX) \arrow[u, "1 \otimes F'"'] \end{tikzcd}$$

in V. ◇

Definition 1.2.11. A V-natural transformation $\theta : F \longrightarrow G$ is called a V-*natural isomorphism* if there exists a V-natural transformation $\theta^{-1} : G \longrightarrow F$ such that the equalities

$$\theta^{-1}\theta = 1_F \quad \text{and} \quad \theta\theta^{-1} = 1_G$$

hold. ◇

The following special cases of horizontal composition will be useful in several computations below.

Lemma 1.2.12 (Whiskering). *Suppose given* V*-categories* B, C, D, *and* E; V*-functors* F, G, H, *and* K; *and a* V*-natural transformation* θ *as in the following diagram:*

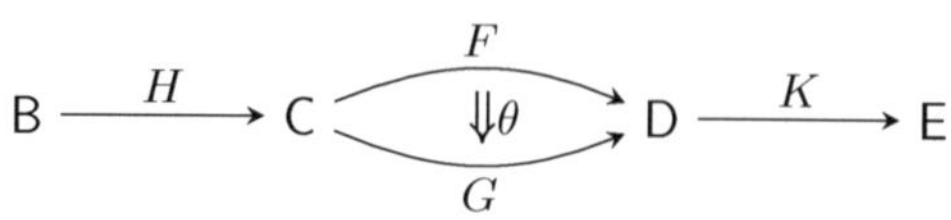

Then for each $Y \in \mathsf{B}$ *and* $X \in \mathsf{C}$ *we have*

$$(\theta * 1_H)_Y = \theta_{HY} \quad \textit{and} \quad (1_K * \theta)_X = (K_{FX,GX}) \circ \theta_X.$$

Proof. For the whiskering of θ with H, consider the following diagram in V, where the top arrow is θ_{HY} and the composite along the remaining boundary is the definition (1.2.10) for $(\theta * 1_H)_Y$.

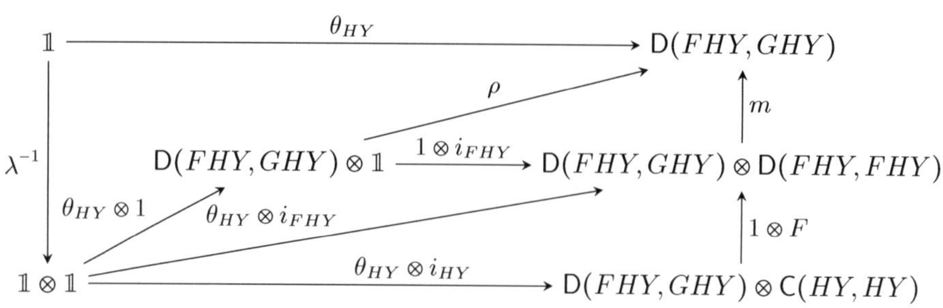

The upper left region commutes by naturality of ρ and the equality $\lambda_{\mathbb{1}} = \rho_{\mathbb{1}}$ (1.1.4). The upper right triangle commutes by the right unity diagram (1.2.3). The lower right triangle commutes by functoriality of $\otimes$ and the identity axiom $F \circ i_{HY} = i_{FHY}$ (1.2.6). The remaining triangle commutes by functoriality of $\otimes$.

For the whiskering of θ with K, consider the following diagram in V, where the top arrow is the composite $(K_{FX,GX}) \circ \theta_X$ and the composite around the remaining boundary is (1.2.10) for $(1_K * \theta)_X$.

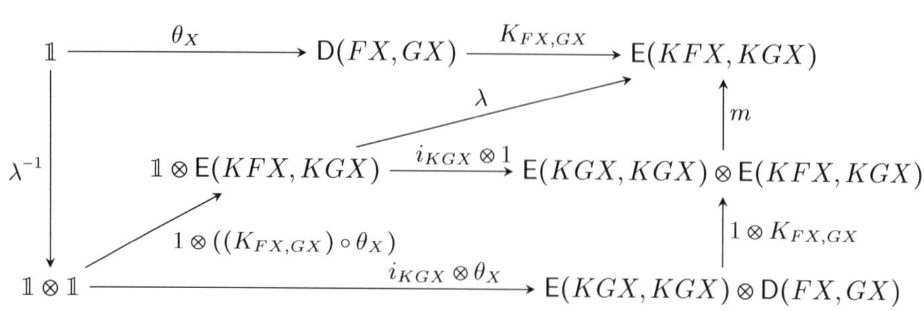

In the above diagram, the upper left region commutes by naturality of λ, the lower region commutes by functoriality of $\otimes$, and the triangular region commutes by the left unity axiom (1.2.3) for m. $\square$

Definition 1.2.13. Suppose V is a monoidal category. The 2-category $\mathsf{V}\text{-}\mathsf{Cat}$ is defined by the following data.

- The objects are small V-categories, 1-cells are V-functors, and 2-cells are V-natural transformations.
- Horizontal composition of 1-cells is given by composition of V-functors.
- Vertical and horizontal composition of 2-cells are given by those of V-natural transformations.
- Identity V-functors are identity 1-cells.
- Identity V-natural transformations are identity 2-cells.

The axioms of a 2-category follow from the defining axioms above. For example, strict associativity of the vertical composition of 2-cells follows from instances of the enriched composition axiom (1.2.5) and the associativity of composition axiom (1.2.2). $\diamond$

Recall from Definition I.6.3.9 the notions of internal adjunction and equivalence in a bicategory. The corresponding notions for $\mathsf{V}\text{-}\mathsf{Cat}$ determine the definitions of V-adjunction and V-equivalence. We give detailed definitions for reference.

Definition 1.2.14. Suppose C and D are V-categories, and $L : \mathsf{C} \longrightarrow \mathsf{D}$ and $R : \mathsf{D} \longrightarrow \mathsf{C}$ are V-functors. A *V-adjunction* $L \dashv R$ consists of

- a V-natural transformation $\eta : 1_{\mathsf{C}} \longrightarrow RL$ called the *unit*, and
- a V-natural transformation $\varepsilon : LR \longrightarrow 1_{\mathsf{D}}$ called the *counit*,

such that the following diagrams commute.

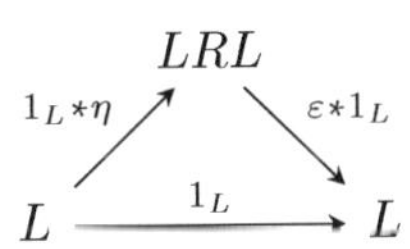

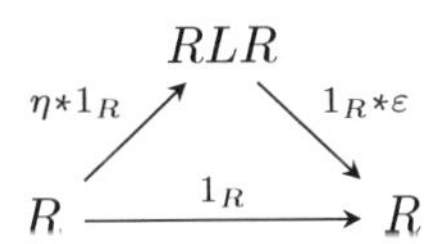

In this case, L is called the *left adjoint*, and R is called the *right adjoint*. $\diamond$

Definition 1.2.15. Suppose $F : \mathsf{C} \longrightarrow \mathsf{D}$ is a V-functor. We call F a *V-equivalence* if there exist

- a V-functor $F' : \mathsf{D} \longrightarrow \mathsf{C}$ and
- V-natural isomorphisms $\eta : 1_{\mathsf{C}} \xrightarrow{\cong} F'F$ and $\varepsilon : FF' \xrightarrow{\cong} 1_{\mathsf{D}}$.

An *adjoint V-equivalence* consists of a V-adjunction $L \dashv R$ such that the unit and counit are V-natural isomorphisms. $\diamond$

Next we define the opposite of a V-category. Unlike the other content in this section, this definition requires that V be braided monoidal

Definition 1.2.16. Suppose $\mathsf{V} = (\mathsf{V}, \otimes, \xi)$ is a braided monoidal category and suppose that C is a V-category. The *opposite V-category*, C^{op}, is defined to have the same objects as C and hom objects

$$\mathsf{C}^{\mathsf{op}}(X,Y) = \mathsf{C}(Y,X).$$

The composition in C^{op} is defined for each triple of objects X, Y, Z in C^{op} as the following composite using the braiding ξ of V and the composition m of C:

$$\begin{array}{ccc} \mathsf{C}^{\mathsf{op}}(Y,Z) \otimes \mathsf{C}^{\mathsf{op}}(X,Y) & & & & \mathsf{C}^{\mathsf{op}}(X,Z) \\ \| & & & & \| \\ \mathsf{C}(Z,Y) \otimes \mathsf{C}(Y,X) & \xrightarrow{\xi} & \mathsf{C}(Y,X) \otimes \mathsf{C}(Z,Y) & \xrightarrow{m} & \mathsf{C}(Z,X) \end{array}$$

The unit of C^{op} is the same as that of C:

$$\mathbb{1} \xrightarrow{i_X} \mathsf{C}(X,X) = \mathsf{C}^{\mathsf{op}}(X,X).$$

We show that these data satisfy the axioms of a V-category in Lemma 1.2.17 below.

◇

Lemma 1.2.17. *In the context of Definition 1.2.16,* C^{op} *is a* V*-category.*

Proof. We show that the axioms for C^{op} follow from the corresponding axioms for C together with braiding properties of ξ. First, the associativity diagram (1.2.2) for C^{op} is the outer diagram below for $W, X, Y, Z \in \mathsf{C}$. In this diagram we use the notation

$$\mathsf{C}^{\mathsf{op}}(W,X) = \mathsf{C}(X,W) = \mathsf{C}_{X,W}$$

and likewise for the other hom objects. We also omit tensor symbols to save space.

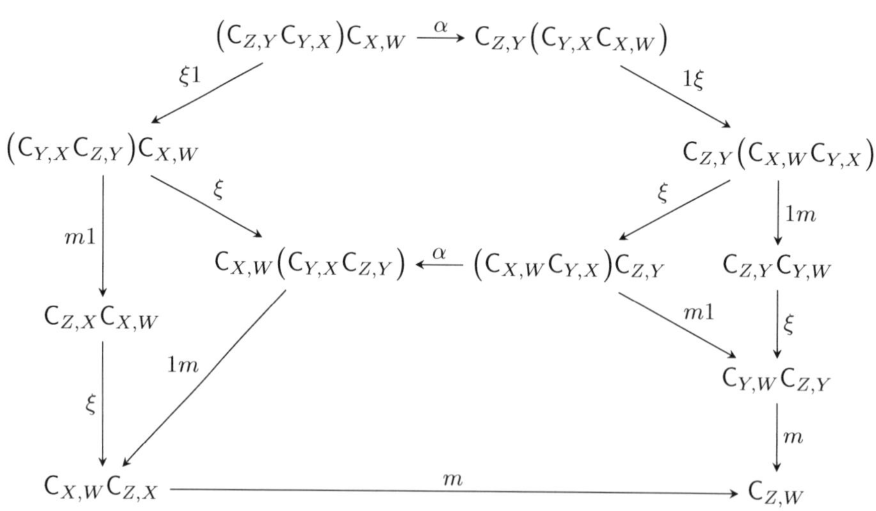

In the above diagram, the upper hexagon commutes by compatibility of ξ with α (1.1.22), the lower pentagon commutes by the associativity diagram (1.2.2) for C, and the other two regions commute by naturality of ξ.

Second, using the same notation, the left and right unity diagrams (1.2.3) for C^{op} are the outer diagrams below.

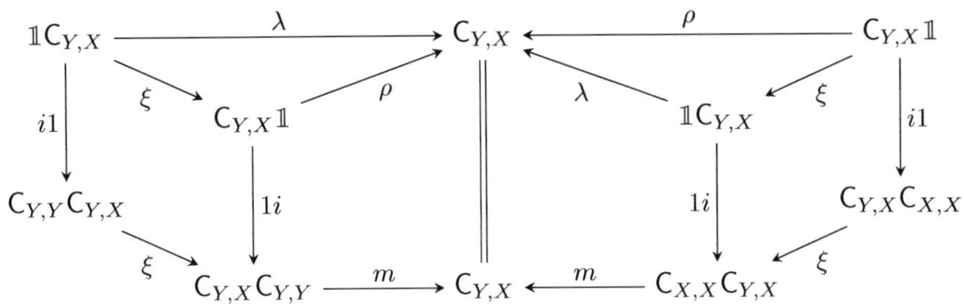

In the above diagram, the top two triangles commute by compatibility of ξ with λ and ρ (1.1.20), the middle two trapezoids commute by the unity diagrams (1.2.3) for C, and the remaining two parallelograms commute by naturality of ξ. $\square$

We will define $(-)^{\mathsf{op}}$ on V-functors and V-natural transformations in Proposition 1.2.19 below. The construction is covariant with respect to horizontal composition, but contravariant with respect to vertical composition of V-natural transformations. We will use the following notation.

Definition 1.2.18. Let $\mathsf{V}\text{-}\mathsf{Cat}^{\mathsf{co}}$ denote the 2-category with the same cells as $\mathsf{V}\text{-}\mathsf{Cat}$, but the direction of 2-cells reversed. So the category of 1- and 2-cells from C to D is given by the opposite category:

$$\mathsf{V}\text{-}\mathsf{Cat}^{\mathsf{co}}(\mathsf{C},\mathsf{D}) = \mathsf{V}\text{-}\mathsf{Cat}(\mathsf{C},\mathsf{D})^{\mathsf{op}}.$$

Identities and horizontal composition in $\mathsf{V}\text{-}\mathsf{Cat}^{\mathsf{co}}$ are given by the opposite functors of those in $\mathsf{V}\text{-}\mathsf{Cat}$. $\diamond$

Proposition 1.2.19. *Suppose* $\mathsf{V} = (\mathsf{V},\otimes,\xi)$ *is a braided monoidal category. Taking opposite* V*-categories defines a 2-functor*

$$(-)^{\mathsf{op}}\colon \mathsf{V}\text{-}\mathsf{Cat} \longrightarrow \mathsf{V}\text{-}\mathsf{Cat}^{\mathsf{co}}.$$

Proof. Continuing from Lemma 1.2.17, we need to define $(-)^{\mathsf{op}}$ for V-functors and V-natural transformations. Given a V-functor F, we define F^{op} to have the same assignment on objects as F and define $F^{\mathsf{op}}_{X,Y} = F_{Y,X}$ as below.

$$\begin{array}{ccc}
\mathsf{C}^{\mathsf{op}}(X,Y) & \xrightarrow{F^{\mathsf{op}}_{X,Y}} & \mathsf{D}^{\mathsf{op}}(FX,FY) \\
\| & & \| \\
\mathsf{C}(Y,X) & \xrightarrow[F_{Y,X}]{} & \mathsf{D}(FY,FX)
\end{array}$$

The composition and identity axioms for F^{op} follow from those for F together with naturality of ξ. For example, the diagram below is the composition axiom for F^{op} at $X,Y,Z \in \mathsf{C}$, with tensor symbols omitted for space.

$$
\begin{array}{ccc}
\mathsf{C}^{\mathsf{op}}(Y,Z)\mathsf{C}^{\mathsf{op}}(X,Y) & \xrightarrow{F^{\mathsf{op}}F^{\mathsf{op}}} & \mathsf{D}^{\mathsf{op}}(FY,FZ)\mathsf{D}^{\mathsf{op}}(FX,FY) \\
\| & & \| \\
\mathsf{C}(Z,Y)\mathsf{C}(Y,X) & \xrightarrow[FF]{} & \mathsf{D}(FZ,FY)\mathsf{D}(FY,FX) \\
\Big\downarrow \xi & & \Big\downarrow \xi \\
\mathsf{C}(Y,X)\mathsf{C}(Z,Y) & \xrightarrow{FF} & \mathsf{D}(FY,FX)\mathsf{D}(FZ,FY) \\
\Big\downarrow m & & \Big\downarrow m \\
\mathsf{C}(Z,X) & \xrightarrow{F} & \mathsf{D}(FZ,FX) \\
\| & & \| \\
\mathsf{C}^{\mathsf{op}}(X,Z) & \xrightarrow[F^{\mathsf{op}}]{} & \mathsf{D}^{\mathsf{op}}(FX,FZ)
\end{array}
$$

Since composition of V-functors is given by composition of assignments on objects and composition in V on hom objects, we have $(GF)^{\mathsf{op}} = G^{\mathsf{op}}F^{\mathsf{op}}$ for V-functors

$$\mathsf{C} \xrightarrow{F} \mathsf{D} \xrightarrow{G} \mathsf{E}.$$

Since the identity of C^{op} has the same components as that of C, we have $(1_{\mathsf{C}})^{\mathsf{op}} = 1_{(\mathsf{C}^{\mathsf{op}})}$.

For V-functors $F, G\colon \mathsf{C} \longrightarrow \mathsf{D}$ and a V-natural transformation $\theta\colon F \longrightarrow G$, the component of $\theta^{\mathsf{op}}\colon G^{\mathsf{op}} \longrightarrow F^{\mathsf{op}}$ at $X \in \mathsf{C}$ is given by θ_X as below.

$$
\mathbb{1} \begin{array}{l} \xrightarrow{\theta^{\mathsf{op}}_X} \mathsf{D}^{\mathsf{op}}(GX,FX) \\ \qquad\qquad\quad \| \\ \xrightarrow[\theta_X]{} \mathsf{D}(FX,GX) \end{array}
$$

The V-naturality axiom for θ^{op} follows from that of θ together with naturality of ξ. Preservation of horizontal composition

$$(\theta' * \theta)^{\mathsf{op}} = (\theta')^{\mathsf{op}} * \theta^{\mathsf{op}},$$

for $F', G'\colon \mathsf{D} \longrightarrow \mathsf{E}$ and $\theta'\colon F' \longrightarrow G'$, follows from the commutative diagram below for each $X \in \mathsf{C}$. The composite along the right, when composed with λ^{-1}, is $(\theta' * \theta)^{\mathsf{op}}_X = (\theta' * \theta)_X$. The composite along the left, when composed with λ^{-1}, is $((\theta')^{\mathsf{op}} * \theta^{\mathsf{op}})_X$. We omit tensor signs to save space and explain commutativity below.

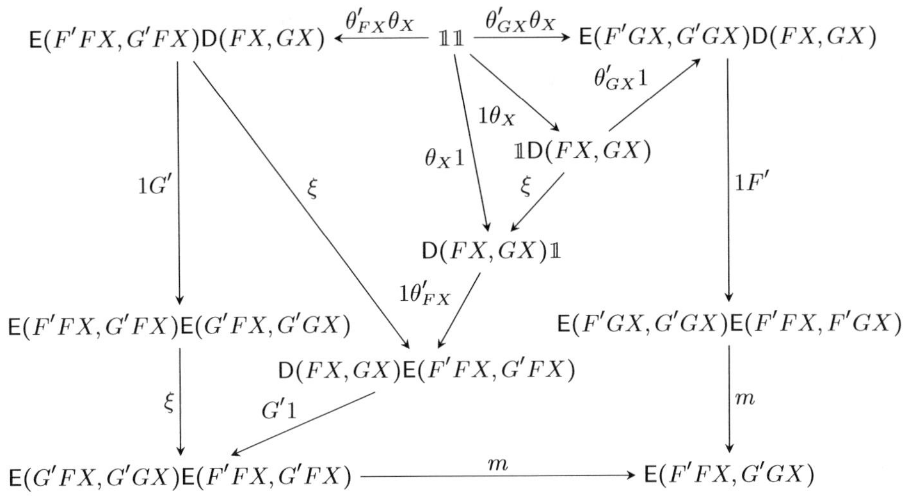

In the above diagram, the top right triangle commutes by functoriality of $\otimes$. The remaining upper/left regions commute by functoriality of $\otimes$ and naturality of ξ, recalling that $\xi_{\mathbb{1},\mathbb{1}} = 1_{\mathbb{1}\mathbb{1}}$ (1.1.21). The lower right region commutes by V-naturality of θ' (1.2.8) at the objects FX and GX, together with the compatibility of ξ with λ and ρ (1.1.20).

Preservation of vertical composition (contravariantly)

$$(\phi\theta)^{\mathsf{op}}_X = (\theta^{\mathsf{op}}\phi^{\mathsf{op}})_X$$

follows simply from naturality of ξ and the definition

$$(\phi\theta)_X = m \circ (\phi_X \otimes \theta_X) \circ \lambda^{-1},$$

again recalling $\xi_{\mathbb{1},\mathbb{1}} = 1_{\mathbb{1}\mathbb{1}}$. □

1.3. The Tensor Product of Enriched Categories

Now we turn to the definition of tensor product for enriched categories. The content of this section, and indeed the rest of this chapter, requires a braiding on the enriching category V. Throughout this section we assume that $\mathsf{V} = (\mathsf{V}, \otimes, \xi)$ is braided monoidal. We use the braiding to define a monoidal product $\otimes$ on V-Cat.

If, moreover, V is symmetric monoidal, then we can define a braiding for $\otimes$ and show that $(\mathsf{V}\text{-}\mathsf{Cat}, \otimes)$ is a symmetric monoidal category. The following table summarizes the relationships.

(1.3.1)

V	V-Cat
monoidal	exists
braided	monoidal
symmetric	symmetric

See Note 1.6.2 for an expansion of this table via the iterated monoidal structures of Chapter II.10.

It will be convenient for our discussion below to introduce the following notation.

Definition 1.3.2. Let ξ_{mid} denote the natural isomorphism whose component at a quadruple of objects X, Y, Z, and W in V is given by the following composite of ξ with associativity isomorphisms.

$$\begin{array}{ccc}
(X \otimes Y) \otimes (Z \otimes W) & \xrightarrow{\xi_{\text{mid}}} & (X \otimes Z) \otimes (Y \otimes W) \\
{\scriptstyle \alpha^{-1}_{X\otimes Y, Z, W}} \downarrow & & \uparrow {\scriptstyle \alpha_{X\otimes Z, Y, W}} \\
((X \otimes Y) \otimes Z) \otimes W & & ((X \otimes Z) \otimes Y) \otimes W \\
{\scriptstyle \alpha_{X,Y,Z} \otimes 1_W} \downarrow & & \uparrow {\scriptstyle \alpha^{-1}_{X,Z,Y} \otimes 1_W} \\
(X \otimes (Y \otimes Z)) \otimes W & \xrightarrow{(1 \otimes \xi) \otimes 1} & (X \otimes (Z \otimes Y)) \otimes W
\end{array}$$

The isomorphisms ξ_{mid} are called *middle four interchange.* By the Braided Coherence Theorem 1.1.38, ξ_{mid} is the unique braided canonical map in V whose underlying braid is s_2 in B_4. ◇

Definition 1.3.3. Suppose that C and D are V-categories with V braided monoidal. The *tensor product* $\mathsf{C} \otimes \mathsf{D}$ consists of the following.

- Objects are given by pairs: $\mathsf{Ob}(\mathsf{C} \otimes \mathsf{D}) = \mathsf{Ob}\,\mathsf{C} \times \mathsf{Ob}\,\mathsf{D}$. Objects are denoted $X \otimes Y$ for $X \in \mathsf{C}$ and $Y \in \mathsf{D}$.
- For objects $X \otimes Y$ and $X' \otimes Y'$, the hom object with domain $X \otimes Y$ and codomain $X' \otimes Y'$ is given by the monoidal product of hom objects
$$(\mathsf{C} \otimes \mathsf{D})(X \otimes Y, X' \otimes Y') = \mathsf{C}(X, X') \otimes \mathsf{D}(Y, Y').$$
- The composition for $X \otimes Y$, $X' \otimes Y'$, and $X'' \otimes Y''$ is given by the following composite in V, where ξ_{mid} is the middle four interchange of Definition 1.3.2:

$$\begin{array}{ccc}
\big(\mathsf{C}(X', X'') \otimes \mathsf{D}(Y', Y'')\big) \otimes \big(\mathsf{C}(X, X') \otimes \mathsf{D}(Y, Y')\big) & & \mathsf{C}(X, X'') \otimes \mathsf{D}(Y, Y'') \\
{\scriptstyle \xi_{\text{mid}}} \searrow & & \nearrow {\scriptstyle m \otimes m} \\
& \big(\mathsf{C}(X', X'') \otimes \mathsf{C}(X, X')\big) \otimes \big(\mathsf{D}(Y', Y'') \otimes \mathsf{D}(Y, Y')\big) &
\end{array}$$

- The identity of $X \otimes Y$ is given by the following composite
$$\mathbb{1} \xrightarrow{\lambda^{-1}} \mathbb{1} \otimes \mathbb{1} \xrightarrow{i_X \otimes i_Y} \mathsf{C}(X, X) \otimes \mathsf{D}(Y, Y).$$
◇

This defines the data of $\mathsf{C} \otimes \mathsf{D}$. The associativity and unity axioms (1.2.2) and (1.2.3), respectively, follow from those for C and D together with coherence for the braiding of V.

Lemma 1.3.4. *In the context of Definition 1.3.3,* $\mathsf{C} \otimes \mathsf{D}$ *is a* V*-category.*

Proof. To verify the associativity diagram (1.2.2), suppose given
$$X_0, X_1, X_2, X_3 \in \mathsf{C} \quad \text{and} \quad Y_0, Y_1, Y_2, Y_3 \in \mathsf{D}.$$
For $i, j \in \{0, 1, 2, 3\}$ we let
$$\mathsf{C}(X_i, X_j) = X_{ij} \quad \text{and} \quad \mathsf{D}(Y_i, Y_j) = Y_{ij}.$$

Then the associativity diagram for $\mathsf{C} \otimes D$ is the following, in which we omit tensor signs to save space. The arrows labeled s_4, s_2, and s_3 are the unique braided canonical morphisms with the given underlying braids.

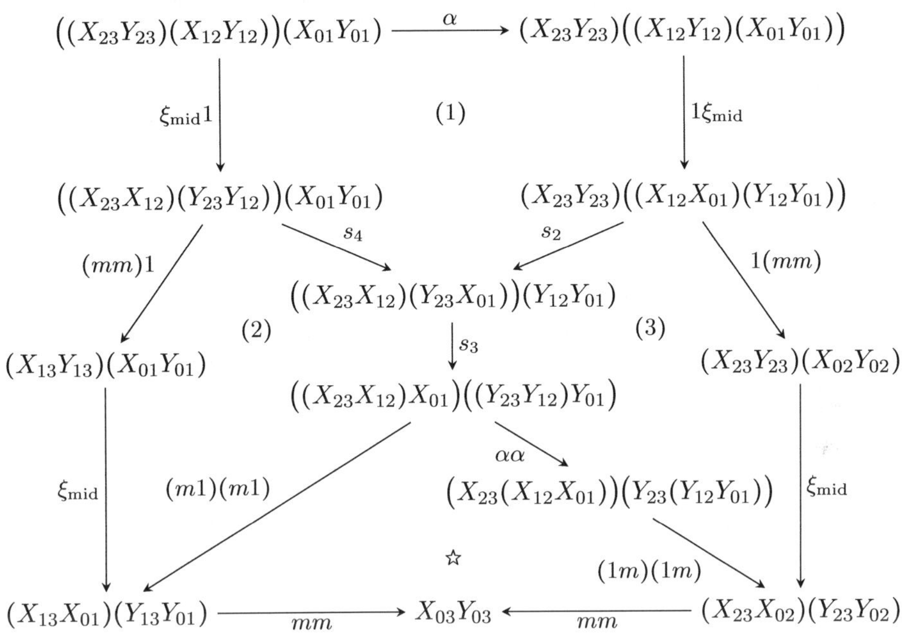

The three numbered regions commute by the Braided Coherence Theorem 1.1.38. The composites around region (1) have the following underlying braid.

$s_4 s_2 = s_2 s_4 \in B_6$

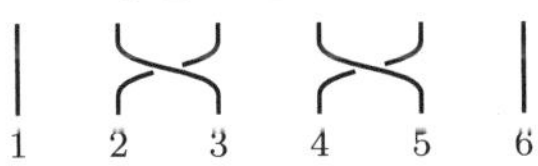

(1)

Composites around regions (2) and (3) have the following underlying braids and are equal by Theorem 1.1.38 together with naturality of the braided canonical maps.

$s_3 s_4 \in B_6$

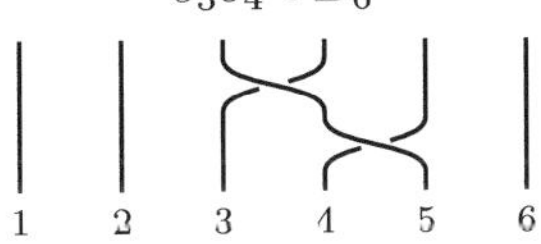

(2)

$s_3 s_2 \in B_6$

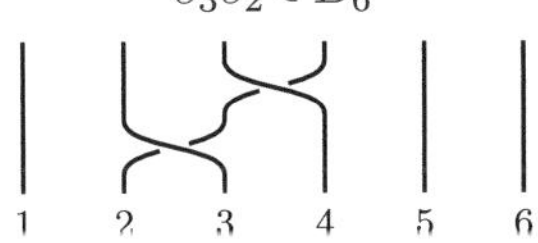

(3)

The remaining region ☆ commutes by associativity in C and D.

For the left unity diagram (1.2.3) we have the following, using the notation as above.

$$\begin{array}{ccccc}
\mathbb{1}(X_{01}Y_{01}) & & \xrightarrow{\quad\lambda\quad} & & X_{01}Y_{01} \\
\Big\downarrow{\scriptstyle \lambda^{-1}(11)} & & & \nearrow{\scriptstyle \lambda\lambda} & \Big\| \\
(\mathbb{1}\mathbb{1})(X_{01}Y_{01}) & \xrightarrow{\xi_{\mathrm{mid}}} & (\mathbb{1}X_{01})(\mathbb{1}Y_{01}) & & \Big\| \\
\Big\downarrow{\scriptstyle (ii)(11)} & & \Big\downarrow{\scriptstyle (i1)(i1)} & \star & \Big\| \\
(X_{11}Y_{11})(X_{01}Y_{01}) & \xrightarrow{\xi_{\mathrm{mid}}} & (X_{11}X_{01})(Y_{11}Y_{01}) & \xrightarrow{mm} & X_{01}Y_{01}
\end{array}$$

In the above diagram, the two unlabeled regions commute by naturality and unity properties of the braiding ξ. The underlying braids of composites around their boundaries are identities. The remaining region labeled ☆ commutes by the left unity in C and D. The right unity diagram for $\mathsf{C} \otimes \mathsf{D}$ is similar. □

Definition 1.3.5. For a braided monoidal category V, the tensor product of V-categories extends to V-functors and V-natural transformations as follows. Suppose that $F, F' : \mathsf{C} \longrightarrow \mathsf{C}'$ and $G, G' : \mathsf{D} \longrightarrow \mathsf{D}'$ are V-functors. Suppose, moreover, that $\theta: F \longrightarrow F'$ and $\omega: G \longrightarrow G'$ are V-natural transformations.

(1) The V-functor $F \otimes G$ is defined on objects by the Cartesian product, so

$$(F \otimes G)(X \otimes Y) = FX \otimes GY.$$

It is defined on hom objects by the monoidal product in V, so for $X, X' \in \mathsf{C}$ and $Y, Y' \in \mathsf{D}$ we have

$$(F \otimes G)_{X \otimes Y, X' \otimes Y'} = (F_{X,X'}) \otimes (G_{Y,Y'}).$$

The composition and identity axioms of Definition 1.2.4 follow from those of F and G separately together with the naturality of the braiding ξ in V.

(2) The V-natural transformation $\theta \otimes \omega$ has components for each $X \otimes Y \in \mathsf{C} \otimes \mathsf{D}$ given by composing the inverse of a unit morphism with θ_X and ω_Y to form the following composite:

$$\mathbb{1} \xrightarrow{\lambda^{-1}} \mathbb{1} \otimes \mathbb{1} \xrightarrow{\theta_X \otimes \omega_Y} \mathsf{C}'(FX, F'X) \otimes \mathsf{D}'(GY, G'Y).$$

The enriched naturality axiom (1.2.8) for $\theta \otimes \omega$ follow from those of θ and ω separately, naturality of the brading ξ, and the unit axiom (1.1.25) for ξ with unit morphisms. ◇

Proposition 1.3.6. *The tensor product is a 2-functor*

$$\otimes : \mathsf{V}\text{-}\mathsf{Cat} \times \mathsf{V}\text{-}\mathsf{Cat} \longrightarrow \mathsf{V}\text{-}\mathsf{Cat}.$$

Proof. We verify functoriality with respect to horizontal composition of V-natural transformations. Functoriality with respect to horizontal composition of 1-cells and vertical composition of 2-cells are similar. Preservation of unit 1- and 2-cells is verified directly from Definition 1.3.5.

Now suppose given V-functors and V-natural transformations as in the diagram below.

$$\begin{tikzcd}[column sep=huge] \mathsf{C} \otimes \mathsf{D} \arrow[r, bend left, "F \otimes G"] \arrow[r, bend right, "H \otimes K"'] & \mathsf{C}' \otimes \mathsf{D}' \arrow[r, bend left, "F' \otimes G'"] \arrow[r, bend right, "H' \otimes K'"'] & \mathsf{C}'' \otimes \mathsf{D}'' \end{tikzcd} \qquad \Downarrow \theta \otimes \omega \qquad \Downarrow \theta' \otimes \omega'$$

We will show that the components of $(\theta' \otimes \omega) * (\theta \otimes \omega)$ and $(\theta' * \theta) \otimes (\omega' * \omega)$ at an object $X \otimes Y \in \mathsf{C} \otimes \mathsf{D}$ are equal. To simplify the diagrams, we will use the following notation:

$$\begin{aligned} \mathsf{C}'(FX, HX) &= X_{00} & \mathsf{D}'(GY, KY) &= Y_{00} \\ \mathsf{C}''(F'FX, F'HX) &= X_{01} & \mathsf{D}''(G'GY, G'KY) &= Y_{01} \\ \mathsf{C}''(F'HX, H'HX) &= X_{12} & \mathsf{D}''(G'KY, K'KY) &= Y_{12} \\ \mathsf{C}''(F'FX, H'HX) &= X_{02} & \mathsf{D}''(G'GY, K'KY) &= Y_{02}. \end{aligned}$$

For example, in this notation the component of $\theta' * \theta$ at X is the composite

$$\mathbb{1} \xrightarrow{\lambda^{-1}} \mathbb{1} \otimes \mathbb{1} \xrightarrow{\theta'_{HX} \otimes \theta_X} X_{12} \otimes X_{00} \xrightarrow{1 \otimes F'} X_{12} \otimes X_{01} \xrightarrow{m} X_{02}.$$

In the diagram below, the composite along the left edge is $(\theta' \otimes \omega') * (\theta \otimes \omega)$ while the composite along the right edge is $(\theta' * \theta) \otimes (\omega' * \omega)$.

$$\begin{tikzcd}[column sep=small] \mathbb{1} \arrow[r, "\lambda^{-1}"] & \mathbb{1}^{\otimes 2} \arrow[r, "\lambda^{-1} \otimes \lambda^{-1}"] & (\mathbb{1}^{\otimes 2}) \otimes (\mathbb{1}^{\otimes 2}) \arrow[dl, "(\theta'_{HX} \otimes \omega'_{KY}) \otimes (\theta_X \otimes \omega_Y)"'] \arrow[dr, "(\theta'_{HX} \otimes \theta_X) \otimes (\omega'_{KY} \otimes \omega_Y)"] & & \\ & (X_{12} \otimes Y_{12}) \otimes (X_{00} \otimes Y_{00}) \arrow[rr, "\xi_{\mathrm{mid}}"] \arrow[d, "(1 \otimes 1) \otimes (F' \otimes G')"'] & & (X_{12} \otimes X_{00}) \otimes (Y_{12} \otimes Y_{00}) \arrow[d, "(1 \otimes F') \otimes (1 \otimes G')"] & \\ & (X_{12} \otimes Y_{12}) \otimes (X_{01} \otimes Y_{01}) \arrow[rr, "\xi_{\mathrm{mid}}"] & & (X_{12} \otimes X_{01}) \otimes (Y_{12} \otimes Y_{01}) \arrow[r, "m \otimes m"] & X_{02} \otimes Y_{02} \end{tikzcd}$$

To see that each subdiagram above commutes, recall that $\xi_{\mathbb{1},\mathbb{1}}$ is the identity on $\mathbb{1} \otimes \mathbb{1}$ (1.1.21). Then both subdiagrams above commute by naturality of ξ. □

We now turn to the unitors, associator, and braiding for $(\mathsf{V}\text{-}\mathsf{Cat}, \otimes)$. Here we simply define the data and show what basic properties they satisfy. We will use these to define monoidal V-categories for general braided monoidal V and then observe, in Theorem 1.5.5, that V Cat is a braided monoidal Cat-category and is symmetric if V is symmetric.

Definition 1.3.7. The *unit* V*-category*, $\mathbb{I}$, is the one-object V-category whose unique hom object is the unit, $\mathbb{1}$, of V. The composition and identity structure morphisms are given, respectively, by the unit and identity morphisms of $\mathbb{1}$. ◇

Definition 1.3.8. The *left unitor*, $\ell^{\otimes}$, and *right unitor*, $r^{\otimes}$, are 2-natural isomorphisms

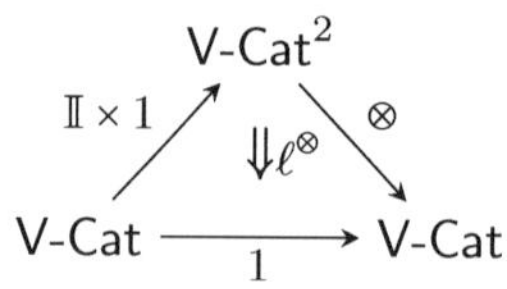

and

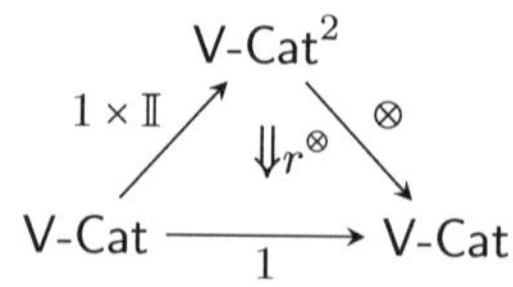

in 2Cat(V-Cat, V-Cat) defined as follows. The unitor components at a V-category C,

$$\mathbb{I}\otimes \mathsf{C} \xrightarrow{\ell^{\otimes}_{\mathsf{C}}} \mathsf{C} \xleftarrow{r^{\otimes}_{\mathsf{C}}} \mathsf{C}\otimes\mathbb{I},$$

are given on objects by the unitors for the Cartesian product and are given on hom objects by the unit isomorphisms

$$\mathbb{1}\otimes \mathsf{C}(X,X') \xrightarrow{\lambda} \mathsf{C}(X,X') \xleftarrow{\rho} \mathsf{C}(X,X')\otimes \mathbb{1}$$

with X and X' objects of C. We show that the components of $\ell^{\otimes}_{\mathsf{C}}$ and $r^{\otimes}_{\mathsf{C}}$ are V-functors in Lemma 1.3.9 below. The naturality of λ and ρ with respect to morphisms in V implies that $\ell^{\otimes}$ and $r^{\otimes}$ are 2-natural with respect to V-functors and V-natural transformations. ◇

Lemma 1.3.9. *In the context of Definition 1.3.8, the components $\ell^{\otimes}_{\mathsf{C}}$ and $r^{\otimes}_{\mathsf{C}}$ are* V*-functors.*

Proof. For $\ell^{\otimes}_{\mathsf{C}}$, the composition axiom (1.2.5) is commutativity of the outer diagram below for each triple $X, X', X'' \in \mathsf{C}$, where we use the notation

$$\mathsf{C}(X,X') = X_{01} \qquad \mathsf{C}(X',X'') = X_{12} \qquad \mathsf{C}(X,X'') = X_{02}.$$

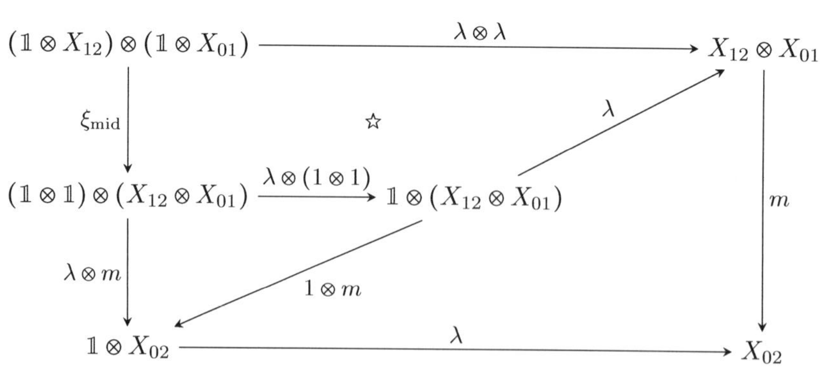

The underlying braids of the two composites around the trapezoid labeled ☆ are both $\mathrm{id}_2 \in B_2$ and therefore ☆ commutes by the Braided Coherence Theorem 1.1.38. The remaining regions commute by functoriality of $\otimes$ and naturality of λ.

The identity axiom (1.2.6) is commutativity of the following diagram for $X \in \mathsf{C}$, which follows from naturality of λ.

$$\begin{array}{ccccc} & & \mathbb{1}\otimes\mathbb{1} & \xrightarrow{1\otimes i_X} & \mathbb{1}\otimes\mathsf{C}(X,X) \\ & \nearrow^{\lambda^{-1}} & & & \downarrow{\scriptstyle\lambda} \\ \mathbb{1} & & \xrightarrow{\quad i_x \quad} & & \mathsf{C}(X,Y) \end{array}$$

The corresponding axioms for $r^{\otimes}_{\mathsf{C}}$ are similar, making use of the right unit isomorphism ρ. □

Definition 1.3.10. The *associator*, $a^{\otimes}$, is a 2-natural isomorphism

$$\begin{array}{ccc} \mathsf{V}\text{-}\mathsf{Cat}^3 & \xrightarrow{\otimes\times 1} & \mathsf{V}\text{-}\mathsf{Cat}^2 \\ {\scriptstyle 1\times\otimes}\downarrow & \overset{a^{\otimes}}{\swarrow} & \downarrow{\scriptstyle\otimes} \\ \mathsf{V}\text{-}\mathsf{Cat}^2 & \xrightarrow[\otimes]{} & \mathsf{V}\text{-}\mathsf{Cat} \end{array}$$

in $2\mathsf{Cat}(\mathsf{V}\text{-}\mathsf{Cat}^3, \mathsf{V}\text{-}\mathsf{Cat})$ defined as follows. For C, D, E in $\mathsf{V}\text{-}\mathsf{Cat}$, the associator component

$$a^{\otimes}_{\mathsf{C},\mathsf{D},\mathsf{E}} : (\mathsf{C}\otimes\mathsf{D})\otimes\mathsf{E} \longrightarrow \mathsf{C}\otimes(\mathsf{D}\otimes\mathsf{E})$$

is given on objects by the associativity isomorphism of the Cartesian product and is given on hom objects by the associativity isomorphism

$$\alpha : \big(\mathsf{C}(X,X')\otimes\mathsf{D}(Y,Y')\big)\otimes\mathsf{E}(Z,Z') \longrightarrow \mathsf{C}(X,X')\otimes\big(\mathsf{D}(Y,Y')\otimes\mathsf{E}(Z,Z')\big)$$

for objects $X, X' \in \mathsf{C}$, $Y, Y' \in \mathsf{D}$, and $Z, Z' \in \mathsf{E}$. The braided monoidal axioms for $(\mathsf{V}, \otimes, \mathbb{1}, \alpha, \lambda, \rho, \xi)$ imply that each component $a^{\otimes}_{\mathsf{C},\mathsf{D},\mathsf{E}}$ is a V-functor. Naturality of α with respect to morphisms in V implies that $a^{\otimes}$ is 2-natural with respect to V-functors and V-natural transformations. ◇

Lemma 1.3.11. *In the context of Definition 1.3.10 the associator components $a^{\otimes}_{\mathsf{C},\mathsf{D},\mathsf{E}}$ are V-functorial.*

Proof. The composition axiom (1.2.5) for $a^{\otimes}_{\mathsf{C},\mathsf{D},\mathsf{E}}$ is commutativity of the outer diagram below for objects $X, X', X'' \in \mathsf{C}$; $Y, Y', Y'' \in \mathsf{D}$; and $Z, Z', Z'' \in \mathsf{E}$. In the below diagram we use the following notation and omit the tensor signs to save space.

$$\begin{array}{lll} \mathsf{C}(X,X') = X_{01} & \mathsf{C}(X',X'') = X_{12} & \mathsf{C}(X,X'') = X_{02} \\ \mathsf{D}(Y,Y') = Y_{01} & \mathsf{D}(Y',Y'') = Y_{12} & \mathsf{D}(Y,Y'') = Y_{02} \\ \mathsf{E}(Z,Z') = Z_{01} & \mathsf{E}(Z',Z'') = Z_{12} & \mathsf{E}(Z,Z'') = Z_{02} \end{array}$$

$$\begin{array}{ccc}
\big((X_{12}Y_{12})Z_{12}\big)\big((X_{01}Y_{01})Z_{01}\big) & \xrightarrow{\alpha\alpha} & \big(X_{12}(Y_{12}Z_{12})\big)\big(X_{01}(Y_{01}Z_{01})\big) \\
\Big\downarrow \xi_{\mathrm{mid}} & & \Big\downarrow \xi_{\mathrm{mid}} \\
\big((X_{12}Y_{12})(X_{01}Y_{01})\big)(Z_{12}Z_{01}) & \text{☆} & (X_{12}X_{01})\big((Y_{12}Z_{12})(Y_{01}Z_{01})\big) \\
\Big\downarrow \xi_{\mathrm{mid}}(11) & & \Big\downarrow (11)\xi_{\mathrm{mid}} \\
\big((X_{12}X_{01})(Y_{12}Y_{01})\big)(Z_{12}Z_{01}) & \xrightarrow{\alpha} & (X_{12}X_{01})\big((Y_{12}Y_{01})(Z_{12}Z_{01})\big) \\
\Big\downarrow (mm)m & & \Big\downarrow m(mm) \\
(X_{02}Y_{02})Z_{02} & \xrightarrow{\alpha} & X_{02}(Y_{02}Z_{02})
\end{array}$$

Commutativity of the region labeled ☆ follows from the Braided Coherence Theorem 1.1.38 because the underlying braids of the boundary composites, shown below, are equal. The remaining rectangle commutes by naturality of α.

$s_2s_4s_3 \in B_6$ $\qquad\qquad$ $s_4s_2s_3 \in B_6$

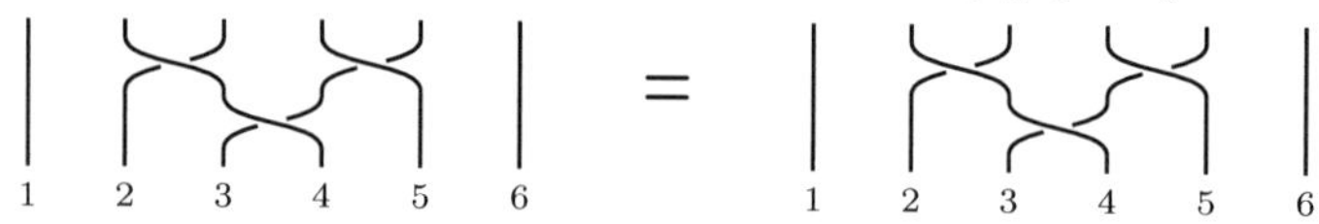

The identity axiom (1.2.6) for $a^{\otimes}_{\mathsf{C},\mathsf{D},\mathsf{E}}$ is similar to that of the unitors, following from unity properties of V (1.1.5) and naturality of α. $\square$

Explanation 1.3.12 (2-Naturality of the Associator and Unitors)**.** The 2-naturality of $a^{\otimes}$ means that the following diagram commutes for V-functors

$$F\colon \mathsf{C} \longrightarrow \mathsf{C}', \quad G\colon \mathsf{D} \longrightarrow \mathsf{D}', \quad \text{and} \quad H\colon \mathsf{E} \longrightarrow \mathsf{E}'.$$

$$\begin{array}{ccc}
(\mathsf{C}\otimes\mathsf{D})\otimes\mathsf{E} & \xrightarrow{(F\otimes G)\otimes H} & (\mathsf{C}'\otimes\mathsf{D}')\otimes\mathsf{E}' \\
\Big\downarrow a^{\otimes} & & \Big\downarrow a^{\otimes} \\
\mathsf{C}\otimes(\mathsf{D}\otimes\mathsf{E}) & \xrightarrow{F\otimes(G\otimes H)} & \mathsf{C}'\otimes(\mathsf{D}'\otimes\mathsf{E}')
\end{array} \tag{1.3.13}$$

The 2-naturality of $\ell^{\otimes}$ and $r^{\otimes}$ yield similar commutative diagrams. $\diamond$

The unitors $\ell^{\otimes}$ and $r^{\otimes}$ as well as the associator $a^{\otimes}$ satisfy the following two important properties generalizing those of monoidal categories.

Lemma 1.3.14 (Middle Unity for $\ell^{\otimes}$ and $r^{\otimes}$)**.** *The following diagram commutes for each pair of* V*-categories* C *and* D*.*

Middle Unity Diagram:

$$\begin{array}{ccc} (\mathsf{C} \otimes \mathbb{I}) \otimes \mathsf{D} & \xrightarrow{a^{\otimes}_{\mathsf{C},\mathbb{I},\mathsf{D}}} & \mathsf{C} \otimes (\mathbb{I} \otimes \mathsf{D}) \\ {\scriptstyle r^{\otimes}_{\mathsf{C}} \otimes 1_{\mathsf{D}}}\downarrow & & \downarrow{\scriptstyle 1_{\mathsf{C}} \otimes \ell^{\otimes}_{\mathsf{D}}} \\ \mathsf{C} \otimes \mathsf{D} & \xrightarrow{1_{\mathsf{C}\otimes\mathsf{D}}} & \mathsf{C} \otimes \mathsf{D} \end{array} \tag{1.3.15}$$

Proof. To verify that the diagram of V-functors commutes, we first observe that it commutes on objects by the middle unity axiom for the Cartesian product. For objects $X, X' \in \mathsf{C}$ and $Y, Y' \in \mathsf{D}$ the middle unity diagram on hom objects is the following.

$$\begin{array}{ccc} (\mathsf{C}(X,X') \otimes \mathbb{1}) \otimes \mathsf{D}(Y,Y') & \xrightarrow{\alpha} & \mathsf{C}(X,X') \otimes (\mathbb{1} \otimes \mathsf{D}(Y,Y')) \\ {\scriptstyle \rho\otimes 1}\downarrow & & \downarrow{\scriptstyle 1\otimes\lambda} \\ \mathsf{C}(X,X') \otimes \mathsf{D}(Y,Y') & \xrightarrow{1} & \mathsf{C}(X,X') \otimes \mathsf{D}(Y,Y') \end{array} \tag{1.3.16}$$

This diagram in V commutes by the middle unity axiom for λ and ρ. □

Lemma 1.3.17 (Pentagon Axiom for $a^{\otimes}$). *The following diagram commutes for each tuple of* V*-categories* A, B, C, *and* D.

Pentagon Diagram:

$$\begin{array}{ccccc} & & (\mathsf{A} \otimes \mathsf{B}) \otimes (\mathsf{C} \otimes \mathsf{D}) & & \\ & \nearrow{\scriptstyle a^{\otimes}_{\mathsf{A}\otimes\mathsf{B},\mathsf{C},\mathsf{D}}} & & \searrow{\scriptstyle a^{\otimes}_{\mathsf{A},\mathsf{B},\mathsf{C}\otimes\mathsf{D}}} & \\ ((\mathsf{A} \otimes \mathsf{B}) \otimes \mathsf{C}) \otimes \mathsf{D} & & & & \mathsf{A} \otimes (\mathsf{B} \otimes (\mathsf{C} \otimes \mathsf{D})) \\ {\scriptstyle a^{\otimes}_{\mathsf{A},\mathsf{B},\mathsf{C}} \otimes 1_{\mathsf{D}}}\downarrow & & & & \uparrow{\scriptstyle 1_{\mathsf{A}} \otimes a^{\otimes}_{\mathsf{B},\mathsf{C},\mathsf{D}}} \\ (\mathsf{A} \otimes (\mathsf{B} \otimes \mathsf{C})) \otimes \mathsf{D} & & \xrightarrow{a^{\otimes}_{\mathsf{A},\mathsf{B}\otimes\mathsf{C},\mathsf{D}}} & & \mathsf{A} \otimes ((\mathsf{B} \otimes \mathsf{C}) \otimes \mathsf{D}) \end{array} \tag{1.3.18}$$

Proof. Our argument that (1.3.18) commutes is similar to the argument of Lemma 1.3.14. First we observe that the diagram commutes on objects by the pentagon axiom for the Cartesian product. Next, we observe that commutativity of the diagram on hom objects is equivalent to the pentagon axiom for those hom objects and the associator in V. □

We also have the following result generalizing the additional unity properties (1.1.4) and (1.1.5).

Lemma 1.3.19. *The following equality in* V-Cat *holds:*

$$\ell^{\otimes}_{\mathbb{I}} = r^{\otimes}_{\mathbb{I}} : \mathbb{I} \otimes \mathbb{I} \longrightarrow \mathbb{I}. \tag{1.3.20}$$

In addition, the following left unity diagram *and* right unity diagram *commute for each pair of* V*-categories* C *and* D.

$$\begin{array}{ccc} (\mathbb{I} \otimes \mathsf{C}) \otimes \mathsf{D} & \xrightarrow{a^{\otimes}_{\mathbb{I},\mathsf{C},\mathsf{D}}} & \mathbb{I} \otimes (\mathsf{C} \otimes \mathsf{D}) \\ {\scriptstyle \ell^{\otimes}_{\mathsf{C}} \otimes 1_{\mathsf{D}}}\searrow & & \swarrow{\scriptstyle \ell^{\otimes}_{\mathsf{C}\otimes\mathsf{D}}} \\ & \mathsf{C} \otimes \mathsf{D} & \end{array} \qquad \begin{array}{ccc} (\mathsf{C} \otimes \mathsf{D}) \otimes \mathbb{I} & \xrightarrow{a^{\otimes}_{\mathsf{C},\mathsf{D},\mathbb{I}}} & \mathsf{C} \otimes (\mathsf{D} \otimes \mathbb{I}) \\ {\scriptstyle r^{\otimes}_{\mathsf{C}\otimes\mathsf{D}}}\searrow & & \swarrow{\scriptstyle 1_{\mathsf{C}} \otimes r^{\otimes}_{\mathsf{D}}} \\ & \mathsf{C} \otimes \mathsf{D} & \end{array} \tag{1.3.21}$$

Proof. We sketch two separate proofs of this result.

(1) Repeat the arguments from Remark I.1.2.5 that establish (1.1.4) and (1.1.5) from the axioms of a monoidal category, using the diagrams (1.3.15) and (1.3.18) instead.
(2) Argue as in Lemmas 1.3.14 and 1.3.17 that the result follows from (1.1.4) and (1.1.5) on objects (for the Cartesian product) and on hom objects (for the monoidal product in V). □

Now we turn to the braiding for the tensor product of V-categories. For simplicity we require that V be symmetric monoidal, although more subtle hypotheses in terms of iterated monoidal structures can be used to obtain more subtle results. See Note 1.6.2 for further comments and references.

Definition 1.3.22. Suppose that (V,ξ) is a symmetric monoidal category. The *braiding*, $\beta^{\otimes}$, is a 2-natural isomorphism

$$\begin{array}{ccc} \mathsf{V}\text{-}\mathsf{Cat}^2 & \xrightarrow{\quad\otimes\quad} & \mathsf{V}\text{-}\mathsf{Cat} \\ & {\scriptstyle\tau}\searrow \quad \Downarrow\beta^{\otimes} \quad \nearrow{\scriptstyle\otimes} & \\ & \mathsf{V}\text{-}\mathsf{Cat}^2 & \end{array}$$

in $\mathsf{2Cat}(\mathsf{V}\text{-}\mathsf{Cat}^2,\mathsf{V}\text{-}\mathsf{Cat})$ defined as follows. For C and D in $\mathsf{V}\text{-}\mathsf{Cat}$, the braiding component

$$\beta^{\otimes}_{\mathsf{C},\mathsf{D}}\colon \mathsf{C}\otimes\mathsf{D} \longrightarrow \mathsf{D}\otimes\mathsf{C}$$

is given on objects by the braiding of the Cartesian product and is given on hom objects by the braiding in V

$$\xi\colon \mathsf{C}(X,X')\otimes\mathsf{D}(Y,Y') \longrightarrow \mathsf{D}(Y,Y')\otimes\mathsf{C}(X,X')$$

for objects $X,X' \in \mathsf{C}$ and $Y,Y' \in \mathsf{D}$. The symmetric monoidal axioms for $(\mathsf{V},\otimes,\mathbb{1},\alpha,\lambda,\rho,\xi)$ imply that each component $\beta^{\otimes}_{\mathsf{C},\mathsf{D}}$ is a V-functor (see Lemma 1.3.23). Naturality of ξ with respect to morphisms in V implies that $\beta^{\otimes}$ is 2-natural with respect to V-functors and V-natural transformations. ◇

Lemma 1.3.23. *In the context of Definition 1.3.22 the braiding components $\beta^{\otimes}_{\mathsf{C},\mathsf{D}}$ are V-functorial.*

Proof. The composition axiom (1.2.5) for $\beta^{\otimes}_{\mathsf{C},\mathsf{D}}$ is commutativity of the outer diagram below for objects $X,X',X'' \in \mathsf{C}$ and $Y,Y',Y'' \in \mathsf{D}$. In the below diagram we use the following notation and omit the tensor signs to save space.

$$\begin{array}{lll} \mathsf{C}(X,X') = X_{01} & \mathsf{C}(X',X'') = X_{12} & \mathsf{C}(X,X'') = X_{02} \\ \mathsf{D}(Y,Y') = Y_{01} & \mathsf{D}(Y',Y'') = Y_{12} & \mathsf{D}(Y,Y'') = Y_{02} \end{array}$$

(1.3.24)
$$\begin{array}{ccc}
(X_{12}Y_{12})(X_{01}Y_{01}) & \xrightarrow{\xi\xi} & (Y_{12}X_{12})(Y_{01}X_{01}) \\
\downarrow{\scriptstyle \xi_{\mathrm{mid}}} & \text{☆} & \downarrow{\scriptstyle \xi_{\mathrm{mid}}} \\
(X_{12}X_{01})(Y_{12}Y_{01}) & \xrightarrow{\xi} & (Y_{12}Y_{01})(X_{12}X_{01}) \\
\downarrow{\scriptstyle mm} & & \downarrow{\scriptstyle mm} \\
X_{02}Y_{02} & \xrightarrow{\xi} & Y_{02}X_{02}
\end{array}$$

In the above diagram, the middle horizontal arrow ξ interchanges $X_{12}X_{01}$ and $Y_{12}Y_{01}$. The composites around the region labeled ☆ have the same underlying permutation and therefore ☆ commutes by the Symmetric Coherence Theorem 1.1.41. The lower rectangle commutes by naturality of ξ.

Checking the identity axiom (1.2.6) for $\beta^{\otimes}_{\mathsf{C},\mathsf{D}}$ is similar to that of the unitors and associator, following from symmetric unity properties of V and naturality of ξ. □

Explanation 1.3.25. In (1.3.24) above, the underlying braids for the two composites around the region labeled ☆ are the following distinct elements of B_4.

$s_2s_1s_3s_2s_2 \in B_4$ $\qquad$ $s_2s_1s_3 \in B_4$

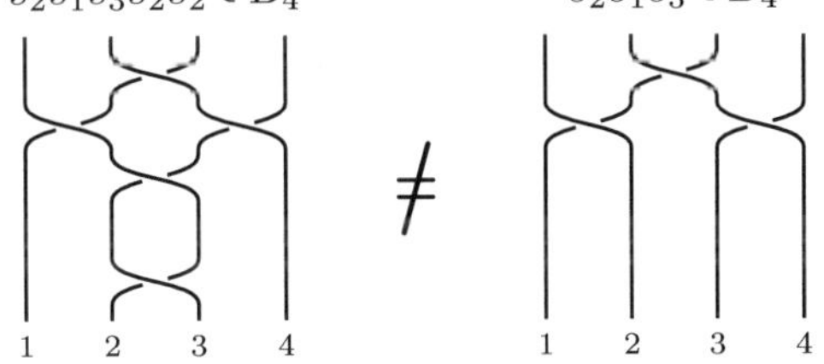

Therefore, if V is merely braided monoidal then the data of $\beta^{\otimes}_{\mathsf{C},\mathsf{D}}$ generally do not form a V-functor. ◇

Explanation 1.3.26 (2-Naturality of the Braiding)**.** The 2-naturality of $\beta^{\otimes}$ means that the following diagram commutes for V-functors

$$F\colon \mathsf{C} \longrightarrow \mathsf{C}' \quad \text{and} \quad G\colon \mathsf{D} \longrightarrow \mathsf{D}'.$$

(1.3.27)
$$\begin{array}{ccc}
\mathsf{C}\otimes\mathsf{D} & \xrightarrow{F\otimes G} & \mathsf{C}'\otimes\mathsf{D}' \\
\downarrow{\scriptstyle \beta^{\otimes}_{\mathsf{C},\mathsf{D}}} & & \downarrow{\scriptstyle \beta^{\otimes}_{\mathsf{C}',\mathsf{D}'}} \\
\mathsf{D}\otimes\mathsf{C} & \xrightarrow{G\otimes F} & \mathsf{D}'\otimes\mathsf{C}'
\end{array}$$

◇

Next we discuss axioms satisfied by $\beta^{\otimes}$.

Convention 1.3.28. Since Definition 1.3.22 assumes that V is symmetric monoidal, that assumption is in effect any time $\beta^{\otimes}$ appears below. ◇

Lemma 1.3.29 (Hexagon Axioms for $\beta^{\otimes}$). *Suppose* V *is symmetric monoidal. The following diagrams commute for each triple of* V*-categories* A, B, *and* C, *with* $a^{-\otimes}$ *denoting the inverse of* $a^{\otimes}$.

Left Hexagon Diagram:

(1.3.30)

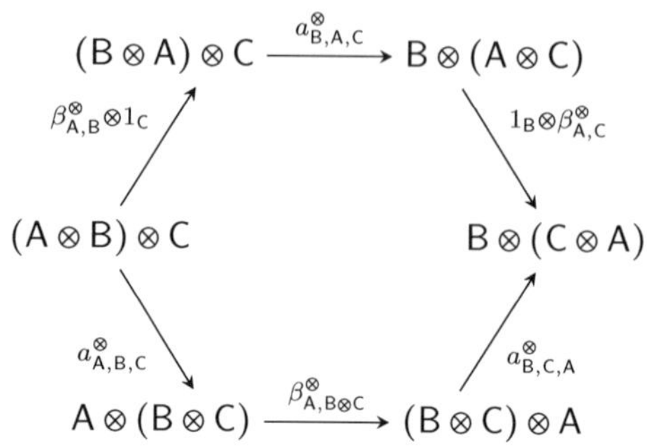

Right Hexagon Diagram:

(1.3.31)

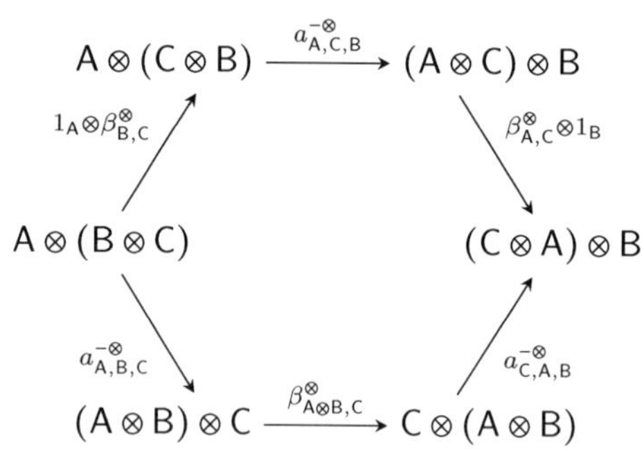

Proof. Our arguments that (1.3.30) and (1.3.31) commute are similar to the arguments of Lemmas 1.3.14 and 1.3.17. Commutativity on objects follows from the hexagon diagrams for the braiding of the Cartesian product, and commutativity on hom objects follows from the hexagon axioms for the braiding of V. □

Lemma 1.3.32 (Symmetry Axiom for $\beta^{\otimes}$). *Suppose* V *is symmetric monoidal. Then the following diagram commutes for each pair of* V*-categories* C *and* D.

Symmetry Diagram:

(1.3.33)

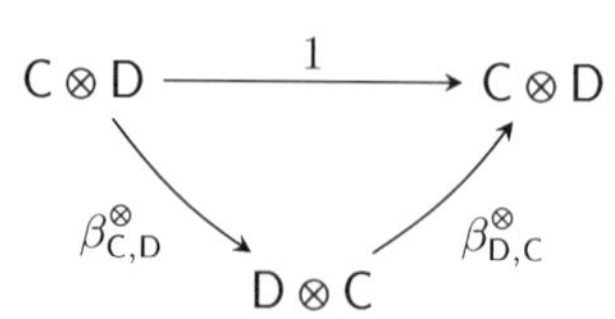

Proof. Similarly to the proofs of Lemmas 1.3.14, 1.3.17, and 1.3.29, commutativity of (1.3.33) follows on objects by the symmetry of the Cartesian product and follows on hom objects by the symmetry axiom for ξ. □

Lemma 1.3.34 (Unit Properties for $\beta^{\otimes}$). *Suppose* V *is symmetric monoidal. The following diagrams commute for each* V*-category* C.

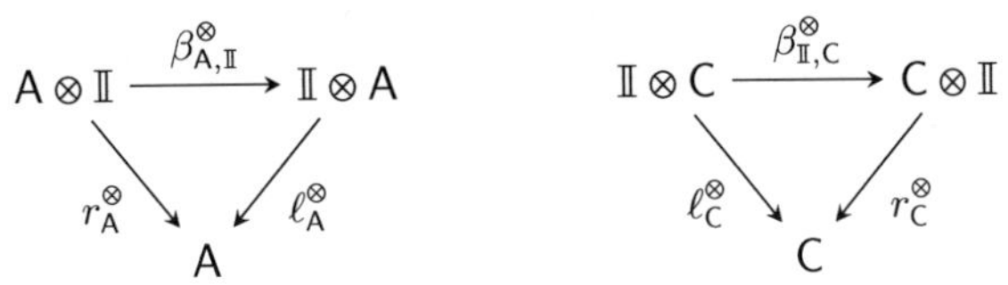

Proof. Commutativity on objects follows from the corresponding properties of the Cartesian product. On hom objects, commutativity follows from the unit diagrams (1.1.20) in V. □

Restricting to the underlying 1-category of V-Cat, the preceding definitions and results of this section prove the following. In fact they prove a stronger Cat-enriched result that we state as Theorem 1.5.5 below, after giving the relevant background in Sections 1.4 and 1.5.

Theorem 1.3.35. *Suppose* $\mathsf{V} = (\mathsf{V}, \otimes, \xi)$ *is a braided monoidal category. Equipped with the data above,* $(\mathsf{V}\text{-}\mathsf{Cat}, \otimes, \mathbb{I}, a^{\otimes}, \ell^{\otimes}, r^{\otimes})$ *is a monoidal category. If* V *is symmetric monoidal then so is* $(\mathsf{V}\text{-}\mathsf{Cat}, \beta^{\otimes})$.

In Explanation 1.3.25 above we explained that the braiding $\beta^{\otimes}$ cannot be defined as a V-functor when V is merely braided monoidal. However, the same assignment on objects can be extended to a V-functor using the opposite V-category construction from Definition 1.2.16. We will make use of this in Example 1.4.27 below.

Definition 1.3.36. Suppose (V, ξ) is a braided monoidal category and suppose $\mathsf{C}, \mathsf{D} \in \mathsf{V}\text{-}\mathsf{Cat}$. Define a V-functor

$$\gamma_{\mathsf{C},\mathsf{D}} \colon \mathsf{C}^{\mathsf{op}} \otimes \mathsf{D}^{\mathsf{op}} \longrightarrow (\mathsf{D} \otimes \mathsf{C})^{\mathsf{op}}$$

with the assignment

$$X \otimes Y \longmapsto Y \otimes X$$

on objects $X \otimes Y \in \mathsf{C}^{\mathsf{op}} \otimes \mathsf{D}^{\mathsf{op}}$. On hom objects we define $(\gamma_{\mathsf{C},\mathsf{D}})_{X \otimes Y, X' \otimes Y'}$ via the braiding ξ below.

$$\begin{array}{ccc}
(\mathsf{C}^{\mathsf{op}} \otimes \mathsf{D}^{\mathsf{op}})(X \otimes Y, X' \otimes Y') & & (\mathsf{D} \otimes \mathsf{C})^{\mathsf{op}}(Y \otimes X, Y' \otimes X') \\
\| & & \| \\
\mathsf{C}^{\mathsf{op}}(X, X') \otimes \mathsf{D}^{\mathsf{op}}(Y, Y') & & (\mathsf{D} \otimes \mathsf{C})(Y' \otimes X', Y \otimes X) \\
\| & & \| \\
\mathsf{C}(X', X) \otimes \mathsf{D}(Y', Y) & \xrightarrow{\quad \xi \quad} & \mathsf{D}(Y', Y) \otimes \mathsf{C}(X', X)
\end{array}$$

We verify the V-functor axioms for $\gamma_{\mathsf{C},\mathsf{D}}$ in Lemma 1.3.37. When clear from context we will omit the subscripts C and D. ◇

Lemma 1.3.37. *In the context of Definition 1.3.36,*

$$\gamma \colon \mathsf{C}^{\mathsf{op}} \otimes \mathsf{D}^{\mathsf{op}} \longrightarrow (\mathsf{D} \otimes \mathsf{C})^{\mathsf{op}}$$

is a V*-functor.*

Proof. To verify the composition axiom (1.2.5) for γ, suppose given

$$X_0, X_1, X_2 \in \mathsf{C} \quad \text{and} \quad Y_0, Y_1, Y_2 \in \mathsf{D}.$$

For $i, j \in \{0,1,2,3\}$ we let

$$\mathsf{C}^{\mathrm{op}}(X_i, X_j) = \overline{X}_{ij} = X_{ji} = \mathsf{C}(X_j, X_i) \quad \text{and} \quad \mathsf{D}^{\mathrm{op}}(Y_i, Y_j) = \overline{Y}_{ij} = Y_{ji} = \mathsf{D}(Y_j, Y_i).$$

Then the composition axiom for γ is the outer diagram below, where we omit tensors to save space.

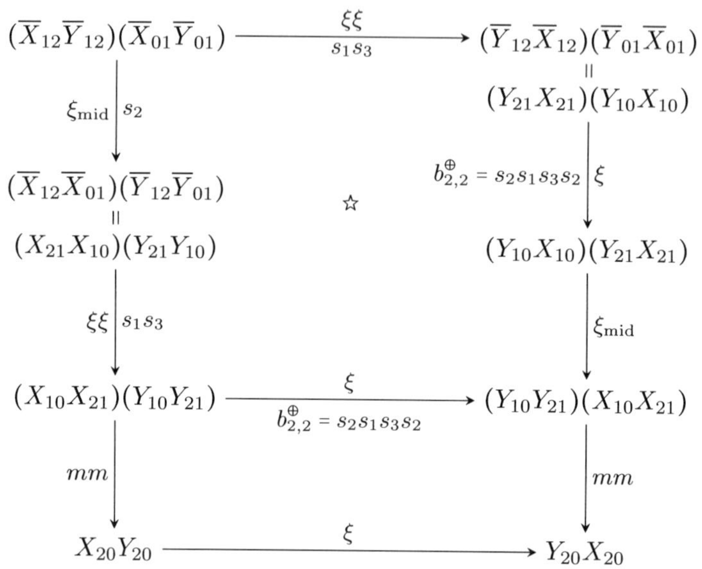

In the diagram above, the vertical composite along the left is the composition for $\mathsf{C}^{\mathrm{op}} \otimes \mathsf{D}^{\mathrm{op}}$, beginning with ξ_{mid} as in Definition 1.3.3. The vertical composite along the right is the composition in $(\mathsf{D} \otimes \mathsf{C})^{\mathrm{op}}$, beginning with the block braid ξ as in Definition 1.2.16. The lower region commutes by naturality of ξ and the upper region ☆ commutes by the Braided Coherence Theorem 1.1.38: each morphism is labeled with its underlying braid, and the two boundary composites are the following.

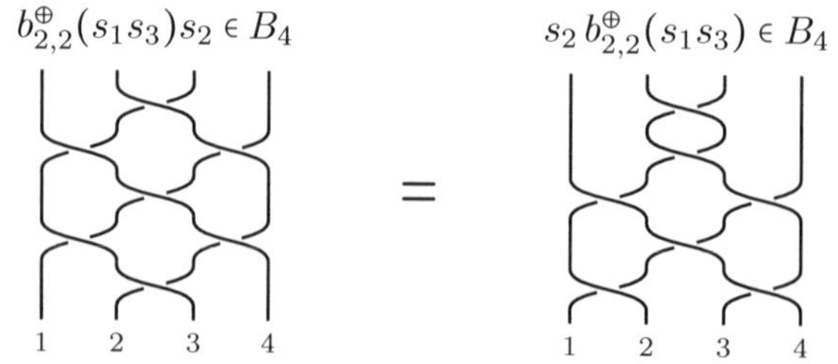

Using the same notation, the identity axiom (1.2.6) for γ is the following.

$$\begin{array}{ccccc} & & & & \overline{X}_{00}\overline{Y}_{00} \\ & & & \overset{ii}{\nearrow} & \| \\ & & & & X_{00}Y_{00} \\ \mathbb{1} & \xrightarrow{\lambda^{-1}} & \mathbb{1}\mathbb{1} & & \downarrow \xi \\ & & & & Y_{00}X_{00} \\ & & & \underset{ii}{\searrow} & \| \\ & & & & \overline{Y}_{00}\overline{X}_{00} \end{array}$$

Recalling that $\xi_{\mathbb{1},\mathbb{1}} = 1_{\mathbb{1}\mathbb{1}}$ (1.1.21), the diagram above commutes by naturality of ξ. □

Explanation 1.3.38. The one-object V-category $\mathbb{I}$ has

$$\mathbb{I}^{\mathsf{op}}(*,*) = \mathbb{I}(*,*) = \mathbb{1}.$$

Since $\xi_{\mathbb{1},\mathbb{1}}$, $\rho_{\mathbb{1}}$, and $\lambda_{\mathbb{1}}$ are identities, the composition laws for $\mathbb{I}$ and its opposite are equal. Therefore, we have

$$\mathbb{I}^{\mathsf{op}} = \mathbb{I}$$

as V-categories. ◇

For future reference, we will need the following properties of γ.

Lemma 1.3.39. *Suppose given* V*-categories*

$$\mathsf{C}, \mathsf{D}, \mathsf{E}, \mathsf{C}_0, \mathsf{C}_1, \mathsf{D}_0, \text{ and } \mathsf{D}_1$$

along with V*-functors*

$$\mathsf{C}_0 \xrightarrow{F} \mathsf{C}_1 \quad \textit{and} \quad \mathsf{D}_0 \xrightarrow{G} \mathsf{D}_1.$$

Then following diagrams of V*-categories and* V*-functors commute.*

(1.3.40)
$$\begin{array}{ccc} \mathsf{C}_0^{\mathsf{op}} \otimes \mathsf{D}_0^{\mathsf{op}} & \xrightarrow{F^{\mathsf{op}} \otimes G^{\mathsf{op}}} & \mathsf{C}_1^{\mathsf{op}} \otimes \mathsf{D}_1^{\mathsf{op}} \\ \downarrow{\scriptstyle \gamma_{\mathsf{C}_0,\mathsf{D}_0}} & & \downarrow{\scriptstyle \gamma_{\mathsf{C}_1,\mathsf{D}_1}} \\ (\mathsf{D}_0 \otimes \mathsf{C}_0)^{\mathsf{op}} & \xrightarrow{(G \otimes F)^{\mathsf{op}}} & (\mathsf{D}_1 \otimes \mathsf{C}_1)^{\mathsf{op}} \end{array}$$

(1.3.41)
$$\begin{array}{ccc} (\mathsf{C}^{\mathsf{op}} \otimes \mathsf{D}^{\mathsf{op}}) \otimes \mathsf{E}^{\mathsf{op}} & \xrightarrow{a^{\otimes}} & \mathsf{C}^{\mathsf{op}} \otimes (\mathsf{D}^{\mathsf{op}} \otimes \mathsf{E}^{\mathsf{op}}) \\ \downarrow{\scriptstyle \gamma_{\mathsf{C},\mathsf{D}} \otimes 1} & & \downarrow{\scriptstyle 1 \otimes \gamma_{\mathsf{D},\mathsf{E}}} \\ (\mathsf{D} \otimes \mathsf{C})^{\mathsf{op}} \otimes \mathsf{E}^{\mathsf{op}} & & \mathsf{C}^{\mathsf{op}} \otimes (\mathsf{E} \otimes \mathsf{D})^{\mathsf{op}} \\ \downarrow{\scriptstyle \gamma_{\mathsf{D}\otimes\mathsf{C},\mathsf{E}}} & & \downarrow{\scriptstyle \gamma_{\mathsf{C},\mathsf{E}\otimes\mathsf{D}}} \\ (\mathsf{E} \otimes (\mathsf{D} \otimes \mathsf{C}))^{\mathsf{op}} & \xleftarrow{(a^{\otimes})^{\mathsf{op}}} & ((\mathsf{E} \otimes \mathsf{D}) \otimes \mathsf{C})^{\mathsf{op}} \end{array}$$

(1.3.42)
$$\begin{array}{ccc} \mathsf{C}^{\mathsf{op}} \otimes \mathbb{I}^{\mathsf{op}} & \xrightarrow{\gamma_{\mathsf{C},\mathbb{I}}} & (\mathbb{I} \otimes \mathsf{C})^{\mathsf{op}} \\ \| & & \\ \mathsf{C}^{\mathsf{op}} \otimes \mathbb{I} & & \downarrow (\ell^{\otimes})^{\mathsf{op}} \\ & \searrow r^{\otimes} & \\ & & \mathsf{C}^{\mathsf{op}} \end{array} \quad \text{and} \quad \begin{array}{ccc} \mathbb{I}^{\mathsf{op}} \otimes \mathsf{C}^{\mathsf{op}} & \xrightarrow{\gamma_{\mathbb{I},\mathsf{C}}} & (\mathsf{C} \otimes \mathbb{I})^{\mathsf{op}} \\ \| & & \\ \mathbb{I} \otimes \mathsf{C}^{\mathsf{op}} & & \downarrow (r^{\otimes})^{\mathsf{op}} \\ & \searrow \ell^{\otimes} & \\ & & \mathsf{C}^{\mathsf{op}} \end{array}$$

Proof. For each of the diagrams, one can check commutativity on objects directly because $(-)^{\mathsf{op}}$ is the identity on objects. On hom objects, one unpacks the various opposites involved and finds that the relevant diagram commutes by the corresponding property of ξ. We give the unpacked diagrams below.

For $X, X' \in \mathsf{C}_0$ and $Y, Y' \in \mathsf{D}_0$, the diagram (1.3.40) at the hom object

$$\mathsf{C}_0^{\mathsf{op}}(X, X') \otimes \mathsf{D}_0^{\mathsf{op}}(Y, Y')$$

is the following. It commutes by naturality of ξ.

$$\begin{array}{ccc} \mathsf{C}_0(X', X) \otimes \mathsf{D}_0(Y', Y) & \xrightarrow{F \otimes G} & \mathsf{C}_1(FX', FX) \otimes \mathsf{D}_1(GY', GY) \\ \downarrow \xi & & \downarrow \xi \\ \mathsf{D}_0(Y', Y) \otimes \mathsf{C}_0(X', X) & \xrightarrow{G \otimes F} & \mathsf{D}_1(GY', GY) \otimes \mathsf{C}_1(FX', FX) \end{array}$$

For $X, X' \in \mathsf{C}$, $Y, Y' \in \mathsf{D}$, and $Z, Z' \in \mathsf{E}$, the diagram (1.3.41) at the hom object

$$(\mathsf{C}^{\mathsf{op}}(X, X') \otimes \mathsf{D}^{\mathsf{op}}(Y, Y')) \otimes \mathsf{E}^{\mathsf{op}}(Z, Z')$$

is the following. It commutes by the hexagon (1.1.22) for ξ and α.

$$\begin{array}{ccc} \big(\mathsf{C}(X', X) \otimes \mathsf{D}(Y', Y)\big) \otimes \mathsf{E}(Z', Z) & \xrightarrow{\alpha} & \mathsf{C}(X', X) \otimes \big(\mathsf{D}(Y', Y) \otimes \mathsf{E}(Z', Z)\big) \\ \downarrow \xi \otimes 1 & & \downarrow 1 \otimes \xi \\ \big(\mathsf{D}(Y', Y) \otimes \mathsf{C}(X', X)\big) \otimes \mathsf{E}(Z', Z) & & \mathsf{C}(X', X) \otimes \big(\mathsf{E}(Z', Z) \otimes \mathsf{D}(Y', Y)\big) \\ \downarrow \xi & & \downarrow \xi \\ \mathsf{E}(Z', Z) \otimes \big(\mathsf{D}(Y', Y) \otimes \mathsf{C}(X', X)\big) & \xleftarrow{\alpha} & \big(\mathsf{E}(Z', Z) \otimes \mathsf{D}(Y', Y)\big) \otimes \mathsf{C}(X', X) \end{array}$$

For $X, X' \in \mathsf{C}$, the diagrams (1.3.42) at the hom object $\mathsf{C}^{\mathsf{op}}(X, X')$ are the following. They commute by the unity properties (1.1.20) for ξ.

$$\begin{array}{ccc} \mathsf{C}(X',X) \otimes \mathbb{1} & \xrightarrow{\xi} & \mathbb{1} \otimes \mathsf{C}(X',X) \\ & {\scriptstyle\rho} \searrow \quad \swarrow {\scriptstyle\lambda} & \\ & \mathsf{C}(X',X) & \end{array} \qquad \begin{array}{ccc} \mathbb{1} \otimes \mathsf{C}(X',X) & \xrightarrow{\xi} & \mathsf{C}(X',X) \otimes \mathbb{1} \\ & {\scriptstyle\lambda} \searrow \quad \swarrow {\scriptstyle\rho} & \\ & \mathsf{C}(X',X) & \end{array}$$

This completes the proof. □

Explanation 1.3.43. One can define an opposite monoidal structure on $\mathsf{V}\text{-}\mathsf{Cat}$ with monoidal product $\otimes^{\tau}$ given by twisting with the braiding τ of $(\mathsf{Cat}, \times)$. Then Lemma 1.3.39 shows that γ provides a monoidal constraint for $(-)^{\mathsf{op}}$, making it a monoidal functor (of 1-categories) from $(\mathsf{V}\text{-}\mathsf{Cat}, \otimes)$ to $(\mathsf{V}\text{-}\mathsf{Cat}, \otimes^{\tau})$. Following Theorem 1.3.35, one can show that γ is 2-natural and one expects these claims will extend to make $(-)^{\mathsf{op}}$ monoidal as a 2-functor to $\mathsf{V}\text{-}\mathsf{Cat}^{\mathsf{co}}$ (with twisted monoidal product). Further development of these points is beyond our current scope. ◇

1.4. Monoidal Enriched Categories

In this section we use the monoidal structure of Section 1.3 to define V-enriched monoidal structures. As in Section 1.3, we assume throughout this section that $(\mathsf{V}, \otimes, \xi)$ is a braided monoidal category. We define monoidal V-categories, V-functors, and V-natural transformations and, under the further assumption that V is symmetric monoidal, we define braided and symmetric variants.

In the definitions below, we will write the tensor product of V-categories as juxtaposition. We will denote the associator, unitors, and braiding of $\mathsf{V}\text{-}\mathsf{Cat}$ with superscripts: $a^{\otimes}$, $\ell^{\otimes}$, $r^{\otimes}$, and $\beta^{\otimes}$. We will use superscripts with a minus sign, $-\otimes$, to denote their inverses.

Motivation 1.4.1 (Monoidal Categories)**.** For a monoidal category C, the associativity and unit isomorphisms are natural transformations described in Definition 1.1.1 via their components. The axioms for these data are given in terms of commuting diagrams of components. They can be rewritten equivalently as equalities of certain composites of natural transformations—that is, as equalities of pasting diagrams in Cat.

This equivalent form of data (as 2-cells) and axioms (as equalities of certain pasting diagrams) yields a definition that can be studied in any monoidal 2-category A. The definitions and results in this section are obtained by applying such a point of view to the 2-category $\mathsf{A} = \mathsf{V}\text{-}\mathsf{Cat}$ with its tensor product. We encourage the reader to observe, for each of the definitions below, how the corresponding diagrams in Cat would define the monoidal structures discussed in Section 1.1. The theory of pseudoalgebras over 2-monads yields an even more general approach to this theory, and we give some references in Note 1.6.3.

However, to minimize the necessary background, our presentation below does not depend on this general theory. Indeed, we will *apply* the material in this section to *define* the 2-dimensional monoidal structure on V-Cat in Section 1.5. Therefore, to avoid circularity, in this section we neither use nor mention general results about 2-dimensional monoidal structure. A reader familiar with the more general theory may recognize many of these results as special cases of more general ones. A reader unfamiliar with the more general theory may take these results as motivating examples. ◇

Definition 1.4.2. A *monoidal* V*-category* is a tuple

$$(\mathsf{K}, \boxtimes, \mathrm{I}^{\boxtimes}, a^{\boxtimes}, \ell^{\boxtimes}, r^{\boxtimes})$$

consisting of the following data.

Base V-category: It has a V-category K called the *base* V*-category*.
Monoidal Composition: It has a V-functor

$$\mathsf{K} \otimes \mathsf{K} \xrightarrow{\boxtimes} \mathsf{K}$$

called the *monoidal composition*.
Monoidal Identity: It has a V-functor

$$\mathbb{I} \xrightarrow{\mathrm{I}^{\boxtimes}} \mathsf{K}$$

called the *monoidal identity*. The image of the unique object in $\mathbb{I}$ is also denoted $\mathrm{I}^{\boxtimes}$ and called the *identity object*.
Monoidal Unitors: It has V-natural isomorphisms

(1.4.3)
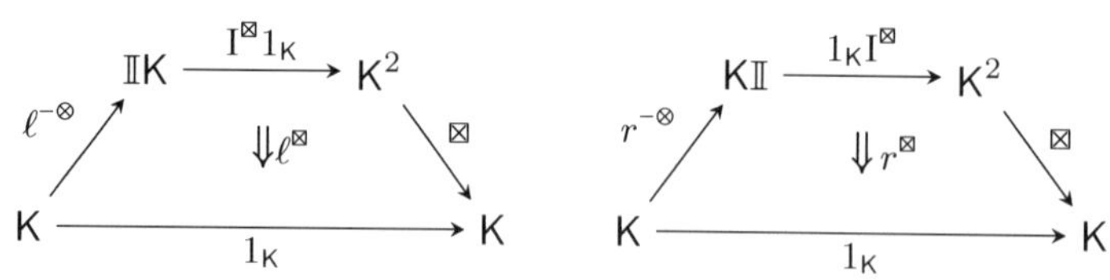

in V-$\mathsf{Cat}(\mathsf{K},\mathsf{K})$, called the *left monoidal unitor* and the *right monoidal unitor*, respectively. Their components at an object $X \in \mathsf{K}$ are, respectively,

$$\mathbb{1} \xrightarrow{\ell^{\boxtimes}_X} \mathsf{K}(\mathrm{I}^{\boxtimes} \boxtimes X, X) \quad \text{and} \quad \mathbb{1} \xrightarrow{r^{\boxtimes}_X} \mathsf{K}(X \boxtimes \mathrm{I}^{\boxtimes}, X).$$

Monoidal Associator: It has a V-natural isomorphism

(1.4.4)

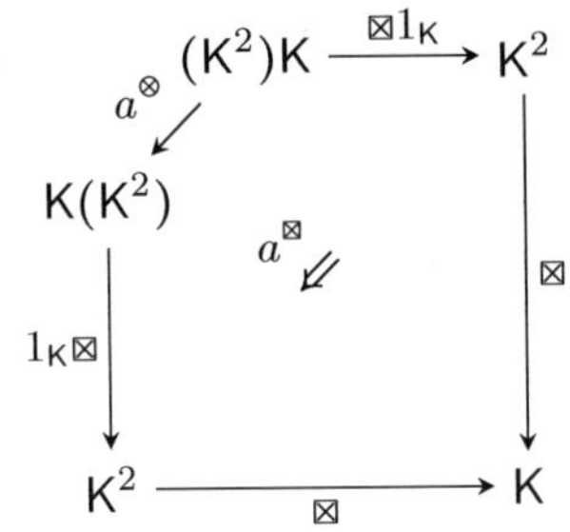

in $\mathsf{V}\text{-}\mathsf{Cat}((\mathsf{K}^2)\mathsf{K},\mathsf{K})$, called the *monoidal associator*. Its component at a triple of objects $X, Y, Z \in \mathsf{K}$ is a morphism in V

$$\mathbb{1} \xrightarrow{a^{\boxtimes}_{X,Y,Z}} \mathsf{K}\big((X \boxtimes Y) \boxtimes Z, X \boxtimes (Y \boxtimes Z)\big).$$

These data are required to satisfy the following two axioms, where we write 1 for the identity V-functor.

Unity Axiom: The composites of the following two *middle unity pasting diagrams* are equal.

(1.4.5)

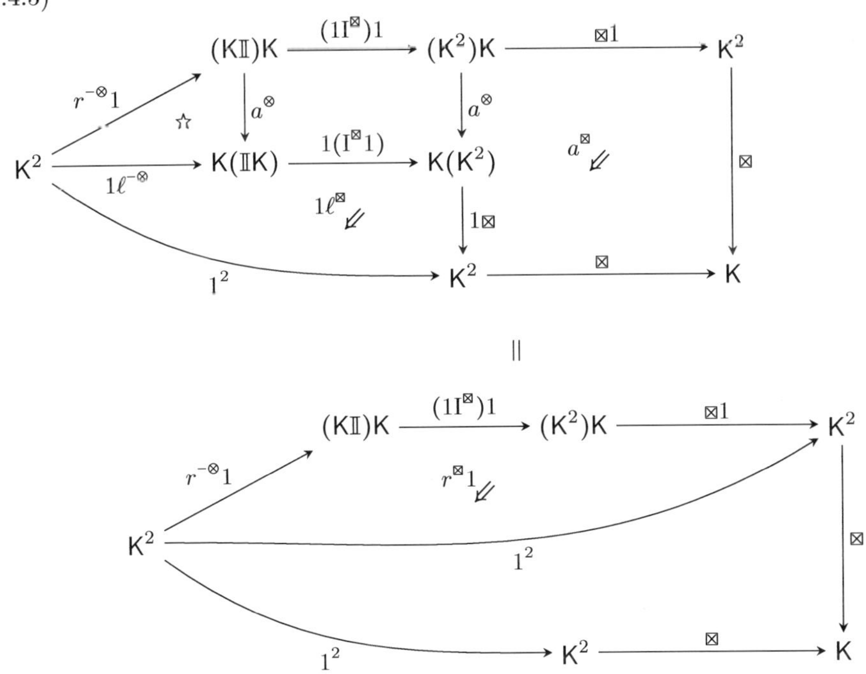

In the first diagram above, the unlabeled rectangle commutes by naturality of $a^{\otimes}$ (1.3.13) and the triangle labeled ☆ commutes by the middle unity for $\ell^{\otimes}$ and $r^{\otimes}$ (1.3.15).

Pentagon Axiom: The composites of the following two *pentagon pasting diagrams* are equal.

(1.4.6)

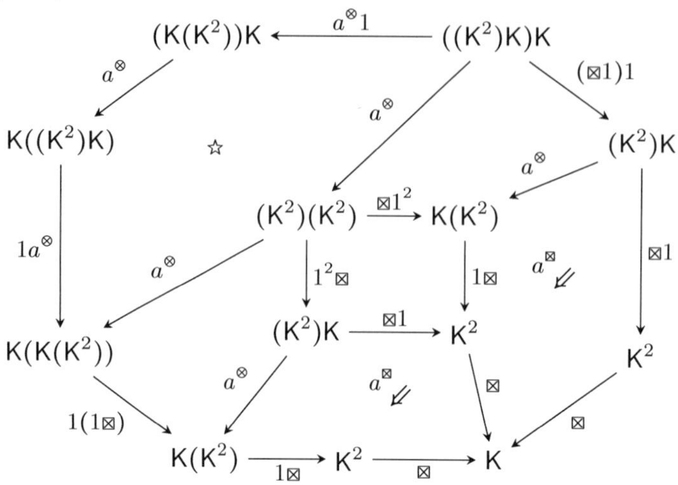

||

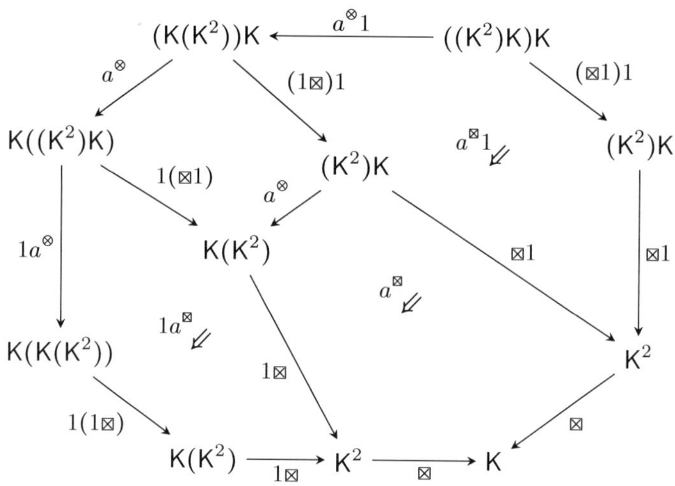

The central square in the first diagram above commutes by 2-functoriality of $\otimes$ in each variable (Proposition 1.3.6). The other unmarked quadrilaterals in the two diagrams above commute by 2-naturality of $a^{\otimes}$ (1.3.13). The pentagon labeled ☆ commutes by the pentagon axiom for $a^{\otimes}$ (1.3.18).

This finishes the definition of a monoidal V-category. We say that a monoidal V-category is *strict* if the monoidal unitors $\ell^{\boxtimes}$ and $r^{\boxtimes}$ along with the monoidal associator $a^{\boxtimes}$ are identity V-natural transformations. ◇

Explanation 1.4.7 (Interchange for a Monoidal V-category)**.** In the context of Definition 1.4.2, the composition axiom for V-functoriality of $\boxtimes$ (Definition 1.2.4) implies that, for all objects

$$X, X', Y, Y', Z, Z' \in \mathsf{K},$$

the following *enriched interchange* diagram in V commutes.

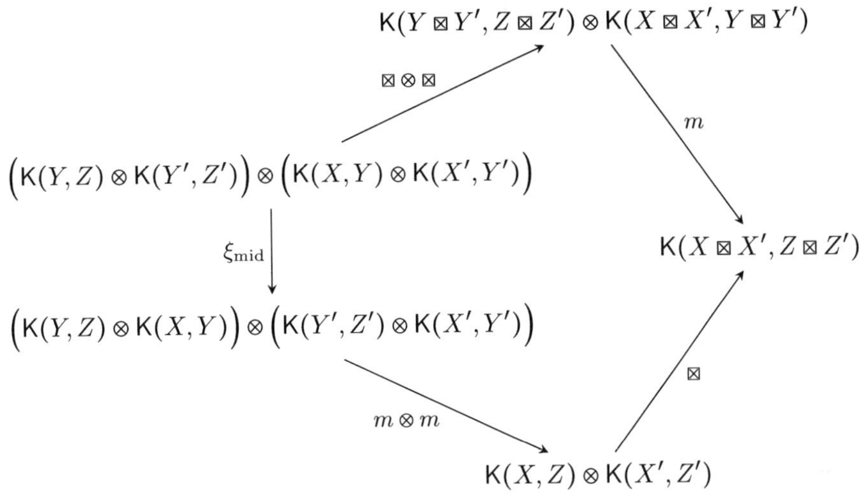

◇

Explanation 1.4.8 (Monoidal V-Category Axiom Components)**.** Each of the axioms in Definition 1.4.2 is an equality of composites of V-natural transformations. In each case one can express the relevant equality via components. For example, the Pentagon Axiom (1.4.6) is equivalent to the following equality of components for each $((X \otimes Y) \otimes Z) \otimes W \in ((\mathsf{K}^2)\mathsf{K})\mathsf{K}$.

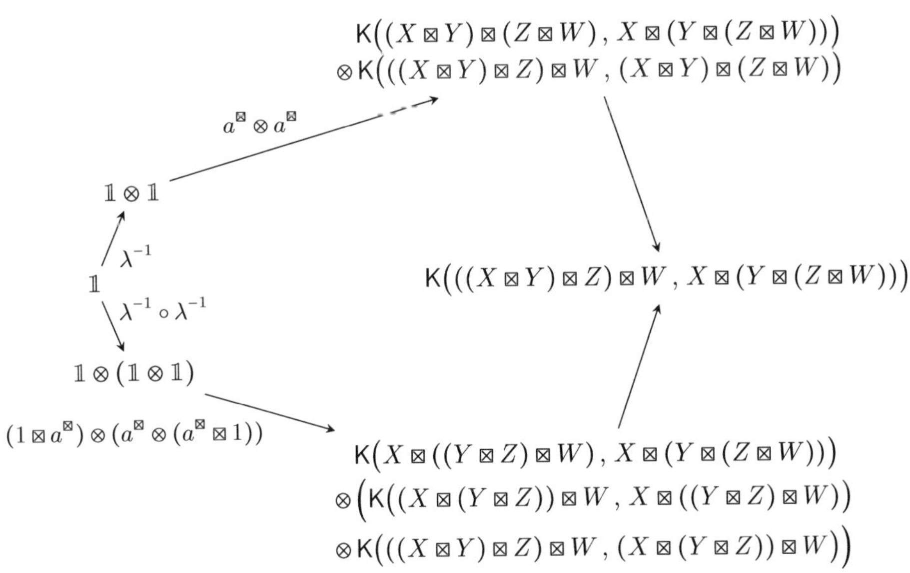

◇

For the definition of braided monoidal V-category, we will need a mate of $a^{\boxtimes}$ defined similarly to the mates of a pentagonator discussed just before Lemma I.6.5.2.

Definition 1.4.9. Suppose K is a monoidal V-category. We let $a_1^{\boxtimes}$ denote the mate of $a^{\boxtimes}$ given by the inverse of $a^{\otimes}$, as shown below.

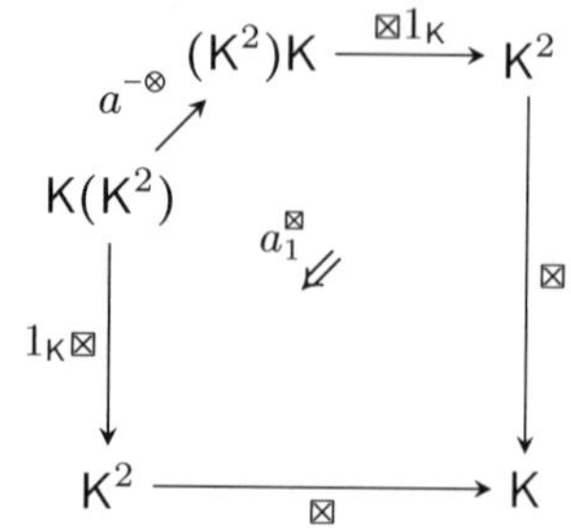

We let $a_1^{-\boxtimes}$ denote the inverse of $a_1^{\boxtimes}$. ◇

Recall from Convention 1.3.28 that we assume V to be symmetric monoidal whenever using the braiding $\beta^{\otimes}$ of $(\mathsf{V}\text{-}\mathsf{Cat}, \otimes)$.

Definition 1.4.10. Suppose V is a symmetric monoidal category. A *braided monoidal* V*-category* is a pair

$$(\mathsf{K}, \beta^{\boxtimes})$$

consisting of the following data.

- K is a monoidal V-category as in Definition 1.4.2.
- $\beta^{\boxtimes}$ is a V-natural isomorphism

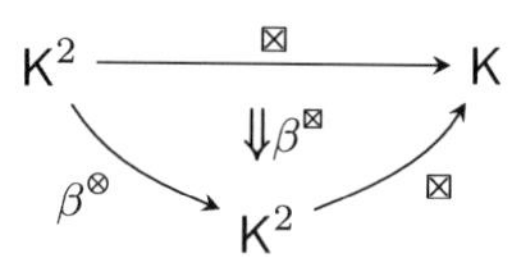

in $\mathsf{V}\text{-}\mathsf{Cat}(\mathsf{K}^2, \mathsf{K})$ called the *braiding* of K.

These data are required to satisfy the following two axioms.

Left Hexagon Axiom: The composites of the following two *left hexagon pasting diagrams* are equal.

(1.4.11)

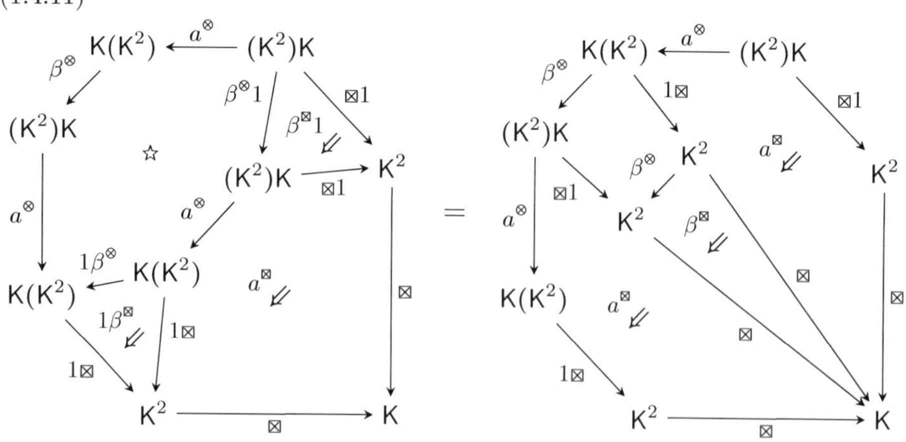

In the diagrams above, the unlabeled quadrilateral commutes by 2-naturality of $\beta^{\otimes}$ (1.3.27) and the hexagon labeled ☆ commutes by the left hexagon axiom for $\beta^{\otimes}$ (1.3.30).

Right Hexagon Axiom: The composites of the following two *right hexagon pasting diagrams* are equal.

(1.4.12)

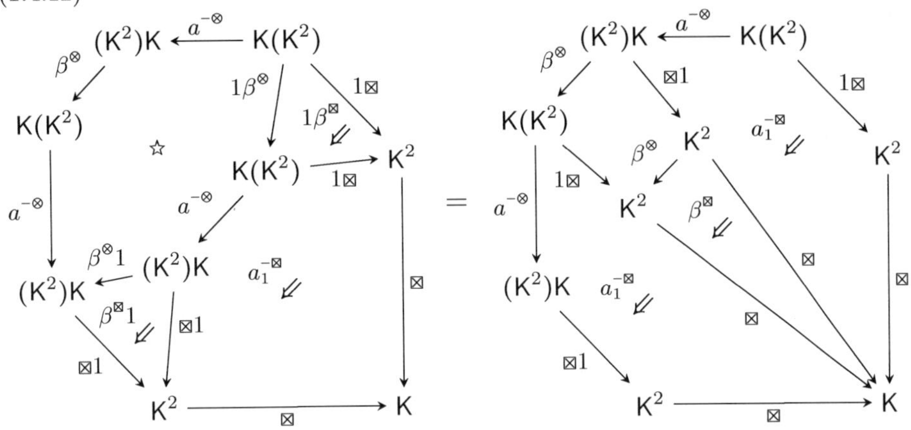

In the diagrams above, the unlabeled quadrilateral commutes by 2-naturality of $\beta^{\otimes}$ (1.3.27) and the hexagon labeled ☆ commutes by the right hexagon axiom for $\beta^{\otimes}$ (1.3.31). The 2-cell isomorphism $a_1^{-\boxtimes}$ is the inverse of $a_1^{\boxtimes}$ described in Definition 1.4.9.

This finishes the definition of a braided monoidal V-category. ◇

Definition 1.4.13. Suppose V is a symmetric monoidal category. A *symmetric monoidal* V *category* is a braided monoidal V category

$$(\mathsf{K}, \beta^{\boxtimes})$$

that satisfies the following axiom.

Symmetry Axiom: The composites of the following two *symmetry pasting diagrams* are equal.

(1.4.14)

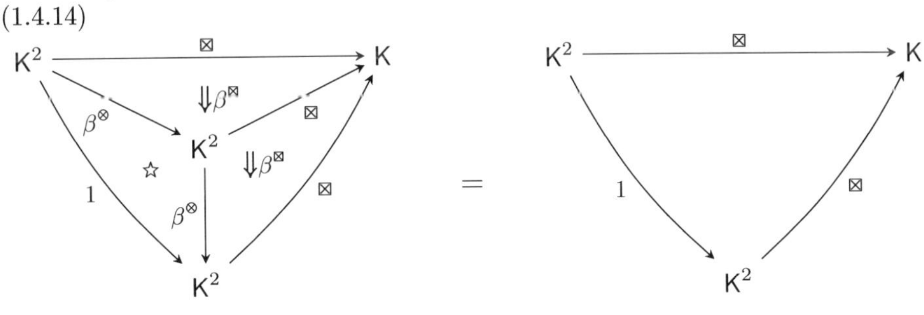

The right hand diagram above indicates the identity V-natural transformation. In the left hand diagram, the triangle labeled ☆ commutes by the symmetry axiom for $\beta^{\otimes}$ (1.3.33). When this axiom holds, $\beta^{\boxtimes}$ is also called the *symmetry* of K. ◇

Next we give the definitions of monoidal V-functors and V-natural transformations. These require only that V is braided monoidal. We will use the following two mates.

Definition 1.4.15. Suppose K is a monoidal V-category. We let $\ell_1^{\boxtimes}$ and $r_1^{\boxtimes}$ denote the mates of $\ell^{\boxtimes}$ and $r^{\boxtimes}$ given, respectively, by using $\ell^{\otimes}$ and $r^{\otimes}$ in place of their inverses, as shown below.

(1.4.16) 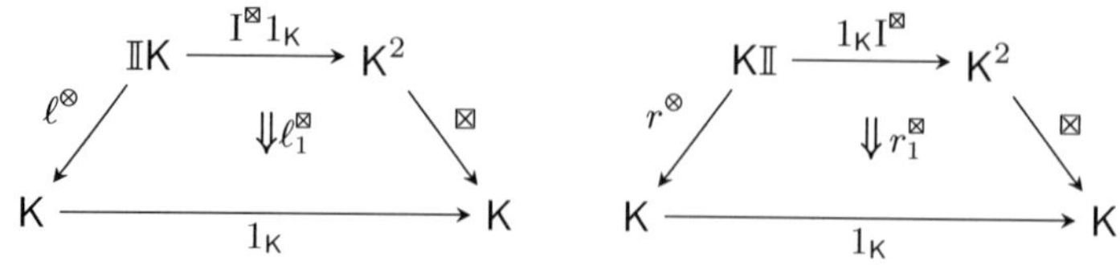

◇

Definition 1.4.17. Suppose K and L are monoidal V-categories. A *monoidal* V-*functor*

$$(F, F^2, F^0)\colon \mathsf{K} \longrightarrow \mathsf{L}$$

consists of

- a V-functor $F\colon \mathsf{K} \longrightarrow \mathsf{L}$;
- a V-natural transformation

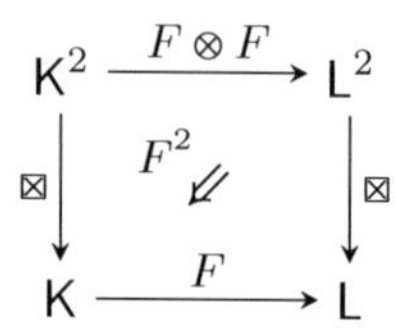

called the *monoidal constraint*; and
- a V-natural transformation

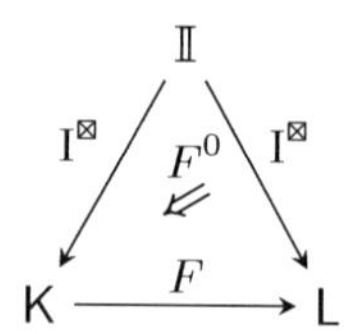

called the *unit constraint.*

These data are required to satisfy the following associativity and unity axioms.

Associativity: The composites of the following two *associativity pasting diagrams* are equal.

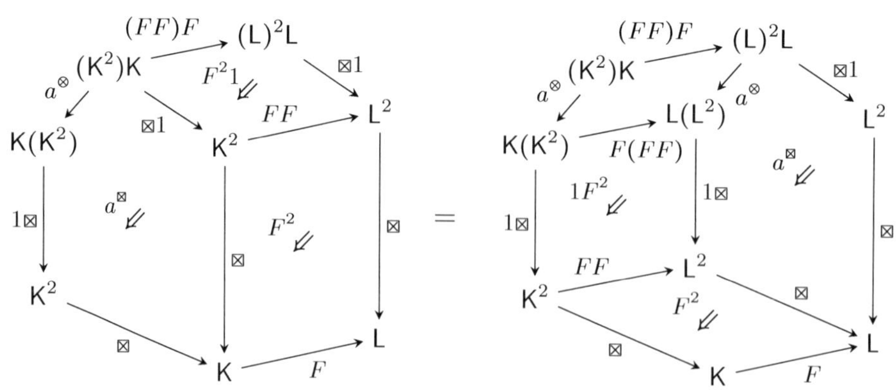

In the right hand diagram above, the unlabeled parallelogram commutes by naturality of $a^{\otimes}$.

Left Unity: The composites of the following two *left unity pasting diagrams* are equal.

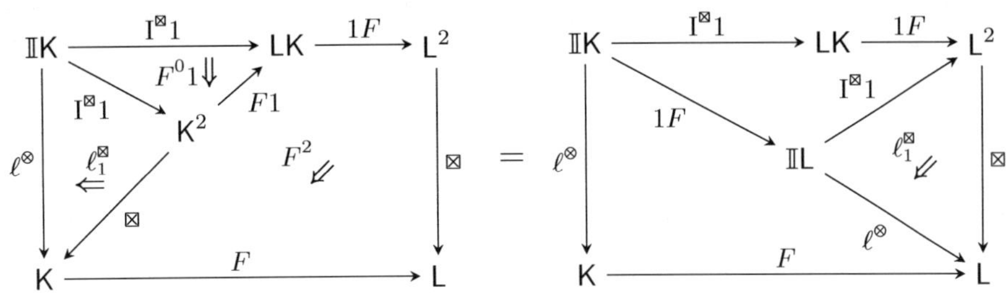

In the right hand diagram above, the lower unlabeled quadrilateral commutes by naturality of $\ell^{\otimes}$ and the upper unlabeled region commutes by 2-functorality of $\otimes$. The 2-cell isomorphisms labeled $\ell_1^{\boxtimes}$ are each the mate of $\ell^{\boxtimes}$ described in Definition 1.4.15.

Right Unity: The composites of the following two *right unity pasting diagrams* are equal.

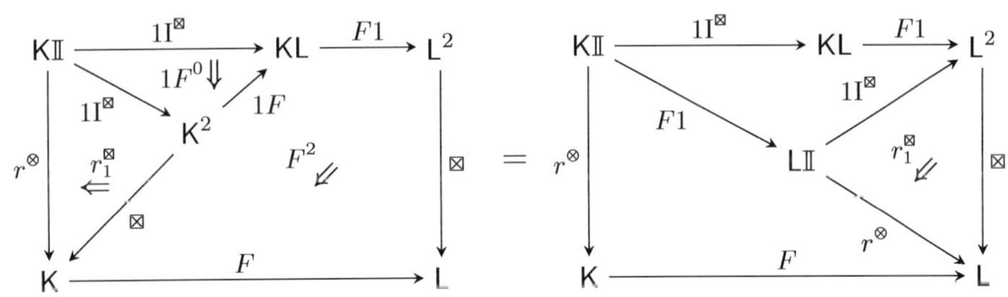

In the right hand diagram above, the lower unlabeled quadrilateral commutes by naturality of $r^{\otimes}$ and the upper unlabeled region commutes by 2-functorality of $\otimes$. The 2-cell isomorphisms labeled $r_1^{\boxtimes}$ are each the mate of $r^{\boxtimes}$ described in Definition 1.4.15.

This finishes the definition of a monoidal V-functor. Moreover, we have the following additional variants.

- A *unital monoidal* V*-functor* is one for which F^0 is invertible.
- A *strictly unital monoidal* V*-functor* is one for which F^0 is an identity.
- A *strong monoidal* V*-functor* is one for which both F^0 and F^2 are invertible.
- A *strict monoidal* V*-functor* is one for which both F^0 and F^2 are identities.

◇

Definition 1.4.18. Suppose K and L are braided monoidal V-categories, with V symmetric monoidal. A *braided monoidal* V-functor

$$(F, F^2, F^0)\colon \mathsf{K} \longrightarrow \mathsf{L}$$

is a monoidal V-functor that satisfies the following axiom.

Braid Axiom: The composites of the following two *braiding pasting diagrams* are equal.

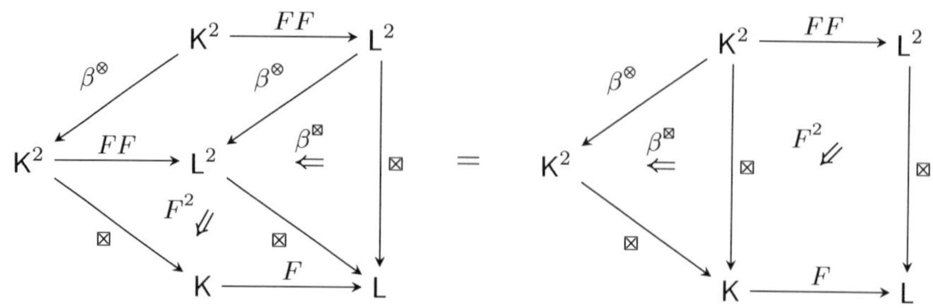

In the left hand diagram above, the unlabeled quadrilateral commutes by naturality of $\beta^{\otimes}$. This finishes the definition of a braided monoidal V-functor.

If K and L are symmetric monoidal V-categories then we say that F is a *symmetric monoidal* V*-functor.* ◇

Definition 1.4.19. For a pair of composable monoidal V-functors

$$\mathsf{K} \xrightarrow{F} \mathsf{L} \xrightarrow{P} \mathsf{M},$$

the *composite* PF is a monoidal V-functor with monoidal and unit constraints given, respectively, by the composites of the following pasting diagrams.

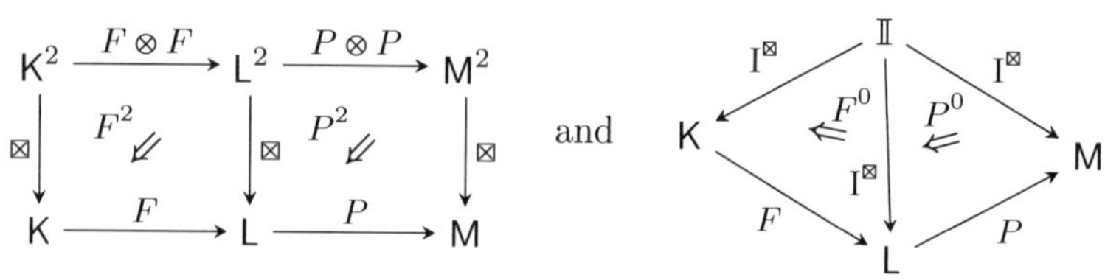

In Proposition 1.4.21 below we verify that (braided) monoidal V-functors are closed under composition. ◇

Lemma 1.4.20. *Suppose* K *is a monoidal* V*-category. The identity* V*-functor* 1_{K} *is monoidal as a* V*-functor. If* K *is braided monoidal (with* V *symmetric) then so is* 1_{K}.

Proof. We take the monoidal and unit constraints of 1_{K} to be identity V-natural transformations. Then the associativity and unity axioms of Definition 1.4.17 are tautologies because horizontal and vertical composition of V-natural transformations are strictly unital. The same holds for the braid axiom of Definition 1.4.18 if K is braided monoidal. □

Proposition 1.4.21. *In the context of Definition 1.4.19, the composite* PF *is a monoidal* V*-functor. If* F *and* P *are braided monoidal* V*-functors (with* V *symmetric monoidal), then so is* PF*. Moreover, composition of monoidal* V*-functors is strictly associative and unital.*

Proof. We verify the associativity axiom of Definition 1.4.17 via the equalities of pasting diagrams below; the unity and (in the braided case) braid axioms are similar.

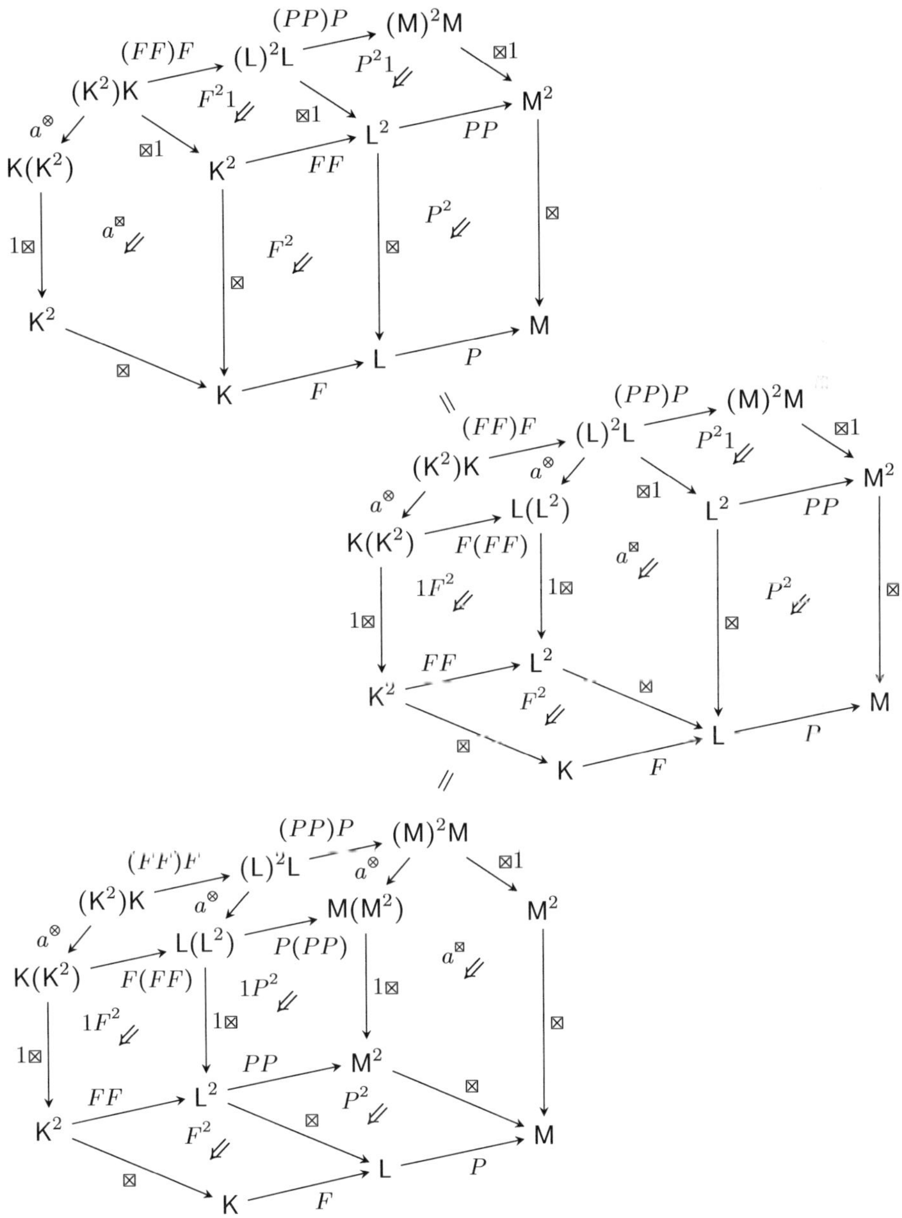

The equalities above follow from the associativity axioms for F and P. This proves the associativity axiom for PF because, by 2-functoriality of $\otimes$ (Proposition 1.3.6), we have the following:

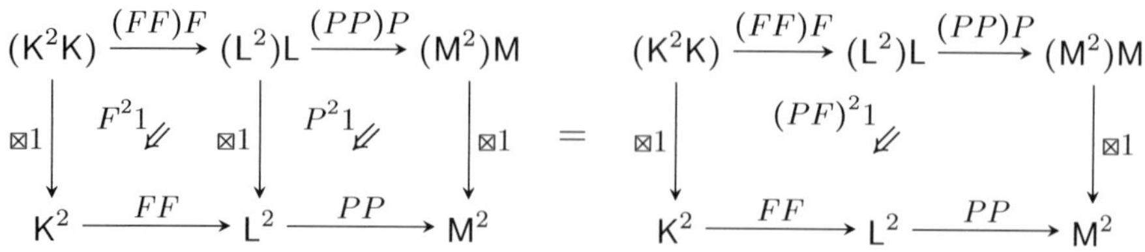

along with a similar equality for $1(PF)^2$.

The assertion that composition of (braided) monoidal V-functors is strictly associative and unital follows from the corresponding statements for underlying V-functors together with strictness of vertical and horizontal composition for V-natural transformations. For example, strictly unital composition implies that the monoidal and unit constraints of $1_{\mathsf{L}} \circ F$ and $F1_{\mathsf{K}}$ are equal to those of F. Similarly, the monoidal and unit constraints of a three-fold composite are given by the three-fold pasting of monoidal and unit constraints. □

Definition 1.4.22. Suppose K and L are monoidal V-categories, and suppose that $F, G\colon \mathsf{K} \longrightarrow \mathsf{L}$ are monoidal V-functors. A *monoidal* V*-natural transformation*

$$\theta\colon F \longrightarrow G$$

is a V-natural transformation of underlying V-functors that satisfies the following two additional axioms.

Monoidal Naturality: The composites of the following two *monoidal naturality pasting diagrams* are equal.

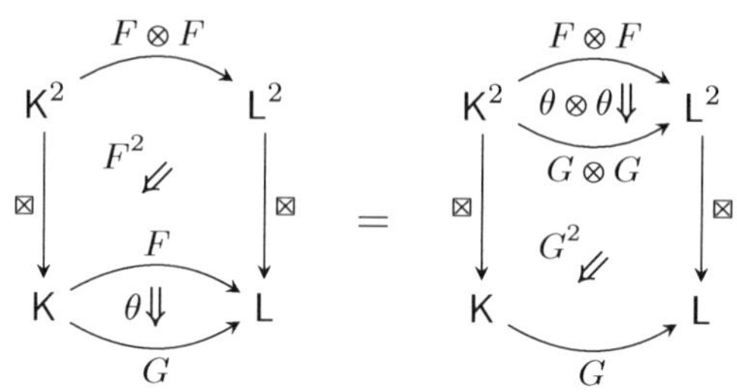

Unit Naturality: The composites of the following two *unit naturality pasting diagrams* are equal.

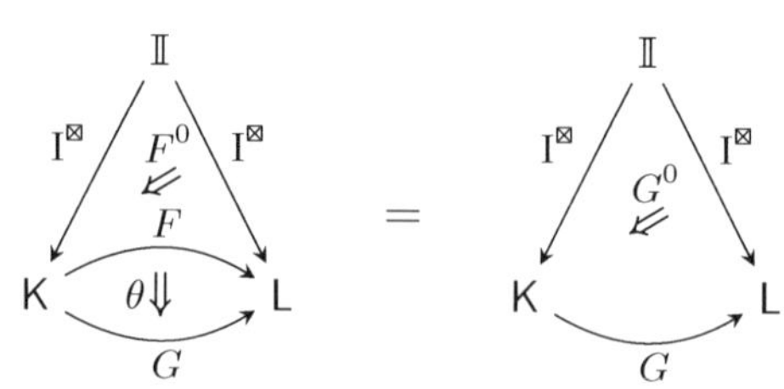

This finishes the definition of a monoidal V-natural transformation. Verification that monoidal V-natural transformations are closed under vertical and horizontal composition is given in Proposition 1.4.24 below. Then identities and composites of monoidal V-natural transformations are defined via underlying V-natural transformations. ◇

Explanation 1.4.23 (Components of monoidal V-natural transformations)**.** Recalling the definitions of horizontal and vertical composition for V-natural transformations (Definition 1.2.9), the monoidal naturality axiom of Definition 1.4.22 is equivalent to commutativity of the following diagram in V for each pair of objects X and Y in K.

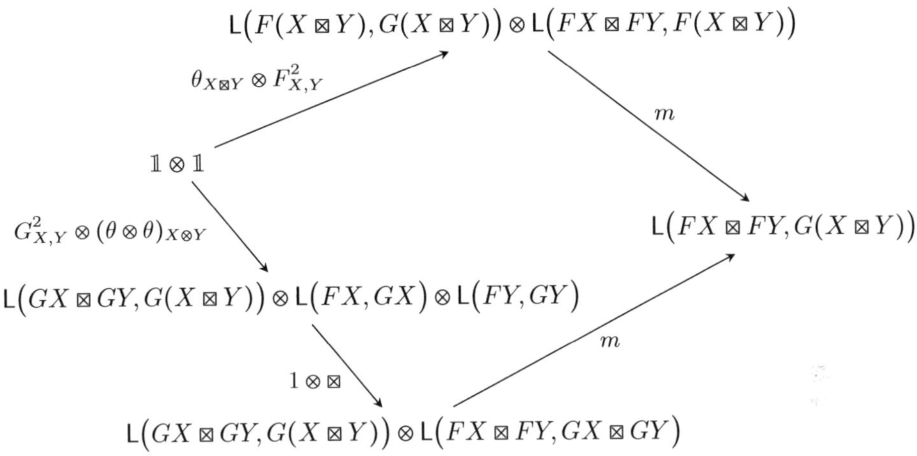

In the diagram above, $(\theta \otimes \theta)_{X \otimes Y}$ is the component of

$$\theta \otimes \theta \colon F \otimes F \longrightarrow G \otimes G$$

at $X \otimes Y$. Following Definition 1.3.5, it is defined to be

$$\mathbb{1} \xrightarrow{\lambda^{-1}} \mathbb{1} \otimes \mathbb{1} \xrightarrow{\theta_X \otimes \theta_Y} \mathsf{L}(FX, GX) \otimes \mathsf{L}(FY, GY)$$

Similarly, the unit naturality axiom of Definition 1.4.22 is equivalent to commutativity of the following diagram in V, with $\mathrm{I} = 1^{\boxtimes}$ denoting the unit objects of K and L.

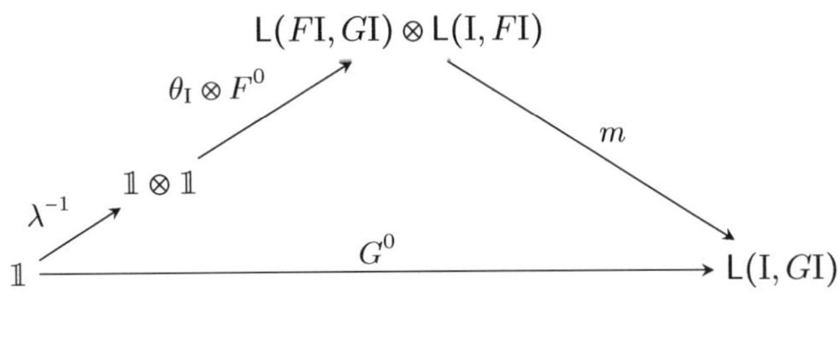

◇

Proposition 1.4.24. *Monoidal* V*-natural transformations are closed under vertical and horizontal composition of underlying* V*-natural transformations.*

Proof. This result follows from 2-functoriality of $\otimes$ (Proposition 1.3.6). We will demonstrate for vertical composition; horizontal composition is similar. Suppose that

$$F, G, H \colon \mathsf{K} \longrightarrow \mathsf{L}$$

are monoidal V-functors between monoidal V-categories. Suppose that

$$\theta \colon F \longrightarrow G \quad \text{and} \quad \phi \colon G \longrightarrow H$$

are monoidal V-natural transformations. Then monoidal naturality of θ and ϕ gives the first two equalities of pasting diagrams below. The third follows by 2-functoriality of $\otimes$; we write $\nu = \phi\theta$ to conserve space.

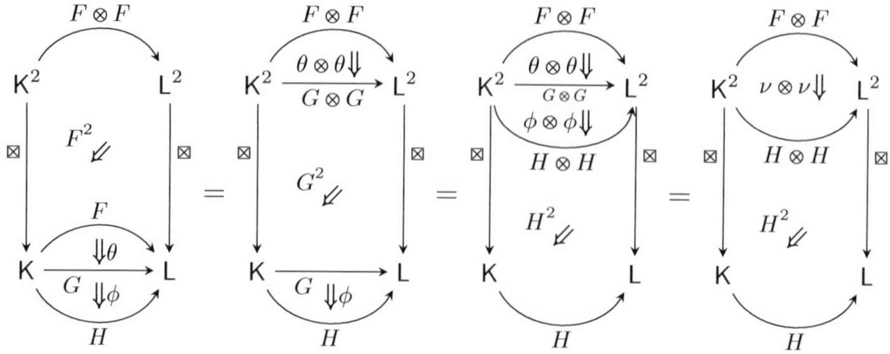

The equalities above show that $\phi\theta$ satisfies the monoidal naturality condition of Definition 1.4.22. The unit naturality condition is similar. □

The results of Propositions 1.4.21 and 1.4.24 show that (braided) monoidal V-functors and monoidal V-natural transformations satisfy the composition laws of a 2-category.

Definition 1.4.25. Suppose $\mathsf{V} = (\mathsf{V}, \otimes, \xi)$ is a braided monoidal category. For items (2) and (3) suppose that V is symmetric. We define the following 2-categories of small monoidal V-categories.

(1) Let V-MCat denote the 2-category whose objects are small monoidal V-categories, 1-cells are monoidal V-functors, and 2-cells are monoidal V-natural transformations.
(2) Let V-BMCat denote the 2-category whose objects are small braided monoidal V-categories, 1-cells are braided monoidal V-functors, and 2-cells are monoidal V-natural transformations.
(3) Let V-SMCat denote the 2-category whose objects are small symmetric monoidal V-categories, 1-cells are symmetric monoidal V-functors, and 2-cells are monoidal V-natural transformations.

These are related by 2-functors where the first recognizes symmetric structure as a special case of the braided analogue, the second forgets the braiding data, and the third forgets all monoidal data:

$$\text{V-SMCat} \longrightarrow \text{V-BMCat} \longrightarrow \text{V-MCat} \longrightarrow \text{V-Cat}. \qquad \diamond$$

Definition 1.4.26. A *monoidal* V*-adjunction* between monoidal V-categories K and L consists of a V-adjunction

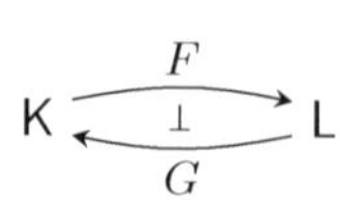

such that F and G are monoidal V-functors and the unit and counit are monoidal V-natural transformations. If K and L are small, this is an internal adjunction in V-MCat. A *monoidal adjoint* V*-equivalence* consists of a monoidal V-adjunction such that the unit and counit are monoidal V-natural isomorphisms. For V symmetric monoidal, a monoidal V-adjunction or V-equivalence is called *braided*, respectively *symmetric*, if F and G are braided, respectively symmetric, monoidal as V-functors. ◇

Next we describe a rotated multiplication on the opposite of a monoidal V-category. Recall from Definition 1.3.36 the V-functor

$$\mathsf{C}^{\mathsf{op}} \otimes \mathsf{D}^{\mathsf{op}} \xrightarrow{\gamma} (\mathsf{D} \otimes \mathsf{C})^{\mathsf{op}}$$

for V-categories C and D. Recall from Proposition 1.2.19 that $(-)^{\mathsf{op}}$ is a 2-functor on V-Cat that is covariant with respect to horizontal composition of 1- and 2-cells, but contravariant with respect to vertical composition of 2-cells.

Example 1.4.27 (Rotated Multiplication)**.** Suppose $\mathsf{V} = (\mathsf{V}, \otimes, \xi)$ is a braided monoidal category and suppose $\mathsf{K} = (\mathsf{K}, \boxtimes, \mathrm{I}, a, \ell, r)$ is a monoidal V-category. The *rotation of* K is defined by the following data.

- The underlying V-category is the opposite, $\overline{\mathsf{K}} = \mathsf{K}^{\mathsf{op}}$.
- The monoidal composition

  $$\overline{\boxtimes}: \overline{\mathsf{K}} \otimes \overline{\mathsf{K}} \longrightarrow \overline{\mathsf{K}}$$

  is defined to be the composite

  $$\mathsf{K}^{\mathsf{op}} \otimes \mathsf{K}^{\mathsf{op}} \xrightarrow{\gamma} (\mathsf{K} \otimes \mathsf{K})^{\mathsf{op}} \xrightarrow{\boxtimes^{\mathsf{op}}} \mathsf{K}^{\mathsf{op}}$$

  using γ (Definition 1.3.36) and the opposite V-functor of $\boxtimes$. Therefore, for objects $X, X', Y, Y' \in \mathsf{K}$ we have $X \overline{\boxtimes} Y = Y \boxtimes X$. The morphism on hom objects is given by the following composite of ξ and $\boxtimes$.

$$\begin{array}{ccccc} \overline{\mathsf{K}}(X, X') \otimes \overline{\mathsf{K}}(Y, Y') & & & & \overline{\mathsf{K}}(X \overline{\boxtimes} Y, X' \overline{\boxtimes} Y') \\ \parallel & & & & \parallel \\ \mathsf{K}(X', X) \otimes \mathsf{K}(Y', Y) & \xrightarrow{\xi} & \mathsf{K}(Y', Y) \otimes \mathsf{K}(X', X) & \xrightarrow{\boxtimes} & \mathsf{K}(Y' \boxtimes X', Y \boxtimes X) \end{array}$$

- The monoidal identity is

  $$\overline{\mathrm{I}} = \mathrm{I}^{\mathsf{op}} : \mathbb{I} = \mathbb{I}^{\mathsf{op}} \longrightarrow \mathsf{K}^{\mathsf{op}}.$$

- The monoidal unitors are given by

  $$\overline{\ell} = r^{-1} : \mathbb{1} \longrightarrow \mathsf{K}(X, X \boxtimes \mathrm{I}) = \overline{\mathsf{K}}(\mathrm{I} \overline{\boxtimes} X, X)$$

  and

  $$\overline{r} = \ell^{-1} : \mathbb{1} \longrightarrow \mathsf{K}(X, \mathrm{I} \boxtimes X) = \overline{\mathsf{K}}(X \overline{\boxtimes} \mathrm{I}, X)$$

  for $X \in \mathsf{K}$.
- The monoidal associator component at $X, Y, Z \in \overline{\mathsf{K}}$ is given by

$$\overline{a}_{X,Y,Z} = a_{Z,Y,X} : \mathbb{1} \longrightarrow \mathsf{K}\big((Z \boxtimes Y) \boxtimes X, Z \boxtimes (Y \boxtimes X)\big) = \overline{\mathsf{K}}\big((X \overline{\boxtimes} Y) \overline{\boxtimes} Z, X \overline{\boxtimes} (Y \overline{\boxtimes} Z)\big)$$

  for $X, Y, Z \in \mathsf{K}$.

To verify that the unitors and associator have the required sources and targets, we use Lemma 1.3.39 to see that the unlabeled regions in each of the following diagrams commute. Recall from Proposition 1.2.19 that $(-)^{\mathsf{op}}$ reverses directions of V-natural transformations. The components of $\overline{r}$, $\overline{\ell}$, and $\overline{a}$ given above are those of the V-natural transformations indicated below.

(1.4.28)

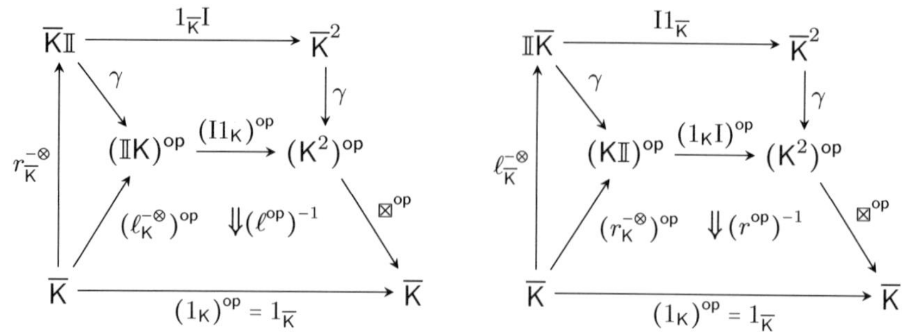

(1.4.29)

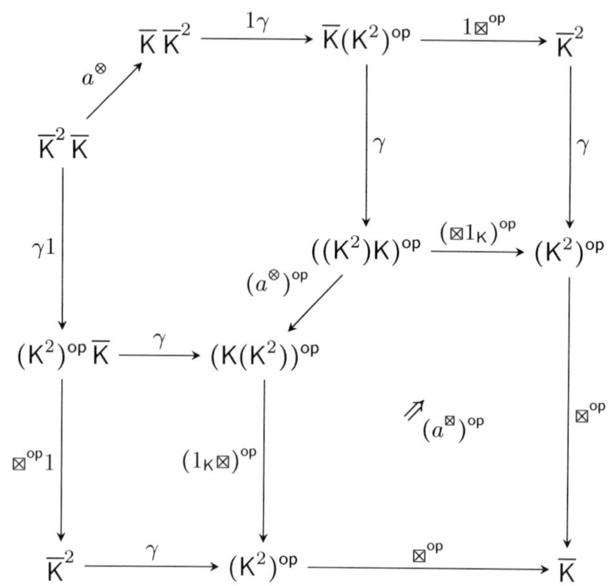

One can use either of two approaches to verify that these data satisfy the unity and associativity axioms of Definition 1.4.2. In one approach, use (2-) naturality of γ to verify that the relevant pasting diagrams reduce to whiskerings of those for K. In the other, consider components as in Explanation 1.4.8 to see that the required equalities for $\overline{\mathsf{K}}$ reduce to corresponding equalities for K. For example, the pentagon axiom for $\overline{a}$ is, for each quadruple of objects $X, Y, Z, W \in \overline{\mathsf{K}}$, an equality of components

$$\begin{aligned} \mathbb{1} \longrightarrow\ & \overline{\mathsf{K}}\Big(\big((X \mathbin{\overline{\boxtimes}} Y) \mathbin{\overline{\boxtimes}} Z\big) \mathbin{\overline{\boxtimes}} W\,,\ X \mathbin{\overline{\boxtimes}} \big(Y \mathbin{\overline{\boxtimes}} (Z \mathbin{\overline{\boxtimes}} W)\big)\Big) \\ &= \mathsf{K}\Big(\big((W \boxtimes Z) \boxtimes Y\big) \boxtimes X\,,\ W \boxtimes \big(Z \boxtimes (Y \boxtimes X)\big)\Big) \end{aligned}$$

The required equality follows from the corresponding equality for a at the quadruple W, Z, Y, X. Similarly, the unity axiom for $\overline{\mathsf{K}}$ at $X, Y \in \overline{\mathsf{K}}$ holds by the unity axiom for K at Y, X. ◇

1.5. Cat-Monoidal 2-Categories

The case $\mathsf{V} = \mathsf{Cat}$, the category of small categories, warrants special attention because the definition of monoidal Cat-category gives an important notion of strict monoidal bicategory. Since the notion of (locally small) 2-category is the same as that of a Cat-enriched category, and each 2-category is a bicategory, the term "monoidal 2-category" might be understood to mean either of the following:

- a monoidal V-category in the case $\mathsf{V} = \mathsf{Cat}$, or
- a monoidal bicategory whose underlying bicategory is a 2-category.

These are distinct notions and therefore to avoid ambiguity we introduce the following terms for the Cat-enriched case.

Definition 1.5.1. Suppose $(\mathsf{V}, \otimes, \mathbb{1}) = (\mathsf{Cat}, \times, \mathbb{1})$ in the context of Definitions 1.4.2, 1.4.10, and 1.4.13. We introduce the following special terminology.

- We say that K is a Cat-*monoidal 2-category* if K is a monoidal Cat-category in the sense of Definition 1.4.2.
- We say that K is a *braided* Cat-*monoidal 2-category* if K is a braided monoidal Cat-category in the sense of Definition 1.4.10.
- We say that K is a *symmetric* Cat-*monoidal 2-category* if K is a symmetric monoidal Cat-category in the sense of Definition 1.4.13. ◇

Next we relate the definition of Cat-monoidal 2-category, along with the braided and symmetric variants, to the 1-categorical analogues.

Explanation 1.5.2 (Relation to Monoidal 1-Categories)**.** Suppose that $(\mathsf{K}, \otimes, \mathbb{I}, a, \ell, r)$ is a Cat-monoidal 2-category.

(1) The underlying 1-categorical data of $(\mathsf{K}, \otimes, \mathbb{I}, a, \ell, r)$ form a monoidal 1-category. The axioms for a monoidal 1-category are obtained by taking components of the diagrams (1.4.5) and (1.4.6) above.

(2) In particular, the components of a, ℓ, and r satisfy the equality

$$\ell_{\mathbb{I}} = r_{\mathbb{I}} \colon \mathbb{I} \otimes \mathbb{I} \longrightarrow \mathbb{I}$$

from (1.1.4) along with the left and right unity diagrams of (1.1.5).

(3) If (K, β) is braided Cat-monoidal, then the underlying 1-category is braided monoidal.

(4) If (K, β) is symmetric Cat-monoidal, then the underlying 1-category is symmetric monoidal. ◇

Next we relate the definition of Cat-monoidal 2-category, along with the braided and symmetric variants, to the definitions of plain, braided, and symmetric monoidal bicategory.

Explanation 1.5.3 (Relation to Monoidal Bicategories)**.** Suppose K is a Cat-monoidal 2-category.

(1) The data of K are those of a monoidal bicategory whose base bicategory is a 2-category and for which the pentagonator π and 2-unitors λ, μ, and ρ are all identities. Commutativity of the pentagon diagram (1.4.6) is equivalent to triviality of π; commutativity of the middle unity diagram (1.4.5) is equivalent to triviality of μ; and the triviality of λ, respectively ρ, is equvialent to commutativity of the left, respectively right, unity diagram from (1.1.5). The non-Abelian 4-cocycle condition (I.6.4.2) along with the

two normalization axioms (I.6.4.3) and (I.6.4.4) then consist entirely of identity 2-cells because $\otimes$ is 2-functorial and a, ℓ, and r are 2-natural.

(2) If, moreover, K is braided Cat-monoidal, then the corresponding bicategory is braided monoidal with hexagonators $R_{-|--}$ and $R_{--|-}$ being identities. The four axioms of Definition I.6.5.3 are equalities of identity 2-cells.

(3) If, moreover, K is symmetric Cat-monoidal, then the corresponding bicategory is symmetric monoidal with syllepsis ν being an identity. The two syllepsis axioms of Definition I.6.5.7 as well as the triple braid axiom of Definition I.6.5.9 are equalities of identity 2-cells. ◇

Example 1.5.4. Equipped with the Cartesian product, $(\mathsf{Cat}, \times, \mathbb{1})$ is a symmetric Cat-monoidal 2-category. The associator, unitors, and braiding are those of the Cartesian product, and are 2-natural by construction. ◇

Theorem 1.5.5. *Suppose* $\mathsf{V} = (\mathsf{V}, \otimes, \xi)$ *is a braided monoidal category. Then the tensor product of* V*-categories makes* $(\mathsf{V}\text{-}\mathsf{Cat}, \otimes, \mathbb{I})$ *a* Cat*-monoidal 2-category. If* V *is symmetric monoidal, then so is* $(\mathsf{V}\text{-}\mathsf{Cat}, \beta^{\otimes})$.

Proof. All of the relevant data and axioms are described in Section 1.3 above. The data are the following.

- The monoidal composition is the tensor product for V-categories; see Definitions 1.3.3 and 1.3.5. It is shown to be a 2-functor in Proposition 1.3.6.
- The identity object is the unit V-category; see Definition 1.3.7.
- The monoidal unitors are $\ell^{\otimes}$ and $r^{\otimes}$; see Definition 1.3.8.
- The monoidal associator is $a^{\otimes}$; see Definition 1.3.10.
- If V is symmetric, the braiding is $\beta^{\otimes}$; see Definition 1.3.22.

The axioms are checked as follows.

- The unity axiom (1.4.5) is verified on components by Lemma 1.3.14.
- The pentagon axiom (1.4.6) is verified on components by Lemma 1.3.17.
- If V is symmetric, the hexagon axioms (1.4.11) and (1.4.12) are verified on components by Lemma 1.3.29.
- If V is symmetric, the symmetry axiom (1.4.14) is verified on components by Lemma 1.3.32. □

Restricting to the underlying 1-categorical data recovers the 1-categorical monoidal structure stated in Theorem 1.3.35.

1.6. Notes

1.6.1 (Enriched Categories)**.** The standard reference for the theory of enriched categories is [**Kel05**]. Many of the basic definitions are developed in [**EK66**], which is itself based on preliminary work referenced there. ◇

1.6.2 (Iterated Monoidal Structure on V and V-Cat)**.** Forcey [**For04**] defines enrichment over V when V has an iterated monoidal structure described in Chapter II.10. The main result [**For04**, Theorem1] shows that if V is n-fold monoidal then V-Cat is $(n-1)$-fold monoidal. The cases $n = 1$ and $n = 2$, along with the relevant Strictification Theorems 1.1.32 and 1.1.39, imply the statements in the top two rows of (1.3.1). Recall from Proposition II.10.1.21 that the case $n = \infty$ corresponds to V being permutative. This explains why our table in (1.3.1) collapses at the symmetric case. ◇

1.6.3 (Enriched Monoidal Categories via 2-Categorical Algebra). Day and Street [**DS97**] develop a theory of pseudomonoids in a Gray monoid and apply it to give definitions of monoidal V-categories (along with the braided and symmetric variants) that correspond, via the tricategorical strictification of [**GPS95, Gur13**], to those of Section 1.4. This approach via pseudomonoids or more general 2-monad theory also appears in the thesis of Cruttwell [**Cru09**]. We have chosen to give more direct definitions to limit the 2-categorical background required for this work. ◇

1.6.4 (Enriched Monoidal Categories for Strict V). For V strict monoidal, work of Kong and Zheng [**KZ18**, Definitions 2.3 and 2.4] gives definitions of (braided) monoidal V-categories in terms of the underlying categories. Theorem 2.5.1 below, which to our knowledge has not previously appeared in the literature, shows that the definitions of [**KZ18**] are special cases of the definitions given here.

Work of Morrison and Penneys [**MP19**] also gives a definition of strict monoidal V-category in the case that V is strict. In that work the authors pose the question of defining non-strict monoidal V-categories and proving coherence theorems. The definitions in this chapter together with the results of Chapter 2 provide positive answers to those questions.

The following table, where V denotes the enriching braided monoidal category, summarizes the various levels of strictness for [**MP19**], [**KZ18**] (where V is denoted B), and this chapter.

	[**MP19**]	[**KZ18**]	1.4
V	strict	strict	non-strict
monoidal V-categories	strict	non-strict	non-strict

◇

CHAPTER 2

Change of Enrichment

This chapter discusses change of enrichment along a monoidal, respectively braided monoidal, respectively symmetric monoidal functor

$$U{:}\,\mathsf{V} \longrightarrow \mathsf{W}.$$

Sections 2.1 through 2.4 describe the general 2-functoriality properties for change of enrichment along U. They show, moreover, that it is compatible with relevant monoidal structures under further assumptions that U, V, and W are braided or symmetric monoidal. The results are summarized in the statements of Theorems 2.2.7, 2.3.9, and 2.4.10.

We have two distinct purposes for covering this material. The first is for Sections 2.5 and 2.6, where change of enrichment along the corepresented functor

$$\mathsf{V}(\mathbb{1},-){:}\,\mathsf{V} \longrightarrow \mathsf{Set}$$

allows us to lift the coherence and strictification theory for ordinary monoidal categories, along with their braided and symmetric variants, to coherence and strictification results for monoidal V-categories. Similar results hold for the corresponding braided and symmetric variants.

Our second purpose for covering this material is for the development of various enriched monoidal categories and functors for our K-theory applications in Part 2. We will apply change of enrichment to develop new enriched monoidal structures from existing ones.

2.1. Change of Enriching Categories

In this section we discuss change of enriching category via a monoidal functor. Throughout this section we assume that

$$U{:}\,\mathsf{V} \longrightarrow \mathsf{W}$$

is a monoidal functor between monoidal categories. Braidings are not required.

Definition 2.1.1. Suppose $(\mathsf{V},\otimes,\mathbb{1})$ and $(\mathsf{W},\otimes,\mathbb{1})$ are monoidal categories and

$$U{:}\,\mathsf{V} \longrightarrow \mathsf{W}$$

is a monoidal functor. Then there is an induced *change of enrichment* 2-functor

$$(-)_U{:}\,\mathsf{V}\text{-}\mathsf{Cat} \longrightarrow \mathsf{W}\text{-}\mathsf{Cat}$$

given as follows.

- For a small V-category C, the W-category C_U has the same objects and has hom objects

$$\mathsf{C}_U(X,Y) = U(\mathsf{C}(X,Y)).$$

The composition and identity of C_U are given by applying U to those of C and composing the monoidal and unit constraints of U, respectively.

- For a V-functor $F\colon \mathsf{C} \longrightarrow \mathsf{D}$, the W-functor F_U is the same assignment on objects and is the induced morphism on hom objects

$$F_U = U(F_{X,Y})\colon U(\mathsf{C}(X,Y)) \longrightarrow U(\mathsf{D}(FX,FY)).$$

- For a V-natural transformation $\theta\colon F \longrightarrow G$, the W-natural transformation θ_U has components given by the following composite in W:

$$\mathbb{1} \xrightarrow{U^0} U\mathbb{1} \xrightarrow{U\theta_X} U(\mathsf{D}(FX,GX)). \qquad \diamond$$

Proposition 2.1.2. *In the context of Definition 2.1.1, $(-)_U$ is a 2-functor.*

Proof. Verifying that $(-)_U$ gives the requisite assignments on 0-, 1-, and 2-cells consists of verifying that each of C_U, F_U, and θ_U satisfy the axioms for W-category, W-functor, and W-natural transformation, respectively. We will check the first, and the other two are similar.

The composition in C_U is defined by the following composite for objects X, Y, and Z in C:
(2.1.3)

$$U(\mathsf{C}(Y,Z)) \otimes U(\mathsf{C}(X,Y)) \xrightarrow{U^2} U(\mathsf{C}(Y,Z) \otimes \mathsf{C}(X,Y)) \xrightarrow{Um} U(\mathsf{C}(X,Z)).$$

The associativity diagram (1.2.2) for this composition is the outer diagram below, where we write $\mathsf{C}_{A,B}$ for the object $\mathsf{C}(A,B)$ to save space.

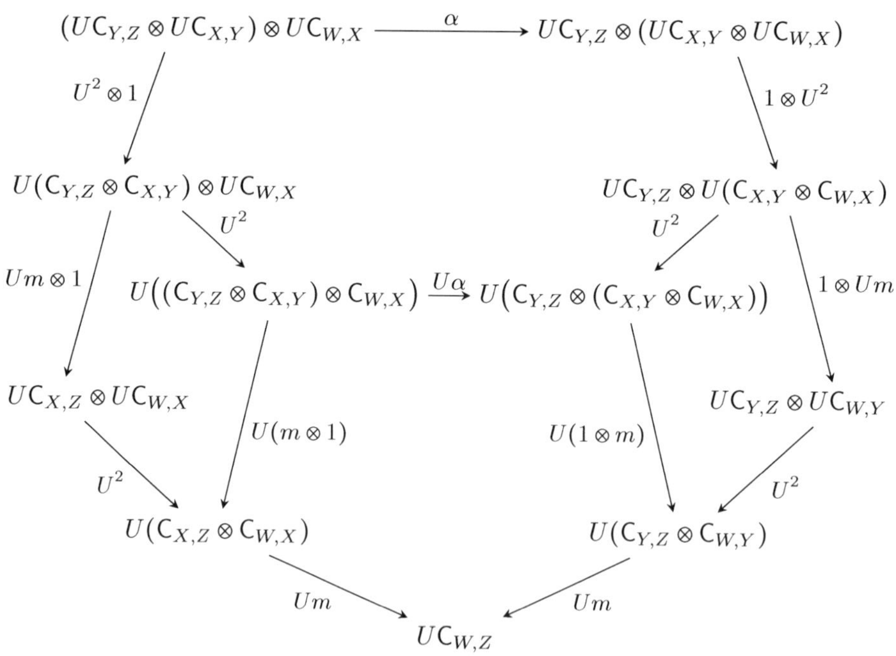

The upper hexagon commutes by the associativity axiom (1.1.9) for U. The lower pentagon commutes by applying U to the associativity diagram (1.2.2) for composition in C. Each of the two quadrilaterals commutes because U preserves identity morphisms and U^2 is natural.

The identity of an object X in C_U is defined by the following composite:

$$\mathbb{1} \xrightarrow{U^0} U\mathbb{1} \xrightarrow{U(i_X)} U(\mathsf{C}(X,X)). \tag{2.1.4}$$

The left side of the unity diagram (1.2.3) for this composition is the outer diagram below; we again abbreviate $\mathsf{C}(A,B)$ as $\mathsf{C}_{A,B}$.

$$\begin{array}{ccccc}
\mathbb{1} \otimes U\mathsf{C}_{X,Y} & & \xrightarrow{\quad\lambda\quad} & & U\mathsf{C}_{X,Y} \\
\downarrow{\scriptstyle U^0 \otimes 1} & & & \nearrow{\scriptstyle U\lambda} & \\
U\mathbb{1} \otimes U\mathsf{C}_{X,Y} & \xrightarrow{U^2} & U(\mathbb{1} \otimes \mathsf{C}_{X,Y}) & & \downarrow{\scriptstyle 1} \\
\downarrow{\scriptstyle Ui_Y \otimes 1} & & \downarrow{\scriptstyle U(i_Y \otimes 1)} & & \\
U\mathsf{C}_{Y,Y} \otimes U\mathsf{C}_{X,Y} & \xrightarrow{U^2} & U(\mathsf{C}_{Y,Y} \otimes \mathsf{C}_{X,Y}) & \xrightarrow{Um} & U\mathsf{C}_{X,Y}
\end{array}$$

The upper quadrilateral commutes by the left unity axiom (1.1.10) for U. The lower left square commutes because U preserves identities and U^2 is natural. The quadrilateral at right commutes by applying U to the left side of the unity diagram (1.2.3) for composition in C. The right side of the unity diagram for composition in C_U is similar.

To verify that $(-)_U$ is a 2-functor, we must verify that it preserves identity 1- and 2-cells, horizontal composition of 1- and 2-cells, and vertical composition of 2-cells. The identity V-functor 1_{C} is given by the identity on objects and identity morphisms on each hom object $\mathsf{C}(X,Y)$. Therefore $(1_{\mathsf{C}})_U = 1_{\mathsf{C}_U}$ because U preserves identity morphisms. For V-functors

$$\mathsf{C} \xrightarrow{F} \mathsf{D} \xrightarrow{G} \mathsf{E},$$

recall from Definition 1.2.4 that the composite GF is given by the composite on objects of C and

$$(GF)_{X,Y} = G_{FX,FY}F_{X,Y}$$

for $X,Y \in \mathsf{C}$. Therefore,

$$(GF)_{U;X,Y} = (G_U \circ F_U)_{X,Y}$$

because U preserves composition. The computations showing that $(-)_U$ preserves horizontal and vertical composition of 2-cells are similar and make use of the monoidal compatibility axioms (1.1.9) and (1.1.10) for U. □

Lemma 2.1.5. *The corepresented functor*

$$\mathsf{V}(\mathbb{1},-)\colon \mathsf{V} \longrightarrow \mathsf{Set}$$

is a monoidal functor with respect to the monoidal product in V *and the Cartesian product in* Set*. If* V *is braided, respectively symmetric, monoidal then* $\mathsf{V}(\mathbb{1},-)$ *is a braided, respectively symmetric, monoidal functor.*

Proof. The monoidal constraint for $\mathsf{V}(\mathbb{1},-)$ is given by functoriality of the monoidal product and composition with a unitor:

$$\mathsf{V}(\mathbb{1},X) \times \mathsf{V}(\mathbb{1},Y) \xrightarrow{-\otimes-} \mathsf{V}(\mathbb{1}\otimes\mathbb{1}, X\otimes Y) \xrightarrow{(\lambda^{-1})^*} \mathsf{V}(\mathbb{1}, X\otimes Y).$$

The unit constraint for $\mathsf{V}(\mathbb{1},-)$ is given by the identity morphism of $\mathbb{1}$, where $*$ denotes the terminal set:

$$* \xrightarrow{1_{\mathbb{1}}} \mathsf{V}(\mathbb{1},\mathbb{1}).$$

The associativity axiom (1.1.9) follows from naturality of α: for morphisms

$$f \in \mathsf{V}(\mathbb{1},X), \quad g \in \mathsf{V}(\mathbb{1},Y), \quad \text{and} \quad h \in \mathsf{V}(\mathbb{1},Z),$$

the following diagram commutes by naturality of α and compatibility with unit isomorphisms (1.1.5).

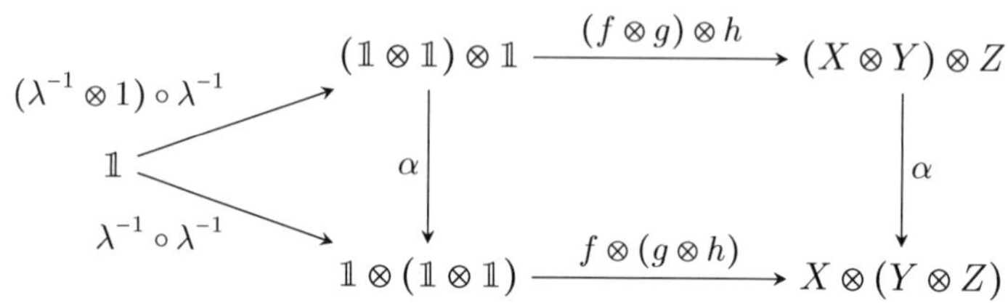

Combining this with the functoriality of $\otimes$ verifies the associativity axiom.

The left and right unity axioms (1.1.10) follow likewise from the naturality of the unitors in V. For example, the left unity axiom follows from commutativity of

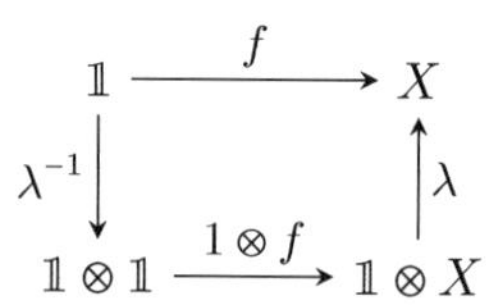

for each $f \in \mathsf{V}(\mathbb{1},X)$. The right unity axiom follows similarly, and uses the equality $\lambda_{\mathbb{1}} = \rho_{\mathbb{1}}$ (1.1.4). This completes the proof that $\mathsf{V}(\mathbb{1},-)$ is a monoidal functor.

If V is braided or symmetric monoidal, then similar analysis shows that the braid compatibility axiom (1.1.18) follows from naturality of ξ and the equality $\xi_{\mathbb{1},\mathbb{1}} = 1_{\mathbb{1}\otimes\mathbb{1}}$ (1.1.21). □

We apply Lemma 2.1.5 together with Proposition 2.1.2 to define the underlying category of a V-category.

Definition 2.1.6. Suppose C is a V-category. The *underlying category of* C is denoted C_0 and defined as follows. The objects of C_0 are those of C, and the morphism sets are defined by

$$\mathsf{C}_0(X,Y) = \mathsf{V}(\mathbb{1},\mathsf{C}(X,Y)).$$

◇

Lemma 2.1.7. *The 2-functor*

$$(-)_0 : \mathsf{V}\text{-}\mathsf{Cat} \longrightarrow \mathsf{Cat}$$

is injective on 2-cells. That is, two parallel V*-natural transformations* θ *and* ω *are equal if and only if their underlying natural transformations* θ_0 *and* ω_0 *are equal.*

Proof. Suppose $F,G\colon \mathsf{C} \longrightarrow \mathsf{D}$ are V-functors and $\theta,\omega\colon F \longrightarrow G$ are V-natural transformations. The component of θ_0 at an object X is defined to be the image of the following composite, where $*$ denotes the terminal set:

$$* \xrightarrow{1_{\mathbb{1}}} \mathsf{V}(\mathbb{1},\mathbb{1}) \xrightarrow{\mathsf{V}(\mathbb{1},\theta_X)} \mathsf{V}(\mathbb{1},\mathsf{D}(FX,GX)).$$

This morphism of sets sends the unique element of $*$ to the composite $\theta_X \circ 1_{\mathbb{1}} \in \mathsf{V}(\mathbb{1}, \mathsf{D}(FX, GX))$. Similarly, we have $(\omega_0)_X = \omega_X \circ 1_{\mathbb{1}}$. The result then follows because $\theta_X = \omega_X$ if and only if $\theta_X \circ 1_{\mathbb{1}} = \omega_X \circ 1_{\mathbb{1}}$. $\square$

Next we define underlying represented, respectively corepresented, functors on the underlying category of a V-category D. We use them to give an alternative condition for V-naturality in Lemma 2.1.11 below.

Definition 2.1.8. Suppose D is a V-category and

$$\theta\colon \mathbb{1} \longrightarrow \mathsf{D}(P,Q)$$

is a morphism in V. Then for each $X \in \mathsf{D}$ we let $\mathsf{D}(X,\theta)$ denote the following composite in V.

$$\begin{array}{ccc} \mathsf{D}(X,P) & \xrightarrow{\quad \mathsf{D}(X,\theta) \quad} & \mathsf{D}(X,Q) \\ {\scriptstyle \lambda^{-1}}\Big\downarrow & & \Big\uparrow{\scriptstyle m} \\ \mathbb{1} \otimes \mathsf{D}(X,P) & \xrightarrow{\quad \theta \otimes 1 \quad} & \mathsf{D}(P,Q) \otimes \mathsf{D}(X,P) \end{array} \tag{2.1.9}$$

Similarly, for each $Y \in \mathsf{D}$ we let $\mathsf{D}(\theta, Y)$ denote the following composite in V.

$$\begin{array}{ccc} \mathsf{D}(Q,Y) & \xrightarrow{\quad \mathsf{D}(\theta,Y) \quad} & \mathsf{D}(P,Y) \\ {\scriptstyle \rho^{-1}}\Big\downarrow & & \Big\uparrow{\scriptstyle m} \\ \mathsf{D}(Q,Y) \otimes \mathbb{1} & \xrightarrow{\quad 1 \otimes \theta \quad} & \mathsf{D}(Q,Y) \otimes \mathsf{D}(P,Q) \end{array} \tag{2.1.10}$$

Recalling the definition of the identity (2.1.4) and composition (2.1.3) in D_0, it follows directly from the axioms for unity (1.2.3) and associativity of composition (1.2.2) that the constructions above define functors

$$\mathsf{D}(X,-)\colon \mathsf{D}_0 \longrightarrow \mathsf{V} \quad \text{and} \quad \mathsf{D}(-,Y)\colon \mathsf{D}_0^{\mathsf{op}} \longrightarrow \mathsf{V}$$

called the *underlying corepresented functor* and *underlying represented functor*, respectively. $\diamond$

Lemma 2.1.11. *Suppose $F, G\colon \mathsf{C} \longrightarrow \mathsf{D}$ are V-functors between V-categories C and D. Suppose that*

$$\theta_X\colon \mathbb{1} \longrightarrow \mathsf{D}(FX, GX)$$

is a collection of morphisms in V for $X \in \mathsf{C}$. The morphisms θ_X are the components of a V-natural transformation θ if and only if the following diagram commutes for each pair of objects X and Y in C.

$$\begin{array}{ccc} \mathsf{C}(X,Y) & \xrightarrow{\quad F \quad} & \mathsf{D}(FX,FY) \\ {\scriptstyle G}\Big\downarrow & & \Big\downarrow{\scriptstyle \mathsf{D}(FX,\theta_Y)} \\ \mathsf{D}(GX,GY) & \xrightarrow{\quad \mathsf{D}(\theta_X, GY) \quad} & \mathsf{D}(FX,GY) \end{array} \tag{2.1.12}$$

Proof. Consider the following diagram in V for each $X, Y \in \mathsf{C}$.

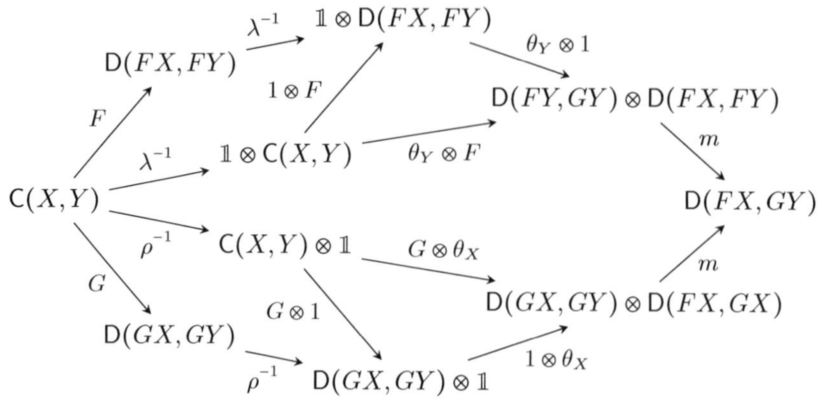

Commutativity of the innermost region in the above diagram is the naturality condition (1.2.8) of Definition 1.2.7 above. On the other hand, by the definitions of $\mathsf{D}(FX, \theta_Y)$ and $\mathsf{D}(\theta_X, GY)$ in Definition 2.1.8, the outermost composites in the above diagram are precisely the morphisms in (2.1.12).

The two quadrilaterals and two triangles in the above diagram commute by naturality of the unitors and functoriality of the monoidal product

$$\otimes : \mathsf{V} \times \mathsf{V} \longrightarrow \mathsf{V},$$

respectively. Therefore, the innermost region in the above diagram commutes if and only if the two composites along the boundary are equal. This proves the result. □

2.2. 2-Functoriality of Change of Enrichment

In this section we show that the assignment from monoidal functors

$$U : \mathsf{V} \longrightarrow \mathsf{W}$$

to 2-functors

$$(-)_U : \mathsf{V}\text{-}\mathsf{Cat} \longrightarrow \mathsf{W}\text{-}\mathsf{Cat}$$

is 2-functorial with respect to monoidal functors and monoidal natural transformations. Throughout this section we work with general monoidal categories, monoidal functors, and monoidal natural transformations; braidings are not assumed.

Next we define the 2-natural transformation, between change of enrichment 2-functors, that is induced by a monoidal natural transformation.

Definition 2.2.1. Suppose V and W are moniodal categories, $U, T : \mathsf{V} \longrightarrow \mathsf{W}$ are moniodal functors, and $\mu : U \longrightarrow T$ is a monoidal natural transformation. For each V-category C let

$$\mathsf{C}_\mu : \mathsf{C}_U \longrightarrow \mathsf{C}_T$$

denote the W-functor that is the identity on objects and whose morphism on hom objects is

$$U\mathsf{C}(X, Y) \xrightarrow{\mu_{\mathsf{C}(X,Y)}} T\mathsf{C}(X, Y),$$

the component of μ at $\mathsf{C}(X, Y)$. The composition and identity axioms of Definition 1.2.4 for C_μ follow from the monoidal axioms (1.1.13) for μ and naturality of μ. ◇

Proposition 2.2.2. *In the context of Definition 2.2.1, the components* C_μ *are 2-natural with respect to* V*-functors and* V*-natural transformations.*

Proof. Suppose given V-categories C and D together with V-functors F and G and a V-natural transformation θ as in the diagram below:

$$\mathsf{C} \underset{G}{\overset{F}{\underset{\Downarrow\theta}{\rightrightarrows}}} \mathsf{D}.$$

Consider the following two diagrams in W-Cat.

$$(2.2.3)\qquad \begin{array}{ccc} \mathsf{C}_U & \underset{G_U}{\overset{F_U}{\underset{\Downarrow\theta_U}{\rightrightarrows}}} & \mathsf{D}_U \\ {\scriptstyle \mathsf{C}_\mu}\downarrow & & \downarrow{\scriptstyle \mathsf{D}_\mu} \\ \mathsf{C}_T & \xrightarrow[G_T]{} & \mathsf{D}_T \end{array} \qquad \begin{array}{ccc} \mathsf{C}_U & \xrightarrow{F_U} & \mathsf{D}_U \\ {\scriptstyle \mathsf{C}_\mu}\downarrow & & \downarrow{\scriptstyle \mathsf{D}_\mu} \\ \mathsf{C}_T & \underset{G_T}{\overset{F_T}{\underset{\Downarrow\theta_T}{\rightrightarrows}}} & \mathsf{D}_T \end{array}$$

The quadrilateral region of each diagram commutes by naturality of μ. For example the diagram involving G becomes the following naturality square for each hom object $\mathsf{C}(X,Y)$.

$$\begin{array}{ccc} U\mathsf{C}(X,Y) & \xrightarrow{UG} & U\mathsf{D}(GX,GY) \\ {\scriptstyle \mu}\downarrow & & \downarrow{\scriptstyle \mu} \\ T\mathsf{C}(X,Y) & \xrightarrow{TG} & T\mathsf{D}(GX,GY) \end{array}$$

To compare the two whiskerings indicated in the two diagrams of (2.2.3), we use the Whiskering Lemma 1.2.12. For each $X \in \mathsf{C}$, the component $(1_{\mathsf{D}_\mu} * \theta_U)_X$ is given, via Lemma 1.2.12, by the upper right composite in the diagram below. Similarly, the component $(\theta_T * 1_{\mathsf{C}_\mu})_X$ is given by the lower composite.

$$\begin{array}{cccc} & U\mathbb{1} & \xrightarrow{U(\theta_X)} & U\mathsf{D}(FX,GX) \\ {\scriptstyle U^0}\nearrow & {\scriptstyle \mu}\downarrow & & \downarrow{\scriptstyle \mu} \\ \mathbb{1} \quad {\scriptstyle T^0}\searrow & T\mathbb{1} & \xrightarrow{T(\theta_X)} & T\mathsf{D}(FX,GX) \end{array}$$

In the above diagram, the triangle commutes by the unit condition (1.1.13) for μ and the square commutes by naturality of μ. Thus the two whiskerings in (2.2.3) are equal. This finishes the verification that $(-)_\mu$ is 2-natural. □

Recall from Definition 1.1.11 the monoidal and unit constraints of a composite of monoidal functors

$$\mathsf{V}_1 \xrightarrow{U_1} \mathsf{V}_2 \xrightarrow{U_2} \mathsf{V}_3.$$

are given, respectively by

$$(U_2U_1)^2 = (U_2(U_1^2)) \circ U_2^2 \quad \text{and} \quad (U_2U_1)^0 = (U_2(U_1^0)) \circ U_2^0.$$

Proposition 2.2.4. *Suppose given monoidal categories* V_1, V_2, *and* V_3 *together with monoidal functors*

$$\mathsf{V}_1 \xrightarrow{U_1} \mathsf{V}_2 \xrightarrow{U_2} \mathsf{V}_3.$$

Then change of enrichment $(-)_{(U_2U_1)}$ *along the composite* U_2U_1 *is equal as a 2-functor to the composite*

$$\mathsf{V}_1\text{-}\mathsf{Cat} \xrightarrow{(-)_{U_1}} \mathsf{V}_2\text{-}\mathsf{Cat} \xrightarrow{(-)_{U_2}} \mathsf{V}_3\text{-}\mathsf{Cat}.$$

Proof. First we describe

$$((-)_{U_1})_{U_2} \quad \text{and} \quad (-)_{(U_2U_1)}$$

on objects of V_1-Cat. For a V_1-category C, the objects of the V_3-categories $(\mathsf{C}_{U_1})_{U_2}$ and $\mathsf{C}_{(U_2U_1)}$ are the same as those of C. Moreover we have an equality of hom objects

$$(\mathsf{C}_{U_1})_{U_2}(X,Y) = U_2(\mathsf{C}_{U_1}(X,Y)) = U_2U_1\mathsf{C}(X,Y) = \mathsf{C}_{(U_2U_1)}(X,Y).$$

To finish checking that $(\mathsf{C}_{U_1})_{U_2}$ and $\mathsf{C}_{(U_2U_1)}$ are equal as V_3-categories, we need to verify that the composition and identity morphisms are equal.

For objects $X, Y, Z \in \mathsf{C}$, the composition morphism for $(\mathsf{C}_{U_1})_{U_2}$ is given by the composite

$$\begin{array}{c} (\mathsf{C}_{U_1})_{U_2}(Y,Z) \otimes (\mathsf{C}_{U_1})_{U_2}(X,Y) \\ \downarrow U_2^2 \\ U_2\big(\mathsf{C}_{U_1}(Y,Z) \otimes \mathsf{C}_{U_1}(X,Y)\big) \\ \downarrow U_2(U_1^2) \\ U_2U_1\big(\mathsf{C}(Y,Z) \otimes \mathsf{C}(X,Y)\big) \xrightarrow{U_2U_1 m} U_2U_1\mathsf{C}(X,Z) \end{array} \tag{2.2.5}$$

because $(U_1 m) \circ U_1^2$ is the composition for C_{U_1} and U_2 is functorial. But since the monoidal constraint of the composite U_2U_1 is precisely $(U_2(U_1^2)) \circ U_2^2$, (2.2.5) also gives the composition morphism for $\mathsf{C}_{(U_2U_1)}$. A similar computation using the unit constraint for (U_2U_1) shows that the identity morphisms of $(\mathsf{C}_{U_1})_{U_2}$ and $\mathsf{C}_{(U_2U_1)}$ are equal. This shows that

$$((-)_{U_1})_{U_2} \quad \text{and} \quad (-)_{(U_2U_1)}$$

give the same assignments on objects of V_1-Cat.

Next we check assignments on 1-cells. For a V_1-functor

$$F: \mathsf{C} \longrightarrow \mathsf{D}$$

the morphisms on hom objects for both $(F_{U_1})_{U_2}$ and $F_{(U_2U_1)}$ are given by

$$U_2U_1\mathsf{C}(X,Y) \xrightarrow{U_2U_1(F_{X,Y})} U_2U_1\mathsf{D}(FX,FY)$$

for $X, Y \in \mathsf{C}$. Therefore, $(F_{U_1})_{U_2} = F_{(U_2U_1)}$.

Finally we check assignments on 2-cells. For a V_1-natural transformation $\theta: F \longrightarrow G$, each of $(\theta_{U_1})_{U_2}$ and $\theta_{(U_2U_1)}$ the component at $X \in C$ given by the composite

$$\mathbb{1} \xrightarrow{U_2^0} U_2\mathbb{1} \xrightarrow{U_2(U_1^0)} U_2U_1\mathbb{1} \xrightarrow{U_2U_1(\theta_X)} U_2U_1\mathsf{D}(FX, GX)$$

because U_2 is functorial and the unit constraint of the composite U_2U_1 is precisely $(U_2(U_1^0)) \circ U_2^0$. This shows that $(\theta_{U_1})_{U_2} = \theta_{(U_2U_1)}$ and completes the proof. □

Recall from Example I.6.1.14 that MCat denotes the 2-category of small monoidal categories, monoidal functors, and monoidal natural transformations.

Motivation 2.2.6 (2-Functoriality and Size). The preceding results of this section can be summarized as a 2-functoriality statement where

$$\begin{aligned} \mathsf{V} &\longmapsto \mathsf{V}\text{-}\mathsf{Cat}, \\ U &\longmapsto (-)_U, \quad \text{and} \\ \mu &\longmapsto (-)_\mu \end{aligned}$$

are the assignments on 0-, 1-, and 2-cells of a 2-functor E out of MCat. However, a nontrivial set-theoretic consideration is necessary. Although we define each V-Cat as a 2-category consisting of small V-categories, V-Cat itself is generally not small. For this reason, the codomain of E must be enlarged. Subtleties of this sort motivated Eilenberg and Mac Lane to introduce the category of "large" categories in the original category theory paper [**EML45**], and that is one resolution available here too.

A similar but technically different approach can be given as follows. The definitions of small monoidal categories, small V-categories, and all other uses of small in this work, implicitly assume a chosen model for set theory, called a *universe* $\mathcal{U}$. The term *set* is reassigned to mean *member of* $\mathcal{U}$, and a category or V-category C called *small* if $\mathsf{Ob}\,\mathsf{C} \in \mathcal{U}$. Let us temporarily call this condition $\mathcal{U}$-small. Grothendieck's *Axiom of Universes* (see Note 2.7.3) implies (by choosing a larger inaccessible cardinal) the existence of a Grothendieck universe $\mathcal{U}'$ such that, for each $\mathcal{U}$-small monoidal category V, and for V-Cat denoting the 2-category of $\mathcal{U}$-small V-categories, we have

$$\mathsf{Ob}(\mathsf{V}\text{-}\mathsf{Cat}) \in \mathcal{U}'.$$

Thus, if 2Cat′ denotes the 2-category of all $\mathcal{U}'$-small 2-categories, then V-Cat is an object of 2Cat′ for each $\mathcal{U}$-small monoidal category V. All of our previous discussion of 2Cat applies to 2Cat′ in the (Grothendieck) universe $\mathcal{U}'$.

We point out that this is only one convenient way of resolving the size subtlety. This one, the Eilenberg-Mac Lane use of "large", and a number of other more nuanced approaches, are outlined in [**Shu∞b**]. We state the next result with a hypothesis about $\mathcal{U}'$ that may follow from any of such approaches. ◇

Theorem 2.2.7. *Suppose $\mathcal{U}'$ is a universe of sets containing* Ob(V-Cat) *for each small monoidal category* V. *Let* 2Cat′ *denote the 2-category of 2-categories that are small with respect to $\mathcal{U}'$. Then change of enrichment provides a 2-functor*

$$E: \mathsf{MCat} \longrightarrow \mathsf{2Cat}'$$

given by

- $E\mathsf{V} = \mathsf{V}\text{-}\mathsf{Cat}$
- $EU = (-)_U$
- $E\mu = (-)_\mu$

for a small monoidal category V*, monoidal functor* U*, and monoidal natural transformation* μ*, respectively.*

Proof. By hypothesis on $\mathcal{U}'$, together with Propositions 2.1.2 and 2.2.2, E is a valid assignment on objects, 1-cells, and 2-cells. Now we verify that E is 2-functorial. To verify that E preserves identities, consider 1_V. Inspection of the definition shows that $(-)_{1_\mathsf{V}}$ is the identity V-functor. Similarly for a monoidal functor U the 2-natural transformation

$$(-)_{1_U} : (-)_U \longrightarrow (-)_U$$

has all identity components and is therefore the identity.

Proposition 2.2.4 shows that E preserves horizontal composition of 1-cells. Now we verify that E preserves horizontal and vertical composition of 2-cells. Suppose given monoidal categories, functors, and natural transformations as in the following diagram:

$$\mathsf{V}_1 \underset{T_1}{\overset{U_1}{\Downarrow\mu}} \mathsf{V}_2 \underset{T_2}{\overset{U_2}{\Downarrow\pi}} \mathsf{V}_3.$$

To verify

$$(E\pi) * (E\mu) = E(\pi * \mu),$$

it suffices to verify that the components at each V_1-category C are equal. Unwinding the definitions, these components are V_3-functors that are identities on objects and whose morphisms on hom objects $\mathsf{C}(X,Y)$ are given by either of the two composites around the square below—the two composites are equal because they are the two equal expressions for the component of $(\pi * \mu)$ at the object $\mathsf{C}(X,Y)$ in V_1.

$$\begin{array}{ccc} U_2U_1\mathsf{C}(X,Y) & \xrightarrow{\pi * U_1} & T_2U_1\mathsf{C}(X,Y) \\ {\scriptstyle U_2 * \mu}\downarrow & & \downarrow{\scriptstyle T_2 * \mu} \\ U_2T_1\mathsf{C}(X,Y) & \xrightarrow{\pi * T_1} & T_2T_1\mathsf{C}(X,Y) \end{array}$$

Lastly we check that E preserves vertical composition of monoidal natural transformations. Suppose given monoidal functors

$$S, T, U : \mathsf{V} \longrightarrow \mathsf{W}$$

and a composable pair of monoidal natural transformations

$$U \xrightarrow{\mu} T \xrightarrow{\nu} S.$$

For each $\mathsf{C} \in \mathsf{V}\text{-}\mathsf{Cat}$, the W-functors

$$\mathsf{C}_U \xrightarrow{\mathsf{C}_\mu} \mathsf{C}_T \xrightarrow{\mathsf{C}_\nu} \mathsf{C}_S \quad \text{and} \quad \mathsf{C}_U \xrightarrow{\mathsf{C}_{(\nu\mu)}} \mathsf{C}_S$$

are identities on objects and have the same morphisms on hom objects

$$U\mathsf{C}(X,Y) \xrightarrow{\mu} T\mathsf{C}(X,Y) \xrightarrow{\nu} S\mathsf{C}(X,Y)$$

because the component of $\nu\mu$ at $\mathsf{C}(X,Y) \in \mathsf{V}$ is the composite of the corresponding components of ν and μ. This completes the proof that E is a 2-functor. □

Explanation 2.2.8. Even though Theorem 2.2.7 is stated for small monoidal categories V, the arguments that change of enrichment E preserves identities and composition of monoidal functors and/or monoidal natural transformations are independent of size. For a statement involving some specific monoidal categories, such as the adjunction result stated in Corollary 2.2.10 below, one can make a size-independent conclusion in one of two ways.

If Grothendieck's *Axiom of Universes* is assumed (see Note 2.7.3), then one can use Theorem 2.2.7 by choosing a sufficiently large universe in which the given monoidal categories are small. In that case, as noted in Motivation 2.2.6, the Axiom of Universes also implies the existence of $\mathcal{U}'$ in the statement of Theorem 2.2.7. In terms of cardinality, these are obtained by taking successively larger inaccessible cardinals.

Alternatively, one can use the preceding results and the arguments given in the proof of Theorem 2.2.7 to show directly that the relevant identities and composites are preserved. This approach is less concise, but avoids the set-theoretic considerations and the assumption of Grothendieck's axiom. ◇

Definition 2.2.9. A *monoidal adjunction* between monoidal categories V and W consists of an adjunction

$$\mathsf{W} \underset{U}{\overset{T}{\rightleftarrows}} \mathsf{V} \quad (T \dashv U)$$

such that T and U are monoidal functors and the unit and counit are monoidal natural transformations. If V and W are small, this is an internal adjunction in MCat. A *monoidal adjoint equivalence* consists of a monoidal adjunction such that the unit and counit are monoidal natural isomorphisms. ◇

Recall, as a special case of Definition I.6.3.9, an adjunction of 2-categories consists of 2-functors F and G

$$\mathsf{A} \underset{G}{\overset{F}{\rightleftarrows}} \mathsf{B} \quad (F \dashv G)$$

together with 2-natural unit and counit satisfying triangle identities. If A and B are small then this is an internal adjunction in $\mathsf{2Cat}$.

Corollary 2.2.10. *Suppose given a monoidal adjunction*

$$\mathsf{W} \underset{U}{\overset{T}{\rightleftarrows}} \mathsf{V}. \quad (T \dashv U)$$

Then change of enrichment induces an adjunction of 2-categories

$$\mathsf{W}\text{-}\mathsf{Cat} \underset{(-)_U}{\overset{(-)_T}{\rightleftarrows}} \mathsf{V}\text{-}\mathsf{Cat}. \quad ((-)_T \dashv (-)_U)$$

If $T \dashv U$ is a monoidal equivalence, then $(-)_T \dashv (-)_U$ is an equivalence of 2-categories.

Proof. For small monoidal categories V and W, the result follows from the statement of Theorem 2.2.7. As discussed in Explanation 2.2.8, one can deduce the result for general V and W by either:

- enlarging to a Grothendieck universe that contains $\mathsf{Ob}\,\mathsf{V}$ and $\mathsf{Ob}\,\mathsf{W}$ (under the assumption of Grothendieck's Axiom of Universes) or
- applying the proofs of Proposition 2.2.4 and Theorem 2.2.7 to T, U, the unit, and the counit of the adjunction. □

The Monoidal Strictification Theorem 1.1.32 is a monoidal equivalence by Lemma 1.1.34 and therefore yields an important special case of Corollary 2.2.10.

Corollary 2.2.11. *Suppose V is a monoidal category and*

$$\mathsf{V}_{\mathsf{st}} \underset{R}{\overset{L}{\rightleftarrows}} \mathsf{V}$$

is a monoidal equivalence with V_{st} strict monoidal. Then there is an equivalence of 2-categories

$$(\mathsf{V}_{\mathsf{st}})\text{-}\mathsf{Cat} \underset{(-)_R}{\overset{(-)_L}{\rightleftarrows}} \mathsf{V}\text{-}\mathsf{Cat}.$$

2.3. Preservation of Enriched Tensor

In this section we discuss change of enrichment along a braided monoidal functor. Throughout, we suppose that $\mathsf{V} = (\mathsf{V}, \otimes, \xi)$ and $\mathsf{W} = (\mathsf{W}, \otimes, \xi)$ are braided monoidal categories and

$$U : \mathsf{V} \longrightarrow \mathsf{W}$$

is a braided monoidal functor. Recall from Definition 2.1.1 the change of enrichment 2-functor induced by U and denoted

$$(-)_U : \mathsf{V}\text{-}\mathsf{Cat} \longrightarrow \mathsf{W}\text{-}\mathsf{Cat}.$$

First we identify how change of enrichment is compatible with the enriched tensor product.

Definition 2.3.1. For V-categories C and D, we define

$$(-)^2_U : \mathsf{C}_U \otimes \mathsf{D}_U \longrightarrow (\mathsf{C} \otimes \mathsf{D})_U$$

to be the identity on objects and given on hom objects by the morphism in W

$$\begin{array}{ccc} (\mathsf{C}_U \otimes \mathsf{D}_U)(X \otimes Y, X' \otimes Y') & & (\mathsf{C} \otimes \mathsf{D})_U(X \otimes Y, X' \otimes Y') \\ \parallel & & \parallel \\ \mathsf{C}_U(X, X') \otimes \mathsf{D}_U(Y, Y') & \xrightarrow{U^2} & (\mathsf{C}(X, X') \otimes \mathsf{D}(Y, Y'))_U \end{array}$$

for $X, X' \in \mathsf{C}$ and $Y, Y' \in \mathsf{D}$. We define

$$(-)^0_U : \mathbb{I} \longrightarrow \mathbb{I}_U$$

as the identity on the unique object and

$$\mathbb{1} \xrightarrow{U^0} U\mathbb{1}$$

on the unique hom object. ◇

Lemma 2.3.2 (W-Functoriality and 2-Naturality of $(-)^2_U$). *The assignment $(-)^2_U$ defines a* W*-functor. The definition of $(-)^2_U$ is 2-natural with respect to*

- V*-functors $\mathsf{C} \longrightarrow \mathsf{C}'$ and $\mathsf{D} \longrightarrow \mathsf{D}'$; and*
- V*-natural transformations thereof.*

Proof. The composition and identity axioms of Definition 1.2.4 require that the following diagrams, respectively, commute for each $X \otimes Y$, $X' \otimes Y'$, and $X'' \otimes Y''$ in $\mathsf{C} \otimes \mathsf{D}$. In the first diagram m_1 denotes the composition in $\mathsf{C}_U \otimes \mathsf{D}_U$ and m_2 denotes the composition in $(\mathsf{C} \otimes \mathsf{D})_U$. In the second diagram i_1 denotes the unit of $\mathsf{C}_U \otimes \mathsf{D}_U$ and i_2 denotes the unit of $(\mathsf{C} \otimes \mathsf{D})_U$.

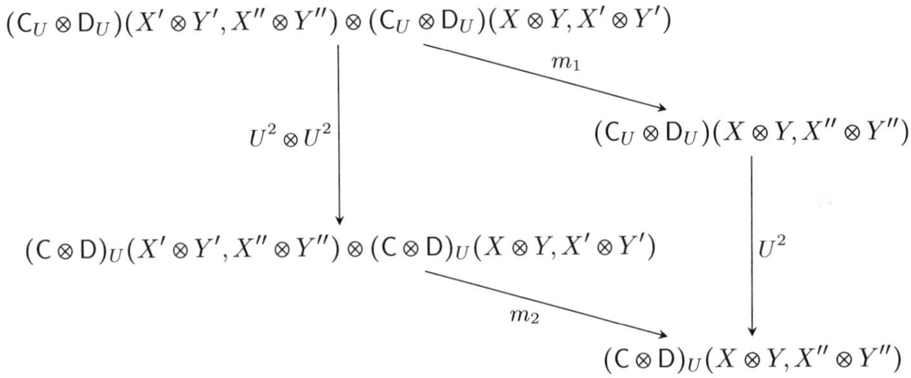

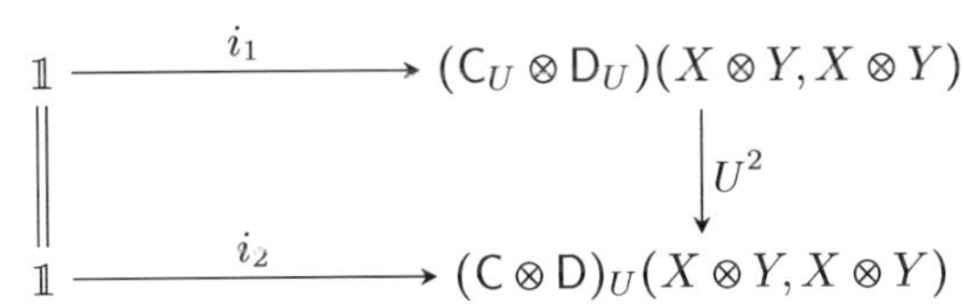

Unpacking the definitions of m_1 and m_2 via Definition 1.3.3 and (2.1.3) shows that the first diagram commutes by the associativity axiom (1.1.9) and the braid axiom (1.1.18) for U. Similarly, unpacking the definitions of i_1 and i_2 via (2.1.4) and Definition 1.3.3 shows that the second diagram commutes by the unity axioms (1.1.10) for U.

To check naturality of $(-)^2_U$, consider W-functors and W-natural transformations as in the diagrams below.

(2.3.3)

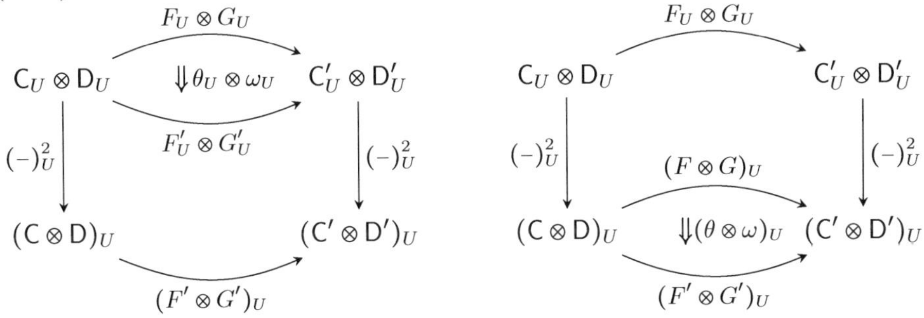

Naturality of $(-)^2_U$ with respect to V-functors, i.e., commutativity of the empty region in each diagram, follows by checking the induced morphisms on hom objects and using naturality of U^2 with respect to morphisms in V. Naturality of $(-)^2_U$ with respect to V-natural transformations, i.e., equality of the indicated whiskerings with $(-)^2_U$, follows by checking components via the Whiskering Lemma 1.2.12. Equality of the relevant components follows from naturality of U^2, functoriality of $\otimes$, naturality of λ in V, and the left unity axiom (1.1.10) for U. □

Lemma 2.3.4 (Associativity Axiom for $(-)_U$). *The following diagram commutes for each triple of* V*-categories* B, C, *and* D.

Associativity Diagram:

$$\begin{array}{ccc}
(\mathsf{B}_U \otimes \mathsf{C}_U) \otimes \mathsf{D}_U & \xrightarrow{\ a^{\otimes}\ } & \mathsf{B}_U \otimes (\mathsf{C}_U \otimes \mathsf{D}_U) \\
{\scriptstyle (-)^2_U \otimes 1} \downarrow & & \downarrow {\scriptstyle 1 \otimes (-)^2_U} \\
(\mathsf{B} \otimes \mathsf{C})_U \otimes \mathsf{D}_U & & \mathsf{B}_U \otimes (\mathsf{C} \otimes \mathsf{D})_U \\
{\scriptstyle (-)^2_U} \downarrow & & \downarrow {\scriptstyle (-)^2_U} \\
((\mathsf{B} \otimes \mathsf{C}) \otimes \mathsf{D})_U & \xrightarrow{\ (a^{\otimes})_U\ } & (\mathsf{B} \otimes (\mathsf{C} \otimes \mathsf{D}))_U
\end{array}$$

Proof. Commutativity of the diagram on objects $(W \otimes X) \otimes Y \in (\mathsf{B} \otimes \mathsf{C}) \otimes \mathsf{D}$ is immediate because $(-)^2_U$ is the identity on objects. For hom objects

$$\big(\mathsf{B}_U(W, W') \otimes \mathsf{C}_U(X, X')\big) \otimes \mathsf{D}_U(Y, Y'),$$

commutativity follows from the definition of $a^{\otimes}$ (Definition 1.3.10) and the associativity axiom (1.1.9) for U. □

Lemma 2.3.5 (Unity Axiom for $(-)_U$). *The following two diagrams commute for each* V*-category* C.

Unity Diagrams:

$$\begin{array}{ccc}
\mathbb{I} \otimes \mathsf{C}_U & \xrightarrow{\ \ell^{\otimes}\ } & \mathsf{C}_U \\
{\scriptstyle (-)^0_U \otimes 1} \downarrow & & \uparrow {\scriptstyle (\ell^{\otimes})_U} \\
\mathbb{I}_U \otimes \mathsf{C}_U & \xrightarrow{\ (-)^2_U\ } & (\mathbb{I} \otimes \mathsf{C})_U
\end{array}
\qquad
\begin{array}{ccc}
\mathsf{C}_U \otimes \mathbb{I} & \xrightarrow{\ r^{\otimes}\ } & \mathsf{C}_U \\
{\scriptstyle 1 \otimes (-)^0_U} \downarrow & & \uparrow {\scriptstyle (r^{\otimes})_U} \\
\mathsf{C}_U \otimes \mathbb{I}_U & \xrightarrow{\ (-)^2_U\ } & (\mathsf{C} \otimes \mathbb{I})_U
\end{array}$$

Proof. Similar to the proof of Lemma 2.3.4, commutativity on objects is immediate and commutativity on hom objects follows from the definitions of $\ell^{\otimes}$ and $r^{\otimes}$ (Definition 1.3.8) together with the unity axioms (1.1.10) for U. □

Lemma 2.3.6 (Braid Axiom for $(-)_U$). *Suppose* V *is symmetric monoidal. The following diagram commutes for each pair of* V*-categories* C *and* D.

Braid Diagram:

$$\begin{array}{ccc} \mathsf{C}_U \otimes \mathsf{D}_U & \xrightarrow{\beta^{\otimes}} & \mathsf{D}_U \otimes \mathsf{C}_U \\ {\scriptstyle (-)^2_U}\downarrow & & \downarrow{\scriptstyle (-)^2_U} \\ (\mathsf{C} \otimes \mathsf{D})_U & \xrightarrow{\beta^{\otimes}_U} & (\mathsf{D} \otimes \mathsf{C})_U \end{array}$$

Proof. Similarly to the proofs of Lemmas 2.3.4 and 2.3.5, commutativity on objects is immediate and commutativity on hom objects follows from the definition of $\beta^{\otimes}$ (Definition 1.3.22) together with the braid axiom (1.1.18) for U. □

Lemmas 2.3.4 through 2.3.6 together verify the axioms of Definitions 1.1.6 and 1.1.17, therefore proving the following result.

Theorem 2.3.7. *Suppose*

$$U \colon \mathsf{V} \longrightarrow \mathsf{W}$$

is a braided monoidal functor between braided monoidal categories. Then the change of enrichment

$$(-)_U \colon \mathsf{V}\text{-}\mathsf{Cat} \longrightarrow \mathsf{W}\text{-}\mathsf{Cat}$$

is a Cat*-monoidal 2-functor with respect to the enriched tensor products. Moreover, if U is a symmetric monoidal functor between symmetric monoidal categories, then $(-)_U$ is symmetric* Cat*-monoidal.*

Theorem 2.3.7 is the first step toward extending Theorem 2.2.7 to the braided and symmetric cases. We will show in Theorem 2.3.9 that change of enrichment provides a 2-functor from braided monoidal categories to Cat-monoidal 2-categories. As discussed in Motivation 2.2.6, this gives a convenient packaging of the relevant statements about preserving units and composition, but at the cost of some set-theoretic subtlety. We will use a similar approach as in Theorem 2.2.7, but applied to the second two of the 2-categories defined as follows.

Definition 2.3.8.

(1) We let BMCat denote the 2-category consisting of
- small braided monoidal categories as objects,
- braided monoidal functors as 1-cells, and
- monoidal natural transformations as 2-cells.

(2) We let SMCat denote the 2-category consisting of
- small symmetric monoidal categories as objects,
- symmetric monoidal functors as 1-cells, and
- monoidal natural transformations as 2-cells.

(3) We let CM2Cat denote the 2-category consisting of
- small Cat-monoidal 2-categories as objects,
- Cat-monoidal 2-functors as 1-cells, and
- Cat-monoidal 2-natural transformations as 2-cells.

(4) We let SCM2Cat denote the 2-category consisting of
- small symmetric Cat-monoidal 2-categories as objects,
- symmetric Cat-monoidal 2-functors as 1-cells, and
- symmetric Cat-monoidal 2-natural transformations as 2-cells.

For verification that these form 2-categories, SMCat is described in Example I.6.1.15. Units and compositions of braided monoidal functors and natural transformations are given by their underlying monoidal counterparts, so the 2-category axioms for BMCat follow from those of MCat (Example I.6.1.14). The 2-categories CM2Cat and SCM2Cat are special cases of V-MCat, respectively V-SMCat in Definition 1.4.25 with V = Cat. ◇

Theorem 2.3.9. *Suppose $\mathcal{U}'$ is a universe of sets containing* Ob(V-Cat) *for each small monoidal category* V*. Let* CM2Cat′ *and* SCM2Cat′ *denote the 2-categories of* Cat*-monoidal, respectively symmetric* Cat*-monoidal, 2-categories that are small with respect to $\mathcal{U}'$. Then change of enrichment provides a 2-functor*

$$E\colon \mathsf{BMCat} \longrightarrow \mathsf{CM2Cat}'$$

given by

- $E\mathsf{V} = \mathsf{V}\text{-}\mathsf{Cat}$
- $EU = (-)_U$
- $E\mu = (-)_\mu$

for a small braided monoidal category V*, braided monoidal functor U, and monoidal natural transformation μ, respectively. Restricted along the inclusion*

$$\mathsf{SMCat} \longrightarrow \mathsf{BMCat},$$

E takes values in SCM2Cat*′.*

Proof. Theorem 2.3.7 shows that the assignment on 1-cells is valid; the assignments on 0- and 2-cells are those of Theorem 2.2.7. Since identities and vertical or horizontal composites of 2-cells in BMCat, respectively CM2Cat′ are those of MCat, respectively 2Cat′, it follows from Theorem 2.2.7 that these are preserved in the braided case.

Given braided monoidal functors

$$\mathsf{V}_1 \xrightarrow{U_1} \mathsf{V}_2 \xrightarrow{U_2} \mathsf{V}_3,$$

Proposition 2.2.4 verifies that

$$((-)_{U_1})_{U_2} \quad \text{and} \quad (-)_{(U_2U_1)}$$

are equal as 2-functors. To complete the proof of the first assertion, we verify that the monoidal and unit constraints are equal. These constraints are 2-natural transformations whose components, at a pair of V_1-categories C and D, are V_3-functors

$$\mathsf{C}_{(U_2U_1)} \otimes \mathsf{D}_{(U_2U_1)} = (\mathsf{C}_{U_1})_{U_2} \otimes (\mathsf{D}_{U_1})_{U_2} \longrightarrow ((\mathsf{C} \otimes \mathsf{D})_{U_1})_{U_2} = (\mathsf{C} \otimes \mathsf{D})_{(U_2U_1)}.$$

Recalling Definition 2.3.1, these V_3-functors are identities on objects. Therefore, it remains only to verify they are equal on hom objects.

By Definition 1.4.19, the monoidal and unit constraints of the composite $((-)_{U_1})_{U_2}$ are given by the composite 2-natural transformations of the following pasting diagrams, where $\mathbb{1}$ denotes the terminal 2-category.

$$\begin{array}{ccccc}
(\mathsf{V}_1\text{-}\mathsf{Cat})^2 & \xrightarrow{(-)_{U_1} \otimes (-)_{U_1}} & (\mathsf{V}_2\text{-}\mathsf{Cat})^2 & \xrightarrow{(-)_{U_2} \otimes (-)_{U_2}} & (\mathsf{V}_3\text{-}\mathsf{Cat})^2 \\
{\scriptstyle\otimes}\downarrow & (-)^2_{U_1} \swarrow & \downarrow{\scriptstyle\otimes} & (-)^2_{U_2} \swarrow & \downarrow{\scriptstyle\otimes} \\
\mathsf{V}_1\text{-}\mathsf{Cat} & \xrightarrow{(-)_{U_1}} & \mathsf{V}_2\text{-}\mathsf{Cat} & \xrightarrow{(-)_{U_2}} & \mathsf{V}_3\text{-}\mathsf{Cat}
\end{array} \tag{2.3.10}$$

$$(2.3.11)\qquad \begin{array}{ccccc} & & \mathbb{1} & & \\ & \swarrow^{\mathbb{I}} & \big\downarrow \mathbb{I} & \searrow^{\mathbb{I}} & \\ \mathsf{V}_1\text{-}\mathsf{Cat} & \overset{(-)^0_{U_1}}{\Leftarrow} & & \overset{(-)^0_{U_2}}{\swarrow\!\!\!\!\Leftarrow} & \mathsf{V}_3\text{-}\mathsf{Cat} \\ & \searrow_{(-)_{U_1}} & \mathsf{V}_2\text{-}\mathsf{Cat} & \nearrow_{(-)_{U_2}} & \end{array}$$

For $X, X' \in \mathsf{C}$ and $Y, Y' \in \mathsf{D}$ the morphism on hom objects given by (2.3.10) (at C, D) is the composite below.

$$\begin{array}{l} (\mathsf{C}_{U_1})_{U_2}(X,X') \otimes (\mathsf{D}_{U_1})_{U_2}(Y,Y') \\ \qquad \big\downarrow U_2^2 \\ U_2\big((\mathsf{C}_{U_1})(X,X') \otimes (\mathsf{D}_{U_1})(Y,Y')\big) \xrightarrow{\ U_2(U_1^2)\ } U_2U_1\big(\mathsf{C}(X,X') \otimes \mathsf{D}(Y,Y')\big) \end{array}$$

This composite is, by Definition 1.1.11, the monoidal constraint of the composite U_2U_1, and therefore is the morphism on hom objects given by $(-)^2_{U_2U_1}$ at C, D. A similar unpacking shows that the unit constraints agree, and this finishes the proof that

$$E\colon \mathsf{BMCat} \longrightarrow \mathsf{CM2Cat}'$$

is a 2-functor.

Recalling Proposition 1.4.21, the composite of symmetric monoidal enriched functors is again symmetric monoidal. Therefore the second assertion, regarding the restriction of E to SMCat, follows from the first assertion and the symmetric cases of Theorems 1.5.5 and 2.3.7. □

Explanation 2.3.12. As in Explanation 2.2.8, we note that one can apply the arguments in the proof of Theorem 2.3.9 to specific cases of interest without assumptions such as the Axiom of Universes. For example, one can do so for the proofs of Corollaries 2.2.10 and 2.2.11 below. ◇

Definition 2.3.13. A *braided monoidal adjunction* between braided monoidal categories V and W consists of an adjunction

$$\mathsf{W} \underset{U}{\overset{T}{\rightleftarrows}} \mathsf{V} \quad (T \dashv U)$$

such that T and U are braided monoidal functors and the unit and counit are monoidal natural transformations. If V and W are small, this is an internal adjunction in BMCat. A *braided monoidal adjoint equivalence* consists of a braided monoidal adjunction such that the unit and counit are monoidal natural isomorphisms.

A braided monoidal adjunction, respectively equivalence, between symmetric monoidal categories is called a *symmetric monoidal adjunction*, respectively *symmetric monoidal equivalence*. ◇

Corollary 2.3.14. *Suppose given a braided monoidal adjunction*

$$\mathsf{W} \underset{U}{\overset{T}{\rightleftarrows}} \mathsf{V}. \quad (T \dashv U)$$

Then change of enrichment induces an adjunction of Cat*-monoidal 2-categories*

$$\mathsf{W}\text{-}\mathsf{Cat} \underset{(-)_U}{\overset{(-)_T}{\rightleftarrows}} \mathsf{V}\text{-}\mathsf{Cat}. \quad (\perp)$$

If $T \dashv U$ is a braided monoidal equivalence, then $(-)_T \dashv (-)_U$ is a Cat*-monoidal equivalence of 2-categories.*

If $T \dashv U$ is a symmetric monoidal adjunction between symmetric monoidal categories, then $(-)_T \dashv (-)_U$ is a symmetric Cat*-monoidal adjunction.*

The Braided Strictification Theorem 1.1.39 is a braided monoidal equivalence by Lemma 1.1.34 and therefore yields an important special case of Corollary 2.3.14.

Corollary 2.3.15. *Suppose* V *is a braided monoidal category and*

$$\mathsf{V}_{\mathsf{st}} \underset{R}{\overset{L}{\rightleftarrows}} \mathsf{V}$$

is a braided monoidal equivalence with V_{st} *strict braided monoidal. Then there is a* Cat*-monoidal equivalence of 2-categories*

$$(\mathsf{V}_{\mathsf{st}})\text{-}\mathsf{Cat} \underset{(-)_R}{\overset{(-)_L}{\rightleftarrows}} \mathsf{V}\text{-}\mathsf{Cat}.$$

The Symmetric Strictification Theorem 1.1.42 is a symmetric monoidal equivalence by Lemma 1.1.34 and therefore yields another important special case of Corollary 2.3.14.

Corollary 2.3.16. *Suppose* V *is a symmetric monoidal category and*

$$\mathsf{V}_{\mathsf{st}} \underset{R}{\overset{L}{\rightleftarrows}} \mathsf{V}$$

is a symmetric monoidal equivalence with V_{st} *strict symmetric monoidal. Then there is a symmetric* Cat*-monoidal equivalence of 2-categories*

$$(\mathsf{V}_{\mathsf{st}})\text{-}\mathsf{Cat} \underset{(-)_R}{\overset{(-)_L}{\rightleftarrows}} \mathsf{V}\text{-}\mathsf{Cat}.$$

2.4. Preservation of Enriched Monoidal Structure

In Theorem 2.3.7 above we showed that change of enrichment along a braided monoidal functor

$$U\colon \mathsf{V} \longrightarrow \mathsf{W}$$

provides a Cat-monoidal 2-functor

$$(-)_U\colon (\mathsf{V}\text{-}\mathsf{Cat}, \otimes) \longrightarrow (\mathsf{W}\text{-}\mathsf{Cat}, \otimes).$$

In this section we show that $(-)_U$ preserves enriched monoidal structure on each monoidal V-category, V-functor, and V-natural transformation. The main result is that $(-)_U$ restricts to 2-functors indicated in the following diagram (with V, W, and U assumed to be symmetric monoidal for the definitions of enriched braided or symmetric monoidal structures). See Theorem 2.4.10 and Explanation 2.4.15.

$$\begin{array}{ccccccc} \mathsf{V\text{-}SMCat} & \longrightarrow & \mathsf{V\text{-}BMCat} & \longrightarrow & \mathsf{V\text{-}MCat} & \longrightarrow & \mathsf{V\text{-}Cat} \\ \downarrow{\scriptstyle (-)_U} & & \downarrow{\scriptstyle (-)_U} & & \downarrow{\scriptstyle (-)_U} & & \downarrow{\scriptstyle (-)_U} \\ \mathsf{W\text{-}SMCat} & \longrightarrow & \mathsf{W\text{-}BMCat} & \longrightarrow & \mathsf{W\text{-}MCat} & \longrightarrow & \mathsf{W\text{-}Cat} \end{array}$$

Again throughout this section V and W are braided monoidal categories and $U\colon \mathsf{V} \longrightarrow \mathsf{W}$ is a braided monoidal functor. We will make use of the following details and conventions.

Convention 2.4.1 (Notation for Monoidal Data)**.** We denote the monoidal products in V-Cat and W-Cat via juxtaposition, and use exponents such as K^2 to denote KK. We also omit superscripts $\otimes$ and let $(\mathbb{I}, a, \ell, r, \beta)$ denote the braided monoidal data of either V-Cat or W-Cat, with context clarifying which we mean. In most cases the data of V-Cat will appear only when considering their images under $(-)_U$, and therefore will appear as $(\mathbb{I}_U, a_U, \ell_U, r_U, \beta_U)$. ◇

Explanation 2.4.2 (2-Functoriality of $(-)_U$)**.** In Definition 2.4.3 below we will use the following three consequences of Proposition 2.1.2, that change of enrichment, $(-)_U$, is a 2-functor.

(1) Change of enrichment strictly preserves identities, so for a V-category K we have $(1_{\mathsf{K}})_U = 1_{(\mathsf{K}_U)}$. We will use 1_U to denote this identity W-functor.
(2) Change of enrichment strictly preserves inverses, so for a V-functor F we have $(F^{-1})_U = (F_U)^{-1}$. We will use this for the images of the unitors, ℓ and r, denoting the inverses of ℓ_U and r_U by ℓ_U^{-1} and r_U^{-1}, respectively.
(3) Change of enrichment strictly preserves horizontal composition. Therefore, if θ is a V-natural transformation whose domain is a composite GF, then the domain of θ_U is $G_U F_U = (GF)_U$. We will apply this to the V-natural transformations in the data of a (braided) monoidal V-category. ◇

Definition 2.4.3. Suppose $\mathsf{K} = (\mathsf{K}, \boxtimes, \mathrm{I}^{\boxtimes}, a^{\boxtimes}, \ell^{\boxtimes}, r^{\boxtimes})$ is a monoidal V-category. Define a monoidal structure

$$(\mathsf{K}_U, \boxtimes', \mathrm{I}', a', \ell', r'),$$

with base W-category K_U and the following additional data. In these definitions we use Convention 2.4.1 for the monoidal data of V-Cat and W-Cat and implicitly use the 2-functoriality of $(-)_U$ as noted in Explanation 2.4.2.

Monoidal Composition: The monoidal composition $\boxtimes'$ is defined to be the composite

$$(\mathsf{K}_U)^2 \xrightarrow{(-)_U^2} (\mathsf{K}^2)_U \xrightarrow{\boxtimes_U} \mathsf{K}_U.$$

Monoidal Identity: The monoidal identity I' is defined to be the composite

$$\mathbb{I} \xrightarrow{(-)_U^0} \mathbb{I}_U \xrightarrow{\mathrm{I}_U^{\boxtimes}} \mathsf{K}_U.$$

Monoidal Unitors: The left unitor, ℓ' is equal to $\ell^{\boxtimes}$ composed with an identity 2-cell as shown in the pasting diagram below. The triangular regions commute by definition, the unlabeled quadrilateral commutes by naturality of

$(-)^2_U$ and the quadrilateral labeled ☆ commutes by the left unity axiom for $(-)_U$, Lemma 2.3.5.

(2.4.4)

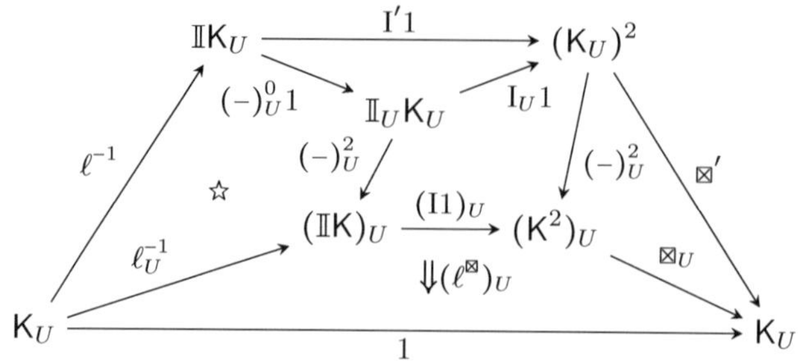

The right unitor, r' is equal to $r^{\boxtimes}$ composed with an identity 2-cell as shown in the pasting diagram below. The triangular regions commute by definition, the unlabeled quadrilateral commutes by naturality of $(-)^2_U$ and the quadrilateral labeled ☆ commutes by the right unity axiom for $(-)_U$, Lemma 2.3.5.

(2.4.5)

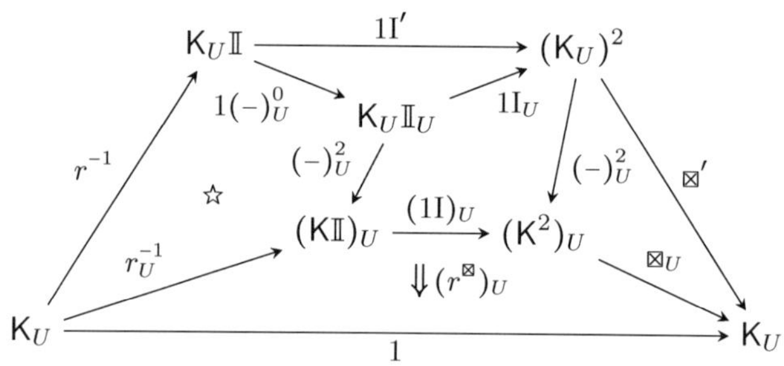

Monoidal Associator: The monoidal associator a' is equal to a whiskering of $a^{\boxtimes}$ composed with identity 2-cells as shown in the pasting diagram below. The triangular regions commute by definition, the unlabeled quadrilaterals commute by naturality of $(-)^2_U$, and the hexagon labeled ☆ commutes by the associativity axiom for $(-)_U$, Lemma 2.3.4.

(2.4.6)

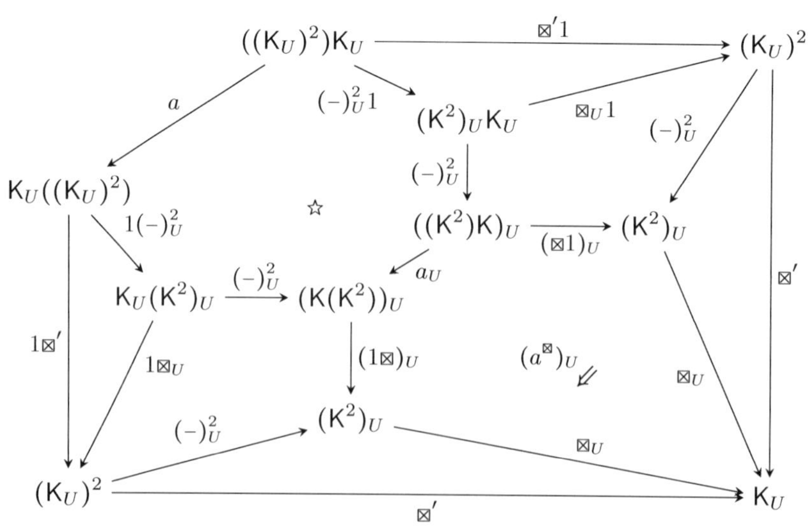

If, moreover, K is braided monoidal with braiding $\beta^{\boxtimes}$ (when V is assumed symmetric), we make the following definition.

Braiding: The braiding β' is equal to a whiskering of $\beta^{\boxtimes}$ composed with identity 2-cells as shown in the pasting diagram below. The triangular regions commute by definition and the quadrilateral labeled ☆ commutes by the braid axiom for $(-)_U$, Lemma 2.3.6.

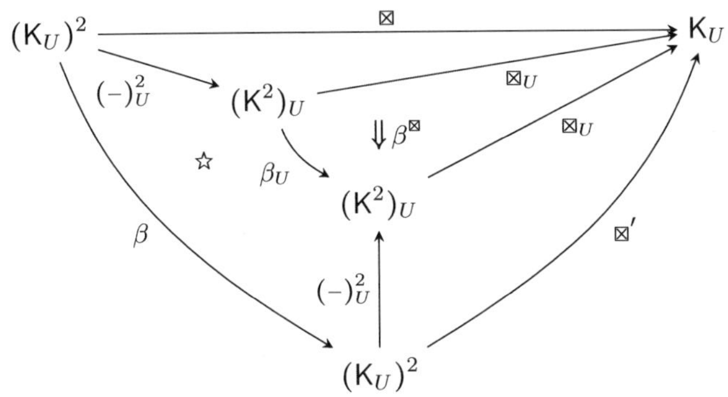

(2.4.7)

This finishes the definition of the data for monoidal structure on K_U. We show that these data satisfy the relevant axioms in Theorem 2.4.10 below. ◇

Definition 2.4.8. Suppose $(F, F^2, F^0)\colon \mathsf{K} \longrightarrow \mathsf{L}$ is a monoidal functor between monoidal V-categories K and L. When K_U and L_U are given the monoidal W-category structure of Definition 2.4.3, then we define a monoidal functor

$$(F_U, (F^2_U)', (F^0_U)')\colon \mathsf{K}_U \longrightarrow \mathsf{L}_U$$

with underlying W-functor F_U and the following monoidal and unit constraints. As above we use Convention 2.4.1 for the monoidal data of V-Cat and W-Cat.

Monoidal Constraint: The monoidal constraint $(F^2_U)'$ is defined to be the wiskering of F^2_U shown below, where the upper square commutes by naturality of $(-)^2_U$.

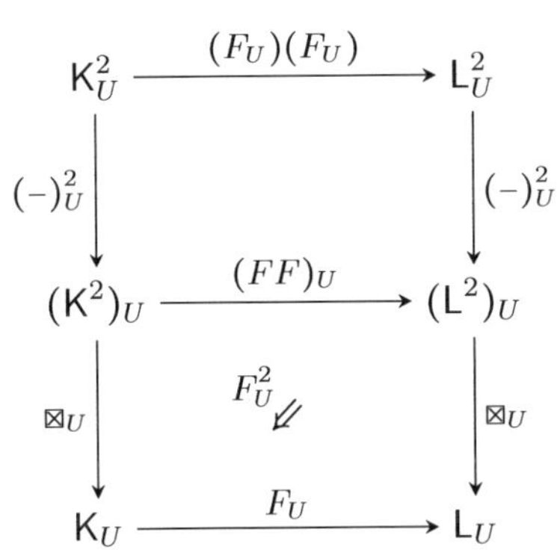

Unit Constraint: The unit constraint $(F_U^0)'$ is defined to be the whiskering of F_U^0 shown below.

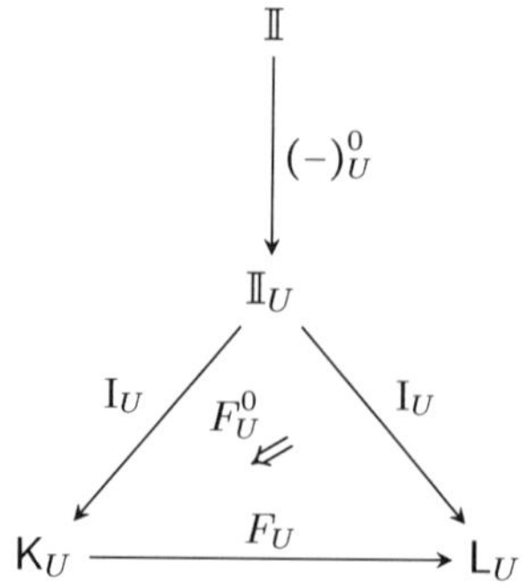

This finishes the definition of $(F_U, (F_U^2)', (F_U^0)')$. We address the relevant axioms in Theorem 2.4.10 below. ◇

Lemma 2.4.9. *Suppose*

$$\mathsf{K} \xrightarrow{F} \mathsf{L} \xrightarrow{P} \mathsf{M}$$

is a composable pair of monoidal V*-functors. Then*

$$(PF)_U = (P_U)(F_U).$$

Proof. The 2-functoriality of $(-)_U$ implies that the asserted equality holds for underlying V-functors; we need to check the monoidal and unit constraints are equal. Recalling Definition 1.4.19, the monoidal and unit constraints of PF are given by pasting those of F and P. Thus the monoidal and unit constraints for $(PF)_U$ are given by the pastings below.

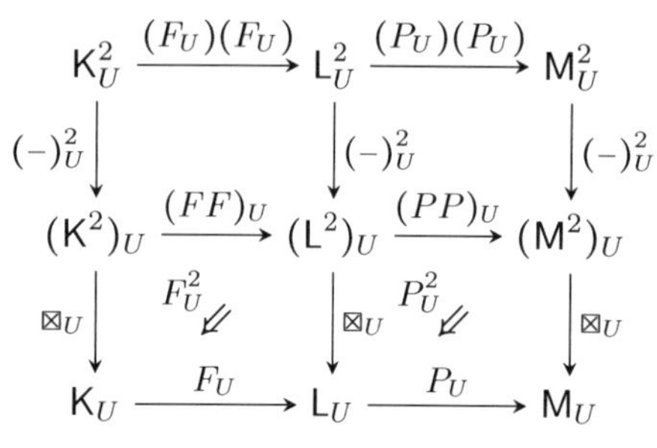

and

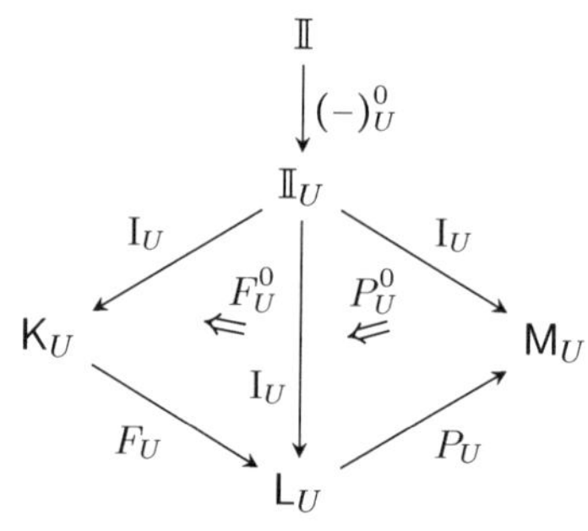

Since (as in any 2-category) horizontal composition of V-natural transformations distributes over vertical composition, the indicated composites are precisely the monoidal and unit constraints of $(P_U)(F_U)$. This completes the proof. □

Theorem 2.4.10. *Suppose*

$$U\colon \mathsf{V} \longrightarrow \mathsf{W}$$

is a braided monoidal functor between braided monoidal categories. In the context of Definitions 2.4.3 and 2.4.8 above, we have the following results. For items (1) *and* (2)*, the braided and symmetric monoidal cases assume moreover that* U*,* V*, and* W *are symmetric monoidal.*

(1) *If* K *is a monoidal, respectively braided monoidal, respectively symmetric monoidal,* V*-category, then* K_U *is a monoidal, respectively braided monoidal, respectively symmetric monoidal,* W*-category.*

(2) *If $F: \mathsf{K} \longrightarrow \mathsf{L}$ is a monoidal, respectively braided monoidal, respectively symmetric monoidal, V-functor between monoidal, respectively braided monoidal, respectively symmetric monoidal, V-categories, then*

$$F_U: \mathsf{K}_U \longrightarrow \mathsf{L}_U$$

is a monoidal, respectively braided monoidal, respectively symmetric monoidal, W-functor.

(3) *If $\theta: F \longrightarrow G$ is a monoidal V-natural transformation between monoidal V-functors F and G, then*

$$\theta_U: F_U \longrightarrow G_U$$

is a monoidal W-natural transformation.

Proof. Each of the three assertions follows by verifying the relevant axioms from Section 1.4. To ease readability of the computations below, we denote the whiskerings of a W-natural transformation θ with the W-functor $(-)^2_U$ as

$$\theta * (-)^2_U \quad \text{and} \quad (-)^2_U * \theta$$

instead of our usual convention using the identity on $(-)^2_U$

$$\theta * 1_{(-)^2_U} \quad \text{and} \quad 1_{(-)^2_U} * \theta.$$

For assertion (1), we begin with the middle unity axiom (1.4.5). Below we show the pasting of a' and $1\ell'$.

(2.4.11)

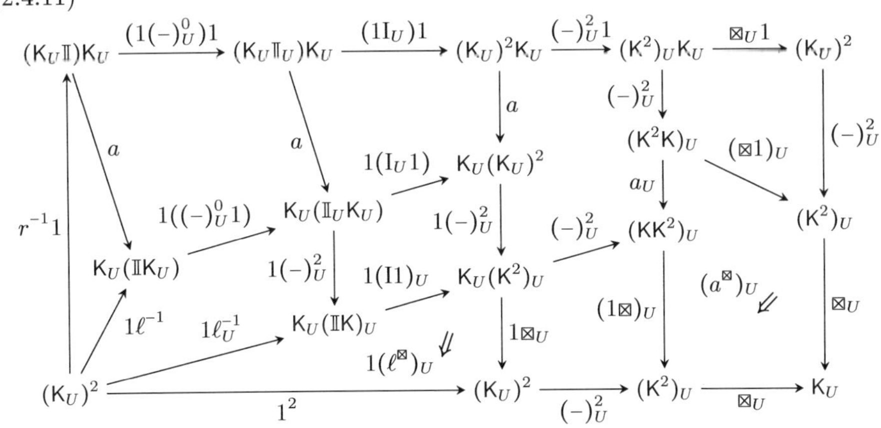

The 2-naturality of $(-)^2_U$ gives the equality

$$(-)^2_U * [1(\ell^{\boxtimes})_U] = (1\ell^{\boxtimes})_U * (-)^2_U,$$

and we use this together with several of the other basic axioms for monoidal structure to show that the composite of (2.4.11) is equal to that of (2.4.12).
(2.4.12)

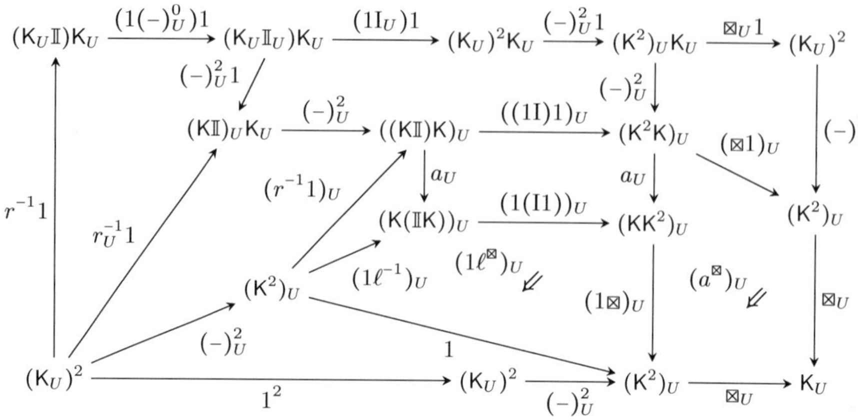

Next we apply $(-)_U$ to the middle unity axiom (1.4.5) for K and use the resulting equality to show that the composite of (2.4.12) is equal to that of (2.4.13).
(2.4.13)

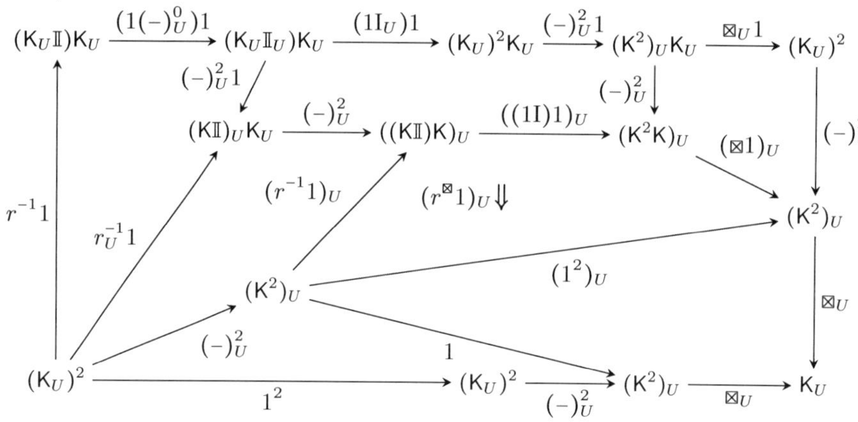

Finally, we again use 2-naturality of $(-)^2_U$, giving

$$(r^{\boxtimes}1_U) * (-)^2_U = (-)^2_U * [(r^{\boxtimes}_U 1)],$$

to conclude that the composite of (2.4.13) is equal to that of (2.4.14) below. This last composite is equal to the whiskering of $r'1$ required for the second half of (1.4.5) for K_U.

(2.4.14)

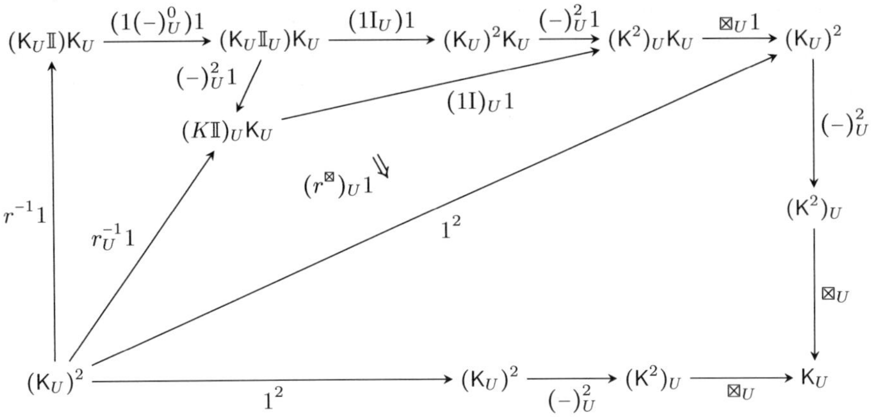

Each of the remaining axioms to be verified for item (1) is an equality of two composites of pasting diagrams, and in each diagram the only nontrivial 2-cells are given by applying $(-)_U$ to corresponding 2-cells of K. The equality then follows as above, using defining properties of (braided/symmetric) monoidal categories and functors, together with the image under $(-)_U$ of the relevant axiom for K.

For example, the pentagon axiom for K_U follows directly from the definitions and the pentagon axiom for K. If K is symmetric, the symmetry axiom for K_U follows likewise from that of K. Similarly, in the case that K is braided, the two hexagon axioms require equalities such as

$$(-)^2_U * (\beta^{\boxtimes}_U 1) = (\beta^{\boxtimes}1)_U * (-)^2_U,$$

that follow from the 2-naturality of $(-)^2_U$ as above. In addition, for the right hexagon axiom (1.4.12) one uses 2-functoriality of $(-)_U$ to see that the mate of a' described in Definition 1.4.9 is equal to an appropriate whiskering of $(a^{\boxtimes}_1)_U$. Thus (again using 2-functoriality of $(-)_U$) the inverse $(a'_1)^{-1}$ is obtained by whiskering $a_1^{-\boxtimes}$.

For assertion (2) the axioms of Definitions 1.4.17 and 1.4.18 are similar but simpler, following from naturality of $(-)^2_U$ and the corresponding axioms for F. For assertion (3) the axioms of Definition 1.4.22 are immediate from the definitions. □

Explanation 2.4.15. Using the notation of Definition 1.4.25, Theorem 2.4.10 proves that $(-)_U$ restricts to 2-functors indicated in the diagram below, with V, W, and U assumed symmetric for the definitions of V-BMCat, W-BMCat, V-SMCat, and W-SMCat.

(2.4.16)

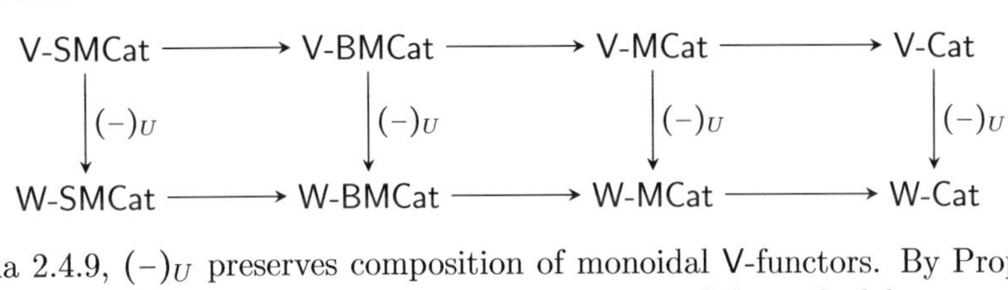

By Lemma 2.4.9, $(-)_U$ preserves composition of monoidal V-functors. By Proposition 1.4.21, composites of braided and symmetric monoidal enriched functors are given by composites of the underlying monoidal enriched functors. By Proposition 1.4.24 composites of monoidal enriched natural transformations are given by composites of the underlying enriched natural transformations. Therefore 2-functoriality of $(-)_U$ on V-Cat implies 2-functoriality of each restriction. ◇

When V is braided monoidal then the corepresented functor $U = \mathsf{V}(\mathbb{1}, -)$ is braided monoidal by Lemma 2.1.5. It is, furthermore, symmetric if V is symmetric. Therefore, by Theorem 2.4.10 we have the following.

Corollary 2.4.17. *Suppose* V *is a braided monoidal category. Then the following statements hold. For items* (1) *and* (2)*, the braided and symmetric monoidal cases assume moreover that* V *is symmetric monoidal.*

(1) *Suppose* K *is a monoidal, respectively braided monoidal, respectively symmetric monoidal,* V*-category. Then the underlying category* K_0 *is a monoidal, respectively braided monoidal, respectively symmetric monoidal, category.*
(2) *Suppose that* $F: \mathsf{K} \longrightarrow \mathsf{L}$ *is a monoidal, respectively braided monoidal, respectively symmetric monoidal,* V*-functor between monoidal, respectively braided monoidal, respectively symmetric monoidal,* V*-categories. Then the underlying functor*

$$F_0: \mathsf{K}_0 \longrightarrow \mathsf{L}_0$$

is monoidal, respectively braided monoidal, respectively symmetric monoidal.
(3) *Suppose that* $\theta: F \longrightarrow G$ *is a monoidal* V*-natural transformation between monoidal* V*-functors. Then the underlying natural transformation*

$$\theta_0: F_0 \longrightarrow G_0$$

is a monoidal natural transformation.

The proof of Theorem 2.4.10 has a technical feature that will be useful to state explicitly. We will use this in the proof of Theorem 2.5.1 below.

Explanation 2.4.18. For each of the axioms (1.4.5), (1.4.6), (1.4.11), (1.4.12), and (1.4.14), let θ denote the composite of the top and/or left pasting diagram for K and ω denote the composite of the bottom and/or right pasting diagram for K. So each axiom is an equality $\theta = \omega$.

Let θ' and ω' denote top/left and bottom/right composites, respectively, in the corresponding axioms for K_U. The monoidal constraint of $(-)_U$ is $(-)^2_U$ and we let $(-)^3_U$ denote the two equal composites

$$(\mathsf{K}_U)^3 \longrightarrow (\mathsf{K}^3)_U$$

given by iterating $(-)^2_U$.

With this notation, the argument for each axiom in the proof of Theorem 2.4.10 consists of two steps:

(1) First, $\theta' = \theta_U * (-)^n_U$ and $\omega' = \omega_U * (-)^n_U$, where $n = 2$ or 3.
(2) Second, $\theta_U * (-)^n_U = \omega_U * (-)^n_U$ by the corresponding axiom for K.

This implies then that $\theta' = \omega'$.

For example, the diagrams (2.4.11) and (2.4.14) above are the pasting diagrams for θ' and ω', respecitevly, in the middle unity axiom. On the other hand, diagrams (2.4.12) and (2.4.13) are simply the pasting diagrams for $\theta_U * (-)^2_U$ and $\omega_U * (-)^2_U$, respectively, with several additional identity 2-cells.

Now we point out that this argument can be used in reverse, under the assumption that $(-)_U$ is faithful on 2-cells. Indeed, if $\theta' = \omega'$, then by item (1) above we conclude $\theta_U * (-)^n_U = \omega_U * (-)^n_U$. Since $(-)^2_U$ is an identity on objects, and hence so is $(-)^3_U$, we apply the Whiskering Lemma 1.2.12 to conclude that the components of θ_U are equal to those of $\theta_U * (-)^n_U$. Similarly, the components of ω_U are equal to those of $\omega_U * (-)^n_U$. Therefore, $\theta_U = \omega_U$. Assuming that $(-)_U$ is injective on 2-cells, then we conclude $\theta = \omega$. ◇

2.5. Coherence of Enriched Monoidal Categories

In this section we extend the coherence results for categories that are monoidal, respectively braided monoidal, respectively symmetric monoidal, to V-enriched counterparts. Throughout we assume $\mathsf{V} = (\mathsf{V}, \otimes, \xi)$ is a braided monoidal category. Recall from Convention 1.3.28 that we assume V to be symmetric monoidal whenever making use of the braiding $\beta^{\otimes}$ for $(\mathsf{V}\text{-}\mathsf{Cat}, \otimes)$. We begin with the following reverse of Corollary 2.4.17, and then use it to lift monoidal coherence results from Cat to $\mathsf{V}\text{-}\mathsf{Cat}$.

Theorem 2.5.1. *Suppose that $\mathsf{V} = (\mathsf{V}, \otimes, \xi)$ is a braided monoidal category and consider the change of enrichment*

$$(-)_0 : \mathsf{V}\text{-}\mathsf{Cat} \longrightarrow \mathsf{Cat}.$$

For the braided and symmetric monoidal cases in the statements below, suppose that V is symmetric monoidal.

(1) *Suppose K is a V-category.*
  (a) *Suppose $(\boxtimes, \mathrm{I}, a, \ell, r)$ is the data of a monoidal V-category structure for K and let $(\boxtimes', \mathrm{I}', a', \ell', r')$ denote the functors and natural transformations given by change of enrichment. Then*

  $$(\mathsf{K}, \boxtimes, \mathrm{I}, a, \ell, r)$$

  is a monoidal V category if and only if

  $$(\mathsf{K}_0, \boxtimes', \mathrm{I}', a', \ell', r')$$

  is a monoidal category.
  (b) *Suppose, moreover, that β is the data of a braiding for a braided monoidal V-category structure on K and let β' denote the natural transformation given by change of enrichment. Then*

  $$(\mathsf{K}, \boxtimes, \mathrm{I}, a, \ell, r, \beta)$$

  is a braided, respectively symmetric, monoidal V-category if and only if

  $$(\mathsf{K}_0, \boxtimes', \mathrm{I}', a', \ell', r', \beta')$$

  is a braided, respectively symmetric, monoidal category.

(2) *Suppose K and L are monoidal, respectively braided monoidal, respectively symmetric monoidal, V-categories and suppose*

$$F : \mathsf{K} \longrightarrow \mathsf{L}$$

is a V-functor. Suppose given V-natural transformations (F^2, F^0) as in the data for F to be a monoidal, respectively braided monoidal, respectively symmetric monoidal, V-functor. Let $((F^2)', (F^0)')$ denote the natural transformations given by change of enrichment. Then

$$(F, F^2, F^0) : \mathsf{K} \longrightarrow \mathsf{L}$$

satisfies the axioms of a monoidal, respectively braided monoidal, respectively symmetric monoidal, V*-functor if and only if*

$$(F_0, (F^2)', (F^0)') \colon \mathsf{K}_0 \longrightarrow \mathsf{L}_0$$

satisfies the axioms of a monoidal, respectively braided monoidal, respectively symmetric monoidal, functor.

(3) *Suppose F and G are monoidal* V*-functors. A* V*-natural transformation*

$$\theta \colon F \longrightarrow G$$

is monoidal V*-natural if and only if the underlying data*

$$\theta_0 \colon F_0 \longrightarrow G_0$$

is a monoidal natural transformation.

Proof. We discuss assertion (1) in the symmetric monoidal case; the same argument applies to the other cases and the other two assertions. One implication is provided by Corollary 2.4.17: if $(\mathsf{K}, \boxtimes, \mathrm{I}, a, \ell, r, \beta)$ is symmetric monoidal in the V-enriched sense, then $(\mathsf{K}_0, \boxtimes', \mathrm{I}', a', \ell', r', \beta')$ satisfies the axioms in Cat for a symmetric monoidal category.

For the reverse implication, suppose that $(\mathsf{K}_0, \boxtimes', \mathrm{I}', a', \ell', r', \beta')$ satisfies the appropriate symmetric monoidal axioms. We then use the technical argument outlined at the end of Explanation 2.4.18 as follows. For each axiom of Definitions 1.4.2, 1.4.10, and 1.4.13, we apply $(-)_0$ to the two sides of the required equality and whisker with a monoidal constraint $(-)_U^2$ or $(-)_U^3$ to obtain two natural transformations that are equal in Cat by the corresponding axiom for K_0. Using Lemma 1.2.12 to check the components of the whiskerings, and noting that each of $(-)_U^n$ is the identity on objects, the images under $(-)_0$ are equal in Cat. We noted in Lemma 2.1.7 above that $(-)_0$ is injective on V-natural transformations and therefore each axiom of Definition 1.4.2 holds in V-Cat. □

As a further corollary of Theorem 2.5.1 and the 2-cell injectivity result of Lemma 2.1.7, we can extend the Coherence Theorems 1.1.31, 1.1.38, and 1.1.41 to give a coherence result (Theorem 2.5.6 below) for monoidal, braided monoidal, and symmetric monoidal V-categories (with V symmetric monoidal in the latter cases). First we extend the definitions of (*permuted*) *word* and (*permuted*) *canonical map*.

Definition 2.5.2. Suppose V is a braided monoidal category (respectively symmetric monoidal in the braided and symmetric cases below) and suppose K is a monoidal V-category. Extending Definitions 1.1.29 and 1.1.30, we say that a V-functor

$$w \colon \mathsf{K}^{\otimes n} \longrightarrow \mathsf{K}$$

is a V-*word* if the underlying functor w_0 composed with the monoidal constraint for $(-)_0$ is a word

$$\mathsf{K}_0^n \cong (\mathsf{K}^n)_0 \xrightarrow{w_0} \mathsf{K}_0.$$

We say that a V-natural isomorphism between V-words of the same length is a *canonical* V*-map* if its underlying natural transformation induces a canonical map.

Moreover, if K is braided, respectively symmetric, monoidal, we extend Definitions 1.1.36 and 1.1.40 to say that $w\sigma$ is a *permuted* V*-word* if the following composite of σ, the monoidal constraint, and w_0 is a permuted word

$$\mathsf{K}_0^n \xrightarrow{\sigma} \mathsf{K}_0^n \cong (\mathsf{K}^n)_0 \xrightarrow{w_0} \mathsf{K}_0.$$

We call this composite the *underlying permuted word.* We say that a V-natural isomorphism is a *braided canonical* V*-map*, respectively *permuted canonical* V*-map* if its underlying natural isomorphism induces a braided, respectively permuted canonical map. Extending Definition 1.1.37, the *underlying braid* of a braided canonical V-map is the underlying braid of its underlying braided canonical map. ◇

Definition 2.5.3. Suppose K is a monoidal V-category and suppose $X=(X_1,\ldots,X_n)$ is a sequence of objects $X_i \in \mathsf{K}$. The *left normalized product* is defined to be

$$(\cdots((X_1 \boxtimes X_2)\boxtimes X_3)\cdots)\boxtimes X_n. \tag{2.5.4}$$

The *right normalized product* is defined to be

$$X_1 \boxtimes (X_2 \boxtimes (X_3 \boxtimes (\cdots(X_{n-1} \boxtimes X_n)\cdots))). \tag{2.5.5}$$

The left and right normalized products of the empty sequence are both defined to be the monoidal identity object $\mathrm{I} \in \mathsf{K}$. ◇

The next result is a coherence theorem for monoidal, braided monoidal, and symmetric monoidal V-categories.

Theorem 2.5.6 (Enriched Monoidal Coherence)**.** *Suppose* V *is braided monoidal (respectively symmetric monoidal in the braided and symmetric cases below) and suppose* K *is a monoidal* V*-category.*

(1) *If* $u, v\colon \mathsf{K}^{\otimes n} \longrightarrow \mathsf{K}$ *are* V*-words of the same length then there exists a unique canonical* V*-map* $u \longrightarrow v$*.*
(2) *Suppose, moreover, that* K *is braided monoidal and* $u\sigma$ *and* $v\tau$ *are permuted* V*-words. If two braided canonical* V*-maps* $u\sigma \longrightarrow v\tau$ *have the same underlying braid, then they are equal.*
(3) *Suppose, moreover, that* K *is symmetric monoidal. If* $u\sigma$ *and* $v\tau$ *are permuted* V*-words of the same length then there exists a unique permuted canonical* V*-map* $u\sigma \longrightarrow v\tau$*.*

Proof. The existence statements follow from the existence statements of Theorems 1.1.31 and 1.1.41 because, by Theorem 2.5.1, the data of the V-enriched monoidal structure corresponds to the data of the underlying monoidal structure. Thus we have V-natural transformations whose underlying natural transformations compose to give any canonical map. The uniqueness statements follow from Lemma 2.1.7 together with the uniqueness statements of Theorems 1.1.31, 1.1.38, and 1.1.41. □

Next we describe an enriched analogue of Epstein's Coherence Theorem 1.1.44 for (symmetric) monoidal functors.

Definition 2.5.7. Suppose $F\colon \mathsf{K} \longrightarrow \mathsf{L}$ is either

- a monoidal V-functor between monoidal V-categories, with V braided monoidal, or
- a symmetric monoidal V-functor between symmetric monoidal V-categories, with V symmetric monoidal.

Recall $F_0\colon \mathsf{K}_0 \longrightarrow \mathsf{L}_0$ denotes the underlying (symmetric) monoidal functor of F.

(1) A V-functor

$$G\colon \mathsf{K}^{\otimes n} \longrightarrow \mathsf{L}$$

is an F-*iterate* if the composite

$$(\mathsf{K}_0^n) \xrightarrow{(-)_0^n} (\mathsf{K}^{\otimes n})_0 \xrightarrow{G_0} \mathsf{L}_0$$

is an F_0-iterate in the sense of Definition 1.1.43 (1).

(2) A V-natural transformation

$$\theta\colon G \longrightarrow H$$

between parallel F-iterates $\mathsf{K}^{\otimes n} \longrightarrow \mathsf{L}$ is an F-*coherent map* if

$$\theta_0 * 1_{(-)_0^n}\colon G_0 \circ (-)_0^n \longrightarrow H_0 \circ (-)_0^n$$

is an F_0-coherent map in the sense of Definition 1.1.43 (2). ◇

Theorem 2.5.8 (Enriched Epstein's Coherence). *In the context of Definition 2.5.7, there exists at most one F-coherent map between any two parallel F-iterates G, $H\colon \mathsf{K}^{\otimes n} \longrightarrow \mathsf{L}$.*

Proof. If θ and ω are F-coherent maps $G \longrightarrow H$, then Epstein's Theorem 1.1.44 implies that we have an equality of F_0-coherent maps

$$\theta_0 * 1_{(-)_0^n} = \omega_0 * 1_{(-)_0^n}.$$

Since $(-)_0^n$ is the identity on objects, the first equality in the Whiskering Lemma 1.2.12 implies that $\theta_0 = \omega_0$. Then the result follows by injectivity of $(-)_0$ on 2-cells (Lemma 2.1.7). □

2.6. Strictification of Enriched Monoidal Categories

Recall from Definition 1.4.2 that a monoidal V-category is called strict if the unitors and associator are identity V-natural transformations. We now use the Enriched Monoidal Coherence Theorem 2.5.6 to extend Mac Lane's Strictification Theorem 1.1.32 to the V-enriched setting. Recall from Definition 1.4.26 that a monoidal adjoint V-equivalence consists of a V-adjoint pair of monoidal V-functors such that the unit and counit are monoidal V-natural isomorphisms.

Theorem 2.6.1 (Enriched Monoidal Strictification). *Suppose V is a braided monoidal category and suppose K is a monoidal V-category. Then there exist*

- *a canonical strict monoidal V-category K_{st} and*
- *a canonical monoidal adjoint V-equivalence*

$$\mathsf{K}_{\mathsf{st}} \underset{i}{\overset{(-)'}{\rightleftarrows}} \mathsf{K}$$

with $(-)'$ unital strong monoidal, i strong monoidal, and $(-)' \circ i = 1_{\mathsf{K}}$.

Proof. The argument here follows the argument of Mac Lane's Strictification Theorem 1.1.32, given in [**ML98**, XI.3 Theorem 1], but uses the Coherence Theorem 2.5.6 in place of the plain monoidal Coherence Theorem 1.1.31.

First we define K_{st} as a V-category. We denote the V-category structure of K with superscripts, such as i^{K} and m^{K}, and use unadorned notation for the structure in K_{st} to be defined. The objects of K_{st} are finite, possibly empty, sequences of objects in K. For such a sequence $X = (X_1, \ldots, X_n)$ we define X' in K to be the left normalized product (2.5.4). In particular, for $X = ()$ the empty tuple we define

$()' = \mathrm{I}^{\mathsf{K}}$, the monoidal unit of K. Given two sequences of objects $X = (X_1, \ldots, X_n)$ and $Y = (Y_1, \ldots, Y_m)$ with $X_i, Y_j \in \mathsf{K}$ we define the hom object of K_{st} as

$$\mathsf{K}_{\mathsf{st}}(X,Y) = \mathsf{K}(X',Y').$$

The identity morphisms and composition in K_{st},

$$\mathbb{1} \xrightarrow{i_X} \mathsf{K}_{\mathsf{st}}(X,X) = \mathsf{K}(X',X')$$

and

$$\mathsf{K}(Y',Z') \otimes \mathsf{K}(X',Y') = \mathsf{K}_{\mathsf{st}}(Y,Z) \otimes \mathsf{K}_{\mathsf{st}}(X,Y) \xrightarrow{m_{X,Y,Z}} \mathsf{K}_{\mathsf{st}}(X,Z) = \mathsf{K}(X',Z'),$$

are likewise defined via the corresponding structure in K:

$$i_X = i^{\mathsf{K}}_{X'} \quad \text{and} \quad m_{X,Y,Z} = m^{\mathsf{K}}_{X',Y',Z'}.$$

This makes K_{st} into a V-category with $(-)'$ a V-functor.

Next we define the monoidal composition

$$\boxtimes \colon \mathsf{K}_{\mathsf{st}} \otimes \mathsf{K}_{\mathsf{st}} \longrightarrow \mathsf{K}_{\mathsf{st}}.$$

The product on objects is given by concatenation of sequences, with the empty sequence as the monoidal unit. Thus the monoidal composition is strictly associative and unital on objects. To define $\boxtimes$ on hom objects, suppose $A, X, B, Y \in \mathsf{K}_{\mathsf{st}}$ are sequences of objects of K. Then we have

$$\mathsf{K}_{\mathsf{st}}(A,X) \otimes \mathsf{K}_{\mathsf{st}}(B,Y) = \mathsf{K}(A',X') \otimes \mathsf{K}(B',Y')$$

and

$$\mathsf{K}_{\mathsf{st}}(A \boxtimes B, X \boxtimes Y) = \mathsf{K}((A \boxtimes B)', (X \boxtimes Y)').$$

By the Coherence Theorem 2.5.6 (1) there is a unique canonical V-map that, on hom objects, induces

$$\mathsf{K}(A' \boxtimes^{\mathsf{K}} B', X' \boxtimes^{\mathsf{K}} Y') \longrightarrow \mathsf{K}((A \boxtimes B)', (X \boxtimes Y)')$$

where $\boxtimes^{\mathsf{K}}$ denotes the monoidal composition in K. Then we define

$$\boxtimes \colon \mathsf{K}_{\mathsf{st}}(A,X) \otimes \mathsf{K}_{\mathsf{st}}(B,Y) \longrightarrow \mathsf{K}_{\mathsf{st}}(A \boxtimes B, X \boxtimes Y)$$

as the composite of this canonical V-map with the monoidal composition in K:

$$\mathsf{K}(A',X') \otimes \mathsf{K}(B',Y') \xrightarrow{\boxtimes^{\mathsf{K}}} \mathsf{K}(A' \boxtimes^{\mathsf{K}} B', X' \boxtimes^{\mathsf{K}} Y') \longrightarrow \mathsf{K}((A \boxtimes B)', (X \boxtimes Y)').$$

For the monoidal axioms of K_{st}, the composites around the boundaries of the unity diagrams (1.4.3) and the associativity diagram (1.4.4) are equal as V-functors by the Coherence Theorem 2.5.6 (1) for K. Therefore, we may define the unitors and associator of K_{st} as identity V-natural transformations. The unity and pentagon axioms for K_{st} then hold because each 2-cell is an identity.

We take the unit constraint $((-)')^0$ to be the identity on

$$()' = \mathrm{I}^{\mathsf{K}}$$

and we take the monoidal constraint $((-)')^2$ to be the V-natural transformation with components

$$\mathbb{1} \longrightarrow \mathsf{K}(X' \boxtimes Y', (X \boxtimes Y)')$$

given by the unique canonical V-maps. To verify that $(-)'$ is monoidal as a V-functor, each axiom of Definition 1.4.17 asserts an equality between two parallel canonical V-maps of K. So they are equal by the uniqueness part of Theorem 2.5.6 (1). Therefore, $(-)'$ is a unital strong monoidal V-functor.

The V-functor

$$i: \mathsf{K} \longrightarrow \mathsf{K}_{\mathsf{st}}$$

is defined by sending each object of K to the length 1 sequence and by the identity on hom objects. Components of its unit and monoidal constraints

$$\mathbb{1} \xrightarrow{i^0} \mathsf{K}_{\mathsf{st}}((),(\mathrm{I}^{\mathsf{K}})) \quad \text{and} \quad \mathbb{1} \xrightarrow{i^2} \mathsf{K}_{\mathsf{st}}((P) \boxtimes (Q),(P \boxtimes^{\mathsf{K}} Q)),$$

with $P, Q \in \mathsf{K}$, are given by the identities of I^{K} and $P \boxtimes^{\mathsf{K}} Q$, respectively. To see that i is a strong monoidal V-functor, one can again appeal to Theorem 2.5.6 (1) for the axioms of Definition 1.4.17, or one can make a direct argument as in Explanation 2.6.2 below.

We then have $(iP)' = P$ for $P \in \mathsf{K}$ and unit of the adjunction $(-)' \dashv i$ given by components

$$\mathbb{1} \xrightarrow{i^{\mathsf{K}}_{X'}} \mathsf{K}(X', X') = \mathsf{K}_{\mathsf{st}}(X, i(X'))$$

for $X \in \mathsf{K}_{\mathsf{st}}$. Thus $(-)' \dashv i$ defines a strong monoidal adjoint V-equivalence. □

Explanation 2.6.2. In the proof of Theorem 2.6.1, one can also give the following direct argument that the V-functor

$$i: \mathsf{K} \longrightarrow \mathsf{K}_{\mathsf{st}}$$

is monoidal.

Associativity: Since i^2 is componentwise an identity morphism and since $\boxtimes$ on hom objects involves a canonical V-map, both composites are equal to a component of the monoidal associator a in K. On the left pasting diagram, this component of a comes from the last 2-cell,

$$1_i * a.$$

On the right pasting diagram, this component of a comes from the middle 2-cell

$$1_{\boxtimes} * (1 \otimes i^2) * 1_{a^{\otimes}}.$$

Left Unity: Both pasting diagrams are equal to an identity morphism in K. The right pasting diagram is the identity because ℓ in K_{st} is the identity. On the left pasting diagram, the first 2-cell

$$1_{\boxtimes} * 1_{(1 \otimes i)} * (i^0 \otimes 1)$$

is given by a component of $(\ell^{\mathsf{K}})^{-1}$ in K and the component of

$$i^2 * 1_{\mathrm{I} \otimes 1}$$

is an identity. The component of the last 2-cell

$$1_i * \ell_1$$

is a component of ℓ^{K} in K. The composite of these is an identity in K.

Right Unity: Similarly, both pasting diagrams are equal to an identity morphism in K. The right pasting diagram is an identity because K_{st} is strict, and the left pasting diagram has identity components given by composing r^{-1} with r.

One way of explaining the asymmetry between the associativity axiom (where both sides are given by generally nontrivial components) and the unity axioms (where both sides are given by identities) is the following. Even though the components of i^0 and i^2 are given by identities in K, they are not identities in K_{st}. Therefore, their whiskerings appearing in the axioms may have non-identity components. Such whiskerings appear on both sides of the associativity axiom, but only on the left side of each unity axiom. ◇

Next we turn to the braided case, extending the Braided Strictification Theorem 1.1.39 to braided monoidal V-categories.

Theorem 2.6.3 (Enriched Braided Strictification). *Suppose V is a symmetric monoidal category and suppose K is a braided monoidal V-category. Then there exist*

- *a canonical braided strict monoidal V-category K_{st} and*
- *a canonical braided monoidal adjoint V-equivalence*

$$\mathsf{K}_{\mathsf{st}} \underset{i}{\overset{(-)'}{\rightleftarrows}} \mathsf{K}$$

with $(-)'$ unital strong braided monoidal, i strong braided monoidal, and $(-)' \circ i = 1_{\mathsf{K}}$.

Proof. The definition of K_{st} and the monoidal adjoint V-equivalence

$$\mathsf{K}_{\mathsf{st}} \underset{i}{\overset{(-)'}{\rightleftarrows}} \mathsf{K}$$

are given as in Theorem 2.6.1. Now using the braiding in K, define the braiding in K_{st} to have components

$$\mathbb{1} \xrightarrow{\beta} \mathsf{K}_{\mathsf{st}}(X \boxtimes Y, Y \boxtimes X) = \mathsf{K}((X \boxtimes Y)', (Y \boxtimes X)')$$

at $X, Y \in \mathsf{K}_{\mathsf{st}}$ given by the composite of the braiding for K and the two canonical V-maps from Theorem 2.5.6 (1) having components

$$\mathbb{1} \longrightarrow \mathsf{K}((X \boxtimes Y)', X' \boxtimes^{\mathsf{K}} Y') \quad \text{and} \quad \mathbb{1} \longrightarrow \mathsf{K}(Y' \boxtimes^{\mathsf{K}} X', (Y \boxtimes X)').$$

For each of the hexagon axioms in K_{st}, (1.4.11) and (1.4.12), both sides have the same underlying braid described in Explanation II.1.2.19. Therefore, the axioms hold by Theorem 2.5.6 (2). Therefore, K_{st} is a braided monoidal V-category.

Now to verify that $(-)'$ satisfies the braid axiom in Definition 1.4.18, note that on both sides each component

$$\mathbb{1} \longrightarrow \mathsf{K}(X' \boxtimes^{\mathsf{K}} Y', (Y \boxtimes X)')$$

has underlying braid given by the same elementary block braid (see Definition II.1.2.3, with m and n given by the lengths of the sequences X and Y). Therefore, the two are equal by Theorem 2.5.6 (2) for K. This shows that $(-)'$ is a strong braided monoidal V-functor.

For i, the underlying braid on both sides of the braid axiom is the generating braid s_1 in B_2. Therefore, i is a strong braided monoidal V-functor. □

Finally we turn to the symmetric case, extending the Symmetric Strictification Theorem 1.1.42 to symmetric monoidal V-categories.

Theorem 2.6.4 (Enriched Symmetric Strictification). *Suppose* V *is a symmetric monoidal category and suppose* K *is a symmetric monoidal* V*-category. Then there exist*

- *a canonical symmetric strict monoidal* V*-category* K_{st} *and*
- *a canonical symmetric monoidal adjoint* V*-equivalence*

$$\mathsf{K}_{\mathsf{st}} \underset{i}{\overset{(-)'}{\rightleftarrows}} \mathsf{K}$$

with $(-)'$ *unital strong symmetric monoidal,* i *strong symmetric monoidal, and* $(-)' \circ i = 1_{\mathsf{K}}$.

Proof. The definition of K_{st} and the adjoint V-equivalence

$$\mathsf{K}_{\mathsf{st}} \underset{i}{\overset{(-)'}{\rightleftarrows}} \mathsf{K}$$

are given as in Theorems 2.6.1 and 2.6.3. The symmetry axiom of Definition 1.4.13 for K_{st} follows from the uniqueness part of Theorem 2.5.6 (3) for K. □

2.7. Notes

2.7.1 (Literature on Change of Enrichment). Change of enrichment along a monoidal functor is discussed in [**EK66**]. The thesis of Cruttwell [**Cru09**] provides some further development: Theorem 2.2.7 above appears as [**Cru09**, Theorem 4.3.2] and Theorem 2.4.10 (1) above appears as [**Cru09**, Theorem 5.7.1]. ◇

2.7.2 (Coherence and Strictification). Kong and Zheng give an enriched strictification theorem [**KZ18**, Proposition 2.4], under the assumption that V is strict. Theorem 2.6.1 generalizes to the case that V is not necessarily strict.

The general theory of pseudoalgebras over 2-monads gives another approach to coherence and strictification results. Starting points for this theory are [**BKP89**] and [**Lac02**]. The thesis of Houston [**Hou07**] discusses pseudomonoids specifically and explains general methods for translating results about monoidal categories to corresponding results for pseudomonoids. ◇

2.7.3 (Axiom of Universes). In Motivation 2.2.6 and Explanation 2.2.8, and in the proofs of Corollaries 2.2.10 and 2.3.14, we refer to Grothendieck's *Axiom of Universes.* This axiom asserts that every set belongs to some Grothendieck universe. See [**JY21**, Section 1.1] for further discussion. The results in this chapter do not depend on the Axiom of Universes, but it gives a convenient way of identifying and precisely navigating the size subtleties that arise in the statements of Theorems 2.2.7 and 2.3.9 and their applications. The survey [**Shu∞b**] describes several other potential approaches. ◇

CHAPTER 3

Self-Enrichment and Enriched Yoneda

Throughout this chapter we assume that V is a symmetric monoidal closed category. The material has three separate but closely related points of focus.

Canonical Self-Enrichment. In Section 3.1 we explain how a symmetric monoidal closed category V is canonically enriched over itself. We denote the resulting V-category as $\underline{\mathsf{V}}$. In Section 3.2 we describe represented V-functors and several consequences of the closed structure on V that will be useful throughout the chapter. In Section 3.3 we apply Theorem 2.5.1 to show that $\underline{\mathsf{V}}$ is symmetric monoidal as a V-category.

Enriched Co/Ends and the Enriched Yoneda Lemma. In Section 3.4 we give a preliminary version of the enriched Yoneda Lemma that we call the Enriched Yoneda Bijection 3.4.12 because it asserts a bijection of certain underlying sets. The Yoneda V-functor is described in Section 3.5 and requires background material on V-enriched coends and ends given there. Then in Section 3.6 we give additional necessary background and prove the full V-Yoneda Lemma 3.6.9.

Enriched Symmetric Monoidal Diagram Categories. In Section 3.7 we apply the previous material to categories of enriched symmetric monoidal functors from a small V-category $\mathcal{D}$ to $\underline{\mathsf{V}}$. Using the V-Yoneda Lemma 3.6.9 we prove an equivalent statement known as the V-Yoneda Density Theorem 3.7.8. With this we show that the category of symmetric monoidal V-functors from $\mathcal{D}$ to $\underline{\mathsf{V}}$ is symmetric monoidal and closed. In Section 3.8 we use change of enrichment to show that the category of enriched diagrams to $\underline{\mathsf{V}}$ is symmetric monoidal as a V-category. The additional section Section 3.9 describes tensor and cotensor structures for enriched categories. The material there will be applied to certain diagram categories in Chapters 4, 7, and 9.

3.1. Self-Enriched Categories

Recall from Definition 1.1.28 that a symmetric monoidal category V is closed if, for each object $X \in \mathsf{V}$, the endofunctor $- \otimes X$ has a right adjoint $[X,-]\colon \mathsf{V} \longrightarrow \mathsf{V}$.

Definition 3.1.1. Suppose V is a symmetric monoidal closed category. For each $X \in \mathsf{V}$ the *evaluation at* X, denoted eval, is the counit

$$[X,-] \otimes X \xrightarrow{\mathsf{eval}} \mathrm{Id}_{\mathsf{V}}.$$

The *coevaluation at* X, denoted coeval, is the unit

$$\mathrm{Id}_{\mathsf{V}} \xrightarrow{\mathsf{coeval}} [X, - \otimes X]. \qquad \diamond$$

Explanation 3.1.2. For reference below, we note two particular consequences of the closed structure on V. Both follow from examining the relevant adjunctions.

(1) The following diagram of sets commutes for each $A, C \in \mathsf{V}$, where the unlabeled isomorphisms are those of the closed structure.

$$\begin{array}{ccc} \mathsf{V}(\mathbb{1},[A,C]) & \xrightarrow{\lambda^*} & \mathsf{V}(\mathbb{1}\otimes\mathbb{1},[A,C]) \\ {\scriptstyle [\lambda,C]_*}\downarrow & & \downarrow{\scriptstyle \cong} \\ \mathsf{V}(\mathbb{1},[\mathbb{1}\otimes A,C]) & \xrightarrow{\cong} & \mathsf{V}(\mathbb{1},[\mathbb{1},[A,C]]) \end{array} \tag{3.1.3}$$

(2) The following diagram in V commutes for each $P, Q, A, C \in \mathsf{V}$, where each of the isomorphisms is given by the closed structure of V.

$$\begin{array}{ccccc} & & [P,[Q\otimes A,C]] & & \\ & \overset{\cong}{\nearrow} & & \overset{\cong}{\searrow} & \\ [P,[Q,[A,C]]] & & & & [P\otimes(Q\otimes A),C] \\ \cong\downarrow & & & & \uparrow\cong \\ [P\otimes Q,[A,C]] & & \xrightarrow{\cong} & & [(P\otimes Q)\otimes A,C] \end{array} \tag{3.1.4}$$

◇

Definition 3.1.5. Suppose V is a symmetric monoidal closed category. We let $\underline{\mathsf{V}}$ denote V equipped with the *canonical self-enrichment* defined as follows.

- The objects are those of V.
- For each pair of objects $X, Y \in \underline{\mathsf{V}}$, the hom object with domain X and codomain Y is $\underline{\mathsf{V}}(X,Y) = [X,Y]$.
- For each triple of objects $X, Y, Z \in \underline{\mathsf{V}}$, the composition

$$m_{X,Y,Z} : [Y,Z] \otimes [X,Y] \longrightarrow [X,Z]$$

is defined as the adjoint to the following composite in V.

$$\begin{array}{ccc} ([Y,Z]\otimes[X,Y])\otimes X & & Z \\ {\scriptstyle \alpha}\downarrow & & \uparrow{\scriptstyle \mathsf{eval}} \\ [Y,Z]\otimes([X,Y]\otimes X) & \xrightarrow{1\otimes\mathsf{eval}} & [Y,Z]\otimes Y \end{array} \tag{3.1.6}$$

- For each object $X \in \underline{\mathsf{V}}$, the identity

$$i_X : \mathbb{1} \longrightarrow [X,X]$$

is defined as the adjoint to the left unit isomorphism in V

$$\mathbb{1} \otimes X \xrightarrow{\lambda} X.$$

Proposition 3.1.11 below shows that these satisfy the axioms of Definition 1.2.1. ◇

Explanation 3.1.7. The multiplication and identity morphisms of Definition 3.1.5 are defined via adjoints. Recall that for an adjuction $F \dashv G$, the adjoints of morphisms

$$FX \xrightarrow{f} Y \quad \text{and} \quad X \xrightarrow{g} GY$$

are the following composites with the unit, respectively counit, of the adjunction:

$$X \longrightarrow GFX \xrightarrow{Gf} GY \quad \text{and} \quad FX \xrightarrow{Fg} FGY \longrightarrow Y.$$

In the context of Definition 3.1.5, this means that the composition of hom objects is given by the following composite.

$$[Y,Z]\otimes[X,Y] \xrightarrow{\text{coeval}} [X,([Y,Z]\otimes[X,Y])\otimes X] \xrightarrow{[X,\alpha]} [X,[Y,Z]\otimes([X,Y]\otimes X)] \xrightarrow{[X,1\otimes\text{eval}]} [X,[Y,Z]\otimes Y] \xrightarrow{[X,\text{eval}]} [X,Y]$$

Similarly, the identity is defined by the following composite:

$$\mathbb{1} \xrightarrow{\text{coeval}} [X,\mathbb{1}\otimes X] \xrightarrow{[X,\lambda]} [X,X].$$

Moreover, m and i are defined as adjoints and adjunction is a bijection, so taking adjoints of m and i, respectively, results in the following two commutative diagrams.

(3.1.8)
$$\begin{array}{ccc}
([Y,Z]\otimes[X,Y])\otimes X & \xrightarrow{m\otimes 1} & [X,Z]\otimes X \\
\downarrow{\scriptstyle \alpha} & & \\
[Y,Z]\otimes([X,Y]\otimes X) & & \downarrow{\scriptstyle \text{eval}} \\
\downarrow{\scriptstyle 1\otimes\text{eval}} & & \\
[Y,Z]\otimes Y & \xrightarrow{\text{eval}} & Z
\end{array}$$

(3.1.9)
$$\begin{array}{ccc}
\mathbb{1}\otimes X & \xrightarrow{i\otimes 1} & [X,X]\otimes X \\
 & \searrow{\scriptstyle \lambda} & \downarrow{\scriptstyle \text{eval}} \\
 & & X
\end{array}$$

We also note, for a morphism $\psi\colon P' \longrightarrow P$ in V, we have the following commutative diagram expressing a compatibility of the composition with the induced morphism $[\psi,Q]\colon[P,Q] \longrightarrow [P',Q]$ for $P,P',Q,R\in\mathsf{V}$.

(3.1.10)
$$\begin{array}{ccc}
[Q,R]\otimes[P,Q] & \xrightarrow{1\otimes[\psi,Q]} & [Q,R]\otimes[P',Q] \\
\downarrow{\scriptstyle m} & & \downarrow{\scriptstyle m} \\
[P,R] & \xrightarrow{[\psi,R]} & [P',R]
\end{array}$$

Commutativity of the above diagram follows by taking adjoints and using the observations above. ◇

Proposition 3.1.11. *Suppose* V *is a symmetric monoidal closed category. Then the closed structure gives* $\underline{\mathsf{V}}$ *the structure of a* V*-category.*

Proof. The data of the canonical self-enrichment is given in Definition 3.1.5. We verify commutativity of the associativity and unity diagrams of Definition 1.2.1 by verifying that their adjoints commute.

To take the adjoint of the associativity diagram (1.2.2) we apply $- \otimes W$ and then compose with the evaluation

$$\mathsf{eval}\colon [W,Z] \otimes W \longrightarrow Z.$$

Doing so yields the outermost composites to Z in the diagram below.

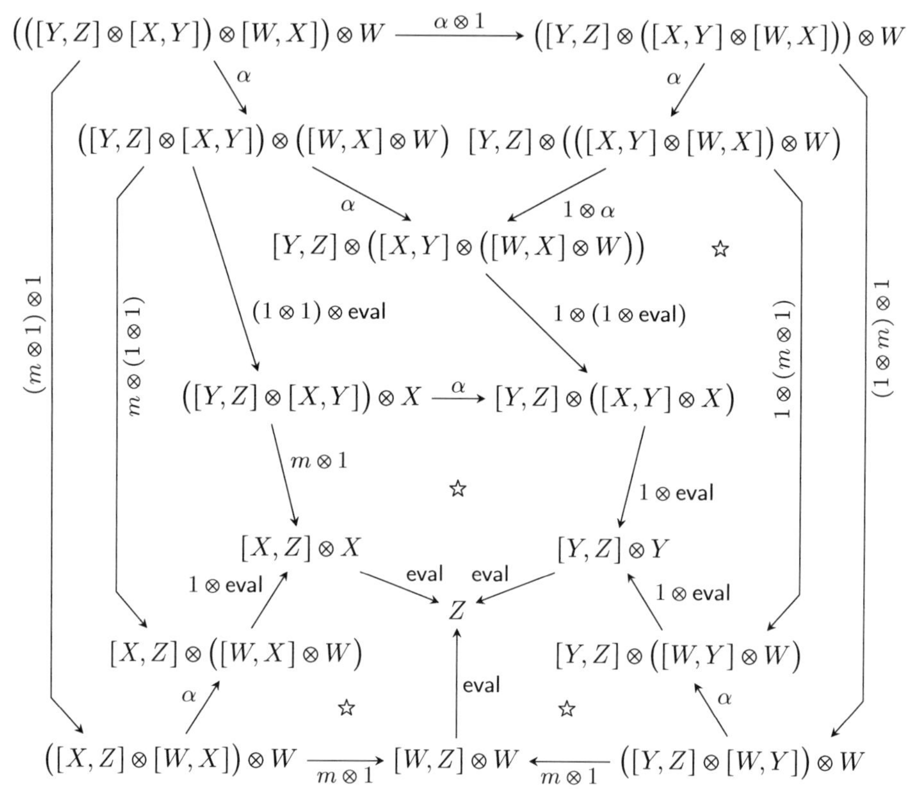

In the diagram above, commutativity of the regions marked ☆ follows from (3.1.8). Commutativity of the remaining subdiagrams follows from naturality of α (three times), the pentagon axiom (1.1.3), and the equality

$$(m \otimes 1) \circ \big((1 \otimes 1) \otimes \mathsf{eval}\big) = m \otimes \mathsf{eval} = (1 \otimes \mathsf{eval}) \circ \big(m \otimes (1 \otimes 1)\big).$$

For the unity diagram (1.2.3), we apply $- \otimes X$ and then compose with the evaluation

$$\mathsf{eval}\colon [X,Y] \otimes X \longrightarrow Y.$$

Doing so yields the outermost composites to Y on both sides of the diagram below.

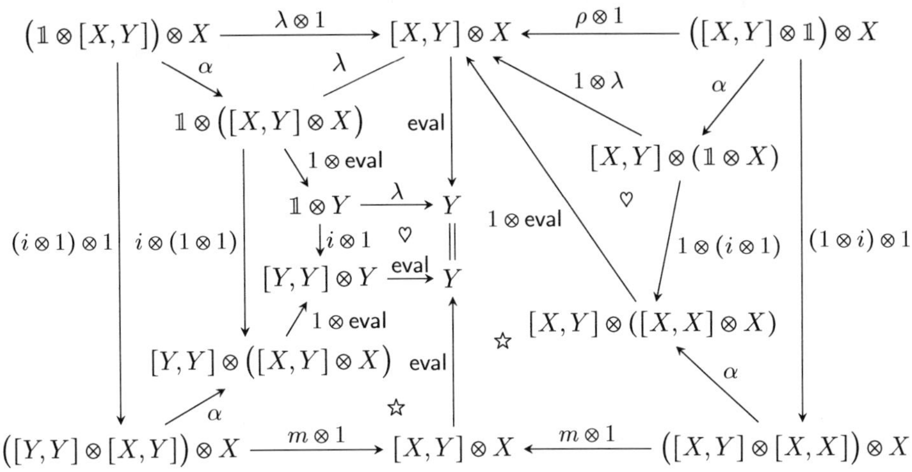

In the diagram above, the unlabeled arrows are composites of an associator with an evaluation. The regions marked ☆ commute by (3.1.8) and the regions marked ♡ commute by (3.1.9). The remaining regions commute by the unity properties (1.1.5) and (1.1.2), naturality of λ, naturality of α (twice), and the equality

$$(1 \otimes \mathsf{eval}) \circ (i \otimes (1 \otimes 1)) = i \otimes \mathsf{eval} = (i \otimes 1) \circ (1 \otimes \mathsf{eval}). \qquad \square$$

3.2. Represented Enriched Functors

In this section we describe corepresented and represented V-functors taking values in $\underline{\mathsf{V}}$. In Lemma 3.2.12 we give a characterization of V-natural transformations between V-functors with codomain $\underline{\mathsf{V}}$.

Definition 3.2.1. Suppose C is a V-category and suppose $X \in \mathsf{C}$. Then $\mathcal{Y}^X = \mathsf{C}(X,-)$ denotes the *corepresented* V*-functor*

$$\mathcal{Y}^X : \mathsf{C} \longrightarrow \underline{\mathsf{V}}$$

whose assignment on objects is $W \longmapsto \mathsf{C}(X,W)$ and whose morphism of hom objects

$$(\mathcal{Y}^X)_{Z,W} : \mathsf{C}(Z,W) \longrightarrow [\mathcal{Y}^X Z, \mathcal{Y}^X W]$$

is defined as the adjoint to composition in C:

$$\mathsf{C}(Z,W) \otimes \mathcal{Y}^X Z = \mathsf{C}(Z,W) \otimes \mathsf{C}(X,Z) \xrightarrow{m} \mathsf{C}(X,W) = \mathcal{Y}^X W$$

for each $Z, W \in \mathsf{C}$. The composition and identity axioms of Definition 1.2.4 follow from the associativity and unity axioms (Definition 1.2.1) of C together with the definitions of composition and identity morphisms in $\underline{\mathsf{V}}$. ◇

For the dual definition, recall from Definition 1.2.16 the opposite of an enriched category.

Definition 3.2.2. Suppose C is a V-category and suppose $Y \in \mathsf{C}$. Then $\mathcal{Y}_Y = \mathsf{C}(-,Y)$ denotes the *represented* V*-functor*

$$\mathcal{Y}_Y : \mathsf{C}^{\mathsf{op}} \longrightarrow \underline{\mathsf{V}}$$

whose assignment on objects is $Z \longmapsto \mathsf{C}(Z,Y)$ and whose morphism of hom objects

$$(\mathcal{Y}_Y)_{Z,W} : \mathsf{C}^{\mathsf{op}}(Z,W) = \mathsf{C}(W,Z) \longrightarrow [\mathcal{Y}_Y Z, \mathcal{Y}_Y W]$$

is the adjoint to composition in C^{op}:

$$\mathsf{C}(W,Z)\otimes\mathcal{Y}_Y Z = \mathsf{C}(W,Z)\otimes\mathsf{C}(Z,Y) \xrightarrow{\xi} \mathsf{C}(Z,Y)\otimes\mathsf{C}(W,Z) \xrightarrow{m} \mathsf{C}(W,Y) = \mathcal{Y}_Y W.$$

Verifying V-functoriality of $\mathcal{Y}_Y$ is similar to that of $\mathcal{Y}^X$. ◇

Definition 3.2.3. Suppose V is a symmetric monoidal closed category and suppose that $\theta\colon P \longrightarrow Q$ is a morphism in V. Let $\theta^\perp$ denote the morphism

$$\mathbb{1} \xrightarrow{\theta^\perp} [P,Q]$$

adjoint to $\theta\circ\lambda$. The correspondence $\theta \leftrightarrow \theta^\perp$ provides an isomorphism of categories between V and $(\underline{\mathsf{V}})_0$, the underlying category of the self-enrichment of V. We call this the *canonical underlying isomorphism* of V. ◇

Explanation 3.2.4. To see that the correspondence $\theta \leftrightarrow \theta^\perp$ in Definition 3.2.3 preserves composition, suppose given a composable pair of morphisms in V

$$P \xrightarrow{\theta} Q \xrightarrow{\omega} R.$$

Then the equality

$$(\omega\theta)^\perp = (\omega^\perp)(\theta^\perp) \tag{3.2.5}$$

is commutativity of the following diagram in V.

$$\begin{array}{ccc} \mathbb{1} & \xrightarrow{(\omega\theta)^\perp} & [P,R] \\ {\scriptstyle\lambda^{-1}}\downarrow & & \uparrow{\scriptstyle m} \\ \mathbb{1}\otimes\mathbb{1} & \xrightarrow{(\omega^\perp)\otimes(\theta^\perp)} & [Q,R]\otimes[P,Q] \end{array}$$

The equality (3.2.5) then follows by considering the adjoints of the composites above, using functoriality of $\otimes$, naturality of λ, and the definition of the composition m. ◇

Recall from Definition 2.1.8 with $\mathsf{D} = \underline{\mathsf{V}}$ we have morphisms

$$\underline{\mathsf{V}}(X,\theta^\perp) = [X,\theta^\perp]\colon [X,P] \longrightarrow [X,Q] \quad\text{and}\quad [Q,Y] \xrightarrow{[\theta^\perp,Y]} [P,Y]. \tag{3.2.6}$$

In Lemmas 3.2.7 and 3.2.8 we show that these morphisms have a simple description via θ.

Lemma 3.2.7. *Suppose $\theta\colon P \longrightarrow Q$ is a morphism in V and X is an object of V. The adjoint of*

$$[X,\theta^\perp]\colon [X,P] \longrightarrow [X,Q]$$

is

$$[X,P]\otimes X \xrightarrow{\mathsf{eval}} P \xrightarrow{\theta} Q.$$

Proof. The result follows by commutativity of the diagram below. Recalling Definition 2.1.8, the adjoint to $[X,\theta^\perp]$ is the composite along the left, bottom, and right of the diagram.

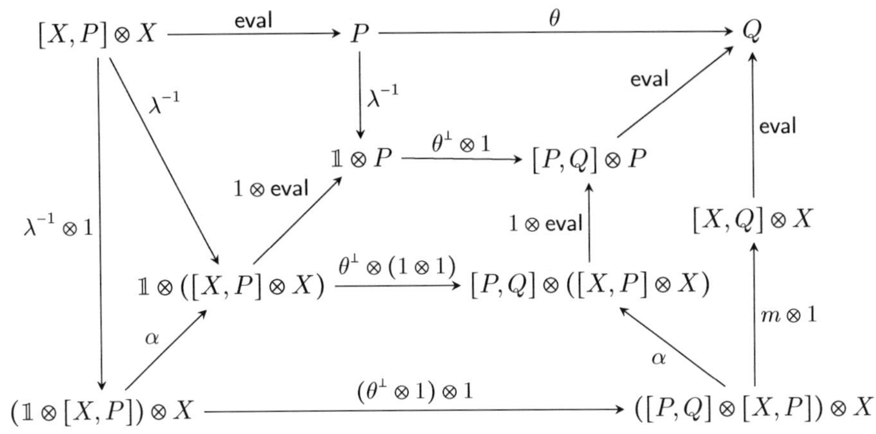

In the above diagram, the upper-right trapezoid commutes by definition of $\theta^\perp$ and the rightmost trapezoid commutes by (3.1.8). The other regions commute by naturality of λ and α, functoriality of $\otimes$, and the left unity diagram (1.1.5). □

Lemma 3.2.8. *Suppose $\theta: P \longrightarrow Q$ is a morphism in V and Y is an object of V. The adjoint of*

$$[\theta^\perp, Y]: [Q, Y] \longrightarrow [P, Y]$$

is

$$[Q, Y] \otimes P \xrightarrow{1 \otimes \theta} [Q, Y] \otimes Q \xrightarrow{\mathsf{eval}} Y.$$

Proof. The result follows by commutativity of the diagram below. Recalling Definition 2.1.8, the adjoint to $[\theta^\perp, Y]$ is the composite along the left, bottom, and right of the diagram.

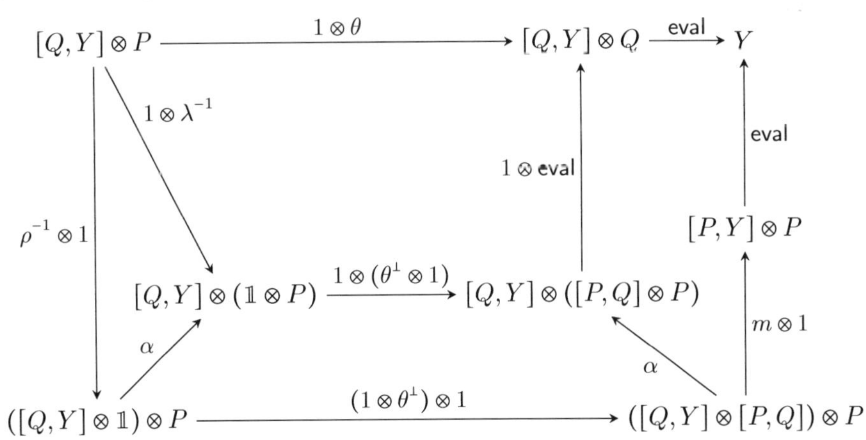

In the above diagram, the upper trapezoid commutes by definition of $\theta^\perp$ and the rightmost trapezoid commutes by (3.1.8). The other regions commute by naturality of α and the middle unity diagram (1.1.2). □

Explanation 3.2.9. The statements of Lemmas 3.2.7 and 3.2.8 will be useful in our work below, but they also allow us to explain the relationship between three subtly different versions of the represented and corepresented functors. Recall we use the notation $\underline{\mathsf{V}}$ to denote V equipped with its self-enrichment as in Definition 3.1.5, and reserve the notation V for the ordinary symmetric monoidal closed category as in Section 1.1.

For $X, Y \in \mathsf{V}$ we have:

(1) the ordinary functors

$$[X,-]: \mathsf{V} \longrightarrow \mathsf{V} \quad \text{and} \quad [-,Y]: \mathsf{V}^{\mathsf{op}} \longrightarrow \mathsf{V}$$

defined via the closed structure on V,

(2) the V-enriched functors

$$\mathcal{Y}^X: \underline{\mathsf{V}} \longrightarrow \underline{\mathsf{V}} \quad \text{and} \quad \mathcal{Y}^Y: \underline{\mathsf{V}}^{\mathsf{op}} \longrightarrow \underline{\mathsf{V}}$$

of Definitions 3.2.1 and 3.2.2 with $\mathsf{C} = \underline{\mathsf{V}}$, and

(3) the ordinary functors

$$[X,(-)^{\perp}]: (\underline{\mathsf{V}})_0 \longrightarrow \mathsf{V} \quad \text{and} \quad [(-)^{\perp}, Y]: (\underline{\mathsf{V}})_0^{\mathsf{op}} \longrightarrow \mathsf{V}$$

defined in (3.2.6) as a special case of Definition 2.1.8.

The functors (1) and (3) correspond under the canonical underlying isomorphism of categories $(\underline{\mathsf{V}})_0 \cong \mathsf{V}$ given by the identity on objects and the correspondence $\theta \leftrightarrow \theta^{\perp}$ from Definition 3.2.3. Indeed for a morphism $\theta: P \longrightarrow Q$ in V we have equalities

$$[X,\theta] = [X,\theta^{\perp}] \quad \text{and} \quad [\theta, Y] = [\theta^{\perp}, Y]$$

because, by Lemma 3.2.7 and Lemma 3.2.8, respectively, the corresponding adjoints are equal.

To compare with (2), we consider the induced functor on $(\underline{\mathsf{V}})_0$. For $\theta: P \longrightarrow Q$ in V the morphism $(\mathcal{Y}^X)_0\theta^{\perp}$ is defined to be the composite

$$\mathbb{1} \xrightarrow{\theta^{\perp}} [P,Q] \xrightarrow{\mathcal{Y}^X} [[X,P],[X,Q]].$$

Using the definition of $\mathcal{Y}^X$ and $[X,\theta^{\perp}]$, one verifies that the adjoint of $(\mathcal{Y}^X)_0\theta^{\perp}$ is equal to the composite

$$\mathbb{1} \otimes [X,P] \xrightarrow{\lambda} [X,P] \xrightarrow{[X,\theta^{\perp}]} [X,Q].$$

With a similar calculation for $(\mathcal{Y}_Y)_0\theta^{\perp}$ we have

$$(\mathcal{Y}^X)_0\theta^{\perp} = [X,\theta^{\perp}]^{\perp} = [X,\theta]^{\perp} \quad \text{and} \quad (\mathcal{Y}_Y)_0\theta^{\perp} = [\theta^{\perp}, Y]^{\perp} = [\theta, Y]^{\perp}. \qquad \diamond$$

In our work below we will use the following characterization of V-naturality for the special case of V-functors with codomain $\underline{\mathsf{V}}$.

Lemma 3.2.10. *Suppose that C is a V-category and that $F, G: \mathsf{C} \longrightarrow \underline{\mathsf{V}}$ are V-functors. Suppose that*

$$\theta_X: FX \longrightarrow GX$$

is a collection of morphisms in V for $X \in \mathsf{C}$. The morphisms

$$\theta_X^{\perp}: \mathbb{1} \longrightarrow [FX, GX]$$

are the components of a V*-natural transformation* $\theta^{\perp}$ *if and only if the following diagram commutes for each pair of objects* X *and* Y *in* C.

$$\begin{array}{ccccc} \mathsf{C}(X,Y)\otimes FX & \xrightarrow{F\otimes 1} & [FX,FY]\otimes FX & \xrightarrow{\mathsf{eval}} & FY \\ {\scriptstyle 1\otimes\theta_X}\big\downarrow & & & & \big\downarrow{\scriptstyle \theta_Y} \\ \mathsf{C}(X,Y)\otimes GX & \xrightarrow{G\otimes 1} & [GX,GY]\otimes GX & \xrightarrow{\mathsf{eval}} & GY \end{array} \tag{3.2.11}$$

Proof. Beginning with the characterization of V-naturality in Lemma 2.1.11, with $\mathsf{D} = \underline{\mathsf{V}}$, we apply $-\otimes FX$ to the diagram (2.1.12) and compose with evaluation. Then Lemmas 3.2.7 and 3.2.8, together with the equality

$$(1\otimes\theta_X)\circ(G\otimes 1) = G\otimes\theta_X = (G\otimes 1)\circ(1\otimes\theta_X),$$

show that the adjoints to $[\theta_X^{\perp}, GY]$ and $[FX, \theta_Y^{\perp}]$ are the composites with $1\otimes\theta_X$ and θ_Y indicated in (3.2.11). This shows that the two composites around the boundary of (3.2.11) are the adjoints of the two composites around (2.1.12). Therefore, for each $X, Y \in \mathsf{C}$, (2.1.12) commutes if and only if (3.2.11) commutes. □

The following immediate applications of Lemma 3.2.10 will be useful in our work below.

Lemma 3.2.12. *Suppose, in the context of Lemma 3.2.10, that each* θ_X *is an isomorphism in* V

$$\theta_X\colon FX \xrightarrow{\cong} GX.$$

Then the components $\theta_X^{\perp}$ *are* V*-natural if and only if the inverse components* $(\theta_X^{-1})^{\perp}$ *are* V*-natural.*

Proof. For each $X, Y \in \mathsf{C}$, the vertical composites in the diagram below are identities.

$$\begin{array}{ccccc} \mathsf{C}(X,Y)\otimes FX & \xrightarrow{F\otimes 1} & [FX,FY]\otimes FX & \xrightarrow{\mathsf{eval}} & FY \\ {\scriptstyle 1\otimes\theta_X}\big\downarrow & & & & \big\downarrow{\scriptstyle \theta_Y} \\ \mathsf{C}(X,Y)\otimes GX & \xrightarrow{G\otimes 1} & [GX,GY]\otimes GX & \xrightarrow{\mathsf{eval}} & GY \\ {\scriptstyle 1\otimes\theta_X^{-1}}\big\downarrow & & & & \big\downarrow{\scriptstyle \theta_Y^{-1}} \\ \mathsf{C}(X,Y)\otimes FX & \xrightarrow{F\otimes 1} & [FX,FY]\otimes FX & \xrightarrow{\mathsf{eval}} & FY \end{array}$$

Therefore, the upper half commutes if and only if the lower half commutes. Hence the result follows from Lemma 3.2.10. □

Lemma 3.2.13. *Suppose* $F\colon \mathsf{C} \longrightarrow \mathsf{D}$ *is a* V*-functor between* V*-categories. Then the morphisms*

$$\mathsf{C}(X,Y) \xrightarrow{F_{X,Y}} \mathsf{D}(FX,FY)$$

for $X, Y \in \mathsf{C}$ *determine* V*-natural transformations*

$$\mathsf{C}(-,W) \xrightarrow{(F_{-,W})^{\perp}} \mathsf{D}(F(-),FW) \quad \text{and} \quad \mathsf{C}(Z,-) \xrightarrow{(F_{Z,-})^{\perp}} \mathsf{D}(FZ,F(-))$$

for $Z, W \in \mathsf{C}$.

Proof. We apply Lemma 3.2.10 to show that both of the V-naturality conditions follow from the composition axiom for F. For each $W \in \mathsf{C}$ consider the V-functors

$$\mathsf{C}(-,W)\colon \mathsf{C}^{\mathsf{op}} \longrightarrow \underline{\mathsf{V}} \quad \text{and} \quad \mathsf{D}(F(-),FW)\colon \mathsf{C}^{\mathsf{op}} \longrightarrow \underline{\mathsf{V}}.$$

The diagram (3.2.11) for $F_{-,W}$ is the outer diagram below, for $X, Y \in \mathsf{C}$, where $m' = m \circ \xi$ is the composition in C^{op}, respectively D^{op}.

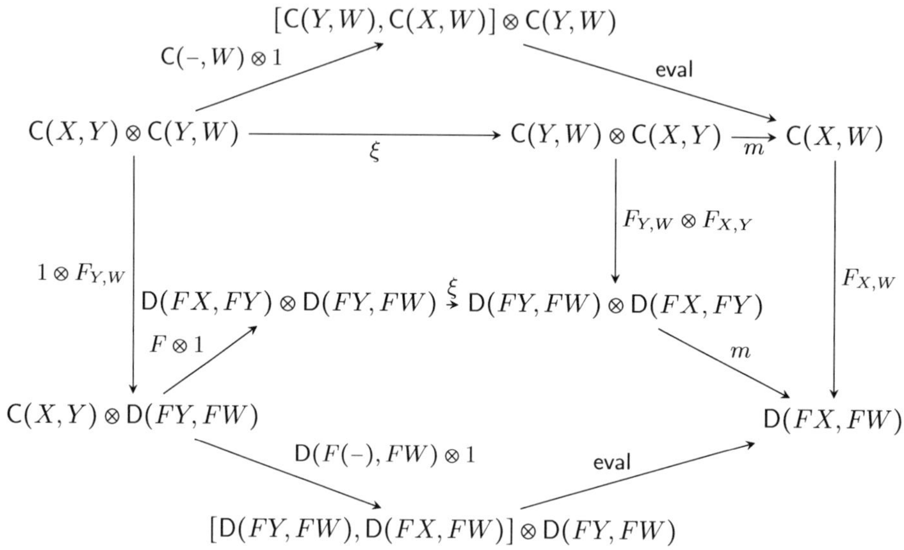

In the above diagram, the upper and lower regions commute by definitions of $\mathsf{C}(-,W)$ and $\mathsf{D}(F(-),FW)$, respectively. The pentagonal region at left commutes by naturality of ξ, and the remaining trapezoid is the composition axiom. Therefore $(F_{-,W})^{\perp}$ is V-natural. The argument for V-naturality of $(F_{Z,-})^{\perp}$ is similar but does not require the symmetry isomorphism ξ. □

In our work below we will need one further result proved using Lemma 3.2.10. For the following, recall from Definition 1.2.14 a V-adjunction $L \dashv R$ is an adjoint pair of 1-cells in V-Cat.

Proposition 3.2.14. *A pair of* V*-functors*

$$\mathsf{C} \underset{R}{\overset{L}{\rightleftarrows}} \mathsf{D}$$

is a V*-adjoint pair with* $L \dashv R$ *if and only if there are isomorphisms in* V

$$\mathsf{D}(LX,Y) \cong \mathsf{C}(X,RY)$$

that are V*-natural with respect to* $X \in \mathsf{C}$ *and* $Y \in \mathsf{D}$.

Proof. For one implication, suppose given unit and counit V-natural transformations

$$\eta\colon 1_{\mathsf{C}} \longrightarrow RL \quad \text{and} \quad \varepsilon\colon LR \longrightarrow 1_{\mathsf{D}}$$

making $L \dashv R$ a V-adjunction. For each $X \in \mathsf{C}$ and $Y \in \mathsf{D}$ we define η^* and ε_* to be the following composites in V, with tensor symbols omitted from the diagrams.

$$
\begin{array}{ccc}
\mathsf{C}(RLX,RY) & \xrightarrow{\eta^*} & \mathsf{C}(X,RY) \\
\downarrow{\scriptstyle \rho^{-1}} & & \nearrow{\scriptstyle m} \\
\mathsf{C}(RLX,RY)\mathbb{1} & & \\
\downarrow{\scriptstyle 1\eta_X} & & \\
\mathsf{C}(RLX,RY)\mathsf{C}(X,RLX) & &
\end{array}
\qquad
\begin{array}{ccc}
\mathsf{D}(LX,LRY) & \xrightarrow{\varepsilon_*} & \mathsf{D}(LX,Y) \\
\downarrow{\scriptstyle \lambda^{-1}} & & \nearrow{\scriptstyle m} \\
\mathbb{1}\mathsf{D}(LX,LRY) & & \\
\downarrow{\scriptstyle \varepsilon_Y 1} & & \\
\mathsf{D}(LRY,Y)\mathsf{D}(LX,LRY) & &
\end{array}
$$

Then we define $\zeta_{X,Y}$ and $\upsilon_{X,Y}$ via the following composites, respectively.

$$
\begin{array}{ccc}
\mathsf{D}(LX,Y) & \xrightarrow{\zeta_{X,Y}} & \mathsf{C}(X,RY) \\
\downarrow{\scriptstyle R_{LX,Y}} & \nearrow{\scriptstyle \eta^*} & \\
\mathsf{C}(RLX,RY) & &
\end{array}
\qquad
\begin{array}{ccc}
\mathsf{C}(X,RY) & \xrightarrow{\upsilon_{X,Y}} & \mathsf{D}(LX,Y) \\
\downarrow{\scriptstyle L_{X,RY}} & \nearrow{\scriptstyle \varepsilon_*} & \\
\mathsf{D}(LX,LRY) & &
\end{array}
$$

To verify that $\zeta^{\perp}$ and $\upsilon^{\perp}$ are V-natural in each variable, first observe by Lemma 3.2.10 that the morphisms $R_{LX,Y}$ and $L_{X,RY}$ define V-natural $R^{\perp}$ and $L^{\perp}$. Then V-naturality of η and ε implies that the composites determining ζ and υ determine V-natural transformations.

The triangle identities of Definition 1.2.14 for η and ε imply that ζ and υ are mutually inverse. For example, the following commutative diagram shows that $\upsilon\zeta$ is equal to the composite $(L\eta^*)\varepsilon^*$, where ε^* is defined dually to ε_*.

$$
\begin{array}{ccccc}
\mathsf{D}(LX,Y) & \xrightarrow{R} & \mathsf{C}(RLX,RY) & \xrightarrow{\eta^*} & \mathsf{C}(X,RY) \\
 & \searrow & \downarrow{\scriptstyle L} & & \downarrow{\scriptstyle L} \\
 & & \mathsf{D}(LRLX,LRY) & \xrightarrow{L\eta^*} & \mathsf{D}(LX,LRY) \\
 & {\scriptstyle \varepsilon^*}\searrow & \downarrow{\scriptstyle \varepsilon_*} & & \downarrow{\scriptstyle \varepsilon_*} \\
 & & \mathsf{D}(LRLX,Y) & \xrightarrow{L\eta^*} & \mathsf{D}(LX,Y)
\end{array}
$$

In the above diagram, the left inner region commutes by the V-naturality condition (1.2.8) for ε. The other regions commute by V-functoriality of L and associativity of composition in D. Then the left triangle identity implies $(L\eta^*)\varepsilon^* = 1$. A similar calculation using V-naturality of η and the right triangle identity shows that the reverse composite $\zeta\upsilon$ is also the identity. This finishes the proof of the first implication.

For the reverse implication, suppose given isomorphisms in V

$$\mathsf{D}(LX,Y) \xrightarrow[\cong]{\zeta_{X,Y}} \mathsf{C}(X,RY)$$

such that $\zeta^{\perp}$ is V-natural with respect to $X \in \mathsf{C}$ and $Y \in \mathsf{D}$. By Lemma 3.2.10, V-naturality of $\zeta^{\perp}$ is equivalent to commutativity of the following diagram in V for each $X, X' \in \mathsf{C}$ and $Y, Y' \in \mathsf{D}$, again with tensor symbols omitted.

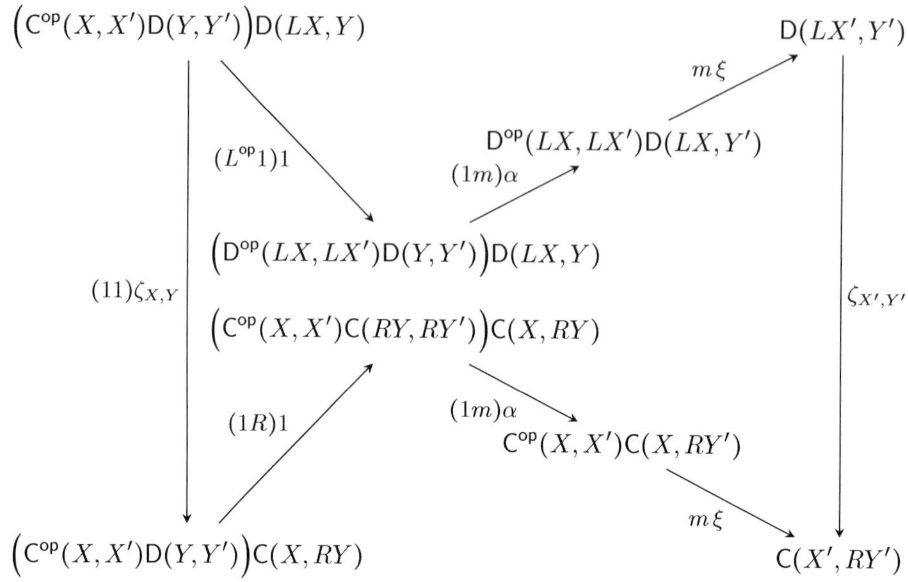

In the above diagram, the upper and lower composites are those that determine the corresponding morphisms in (3.2.11) for $\mathsf{D}(L(-),-)$ and $\mathsf{C}(-,R(-))$.

This V-naturality condition implies that the following two diagrams commute for each $X, X' \in \mathsf{C}$ and $Y, Y' \in \mathsf{D}$, again with tensor symbols omitted. The first follows from the special case $X = X'$ and the second follows from the special case $Y = Y'$.

(3.2.15)

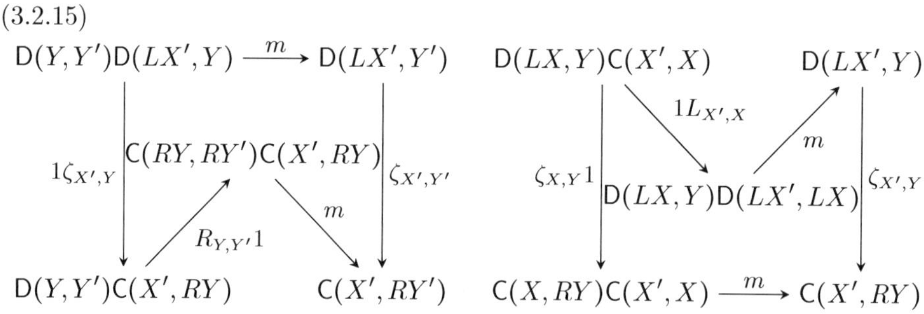

Now we define components of a unit η and counit ε via the following composites, respectively.

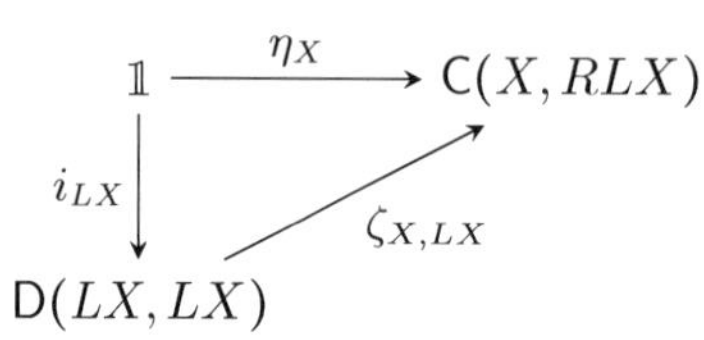

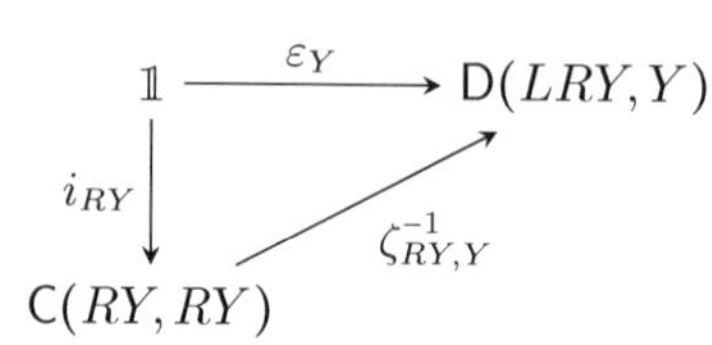

We use (3.2.15) to show that η and ε are V-natural. The V-naturality condition (1.2.8) for η at $X, X' \in \mathsf{C}$ and $Y, Y' \in \mathsf{D}$ is the outer diagram below, with tensor symbols omitted.

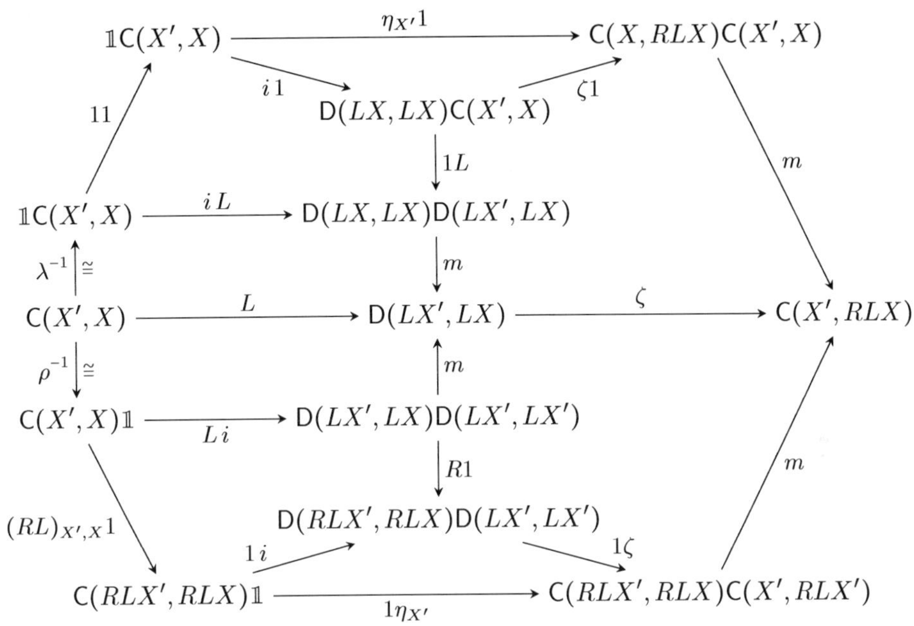

In the above diagram, the lower right region commutes by the first diagram of (3.2.15), with $Y = LX'$ and $Y' = LX$. The upper right region commutes by the second diagram of (3.2.15), with $Y = LX$. The other regions of the above diagram commute by definition of η, unity properties, and V-functoriality of L and R. The argument for V naturality of ε is similar, again reducing to special cases of (3.2.15).

The triangle identities for η and ε follow from the definitions—notably the invertibility of each $\zeta_{X,Y}$. One uses the formula for composition of natural transformations (Definition 1.2.9) and again reduces to special cases of (3.2.15) to show that each of the relevant components is an identity. □

3.3. Self-Enriched Symmetric Monoidal Categories

In this section we show that the canonical self-enrichment of a symmetric monoidal closed category V extends to an enrichment of the data $\otimes$, $\mathbb{1}$, α, λ, ρ, and ξ over V. In Theorem 3.3.2 we prove that $\underline{\mathsf{V}}$ is a symmetric monoidal V-category.

Definition 3.3.1. We extend the monoidal product and monoidal unit of V to V-functors

$$\boxtimes\colon \underline{\mathsf{V}} \otimes \underline{\mathsf{V}} \longrightarrow \underline{\mathsf{V}} \quad \text{and} \quad \mathrm{I}\colon \mathbb{I} \longrightarrow \underline{\mathsf{V}}$$

with the same assignments on objects, so

$$X \boxtimes Y = X \otimes Y \quad \text{and} \quad \mathrm{I}* = \mathbb{1}.$$

We use the notation $\boxtimes$ and I to help distinguish between the different roles that $\otimes$ and $\mathbb{1}$ play in this definition.

For a pair of objects $(X,X'),(Y,Y') \in \underline{\mathsf{V}} \otimes \underline{\mathsf{V}}$, the morphism

$$\boxtimes_{(X,X'),(Y,Y')} \colon [X,Y] \otimes [X',Y'] \longrightarrow [X \boxtimes X', Y \boxtimes Y']$$

is defined as the adjoint to the following composite, where ξ_{mid} is the middle four interchange of Definition 1.3.2.

$$\begin{array}{ccc} ([X,Y] \otimes [X',Y']) \otimes (X \otimes X') & \xrightarrow{\xi_{\text{mid}}} & ([X,Y] \otimes X) \otimes ([X',Y'] \otimes X') \\ & & \downarrow \mathsf{eval} \otimes \mathsf{eval} \\ & & Y \otimes Y' \end{array}$$

The morphism

$$\mathbb{1} \longrightarrow [\mathbb{1}, \mathbb{1}]$$

is defined as the adjoint to

$$\mathbb{1} \otimes \mathbb{1} \xrightarrow{\lambda_{\mathbb{1}} = \rho_{\mathbb{1}}} \mathbb{1}.$$

For both $\boxtimes$ and I, the composition and identity axioms of Definition 1.2.4 follow by verifying that the relevant adjoints are equal. ◇

Theorem 3.3.2. *Suppose* $\mathsf{V} = (\mathsf{V}, \otimes, \mathbb{1}, \alpha, \lambda, \rho, \xi)$ *is a symmetric monoidal closed category. Then* $\underline{\mathsf{V}}$ *is symmetric monoidal as a* V*-category.*

Proof. Definition 3.3.1 extends $\otimes$ and $\mathbb{1}$ to V-functors denoted $\boxtimes$ and I, respectively. Using the notation of Definition 3.2.3, the natural transformations α, λ, ρ, and ξ define the following adjoint components for $X, Y, Z \in \mathsf{V}$.

$$\begin{aligned} \alpha^{\perp}_{X,Y,Z} \colon \mathbb{1} &\longrightarrow [(X \boxtimes Y) \boxtimes Z, X \boxtimes (Y \boxtimes Z)] \\ \lambda^{\perp}_X \colon \mathbb{1} &\longrightarrow [\mathbb{1} \boxtimes X, X] \\ \rho^{\perp}_X \colon \mathbb{1} &\longrightarrow [X \boxtimes \mathbb{1}, X] \\ \xi^{\perp}_{X,Y} \colon \mathbb{1} &\longrightarrow [X \boxtimes Y, Y \boxtimes X]. \end{aligned}$$

We use Lemma 3.2.10 to verify V-naturality for each collection of components. We then verify that these satisfy the axioms of Definitions 1.4.2, 1.4.10, and 1.4.13. As in Section 1.4, we will write the monoidal product in $\mathsf{V}\text{-}\mathsf{Cat}$ as juxtaposition throughout this proof.

For $\alpha^{\perp}$, we have $\mathsf{C} = (\mathsf{V}^2)\mathsf{V}$, with $F = \boxtimes \circ (\boxtimes 1)$ and $G = \boxtimes \circ (1 \boxtimes)$. Recall from Definition 3.3.1 that

$$\boxtimes \colon [X,Y] \otimes [X',Y'] \longrightarrow [X \boxtimes X', Y \boxtimes Y']$$

is adjoint to a composite of ξ_{mid} with $\mathsf{eval} \otimes \mathsf{eval}$. Using functoriality of $\otimes$, the morphisms of hom objects defined by F are adjoint to

$$\begin{array}{c} (([X,Y] \otimes [X',Y']) \otimes [X'',Y'']) \otimes ((X \otimes X') \otimes X'') \\ \downarrow \\ (([X,Y] \otimes X) \otimes ([X',Y'] \otimes X')) \otimes ([X'',Y''] \otimes X'') \\ \downarrow (\mathsf{eval}^{\otimes 2}) \otimes \mathsf{eval} \\ (Y \otimes Y') \otimes Y'' \end{array}$$

where the first unlabeled morphism is a permutation of terms given by α and ξ. Similarly, the morphisms of hom objects defined by G are adjoint to

$$\begin{array}{c}([X,Y]\otimes([X',Y']\otimes[X'',Y'']))\otimes(X\otimes(X'\otimes X''))\\ \big\downarrow\\ ([X,Y]\otimes X)\otimes(([X',Y']\otimes X')\otimes([X'',Y'']\otimes X''))\\ \big\downarrow{\scriptstyle \mathsf{eval}\otimes(\mathsf{eval}^{\otimes 2})}\\ Y\otimes(Y'\otimes Y'')\end{array}$$

where the first unlabeled morphism is a permutation of terms given by α and ξ. Now we apply Lemma 3.2.10. Since the morphisms along the top and bottom of (3.2.11) are the adjoints of F and G, respectively, then $\alpha^{\perp}$ is V-natural if and only if the boundary of the following diagram commutes, where the two unlabeled morphisms are permutations of terms given by α and ξ.

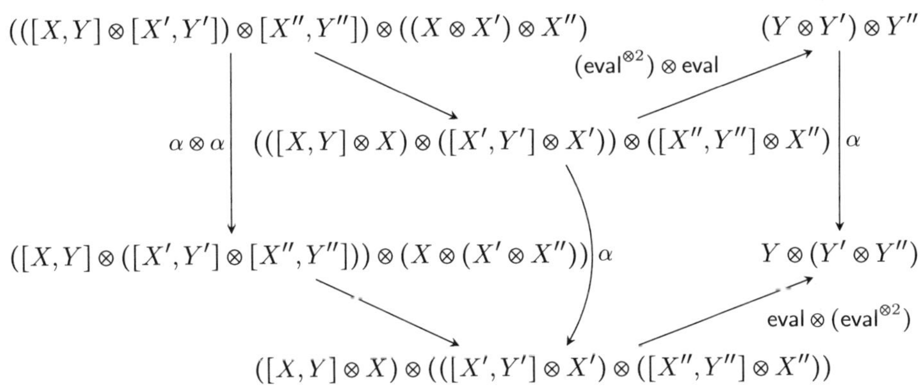

The left half of the above diagram commutes by the Symmetric Coherence Theorem 1.1.41. The right half commutes by naturality of α.

Verifying V-naturality for $\lambda^{\perp}$, $\rho^{\perp}$, and $\xi^{\perp}$ follows the same structure: In each case the relevant functors F and G are defined via their adjoints, which involve canonical morphisms and evaluation. In each case, the corresponding version of (3.2.11) has one half that commutes by coherence, and another half that commutes by naturality.

The result then follows by Theorem 2.5.1: the symmetric monoidal axioms for $\underline{\mathsf{V}}$ as a V-category hold if and only if the corresponding axioms hold for V as a plain category. □

Definition 3.3.3. Suppose V and W are symmetric monoidal closed categories and suppose $U\colon \mathsf{V} \longrightarrow \mathsf{W}$ is a symmetric monoidal functor. For each $X, Y \in \mathsf{V}$, let

$$U_{X,Y}\colon U[X,Y] \longrightarrow [UX, UY]$$

be the morphism in W adjoint to the composite

$$U[X,Y]\otimes UX \xrightarrow{U^2} U([X,Y]\otimes X) \xrightarrow{U\mathsf{eval}} UY.$$

The *standard enrichment* of U is the symmetric monoidal W-functor

$$U : \underline{\mathsf{V}}_U \longrightarrow \underline{\mathsf{W}}$$

defined by the same assignment on objects as the underlying functor U and defined on hom objects by the morphisms $U_{X,Y}$. This finishes the definition of U as an assignment on objects and a collection of morphisms between hom objects. We verify that these define a symmetric monoidal W-functor in Proposition 3.3.4. ◇

Proposition 3.3.4. *In the context of Definition 3.3.3, the standard enrichment of U is a symmetric monoidal* W*-functor.*

Proof. This proof has three parts. The first is to check the composition and identity axioms of Definition 1.2.4 to verify that U is a W-functor. Each of these is verified by checking that its adjoint commutes. For the composition axiom, with $X, Y, Z \in \mathsf{V}$, taking adjoints and using the definitions of $U_{X,Z}$ and the composition in $\underline{\mathsf{W}}$ yields the outer composites in the following diagram. In this diagram, and the other diagrams in this proof, we abbreviate the monoidal products in V and W as concatenation and use subscripts A_U to denote UA for $A \in \mathsf{V}$.

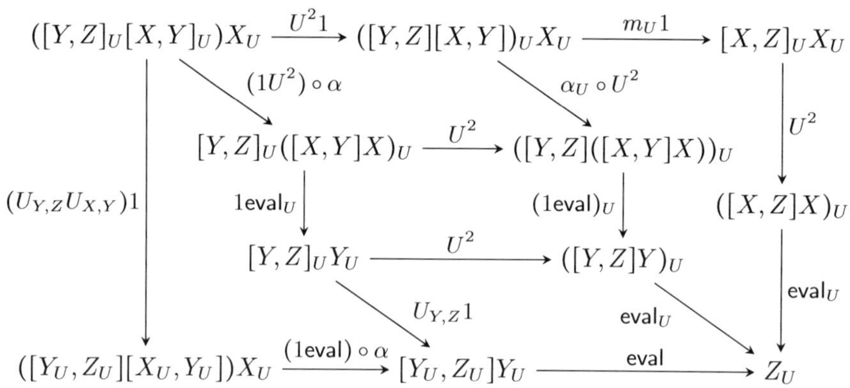

Commutativity of the subdiagrams above follows from the associativity axiom (1.1.9) for U; naturality of U^2; and the definitions of $U_{X,Y}$, $U_{Y,Z}$, and the composition in $\underline{\mathsf{V}}$. A similar adjoint diagram verifies the identity axiom for U.

For the second part of this proof, we use Lemma 3.2.10 to verify that the adjoints $(U^0)^\perp$ and $(U^2)^\perp$ are W-natural. These provide the enriched monoidal and unit constraints for U.

The component of $(U^2)^\perp$ at an object $X \otimes X' \in \mathsf{V} \otimes \mathsf{V}$ is a morphism in W

$$(U^2)^\perp_{X \otimes X'} : \mathbb{1} \longrightarrow [(UX) \otimes (UX'), U(X \otimes X')].$$

For each $X \otimes X'$ and $Y \otimes Y'$ in $\mathsf{V} \otimes \mathsf{V}$, the relevant version of (3.2.11) is the following diagram.

(3.3.5)

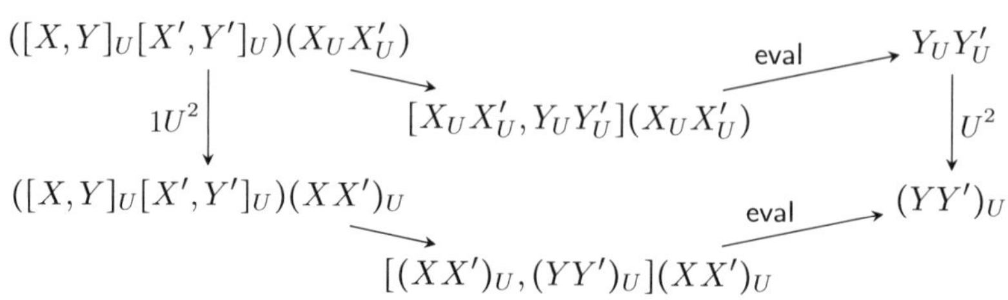

In the above diagram, the top unlabeled morphism is the identity on $(X_U X'_U)$ and the following composite on the other factor:

$$[X,Y]_U[X',Y']_U \xrightarrow{U_{X,Y}U_{X',Y'}} [X_U,Y_U][X'_U,Y'_U] \xrightarrow{\boxtimes} [X_U X'_U, Y_U Y'_U].$$

The bottom unlabeled morphism in (3.3.5) is the identity on $(XX')_U$ and the following composite on the other factor:

$$[X,Y]_U[X',Y']_U \xrightarrow{U^2} ([X,Y][X'Y'])_U \xrightarrow{\boxtimes_U} [XX',YY']_U \xrightarrow{U_{XX',YY'}} [(XX')_U,(YY')_U].$$

Commutativity of (3.3.5) follows from Epstein's Coherence Theorem 1.1.44 together with the naturality of U^2.

The component of $(U^0)^\perp$ at the unique object of $\mathbb{I}$ is a morphism in W

$$\mathbb{1} \longrightarrow [\mathbb{1}, \mathbb{1}_U],$$

where $\mathbb{1}$ denotes the monoidal unit of either V or W, distinguished by context. Recall from Definition 3.3.1 that the unit for self-enriched monoidal structure is determined by the adjoint of $\lambda_{\mathbb{1}} = \rho_{\mathbb{1}}$. Let

$$\widetilde{\lambda}\colon \mathbb{1} \longrightarrow [\mathbb{1}, \mathbb{1}]$$

denote this adjoint of $\lambda_{\mathbb{1}}$ in both V and W. With this notation, and recalling the definition of the monoidal identity for $\underline{\mathsf{V}}_U$ from Definition 2.4.3, the version of (3.2.11) for W-naturality of $(U^0)^\perp$ is commutativity of the outer diagram below.

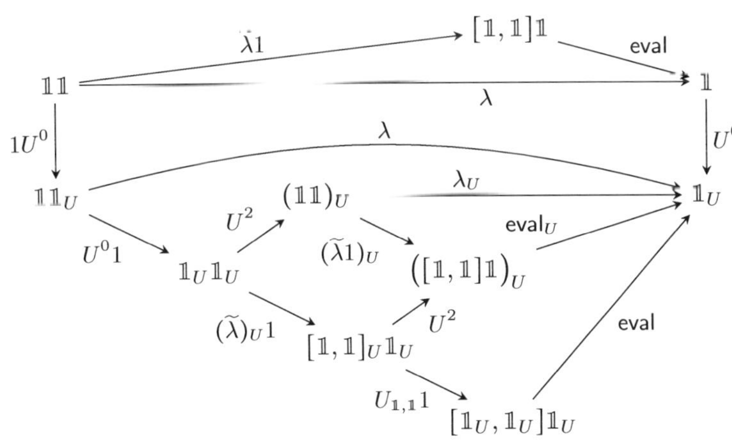

In the above diagram, the two triangular regions involving $\widetilde{\lambda}$ commute by definition. The upper central region commutes by naturality of λ. The two lower quadrilateral regions commute by naturality of U^2 and definition of $U_{\mathbb{1},\mathbb{1}}$, respectively. The remaining region, involving U^0, U^2, and λ, is the left unity diagram (1.1.10) for U.

The third and final part of this proof consists of checking the symmetric monoidal axioms of Definitions 1.4.17 and 1.4.18. This is similar to the verification of axioms in the proof of Theorem 3.3.2. We check the axioms on components, and in each case taking adjoints yields the corresponding axiom for (U, U^2, U^0).

To illustrate, the associativity axiom of Definition 1.4.17 is the following equality of components for each

$$(X \otimes Y) \otimes Z \in (\mathsf{V}^{\otimes 2}) \otimes \mathsf{V}.$$

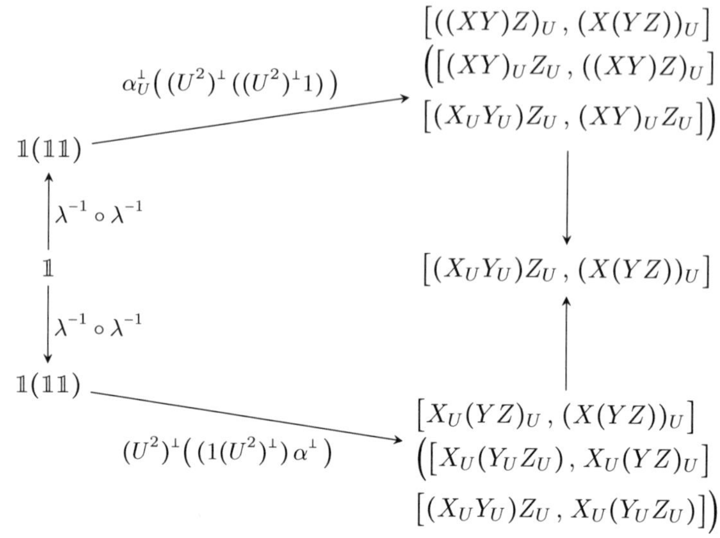

The adjoints of the two composites in the above diagram are two morphisms

$$(X_U Y_U)Z_U \longrightarrow (X(YZ))_U,$$

and it follows from the definitions of $\alpha^\perp$ and $(U^2)^\perp$ that they are the two composites in the associativity axiom (1.1.9) for U. The two unity axioms of Definition 1.4.17 and the braid axiom of Definition 1.4.18 follow in the same way. $\square$

3.4. Enriched Yoneda Bijection

In this section we prove the V-Yoneda Bijection 3.4.12. This is a preliminary version that will be used in the proof of the full V-Yoneda Lemma 3.6.9. We assume throughout that V is symmetric monoidal closed and that C is a small V-category. Recall from Definitions 3.2.1 and 3.2.2 the corepresented and represented V-functors

$$\mathcal{Y}^X = \mathsf{C}(X,-)\colon \mathsf{C} \longrightarrow \underline{\mathsf{V}} \quad \text{and} \quad \mathcal{Y}_Y = \mathsf{C}(-,Y)\colon \mathsf{C}^{\mathsf{op}} \longrightarrow \underline{\mathsf{V}}$$

for $X, Y \in \mathsf{C}$.

Definition 3.4.1. Suppose $W \in \mathsf{V}$. We let κ_W denote the composite shown below.

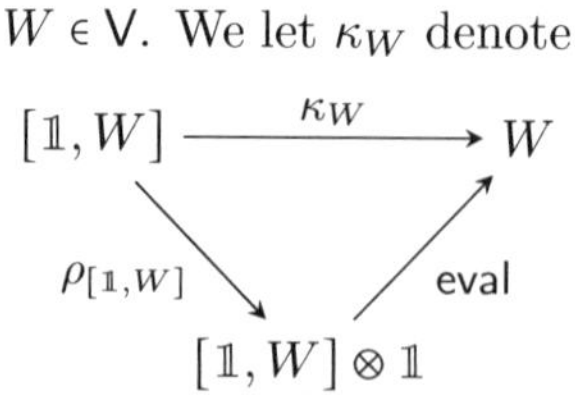

Using the triangle identities for the closed monoidal structure of V, one verifies that κ_W is an isomorphism with inverse given by

$$W \xrightarrow{\;\mathsf{coeval}\;} [\mathbb{1}, W \otimes \mathbb{1}] \xrightarrow{\;[\mathbb{1},\rho]\;} [\mathbb{1}, W].$$

When clear from context we will often omit the subscript W. $\diamond$

Explanation 3.4.2. Checking adjoints shows that the following two diagrams involving κ commute for $Z, W \in \mathsf{V}$. The first uses unity properties in V, and the second follows from the first along with additional unity properties, naturality of the monoidal structure morphisms, and the commutativity of (3.1.8) relating the enriched composition m with the counit eval in V.

(3.4.3)
$$\begin{array}{ccc} [\mathbb{1}, Z] \otimes \mathbb{1} & \xrightarrow{\kappa_Z \otimes 1} & Z \otimes \mathbb{1} \\ & \searrow^{\mathsf{eval}} & \downarrow \rho_Z \\ & & Z \end{array} \qquad \begin{array}{ccc} [Z,W] \otimes [\mathbb{1}, Z] & \xrightarrow{1 \otimes \kappa_Z} & [Z,W] \otimes Z \\ \downarrow m & & \downarrow \mathsf{eval} \\ [\mathbb{1}, W] & \xrightarrow{\kappa_W} & W \end{array}$$

◇

Next we record some notation to be used in our discussion of the V-enriched Yoneda lemmas.

Definition 3.4.4. Suppose C is a small V-category and $F, G : \mathsf{C} \longrightarrow \underline{\mathsf{V}}$ are V-functors.

- We let $\mathsf{V}\text{-nat}(F, G)$ denote the collection of V-natural transformations $F \longrightarrow G$. We note that $\mathsf{V}\text{-nat}(F, G)$ is a set because the components of a V-natural transformation are indexed by objects of C.
- For objects $X, Y \in \mathsf{V}$, recall that $\mathsf{V}(X, Y)$ denotes the *set* of morphisms $X \longrightarrow Y$ in the category V, while $\underline{\mathsf{V}}(X, Y) = [X, Y]$ denotes the V-object of morphisms for the standard self-enrichment of V.
- Recall that $\mathcal{Y}^X = \mathsf{C}(X, -)$ denotes the corepresented V-functor $\mathsf{C} \longrightarrow \underline{\mathsf{V}}$ for $X \in \mathsf{C}$ (Definition 3.2.1).
- Given a morphism $\eta \in \mathsf{V}(\mathbb{1}, FX)$ for $X \in \mathsf{C}$, define $\overline{\eta}_Y$ for each $Y \in \mathsf{V}$ as the composite in the following diagram.

(3.4.5)
$$\begin{array}{ccc} \mathsf{C}(X, Y) & \xrightarrow{\overline{\eta}_Y} & FY \\ \downarrow F & & \uparrow \kappa_{FY} \\ [FX, FY] & \xrightarrow{[\eta, FY]} & [\mathbb{1}, FY] \end{array}$$

Using the notation of Definition 3.2.3, let

(3.4.6)
$$\widetilde{\eta}_Y = (\overline{\eta}_Y)^{\perp} : \mathbb{1} \longrightarrow [\mathsf{C}(X, Y), FY].$$

By Lemma 3.2.10, using the V-functoriality of F and $\mathcal{Y}^X$, one verifies that the components $\widetilde{\eta}_Y$ define a V-natural transformation

$$\widetilde{\eta} : \mathcal{Y}^X \longrightarrow F.$$

- Given $X \in \mathsf{C}$ and a V-natural transformation $\theta : \mathcal{Y}^X \longrightarrow F$ we define $\theta_0 : \mathbb{1} \longrightarrow FX$ as the composite in the following diagram.

(3.4.7)
$$\begin{array}{ccc} \mathbb{1} & \xrightarrow{\theta_0} & FX \\ \downarrow \theta_X & & \uparrow \kappa_{FX} \\ [\mathsf{C}(X, X), FX] & \xrightarrow{[i_X, FX]} & [\mathbb{1}, FX] \end{array}$$

◇

Explanation 3.4.8. Our proof of Theorem 3.4.12 will rely on several observations, recorded as commutativity of the diagrams below. In each case one verifies the equality by taking adjoints and using

- basic properties of the closed monoidal structure on V (Definition 1.1.1, (1.1.4), and (1.1.5)),
- definitions and basic properties of the canonical self-enrichment (Definition 3.1.5 and Explanation 3.1.7),
- definition of the corepresented functor $\mathcal{Y}^X$ (Definition 3.2.1), and
- properties of the isomorphisms κ (Explanation 3.4.2).

Thus the following diagrams commute for $X, Y \in \mathsf{C}$.

(3.4.9)
$$\begin{array}{ccc} \mathbb{1} & \xrightarrow{\ \eta\ } & FX \\ {\scriptstyle i_{FX}}\big\downarrow & & \big\uparrow{\scriptstyle \kappa_{FX}} \\ [FX,FX] & \xrightarrow{\ [\eta,FX]\ } & [\mathbb{1},FX] \end{array}$$

(3.4.10)
$$\begin{array}{ccc} \mathcal{Y}^X Y \otimes \mathcal{Y}^X X & & \\ {\scriptstyle \mathcal{Y}^X \otimes 1}\big\downarrow & \searrow^{m} & \\ & & \mathcal{Y}^X Y \\ & \nearrow_{\mathsf{eval}} & \\ [\mathcal{Y}^X X, \mathcal{Y}^X Y] \otimes \mathcal{Y}^X X & & \end{array}$$

(3.4.11)
$$\begin{array}{ccccc} [\mathcal{Y}^X Y, W] \otimes \mathsf{C}(X,Y) & & \xrightarrow{\ \mathsf{eval}\ } & & W \\ {\scriptstyle 1 \otimes \mathcal{Y}^X}\big\downarrow & & & & \big\uparrow{\scriptstyle \kappa_W} \\ [\mathcal{Y}^X Y, W] \otimes [\mathcal{Y}^X X, \mathcal{Y}^X Y] & \xrightarrow{\ m\ } & [\mathcal{Y}^X X, W] & \xrightarrow{\ [i_X, W]\ } & [\mathbb{1}, W] \end{array}$$

◇

Theorem 3.4.12 (V-Yoneda Bijection). *Suppose* V *is a symmetric monoidal closed category and* C *is a small* V*-category. For each* V*-functor* $F: \mathsf{C} \longrightarrow \underline{\mathsf{V}}$ *there is a bijection of sets*

(3.4.13)
$$\mathsf{V}\text{-nat}(\mathcal{Y}^X, F) \cong \mathsf{V}(\mathbb{1}, FX)$$

induced by the assignments

$$\theta \longmapsto \theta_0 \quad \textit{and} \quad \eta \longmapsto \widetilde{\eta}$$

of Definition 3.4.4 for $\theta \in \mathsf{V}\text{-nat}(\mathcal{Y}^X, F)$ *and* $\eta \in \mathsf{V}(\mathbb{1}, FX)$*. Dually, for each* V*-functor* $G: \mathsf{C}^{\mathsf{op}} \longrightarrow \underline{\mathsf{V}}$ *there is a bijection of sets*

(3.4.14)
$$\mathsf{V}\text{-nat}(\mathcal{Y}_Y, G) \cong \mathsf{V}(\mathbb{1}, GY).$$

Proof. For the second assertion, we note that

$$\mathcal{Y}_Y = \mathsf{C}(-, Y) = \mathsf{C}^{\mathsf{op}}(Y, -)$$

as V-functors. Therefore, the second follows from applying the first to C^{op}.

To prove the first assertion, suppose given $\eta \in \mathsf{V}(\mathbb{1}, FX)$. Then $\widetilde{\eta}_0$ is the composite

$$\mathbb{1} \xrightarrow{\widetilde{\eta}_X} [\mathsf{C}(X,X), FX] \xrightarrow{[i_X, FX]} [\mathbb{1}, FX] \xrightarrow{\kappa} FX$$

and $\widetilde{\eta}_X = (\overline{\eta}_X)^{\perp}$. To see that $\widetilde{\eta}_0 = \eta$, we first use naturality of ρ to obtain commutative rectangles along the top of the diagram below.

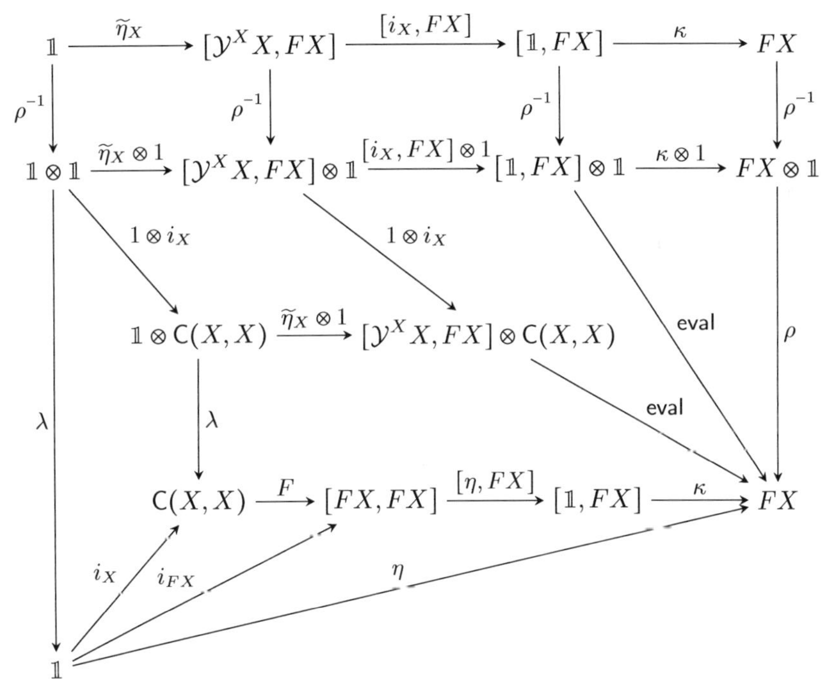

In the above diagram, the left vertical composite is the identity because $\lambda_{\mathbb{1}} = \rho_{\mathbb{1}}$ (1.1.4). Then we obtain the remaining commutative regions using the following, respectively: naturality of λ, functoriality of $\otimes$, the adjoint of $[i_X, FX]$, the triangle (3.4.3), the definition of $\widetilde{\eta}_X = (\overline{\eta}_X)^{\perp}$, the identity axiom for F (Definition 1.2.4), and (3.4.9).

Next, suppose given $\theta \in \mathsf{V}\text{-nat}(\mathcal{Y}^X, F)$. Then $\widetilde{(\theta_0)}$ is a V-natural transformation whose component at $Y \in \mathsf{C}$ is $\overline{(\theta_0)}_Y^{\perp}$ with $\overline{(\theta_0)}_Y$ given by the composite

$$\mathsf{C}(X,Y) \xrightarrow{F} [FX, FY] \xrightarrow{[\theta_0, FY]} [\mathbb{1}, FY] \xrightarrow{\kappa} FY.$$

To see that $\widetilde{(\theta_0)}_Y = \theta_Y$ for each $Y \in \mathsf{C}$, we use the diagram below to show that the adjoint of θ_Y, along the top edge, is equal to $\overline{(\theta_0)}_Y \circ \lambda$, along the left and bottom edges.

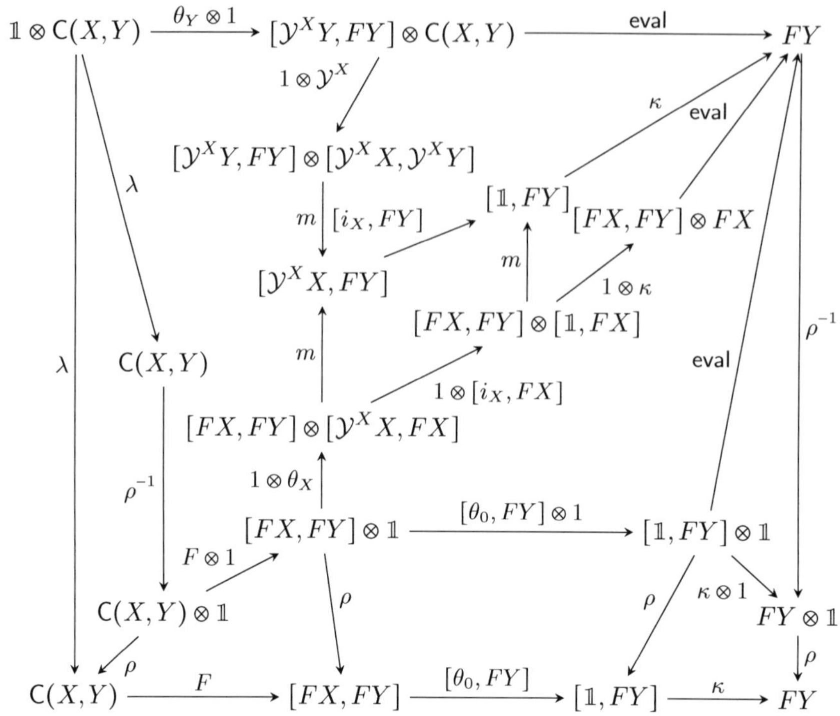

In the above diagram, each of the quadrilaterals along the bottom commutes by naturality of ρ and the trapezoid at left commutes by invertibility of ρ. Commutativity of the remaing regions is a consequence of the following, respectively: the V-naturality of θ (1.2.8), compatibility of $\mathcal{Y}^X$ with m (3.4.11), compatibility of m with morphisms in V (3.1.10), the square (3.4.3), the definition of θ_0, and the triangle (3.4.3). $\square$

The following application of Theorem 3.4.12 will be useful in Section 3.9 below.

Proposition 3.4.15. *Suppose* D *and* C *are small* V*-categories and suppose given* V*-functors*

$$\mathsf{D} \underset{G}{\overset{F}{\rightleftarrows}} \mathsf{C}.$$

Then the underlying functors form an adjoint pair $F_0 \dashv G_0$ *if and only if* $F \dashv G$ *is a* V*-adjoint pair.*

Proof. One implication follows directly from 2-functoriality of

$$(-)_0 \colon \mathsf{V}\text{-}\mathsf{Cat} \longrightarrow \mathsf{Cat},$$

proved in Proposition 2.1.2 and Lemma 2.1.5.

For the reverse implication, suppose given V-functors F and G such that $F_0 \dashv G_0$ is an adjoint pair. Then we have unit and counit with components

$$\eta_X \in \mathsf{D}_0(X, GFX) = \mathsf{V}(\mathbb{1}, \mathsf{D}(X, GFX)) \quad \text{and}$$
$$\varepsilon_Y \in \mathsf{C}_0(FGY, Y) = \mathsf{V}(\mathbb{1}, \mathsf{C}(FGY, Y))$$

for each $X \in \mathsf{D}$ and $Y \in \mathsf{C}$. By the Yoneda Bijection 3.4.12 we have

$$\mathsf{V}\text{-nat}(\mathsf{C}(FX,-),\mathsf{D}(X,G(-))) \cong \mathsf{V}(\mathbb{1},\mathsf{D}(X,GFX)) \quad \text{and}$$
$$\mathsf{V}\text{-nat}(\mathsf{D}(-,GY),\mathsf{C}(F(-),Y) \cong \mathsf{V}(\mathbb{1},\mathsf{C}(FGY,Y)).$$

Recalling the proof of Theorem 3.4.12, the morphisms η_X and ε_Y therefore induce components

$$\mathsf{C}(FX,Y) \xrightarrow{\overline{\eta}_{X,Y}} \mathsf{D}(X,GY) \quad \text{and} \quad \mathsf{D}(X,GY) \xrightarrow{\overline{\varepsilon}_{X,Y}} \mathsf{C}(FX,Y) \tag{3.4.16}$$

such that the $\overline{\eta}^{\perp}$ is V-natural with respect to $Y \in \mathsf{V}$ for each $X \in \mathsf{V}$ and $\overline{\varepsilon}^{\perp}$ is V-natural with respect to $X \in \mathsf{V}$ for each $Y \in \mathsf{V}$. The triangle identities for $F_0 \dashv G_0$ imply that these components are mutually inverse isomorphisms in V.

For V-naturality of each morphism with respect to the other variable, we recall from Lemma 3.2.12 that a collection of isomorphisms in V determines a V-natural transformation if and only if its collection of inverse morphisms does so. Therefore, the V-naturality of $\overline{\eta}^{\perp}$ with respect to Y implies that of $\overline{\varepsilon}^{\perp}$ and the V-naturality of $\overline{\varepsilon}^{\perp}$ with respect to X implies that of $\overline{\eta}^{\perp}$. Lastly, by Proposition 3.2.14 the V-naturality of isomorphisms (3.4.16) implies that $F \dashv G$ is a V-adjunction. □

3.5. Enriched Ends and Internal Mapping Objects

The statement of the full V-enriched Yoneda Lemma 3.6.9 requires the following V-enriched notion of coends and ends generalizing the plain categorical notion from Definition I.1.1.16. Proposition 3.5.5 shows that all V-coends, respectively V-ends, exist if V is cocomplete, respectively complete.

Definition 3.5.1. Suppose A is a V-category and suppose $F: \mathsf{A}^{\mathsf{op}} \otimes \mathsf{A} \longrightarrow \underline{\mathsf{V}}$ is a V-functor.

(1) A V-*cowedge* of F is a pair (X,ζ) consisting of
  - an object $X \in \underline{\mathsf{V}}$ and
  - morphisms $\zeta_a: F(a,a) \longrightarrow X$ in V for $a \in \mathsf{A}$

  such that the following diagram in V commutes for all objects $a, a' \in \mathsf{A}$.

$$\begin{array}{ccc} \mathsf{A}(a,a') & \xrightarrow{F(-,a)} & [F(a',a),F(a,a)] \\ {\scriptstyle F(a',-)}\downarrow & & \downarrow{\scriptstyle [F(a',a),\zeta_a]} \\ [F(a',a),F(a',a')] & \xrightarrow{[F(a',a),\zeta_{a'}]} & [F(a',a),X] \end{array} \tag{3.5.2}$$

(2) A V-*coend* of F is an initial V-cowedge and is denoted by the pair

$$\left(\int^{a\in\mathsf{A}} F(a,a),\omega\right)$$

  or simply by $\int^{a\in\mathsf{A}} F(a,a)$ with ω implicit.

A V-*end* of F is the dual notion given as follows.

(3) A V-*wedge* of F is a pair (Y,δ) consisting of
  - an object $Y \in \underline{\mathsf{V}}$ and
  - morphisms $\delta_a: Y \longrightarrow F(a,a)$ in V for $a \in \mathsf{A}$

such that the following diagram in V commutes for all objects $a, a' \in \mathsf{A}$.

(3.5.3)
$$\begin{array}{ccc} \mathsf{A}(a,a') & \xrightarrow{\quad F(a,-) \quad} & [F(a,a), F(a,a')] \\ {\scriptstyle F(-,a')}\downarrow & & \downarrow{\scriptstyle [\delta_a, F(a,a')]} \\ [F(a',a'), F(a,a')] & \xrightarrow{\quad [\delta_{a'}, F(a,a')] \quad} & [Y, F(a,a')] \end{array}$$

(4) A V-*end* of F is a terminal V-wedge and is denoted by the pair

$$\Big(\int_{a\in\mathsf{A}} F(a,a), \sigma\Big)$$

or simply by $\int_{a\in\mathsf{A}} F(a,a)$ with σ implicit.

When clear from context, we will omit the ambient category from the superscript or subscript. ◇

Explanation 3.5.4 (Universal Properties of Enriched Coends and Ends)**.** The definition of a V-coend for F as an initial V-cowedge means that it satisfies the following universal property. Given any V-cowedge with components ζ_a as shown below, there exists a unique morphism in V such that the following diagram commutes for all $a \in \mathsf{A}$.

$$\begin{array}{ccc} F(a,a) & & \\ {\scriptstyle \omega_a}\downarrow & \searrow^{\zeta_a} & \\ \int^a F(a,a) & \overset{\exists !}{\dashrightarrow} & X \end{array}$$

Dually, a V-end for F satisfies the following universal property. Given any V-wedge with components δ_a as shown below, there exists a unique morphism in V such that the following diagram commutes for all $a \in \mathsf{A}$.

$$\begin{array}{ccc} Y & \overset{\exists !}{\dashrightarrow} & \int_a F(a,a) \\ & \searrow_{\delta_a} & \downarrow{\scriptstyle \sigma_a} \\ & & F(a,a) \end{array}$$

◇

Proposition 3.5.5. *Suppose given a* V*-functor*

$$F\colon \mathsf{A}^{\mathsf{op}} \otimes \mathsf{A} \longrightarrow \underline{\mathsf{V}}$$

as in Definition 3.5.1. If V *is cocomplete and* A *is small, then the* V*-coend of* F *exists and is given by a coequalizer in* V

(3.5.6)
$$\int^{a\in\mathsf{A}} F(a,a) \cong \mathsf{coeq}\Big(\coprod_{a,a'\in\mathsf{A}} \mathsf{A}(a,a') \otimes F(a',a) \rightrightarrows \coprod_{a\in\mathsf{A}} F(a,a) \Big).$$

Dually, if V *is complete and* A *is small, then the* V*-end of* F *exists and is given by an equalizer in* V

$$\int_{a\in\mathsf{A}} F(a,a) \cong \mathsf{eq}\Big(\prod_{a\in\mathsf{A}} F(a,a) \rightrightarrows \prod_{a,a'\in\mathsf{A}} [\mathsf{A}(a,a'),F(a,a')]\Big). \tag{3.5.7}$$

Proof. The symmetric monoidal closed structure of V gives isomorphisms

$$\mathsf{V}(P,[Q,R]) \cong \mathsf{V}(P\otimes Q,R) \xrightarrow[\cong]{\xi^*} \mathsf{V}(Q\otimes P,R) \cong \mathsf{V}(Q,[P,R]) \tag{3.5.8}$$

for each $P,Q,R \in \mathsf{V}$. Using the first bijection of (3.5.8), the V-cowedge condition (3.5.2) for components ζ_a is equivalent to commutativity of the following diagram for each $a,a' \in \mathsf{A}$.

$$\begin{array}{ccc} \mathsf{A}(a,a')\otimes F(a',a) & \longrightarrow & F(a,a) \\ \big\downarrow & & \big\downarrow{\scriptstyle \zeta_a} \\ F(a',a') & \xrightarrow{\zeta_{a'}} & X \end{array}$$

In the above diagram, the unlabeled horizontal, respectively vertical, morphism corresponds to $F(-,a)$, respectively $F(a',-)$ under the first bijection of (3.5.8). Therefore the coend of F can be expressed as the indicated coequalizer.

Dually, using the composite bijection of (3.5.8), the V-wedge condition (3.5.3) for components δ_a is equivalent to commutativity of the following diagram for each $a,a' \in \mathsf{A}$.

$$\begin{array}{ccc} Y & \xrightarrow{\delta_a} & F(a,a) \\ {\scriptstyle \delta_{a'}}\big\downarrow & & \big\downarrow \\ F(a',a') & \longrightarrow & [\mathsf{A}(a,a'),F(a,a')] \end{array}$$

In the above diagram, the unlabeled vertical, respectively horizontal, morphism corresponds to $F(a,-)$, respectively $F(-,a')$ under the bijection (3.5.8). □

Explanation 3.5.9. In the context of Definition 3.5.1, a V-coend $\int^a F(a,a)$ is called a *coend in* $\underline{\mathsf{V}}$ if, for each $b \in \mathsf{V}$, the morphisms $[\omega_a, b]$ induce an isomorphism

$$\Big[\int^a F(a,a), b\Big] \xrightarrow{\cong} \int_a \big[F(a,a), b\big]. \tag{3.5.10}$$

Similarly, a V-end $\int_a F(a,a)$ is called an *end in* $\underline{\mathsf{V}}$ if, for each $b \in \mathsf{V}$, the morphisms $[b,\sigma_a]$ induce an isomorphism

$$\Big[b, \int_a F(a,a)\Big] \xrightarrow{\cong} \int_a \big[b, F(a,a)\big]. \tag{3.5.11}$$

If, as we assume throughout this chapter, V is symmetric monoidal closed, then (3.5.10) and (3.5.11) are necessarily isomorphisms and therefore every V-end, respectively V-coend, is an end in $\underline{\mathsf{V}}$, respectively coend in $\underline{\mathsf{V}}$. Therefore, we will not have need for this alternate terminology. ◇

Definition 3.5.12. Suppose $F, G\colon \mathsf{C} \longrightarrow \underline{\mathsf{V}}$ are V-functors with V complete. We define the *mapping object* as the V-end

$$\mathsf{Map}(F,G) = \int_{Y\in\mathsf{C}} \big[FY, GY\big] \tag{3.5.13}$$

with V-wedge components

$$\mathsf{Map}(F,G) \xrightarrow{\sigma_Z} [FZ, GZ]$$

for each $Z \in \mathsf{C}$. ◇

Lemma 3.5.14. *In the context of Definition 3.5.12, the underlying set*

$$\mathsf{V}\big(\mathbb{1}, \mathsf{Map}(F,G)\big)$$

is naturally isomorphic to the set of V*-natural transformations* $F \longrightarrow G$.

Proof. By universality of the end, each morphism in V from $\mathbb{1}$ to $\mathsf{Map}(F,G)$ is uniquely determined by a V-wedge with components

$$\mathbb{1} \xrightarrow{\theta_X} [FX, GX]$$

such that the following diagram commutes for each $X, Y \in \mathsf{C}$.

$$\begin{array}{ccc}
\mathsf{C}(X,Y) & \xrightarrow{[FX,-]} & [[FX,GX],[FX,GY]] \\
{\scriptstyle [-,GY]}\downarrow & & \downarrow{\scriptstyle [\theta_X,[FX,GY]]} \\
[[FY,GY],[FX,GY]] & \xrightarrow{[\theta_Y,[FX,GY]]} & [\mathbb{1},[FX,GY]]
\end{array}$$

Taking adjoints and unpacking the definitions of the morphisms involved, commutativity of the diagram above is equivalent to commutativity of the defining diagram (1.2.8) for V-naturality of θ. □

Definition 3.5.15. Suppose $\theta\colon P \longrightarrow [Q,R]$ is a morphism in V. The *transform* of θ is

$$\theta^{\#}\colon Q \longrightarrow [P,R]$$

defined as the adjoint of

$$Q \otimes P \xrightarrow{\xi} P \otimes Q \xrightarrow{\theta\otimes 1} [Q,R]\otimes Q \xrightarrow{\mathsf{eval}} R.$$ ◇

The following result will be useful in our discussion of the Enriched Yoneda Lemma 3.6.9 below.

Lemma 3.5.16. *Suppose* $F\colon \mathsf{C} \longrightarrow \underline{\mathsf{V}}$ *is a* V*-functor, suppose* $X \in \mathsf{C}$*, and suppose*

$$\theta_Y\colon \mathcal{Y}^X Y \longrightarrow [Q, FY]$$

is a morphism in V *for each* $Y \in \mathsf{C}$*. The transforms*

$$\theta_Y^{\#}\colon Q \longrightarrow [\mathcal{Y}^X Y, FY]$$

form a wedge (Q,θ) *if and only if the morphisms*

$$\theta_Y^{\perp}\colon \mathbb{1} \longrightarrow [\mathcal{Y}^X Y, [Q, FY]]$$

adjoint to $\theta_Y \circ \lambda$ *are components of a* V*-natural transformation* $\mathcal{Y}^X \longrightarrow [Q, F(-)]$.

Proof. The V-naturality condition for $\theta^{\perp}_{Y}$ given in Lemma 3.2.10 is that the diagram below commutes for each $Y \in \mathsf{C}$.
(3.5.17)

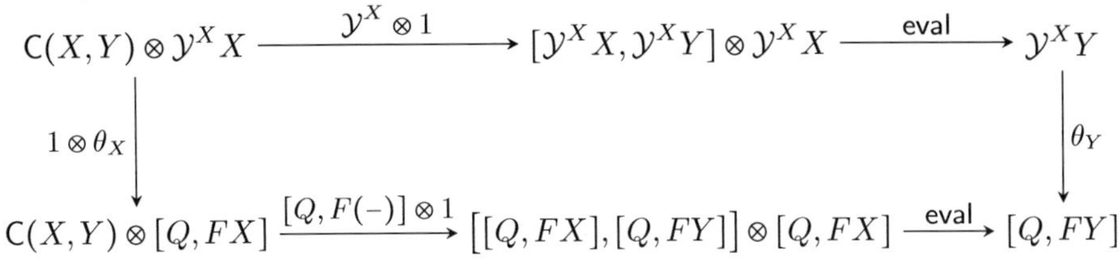

To explain the V-wedge condition (3.5.3) for $\theta^{\#}$, we let $\overline{\theta}_Y$ denote the adjoint of θ_Y, given by the following composite.

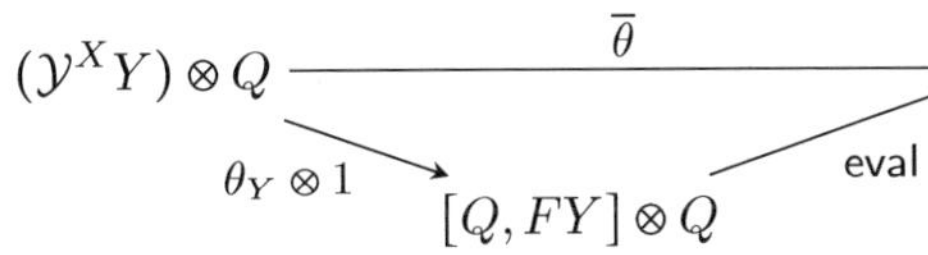

The V-wedge condition (3.5.3) for $\theta^{\#}$ is that the outer diagram below commutes for each $Y \in \mathsf{C}$. In this diagram the unlabeled isomorphisms are given by the monoidal closed structure of V.
(3.5.18)

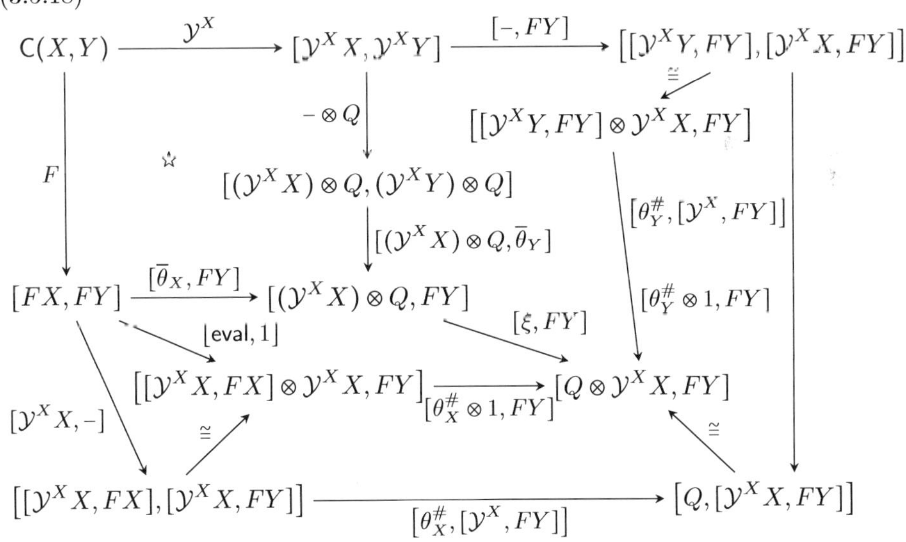

In the above diagram, each region except the one labeled ☆ commutes as a consequence of the symmetric monoidal closed structure of V and the definitions of the relevant morphisms. Therefore $\theta^{\#}$ satisfies the V-wedge condition if and only if the region ☆ commutes.

The adjoints of the two composites around the boundary of ☆ are given by the composites around the boundary of the diagram below. In this below diagram, each region except the one labeled ♡ commutes as a consequence of the symmetric monoidal closed structure of V and the definitions of the relevant morphisms. Therefore, the composites around the boundary of the diagram below are equal if and only if the composites around the boundary of the region ♡ are equal.

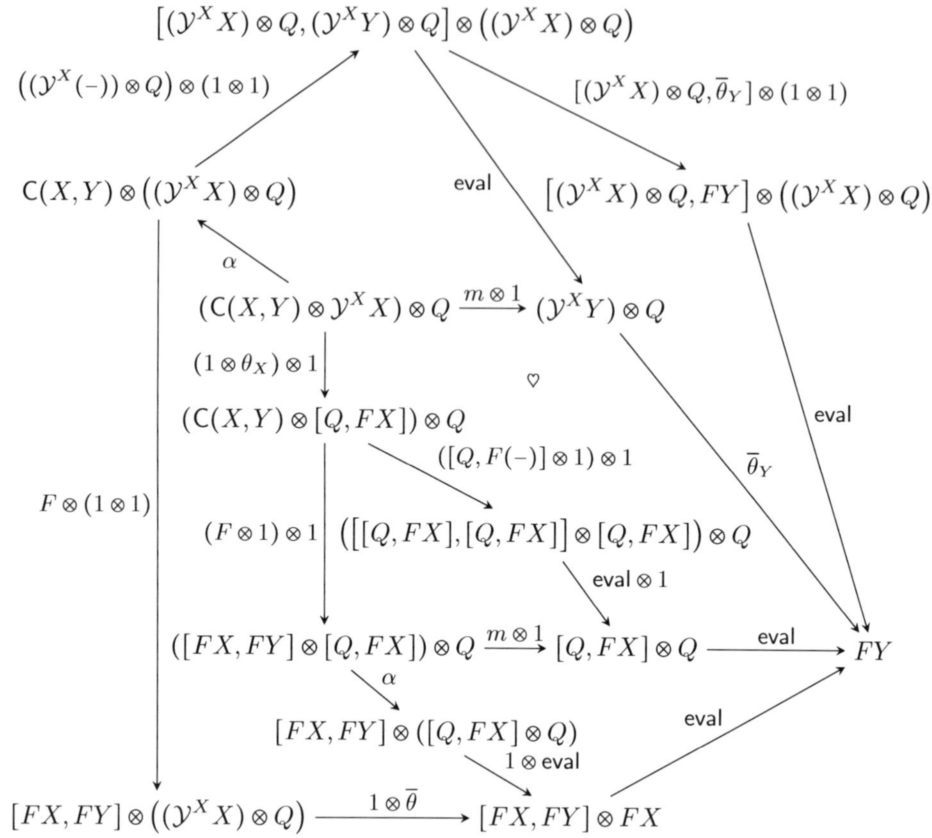

Noting that the composite along the top of (3.5.17) is the composition m in C, the composites around the boundary of the region labeled ♡ in the diagram above are the adjoints of the composites around the boundary of (3.5.17). Thus for each $Y \in \mathsf{C}$ the V-naturality diagram (3.5.17) commutes if and only if the wedge diagram (3.5.18) does so. □

Now we show that the mapping objects give a V-category of V-functors $\mathsf{C} \longrightarrow \underline{\mathsf{V}}$.

Definition 3.5.19. Suppose C is a small V-category with V complete. We define the *mapping* V*-category* $\mathsf{C}\text{-}\mathsf{V}$ to have objects given by V-functors $\mathsf{C} \longrightarrow \underline{\mathsf{V}}$ and hom objects

$$\mathsf{C}\text{-}\mathsf{V}(F,G) = \mathsf{Map}(F,G)$$

for V-functors F and G.

For each $F \in \mathsf{C}\text{-}\mathsf{V}$, the identity i_F is the universal arrow induced by $\theta_Z = i_{FZ}$ for each $Z \in \mathsf{C}$.

(3.5.20)

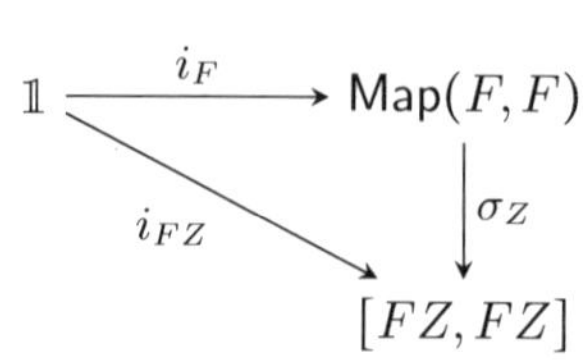

For $F, G, H \in \mathsf{C}\text{-}\mathsf{V}$, the composition morphism $m = m^{\mathsf{Map}}$ is the universal arrow induced by the composite $m \circ (\sigma \otimes \sigma)$ below.

$$
\begin{array}{ccc}
\mathsf{Map}(G,H) \otimes \mathsf{Map}(F,G) & \xrightarrow{\quad m^{\mathsf{Map}} \quad} & \mathsf{Map}(F,H) \\
{\scriptstyle \sigma_Z \otimes \sigma_Z} \big\downarrow & & \big\downarrow {\scriptstyle \sigma_Z} \\
[GZ, HZ] \otimes [FZ, GZ] & \xrightarrow{\quad m \quad} & [FZ, HZ]
\end{array} \tag{3.5.21}
$$

In Lemma 3.5.24 we show that the unity and associativity axioms hold for $\mathsf{C}\text{-}\mathsf{V}$, making it a V-category. Lemma 3.5.14 above shows that the underlying category of $\mathsf{C}\text{-}\mathsf{V}$ is isomorphic to the plain category $\mathsf{V}\text{-}\mathsf{Cat}(\mathsf{C}, \underline{\mathsf{V}})$. ◇

Explanation 3.5.22. In (3.5.21) above, the adjoint of $\sigma_Z \circ m^{\mathsf{Map}}$ is equal to the lower left composite below.

$$
\begin{array}{ccc}
\big(\mathsf{Map}(G,H) \otimes \mathsf{Map}(F,G)\big) \otimes FZ & & \\
{\scriptstyle (\sigma_Z \otimes \sigma_Z) \otimes 1} \big\downarrow & & \\
\big([GZ,HZ] \otimes [FZ,GZ]\big) \otimes FZ & \xrightarrow{\quad \mathsf{eval} \circ (m \otimes 1) \quad} & HZ \\
{\scriptstyle \alpha} \big\downarrow & & \big\uparrow {\scriptstyle \mathsf{eval}} \\
[GZ,HZ] \otimes \big([FZ,GZ] \otimes FZ\big) & \xrightarrow[\quad 1 \otimes \mathsf{eval} \quad]{} & [GZ,HZ] \otimes GZ
\end{array} \tag{3.5.23}
$$

The rectangle commutes by definition of the composition m in $\underline{\mathsf{V}}$ via (3.1.6). ◇

Lemma 3.5.24. *In the context of Definition 3.5.19,* $\mathsf{C}\text{-}\mathsf{V}$ *is a* V*-category.*

Proof. We begin with the left unity diagram (1.2.3). By the universal property of V-ends, it suffices to show that the following diagram commutes for each $Z \in \mathsf{C}$. Here and throughout the rest of this proof we abbreviate $\mathsf{Map}(-,-)$ as $\mathsf{M}(-,-)$ and omit the tensor symbols in V.

$$
\begin{array}{ccccc}
\mathbb{1}\mathsf{M}(F,G) & \xrightarrow{\ \lambda\ } & \mathsf{M}(F,G) & & \\
 & & & \searrow^{\sigma_Z} & \\
{\scriptstyle i_G 1} \big\downarrow & & & & [FZ, GZ] \\
 & & & \nearrow_{\sigma_Z} & \\
\mathsf{M}(G,G)\mathsf{M}(F,G) & \xrightarrow{\ m^{\mathsf{Map}}\ } & \mathsf{M}(F,G) & &
\end{array}
$$

To see that the above diagram commutes, we verify that the following adjoint diagram commutes, where we have used (3.5.23) for the adjoint of $\sigma_Z \circ m^{\mathsf{Map}}$.

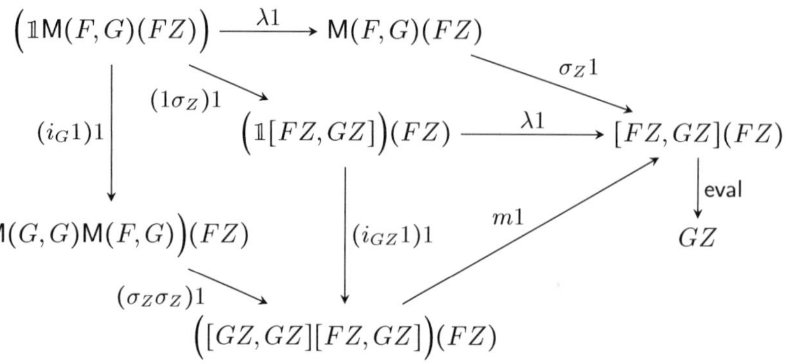

In the above diagram, the parallelogram at left commutes by the definition of i_G (3.5.20), the upper parallelogram commutes by naturality of λ, and the remaining triangle commutes by the left unity diagram for $\underline{\mathsf{V}}$. This verifies that the left unity diagram for C-V commutes. The right unity diagram for C-V is similar, using the right unity of $\underline{\mathsf{V}}$.

Now we turn to the associativity diagram (1.2.2) for m^{Map}. Again using universality of V-ends, it suffices to show that the following diagram commutes for $E, F, G, H \in$ C-V.

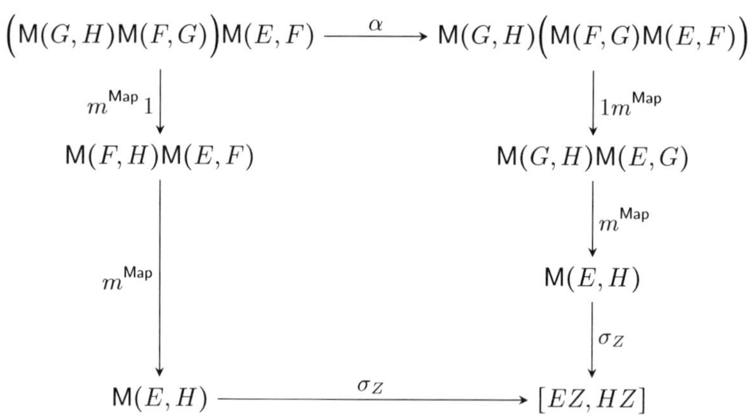

To see that the above diagram commutes, we verify that the following adjoint diagram commutes, where we have used (3.5.23) twice for adjoints of $\sigma_Z \circ m^{\mathsf{Map}}$ and abbreviated

$$L = \Big(\big([GZ,HZ][FZ,GZ]\big)[EZ,FZ]\Big)(EZ)$$

and

$$R = \Big([GZ,HZ]\big([FZ,GZ][EZ,FZ]\big)\Big)(EZ).$$

$$\begin{array}{ccc}
\Big(\big(\mathsf{M}(G,H)\mathsf{M}(F,G)\big)\mathsf{M}(E,F)\Big)(EZ) & \xrightarrow{\alpha 1} & \Big(\mathsf{M}(G,H)\big(\mathsf{M}(F,G)\mathsf{M}(E,F)\big)\Big)(EZ) \\
\swarrow (m^{\mathsf{Map}}1)1 \quad \searrow ((\sigma_Z\sigma_Z)\sigma_Z)1 & & (\sigma_Z(\sigma_Z\sigma_Z))1 \swarrow \quad \searrow (1m^{\mathsf{Map}})1 \\
\big(\mathsf{M}(F,H)\mathsf{M}(E,F)\big)(EZ) \qquad L & \xrightarrow[\alpha 1]{} & R \qquad \big(\mathsf{M}(G,H)\mathsf{M}(E,G)\big)(EZ) \\
(\sigma_Z\sigma_Z)1 \searrow \quad \swarrow (m1)1 & & (1m)1 \searrow \quad \swarrow (\sigma_Z\sigma_Z)1 \\
\big([FZ,HZ][EZ,FZ]\big)(EZ) & & \big([GZ,HZ][EZ,GZ]\big)(EZ) \\
m1 \searrow & & \swarrow m1 \\
& [EZ,HZ](EZ) \xrightarrow{\mathsf{eval}} HZ &
\end{array}$$

In the above diagram, the quadrilaterals at left and right commute by definition of m^{Map}. The upper middle region commutes by naturality of α and the lower middle region commutes by associativity for m in $\underline{\mathsf{V}}$. This completes the proof that C-V is a V-category. $\square$

Definition 3.5.25 (Yoneda V-functor)**.** Suppose C is a small V-category with V complete. Define the *Yoneda* V*-functor*

$$\mathcal{Y}^{(-)}\colon \mathsf{C} \longrightarrow (\mathsf{C}\ \mathsf{V})^{\mathsf{op}}$$

by

$$X \longmapsto \mathcal{Y}^X$$

on objects and the following universal arrow on hom objects:

$$\text{(3.5.26)} \qquad \begin{array}{ccc}
\mathsf{C}(X,Y) & \xrightarrow{(\mathcal{Y}^{(-)})_{X,Y}} & \mathsf{Map}(\mathcal{Y}^Y,\mathcal{Y}^X) \\
& (\mathcal{Y}_Z)_{X,Y} \searrow & \downarrow \sigma_Z \\
& & [\mathsf{C}(Y,Z),\mathsf{C}(X,Z)]
\end{array}$$

We show in Lemma 3.5.29 that the identity and composition axioms hold for $\mathcal{Y}^{(-)}$, making it a V-functor. We show in Corollary 3.6.11 that $\mathcal{Y}^{(-)}$ induces isomorphisms on hom objects. For this reason, $\mathcal{Y}^{(-)}$ is also called the V*-Yoneda embedding.* ◇

Explanation 3.5.27. In the context of Definition 3.5.25 above, the adjoint of $\sigma_Z \circ (\mathcal{Y}^{(-)})_{X,Y}$ is the adjoint of $(\mathcal{Y}_Z)_{X,Y}$. Recalling Definition 3.2.2, it is the below composite of the symmetry ξ in V and the composition m of C:

$$\text{(3.5.28)} \qquad \mathsf{C}(X,Y) \otimes \mathsf{C}(Y,Z) \xrightarrow{\xi} \mathsf{C}(Y,Z) \otimes \mathsf{C}(X,Y) \xrightarrow{m} \mathsf{C}(X,Z). \qquad ◇$$

Lemma 3.5.29. *In the context of Definition 3.5.25, $\mathcal{Y}^{(\ )}$ is a* V*-functor.*

Proof. Throughout this proof we will omit the tensor symbols in V and we will use the following abbreviations for $X, Y \in \mathsf{C}$:

$$\mathsf{M} = \mathsf{Map} \quad \text{and} \quad \mathsf{C}_{X,Y} = \mathsf{C}(X,Y).$$

Now we begin with the identity axiom (1.2.6). By universality of V-ends, it suffices to show that the following diagram commutes for $Y, Z \in \mathsf{C}$.

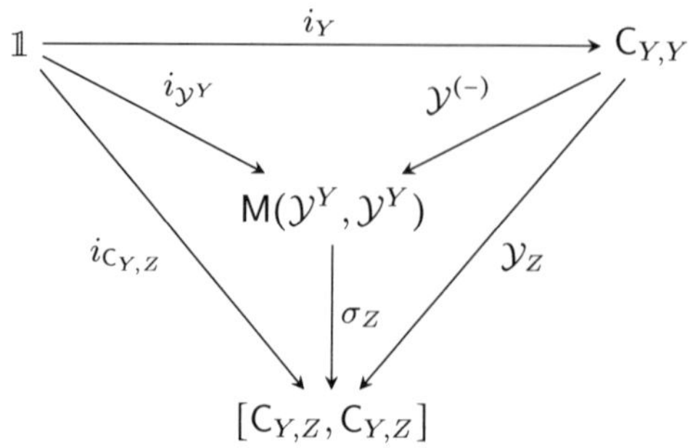

Taking adjoints, commutativity of the above diagram is equivalent to that of the outer diagram below. We have used (3.5.28) for the adjoint of $\mathcal{Y}_Z$.

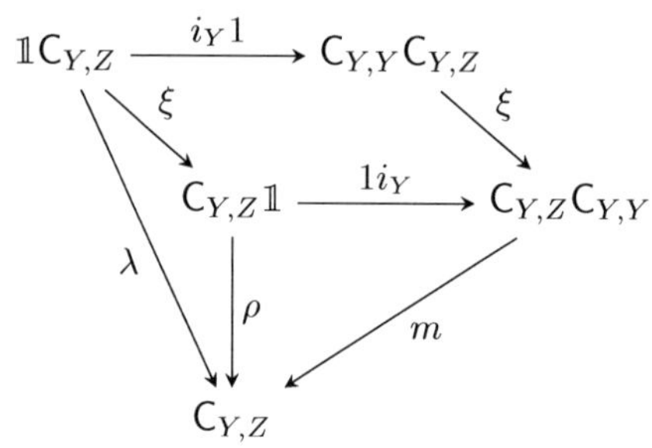

In the above diagram, the upper parallelogram commutes by naturality of ξ. The triangle at left commutes by symmetry of units (1.1.20) and the triangle at right commutes by the right unity axiom for composition m in C.

Now we turn to the composition axiom (1.2.5). By universality of V-ends, it suffices to verify that the following diagram commutes for $W, X, Y, Z \in \mathsf{C}$. In the below diagram, $(m^{\mathsf{Map}})'$ denotes the composition in $(\mathsf{C}\text{-}\mathsf{V})^{\mathsf{op}}$ and is defined via the lower triangle.

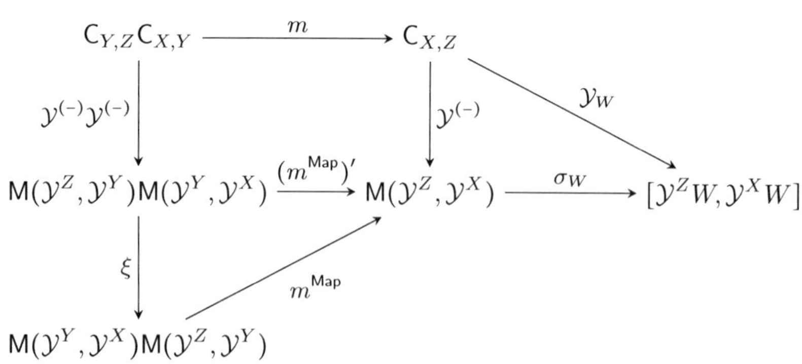

Taking adjoints, commutativity of the diagram above is equivalent to that of the outer diagram below. We have used (3.5.28) for the adjoint of $\mathcal{Y}_Z$ and the lower composite of (3.5.23) for the adjoint of $\sigma_W \circ m^{\mathsf{Map}}$.

(3.5.30)

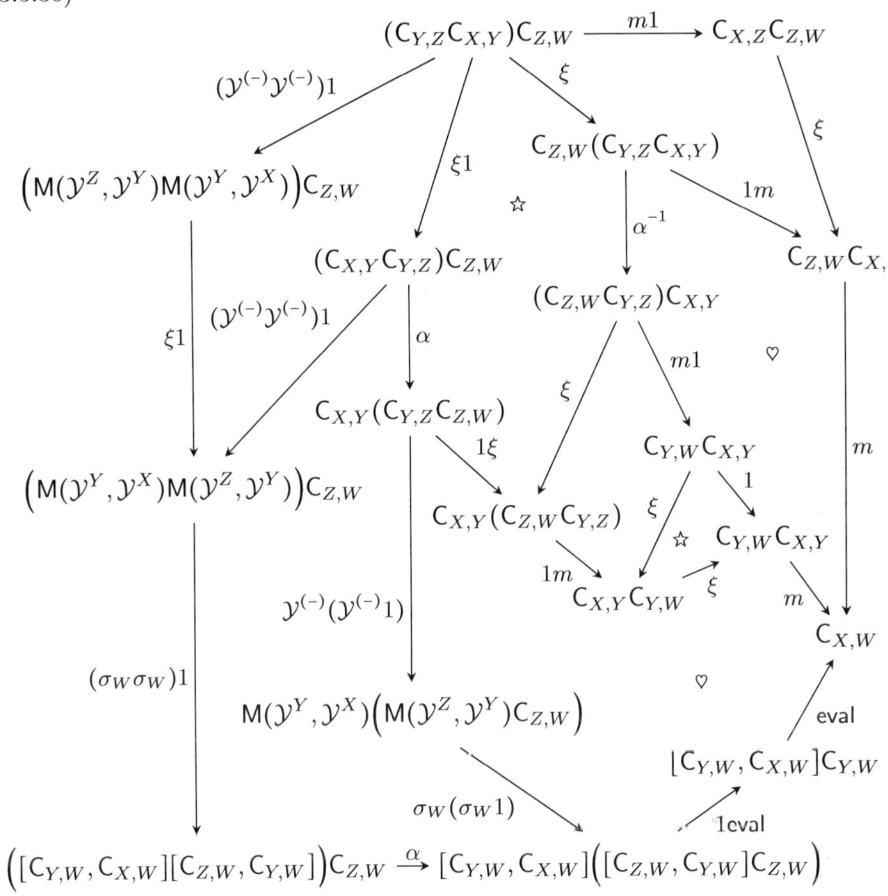

In the above diagram, the four unlabeled regions commute by naturality of α (once) and ξ (three times). The two regions labeled ♡ commute, respectively, by the definition of $\mathcal{Y}^{(-)}$ and by associativity of the composition m in C. The two regions labeled ☆ commute by the symmetry axiom (1.1.24) for ξ and the Symmetric Coherence Theorem 1.1.41 for V. □

Explanation 3.5.31. We note that the composition axiom for $\mathcal{Y}^{(-)}$ depends heavily on the underlying assumption that V is symmetric monoidal. The two regions marked ☆ in (3.5.30) do not commute when V is merely braided monoidal. ◇

Definition 3.5.32. Suppose $F\colon \mathsf{C} \longrightarrow \underline{\mathsf{V}}$ is a V-functor with V complete. Define

$$\mathsf{Map}(-,F)\colon (\mathsf{C}\text{-}\mathsf{V})^{\mathsf{op}} \longrightarrow \underline{\mathsf{V}}$$

by

$$G \longmapsto \mathsf{Map}(G,F) = \int_{Z\in\mathsf{C}} [GZ,FZ]$$

on objects and the following universal arrow on hom objects:

$$
\begin{array}{ccc}
\mathsf{Map}(G,H) & \xrightarrow{\ \left(\mathsf{Map}(-,F)\right)_{G,H}\ } & [\mathsf{Map}(H,F),\mathsf{Map}(G,F)] \\
{\scriptstyle \sigma_W}\downarrow & & \downarrow{\scriptstyle [\mathsf{Map}(H,F),\sigma_W]} \\
[GW,HW] & & [\mathsf{Map}(H,F),[GW,FW]] \\
& {\scriptstyle [-,FW]}\searrow \quad \nearrow{\scriptstyle [\sigma_W,[GW,FW]]} & \\
& [[HW,FW],[GW,FW]] &
\end{array}
\tag{3.5.33}
$$

As we describe in Explanation 3.5.34 below, $\mathsf{Map}(-,F)$ is the represented V-functor $\mathcal{Y}_F$ for the V-category $\mathsf{C}\text{-}\mathsf{V}$. ◇

Explanation 3.5.34 (V-Functoriality of $\mathsf{Map}(-,F)$)**.** In the context of Definition 3.5.32 above, the adjoint of (3.5.33) is given by the following diagram, where the unlabeled top arrow is the adjoint of $\left(\mathsf{Map}(-,F)\right)_{G,H}$.

$$
\begin{array}{ccc}
\mathsf{Map}(G,H)\otimes\mathsf{Map}(H,F) & \longrightarrow & \mathsf{Map}(G,F) \\
& & \quad\searrow{\scriptstyle \sigma_W} \\
{\scriptstyle \sigma_W\otimes\sigma_W}\downarrow & & \qquad [GW,FW] \\
& & \quad\nearrow{\scriptstyle m} \\
[GW,HW]\otimes[HW,FW] & \xrightarrow{\ \xi\ } & [HW,FW]\otimes[GW,HW]
\end{array}
\tag{3.5.35}
$$

Recalling (3.5.21), this shows that $\left(\mathsf{Map}(-,F)\right)_{G,H}$ is adjoint to

$$\left(m^{\mathsf{Map}}\right)' = m^{\mathsf{Map}}\circ\xi,$$

the composition in $(\mathsf{C}\text{-}\mathsf{V})^{\mathsf{op}}$. Therefore, $\mathsf{Map}(-,F)$ is the V-functor represented by F. ◇

Now Lemma 3.5.29 and V-functoriality of $\mathsf{Map}(-,F)$ give the following.

Proposition 3.5.36. *For each* V*-functor* $F\colon\mathsf{C}\longrightarrow\underline{\mathsf{V}}$,

$$\mathsf{Map}(\mathcal{Y}^{(-)},F)\colon\mathsf{C}\longrightarrow\underline{\mathsf{V}}$$

is a V*-functor.*

Explanation 3.5.37. The composite $\mathsf{Map}(\mathcal{Y}^{(-)},F)$ in Proposition 3.5.36 is given on objects by the assignment

$$X\longmapsto\mathsf{Map}(\mathcal{Y}^X,F)=\int_{Z\in\mathsf{C}}[\mathcal{Y}^XZ,FZ]$$

for $X \in \mathsf{C}$. The morphism on hom objects $\mathsf{C}(X,Y)$ is the top horizontal morphism in the following diagram, induced by the other morphisms.

$$
\begin{tikzcd}
\mathsf{C}(X,Y) \ar[rr, "\mathsf{Map}(\mathcal{Y}^{(-)},F)"] \ar[d, "\mathsf{C}(-,Z)"'] \ar[dr, "\mathcal{Y}^{(-)}"] & & {[\mathsf{Map}(\mathcal{Y}^X,F),\mathsf{Map}(\mathcal{Y}^Y,F)]} \ar[d, "{[\mathsf{Map}(\mathcal{Y}^X,F),\sigma_Z]}"] \\
& \mathsf{Map}(\mathcal{Y}^Y,\mathcal{Y}^X) \ar[ur, "\mathsf{Map}(-,F)"'] \ar[dl, "\sigma_Z"] & \\
{[\mathcal{Y}^Y Z,\mathcal{Y}^X Z]} \ar[dr, "{[-,FZ]}"'] & & {\big[\mathsf{Map}(\mathcal{Y}^X,F),[\mathcal{Y}^Y Z,FZ]\big]} \\
& {\big[[\mathcal{Y}^X Z,FZ],[\mathcal{Y}^Y Z,FZ]\big]} \ar[ur, "{\big[\sigma_Z,[\mathcal{Y}^Y Z,FZ]\big]}"'] &
\end{tikzcd}
\tag{3.5.38}
$$

In the above diagram, the upper triangular region is the definition of the composite V-functor on hom objects, and the remaining two regions are the definitions (3.5.26) and (3.5.33) of $\mathcal{Y}^{(-)}$ and $\mathsf{Map}(-,F)$, respectively. $\diamond$

3.6. Enriched Yoneda Lemma

Throughout this section we assume V is a symmetric monoidal closed category that is complete. The main purpose of this section is to prove the V-Yoneda Lemma 3.6.9.

Definition 3.6.1. For each V-functor $F\colon \mathsf{C} \longrightarrow \underline{\mathsf{V}}$ and each $X \in \mathsf{C}$ let

$$FX \xrightarrow{\phi_X} \mathsf{Map}(\mathcal{Y}^X,F)$$

denote the morphism in V induced by morphisms

$$FX \xrightarrow{F^{\#}_{X,Z}} [\mathcal{Y}^X Z,FZ]$$

that are the transforms of

$$\mathcal{Y}^X Z = \mathsf{C}(X,Z) \xrightarrow{F_{X,Z}} [FX,FZ].$$

We show in Lemma 3.6.5 below that the ϕ_X determine a V-natural transformation

$$\phi^{\perp}\colon F \longrightarrow \mathsf{Map}(\mathcal{Y}^{(-)},F). \qquad \diamond$$

Explanation 3.6.2 (The morphisms ϕ_X)**.** The composition axiom for F (1.2.5) together with Lemma 3.2.10 shows that the morphisms

$$F^{\perp}_{X,Z}\colon \mathbb{1} \longrightarrow [\mathsf{C}(X,Z),[FX,FZ]]$$

determine a V-natural transformation $\mathsf{C}(-,-) \longrightarrow [F(-),F(-)]$. In particular, for fixed X, Lemma 3.5.16 shows that the morphisms $F^{\#}_{X,Z}$ for $Z \in \mathsf{C}$ satisfy the V-wedge condition (3.5.3), thus inducing the morphisms

$$FX \xrightarrow{\phi_X} \int_{Z\in\mathsf{C}} \big[\mathcal{Y}^X Z,FZ\big] = \mathsf{Map}(\mathcal{Y}^X,F)$$

in V. $\diamond$

We will use the following result with $H = \mathcal{Y}_Z$ to show that each $F^{\#}_{-,Z}$ determines a V-natural transformation. Then we use that result to show that $\phi^{\perp}$ is V-natural.

Lemma 3.6.3. *Suppose $F: \mathsf{C} \longrightarrow \underline{\mathsf{V}}$ and $H: \mathsf{C}^{\mathsf{op}} \longrightarrow \underline{\mathsf{V}}$ are V-functors and suppose $Q \in \mathsf{V}$. A collection of morphisms*

$$\theta_X : FX \longrightarrow [HX, Q]$$

for $X \in \mathsf{C}$ determines a V-natural transformation $\theta^{\perp}$ if and only if the transforms

$$\theta^{\#}_X : HX \longrightarrow [FX, Q]$$

determine a V-natural transformation $(\theta^{\#})^{\perp}$.

Proof. We apply Lemma 3.2.10. Throughout this proof we omit tensor symbols and use $[-, Q]$ to denote the V-enriched functor from $\underline{\mathsf{V}}^{\mathsf{op}}$ to $\underline{\mathsf{V}}$ represented by Q. The diagram (3.2.11) for $\theta^{\perp}$ is the following, for $X, Y \in \mathsf{C}$.

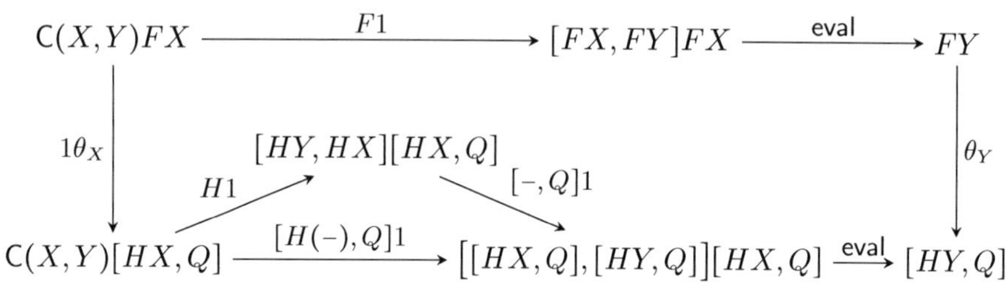

The diagram (3.2.11) for $(\theta^{\#})^{\perp}$ is the following, for $X, Y \in \mathsf{C}$.

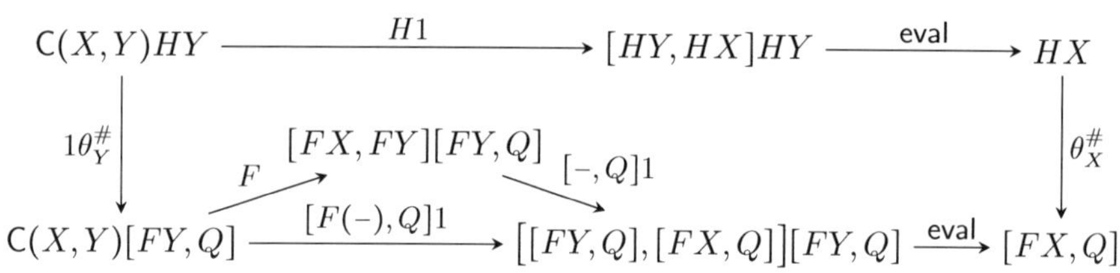

Now let

$$\overline{\theta}_X : FX \otimes HX \longrightarrow Q$$

denote the morphism adjoint to θ_X for each $X \in \mathsf{C}$, so that $\overline{\theta}_X \xi$ is the adjoint of $\theta^{\#}_X$. Then taking adjoints and using naturality of ξ and α together with the definition of the represented V-functors $[-, Q]$ (Definition 3.2.2), we see that each of the diagrams above commutes if and only if the following diagram commutes.

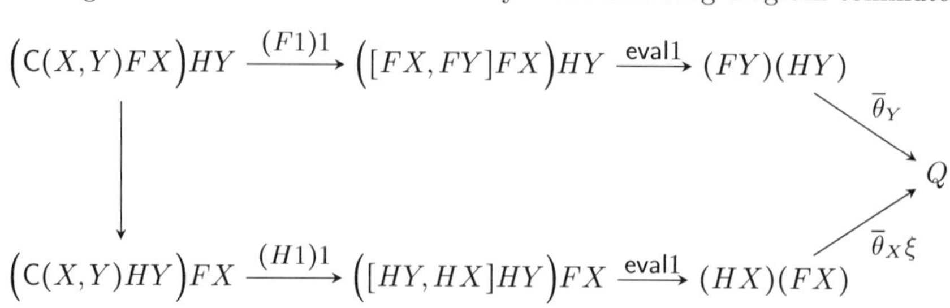

Therefore, $\theta^{\perp}$ is V-natural if and only if $(\theta^{\#})^{\perp}$ is V-natural. □

Lemma 3.6.4. *For each* V*-functor* $F{:}\,\mathsf{C} \longrightarrow \underline{\mathsf{V}}$ *and each* $Z \in \mathsf{C}$*, the transforms* $F^{\#}_{X,Z}$ *determine a* V*-natural transformation*

$$(F^{\#}_{-,Z})^{\perp}{:}\,F \longrightarrow [\mathcal{Y}_Z, FZ]$$

Proof. By Lemma 3.2.13, the composition axiom for F implies that $F_{-,Z}$ determines a V-natural transformation

$$F^{\perp}_{-,Z}{:}\,\mathcal{Y}_Z \longrightarrow [F(-), FZ].$$

Then the result follows from Lemma 3.6.3. □

Lemma 3.6.5. *In the context of Definition 3.6.1, the morphisms*

$$\phi_X{:}\,FX \longrightarrow \mathsf{Map}(\mathcal{Y}^X, F)$$

for $X \in \mathsf{C}$ *are components of a* V*-natural transformation*

$$\phi^{\perp}{:}\,F \longrightarrow \mathsf{Map}(\mathcal{Y}^{(-)}, F).$$

Proof. We apply Lemma 3.2.10. We omit tensor symbols and abbreviate $\mathsf{M} = \mathsf{Map}$ in this proof. The diagram (3.2.11) for ϕ is the upper region labeled ☆ in the diagram below. By universality of V-ends it suffices to show that the composites around ☆ are equal after composed with σ_Z for each $Z \in \mathsf{C}$.

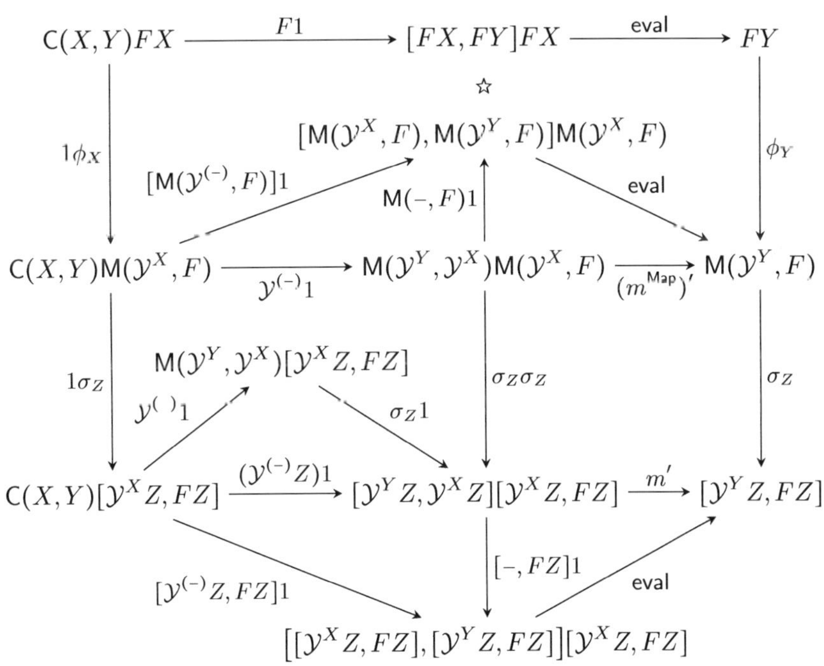

In the above diagram, the two triangular regions below ☆ commute by definition, where $(m^{\mathsf{Map}})' = m^{\mathsf{Map}} \circ \xi$ is the composition in $(\mathsf{C}\text{-}\mathsf{V})^{\mathsf{op}}$. We also let $m' = m \circ \xi$. The remaining regions, from left to right and top to bottom, commute by functoriality of ⊗, definition of m^{Map} (3.5.21) and naturality of ξ, definition of $\mathcal{Y}^{(-)}$ on hom objects (3.5.26), and definition of $[\mathcal{Y}^{(-)}Z, FZ]$.

Lastly, by Definition 3.6.1 the components of ϕ are determined by the equalities

$$F^{\#}_{X,Z} = \sigma_Z \circ \phi_X \quad \text{and} \quad F^{\#}_{Y,Z} = \sigma_Z \circ \phi_Y.$$

Therefore, the outer diagram above commutes by naturality of $(F^{\#}_{-,Z})^{\perp}$, proved in Lemma 3.6.4. This implies that the composites around ☆ are equal after composing with each σ_Z, and thus completes the proof that $\phi^{\perp}$ is V-natural. □

Lemma 3.6.6. *Suppose that f, g, and h are morphisms in* V *as shown at left below. Then the diagram at left commutes if and only if the diagram at right commutes.*

(3.6.7)

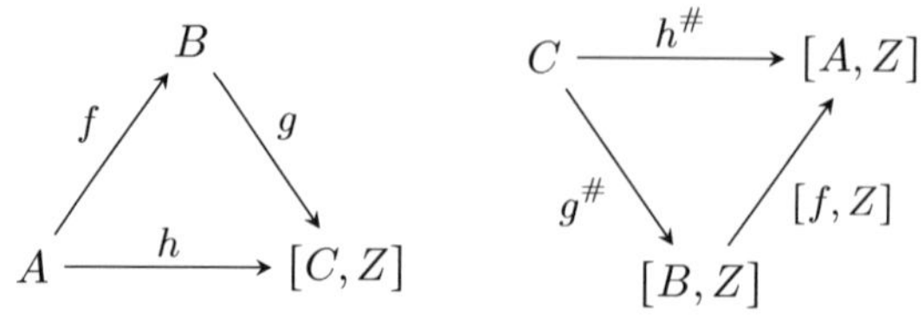

Proof. By taking adjoints one verifies that each of the two traingles in (3.6.7) commutes if and only if the triangle below commutes, where the unlabeled arrows are the adjoints of g and h, respectively.

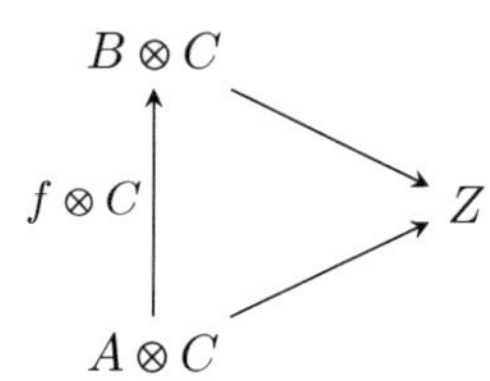

□

Lemma 3.6.8. *Suppose $\eta\colon \mathbb{1} \longrightarrow [Q, A]$ is a morphism in* V *and let η' denote the composite*

$$Q \xrightarrow{\eta^{\#}} [\mathbb{1}, A] \xrightarrow[\cong]{\kappa} A.$$

Then the following diagram in V *commutes.*

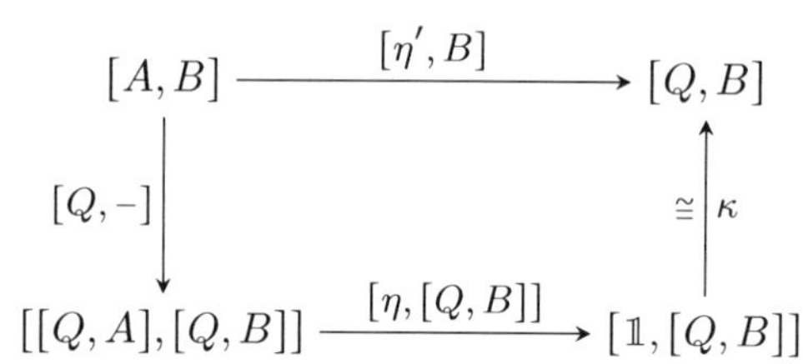

Proof. Taking adjoints, the diagram in the statement commutes if and only if the outer diagram below commutes.

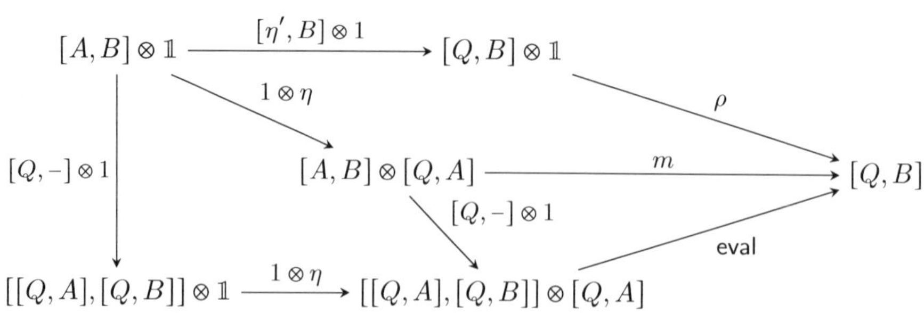

In the above diagram, commutativity of the region at left follows from functoriality of ⊗. Commutativity of the lower right triangle follows from the definition of $[Q,-]$. Commutativity of the remaining region follows from another adjoint diagram argument comparing the corresponding morphisms

$$([A,B]\otimes \mathbb{1})\otimes Q \longrightarrow B. \qquad \square$$

Now we use the V-Yoneda Bijection 3.4.12 to prove the following.

Theorem 3.6.9 (V-Yoneda Lemma). *Suppose* V *is a symmetric monoidal closed category that is complete. Suppose* C *is a small* V*-category and* $F\colon \mathsf{C} \longrightarrow \underline{\mathsf{V}}$ *is a* V*-functor. Then*

$$FX \xrightarrow{\phi_X} \mathsf{Map}(\mathcal{Y}^X, F)$$

is an isomorphism for each $X \in \mathsf{C}$*. These provide the components of a* V*-natural isomorphism*

$$\phi^{\perp}\colon F \longrightarrow \mathsf{Map}(\mathcal{Y}^{(-)}, F).$$

Proof. The definition of ϕ is given in Definition 3.6.1 above. The V-naturality of $\phi^{\perp}$ is proved in Lemma 3.6.5 above. Now we show that each ϕ_X is an isomorphism in V by showing that FX satisfies the universal property of the end $\mathsf{Map}(\mathcal{Y}^X, F)$.

To do so, suppose that

$$Q \xrightarrow{\theta_Z^{\#}} [\mathcal{Y}^X Z, FZ]$$

is a collection of morphisms for $Z \in \mathsf{V}$ that are transforms of

$$\mathcal{Y}^X Z \xrightarrow{\theta_Z} [Q, FZ].$$

By Lemma 3.5.16 the transforms $\theta^{\#}$ satisfy the V-wedge condition (3.5.3) if and only if $\theta^{\perp}$ are the components of a V-natural transformation $\mathcal{Y}^X \longrightarrow [Q, F(-)]$. In this case, the V-Yoneda Bijection of Theorem 3.4.12 implies that there is a unique morphism in V

$$\eta\colon \mathbb{1} \longrightarrow [Q, FX]$$

such that the outer square of the diagram below commutes for each $Y \in \mathsf{C}$.

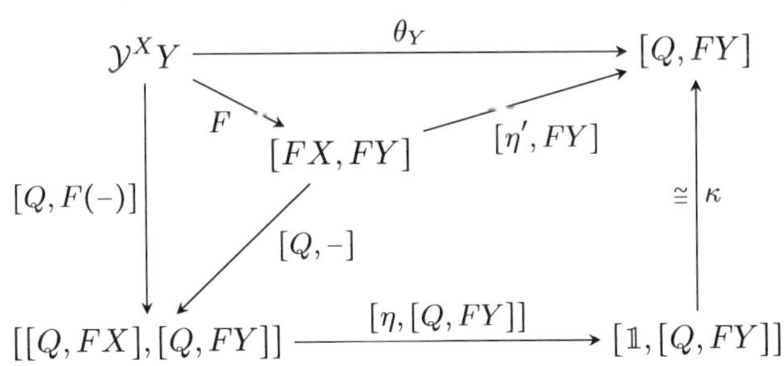

In the above diagram, the lower right quadrilateral commutes by Lemma 3.6.8 and therefore the upper triangular region also commutes. By Lemma 3.6.6 the commutativity of that upper triangular region is equivalent to commutativity of the triangle below.

(3.6.10)

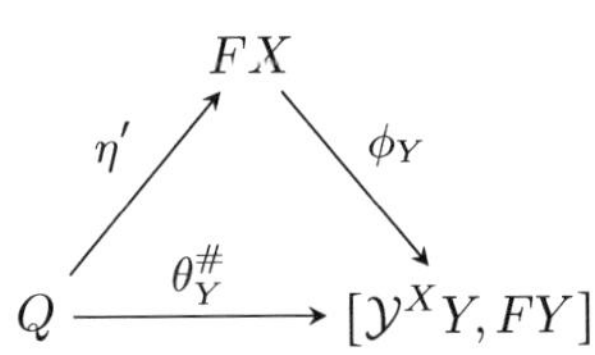

This argument shows that uniqueness of η' making (3.6.10) commute is equivalent to uniquess of η in the V-Yoneda Bijection of Theorem 3.4.12. Therefore, (FX,ϕ) satisfies the universal property of the end defining $\mathsf{Map}(\mathcal{Y}^X,F)$. □

Corollary 3.6.11 (V-Yoneda Embedding Theorem). *The Yoneda* V*-functor induces isomorphisms in* V

$$\mathsf{C}(X,Y) \xrightarrow{(\mathcal{Y}^{(-)})_{X,Y}} \mathsf{Map}(\mathcal{Y}^Y,\mathcal{Y}^X)$$

for $X,Y \in \mathsf{C}$.

Proof. Taking $F = \mathcal{Y}^X$ in Theorem 3.6.9 (interchanging X and Y), we have

$$\phi_Y = \mathcal{Y}^{(-)}_{X,Y}$$

because the transform

$$(\mathcal{Y}^X)^{\#}_{Y,Z}: \mathcal{Y}^X Y \longrightarrow [\mathcal{Y}^Y Z, \mathcal{Y}^X Z]$$

in Definition 3.6.1 is the same as $(\mathcal{Y}_Z)_{X,Y}$ in (3.5.26). Therefore, the result follows from Theorem 3.6.9. □

Definition 3.6.12. Suppose $F,G\colon \mathsf{C} \longrightarrow \underline{\mathsf{V}}$ are V-functors with V complete and suppose

$$\theta\colon F \longrightarrow G$$

is a V-natural transformation. We define

$$\mathsf{Map}(\mathcal{Y}^{(-)},F) \xrightarrow{\mathsf{Map}(\mathcal{Y}^{(-)},\theta)} \mathsf{Map}(\mathcal{Y}^{(-)},G)$$

to be the V-natural transformation with component $\mathsf{Map}(\mathcal{Y}^X,\theta)$ at $X \in \mathsf{C}$ determined by the following diagram for $Z \in \mathsf{C}$.

(3.6.13)
$$\begin{array}{ccc}
\mathsf{Map}(\mathcal{Y}^X,F) & \xrightarrow{\mathsf{Map}(\mathcal{Y}^X,\theta)} & \mathsf{Map}(\mathcal{Y}^X,G) \\
\downarrow{\scriptstyle \sigma_Z} & & \searrow{\scriptstyle \sigma_Z} \\
[\mathcal{Y}^X Z, FZ] & & [\mathcal{Y}^X Z, GZ] \\
\downarrow{\scriptstyle [\mathcal{Y}^X Z,\lambda^{-1}]} & & \nearrow{\scriptstyle [\mathcal{Y}^X Z,\mathsf{eval}]} \\
[\mathcal{Y}^X Z, \mathbb{1} \otimes (FZ)] & \xrightarrow{[\mathcal{Y}^X Z,\theta_Z 1]} & [\mathcal{Y}^X Z, [FZ,GZ] \otimes (FZ)]
\end{array}$$

That is, we define

$$\mathsf{Map}(\mathcal{Y}^{(-)},\theta)_X = \big(\mathsf{Map}(\mathcal{Y}^X,\theta)\big)^{\perp}.$$

We show in Lemma 3.6.14 that $\mathsf{Map}(\mathcal{Y}^{(-)},\theta)$ is V-natural. ◇

Lemma 3.6.14. *In the context of Definition 3.6.12,*

(1) $\mathsf{Map}(\mathcal{Y}^{(-)}, \theta)$ *is a* V*-natural transformation and*
(2) *the following diagram of* V*-natural transformations is commutative, where* $\phi^{\perp}$ *denotes the* V*-natural transformations of Definition 3.6.1 for both* F *and* G*, respectively.*

$$\begin{CD} F @>{\phi^{\perp}}>> \mathsf{Map}(\mathcal{Y}^{(-)}, F) \\ @V{\theta}VV @VV{\mathsf{Map}(\mathcal{Y}^{(-)}, \theta)}V \\ G @>{\phi^{\perp}}>> \mathsf{Map}(\mathcal{Y}^{(-)}, G) \end{CD} \tag{3.6.15}$$

Proof. It suffices, for both assertions in the statement, to show that (3.6.15) is componentwise commutative. Then V-naturality of

$$\mathsf{Map}(\mathcal{Y}^{(-)}, \theta) = (\phi^{\perp})\,\theta\,(\phi^{\perp})^{-1}$$

follows from V-naturality of θ and of each $\phi^{\perp}$, proved in Lemma 3.6.5.

For each $X \in \mathsf{C}$, let ω_X denote the composite

$$FX \xrightarrow{\lambda^{-1}} \mathbb{1} \otimes FX \xrightarrow{\theta_X \otimes 1} [FX, GX] \otimes FX \xrightarrow{\mathsf{eval}} GX,$$

so that $\theta_X = \omega_X^{\perp}$. Then by functoriality of $(-)^{\perp}$ (3.2.5) we have

$$(\phi_X^{\perp})\,\theta_X\,(\phi_X^{\perp})^{-1} = (\phi_X \omega_X \phi_X^{-1})^{\perp}.$$

Therefore, using the universality of V-ends, the component of (3.6.15) at X commutes if and only if the following diagram commutes for each $Z \in \mathsf{C}$. Here and throughout the rest of this proof we omit tensor symbols. The triangles below determine the morphisms ϕ_X (Definition 3.6.1)

$$\begin{array}{ccccc} & \nearrow^{\phi_X} & \mathsf{Map}(\mathcal{Y}^X, F) & \searrow^{\sigma_Z} & \\ FX & & \xrightarrow{\quad F^{\#}_{X,Z} \quad} & & [\mathcal{Y}^X Z, FZ] \\ \Big\downarrow{\scriptstyle \omega_X} & & & & \Big\downarrow{\scriptstyle [\mathcal{Y}^X Z, \omega_Z]} \\ GX & & \xrightarrow{\quad G^{\#}_{X,Z} \quad} & & [\mathcal{Y}^X Z, GZ] \\ & \searrow_{\phi_X} & \mathsf{Map}(\mathcal{Y}^X, G) & \nearrow_{\sigma_Z} & \end{array}$$

By taking adjoints of the composites in the inner rectangle above—using Definition 3.5.15 for the adjoints of the transforms $F^{\#}$ and $G^{\#}$—we obtain the rectangle (3.2.11) that determines V-naturality of $\omega^{\perp} = \theta$. Therefore, V-naturality of θ implies that the above diagram commutes for each $Z \in \mathsf{C}$. Hence (3.6.15) commutes componentwise. □

Explanation 3.6.16. The commutativity of (3.6.15) also proves that the assignment

$$\theta \longmapsto \mathsf{Map}(\mathcal{Y}^{(-)}, \theta)$$

is functorial with respect to identities and vertical composites. Therefore, we have a functor of categories

$$\mathsf{V}\text{-}\mathsf{Cat}(\mathsf{C}, \underline{\mathsf{V}}) \xrightarrow{\mathsf{Map}(\mathcal{Y}^{(-)}, -)} \mathsf{V}\text{-}\mathsf{Cat}(\mathsf{C}, \underline{\mathsf{V}}).$$

The V-Yoneda Lemma 3.6.9 together with (3.6.15) imply that this functor is naturally isomorphic to the identity functor. ◇

3.7. Symmetric Monoidal Diagram Categories

Throughout this section we let V denote a symmetric monoidal closed category that is complete and cocomplete. We let $\mathcal{D}$ denote a small symmetric monoidal V-category. The purpose of this section is to describe a symmetric monoidal closed structure on the category of V-functors from $\mathcal{D}$ to $\underline{\mathsf{V}}$. Throughout this section, recall from Theorem 3.3.2 that the monoidal data of V are V-natural. These, along with the V-natural monoidal structure of $\mathcal{D}$, will be used as part of the V-natural monoidal structure on the V-functor category.

Definition 3.7.1. Suppose C is a V-category and $\mathcal{D}$ is a small V-category. A V-functor $\mathcal{D} \longrightarrow \mathsf{C}$ is called a *$\mathcal{D}$-shaped diagram* in C. The category of V-functors and V-natural transformations is denoted $\mathcal{D}$-C. In the case $\mathsf{C} = \underline{\mathsf{V}}$ with its canonical self-enrichment, we will write

$$\mathcal{D}\text{-}\mathsf{V} = \mathsf{V}\text{-}\mathsf{Cat}(\mathcal{D}, \underline{\mathsf{V}}).$$

In this context we will often use lowercase letters $a, b, c, \ldots \in \mathcal{D}$ for objects of $\mathcal{D}$, uppercase letters $X, Y, Z, \ldots \in \mathcal{D}$-$\mathsf{C}$ for $\mathcal{D}$-shaped diagrams, and subscripts X_a for the value of a diagram X at an object a. ◇

Explanation 3.7.2. In Definition 3.5.19 and Lemma 3.5.24, with $\mathcal{D} = \mathsf{C}$, we used the notation $\mathcal{D}$-V for the V-enriched category of $\mathcal{D}$-shaped diagrams in $\underline{\mathsf{V}}$ with hom objects given by the mapping objects $\mathsf{Map}(-,-)$ (Definition 3.5.12). By Lemma 3.5.14, the underlying category of this V-enriched category is isomorphic to the category of V-functors and V-natural transformations denoted $\mathcal{D}$-V in Definition 3.7.1. Throughout this section we will let $\mathcal{D}$-V denote the ordinary category and define a symmetric monoidal closed structure for it. In Explanation 3.8.3 we will relate this back to the V-enriched structure of Definition 3.5.19 and Lemma 3.5.24. ◇

Definition 3.7.3. Suppose that $(\mathsf{V}, \otimes)$ is a symmetric monoidal closed category that is complete and cocomplete. Suppose that $\mathcal{D}$ is small symmetric monoidal V-category with product denoted $\boxdot$. Suppose given $\mathcal{D}$-shaped diagrams $X, Y \in \mathcal{D}$-V. The *Day convolution product* of X and Y is denoted $X \otimes Y$ and is defined as the V-coend

$$X \otimes Y = \int^{(a,b)\in\mathcal{D}\otimes\mathcal{D}} \mathcal{D}(a \boxdot b, -) \otimes \big((X_a) \otimes (Y_b)\big). \tag{3.7.4}$$

The *hom diagram* from X to Y is denoted $\mathsf{Hom}_{\mathcal{D}}(X,Y)$ and is defined as the V-end

$$\mathsf{Hom}_{\mathcal{D}}(X,Y) = \int_{c\in\mathcal{D}} \Big[\int^{b\in\mathcal{D}} \mathcal{D}(- \boxdot b, c) \otimes X_b, Y_c \Big] \tag{3.7.5}$$

$$\cong \int_{(b,c)\in\mathcal{D}\otimes\mathcal{D}} \Big[\mathcal{D}(- \boxdot b, c) \otimes X_b, Y_c \Big] \tag{3.7.6}$$

with the isomorphism following from (3.5.10). The assumption that V is complete and cocomplete implies that the objects

$$(X \otimes Y)_a \in \mathsf{V} \quad \text{and} \quad \mathsf{Hom}_{\mathcal{D}}(X,Y)_a \in \mathsf{V}$$

exist for all diagrams X and Y in $\mathcal{D}$-V and all $a \in \mathcal{D}$. The V-functoriality of

$$X \otimes Y : \mathcal{D} \longrightarrow \underline{\mathsf{V}} \quad \text{and} \quad \mathsf{Hom}_{\mathcal{D}}(X,Y) : \mathcal{D} \longrightarrow \underline{\mathsf{V}}$$

follows from universality of the (co)ends along with V-functoriality of the monoidal products and internal homs in $\underline{\mathsf{V}}$ and $\mathcal{D}$. ◇

In Proposition 3.7.19 below we give the definitions of $\otimes$ and $\mathsf{Hom}_{\mathcal{D}}$ on morphisms of $\mathcal{D}$-V—that is, V-natural transformations—and show that they are functors

$$\otimes\colon \mathcal{D}\text{-}\mathsf{V} \times \mathcal{D}\text{-}\mathsf{V} \longrightarrow \mathcal{D}\text{-}\mathsf{V} \quad \text{and} \quad \mathsf{Hom}_{\mathcal{D}}\colon \mathcal{D}\text{-}\mathsf{V}^{\mathsf{op}} \times \mathcal{D}\text{-}\mathsf{V} \longrightarrow \mathcal{D}\text{-}\mathsf{V}.$$

We show that $\otimes$ and $\mathsf{Hom}_{\mathcal{D}}$ are adjoints in Proposition 3.7.21. Finally in Theorem 3.7.22 we show that these give a symmetric monoidal closed structure on $\mathcal{D}$-V.

To express the associativity and unit isomorphisms for Day convolution, we will need some introductory definitions and results related to the V-Yoneda Lemma 3.6.9. For $X \in \mathcal{D}$-V, it follows from V-naturality of the morphisms

$$\mathcal{D}(x,s) \xrightarrow{X_{x,s}} [X_x, X_s].$$

and Lemma 3.5.16 that the adjoints

$$\mathcal{D}(x,s) \otimes X_x \xrightarrow{\overline{X}_{x,s}} X_s$$

satisfy the V-cowedge condition (3.5.2).

Definition 3.7.7. For each V-functor $X\colon \mathcal{D} \longrightarrow \underline{\mathsf{V}}$ and each $s \in \mathcal{D}$, let

$$\int^{x \in \mathcal{D}} \mathcal{D}(x,s) \otimes X_x \xrightarrow{\psi_s} X_s$$

denote the morphism induced by the adjoints of

$$\mathcal{D}(x,s) \xrightarrow{X_{x,s}} [X_x, X_s].$$ ◇

Theorem 3.7.8 (V-Yoneda Density). *Suppose $\mathcal{D}$ is a small V-category and*

$$X\colon \mathcal{D} \longrightarrow \underline{\mathsf{V}}$$

is a V-functor. Then for each $s \in \mathcal{D}$

$$\int^{x} \mathcal{D}(x,s) \otimes X_x \xrightarrow{\psi_s} X_s$$

is an isomorphism. These provide components of a V-natural isomorphism

$$\int^{x} \mathcal{D}(x,-) \otimes X_x \xrightarrow{\psi^{\perp}} X.$$

Proof. The V-naturality of $\psi^{\perp}$ follows via Lemma 3.2.10 from V-naturality of the morphisms $(X_{x,s})^{\perp}$. To prove that each ψ_s is an isomorphism in V, it suffices to prove that the induced morphisms

$$[X_s, Z] \xrightarrow{[\psi_s, Z]} \Big[\int^{x \in \mathcal{D}} \mathcal{D}(x,s) \otimes X_x, Z\Big]$$

are isomorphisms for each $Z \in \mathsf{V}$. First by (3.5.10) we have

$$\Big[\int^{x \in \mathcal{D}} \mathcal{D}(x,s) \otimes X_x, Z\Big] \cong \int_{x \in \mathcal{D}} \big[\mathcal{D}(x,s) \otimes X_x, Z\big]$$

and then using the closed structure we have

$$\big[\mathcal{D}(x,s) \otimes X_x, Z\big] \cong \big[\mathcal{D}(x,s), [X_x, Z]\big]$$

for each $x \in \mathcal{D}$. The composite morphisms

$$[X_s, Z] \longrightarrow \int_{x \in \mathcal{D}} \Big[\mathcal{D}(x,s), [X_x, Z]\Big] = \int_{x \in \mathcal{D}^{\mathrm{op}}} \Big[\mathcal{D}^{\mathrm{op}}(s,x), [X_x, Z]\Big]$$

are those induced by the structure morphisms of the V-functor

$$[X_{(-)}, Z] \colon \mathcal{D}^{\mathrm{op}} \longrightarrow \underline{\mathsf{V}}.$$

Therefore, the V-Yoneda Lemma 3.6.9 shows that these morphisms, and hence also the morphisms $[\psi_s, Z]$, are isomorphisms for each $Z \in \mathsf{V}$. □

Explanation 3.7.9. In the context of Theorem 3.7.8, the V-Yoneda Lemma 3.6.9 can be written in the following equivalent form. For each $s \in \mathcal{D}$

$$X_s \xrightarrow{\phi_s} \int_x [\mathcal{D}(s,x), X_x]$$

is an isomorphism. These provide components of a V-natural isomorphism

$$X \xrightarrow{\phi^{\perp}} \int_x [\mathcal{D}(-,x), X_x] \tag{3.7.10}$$ ◇

As an immediate application of Theorem 3.7.8, we have the following.

Explanation 3.7.11. In the context of Definition 3.7.3 we have a sequence of V-natural isomorphisms given by the isomorphism (3.5.10), interchanging order of ends, the symmetric monoidal closed structure of V, the isomorphism (3.5.11), and the V-Yoneda Density Theorem 3.7.8.

$$\begin{aligned}
\mathsf{Hom}_{\mathcal{D}}(X,Y) &= \int_c \Big[\int^b \mathcal{D}(- \boxdot b, c) \otimes X_b, Y_c\Big] \\
&\cong \int_c \int_b \Big[\mathcal{D}(- \boxdot b, c) \otimes X_b, Y_c\Big] \\
&\cong \int_b \int_c \Big[X_b, [\mathcal{D}(- \boxdot b, c), Y_c]\Big] \\
&\cong \int_b \Big[X_b, \int_c [\mathcal{D}(- \boxdot b, c), Y_c]\Big] \\
&\cong \int_{b \in \mathcal{D}} \Big[X_b, Y_{- \boxdot b}\Big].
\end{aligned} \tag{3.7.12}$$

With this latter formulation one can see more directly that we have

$$\mathsf{Map}(X,Y) \cong \mathsf{Hom}_{\mathcal{D}}(X,Y)_e \tag{3.7.13}$$

where $e \in \mathcal{D}$ is the monoidal unit. ◇

We will use Theorem 3.7.8 as stated, along with two special cases defined as follows.

Definition 3.7.14. For each triple $a, b, c \in \mathcal{D}$, let

$$\int^x \mathcal{D}(x \boxdot b, c) \otimes \mathcal{D}(a, x) \xrightarrow[\cong]{\gamma_{a,c;b}} \mathcal{D}(a \boxdot b, c) \tag{3.7.15}$$

denote the morphisms induced by

$$\mathcal{D}(x \boxdot b, c) \otimes \mathcal{D}(a,x) \xrightarrow{1 \otimes (- \boxdot b)} \mathcal{D}(x \boxdot b, c) \otimes \mathcal{D}(a \boxdot b, x \boxdot b) \xrightarrow{m} \mathcal{D}(a \boxdot b, c).$$

Similarly, let

$$\int^x \mathcal{D}(b \boxdot x, c) \otimes \mathcal{D}(a, x) \xrightarrow[\cong]{\gamma_{b;a,c}} \mathcal{D}(b \boxdot a, c) \tag{3.7.16}$$

denote the morphisms induced by

$$\mathcal{D}(b \boxdot x, c) \otimes \mathcal{D}(a, x) \xrightarrow{1 \otimes (b \boxdot -)} \mathcal{D}(b \boxdot x, c) \otimes \mathcal{D}(b \boxdot a, b \boxdot x) \xrightarrow{m} \mathcal{D}(b \boxdot a, c).$$

Applying Theorem 3.7.8 with $X_s = \mathcal{D}(s \boxdot b, c)$ and $X_s = \mathcal{D}(b \boxdot s, c)$, respectively, together with an appropriate symmetry isomorphism ξ shows that the $\gamma_{a,c;b}$ and $\gamma_{b;a,c}$ are isomorphisms. ◇

Using the isomorphisms given above, we define the following V-natural isomorphisms of diagrams. These will define the components of the associativity and unit isomorphisms for the Day convolution product in Theorem 3.7.22.

Definition 3.7.17. Suppose $\mathcal{D}$ is a small symmetric monoidal V-category with monoidal unit e. Suppose X, Y, and Z are V-functors $\mathcal{D} \longrightarrow \underline{\mathsf{V}}$. Define the *unit diagram* J to be the V-functor

$$J = \mathcal{Y}^e = \mathcal{D}(e, -): \mathcal{D} \longrightarrow \underline{\mathsf{V}}. \tag{3.7.18}$$

Define V-natural isomorphisms

$$\begin{aligned}
\alpha_{X,Y,Z}&: (X \otimes Y) \otimes Z \longrightarrow X \otimes (Y \otimes Z)\\
\lambda_X&: J \otimes X \longrightarrow X\\
\rho_X&: X \otimes J \longrightarrow X\\
\xi_{X,Y}&: X \otimes Y \longrightarrow Y \otimes X
\end{aligned}$$

as follows.

Associativity: The *associativity isomorphsm* $\alpha_{X,Y,Z}$ is the V-natural isomorphism with components determined by the following composite, explained below, for each $s \in \mathcal{D}$:

$$\begin{aligned}
[(X \otimes Y) \otimes Z]_s &= \int^{x,r} \mathcal{D}(x \boxdot r, s) \otimes (X \otimes Y)_x \otimes Z_r\\
&= \int^{x,r} \mathcal{D}(x \boxdot r, s) \otimes \Big(\Big(\int^{p,q} \mathcal{D}(p \boxdot q, x) \otimes (X_p \otimes Y_q)\Big) \otimes Z_r\Big)\\
&\cong \int^{p,q,r} \Big(\int^{x} \mathcal{D}(x \boxdot r, s) \otimes \mathcal{D}(p \boxdot q, x)\Big) \otimes \Big((X_p \otimes Y_q) \otimes Z_r\Big)\\
&\cong \int^{p,q,r} \Big(\int^{x} \mathcal{D}(p \boxdot x, s) \otimes \mathcal{D}(q \boxdot r, x)\Big) \otimes \Big(X_p \otimes (Y_q \otimes Z_r)\Big)\\
&\cong \int^{p,x} (\mathcal{D}(p \boxdot x, s) \otimes X_p) \otimes \Big(\int^{q,r} \mathcal{D}(q \boxdot r, x) \otimes (Y_q \otimes Z_r)\Big)\\
&= [X \otimes (Y \otimes Z)]_s.
\end{aligned}$$

The first isomorphism above is given by associativity isomorphisms in V along with interchanging order of coends. The middle isomorphism is given by associativity

$$(X_p \otimes Y_q) \otimes Z_r \cong X_p \otimes (Y_q \otimes Z_r)$$

in V together with the morphisms $\gamma_{(p\boxdot q),s;r}$ and $\gamma^{-1}_{p;(q\boxdot r),s}$ from Definition 3.7.14 as well as the isomorphism

$$\mathcal{D}(p \boxdot (q \boxdot r), s) \cong \mathcal{D}((p \boxdot q) \boxdot r, s)$$

induced by associativity in $\mathcal{D}$. The third isomorphism is similar to the first, but also uses a symmetry isomorphism in V to permute X_p with $\mathcal{D}(q \boxdot r, x)$.

Left Unit: The *left unit isomorphism* λ_X is the V-natural isomorphism with components determined by the following composite of isomorphisms, explained below, for each $s \in \mathcal{D}$:

$$\begin{aligned}(J \otimes X)_s &= \int^{p,q} \mathcal{D}(p \boxdot q, s) \otimes (\mathcal{D}(e,p) \otimes X_q)\\ &\cong \int^q \Big(\int^p \mathcal{D}(p \boxdot q, s) \otimes \mathcal{D}(e,p) \Big) \otimes X_q\\ &\cong \int^q \mathcal{D}(q,s) \otimes X_q\\ &\cong X_s.\end{aligned}$$

The first isomorphism is given by associativity in V and interchanging order of limits. The second isomorphism is given by $\gamma_{e,s;q}$ (3.7.15) and the left unit isomorphism of $\mathcal{D}$. The final isomorphism is the V-Yoneda Density Theorem Theorem 3.7.8 for X.

Right Unit: The *right unit isomorphism* ρ_X is the V-natural isomorphism with components determined by the following composite for each $s \in \mathcal{D}$ defined similarly to those for λ_X:

$$\begin{aligned}(X \otimes J)_s &= \int^{q,p} \mathcal{D}(q \boxdot p, s) \otimes (X_q \otimes \mathcal{D}(e,p))\\ &\cong \int^q \Big(\int^p \mathcal{D}(q \boxdot p, s) \otimes \mathcal{D}(e,p) \Big) \otimes X_q\\ &\cong \int^q \mathcal{D}(q,s) \otimes X_q\\ &\cong X_s.\end{aligned}$$

The only significant difference between ρ_X and λ_X is that the first isomorphism above also requires use of the symmetry isomorphism in V

$$X_q \otimes \mathcal{D}(e,p) \xrightarrow[\cong]{\xi} \mathcal{D}(e,p) \otimes X_q.$$

Symmetry: The *symmetry isomorphism* $\xi_{X,Y}$ is the V-natural isomorphism with components determined by the following isomorphism for each $s \in \mathcal{D}$.

$$\begin{aligned}(X \otimes Y)_s &= \int^{p,q} \mathcal{D}(p \boxdot q, s) \otimes (X_p \otimes Y_q)\\ &\cong \int^{q,p} \mathcal{D}(q \boxdot p, s) \otimes (Y_q \otimes X_p)\\ &= (Y \otimes X)_s\end{aligned}$$

This isomorphism is induced by interchanging order of limits along with the symmetry isomorphisms

$$X_p \otimes Y_q \cong Y_q \otimes X_p \quad \text{and} \quad \mathcal{D}(p \boxdot q, s) \cong \mathcal{D}(q \boxdot p, s)$$

of V and $\mathcal{D}$, respectively. ◇

Now we give a series of results aiming to show that $\mathcal{D}$-V has a symmetric monoidal closed structure.

Proposition 3.7.19. *The convolution product and hom diagram of Definition 3.7.3 are functorial with respect to* V*-natural transformations.*

Proof. Suppose given V-natural transformations

$$\theta\colon X \longrightarrow X' \quad \text{and} \quad \eta\colon Y \longrightarrow Y'$$

for $X, X', Y, Y' \in \mathcal{D}\text{-}\mathsf{V}$. Then

$$\theta \otimes \eta \colon X \otimes Y \longrightarrow X' \otimes Y'$$

has components $(\theta \otimes \eta)_c$ for $c \in \mathcal{D}$ defined as

$$\int^{a,b} \mathcal{D}(a \boxdot b, c) \otimes (X_a \otimes Y_b) \xrightarrow{\int^{a,b} 1 \otimes (\theta_a \otimes \eta_b)} \int^{a,b} \mathcal{D}(a \boxdot b, c) \otimes (X'_a \otimes Y'_b).$$

Similarly,

$$\mathsf{Hom}_{\mathcal{D}}(\theta, \eta) \colon \mathsf{Hom}_{\mathcal{D}}(X', Y) \longrightarrow \mathsf{Hom}_{\mathcal{D}}(X, Y')$$

has components $\mathsf{Hom}_{\mathcal{D}}(\theta, \eta)_a$ for $a \in \mathcal{D}$ defined, via (3.7.12), as

$$\mathsf{Hom}_{\mathcal{D}}(X', Y)_a \cong \int_b [X'_b, Y_{a \boxdot b}] \xrightarrow{\int_b [\theta_b, \eta_{a \boxdot b}]} \int_b [X_b, Y'_{a \boxdot b}] \cong \mathsf{Hom}_{\mathcal{D}}(X, Y')_a.$$

To verify that these components $(\theta \otimes \eta)_c$ and $\mathsf{Hom}_{\mathcal{D}}(\theta, \eta)_a$ define V-natural transformations, we use the alternative characterization given in Lemma 2.1.11. For $\theta \otimes \eta$, this means we must verify the following version of (2.1.12) commutes for each pair of objects $c, c' \in \mathcal{D}$.

$$\begin{array}{ccc}
\mathcal{D}(c, c') & \xrightarrow{(X \otimes Y)_{c,c'}} & [(X \otimes Y)_c, (X \otimes Y)_{c'}] \\
\Big\downarrow {\scriptstyle (X' \otimes Y')_{c,c'}} & & \Big\downarrow {\scriptstyle [(X \otimes Y)_c, (\theta \otimes \eta)_{c'}]} \\
[(X' \otimes Y')_c, (X' \otimes Y')_{c'}] & \xrightarrow{[(\theta \otimes \eta)_c, (X' \otimes Y')_{c'}]} & [(X \otimes Y)_c, (X' \otimes Y')_{c'}]
\end{array}$$

The adjoint of the above diagram is the outer diagram below. The unlabeled morphisms are given by associators in V and universality of the coends; the morphisms m_* are induced by the composition of $\mathcal{D}$. To save space the monoidal products $\boxdot$ and $\otimes$ are denoted by juxtaposition.
(3.7.20)

$$\begin{array}{ccccc}
 & & \int^{a,b} \Big(\mathcal{D}(c,c')\mathcal{D}(ab,c)\Big)(X_a Y_b) & & \\
 & \nearrow & & \searrow {\scriptstyle m_*} & \\
\mathcal{D}(c,c')\Big(\int^{a,b} \mathcal{D}(ab,c)(X_a Y_b)\Big) & & \Big\downarrow {\scriptstyle \int^{a,b} (11)(\theta_a \eta_b)} & & \int^{a,b} \mathcal{D}(ab,c')(X_a Y_b) \\
\Big\downarrow {\scriptstyle 1\left(\int^{a,b} 1(\theta_a \eta_b)\right)} & & \int^{a,b} \Big(\mathcal{D}(c,c')\mathcal{D}(ab,c)\Big)(X'_a Y'_b) & & \Big\downarrow {\scriptstyle \int^{a,b} 1(\theta_a \eta_b)} \\
 & \nearrow & & \searrow {\scriptstyle m_*} & \\
\mathcal{D}(c,c')\Big(\int^{a,b} \mathcal{D}(ab,c)(X'_a Y'_b)\Big) & & & & \int^{a,b} \mathcal{D}(ab,c')(X'_a Y'_b)
\end{array}$$

The above diagram commutes by functoriality of $\otimes$ and naturality of the monoidal structure and coend morphisms. This shows that $\theta \otimes \eta$ is V-natural. The functoriality of $\otimes$ with respect to identities and composition of V-natural transformations follows from functoriality of $\otimes$ and universality of the coends.

A similar analysis shows that $\mathsf{Hom}_{\mathcal{D}}(\theta, \eta)$ is V-natural: the adjoint of the diagram (2.1.12) for $\mathsf{Hom}_{\mathcal{D}}$ is the following, explained below.

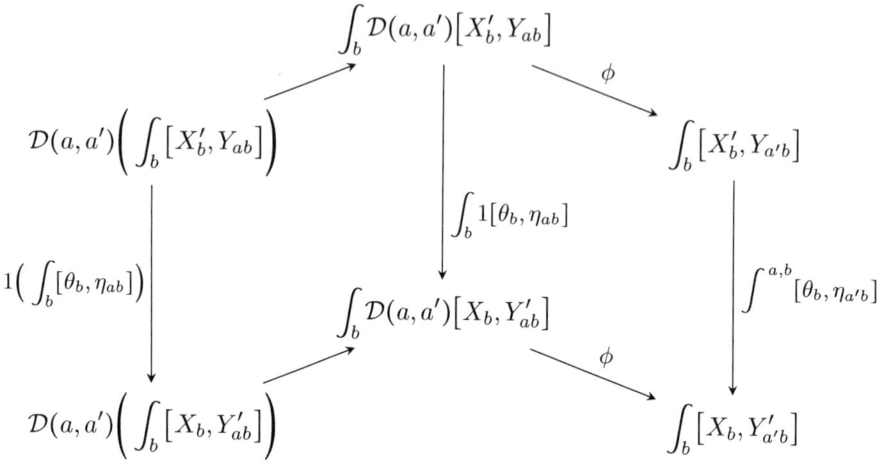

In the above diagram, the unlabeled morphisms are given by universality of the ends and the morphisms labeled ϕ are induced by the composites

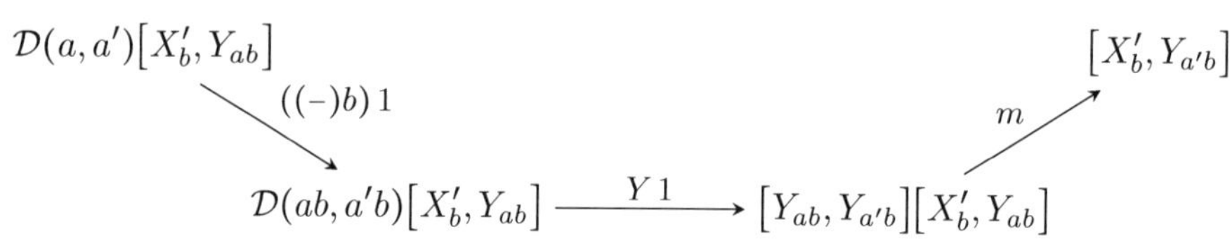

for each $b \in \mathcal{D}$. Commutativity of (3.7.20), and hence V-naturality of $\mathsf{Hom}_{\mathcal{D}}(\theta, \eta)$, follows as for $\theta \otimes \eta$. Then functoriality of $\mathsf{Hom}_{\mathcal{D}}$ with respect to θ (contravariant) and η (covariant) follows from functoriality of $[-,-]$ and universality of the ends. $\square$

Proposition 3.7.21. *The convolution product and hom diagram of Definition 3.7.3 are adjoint: for each $Y \in \mathcal{D}$-V there is an adjunction*

$$- \otimes Y \dashv \mathsf{Hom}_{\mathcal{D}}(Y, -).$$

Proof. We will show the following isomorphism in V for each triple of diagrams $X, Y, Z \in \mathcal{D}$-V:

$$\mathsf{Map}\big(X \otimes Y, Z\big) \cong \mathsf{Map}\big(X, \mathsf{Hom}_{\mathcal{D}}(Y, Z)\big).$$

The isomorphism of underlying sets of V-natural transformations then follows from Lemma 3.5.14.

Beginning with the definitions of Map (3.5.13) and $\otimes$ (3.7.4) we have the following sequence of isomorphisms, explained below:

$$\begin{aligned}\mathsf{Map}\big(X\otimes Y, Z\big) &= \int_c \big[(X\otimes Y)_c\,,\, Z_c\big] = \int_c \Big[\int^{a,b} \mathcal{D}(a\boxdot b, c)\otimes (X_a\otimes Y_b)\,,\, Z_c\Big]\\ &\cong \int_{c,a} \Big[X_a \otimes \Big(\int^b \mathcal{D}(a\boxdot b,c)\otimes Y_b\Big)\,,\, Z_c\Big]\\ &\cong \int_{a,c} \Big[X_a\,,\, \Big[\int^b \mathcal{D}(a\boxdot b,c)\otimes Y_b\,,\, Z_c\Big]\Big]\\ &\cong \int_a \Big[X_a\,,\, \int_c \Big[\int^b \mathcal{D}(a\boxdot b,c)\otimes Y_b\,,\, Z_c\Big]\Big]\\ &= \int_a \Big[X_a\,,\, \mathsf{Hom}_{\mathcal{D}}(Y,Z)_a\Big] = \mathsf{Map}\big(X, \mathsf{Hom}_{\mathcal{D}}(Y,Z)\big).\end{aligned}$$

After using the definitions in the first line, the middle three lines are given by:

(1) the isomorphism (3.5.10) and then associativity and symmetry in V followed by commuting $X_a \otimes -$ with coends;
(2) the closed structure of V and interchanging order of ends; and
(3) the isomorphism (3.5.11).

The final line is given by definition of the hom diagram (3.7.5) and of the mapping object. □

Theorem 3.7.22 ([**Day70**]). *Suppose* V *is a symmetric monoidal closed category that is complete and cocomplete. Suppose $\mathcal{D}$ is a small symmetric monoidal* V*-category. Then $\mathcal{D}$-*V *is a symmetric monoidal closed category with the following data.*

- *The monoidal product is given by the Day convolution product* (3.7.4).
- *The closed structure is given by the hom diagram* (3.7.5).
- *The monoidal unit is given by the diagram $J = \mathcal{Y}^e$* (3.7.18), *where e denotes the monoidal unit of $\mathcal{D}$.*
- *The unit, associativity, and symmetry isomorphisms are given by the components in Definition 3.7.17.*

Proof. Functoriality of the monoidal product and internal hom are discussed in Proposition 3.7.19. The symmetric monoidal axioms of Definitions 1.1.1 and 1.1.23 follow from

- the universal properties of (co)ends,
- symmetric monoidal axioms for $\mathcal{D}$ and V, and
- functoriality of the monoidal structures for $\mathcal{D}$ and V.

As an example, we prove the pentagon axiom (1.1.3) for $\mathcal{D}$-V. For $X, Y, Z, W \in \mathcal{D}$-V we compute below each of the five vertices of the pentagon. In each case, first we use the definition of $\otimes$ (3.7.4), second we interchange coends and use associativity along with (in all but the first case) symmetry in V, and third we use isomorphisms

γ from Definition 3.7.14. To save space, we denote $\boxdot$ and $\otimes$ by juxtaposition.

(3.7.23)

$$\begin{aligned}\Big[\big((XY)Z\big)W\Big]_t &= \int^{x,s} \mathcal{D}(xs,t)\bigg(\int^{y,r} \mathcal{D}(yr,x)\bigg(\bigg(\int^{p,q} \mathcal{D}(pq,y)(X_pY_q)\bigg)Z_r\bigg)W_s\bigg)\\ &\cong \int^{p,q,r,s}\bigg(\int^{x,y} \mathcal{D}(xs,t)\big(\mathcal{D}(yr,x)\mathcal{D}(pq,y)\big)\bigg)\Big(\big((X_pY_q)Z_r\big)W_s\Big)\\ &\cong \int^{p,q,r,s} \mathcal{D}\big(((pq)r)s,t\big)\Big(\big((X_pY_q)Z_r\big)W_s\Big)\end{aligned}$$

(3.7.24)

$$\begin{aligned}\big[(XY)(ZW)\big]_t &= \int^{x,y} \mathcal{D}(xy,t)\bigg(\bigg(\int^{p,q} \mathcal{D}(pq,x)(X_pY_q)\bigg)\bigg(\int^{r,s} \mathcal{D}(rs,y)(Z_rW_s)\bigg)\bigg)\\ &\cong \int^{p,q,r,s}\bigg(\int^{x,y} \mathcal{D}(xy,t)\big(\mathcal{D}(pq,x)\mathcal{D}(rs,y)\big)\bigg)\big((X_pY_q)(Z_rW_s)\big)\\ &\cong \int^{p,q,r,s} \mathcal{D}\big((pq)(rs),t\big)\big((X_pY_q)(Z_rW_s)\big)\end{aligned}$$

(3.7.25)

$$\begin{aligned}\big[X(Y(ZW))\big]_t &= \int^{p,x} \mathcal{D}(px,t)\bigg(X_p\bigg(\int^{q,y} \mathcal{D}(qy,x)\bigg(Y_q\bigg(\int^{r,s} \mathcal{D}(rs,y)(Z_rW_s)\bigg)\bigg)\bigg)\bigg)\\ &\cong \int^{p,q,r,s}\bigg(\int^{x,y} \mathcal{D}(px,t)\big(\mathcal{D}(qy,x)\mathcal{D}(rs,y)\big)\bigg)\Big(X_p\big(Y_q(Z_rW_s)\big)\Big)\\ &\cong \int^{p,q,r,s} \mathcal{D}\big(p(q(rs)),t\big)\Big(X_p\big(Y_q(Z_rW_s)\big)\Big)\end{aligned}$$

(3.7.26)

$$\begin{aligned}\big[(X(YZ))W\big]_t &= \int^{x,s} \mathcal{D}(xs,t)\bigg(\bigg(\int^{p,y} \mathcal{D}(py,x)\bigg(X_p\bigg(\int^{q,r} \mathcal{D}(qr,y)(Y_qZ_r)\bigg)\bigg)\bigg)W_s\bigg)\\ &\cong \int^{p,q,r,s}\bigg(\int^{x,y} \mathcal{D}(xs,t)\big(\mathcal{D}(py,x)\mathcal{D}(qr,y)\big)\bigg)\Big(\big(X_p(Y_qZ_r)\big)W_s\Big)\\ &\cong \int^{p,q,r,s} \mathcal{D}\big((p(qr))s,t\big)\Big(\big(X_p(Y_qZ_r)\big)W_s\Big)\end{aligned}$$

(3.7.27)

$$\begin{aligned}\big[X((YZ)W)\big]_t &= \int^{p,x} \mathcal{D}(px,t)\bigg(X_p\bigg(\int^{y,s} \mathcal{D}(ys,x)\bigg(\bigg(\int^{q,r} \mathcal{D}(qr,y)(Y_qZ_r)\bigg)W_s\bigg)\bigg)\bigg)\\ &\cong \int^{p,q,r,s}\bigg(\int^{x,y} \mathcal{D}(px,t)\big(\mathcal{D}(ys,x)\mathcal{D}(qr,y)\big)\bigg)\Big(X_p\big((Y_qZ_r)W_s\big)\Big)\\ &\cong \int^{p,q,r,s} \mathcal{D}\big(p((qr)s),t\big)\Big(X_p\big((Y_qZ_r)W_s\big)\Big)\end{aligned}$$

Then, by the universal property of coends, the pentagon axiom for each quadruple $X, Y, Z, W \in \mathcal{D}\text{-}\mathsf{V}$ follows from:

(1) the pentagon axiom for $\mathcal{D}$ (involving the components at p, q, r, and s),

(2) the pentagon axiom for V (involving the components at X_p, Y_q, Z_r, and W_s), and
(3) the functoriality of $\otimes$ in V.

The other symmetric monoidal category axioms for $\mathcal{D}$-V are proved in similar ways. The closed structure is provided by the $\otimes$-$\mathsf{Hom}_{\mathcal{D}}$ adjunction are given in Proposition 3.7.21. $\square$

Recall the concept of a symmetric monoidal V-functor from Definitions 1.4.17 and 1.4.18. We end this section with the following change-of-shape theorem.

Theorem 3.7.28. *Suppose* V *is a symmetric monoidal closed category that is complete and cocomplete. Suppose* $F\colon \mathcal{D} \longrightarrow \mathcal{E}$ *is a symmetric monoidal* V*-functor between small symmetric monoidal* V*-categories. Then precomposition with* F *is part of a symmetric monoidal functor*

$$\mathcal{E}\text{-}\mathsf{V} \xrightarrow{\ F^*\ } \mathcal{D}\text{-}\mathsf{V}.$$

Proof. Since F is a V-functor, precomposition with it yields a well-defined V-functor, $F^* = - \circ F$. The unit constraint of F^* is the following composite of V-natural transformations, with e and e' the monoidal units in, respectively, $\mathcal{D}$ and $\mathcal{E}$.

$$\begin{array}{ccc}
J^{\mathcal{D}} = \mathcal{D}(e,-) & \xrightarrow{\ (F^*)^0\ } & \mathcal{E}\big(e',F(-)\big) = J^{\mathcal{E}} \circ F \\
 & {\scriptstyle F}\searrow \qquad \nearrow{\scriptstyle (F^0)^*} & \\
 & \mathcal{E}\big(Fe,F(-)\big) &
\end{array}$$

Here $(F^0)^*$ is induced by the monoidal constraint F^0 of the monoidal V-functor (F,F^2,F^0).

The monoidal constraint $(F^*)^2$ of F^* is defined as follows. For $X,Y \in \mathcal{E}$-V, the (X,Y) component of $(F^*)^2$ is defined by the following commutative diagram in $\mathcal{D}$-V, with the monoidal products in $\mathcal{D}$, $\mathcal{E}$, and V denoted by juxtaposition.

$$\begin{array}{ccc}
F^*X \otimes F^*Y & \xrightarrow{\qquad (F^*)^2_{X,Y} \qquad} & F^*(X\otimes Y) \\
\| & & \| \\
\int^{(a,b)\in\mathcal{D}\otimes\mathcal{D}} \mathcal{D}(ab,-)(X_{Fa}Y_{Fb}) & & \int^{(c,d)\in\mathcal{E}\otimes\mathcal{E}} \mathcal{E}\big(cd,F(-)\big)(X_cY_d) \\
{\scriptstyle \omega_{a,b}}\uparrow & & \uparrow{\scriptstyle \omega_{Fa,Fb}} \\
\mathcal{D}(ab,-)(X_{Fa}Y_{Fb}) & & \mathcal{E}\big((Fa)(Fb),F(-)\big)(X_{Fa}Y_{Fb}) \\
 & {\scriptstyle F\otimes 1}\searrow \qquad \nearrow{\scriptstyle (F^2_{a,b})^*\otimes 1} & \\
 & \mathcal{E}\big(F(ab),F(-)\big)(X_{Fa}Y_{Fb}) &
\end{array}$$

The two vertical morphisms ω are V-coend structure morphisms. The naturality of $(F^*)^2$ follows from the V-naturality of F^2. The axioms in Definitions 1.1.6 and 1.1.17 for $\big(F^*,(F^*)^2,(F^*)^0\big)$ to be a symmetric monoidal functor follow from the corresponding axioms for the symmetric monoidal V-functor (F,F^2,F^0) in Definitions 1.4.17 and 1.4.18 $\square$

3.8. Enriched Diagram Categories

Throughout this section we continue to let V denote a symmetric monoidal closed category that is complete and cocomplete.

Lemma 3.8.1. *In the context of Definition 3.7.17, evaluation at e defines a symmetric monoidal functor*

$$\mathsf{ev}_e \colon \mathcal{D}\text{-}\mathsf{V} \longrightarrow \mathsf{V}.$$

Proof. For $X, Y \in \mathcal{D}\text{-}\mathsf{V}$ we define $\mathsf{ev}_e X = X_e$ and for a natural transformation of V-functors we let ev_e take the component at e. The monoidal constraint

$$X_e \otimes Y_e \xrightarrow{\mathsf{ev}_e^2} (X \otimes Y)_e = \int^{p,q} \mathcal{D}(p \boxdot q, e) \otimes \big(X_p \otimes Y_q\big)$$

is given by the following composite with the structure morphism $\omega_{(e,e)}$ for the coend:

$$X_e \otimes Y_e \longrightarrow \mathcal{D}(e \boxdot e, e) \otimes \big(X_e \otimes Y_e\big) \xrightarrow{\omega_{(e,e)}} \int^{p,q} \mathcal{D}(p \boxdot q, e) \otimes \big(X_p \otimes Y_q\big). \tag{3.8.2}$$

The unlabeled morphism above is given by the inverse of the left unitor and tensoring with the composite

$$\mathbb{1} \xrightarrow{i_e} \mathcal{D}(e,e) \xrightarrow{\cong} \mathcal{D}(e \boxdot e, e)$$

induced by the identity and unitors for e in $\mathcal{D}$.

Each of the symmetric monoidal axioms for ev_e follows from universality of the coends and a corresponding commutativity of a diagram of components at e. For example, the symmetry axiom follows from commutativity of

$$\begin{array}{ccc} X_e \otimes Y_e & \xrightarrow{\quad\xi\quad} & Y_e \otimes X_e \\ \downarrow & & \downarrow \\ \mathcal{D}(e \boxdot e, e) \otimes \big(X_e \otimes Y_e\big) & \xrightarrow{\cong} & \mathcal{D}(e \boxdot e, e) \otimes \big(Y_e \otimes X_e\big) \end{array}$$

In the above diagram, the vertical morphisms are given as in (3.8.2) and the unlabeled horizontal morphism is induced by the symmetry on $e \boxdot e$ (which is the identity) and on $X_e \otimes Y_e$. The diagram therefore commutes by naturality of the left unit isomorphism. The associativity and unity axioms for ev_e follow similarly. $\square$

For a symmetric monoidal functor $U \colon \mathsf{V} \longrightarrow \mathsf{W}$, recall the change of enrichment 2-functor $(-)_U \colon \mathsf{V}\text{-}\mathsf{Cat} \longrightarrow \mathsf{W}\text{-}\mathsf{Cat}$ from Definition 2.1.1.

Explanation 3.8.3 (Various Enrichments for $\mathcal{D}\text{-}\mathsf{V}$)**.** We now have three interpretations of the notation $\mathcal{D}\text{-}\mathsf{V}$:

(1) In the notation of Definition 3.7.1 and Theorem 3.7.22, $\mathcal{D}\text{-}\mathsf{V}$ denotes the symmetric monoidal closed category of V-functors and V-natural transformations from $\mathcal{D}$ to $\underline{\mathsf{V}}$.

(2) Via the canonical self-enrichment, $\mathcal{D}\text{-}\mathsf{V}$ is symmetric monoidal as a $\mathcal{D}\text{-}\mathsf{V}$-category with hom objects given by (3.7.12)

$$\mathsf{Hom}(X,Y) \cong \int_{b \in \mathcal{D}} [X_b, Y_{-\boxdot b}]$$

for $X, Y \in \mathcal{D}\text{-}\mathsf{V}$.

(3) In the notation of Definition 3.5.19 and Lemma 3.5.24, $\mathcal{D}$-V denotes the V-enriched category with

$$\mathcal{D}\text{-}\mathsf{V}(X, Y) = \mathsf{Map}(X, Y).$$

We discussed items (1) and (3) in Explanation 3.7.2 above: by Lemma 3.5.14, (1) is the underlying category of (3). Now we observe, by (3.7.13), that (3) is obtained from (2) by change of enrichment along the symmetric monoidal functor ev_e. This last observation shows that (3) is symmetric monoidal as a V-category by Theorem 2.4.10. ◇

Next we show that change of enrichment along a symmetric monoidal functor $U: \mathsf{V} \longrightarrow \mathsf{W}$ defines a symmetric monoidal functor on diagram categories. Recall from Proposition 3.3.4 that the standard enrichment of U defines a symmetric monoidal W-functor $\mathsf{V}_U \longrightarrow \mathsf{W}$.

Proposition 3.8.4. *Suppose* V *and* W *are complete and cocomplete symmetric monoidal closed categories. Suppose* $U: \mathsf{V} \longrightarrow \mathsf{W}$ *is a symmetric monoidal functor. Then change of enrichment and composition with the standard enrichment of* U *defines a symmetric monoidal functor*

$$U_*: \mathcal{D}\text{-}\mathsf{V} \longrightarrow \mathcal{D}_U\text{-}\mathsf{W}$$

Proof. The underlying functor U_* is given by

$$\mathsf{V}\text{-}\mathsf{Cat}(\mathcal{D}, \mathsf{V}) \xrightarrow{(-)_U} \mathsf{W}\text{-}\mathsf{Cat}(\mathcal{D}_U, \mathsf{V}_U) \xrightarrow{U \circ -} \mathsf{W}\text{-}\mathsf{Cat}(\mathcal{D}_U, \mathsf{W}).$$

For $X \in \mathcal{D}$-V, the W-enriched diagram U_*X sends $p \in \mathcal{D}_U$ to $(U_*X)_p = U(X_p)$. In our work below we will omit parentheses from such expressions. For $p, q \in \mathcal{D}$ the morphism on hom objects

$$(U_*X)_{p,q}: U\mathcal{D}(p,q) \longrightarrow [UX_p, UX_q]$$

is the morphism in W adjoint to the following composite:

$$\begin{array}{ccc} U(\mathcal{D}(p,q) \otimes X_p) & \xrightarrow{U(X_{p,q} \otimes 1)} & U([X_p, X_q] \otimes X_p) \\ {\scriptstyle U^2} \nearrow & & \searrow {\scriptstyle U\mathsf{eval}} \\ U\mathcal{D}(p,q) \otimes UX_p & & UX_q. \end{array}$$

For $X, Y \in \mathcal{D}$-V the monoidal constraint

$$(U_*X) \otimes (U_*Y) \xrightarrow{U_*^2} U_*(X \otimes Y)$$

is given by components for each $s \in \mathcal{D}$ via the following composite. The first morphism below uses the monoidal constraint of U (twice) and the second uses the universal property of the coend.

$$\begin{array}{ccc}
\big((U_*X)\otimes(U_*Y)\big)_s & \xrightarrow{\quad U_*^2 \quad} & U_*(X\otimes Y)_s \\
\| & & \| \\
\int^{p,q}\mathcal{D}_U(p\boxdot q,s)\otimes\big((UX_p)\otimes(UY_q)\big) & & U\Big(\int^{p,q}\mathcal{D}(p\boxdot q,s)\otimes(X_p\otimes Y_q)\Big) \\
& \searrow \qquad \nearrow & \\
& \int^{p,q}U\big(\mathcal{D}(p\boxdot q,s)\otimes(X_p\otimes Y_q)\big) &
\end{array}$$

We take the unit constraint U_*^0 to be the identity because

$$U_*J = U\mathcal{D}(e,-)$$

is the monoidal unit of $\mathcal{D}_U$-W.

The symmetric monoidal functor axioms for U_* follow from those of U together with universality of coends. We illustrate this with the left unity axiom (1.1.10). The right unity, associativity, and symmetry axioms for U_* follow similarly.

The left unity axiom for U_* requires that the following diagram commute for each $s\in\mathcal{D}$.

$$\begin{array}{ccc}
\big(UJ\otimes U_*X\big)_s & \xrightarrow{\quad \lambda_{U_X X} \quad} & U_*X_s \\
{\scriptstyle 1}\downarrow & & \uparrow{\scriptstyle U_*\lambda_X} \\
\int^{p,q}U\mathcal{D}(p\boxdot q,s)\otimes\big(U\mathcal{D}(e,p)\otimes UX_q\big) & \xrightarrow{U_*^2} & U\Big(\int^{p,q}\mathcal{D}(p\boxdot q,s)\otimes(\mathcal{D}(e,p)\otimes X_q)\Big)
\end{array}$$

Commutativity of the above diagram follows from commutativity of the following two diagrams for $p,q,s\in\mathcal{D}$, which we explain below.

(3.8.5)
$$\begin{array}{ccc}
U\mathcal{D}(p\boxdot q,s)\otimes U\mathcal{D}(e,p) & \xrightarrow{\quad \gamma \quad} & U\mathcal{D}(e\boxdot q,s) \\
{\scriptstyle U^2}\downarrow & \nearrow{\scriptstyle U\gamma} & \\
U\big(\mathcal{D}(p\boxdot q,s)\otimes\mathcal{D}(e,p)\big) & &
\end{array}$$

(3.8.6)
$$\begin{array}{ccc}
\int^{q}U\mathcal{D}(q,s)\otimes UX_q & \xrightarrow{\quad \cong \quad} & UX_s \\
\downarrow & & \uparrow{\scriptstyle\cong} \\
\int^{q}U\big(\mathcal{D}(q,s)\otimes X_q\big) & \xrightarrow{\quad \exists! \quad} & U\Big(\int^{q}\mathcal{D}(q,s)\otimes X_q\Big)
\end{array}$$

The diagram (3.8.5) compares the morphisms γ of Definition 3.7.14 for $\mathcal{D}_U$ and $\mathcal{D}$. Commutativity follows from the definition of composition in $\mathcal{D}_U$ via U^2. The diagram (3.8.6) compares the isomorphisms of the V-Yoneda Density Theorem 3.7.8 for UX and X. Commutativity follows from universality of the relevant coends. □

One family of important special cases is those arising from ordinary symmetric monoidal categories $\mathcal{D}$ via the following change of enrichment.

Definition 3.8.7. Let

$$F_{\mathbb{1}} \colon (\mathsf{Set}, \times, *) \longrightarrow (\mathsf{V}, \otimes, \mathbb{1}) \tag{3.8.8}$$

denote the strictly unital strong symmetric monoidal functor that sends a set X to $\coprod_X \mathbb{1}$. The unit constraint is $1_{\mathbb{1}}$. The monoidal constraint

$$FX \otimes FY \xrightarrow{F^2_{X,Y}} F(X \times Y)$$

is the following composite

$$\Big(\coprod_X \mathbb{1}\Big) \otimes \Big(\coprod_Y \mathbb{1}\Big) \xrightarrow{\cong} \coprod_{X\times Y} (\mathbb{1} \otimes \mathbb{1}) \xrightarrow{\cong} \coprod_{X\times Y} \mathbb{1}.$$

In the above composite, the first morphism is the canonical isomorphism commuting $\otimes$ with small coproducts (since V is symmetric monoidal closed) and the second isomorphism is given by the left unit isomorphism in V. ◇

Definition 3.8.9. Suppose $(\mathcal{D}, \boxdot)$ is a small symmetric monoidal category. The *unitary enrichment* of $\mathcal{D}$ over V is denoted $\mathcal{D}_{\mathbb{1}}$ and is defined as the change of enrichment for $\mathcal{D}$, in the sense of Definition 2.1.1 and Proposition 2.1.2, along the functor $F_{\mathbb{1}}$ of (3.8.8) above.

So $\mathcal{D}_{\mathbb{1}}$ has the same objects as $\mathcal{D}$ and hom objects given by the coproduct in V

$$\mathcal{D}_{\mathbb{1}}(b,c) = \coprod_{p \in \mathcal{D}(b,c)} \mathbb{1}. \tag{3.8.10}$$

By Theorem 2.4.10, $\mathcal{D}_{\mathbb{1}}$ is a symmetric monoidal V-category. ◇

Recall from Lemma 2.1.5 the corepresented functor

$$\mathsf{V}(\mathbb{1}, -) \colon \mathsf{V} \longrightarrow \mathsf{Set}$$

is symmetric monoidal.

Lemma 3.8.11. *In the context of Definition 3.8.7, the corepresented functor*

$$\mathsf{V}(\mathbb{1}, -) \colon \mathsf{V} \longrightarrow \mathsf{Set}$$

is right adjoint to $F_{\mathbb{1}}$.

Proof. The unit has component at a set X

$$\eta_X \colon X \longrightarrow \mathsf{V}\Big(\mathbb{1}, \coprod_X \mathbb{1}\Big)$$

which sends an element $i \in X$ to

$$\kappa_i \colon \mathbb{1} \longrightarrow \coprod_X \mathbb{1},$$

the canonical inclusion at summand i. The counit has component at an object A of V

$$\varepsilon_A \colon F_{\mathbb{1}}\big(\mathsf{V}(\mathbb{1}, A)\big) = \coprod_{\mathsf{V}(\mathbb{1},A)} \mathbb{1} \longrightarrow A$$

induced, for each summand $f \in \mathsf{V}(\mathbb{1}, A)$, by f.

The right triangle identity A in V is the composite

$$\mathsf{V}(\mathbb{1},A) \longrightarrow \mathsf{V}(\mathbb{1}, \coprod_{\mathsf{V}(\mathbb{1},A)} \mathbb{1}) \longrightarrow \mathsf{V}(\mathbb{1},A).$$

This composite is the identity because each $f \in \mathsf{V}(\mathbb{1},A)$ is sent to the composite

$$\mathbb{1} \xrightarrow{\kappa_f} \coprod_{\mathsf{V}(\mathbb{1},A)} \mathbb{1} \xrightarrow{\varepsilon_A} A$$

that, by definition of ε, is f. The left triangle identity for X in Set is the composite

$$\coprod_X \mathbb{1} \longrightarrow \coprod_{\mathsf{V}(\mathbb{1},\coprod_X \mathbb{1})} \mathbb{1} \longrightarrow \coprod_X \mathbb{1}.$$

This composite is the identity because the summand $\mathbb{1}$ at $i \in X$ maps via the identity on $\mathbb{1}$ to the summand at κ_i in the middle term and, by definition of ε, this summand maps via κ_i to $\coprod_X \mathbb{1}$. □

Proposition 3.8.12. *In the context of Definition 3.8.9, there is an isomorphism of categories*

$$\mathsf{Cat}(\mathcal{D},\mathsf{V}) \cong \mathsf{V}\text{-}\mathsf{Cat}(\mathcal{D}_{\mathbb{1}},\underline{\mathsf{V}}) = (\mathcal{D}_{\mathbb{1}})\text{-}\mathsf{V}. \tag{3.8.13}$$

Proof. This is the isomorphism on hom categories of the change of enrichment adjunction induced by

$$F_{\mathbb{1}} \dashv \mathsf{V}(\mathbb{1},-)$$

of Lemma 3.8.11. More explicitly, a functor of categories

$$X\colon \mathcal{D} \longrightarrow \mathsf{V}$$

corresponds to a V-functor

$$X_{\mathbb{1}}\colon \mathcal{D}_{\mathbb{1}} \longrightarrow \underline{\mathsf{V}}$$

with the same assignment on objects and with

$$\mathcal{D}_{\mathbb{1}}(a,b) = \coprod_{p\in\mathcal{D}(a,b)} \mathbb{1} \xrightarrow{(X_{\mathbb{1}})_{a,b}} [X_a, X_b]$$

given by the bijection

$$\mathsf{V}(\mathbb{1},[X_a,X_b]) \cong \mathsf{V}(X_a,X_b)$$

for each pair of objects a and b in $\mathcal{D}$. The bijection of natural and V-natural transformations is similar. □

Example 3.8.14 (Diagrams in Cat)**.** For this example, suppose

$$(\mathsf{V},\otimes,\mathbb{1}) = (\mathsf{Cat},\times,\mathbb{1}),$$

the category of small categories with the Cartesian product and terminal category as its unit. Suppose $\mathcal{D}$ is an ordinary symmetric monoidal 1-category, regarded as a symmetric Cat-monoidal 2-category with identity 2-cells. Then $\mathcal{D}$ has the unitary enrichment over Cat. Suppose given diagrams

$$X, Y\colon \mathcal{D} \longrightarrow \mathsf{Cat}.$$

We have the following simplification in this case, for objects b and c in $\mathcal{D}$:

$$\Big[\mathcal{D}(b,c),[X_b,Y_c]\Big] \cong \Big[\coprod_{p\in\mathcal{D}(b,c)} X_b, Y_c\Big] \cong \prod_{p\in\mathcal{D}(b,c)} [X_b,Y_c] \tag{3.8.15}$$

where the coproduct and product are indexed over morphisms in $\mathcal{D}$ and $[-,-]$ denotes the category of functors and natural transformations—the closed structure of Cat.

Recall from (3.7.13) we have

$$\mathsf{Hom}_{\mathcal{D}}(X,Y)_e \cong \mathsf{Map}(X,Y).$$

Using the simplification (3.8.15) and the equalizer formula (3.5.7), we have

$$\mathsf{Map}(X,Y) \cong \mathsf{eq}\Big(\prod_b [X_b,Y_b] \rightrightarrows \prod_{p:b \to c} [X_b,Y_c]\Big). \tag{3.8.16}$$

The two arrows in the equalizer are given, for each component $p \in \mathcal{D}(b,c)$, by pre- and post-composition with X_p and Y_p. Therefore a tuple of morphisms

$$\alpha_b \colon X_b \longrightarrow Y_b$$

determines an object of $\mathsf{Map}(X,Y)$ if and only if the following diagram in Cat commutes for each morphism p.

$$\begin{array}{ccc} X_b & \xrightarrow{X_p} & X_c \\ {\scriptstyle \alpha_b}\downarrow & & \downarrow{\scriptstyle \alpha_c} \\ Y_b & \xrightarrow{Y_p} & Y_c \end{array} \tag{3.8.17}$$

This is precisely the 1-cell naturality condition (I.6.2.24) for α to define a 2-natural transformation from X to Y. The 2-cell naturality condition (I.6.2.25) is trivially satisfied because $\mathcal{D}$ has only identity 2-cells. Similarly, the morphisms in $\mathsf{Map}(X,Y)$ correspond to tuples of natural transformations, Γ_b, such that the whiskerings with X_p or Y_p are equal. This is precisely the modification axiom (I.6.3.2) for Γ to be a modification between 2-natural transformations. ◇

3.9. Tensored and Cotensored Enriched Categories

In this section we define and prove basic results about tensor and cotensor structures.

Definition 3.9.1 (Tensored and Cotensored)**.** Suppose V is a symmetric monoidal closed category and suppose C is a V-category.

Tensored: We say that C is *tensored over* V if C is equipped, for each $X \in \mathsf{C}$ and $A \in \mathsf{V}$, with an object $X \otimes A \in \mathsf{C}$ together with isomorphisms in V

$$\mathsf{C}(X \otimes A, Y) \cong [A, \mathsf{C}(X,Y)]$$

for all $Y \in \mathsf{C}$.

Cotensored: We say that C is *cotensored over* V if C is equipped, for each $X \in \mathsf{C}$ and $A \in \mathsf{V}$, with an object $X^A \in \mathsf{C}$ together with isomorphisms in V

$$\mathsf{C}(Y, X^A) \cong [A, \mathsf{C}(Y,X)]$$

for all $Y \in \mathsf{C}$. ◇

Example 3.9.2. The closed structure for V makes the self-enriched V-category $\underline{\mathsf{V}}$ tensored and cotensored over V with tensor given by the monoidal product and cotensor given by the closed structure. Isomorphisms

$$[X \otimes Y, Z] \cong [X, [Y,Z]]$$

for X, Y, and Z in V follow from the Yoneda Lemma, the closed structure, and the associativity in V. Indeed, for $W \in \mathsf{V}$ we have

$$\begin{aligned} \mathsf{V}(W,[X\otimes Y,Z]) \cong \mathsf{V}(W\otimes(X\otimes Y),Z) \xrightarrow[\cong]{\alpha^*} & \mathsf{V}((W\otimes X)\otimes Y,Z) \\ & \cong \mathsf{V}(W\otimes X,[Y,Z]) \\ & \cong \mathsf{V}(W,[X,[Y,Z]]). \end{aligned}$$

Then the tensor and cotensor isomorphisms of Definition 3.9.1 follow from the symmetry of V. ◇

Although Definition 3.9.1 does not assert any particular compatibility of the tensor/cotensor isomorphisms, the following result uses Proposition 3.4.15 to show that the tensor and cotensor determine V-adjunctions.

Proposition 3.9.3. *In the context of Definition 3.9.1, for each X and Y in V there are unique extensions of the tensor and cotensor to V-functors*

$$\underline{\mathsf{V}} \xrightarrow{X\otimes -} \mathsf{C} \quad \textit{and} \quad \underline{\mathsf{V}} \xrightarrow{Y^{(-)}} \mathsf{C}^{\mathsf{op}},$$

respectively, such that the following are V-adjunctions

$$(X\otimes -) \dashv \mathcal{Y}^X \quad \textit{and} \quad Y^{(-)} \dashv \mathcal{Y}_Y.$$

Proof. We first note that, by the Yoneda Lemma for ordinary categories, the tensors and cotensors on underlying categories extend uniquely to bifunctors

$$\mathsf{C}_0 \times \mathsf{V} \xrightarrow{-\otimes -} \mathsf{C}_0 \quad \text{and} \quad \mathsf{C}_0 \times \mathsf{V}^{\mathsf{op}} \xrightarrow{(-)^{(-)}} \mathsf{C}_0$$

such that for each $X, Y \in \mathsf{V}$ there are adjunctions

$$\mathsf{V} \underset{\mathsf{C}(X,-)}{\overset{X\otimes -}{\underset{\perp}{\rightleftarrows}}} \mathsf{C}_0 \quad \text{and} \quad \mathsf{V} \underset{\mathsf{C}(-,Y)}{\overset{Y^{(-)}}{\underset{\perp}{\rightleftarrows}}} \mathsf{C}_0^{\mathsf{op}}$$

of underlying categories. It suffices to show that the underlying functors $X\otimes -$ and $Y^{(-)}$ can be extended to V-functors. Then the result follows from Proposition 3.4.15.

We will give the argument for $X\otimes -$; the argument for $Y^{(-)}$ is similar and formally dual. To begin, we first observe the following details for $X\otimes -$. Both follow from the details noted in Explanation 3.1.2. Here and throughout the rest of the proof we omit tensor symbols to save space.

For each $Y \in \mathsf{C}$ and $P, A \in \mathsf{V}$, the tensor and closed structures give isomorphisms in V

$$\mathsf{C}(X(PA),Y) \cong [PA,\mathsf{C}(X,Y)] \cong [P,[A,C(X,Y)]] \cong [P,\mathsf{C}(XA,Y)] \cong \mathsf{C}((XA)P,Y).$$

The induced bijection of morphisms in C_0

$$\mathsf{C}_0(X(PA),Y) \cong \mathsf{C}_0((XA)P,Y) \tag{3.9.4}$$

is natural with respect to $Y \in \mathsf{C}_0$ and $P, A \in \mathsf{V}$ by naturality of the underlying adjunction. Therefore, by the Yoneda Lemma for ordinary categories, the bijection (3.9.4) is induced by an isomorphism $w = w_{P,A}$ in C_0

$$(XA)P \xrightarrow[\cong]{w_{P,A}} X(PA)$$

that is natural with respect to $P, A \in \mathsf{V}$. Unwinding the relevant adjunctions and using $C = \mathsf{C}(X, Y)$ in (3.1.4), the following diagram in C_0 commutes for each $P, Q, A \in \mathsf{V}$.

(3.9.5)
$$\begin{array}{c} ((XA)Q)P \xrightarrow{w_{Q,A} \otimes P} (X(QA))P \xrightarrow{w_{P,QA}} X(P(QA)) \\ ((XA)Q)P \xrightarrow{w_{P,Q}} (XA)(PQ) \xrightarrow{w_{PQ,A}} X((PQ)A) \xrightarrow{X\alpha_{P,Q,A}} X(P(QA)) \end{array}$$

Similarly, there is a morphism
$$j_X \in \mathsf{C}_0(X\mathbb{1}, X)$$
corresponding, under the following isomorphisms of the closed and tensor structures, to $1_X \in \mathsf{C}_0(X, X)$:
$$\mathsf{V}(\mathbb{1}, \mathsf{C}(X,X)) \xrightarrow{\lambda^*} \mathsf{V}(\mathbb{1}\mathbb{1}, \mathsf{C}(X,X)) \xrightarrow{\cong} \mathsf{V}(\mathbb{1}, [\mathbb{1}, \mathsf{C}(X,X)]) \xrightarrow{\cong} \mathsf{V}(\mathbb{1}, \mathsf{C}(X\mathbb{1}, X)).$$
Unwinding the relevant adjunctions and using $C = \mathsf{C}(X, Y)$ in (3.1.3), the following diagram in C_0 commutes for each $A \in \mathsf{V}$.

(3.9.6)
$$(XA)\mathbb{1} \xrightarrow{w_{\mathbb{1},A}} X(\mathbb{1}A) \xrightarrow{X\lambda} XA \qquad = \qquad (XA)\mathbb{1} \xrightarrow{j_{XA}} XA$$

Now we give the definition of $X \otimes -$ on hom objects. For each $A, A', P \in \mathsf{V}$ we have

(3.9.7)
$$\begin{aligned} \mathsf{V}(P, [A, A']) \cong \mathsf{V}(PA, A') &\longrightarrow \mathsf{V}(\mathbb{1}, \mathsf{C}(X(PA), XA')) \cong \mathsf{V}(\mathbb{1}, \mathsf{C}((XA)P, XA')) \\ &\cong \mathsf{V}(\mathbb{1}, [P, \mathsf{C}(XA, XA')]) \cong \mathsf{V}(P, \mathsf{C}(XA, XA')) \end{aligned}$$

where:

- the first and last isomorphisms are given by the closed structure and unit isomorphism of V,
- the arrow on the first row is given by the underlying functor $X \otimes -$,
- the second isomorphism in the first row is given by precomposition with $w_{P,A}$, and
- the first isomorphism on the second row is given by the tensor structure of C.

Taking $P = [A, A']$, the identity $1_{[A,A']}$ corresponds to a morphism in V that we take as the definition of $X \otimes -$ on hom objects:
$$[A, A'] \xrightarrow{(X \otimes -)_{A,A'}} \mathsf{C}(XA, XA').$$
Now we verify the axioms of Definition 1.2.4 for $X \otimes -$.

The composition axiom (1.2.5) follows from commutativity of the below outer diagram for each $P, Q, A, A', A'' \in \mathsf{V}$. Following the diagram we explain commutativity of each region.

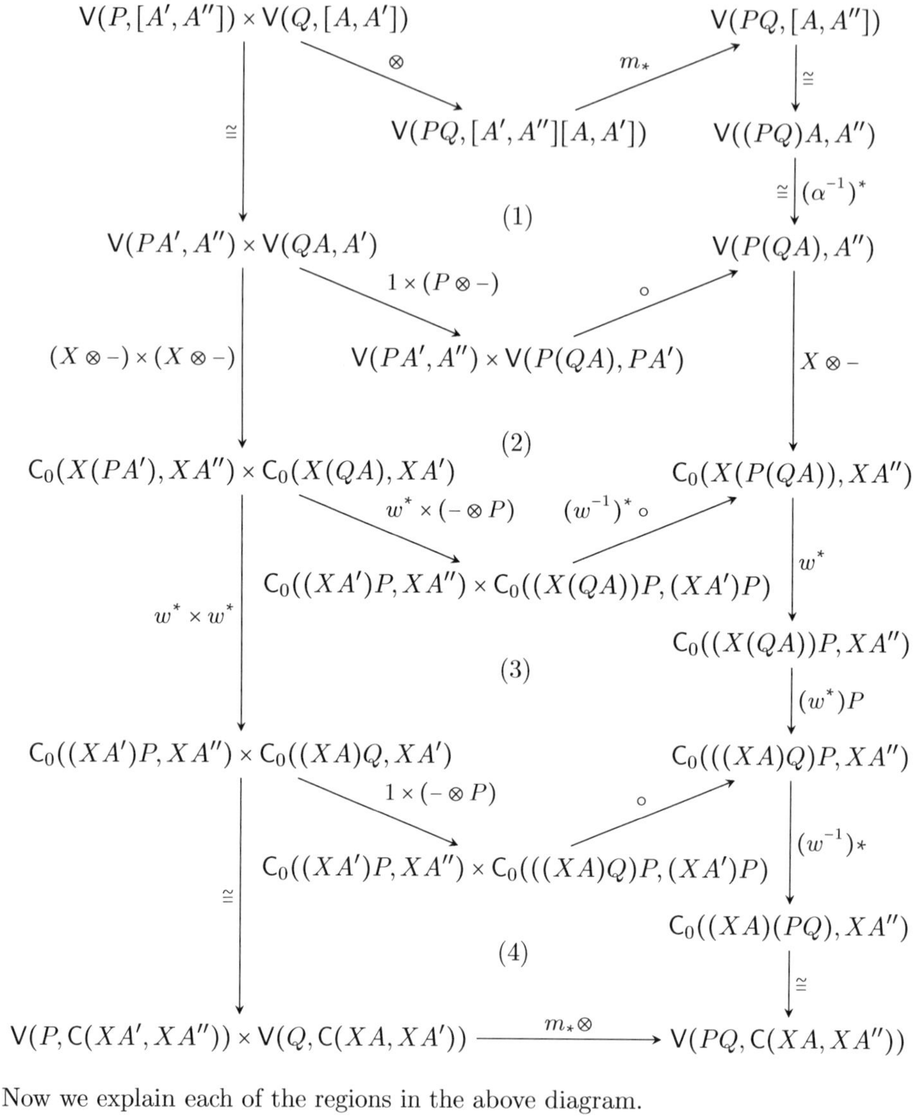

Now we explain each of the regions in the above diagram.

(1) The unlabeled vertical isomorphisms are given by the closed structure of V. The region commutes by functoriality of the monoidal product in V.

(2) The lower right morphism is given by composition in C_0 and precomposition with w^{-1}. Unwinding the definitions, commutativity of this region follows from the naturality noted in (3.9.4) for morphisms $QA \longrightarrow A'$ in V.

(3) Unwinding the definitions, commutativity of this diagram follows from underlying functoriality of $- \otimes P$ with respect to morphisms in C_0.

(4) The unlabeled vertical isomorphisms are from the closed structure and unit isomorphisms of V. The lower horizontal morphism is the monoidal

product of V followed by the composition in C. Commutativity of this diagram follows the same argument as (1), using underlying functoriality of $- \otimes P$.

Lastly, we observe that the vertical composite along the right is equal to $(X \otimes -)_{A,A''}$ by the compatibility of w with α noted in (3.9.5) above.

The identity axiom (1.2.6) for $(X \otimes -)$ follows from commutativity of the diagram below by evaluating at $i_A \in \mathsf{V}(\mathbb{1}, [A, A])$, as we explain below.

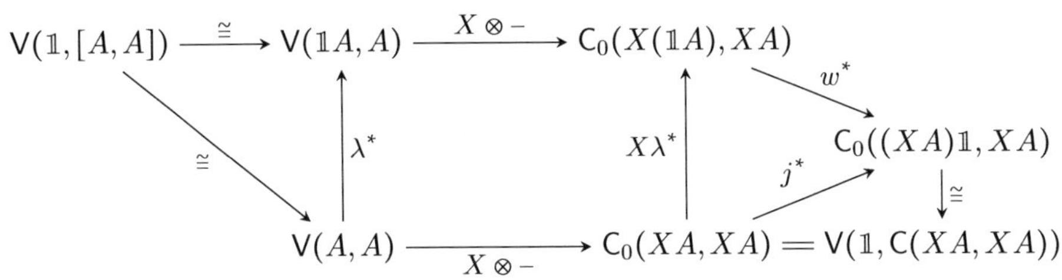

In the above diagram, the top and right composite is (3.9.7) with $A' = A$ and $P = \mathbb{1}$, and therefore the image of i_A under this composite is

$$\mathbb{1} \xrightarrow{i_A} [A, A] \xrightarrow{(X \otimes -)_{A,A}} \mathsf{C}(XA, XA).$$

The left and bottom composite is i_{XA} because the underlying functor $X \otimes -$ preserves identity morphisms. The middle rectangle commutes by underlying functoriality of $X \otimes -$. The triangle involving $X\lambda$, w, and j commutes by (3.9.6) above. The remaining triangles commute by definition. This completes the proof that $X \otimes -$ extends to a V-functor determined by the underlying tensor adjunction. □

Recall from Definition 2.1.1 that for V-category C and a monoidal functor

$$U : \mathsf{V} \longrightarrow \mathsf{W}$$

we have the W-category C_U with hom objects $\mathsf{C}_U(X, Y) = U\mathsf{C}(X, Y)$ for $X, Y \in \mathsf{C}$. The following result generalizes the change of enrichment for change of tensors and cotensors as well.

Theorem 3.9.8. *Suppose* V *and* W *are symmetric monoidal closed categories and suppose*

$$\mathsf{W} \underset{U}{\overset{F}{\rightleftarrows}} \mathsf{V} \quad (F \dashv U)$$

is an adjunction of monoidal functors, with F^2 invertible. If C *is enriched, tensored, and cotensored over* V, *then* C_U *is enriched, tensored, and cotensored over* W *with*

- $X \otimes A = X \otimes FA$ *and*
- $X^A = X^{FA}$

for $X \in \mathsf{C}$ *and* $A \in \mathsf{W}$.

Proof. The assertion that C_U is a W-category is proved in Proposition 2.1.2. To show that C_U is tensored over W it suffices, by the Yoneda Lemma for ordinary categories, to show, for each $X, Y \in \mathsf{C}$ and $A \in \mathsf{V}$, there are bijections of sets

$$\mathsf{W}(B, \mathsf{C}_U(X \otimes FA, Y)) \cong \mathsf{W}(B, [A, \mathsf{C}_U(X, Y)])$$

that are natural with respect to $B \in \mathsf{W}$. This we show by the following sequence of natural bijections:

$$\begin{aligned}
\mathsf{W}(B, \mathsf{C}_U(X \otimes FA, Y)) &= \mathsf{W}(B, U\mathsf{C}(X \otimes FA, Y)) \\
&\cong \mathsf{V}(FB, \mathsf{C}(X \otimes FA, Y)) && \text{since } F \dashv U, \\
&\cong \mathsf{V}(FB, [FA, \mathsf{C}(X,Y)]) && \text{since } \mathsf{C} \text{ tensored over } \mathsf{V}, \\
&\cong \mathsf{V}(FB \otimes FA, \mathsf{C}(X,Y)) && \text{since } \mathsf{V} \text{ closed monoidal}, \\
&\cong \mathsf{V}(F(B \otimes A), \mathsf{C}(X,Y)) && \text{since } F^2 \text{ invertible}, \\
&\cong \mathsf{W}(B \otimes A, \mathsf{C}_U(X,Y)) && \text{since } F \dashv U, \\
&\cong \mathsf{W}(B, [A, \mathsf{C}_U(X,Y)]) && \text{since } \mathsf{W} \text{ closed monoidal}.
\end{aligned}$$

A dual argument shows that C_U is cotensored over W. We have the following sequence of natural bijections:

$$\begin{aligned}
\mathsf{W}(B, \mathsf{C}_U(X, Y^{FA})) &= \mathsf{W}(B, U\mathsf{C}(X, Y^{FA})) \\
&\cong \mathsf{V}(FB, \mathsf{C}(X, Y^{FA})) && \text{since } F \dashv U, \\
&\cong \mathsf{V}(FB, [FA, \mathsf{C}(X,Y)]) && \text{since } \mathsf{C} \text{ cotensored over } \mathsf{V}, \\
&\cong \mathsf{V}(FB \otimes FA, \mathsf{C}(X,Y)) && \text{since } \mathsf{V} \text{ closed monoidal}, \\
&\cong \mathsf{V}(F(B \otimes A), \mathsf{C}(X,Y)) && \text{since } F^2 \text{ invertible}, \\
&\cong \mathsf{W}(B \otimes A, U\mathsf{C}(X,Y)) && \text{since } F \dashv U, \\
&\cong \mathsf{W}(B, [A, U\mathsf{C}(X,Y)]) && \text{since } \mathsf{W} \text{ closed monoidal}.
\end{aligned}$$

This completes the proof. □

Example 3.9.9. In the context of Theorem 3.9.8, the special case $\mathsf{C} = \underline{\mathsf{V}}$ shows that $\underline{\mathsf{V}}_U$ is enriched, tensored, and cotensored over W. ◇

One of our important applications of Theorem 3.9.8 will be in the following context.

- Suppose V is a complete and cocomplete symmetric monoidal closed category.
- Suppose $\mathcal{D}$ is a small symmetric monoidal V-category with monoidal unit e.

In this context, recalling Definition 3.7.1 and Theorem 3.7.22, we have a symmetric monoidal closed category $\mathcal{D}$-V with monoidal product given by Day convolution and monoidal unit given by the V-functor

$$J = \mathcal{Y}^e = \mathcal{D}(e, -) : \mathcal{D} \longrightarrow \underline{\mathsf{V}}.$$

In this context, by Lemma 3.8.1, there is a symmetric monoidal functor

$$\mathsf{ev}_e : \mathcal{D}\text{-}\mathsf{V} \longrightarrow \mathsf{V}$$

given by evaluation at e. Now we define a left adjoint for ev_e.

Definition 3.9.10. Suppose $(\mathsf{V}, \otimes, \mathbb{1})$ is a complete and cocomplete symmetric monoidal closed category. Suppose $(\mathcal{D}, \boxdot, e)$ is a small symmetric monoidal V-category. Define a strong symmetric monoidal functor

$$L_e \colon \mathsf{V} \longrightarrow \mathcal{D}\text{-}\mathsf{V}$$

as follows. For each object A of V, let

$$L_e A = A \otimes J,$$

the composite of V-functors

$$\mathcal{D} \xrightarrow{\ J\ } \underline{\mathsf{V}} \xrightarrow{\ A \otimes -\ } \underline{\mathsf{V}}. \tag{3.9.11}$$

For morphisms $A \longrightarrow B$ in V we have corresponding V-natural transformations

$$A \otimes - \longrightarrow B \otimes -.$$

Whiskering with J defines L_e on morphisms. Functoriality of L_e follows because whiskering is functorial with respect to vertical composition of 2-cells.

Recalling the symmetric monoidal data for $\underline{\mathsf{V}}$ from Definition 3.3.1 and Theorem 3.3.2, the monoidal unit for $\underline{\mathsf{V}}$ is given by $\mathbb{1}$ and provides a V-natural isomorphism

$$1_{\underline{\mathsf{V}}} \cong \mathbb{1} \otimes -$$

which then provides

$$L_e^0 \colon J \xrightarrow{\ \cong\ } \mathbb{1} \otimes J = L_e \mathbb{1}.$$

The monoidal constraint L_e^2 is given by the composite of V-natural isomorphisms indicated by the following pasting diagram involving

- functoriality of the tensor product of V-categories with respect to composition of V-functors (Proposition 1.3.6),
- the unit isomorphism λ for J, from Definition 3.7.17, and
- the unique permuted canonical V-map provided by the Enriched Monoidal Coherence Theorem 2.5.6.

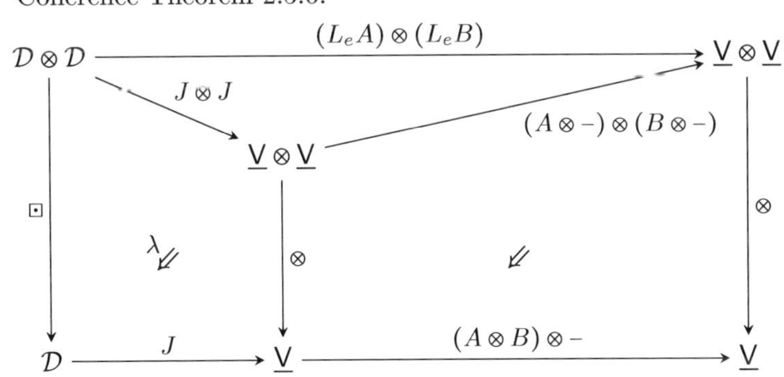

◇

Lemma 3.9.12. *In the context of Definition 3.9.10, L_e is a strong symmetric monoidal functor and provides an adjunction*

$$\mathsf{V} \underset{\mathsf{ev}_e}{\overset{L_e}{\rightleftarrows}} \mathcal{D}\text{-}\mathsf{V} \qquad (L_e \dashv \mathsf{ev}_e) \tag{3.9.13}$$

where ev_e is evaluation at the monoidal unit e, as in Lemma 3.8.1.

Proof. The associativity axiom (1.1.9) and unity axiom (1.1.10) for L_e hold by Enriched Monoidal Coherence Theorem 2.5.6 for $\underline{\mathsf{V}}$ and ordinary Monoidal Coherence Theorem 1.1.31 for $\mathcal{D}$-V. The symmetry axiom (1.1.18) for L_e follows similarly, using the symmetric case of Theorem 2.5.6.

For A in V we have unit components η_A given by the composite

$$A \xrightarrow[\cong]{\rho^{-1}} A \otimes \mathbb{1} \xrightarrow{1 \otimes 1_e} A \otimes \mathcal{D}(e,e) = (\mathsf{ev}_e L_e)A.$$

Naturality of η follows from naturality of ρ and functoriality of the monoidal product in V.

For $X \in \mathcal{D}$-V we have counit components

$$\varepsilon_X : (L_e \mathsf{ev}_e)X = X_e \otimes J \longrightarrow X$$

defined, as V-natural transformations, via components

$$\mathbb{1} \longrightarrow [\mathcal{D}(e,a), [X_e, X_a]] \xrightarrow{\cong} [X_e \otimes \mathcal{D}(e,a), X_a]$$

where the first morphism is the adjoint of $X_{e,a}$ and the second uses the symmetry and closed monoidal structure of V. The V-naturality of each component ε_X follows from functoriality of X. The naturality of ε with respect to morphisms (V-natural transformations) θ from X to Y in $\mathcal{D}$-V is the commutativity, for each a in $\mathcal{D}$, of the following diagram in V.

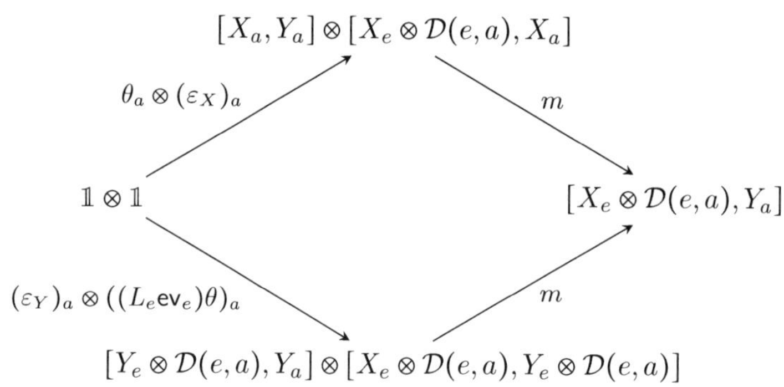

Taking adjoints, commutativity of the diagram above follows from V-naturality of θ.

The left triangle identity for A in V is the composite of V-natural transformations

$$A \otimes J \xrightarrow{\eta_A \otimes J} (A \otimes \mathcal{D}(e,e)) \otimes J \xrightarrow{\varepsilon} A \otimes J \tag{3.9.14}$$

whose component at each a in $\mathcal{D}$ is adjoint to the composite

$$A \otimes \mathcal{D}(e,a) \longrightarrow (A \otimes \mathcal{D}(e,e)) \otimes \mathcal{D}(e,a) \longrightarrow A \otimes \mathcal{D}(e,a)$$

given by composition with 1_e. Therefore (3.9.14) is the identity. The right triangle identity for X in $\mathcal{D}$-V is the composite

$$X_e \xrightarrow{\eta_{X_e}} X_e \otimes \mathcal{D}(e,e) \xrightarrow{(\varepsilon_x)_e} X_e$$

also given by composition with 1_e. This completes the proof. □

Applying Theorem 3.9.8 to the adjunction of (3.9.13) then gives the following.

Corollary 3.9.15. *Suppose* V *is a complete and cocomplete symmetric monoidal closed category. Suppose* $\mathcal{D}$ *is a small symmetric monoidal* V*-category with monoidal unit* e*. Then* $\mathcal{D}$-V *is symmetric monoidal closed and, by changing enrichment along* ev_e*, is enriched, tensored, and cotensored over* V*.*

Explanation 3.9.16. In the context of Corollary 3.9.15, one can give a separate, more direct, explanation for the tensoring and cotensoring of $\mathcal{D}$-V over V. Note that the following explanation does not depend on the monoidal structure of $\mathcal{D}$. Recall from Explanation 3.8.3 that we regard $\mathcal{D}$-V as a V-enriched category with hom objects

$$\mathsf{Map}(X,Y) = \int_b [X_b, Y_b]$$

for X and Y in $\mathcal{D}$-V. Then for A in V we define the tensor and cotensor via composition of V-functors

$$X \otimes A : \mathcal{D} \xrightarrow{X} \underline{\mathsf{V}} \xrightarrow{-\otimes A} \underline{\mathsf{V}}$$

and

$$Y^A : \mathcal{D} \xrightarrow{Y} \underline{\mathsf{V}} \xrightarrow{[A,-]} \underline{\mathsf{V}}$$

generalizing the definition of L_e from Definition 3.9.10.

Then for the isomorphisms making $\mathcal{D}$-V tensored over V we have the following, using the closed structure of V and (3.5.11)

$$\begin{aligned}
\mathsf{Map}(X \otimes A, Y) &= \int_b [(X \otimes A)_b, Y_b] \\
&= \int_b [X_b \otimes A, Y_b] \\
&\cong \int_b [A, [X_b, Y_b]] \\
&\cong \Big[A, \int_b [X_b, Y_b]\Big] \\
&= [A, \mathsf{Map}(X,Y)].
\end{aligned}$$

Similarly, for the cotensor isomorphisms we have

$$\begin{aligned}
\mathsf{Map}(X, Y^A) &= \int_b [X_b, (Y^A)_b] \\
&= \int_b [X_b, [A, Y_b]] \\
&\cong \int_b [A, [X_b, Y_b]] \\
&\cong [A, \mathsf{Map}(X,Y)].
\end{aligned}$$

◇

3.10. Notes

3.10.1 (Self-Enrichment)**.** Kelly shows a version of Proposition 3.1.11 in [**Kel05**, Section 1.6]. Our approach is more elementary and direct than that of [**Kel05**], which is in turn a simplification of general coherence methods of [**KML71a, KML71b, KML72**]. ◇

3.10.2 (Enriched Yoneda Results)**.** The V-Yoneda Bijection of Theorem 3.4.12 is often called the *Weak Yoneda Lemma.* The V-Yoneda Density Theorem 3.7.8 is often called simply the Density Theorem or the Yoneda Lemma. It is equivalent to the V-Yoneda Lemma 3.6.9. For further discussion of these results we refer the reader to [**DK69**], [**Kel05**, Sections 1.9, 2.4, and 5.1], [**Bor94b**, Section 6.3], [**Lor21**, Section 4.3.2]. ◇

3.10.3 (Simplifications for Enriched Co/Ends in V)**.** Our treatment of V-coends and V-ends in Section 3.5 involves several simplifications that do not hold for the more general theory of enriched co/ends.

- When V is symmetric monoidal closed, V-enriched co/limits in $\underline{V}$ (in the conical, unweighted sense) are canonically isomorphic to those in the underlying category V_0. Thus co/completeness of V_0 in the unenriched sense implies enriched co/completeness of V (with respect to conical co/limits). See [**DK69, Kel05**] for further explanation.
- The generalization of the isomorphisms (3.5.10) and (3.5.11) provides the more useful notion of co/end in a general V-category C. See [**DK69, Day70**].
- Our existence result Proposition 3.5.5 is a special case of that given in [**DK69**].
- The theory of V-enriched *weighted co/limits* extends the theory of conical co/limits and is defined in terms of certain co/ends. These more general constructions are necessary for general V-enriched category theory, and we refer the reader to [**Kel05, Lor21**] for further explanation. ◇

3.10.4 (Enriched Diagram Categories)**.** Our discussion of enriched diagram categories in Section 3.7 follows [**DK69, Day70**]. A different sketch of the main ideas around Day's convolution product, under slightly simpler assumptions, is given in [**Ric20**]. The thesis of Corner [**Cor16**] gives a review of Day convolution and an extension to monoidal bicategories. Day-Street [**DS97**] gives a more general treatment of the convolution product and hom diagram as part of a more general theory of V-*modules*. ◇

3.10.5 (Enriched Tensors and Cotensors)**.** For further discussion of enriched tensors and cotensors we refer the reader to [**Kel69, Bor94b, Kel05, Rie14**]. In the literature, the term *copowered*, respectively *powered*, is sometimes used in place of the term tensored, respectively cotensored.

Our Definition 3.9.1 differs slightly from the literature in that the tensor and cotensor isomorphisms are not assumed to be V-natural. Thus our conclusion in Proposition 3.9.3 does not have all the features of an *enriched 2-variable adjunction* as in [**Rie14**, Remark 3.7.4] or [**Shu∞a**]. Theorem 3.9.8 and Corollary 3.9.15, which will be used in our work below, only establish tensor and cotensor structures in the sense of Definition 3.9.1. ◇

CHAPTER 4

Pointed Objects, Smash Products, and Pointed Homs

Throughout this chapter we suppose C is a category with a terminal object T and discuss objects X of C that are pointed by a morphism

$$\iota^X : T \longrightarrow X.$$

The category of pointed objects and morphisms, C_*, is the category under T. With the exception of some preliminary definitions in Section 4.1, we assume throughout that C is complete and cocomplete with a symmetric monoidal closed structure

$$(\mathsf{C}, \otimes, \mathbb{1}, \alpha, \lambda, \rho, \xi, \mathsf{Hom}).$$

In Section 4.1 we define the smash product of pointed objects and show that it induces a symmetric monoidal product on C_*. In Section 4.2 we define the pointed hom and show that it is adjoint to the smash product. The following examples of pointed categories will be essential parts for our algebraic K-theory work in Part 2.

- With $\mathsf{C} = \mathsf{Cat}$ and $T = \mathbb{1}$, the terminal category, Cat_* is the category of pointed categories and pointed functors.
- With $\mathsf{C} = \mathsf{PermCat}$ and $T = \mathbb{1}$ with its trivial monoidal product, every permutative category has a canonical basepoint given by its monoidal unit. For symmetric monoidal functors, preserving this basepoint is precisely the strictly unital condition.
- In Section 5.6 we describe the Boardman-Vogt tensor product for multicategories and its associated smash product.
- With $\mathsf{C} = \mathsf{Set}$ and $T = *$, the one-point set, Set_* is the category of pointed sets and pointed functions.
- In Section 7.1 we describe the category of pointed simplicial sets, sSet_*. Symmetric spectra, defined in Section 7.4, are certain sequences of pointed simplicial sets.

In many—but not all—of these applications, $\otimes$ is the Cartesian product in C and the monoidal unit is a terminal object so we have $\mathbb{1} = T$. The important exception is $\mathsf{C} = \mathsf{Multicat}$, the category of small multicategories discussed in Chapters 5 and 6.

In Section 4.3 below we discuss categories of pointed diagrams. Theorem 4.3.37 gives a summary statement. The development of K-theory for small permutative categories, given in Chapters 8, 9, and 10, will make use of certain pointed diagram categories and their symmetric monoidal closed structure.

4.1. Smash Products

We begin by defining categories of pointed objects. We show that completeness and cocompleteness are inherited from the ambient category C, and then we go on to define the smash product of pointed objects.

Definition 4.1.1. Suppose C is a category with terminal object T. We let C_* denote the category under T. Its objects are morphisms

$$\iota^X : T \longrightarrow X \in \mathsf{C},$$

which are called *pointed objects* with ι^X the *basepoint*. The morphisms of C_*, which are called *pointed morphisms*, are those morphisms in C that preserve the structure morphisms ι. ◇

Definition 4.1.2. Suppose C is a category with a terminal object T and with coproducts. If X is an object of C, then $X_+ = X \coprod T$ denotes the pointed object given by the structure morphism of the coproduct

$$T \longrightarrow X \coprod T.$$

This defines a functor $\mathsf{C} \longrightarrow \mathsf{C}_*$ called *adjoining a disjoint basepoint.* ◇

Proposition 4.1.3. *In the context of Definition 4.1.2, there is an adjunction of categories*

$$\mathsf{C} \underset{U}{\overset{(-)_+}{\rightleftarrows}} \mathsf{C}_*$$

with left adjoint $(-)_+$ given by adjoining a disjoint basepoint and right adjoint U given by forgetting basepoints.

Proof. The unit of the adjunction is the coproduct summand inclusion away from T. The component of the counit at a pointed object Y is

$$(UY)_+ \longrightarrow Y \quad \text{for} \quad Y \in \mathsf{C}_*$$

given by the identity on Y and the unique pointed morphism from the disjoint basepoint to Y. The triangle identities are the following composites for $X \in \mathsf{C}$ and $Y \in \mathsf{C}_*$:

$$X_+ \longrightarrow (U(X_+))_+ \longrightarrow X_+ \quad \text{and} \quad UY \longrightarrow U((UY)_+) \longrightarrow UY. \qquad \square$$

Definition 4.1.4. Suppose C is a category with terminal object T. For pointed objects X and Y the *wedge product* $X \vee Y$ is the pushout in C of the span

$$X \xleftarrow{\iota^X} T \xrightarrow{\iota^Y} Y.$$

Equivalently, it is the coequalizer of the two induced morphisms

$$T \underset{\iota^Y}{\overset{\iota^X}{\rightrightarrows}} X \coprod Y \dashrightarrow X \vee Y.$$

The wedge product is sometimes also called the *wedge sum.* ◇

Theorem 4.1.5. *Suppose C is a category with terminal object T. If C is complete and cocomplete, then so is C_*.*

Proof. The coproduct of pointed objects is given by the wedge product. Coequalizers of pointed objects are given by coequalizers in C equipped with the induced basepoint. Similarly, products and equalizers in C_* are given in C and equipped with the basepoints induced by these constructions. $\square$

Definition 4.1.6 (Smash Product). Suppose that $\mathsf{C} = (\mathsf{C}, \otimes, \mathbb{1})$ is a monoidal category with terminal object T and with coproducts. Then the *smash product* $X \wedge Y$ is the following pushout in C.

$$\begin{array}{ccc} (X \otimes T) \coprod (T \otimes Y) & \xrightarrow{(1_X \otimes \iota^Y) \coprod (\iota^X \otimes 1_Y)} & X \otimes Y \\ \downarrow & & \downarrow \\ T & \longrightarrow & X \wedge Y \end{array} \tag{4.1.7}$$

The *smash unit* E is defined by adjoining a disjoint basepoint to the unit of C:

$$E = \mathbb{1}_+ = \mathbb{1} \coprod T.$$

In Theorem 4.1.8 we show that $(\mathsf{C}_*, \wedge, E)$ is a symmetric monoidal category when $(\mathsf{C}, \otimes)$ is symmetric monoidal closed. In Section 4.2 we show, furthermore, that under these hypotheses $\wedge$ also has an adjoint. ◇

Beginning now, and continuing through Section 4.2, we suppose that $(\mathsf{C}, \otimes, \mathsf{Hom})$ is a complete and cocomplete symmetric monoidal closed category with terminal object T.

Theorem 4.1.8. *Suppose*

$$\mathsf{C} = (\mathsf{C}, \otimes, \mathbb{1}, \alpha, \lambda, \rho, \xi, \mathsf{Hom})$$

is a symmetric monoidal closed category that is complete and cocomplete with terminal object T. Then $(\mathsf{C}_*, \wedge, E)$ is a complete and cocomplete symmetric monoidal category.

Proof. Theorem 4.1.5 shows that C_* is complete and cocomplete. Now we show that the smash product defines a symmetric monoidal product with unit

$$E = \mathbb{1}_+ = \mathbb{1} \coprod T.$$

Because C is symmetric monoidal closed, the functors

$$X \otimes - : \mathsf{C} \longrightarrow \mathsf{C} \quad \text{and} \quad - \otimes Z : \mathsf{C} \longrightarrow \mathsf{C}$$

for $X, Z \in \mathsf{C}$ are left adjoints. Throughout this proof we will use the fact that these left adjoints preserve small colimits.

To construct the associativity isomorphism $\alpha^\wedge$ for the smash product, suppose (X, ι^X), (Y, ι^Y), and (Z, ι^Z) are pointed objects. We write

$$X \otimes Y \xrightarrow{\pi_{X,Y}} X \wedge Y \in \mathsf{C}$$

for the right vertical morphism in the pushout (4.1.7). Omitting $\otimes$ to save space, first observe that there is a pushout

$$\begin{array}{ccc} (XY)T \coprod (XT)Z \coprod (TY)Z & \longrightarrow & (XY)Z \\ & & \downarrow{\scriptstyle \pi_{X,Y} 1} \\ \downarrow & & (X \wedge Y)Z \\ & & \downarrow{\scriptstyle \pi_{X \wedge Y, Z}} \\ T & \longrightarrow & (X \wedge Y) \wedge Z \end{array} \tag{4.1.9}$$

in C with the top horizontal arrow induced by the basepoints of X, Y, and Z. In more detail, consider the following diagram in C.

(4.1.10)
$$\begin{array}{ccc} (XY)T \amalg (XT)Z \amalg (TY)Z & \longrightarrow & (XY)Z \\ \omega \downarrow & & \downarrow \pi_{X,Y}1 \\ (X \wedge Y)T \amalg TZ & \longrightarrow & (X \wedge Y)Z \\ \downarrow & & \downarrow \pi_{X\wedge Y,Z} \\ T & \longrightarrow & (X \wedge Y) \wedge Z \end{array}$$

- The bottom square in (4.1.10) is a pushout by the definition (4.1.7) of the smash product.
- The top square in (4.1.10) is obtained by first applying $- \otimes Z$ to the pushout (4.1.7) defining $X \wedge Y$. This yields the pushout

(4.1.11)
$$\begin{array}{ccc} (XT)Z \amalg (TY)Z & \longrightarrow & (XY)Z \\ \tau \downarrow & & \downarrow \pi_{X,Y}1 \\ TZ & \longrightarrow & (X \wedge Y)Z \end{array}$$

because $- \otimes Z$ is a left adjoint, which preserves all small colimits. The top left vertical arrow in (4.1.10) is the coproduct

$$\omega = (\pi_{X,Y} \otimes 1_T) \amalg \tau.$$

Since (4.1.11) is a pushout, so is the top square in (4.1.10) by the functoriality of $\otimes$ in C.

Therefore, the outer diagram in (4.1.10), which is (4.1.9), is a pushout.

A similar argument using $X \otimes -$ yields the following pushout in C with the right normalized bracketing:

(4.1.12)
$$\begin{array}{ccc} X(YT) \amalg X(TZ) \amalg T(YZ) & \longrightarrow & X(YZ) \\ & & \downarrow 1\pi_{Y,Z} \\ \downarrow & & X(Y \wedge Z) \\ & & \downarrow \pi_{X,Y\wedge Z} \\ T & \longrightarrow & X \wedge (Y \wedge Z) \end{array}$$

Now we consider the following morphism from the pushout (4.1.9), shown in the back square below, to the pushout (4.1.12), shown in the front square below.

(4.1.13)
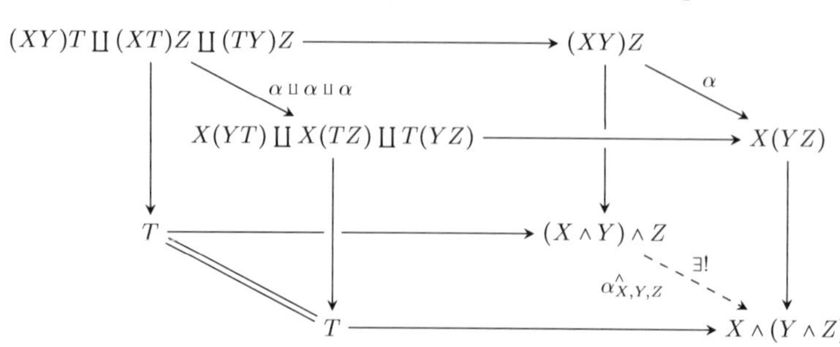

The left face in (4.1.13) commutes because T is a terminal object. The top face commutes by the naturality of α. The universal property of pushouts implies that there exists a unique dotted morphism $\alpha^{\wedge}_{X,Y,Z}$ that makes the entire cube commutative. Since each α is an isomorphism, so is $\alpha^{\wedge}_{X,Y,Z}$. We define this as the associativity isomorphism $\alpha^{\wedge}$ for the pointed objects X, Y, and Z. The naturality of $\alpha^{\wedge}$ follows from the naturality of α and the universal property of pushouts.

The pentagon axiom (1.1.3) for $\alpha^{\wedge}$ follows from that of α and the universal property of pushouts. More precisely, for pointed objects W, X, Y, and Z, we write

$$V_l = ((WX)Y)T \amalg ((WX)T)Z \amalg ((WT)Y)Z \amalg ((TX)Y)Z$$
$$V_r = W(X(YT)) \amalg W(X(TZ)) \amalg W(T(YZ)) \amalg T(X(YZ))$$

and consider the following cube.

(4.1.14)

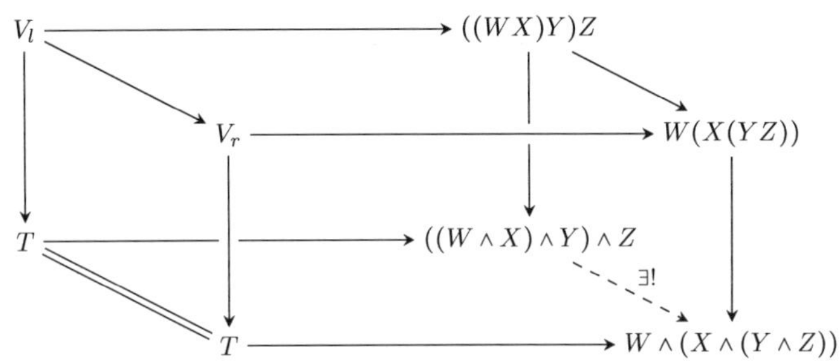

In the top face in (4.1.14), the morphism $V_l \longrightarrow V_r$ from the back pushout to the front pushout is a coproduct of four morphisms, each being either composite in the pentagon diagram (1.1.3) in C. The top right morphism

$$((W \otimes X) \otimes Y) \otimes Z \longrightarrow W \otimes (X \otimes (Y \otimes Z))$$

is either composite in the pentagon diagram in C for W, X, Y, and Z. The universal property of pushouts implies that there is a unique dotted arrow that renders the cube commutative, which proves the pentagon axiom for $\alpha^{\wedge}$.

To define the left unit isomorphism $\lambda^{\wedge}$ for the smash product, for a pointed object X, consider the following diagram in C.

(4.1.15)

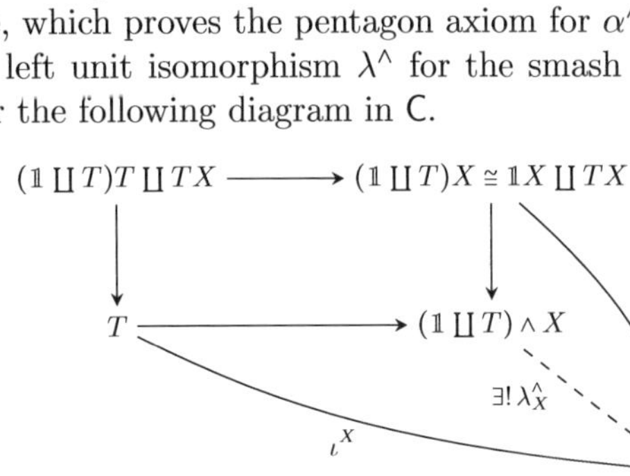

In (4.1.15), the top left square is a pushout by the definition (4.1.7) of the smash product. The morphism ℓ is induced by the left unit isomorphism

$$\mathbb{1} \otimes X \xrightarrow{\lambda_X} X$$

in C and the composite

$$T \otimes X \longrightarrow T \xrightarrow{\iota^X} X.$$

The outer diagram in (4.1.15) is commutative, so there is a unique dotted arrow $\lambda^{\wedge}_X$ that renders the entire diagram commutative. Its naturality follows from that of λ and the universal property of pushouts. Since λ is an isomorphism, $\lambda^{\wedge}_X$ is an isomorphism, with inverse the composite

$$X \xrightarrow{\lambda^{-1}} \mathbb{1} \otimes X \longrightarrow (\mathbb{1} \textstyle\coprod T) \otimes X \longrightarrow (\mathbb{1} \coprod T) \wedge X.$$

The right unit isomorphism $\rho^{\wedge}$ for the smash product is defined similarly by the right unit isomorphism ρ in C and the pushout (4.1.7) that defines $\wedge$.

With the smash unit

$$E = \mathbb{1} \textstyle\coprod T,$$

the unity axiom (1.1.2) for the smash product asserts the commutativity of the following triangle in C for pointed objects X and Y.

(4.1.16)
$$\begin{array}{ccc} (X \wedge E) \wedge Y & \xrightarrow{\alpha^{\wedge}_{X,E,Y}} & X \wedge (E \wedge Y) \\ & {\scriptstyle \rho^{\wedge}_X \wedge 1_Y} \searrow \quad \swarrow {\scriptstyle 1_X \wedge \lambda^{\wedge}_Y} & \\ & X \wedge Y & \end{array}$$

Similar to the above proof of the pentagon axiom, the proof of the unity axiom for the smash product is a cube argument that reduces to the unity axiom for $\otimes$. More precisely, we consider the following morphism of spans in C, with the $\otimes$ symbol omitted to save space.

(4.1.17)
$$\begin{array}{ccccc} (XE)T \coprod (XT)Y \coprod (TE)Y & & \longrightarrow & & (XE)Y \\ \downarrow & \searrow^{f} & & & \searrow^{g} \\ T & & XT \coprod TY & \longrightarrow & XY \\ & \diagdown\!\!\diagdown & \downarrow & & \\ & & T & & \end{array}$$

Using the fact that $- \otimes E$ commutes with small colimits, the morphisms f and g in (4.1.17) are defined by the following morphisms, with t the unique morphism to the terminal object T and ρ the right unit isomorphism in C.

(4.1.18)
$$\begin{aligned} (XE)T \cong (X\mathbb{1})T \textstyle\coprod (XT)T &\xrightarrow{(\rho 1_T,\, (\iota^X \circ t) 1_T)} XT \\ (XT)Y &\xrightarrow{\;\; t1_Y \;\;} TY \\ (TE)Y &\xrightarrow{\;\; t1_Y \;\;} TY \\ (XE)Y \cong (X\mathbb{1})Y \textstyle\coprod (XT)Y &\xrightarrow{(\rho 1_Y,\, (\iota^X \circ t) 1_Y)} XY \end{aligned}$$

Since (4.1.17) is commutative, there is a unique induced morphism from the back pushout to the front pushout, namely,

$$(X \wedge E) \wedge Y \xrightarrow{\rho^{\wedge}_X \wedge 1_Y} X \wedge Y$$

in the triangle (4.1.16).

On the other hand, the composite

$$(X \wedge E) \wedge Y \xrightarrow{(1_X \wedge \lambda^\wedge_Y)\circ \alpha^\wedge_{X,E,Y}} X \wedge Y$$

in (4.1.16) is induced by the following morphisms, instead of those in (4.1.18), from the back span to the front span in (4.1.17).

$$\begin{array}{lllll}
(XE)T & \xrightarrow{\alpha} & X(ET) & \xrightarrow{1_X t} & XT \\
(XT)Y & \xrightarrow{\alpha} & X(TY) & \xrightarrow{1_X t} & XT \\
(T\mathbb{1})Y & \xrightarrow{\alpha} & T(\mathbb{1}Y) & \xrightarrow{1_T \lambda} & TY \\
(TT)Y & \xrightarrow{\alpha} & T(TY) & \xrightarrow{1_T(\iota^Y \circ t)} & TY \\
(X\mathbb{1})Y & \xrightarrow{\alpha} & X(\mathbb{1}Y) & \xrightarrow{1_X \lambda} & XY \\
(XT)Y & \xrightarrow{\alpha} & X(TY) & \xrightarrow{1_X(\iota^Y \circ t)} & XY
\end{array}$$

To see that the two induced morphisms in (4.1.16) are equal, by the universal property of the back pushout in (4.1.17), we must check that the two composites

$$(XE)Y \cong (X\mathbb{1})Y \amalg (XT)Y \underset{(1_X\lambda_Y,\, 1_X(\iota^Y \circ t))\circ(\alpha \amalg \alpha)}{\overset{(\rho_X 1_Y,\, (\iota^X \circ t)1_Y)}{\rightrightarrows}} XY \xrightarrow{\pi_{X,Y}} X \wedge Y \tag{4.1.19}$$

are equal. On the summand $(X\mathbb{1})Y$, the equality

$$\rho_X \otimes 1_Y = (1_X \otimes \lambda_Y)\circ \alpha \colon (X \otimes \mathbb{1}) \otimes Y \longrightarrow X \otimes Y$$

holds by the unity axiom in C. On the other summand $(XT)Y$, the two composites in (4.1.19) form the outer diagram below.

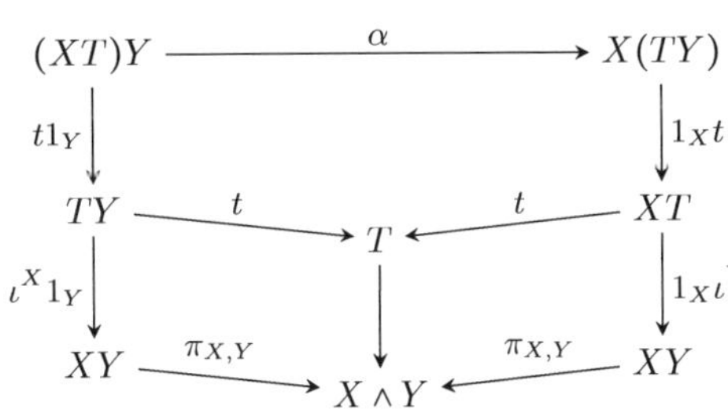

In the above diagram, the top pentagon commutes because T is a terminal object. The two lower parallelograms commute by the pushout definition of $X \wedge Y$. This proves the unity axiom for the smash product.

The symmetry isomorphism

$$X \wedge Y \xrightarrow{\xi^\wedge_{X,Y}} Y \wedge X$$

for the smash product is defined as the unique induced morphism from the pushout (4.1.7) defining $X \wedge Y$ to the pushout defining $Y \wedge X$, using the symmetry isomorphism $\xi_{X,Y}$ in C. Similar to the previous two paragraphs, the symmetry axiom (1.1.24) and the unit axiom (1.1.25) for $\wedge$ follow from those in C. In the presence of the symmetry axiom, the hexagon axiom (1.1.26) for $\wedge$ is equivalent to the left hexagon axiom in (1.1.16), which asserts the commutativity of the following diagram for pointed objects X, Y, and Z.

$$
\begin{array}{ccc}
(Y \wedge X) \wedge Z & \xrightarrow{\alpha^\wedge} & Y \wedge (X \wedge Z) \\
{\scriptstyle \xi^\wedge \wedge 1} \nearrow & & \searrow {\scriptstyle 1 \wedge \xi^\wedge} \\
(X \wedge Y) \wedge Z & & Y \wedge (Z \wedge X) \\
{\scriptstyle \alpha^\wedge} \searrow & & \nearrow {\scriptstyle \alpha^\wedge} \\
X \wedge (Y \wedge Z) & \xrightarrow{\xi^\wedge} & (Y \wedge Z) \wedge X
\end{array} \tag{4.1.20}
$$

To prove the commutativity of (4.1.20), we again use a cube argument and consider the following morphisms of spans in C.

(4.1.21)

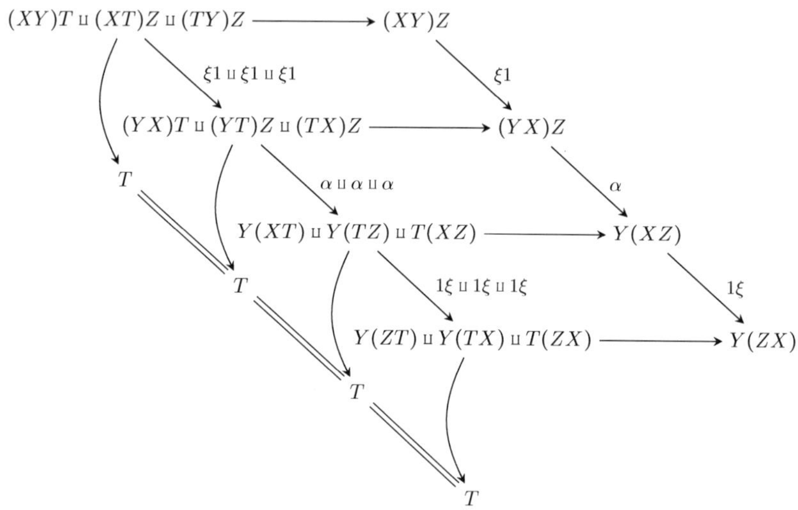

In (4.1.21), the top middle face is commutative by the naturality of α. The other two top faces are commutative by the naturality of ξ. So the diagram (4.1.21) is commutative. The unique induced morphism from the back pushout to the front pushout is the top composite in the left hexagon diagram (4.1.20).

Moreover, the left hexagon axiom in C is an equality

$$1\xi \circ \alpha \circ \xi 1 = \alpha \circ \xi \circ \alpha, \tag{4.1.22}$$

with the left-hand side the composite of the indicated morphisms in (4.1.21). Similarly, the left hexagon axiom in C yields the equality

$$
\begin{aligned}
&(1\xi \amalg 1\xi \amalg 1\xi) \circ (\alpha \amalg \alpha \amalg \alpha) \circ (\xi 1 \amalg \xi 1 \amalg \xi 1) \\
&= (\alpha \amalg \alpha \amalg \alpha) \circ (\xi \amalg \xi \amalg \xi) \circ (\alpha \amalg \alpha \amalg \alpha),
\end{aligned} \tag{4.1.23}
$$

with the top line the composite of the indicated morphisms in (4.1.21). Using

- the right-hand side of (4.1.22) and
- the bottom line in (4.1.23)

in the diagram (4.1.21), the unique induced morphism from the back pushout to the front pushout is the bottom composite in the left hexagon diagram (4.1.20). Therefore, (4.1.20) is commutative. □

4.2. Pointed Homs

We remind the reader that $\mathsf{C} = (\mathsf{C}, \otimes, \mathsf{Hom})$ is assumed to be a complete and cocomplete symmetric monoidal closed category with terminal object T. In Section 4.1 we defined the smash product and showed that it defines a symmetric monoidal product $\wedge$ on C_*, with unit $E = \mathbb{1} \coprod T$. Now we define a closed structure for $(\mathsf{C}_*, \wedge, E)$

Definition 4.2.1. For pointed objects X and Y in C_*, we define the *pointed hom* as the following pullback in C.

$$\begin{array}{ccc} \mathsf{Hom}_*(X,Y) & \longrightarrow & T \\ \big\downarrow & & \big\downarrow \\ \mathsf{Hom}(X,Y) & \longrightarrow & \mathsf{Hom}(T,Y) \end{array} \tag{4.2.2}$$

The composite

$$T \cong \mathsf{Hom}(X,T) \longrightarrow \mathsf{Hom}(X,Y) \longrightarrow \mathsf{Hom}(T,Y)$$

induced by the structure morphisms for X and Y is equal to the vertical morphism in the diagram, and therefore induces a canonical structure morphism $T \longrightarrow \mathsf{Hom}_*(X,Y)$ making $\mathsf{Hom}_*(X,Y)$ a pointed object. ◇

Theorem 4.2.3. *Suppose*

$$\mathsf{C} = (\mathsf{C}, \otimes, \mathbb{1}, \alpha, \lambda, \rho, \xi, \mathsf{Hom})$$

is a symmetric monoidal closed category that is complete and cocomplete with terminal object T. Then

$$(\mathsf{C}_*, \wedge, E, \mathsf{Hom}_*)$$

is a complete and cocomplete symmetric monoidal closed category.

Proof. Theorem 4.1.5 shows that C_* is complete and cocomplete. Theorem 4.1.8 shows that $(\mathsf{C}_*, \wedge)$ is symmetric monoidal with unit E. Now we show that Hom_* is adjoint to $\wedge$.

We will show

$$\mathsf{Hom}_*(X \wedge Y, Z) \cong \mathsf{Hom}_*(X, \mathsf{Hom}_*(Y,Z))$$

by showing that the right hand side satisfies the universal property of the pullback defining the left hand side. To do so, first note that the functors

$$\mathsf{Hom}(X,-)\colon \mathsf{C} \longrightarrow \mathsf{C} \quad \text{and} \quad \mathsf{Hom}(-,Z)\colon \mathsf{C}^{\mathsf{op}} \longrightarrow \mathsf{C}$$

both preserve small limits by the $\otimes$-Hom adjunction. That is, $\mathsf{Hom}(X,-)$ preserves small limits in C, and $\mathsf{Hom}(-,Z)$ converts small colimits in C (limits in C^{op}) to limits in C.

In the diagram below, we apply $\mathsf{Hom}(-, Z)$ to the pushout diagram (4.1.7) defining $X \wedge Y$. This yields a pullback in C shown at right. Next we observe that the definition of $\mathsf{Hom}_*(X \wedge Y, Z)$ makes it a pullback shown at left.

(4.2.4)

$$\begin{array}{ccccc}
\mathsf{Hom}_*(X \wedge Y, Z) & \longrightarrow & \mathsf{Hom}(X \wedge Y, Z) & \longrightarrow & \mathsf{Hom}(X \otimes Y, Z) \\
\downarrow & & \downarrow & & \downarrow \\
T & \longrightarrow & \mathsf{Hom}(T, Z) & \longrightarrow & \mathsf{Hom}((X \otimes T) \amalg (T \otimes Y), Z)
\end{array}$$

We will show that $\mathsf{Hom}_*(X, \mathsf{Hom}_*(Y, Z))$ is also a pullback of the outer rectangle, via a sequence of isomorphisms that are natural in X, Y, and Z.

To begin, consider the following diagram in C. The horizontal morphisms are given by the unique morphisms to T or by pulling back along the basepoint of X. The other morphisms are described below.

(4.2.5)

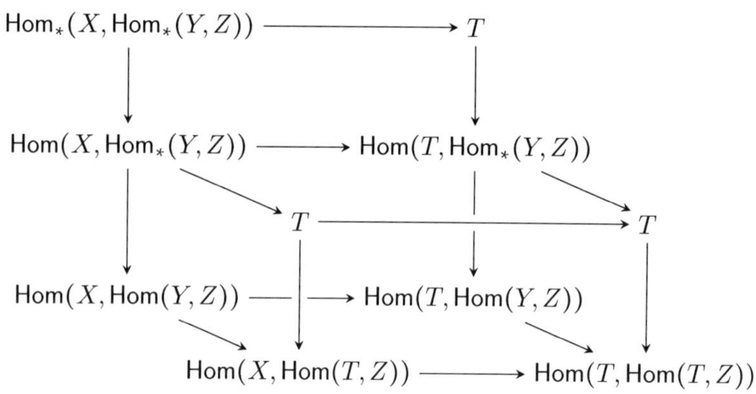

The upper square of (4.2.5) is the pullback defining $\mathsf{Hom}_*(X, \mathsf{Hom}_*(Y, Z))$, and the other two squares involving a single vertex T (that is, the left vertical and right vertical squares) are given by applying $\mathsf{Hom}(X, -)$, respectively $\mathsf{Hom}(T, -)$ to the pullback squares defining $\mathsf{Hom}_*(Y, Z)$. Thus, because $\mathsf{Hom}(X, -)$ and $\mathsf{Hom}(T, -)$ are right adjoints, each of the squares involving a single vertex T is a pullback.

Now we use the closed structure of C to replace four lower vertices of (4.2.5) with objects of the form $\mathsf{Hom}(- \otimes -, Z)$ and recognize the right vertical composite

$$T \longrightarrow \mathsf{Hom}(T, \mathsf{Hom}_*(Y, Z)) \longrightarrow \mathsf{Hom}(T \otimes Y, Z)$$

as equal to the composite

$$T \cong \mathsf{Hom}(T \otimes Y, T) \longrightarrow \mathsf{Hom}(T \otimes Y, Z).$$

Using this equality and rearranging (4.2.5), we have $\mathsf{Hom}_*(X, \mathsf{Hom}_*(Y,Z))$ as a limit in the following diagram in C.

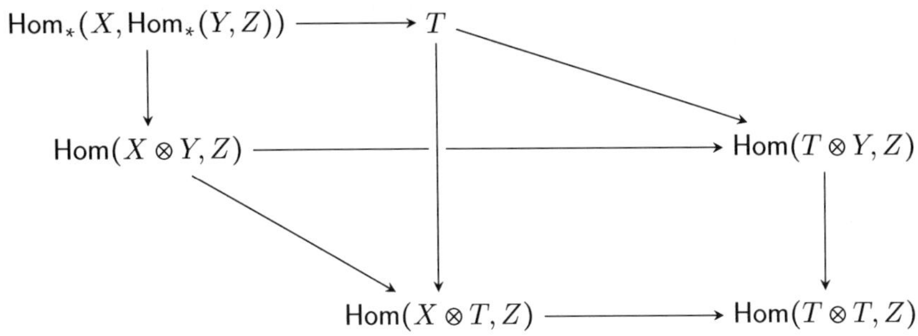

The vertex $\mathsf{Hom}(T \otimes T, Z)$ is redundant, and we replace the two arrows out of T, respectively $\mathsf{Hom}(X \otimes Y, Z)$, with the corresponding arrow into

$$\mathsf{Hom}(X \otimes T, Z) \times \mathsf{Hom}(T \otimes Y, Z) \cong \mathsf{Hom}\big((X \otimes T) \textstyle\coprod (T \otimes Y), Z \big).$$

This exhibits $\mathsf{Hom}_*(X, \mathsf{Hom}_*(Y,Z))$ as the pullback below, which is the outer pullback of (4.2.4).

$$\begin{array}{ccc}
\mathsf{Hom}_*(X, \mathsf{Hom}_*(Y,Z)) & \longrightarrow & T \\
\downarrow & & \downarrow \\
\mathsf{Hom}(X \otimes Y, Z) & \longrightarrow & \mathsf{Hom}((X \otimes T) \coprod (T \otimes Y), Z)
\end{array}$$

Therefore

$$\mathsf{Hom}_*(X, \mathsf{Hom}_*(Y,Z)) \cong \mathsf{Hom}_*(X \wedge Y, Z),$$

naturally in X, Y, and Z. □

4.3. Pointed Diagram Categories

In this section we suppose that

$$\mathsf{V} = (\mathsf{V}, \otimes, \mathbb{1}, \alpha, \lambda, \rho, [-,-], \mathrm{T})$$

is a complete and cocomplete symmetric monoidal closed category with chosen terminal object T, sometimes denoted T^{V}. Recalling the smash product and internal hom constructions of Sections 4.1 and 4.2, we have a symmetric monoidal closed category

$$(\mathsf{V}_*, \wedge, E, [-,-]_*)$$

where $E = \mathbb{1} \coprod \mathrm{T}$ is the monoidal unit and $[-,-]_*$ denotes the pointed hom of Definition 4.2.1.

Convention 4.3.1. To avoid ambiguity with our notation for diagram categories, throughout this section we write

$$\mathsf{VCat} \quad \text{and} \quad \mathsf{V}_*\mathsf{Cat} \tag{4.3.2}$$

for the 2-categories of V-enriched, respectively V_*-enriched, categories, functors and natural transformations. ◇

Definition 4.3.3. Suppose $(\mathcal{D}, \boxdot, e)$ is a symmetric monoidal category. We say that T is a *null object* for $\mathcal{D}$ if

- T is both initial and terminal in $\mathcal{D}$, and
- there are natural isomorphisms

$$a \boxdot \mathrm{T} \cong \mathrm{T} \cong \mathrm{T} \boxdot a$$

for objects a in $\mathcal{D}$. ◇

We will use the following notation throughout this section.

Definition 4.3.4. Let $(\mathsf{Set}_*, \wedge, \underline{1}, *)$ denote the category of pointed finite sets with its smash product $\wedge$, unit $\underline{1} = \{0, 1\}$, with basepoint element 0, and terminal 1-point set $*$. ◇

The main result of this section, Theorem 4.3.19, gives a comparison between pointed diagrams from $(\mathcal{D}, \mathrm{T}^{\mathcal{D}})$ to V_* with enriched diagrams out of a certain V_*-enriched category $\widehat{\mathcal{D}}$ described in Definition 4.3.11. To prepare for this, we first give the following general comparison between

- pointed functors between categories whose basepoints are both initial and terminal, and
- Set_*-enriched functors between such categories.

Lemma 4.3.5. *Suppose* $(\mathsf{B}, \mathrm{T}^{\mathsf{B}})$ *and* $(\mathsf{C}, \mathrm{T}^{\mathsf{C}})$ *are pointed categories in which the basepoints* T^{B} *and* T^{C} *are both initial and terminal.*

(1) *Taking zero morphisms as basepoints of their hom sets, both* B *and* C *are enriched categories over* Set_*.
(2) *There is an equivalence of categories*

$$\mathsf{Cat}_*((\mathsf{B}, \mathrm{T}^{\mathsf{B}}), (\mathsf{C}, \mathrm{T}^{\mathsf{C}})) \simeq (\mathsf{Set}_*\mathsf{Cat})(\mathsf{B}, \mathsf{C}), \tag{4.3.6}$$

where the right hand side denotes the category of Set_**-enriched functors and natural transformations.*

Proof. For the first assertion, the composition and identities of B and C are Set_*-enriched because composition with a zero morphism is zero. For the second assertion, first observe that a pointed functor

$$H: (\mathsf{B}, \mathrm{T}^{\mathsf{B}}) \longrightarrow (\mathsf{C}, \mathrm{T}^{\mathsf{C}})$$

preserves the zero morphisms and is therefore Set_*-enriched. Pointed natural transformations are likewise Set_*-enriched. This provides a functor

$$\mathsf{Cat}_*((\mathsf{B}, \mathrm{T}^{\mathsf{B}}), (\mathsf{C}, \mathrm{T}^{\mathsf{C}})) \longrightarrow (\mathsf{Set}_*\mathsf{Cat})(\mathsf{B}, \mathsf{C})$$

For the reverse direction, the forgetful functor

$$U: \mathsf{Set}_* \longrightarrow \mathsf{Set},$$

which is the identity on objects and morphisms, is symmetric monoidal with unit constraint the inclusion of the one point set at the non-basepoint element of $\underline{1}$, the monoidal unit for Set_*. The monoidal constraint

$$(UX) \times (UY) = X \times Y \xrightarrow{U^2_{X,Y}} X \wedge Y$$

for pointed sets X and Y is given by the structure morphism of the pushout (4.1.7) defining $X \wedge Y$. Changing enrichment along U provides a forgetful functor

$$(\mathsf{Set}_*\mathsf{Cat})(\mathsf{B},\mathsf{C}) \longrightarrow \mathsf{Cat}(\mathsf{B},\mathsf{C}).$$

from the category of Set_*-enriched functors and natural transformations to the underlying category of ordinary Set-enriched functors and natural transformations.

We show that Set_*-enriched functors preserve initial and terminal objects up to unique isomorphism, and then use these isomorphisms to provide strictly pointed functors. To make this explicit, let $0_{Z,W}$ denote the zero morphism

$$Z \longrightarrow \top \longrightarrow W$$

factoring through the respective basepoint object for each pair of objects Z and W objects in either B or C. Then we note the following.

- If F is a Set_*-enriched functor from B to C, then preserving the basepoint of each hom set means
$$F(0_{X,Y}) = 0_{FX,FY}$$
for each pair of objects X and Y in B.
- For any object Z of B or C, we have
$$0_{Z,Z} = 1_Z$$
if and only if $Z \cong \top$, the respective basepoint object. This isomorphism is necessarily unique.
- By the identity axiom (1.2.6) for a Set_*-enriched functor $F \colon \mathsf{B} \longrightarrow \mathsf{C}$, we have
$$1_{F(\top^{\mathsf{B}})} = F(1_{\top^{\mathsf{B}}}).$$

Combining these observations, we have

$$1_{F(\top^{\mathsf{B}})} = F(1_{\top^{\mathsf{B}}}) = F(0_{\top^{\mathsf{B}},\top^{\mathsf{B}}}) = 0_{F(\top^{\mathsf{B}}),F(\top^{\mathsf{B}})}$$

and, therefore, a unique isomorphism

$$F(\top^{\mathsf{B}}) \cong \top^{\mathsf{C}}. \tag{4.3.7}$$

Since $\top^{\mathsf{B}}$ and $\top^{\mathsf{C}}$ are both initial and terminal, we can change the definition of F on objects to

$$F'X = \begin{cases} FX & \text{if } \; X \neq \top^{\mathsf{B}} \\ \top^{\mathsf{C}} & \text{if } \; X = \top^{\mathsf{B}}. \end{cases}$$

and define F' on morphisms via F and composition with (4.3.7) for the (necessarily unique) morphisms to or from $\top^{\mathsf{B}}$ in B. The functoriality of F and uniqueness of (4.3.7) ensure that F' remains functorial. We likewise replace a Set_*-enriched natural transformation

$$\alpha \colon F \longrightarrow G$$

with a pointed natural transformation

$$\alpha' \colon F' \longrightarrow G'$$

by changing the component at $\top^{\mathsf{B}}$ to be the identity of $\top^{\mathsf{C}}$. Naturality of α' follows from that of α and the uniqueness of (4.3.7).

For a pointed functor

$$H \colon (\mathsf{B}, \top^{\mathsf{B}}) \longrightarrow (\mathsf{C}, \top^{\mathsf{C}}),$$

we have $H' = H$. For a general Set_*-enriched functor F as above, there is a natural isomorphism

$$\varepsilon\colon F' \xrightarrow{\cong} F$$

whose only potential non-identity component is the isomorphism (4.3.7). Therefore we have $\varepsilon' = 1_{F'}$. This proves the triangle identities for the second assertion. □

Recall from Definition 3.8.9 the unitary enrichment of a symmetric monoidal category over V. Now we introduce a pointed variant over V_*.

Definition 4.3.8. For a pointed set X, we let $X^\flat$ denote the *punctured set*

$$X^\flat = X \setminus \{*\}$$

consisting of all elements of X except its basepoint. The symbol $\flat$ is the musical flat symbol, and $X^\flat$ may be read as *X-flat* or *X-punctured,* with the mnemonic that a punctured tire is flat. ◇

Definition 4.3.9. Define a pointed and strictly unital strong symmetric monoidal functor

$$\widehat{F}\colon (\mathsf{Set}_*, \wedge, \underline{1}, *) \longrightarrow (\mathsf{V}_*, \wedge, E, \mathrm{T}) \tag{4.3.10}$$

by

$$\widehat{F}X = \bigvee_{i \in X^\flat} E \quad \text{for} \quad X \in \mathsf{Set}_*$$

where we take T for the wedge over an empty indexing set. For a morphism of pointed sets $f\colon X \longrightarrow Y$, the induced morphism

$$\bigvee_{i \in X^\flat} E \longrightarrow \bigvee_{i \in Y^\flat} E$$

has summand at $i \in X^\flat$ given by the identity 1_E if $f(i)$ is not the basepoint of Y (and, therefore, $f(i) \in Y^\flat$) or by the zero morphism to the target (factoring through the basepoint of the wedge) if $f(i) = *$.

The unit constraint $\widehat{F}^0$ is 1_E. The monoidal constraint

$$\widehat{F}X \wedge \widehat{F}Y \xrightarrow{\widehat{F}^2_{X,Y}} \widehat{F}(X \wedge Y)$$

is given by the following composite

$$\Big(\bigvee_{i \in X^\flat} E\Big) \wedge \Big(\bigvee_{j \in Y^\flat} E\Big) \xrightarrow{\cong} \bigvee_{(i,j) \in X^\flat \times Y^\flat} E \wedge E \xrightarrow{\cong} \bigvee_{(i,j) \in (X \wedge Y)^\flat} E.$$

In the above composite, the first morphism is the canonical isomorphism commuting $\wedge$ with small coproducts (since V_* is symmetric monoidal closed) and the second morphism is given by the bijection of indexing sets

$$X^\flat \times Y^\flat \cong (X \wedge Y)^\flat$$

and the unit isomorphisms of E. ◇

Definition 4.3.11. Suppose $(\mathcal{D}, \boxdot, e, \mathrm{T}^{\mathcal{D}})$ is a small symmetric monoidal category with null object $\mathrm{T}^{\mathcal{D}}$ and with its canonical enrichment over Set_* as in Lemma 4.3.5 (1).

Suppose T^{V} is a chosen terminal object of V. The *pointed unitary enrichment* of $\mathcal{D}$ over V_* is denoted $\widehat{\mathcal{D}}$ and is defined as the change of enrichment for $\mathcal{D}$, in the sense of Definition 2.1.1 and Proposition 2.1.2, along the functor $\widehat{F}$ of (4.3.10) above.

So $\widehat{\mathcal{D}}$ has the same objects as $\mathcal{D}$ and has hom objects given by the coproduct in V_*

$$\widehat{\mathcal{D}}(a,b) = \bigvee_{p\in(\mathcal{D}(a,b))^\flat} E \tag{4.3.12}$$

where we take T^{V} for the empty wedge. By Theorem 2.4.10, $\widehat{\mathcal{D}}$ is a symmetric monoidal V_*-category. ◇

Explanation 4.3.13. Recalling the unitary enrichment of $\mathcal{D}$ over E, the monoidal unit of V_*, Proposition 3.8.12 provides an isomorphism of categories

$$\mathsf{Cat}(\mathcal{D},\mathsf{V}_*) \cong \mathsf{V}_*\mathsf{Cat}(\mathcal{D}_E,\underline{\mathsf{V}_*}), \tag{4.3.14}$$

where $\mathsf{V}_*\mathsf{Cat}$ is the 2-category of V_*-categories, V_*-functors, and V_*-natural transformations. When $\mathcal{D}$ has a null object as its basepoint, as in Definition 4.3.11, then there is also a V_*-functor

$$\mathcal{D}_E \longrightarrow \widehat{\mathcal{D}} \tag{4.3.15}$$

induced, as in Definition 2.2.1, by a monoidal natural transformation

$$F_E \longrightarrow \widehat{F}$$

whose component at a pointed set X is the morphism

$$\bigvee_{i\in X} E \longrightarrow \bigvee_{i\in X^\flat} E$$

given by the identity 1_E for each non-basepoint summand and by the zero morphism for the basepoint summand. Restriction along the V_*-functor (4.3.15) provides a 2-functor

$$\mathsf{V}_*\mathsf{Cat}(\widehat{\mathcal{D}},\underline{\mathsf{V}_*}) \longrightarrow \mathsf{V}_*\mathsf{Cat}(\mathcal{D}_E,\underline{\mathsf{V}_*}). \tag{4.3.16}$$

We show in Theorem 4.3.19 below that this corresponds under the isomorphism (4.3.14) to the forgetful inclusion of pointed functors among all functors. ◇

Example 4.3.17. In the case $\mathsf{V}_* = \mathsf{Set}_*$, the enrichment $\widehat{\mathcal{D}}$ of Definition 4.3.11 is the canonical Set_*-enrichment described in Lemma 4.3.5 (1). ◇

For the remainder of this section we will have the following standing assumptions.

Convention 4.3.18 (Context for Pointed Diagram Categories).

(1) Suppose $(\mathsf{V},\otimes,\mathbb{1},\mathrm{T}^{\mathsf{V}})$ is a complete and cocomplete symmetric monoidal closed category with chosen terminal object T^{V}.

(2) Then by Theorem 4.2.3 $(\mathsf{V}_*,\wedge,E,[-,-]_*)$ is a complete and cocomplete symmetric monoidal closed category with monoidal smash product $\wedge$, monoidal unit

$$E = \mathbb{1} \coprod \mathrm{T}^{\mathsf{V}}$$

and pointed hom $[-,-]_*$.

(3) Suppose $(\mathcal{D},\boxdot,e,\mathrm{T}^{\mathcal{D}})$ is a small pointed symmetric monoidal category with chosen null object $\mathrm{T}^{\mathcal{D}}$ as basepoint. ◇

Theorem 4.3.19. *In the context of Convention 4.3.18 there is an equivalence of categories*

$$\mathsf{Cat}_*((\mathcal{D},\mathrm{T}^{\mathcal{D}}),(\mathsf{V}_*,\mathrm{T}^{\mathsf{V}})) \xrightarrow{\ \simeq\ } \mathsf{V}_*\mathsf{Cat}(\widehat{\mathcal{D}},\underline{\mathsf{V}_*}). \tag{4.3.20}$$

Moreover, this equivalence commutes with the isomorphism (4.3.14) *in the sense of the following commutative diagram, where the vertical functors are given by the forgetful inclusion and by* (4.3.16).

(4.3.21)
$$\begin{array}{ccc}
\mathsf{Cat}_*((\mathcal{D},\mathrm{T}^{\mathcal{D}}),(\mathsf{V}_*,\mathrm{T}^{\mathsf{V}})) & \xrightarrow{\simeq} & \mathsf{V}_*\mathsf{Cat}(\widehat{\mathcal{D}},\underline{\mathsf{V}_*}) \\
\downarrow & & \downarrow \\
\mathsf{Cat}(\mathcal{D},\mathsf{V}_*) & \xrightarrow{\cong} & \mathsf{V}_*\mathsf{Cat}(\mathcal{D}_E,\underline{\mathsf{V}_*})
\end{array}$$

Proof. As noted in the proof of Proposition 3.8.12, the lower isomorphism of (4.3.21) is the isomorphism on hom categories for the change of enrichment adjunction induced by

$$F_E \dashv \mathsf{V}_*(E,-).$$

The right adjoint $\mathsf{V}_*(E,-)$ factors through Set_* as

$$\mathsf{V}_* \xrightarrow{\widehat{U}} \mathsf{Set}_* \longrightarrow \mathsf{Set}$$

where the second functor forgets the basepoint and, for each A in V_* we let $\widehat{U}A$ denote $\mathsf{V}_*(E,A)$ equipped with the zero basepoint.

As in Lemma 3.8.11, $\widehat{U}$ is right adjoint to $\widehat{F}$ with natural isomorphisms

$$\mathsf{V}_*\Big(\bigvee_{X^\flat} E, A\Big) \cong \prod_{X^\flat} \mathsf{V}_*(E,A) \cong \mathsf{Set}_*(X,\mathsf{V}_*(E,A))$$

for each A in V_*. Thus we have the following serially-commuting diagram of adjunctions, where the vertical functors are given by the forgetful inclusions.

$$\begin{array}{ccc}
\mathsf{Set}_* & \underset{\widehat{U}}{\overset{\widehat{F}}{\rightleftarrows}}\ (\perp) & \mathsf{V}_* \\
\downarrow & & \downarrow \\
\mathsf{Set} & \underset{\mathsf{V}_*(E,-)}{\overset{F_E}{\rightleftarrows}}\ (\perp) & \mathsf{V}
\end{array}$$

In the above diagram, the lower adjunction induces the isomorphism (4.3.14) shown along the bottom edge of (4.3.21). The upper adjunction induces an isomorphism

(4.3.22)
$$(\mathsf{Set}_*\mathsf{Cat})(\mathcal{D},\mathsf{V}_*) \cong \mathsf{V}_*\mathsf{Cat}(\widehat{\mathcal{D}},\underline{\mathsf{V}_*})$$

between V_*-enriched diagrams and Set_*-enriched diagrams. We compose this isomorphism with the equivalence of Lemma 4.3.5 (2) to obtain (4.3.20). The diagram (4.3.21) commutes by commutativity of the adjunctions above. □

Continuing in the context of Convention 4.3.18, the Day convolution and hom diagram for enriched diagrams from $\widehat{\mathcal{D}}$ to $\underline{\mathsf{V}_*}$ provide a symmetric monoidal closed structure for pointed diagrams from $(\mathcal{D},\mathrm{T}^{\mathcal{D}})$ to $(\mathsf{V}_*,\mathrm{T}^{\mathsf{V}})$. We record these here for later reference, using $\widehat{\mathcal{D}} \wedge \widehat{\mathcal{D}}$ to denote the monoidal product as V_*-categories, called the tensor product in Definition 1.3.3.

Definition 4.3.23. In the context of Convention 4.3.18, suppose given pointed diagrams

$$X, Y \in \mathsf{Cat}_*((\mathcal{D}, \mathrm{T}^{\mathcal{D}}), (\mathsf{V}_*, \mathrm{T}^{\mathsf{V}})).$$

We will use the notation

$$\mathcal{D}^\flat(x,y) = \big(\mathcal{D}(x,y)\big)^\flat \quad \text{for} \quad x, y \in \mathcal{D}.$$

The Day convolution product for X and Y is the V_*-coend

$$X \wedge Y = \int^{(a,b) \in \widehat{\mathcal{D}} \wedge \widehat{\mathcal{D}}} \bigvee_{\mathcal{D}^\flat(a \boxdot b, -)} X_a \wedge Y_b \tag{4.3.24}$$

where we remind the reader that an empty wedge is T^{V} by choice in the definition of $\widehat{F}$ (Definition 4.3.9) and $\widehat{\mathcal{D}}$ (4.3.12). If the input object is $\mathrm{T}^{\mathcal{D}}$, we choose T^{V} for the coend. This definition is naturally isomorphic to the Day convolution (3.7.4) for $\widehat{\mathcal{D}}$ and V_*, via the isomorphisms

$$\widehat{\mathcal{D}}(a \boxdot b, c) \wedge (X_a \wedge Y_b) = \bigvee_{\mathcal{D}^\flat(a \boxdot b, c)} E \wedge (X_a \wedge Y_b) \cong \bigvee_{\mathcal{D}^\flat(a \boxdot b, c)} X_a \wedge Y_b$$

induced by left unit isomorphisms in V_*, for a, b, and c in $\mathcal{D}$.

The hom diagram from X to Y is the V_*-end

$$\mathsf{Hom}_{\mathcal{D}_*}(X,Y) = \int_{(b,c) \in \widehat{\mathcal{D}} \wedge \widehat{\mathcal{D}}} \Big[\bigvee_{\mathcal{D}^\flat(- \boxdot b; c)} X_b, Y_c \Big]_* \tag{4.3.25}$$

$$\cong \int_{b \in \widehat{\mathcal{D}}} [X_b, Y_{- \boxdot b}]_* \tag{4.3.26}$$

where the isomorphism is from (3.7.12). We again recall that an empty wedge is T^{V} by choice in the definition of $\widehat{F}$ (Definition 4.3.9). When the input object is $\mathrm{T}^{\mathcal{D}}$, we choose T^{V} for the end. This definition is naturally isomorphic to the hom diagram (3.7.5) for $\widehat{\mathcal{D}}$ and V_*, via the isomorphisms

$$\big[\widehat{\mathcal{D}}(a \boxdot b, c) \wedge X_b, Y_c\big]_* = \Big[\bigvee_{\mathcal{D}^\flat(a \boxdot b, c)} E \wedge X_b, Y_c \Big]_* \cong \Big[\bigvee_{\mathcal{D}^\flat(a \boxdot b, c)} X_b, Y_c \Big]_*$$

induced by the left unit isomorphisms in V_*, for a, b, and c in $\mathcal{D}$.

The mapping object from X to Y is the V_*-end

$$\mathsf{Map}_{\mathcal{D}_*}(X,Y) = \int_{b \in \widehat{\mathcal{D}}} [X_b, Y_b]_* \cong \mathsf{Hom}_{\mathcal{D}_*}(X,Y)_e \tag{4.3.27}$$

where the isomorphism follows from (3.7.12) and the unit isomorphisms in $\mathcal{D}$.

The monoidal unit diagram is the V_*-functor

$$J(-) = \bigvee_{\mathcal{D}^\flat(e,-)} E \colon \widehat{\mathcal{D}} \longrightarrow \underline{\mathsf{V}_*} \tag{4.3.28}$$

where, as above, we take the empty wedge to be T^{V}. ◇

Explanation 4.3.29. The Day convolution, hom diagram, and monoidal unit defined in Definition 4.3.23 are all defined as V_*-enriched functors

$$\widehat{\mathcal{D}} \longrightarrow \underline{\mathsf{V}_*}.$$

Equivalently, using the isomorphism (4.3.22) from the proof of Theorem 4.3.19, these are Set_*-enriched functors

$$\mathcal{D} \longrightarrow \mathsf{V}_*.$$

Our choices of initial and terminal object when the inputs are $\mathrm{T}^{\mathcal{D}}$ make each of these a pointed functor

$$(\mathcal{D}, \mathrm{T}^{\mathcal{D}}) \longrightarrow (\mathsf{V}_*, \mathrm{T}^{\mathsf{V}}).$$

Moreover, the associativity, unit, and symmetry isomorphisms of Definition 3.7.17 are all pointed natural transformations with these definitions. We will use the notation of Definition 4.3.23 for any of these three contexts. ◇

Combining Definition 4.3.23 and Explanation 4.3.29, we have the following corollary of Theorems 3.7.22 and 4.3.19.

Corollary 4.3.30. *In the context of Convention 4.3.18, the Day convolution, monoidal unit, and hom diagram for pointed functors*

$$(\mathcal{D}, \mathrm{T}^{\mathcal{D}}) \longrightarrow (\mathsf{V}_*, \mathrm{T}^{\mathsf{V}})$$

provide a symmetric monoidal closed structure for

$$\mathsf{Cat}_*((\mathcal{D}, \mathrm{T}^{\mathcal{D}}), (\mathsf{V}_*, \mathrm{T}^{\mathsf{V}})).$$

This makes the functors in the equivalence (4.3.20) *into strong symmetric monoidal functors.*

Recall from Definition 3.7.1 we have the notation

$$(\widehat{\mathcal{D}})\text{-}(\mathsf{V}_*) = \mathsf{V}_*\mathsf{Cat}(\widehat{\mathcal{D}}, \underline{\mathsf{V}_*}) \tag{4.3.31}$$

for the category of enriched $\widehat{\mathcal{D}}$-shaped diagrams in the self-enrichment of V_*.

Definition 4.3.32. In the context of Convention 4.3.18, we let

$$\mathcal{D}_*\text{-}\mathsf{V} = \mathsf{Cat}_*\big((\mathcal{D}, \mathrm{T}^{\mathcal{D}}), (\mathsf{V}_*, \mathrm{T}^{\mathsf{V}})\big) \tag{4.3.33}$$

◇

Explanation 4.3.34. In the notation of Definition 4.3.32, Theorem 4.3.19 gives an equivalence of categories

$$\mathcal{D}_*\text{-}\mathsf{V} \simeq (\widehat{\mathcal{D}})\text{-}(\mathsf{V}_*) \tag{4.3.35}$$

and Definition 4.3.23, together with Explanation 4.3.29, uses this equivalence to define

$$(\mathcal{D}_*\text{-}\mathsf{V}, \wedge, J, \mathsf{Hom}_{\mathcal{D}_*})$$

as a symmetric monoidal closed category such that (4.3.35) is an equivalence via strong symmetric monoidal functors. ◇

As an application of Lemmas 3.8.1 and 3.9.12 together with the equivalence of categories from Theorem 4.3.19, evaluation at the monoidal unit of $\widehat{\mathcal{D}}$ defines a symmetric monoidal functor ev_e with left adjoint we again denote L_e. The left adjoint

$$L_e \colon \mathsf{V}_* \rightleftarrows \mathcal{D}_*\text{-}\mathsf{V} \colon \mathsf{ev}_e \tag{4.3.36}$$

is strong symmetric monoidal by Lemma 3.9.12 and Corollary 4.3.30.

Theorem 4.3.37. *Suppose* $(\mathsf{V}, \otimes, \mathbb{1}, \mathrm{T}^{\mathsf{V}})$ *is a complete and cocomplete symmetric monoidal closed category with chosen terminal object* T^{V}*. Suppose* $(\mathcal{D}, \boxdot, e, \mathrm{T}^{\mathcal{D}})$ *is a small symmetric monoidal category with chosen null object* $\mathrm{T}^{\mathcal{D}}$*. Then the category of pointed diagrams*

$$\mathcal{D}_*\text{-}\mathsf{V} = \mathsf{Cat}_*\big((\mathcal{D}, \mathrm{T}^{\mathcal{D}}), (\mathsf{V}_*, \mathrm{T}^{\mathsf{V}})\big)$$

is a complete and cocomplete symmetric monoidal closed category with

- *monoidal product given by the Day convolution* $\wedge$,
- *internal hom given by* $\mathsf{Hom}_{\mathcal{D}_*}$, *and*
- *monoidal unit* J.

Moreover, the adjunction (L_e, ev_e) *makes the pointed diagram category enriched, tensored, and cotensored over* V_* *with mapping objects given by* $\mathsf{Map}_{\mathcal{D}_*}$.

Proof. Limits and colimits of pointed diagrams are computed objectwise, with functoriality determined by the universal properties. The symmetric monoidal closed structure is from Definition 4.3.23 and Explanation 4.3.29 together with Theorems 3.7.22 and 4.3.19. The tensor and cotensor over V_* follows from the adjunction (4.3.36) and Corollary 3.9.15 or from Explanation 3.9.16 and the equivalence of Theorem 4.3.19 directly. □

4.4. Notes

4.4.1 (Smash and Pointed Hom)**.** Our proof of the $\wedge$-Hom_* adjunction in Theorem 4.2.3 follows that of [**EM09**, Lemma 4.20]. An alternate proof of Theorem 4.1.8, using the Yoneda Lemma and the $\wedge$-Hom_* adjunction from the proof of Theorem 4.2.3 instead of our direct pushout arguments, is sketched in the proofs of [**EM09**, Theorem 4.11 and Lemma 4.20]. ◇

4.4.2 (Choices for Pointed Convolution, Hom, and Monoidal Unit Diagrams)**.** In Definition 4.3.23 we emphasize choosing T^{V}, instead of some other initial and terminal object, for the empty wedges and constant co/ends when the input object is $\mathrm{T}^{\mathcal{D}}$. These choices make the convolution, hom, and monoidal unit diagrams automatically pointed diagrams.

Without these conventions, the respective diagrams will still be enriched in Set_* and one can alternatively apply the strictification $(-)'$ defined in the proof of Lemma 4.3.5 to obtain pointed convolution, hom, and monoidal unit diagrams. ◇

CHAPTER 5

Multicategories

In this chapter and the next we recall parts of the theory of multicategories. We will use this in Chapters 8, 9, and 10 to define the K-theory functors K^{Se} (Segal) and K^{EM} (Elmendorf-Mandell) that construct symmetric spectra from small permutative categories. We will show that K^{EM} preserves Cat-enrichment, as defined in Chapter 6, and then use this in Part 2, Chapters 11, 12, and 13, to transport E_n-monoidal structures on small permutative categories to E_n-monoidal structures on the corresponding symmetric spectra.

In Section 5.1 we give the basic definitions and define the 2-category of small multicategories, denoted $\mathsf{Multicat}$. In Section 5.2 we discuss the Cartesian product of multicategories. Then we define pointed multicategories in Section 5.3 and explain the canonical inclusions of permutative categories as (pointed) multicategories. Our K-theory constructions in Part 2 will make use of these inclusions.

The remaining sections are devoted to showing that $\mathsf{Multicat}$ is symmetric monoidal closed. In Section 5.4 we review the general theory of limits and colimits for algebras over monads. We apply this in Section 5.5 to show that $\mathsf{Multicat}$ is complete and cocomplete. Then in Sections 5.6 and 5.7 we use certain colimits and limits to develop the tensor product and internal hom for $\mathsf{Multicat}$. Applying the general theory of Chapter 4, these descend to the category of small pointed multicategories, giving a smash product and pointed hom.

5.1. The 2-Category of Multicategories

We begin with some notation and then give the definition of multicategory.

Definition 5.1.1. Suppose C is a class.

(1) Denote by
$$\mathsf{Prof}(C) = \coprod_{n \geq 0} C^{\times n}$$
the class of finite ordered sequences of elements in C. An element in $\mathsf{Prof}(C)$ is called a *C-profile.*

(2) A typical C-profile of length $n = \operatorname{len}\langle c \rangle$ is denoted by $\langle c \rangle = (c_1, \ldots, c_n) \in C^{\times n}$ or by $\langle c_i \rangle_i$ to indicate the indexing variable. The empty C-profile is denoted by $\langle \rangle$.

(3) We let $\oplus$ denote the concatenation of profiles, and note that $\oplus$ is an associative binary operation with unit given by the empty tuple $\langle \rangle$.

(4) An element in $\mathsf{Prof}(C) \times C$ is denoted as $\big(\langle c \rangle ; c'\big)$ with $c' \in C$ and $\langle c \rangle \in \mathsf{Prof}(C)$. ◇

Now we come to the definition of multicategory. In the literature, what we call a multicategory in the next definition is sometimes called a *symmetric multicategory* or a *colored operad*, with the term multicategory reserved for the non-symmetric version. Since we only consider the version with symmetric group action, we simply call them multicategories.

Definition 5.1.2. A *multicategory* $(\mathsf{M}, \gamma, 1)$ consists of the following data.

- M is equipped with a class $\mathsf{Ob}\,\mathsf{M}$ of *objects*. We abbreviate $\mathsf{Prof}(\mathsf{Ob}\,\mathsf{M})$ as $\mathsf{Prof}(\mathsf{M})$.
- For $c' \in \mathsf{Ob}\,\mathsf{M}$ and $\langle c \rangle = (c_1, \ldots, c_n) \in \mathsf{Prof}(\mathsf{M})$, M is equipped with a set
$$\mathsf{M}\big(\langle c \rangle\,;c'\big) = \mathsf{M}(c_1, \ldots, c_n\,;c')$$
of *n-ary operations* with *input profile* $\langle c \rangle$ and *output* c'.
- For $\big(\langle c \rangle\,;c'\big) \in \mathsf{Prof}(\mathsf{M}) \times \mathsf{Ob}\,\mathsf{M}$ as above and a permutation $\sigma \in \Sigma_n$, M is equipped with a bijection
$$\mathsf{M}\big(\langle c \rangle\,;c'\big) \xrightarrow[\cong]{\sigma} \mathsf{M}\big(\langle c \rangle\sigma\,;c'\big),$$
called the *right action* or the *symmetric group action*, in which
$$\langle c \rangle\sigma = \big(c_{\sigma(1)}, \ldots, c_{\sigma(n)}\big)$$
is the right permutation of $\langle c \rangle$ by σ.
- For $c \in \mathsf{Ob}\,\mathsf{M}$, M is equipped with an element
$$1_c \in \mathsf{M}(c\,;c),$$
called the *c-colored unit.*
- For $c'' \in \mathsf{Ob}\,\mathsf{M}$, $\langle c' \rangle = (c'_1, \ldots, c'_n) \in \mathsf{Prof}(\mathsf{M})$, and $\langle c_j \rangle = (c_{j,1}, \ldots, c_{j,k_j}) \in \mathsf{Prof}(\mathsf{M})$ for each $j \in \{1, \ldots, n\}$, let $\langle c \rangle = \oplus_j \langle c_j \rangle \in \mathsf{Prof}(\mathsf{M})$ be the concatenation of the $\langle c_j \rangle$. Then M is equipped with a map
$$\mathsf{M}\big(\langle c' \rangle\,;c''\big) \times \prod_{j=1}^{n} \mathsf{M}\big(\langle c_j \rangle\,;c'_j\big) \xrightarrow{\ \gamma\ } \mathsf{M}\big(\langle c \rangle\,;c''\big) \tag{5.1.3}$$
called the *composition.*

These data are required to satisfy the following axioms.

Symmetric Group Action: For $\big(\langle c \rangle\,;c'\big) \in \mathsf{Prof}(\mathsf{M}) \times \mathsf{Ob}\,\mathsf{M}$ with $n = \mathrm{len}\langle c \rangle$ and $\sigma, \tau \in \Sigma_n$, the diagram
$$\begin{array}{ccc} \mathsf{M}\big(\langle c \rangle\,;c'\big) & \xrightarrow{\ \sigma\ } & \mathsf{M}\big(\langle c \rangle\sigma\,;c'\big) \\ & {}_{\sigma\tau}\searrow & \downarrow{\scriptstyle \tau} \\ & & \mathsf{M}\big(\langle c \rangle\sigma\tau\,;c'\big) \end{array}$$
is commutative. Moreover, the identity permutation in Σ_n acts as the identity map on $\mathsf{M}\big(\langle c \rangle\,;c'\big)$.

Associativity: Suppose given
- $c''' \in \mathsf{Ob}\,\mathsf{M}$,
- $\langle c'' \rangle = (c''_1, \ldots, c''_n) \in \mathsf{Prof}(\mathsf{M})$,
- $\langle c'_j \rangle = (c'_{j,1}, \ldots, c'_{j,k_j}) \in \mathsf{Prof}(\mathsf{M})$ for each $j \in \{1, \ldots, n\}$, and
- $\langle c_{j,i} \rangle = (c_{j,i,1}, \ldots, c_{j,i,\ell_{j,i}}) \in \mathsf{Prof}(\mathsf{M})$ for each $j \in \{1, \ldots, n\}$ and each $i \in \{1, \ldots, k_j\}$,

such that $k_j = \operatorname{len}\langle c_j'\rangle > 0$ for at least one j. For each j, let $\langle c_j\rangle = \oplus_{i=1}^{k_j}\langle c_{j,i}\rangle$ denote the concatenation of the $\langle c_{j,i}\rangle$. Let $\langle c\rangle = \oplus_{j=1}^{n}\langle c_j\rangle$ denote the concatenation of the $\langle c_j\rangle$. Let $\langle c'\rangle = \oplus_{j=1}^{n}\langle c_j'\rangle$ denote the concatenation of the $\langle c_j'\rangle$.

Then the *associativity diagram* below commutes.

(5.1.4)
$$\begin{array}{c}
\mathsf{M}(\langle c''\rangle;c''') \times \Big[\prod_{j=1}^{n}\mathsf{M}(\langle c_j'\rangle;c_j'')\Big] \times \prod_{j=1}^{n}\Big[\prod_{i=1}^{k_j}\mathsf{M}(\langle c_{j,i}\rangle;c_{j,i}')\Big] \xrightarrow{(\gamma,1)} \mathsf{M}(\langle c'\rangle;c''') \times \prod_{j=1}^{n}\Big[\prod_{i=1}^{k_j}\mathsf{M}(\langle c_{j,i}\rangle;c_{j,i}')\Big] \xrightarrow{\gamma} \mathsf{M}(\langle c\rangle;c''')\\
\text{permute}\ \Big\downarrow\cong\\
\mathsf{M}(\langle c''\rangle;c''') \times \prod_{j=1}^{n}\Big[\mathsf{M}(\langle c_j'\rangle;c_j'') \times \prod_{i=1}^{k_j}\mathsf{M}(\langle c_{j,i}\rangle;c_{j,i}')\Big] \xrightarrow{(1,\prod_j\gamma)} \mathsf{M}(\langle c''\rangle;c''') \times \prod_{j=1}^{n}\mathsf{M}(\langle c_j\rangle;c_j'') \xrightarrow{\gamma} \mathsf{M}(\langle c\rangle;c''')
\end{array}$$

Unity: Suppose $c' \in \operatorname{Ob}\mathsf{M}$.

(1) If $\langle c\rangle = (c_1,\ldots,c_n) \in \mathsf{Prof}(\mathsf{M})$ has length $n \geq 1$, then the following *right unity diagram* is commutative. Here $\{*\}$ is the one-point set, and $\{*\}^n$ is its n-fold product.

(5.1.5)
$$\begin{array}{ccc}
\mathsf{M}(\langle c\rangle;c') \times \{*\}^n & \xrightarrow{\cong} & \mathsf{M}(\langle c\rangle;c')\\
{\scriptstyle (1,\sqcap 1_{c_j})}\big\downarrow & & \big\downarrow{\scriptstyle 1}\\
\mathsf{M}(\langle c\rangle;c') \times \prod_{j=1}^{n}\mathsf{M}(c_j;c_j) & \xrightarrow{\gamma} & \mathsf{M}(\langle c\rangle;c')
\end{array}$$

(2) For any $\langle c\rangle \in \mathsf{Prof}(\mathsf{M})$, the *left unity diagram* below is commutative.

(5.1.6)
$$\begin{array}{ccc}
\{*\} \times \mathsf{M}(\langle c\rangle;c') & \xrightarrow{\cong} & \mathsf{M}(\langle c\rangle;c')\\
{\scriptstyle (1_{c'},1)}\big\downarrow & & \big\downarrow{\scriptstyle 1}\\
\mathsf{M}(c';c') \times \mathsf{M}(\langle c\rangle;c') & \xrightarrow{\gamma} & \mathsf{M}(\langle c\rangle;c')
\end{array}$$

Equivariance: Suppose that in the definition of γ (5.1.3), $\operatorname{len}\langle c_j\rangle = k_j \geq 0$.

(1) For each $\sigma \in \Sigma_n$, the following *top equivariance diagram* is commutative.

(5.1.7)

$$\begin{array}{ccc} \mathsf{M}\big(\langle c'\rangle; c''\big) \times \prod_{j=1}^{n} \mathsf{M}\big(\langle c_j\rangle; c'_j\big) & \xrightarrow{(\sigma,\sigma^{-1})} & \mathsf{M}\big(\langle c'\rangle\sigma; c''\big) \times \prod_{j=1}^{n} \mathsf{M}\big(\langle c_{\sigma(j)}\rangle; c'_{\sigma(j)}\big) \\ \gamma\downarrow & & \downarrow\gamma \\ \mathsf{M}\big(\langle c_1\rangle, \ldots, \langle c_n\rangle; c''\big) & \xrightarrow{\sigma\langle k_{\sigma(1)},\ldots,k_{\sigma(n)}\rangle} & \mathsf{M}\big(\langle c_{\sigma(1)}\rangle, \ldots, \langle c_{\sigma(n)}\rangle; c''\big) \end{array}$$

Here $\sigma\langle k_{\sigma(1)}, \ldots, k_{\sigma(n)}\rangle \in \Sigma_{k_1+\cdots+k_n}$ is right action of the block permutation (II.1.1.19) that permutes the n consecutive blocks of lengths $k_{\sigma(1)}, \ldots, k_{\sigma(n)}$ as σ permutes $\{1, \ldots, n\}$, leaving the relative order within each block unchanged.

(2) Given permutations $\tau_j \in \Sigma_{k_j}$ for $1 \leq j \leq n$, the following *bottom equivariance diagram* is commutative.

$$\begin{array}{ccc} \mathsf{M}\big(\langle c'\rangle; c''\big) \times \prod_{j=1}^{n} \mathsf{M}\big(\langle c_j\rangle; c'_j\big) & \xrightarrow{(1,\prod\tau_j)} & \mathsf{M}\big(\langle c'\rangle; c''\big) \times \prod_{j=1}^{n} \mathsf{M}\big(\langle c_j\rangle\tau_j; c'_j\big) \\ \gamma\downarrow & & \downarrow\gamma \\ \mathsf{M}\big(\langle c_1\rangle, \ldots, \langle c_n\rangle; c''\big) & \xrightarrow{\tau_1\times\cdots\times\tau_n} & \mathsf{M}\big(\langle c_1\rangle\tau_1, \ldots, \langle c_n\rangle\tau_n; c''\big) \end{array} \tag{5.1.8}$$

Here $\tau_1 \times \cdots \times \tau_n \in \Sigma_{k_1+\cdots+k_n}$ is the block sum (II.1.1.8) given by the image of $(\tau_1, \ldots, \tau_n)$ under the canonical inclusion

$$\Sigma_{k_1} \times \cdots \times \Sigma_{k_n} \longrightarrow \Sigma_{k_1+\cdots+k_n}.$$

This finishes the definition of a multicategory.

Moreover:

- A multicategory with only one object is an *operad.*
- If M is an operad, then its set of n-ary operations is denoted by M_n.
- A multicategory is *small* if its class of objects is a set.
- We will denote the image of an operation $\phi \in \mathsf{M}\big(\langle c\rangle; c'\big)$ under the right action of a permutation σ as $\phi \cdot \sigma \in \mathsf{M}\big(\langle c\rangle\sigma; c'\big)$.
- We will denote the composition of operations $\phi' \in \mathsf{M}\big(\langle c'\rangle; c''\big)$ and $\phi_j \in \mathsf{M}\big(\langle c_j\rangle; c'_j\big)$ with juxtaposition or $\circ$, as in the following:

$$\gamma(\phi', (\phi_1, \ldots, \phi_n)) = \phi' \circ (\phi_1, \ldots, \phi_n) = (\phi')(\phi_1, \ldots, \phi_n). \qquad \diamond$$

Example 5.1.9 (Unary Multicategories)**.** Suppose M is a category. Then there is a multicategory M^u whose object class is $\operatorname{Ob}\mathsf{M}$ and whose operations are given by

$$\mathsf{M}^u\big(\langle c\rangle; c'\big) = \begin{cases} \mathsf{M}(c, c') & \text{if } \langle c\rangle = c \in \operatorname{Ob}\mathsf{M}, \\ \varnothing & \text{otherwise.} \end{cases}$$

In other words, M^u only has unary operations, which are the morphisms in M. The colored units and composition in M^u are given by those in M. There are no non-trivial symmetric group actions, and the multicategory axioms reduce to the axioms of a category.

Conversely, if M is a multicategory then there is an underlying category, $r\mathsf{M}$ consisting of the same objects and morphisms given by the unary operations of M. ◇

Example 5.1.10 (Endomorphism Operad). Suppose X is an object of a permutative category $(\mathsf{C},\oplus,e)$. Then there is a multicategory $\mathsf{End}(X)$ called the *endomorphism operad* with a single object and with n-ary operations

$$\mathsf{End}(X)_n = \mathsf{C}(X^{\oplus n}, X).$$

One detail of this example that will be important for our work below is the composition with nullary operations $\mathsf{End}(X)_0 = \mathsf{C}(e,X)$. For example, the composition

$$\mathsf{End}(X)_2 \times \big(\mathsf{End}(X)_0 \times \mathsf{End}(X)_1\big) \longrightarrow \mathsf{End}(X)_1$$

is given by the following composite for $f_i \in \mathsf{End}(X)_i$:

$$X \longrightarrow e \oplus X \xrightarrow{f_0 \oplus f_1} X \oplus X \xrightarrow{f_2} X.$$

The unlabeled morphism is given by (the inverse of) a unit isomorphism, which is an identity because C is permutative. We will recall this observation in the proof of Proposition 8.5.4 below. ◇

Example 5.1.11 (Endomorphism Multicategory). Suppose $(\mathsf{C},\oplus,e)$ is a small permutative category. Then Example 5.1.10 generalizes to define the *endomorphism multicategory* $\mathsf{End}(\mathsf{C})$ with object set given by $\mathsf{Ob}\,\mathsf{C}$ and with

$$\mathsf{End}(\mathsf{C})\big(\langle X\rangle\,;Y\big) = \mathsf{C}(X_1 \oplus \cdots \oplus X_n, Y)$$

for $Y \in \mathsf{Ob}\,\mathsf{C}$ and $\langle X\rangle = (X_1,\cdots,X_n) \in (\mathsf{Ob}\,\mathsf{C})^{\times n}$. We note as in Example 5.1.10 that compositions involving the empty tuple are defined via the (strict) unit isomorphisms of C. ◇

Definition 5.1.12. A *multifunctor* $F : \mathsf{M} \longrightarrow \mathsf{N}$ between multicategories M and N consists of the following data:

- an assignment
$$F : \mathsf{Ob}\,\mathsf{M} \longrightarrow \mathsf{Ob}\,\mathsf{N},$$
where $\mathsf{Ob}\,\mathsf{M}$ and $\mathsf{Ob}\,\mathsf{N}$ are the classes of objects of M and N, respectively, and
- for each $\big(\langle c\rangle\,;c'\big) \in \mathsf{Prof}(\mathsf{M}) \times \mathsf{Ob}\,\mathsf{M}$ with $\langle c\rangle = (c_1,\ldots,c_n)$, a function
$$F : \mathsf{M}\big(\langle c\rangle\,;c'\big) \longrightarrow \mathsf{N}\big(F\langle c\rangle\,;Fc'\big),$$
where $F\langle c\rangle = (Fc_1,\ldots,Fc_n)$.

These data are required to preserve the symmetric group action, the colored units, and the composition in the following sense.

Symmetric Group Action: For each $\big(\langle c\rangle\,;c'\big)$ as above and each permutation $\sigma \in \Sigma_n$, the following diagram is commutative.

$$\begin{array}{ccc} \mathsf{M}\big(\langle c\rangle\,;c'\big) & \xrightarrow{F} & \mathsf{N}\big(F\langle c\rangle\,;Fc'\big) \\ {\scriptstyle\sigma}\downarrow{\scriptstyle\cong} & & {\scriptstyle\sigma}\downarrow{\scriptstyle\cong} \\ \mathsf{M}\big(\langle c\rangle\sigma\,;c'\big) & \xrightarrow{F} & \mathsf{N}\big(F\langle c\rangle\sigma\,;c'\big) \end{array} \tag{5.1.13}$$

Units: For each $c \in \mathsf{Ob}\,\mathsf{M}$, the following equality holds:

$$F1_c = 1_{Fc} \in \mathsf{N}\big(Fc\,;Fc\big). \tag{5.1.14}$$

Composition: For c'', $\langle c_j'\rangle$, and $\langle c\rangle = \oplus_j \langle c_j\rangle$ as in the definition of γ (5.1.3), the following diagram is commutative.

$$\begin{array}{ccc} \mathsf{M}\big(\langle c'\rangle\,;c''\big) \times \prod\limits_{j=1}^{n} \mathsf{M}\big(\langle c_j\rangle\,;c_j'\big) & \xrightarrow{(F,\prod F)} & \mathsf{N}\big(F\langle c'\rangle\,;Fc''\big) \times \prod\limits_{j=1}^{n} \mathsf{N}\big(F\langle c_j\rangle\,;Fc_j'\big) \\ {\scriptstyle\gamma}\big\downarrow & & \big\downarrow{\scriptstyle\gamma} \\ \mathsf{M}\big(\langle c\rangle\,;c''\big) & \xrightarrow{\quad F \quad} & \mathsf{N}\big(F\langle c\rangle\,;Fc''\big) \end{array} \tag{5.1.15}$$

This finishes the definition of a multifunctor.

Moreover:

(1) For another multifunctor $G : \mathsf{N} \longrightarrow \mathsf{P}$ between multicategories, where P has object class $\mathsf{Ob}\,\mathsf{P}$, the *composition* $GF : \mathsf{M} \longrightarrow \mathsf{P}$ is the multifunctor defined by composing the assignments on objects

$$\mathsf{Ob}\,\mathsf{M} \xrightarrow{F} \mathsf{Ob}\,\mathsf{N} \xrightarrow{G} \mathsf{Ob}\,\mathsf{P}$$

and the functions on n-ary operations

$$\mathsf{M}\big(\langle c\rangle\,;c'\big) \xrightarrow{F} \mathsf{N}\big(F\langle c\rangle\,;Fc'\big) \xrightarrow{G} \mathsf{P}\big(GF\langle c\rangle\,;GFc'\big).$$

(2) The *identity multifunctor* $1_{\mathsf{M}} : \mathsf{M} \longrightarrow \mathsf{M}$ is defined by the identity assignment on objects and the identity function on n-ary operations.

(3) An *operad morphism* is a multifunctor between two multicategories with one object. ◇

Lemma 5.1.16. *Composition of multifunctors is well defined, associative, and unital with respect to the identity multifunctors.*

Proof. For multifunctors $F : \mathsf{M} \longrightarrow \mathsf{N}$ and $G : \mathsf{N} \longrightarrow \mathsf{P}$, GF preserves the compositions in M and P because the following diagram is commutative.

$$\begin{array}{cccccc} & \mathsf{M}\big(\langle c'\rangle\,;c''\big) \times \prod\limits_{j=1}^{n} \mathsf{M}\big(\langle c_j\rangle\,;c_j'\big) & \xrightarrow{\quad\gamma\quad} & \mathsf{M}\big(\langle c\rangle\,;c''\big) & \\ & {\scriptstyle(F,\prod F)}\big\downarrow & & \big\downarrow{\scriptstyle F} & \\ {\scriptstyle(GF,\prod GF)} & \mathsf{N}\big(F\langle c'\rangle\,;Fc''\big) \times \prod\limits_{j=1}^{n} \mathsf{N}\big(F\langle c_j\rangle\,;Fc_j'\big) & \xrightarrow{\quad\gamma\quad} & \mathsf{N}\big(F\langle c\rangle\,;Fc''\big) & {\scriptstyle GF} \\ & {\scriptstyle(G,\prod G)}\big\downarrow & & \big\downarrow{\scriptstyle G} & \\ & \mathsf{P}\big(GF\langle c'\rangle\,;GFc''\big) \times \prod\limits_{j=1}^{n} \mathsf{P}\big(GF\langle c_j\rangle\,;GFc_j'\big) & \xrightarrow{\gamma} & \mathsf{P}\big(GF\langle c\rangle\,;GFc''\big) & \end{array}$$

Similarly, GF preserves the symmetric group actions and the colored units in M and P, so it is a multifunctor. Composition of multifunctors is associative and unital because composition of functions is associative and unital. □

Natural transformations and their horizontal and vertical compositions also have direct generalizations to multicategories.

Definition 5.1.17. Suppose $F, G : \mathsf{M} \longrightarrow \mathsf{N}$ are multifunctors as in Definition 5.1.12. A *multinatural transformation* $\alpha : F \longrightarrow G$ consists of unary operations

$$\alpha_c \in \mathsf{N}\big(Fc\,;Gc\big) \quad \text{for} \quad c \in \mathsf{Ob}\,\mathsf{M}$$

such that, for each n-ary operation $p \in \mathsf{M}(\langle c\rangle; c')$ with $\langle c\rangle = (c_1, \ldots, c_n)$, the following *naturality condition* holds, with composition taken in N:

$$(Gp) \circ (\alpha_{c_1}, \ldots, \alpha_{c_n}) = \alpha_{c'} \circ (Fp) \in \mathsf{N}(F\langle c\rangle; Gc').$$

- Each α_c is called a *component* of α.
- The *identity multinatural transformation* $1_F : F \longrightarrow F$ has components

$$(1_F)_c = 1_{Fc} \in \mathsf{N}(Fc; Fc) \quad \text{for} \quad c \in \mathsf{Ob}\,\mathsf{M}. \qquad \diamond$$

Definition 5.1.18. Suppose $\alpha : F \longrightarrow G$ is a multinatural transformation between multifunctors as in Definition 5.1.17.

(1) Suppose $\beta : G \longrightarrow H$ is a multinatural transformation for a multifunctor $H : \mathsf{M} \longrightarrow \mathsf{N}$. The *vertical composition*

$$\beta\alpha : F \longrightarrow H$$

is the multinatural transformation with components given by composition in N

$$(\beta\alpha)_c = \beta_c \circ \alpha_c \in \mathsf{N}(Fc; Hc) \quad \text{for} \quad c \in \mathsf{Ob}\,\mathsf{M}.$$

(2) Suppose $\alpha' : F' \longrightarrow G'$ is a multinatural transformation for multifunctors $F', G' : \mathsf{N} \longrightarrow \mathsf{P}$. The *horizontal composition*

$$\alpha' * \alpha : F'F \longrightarrow G'G$$

is the multinatural transformation with components given by composition in P

$$(\alpha' * \alpha)_c = \alpha'_{Gc} \circ (F'\alpha_c) = (G'\alpha_c) \circ (\alpha'_{Fc}) \in \mathsf{P}(F'Fc; G'Gc)$$

for each object $c \in \mathsf{Ob}\,\mathsf{M}$, in which the second equality follows from the naturality of α'. $\diamond$

Example 5.1.19 (Functors and Natural Transformations)**.** A functor $F : \mathsf{M} \longrightarrow \mathsf{N}$ between categories is also a multifunctor when M and N are regarded as multicategories with only unary operations as in Example 5.1.9. Similarly, a natural transformation $\alpha : F \longrightarrow F'$ between functors is a multinatural transformation when F and F' are regarded as multifunctors. $\diamond$

Theorem 5.1.20. *There is a 2-category* $\mathsf{Multicat}$ *consisting of the following data.*

- *Its objects are small multicategories.*
- *For small multicategories* M *and* N*, the hom category* $\mathsf{Multicat}(\mathsf{M}, \mathsf{N})$ *has:*
  - *multifunctors* $\mathsf{M} \longrightarrow \mathsf{N}$ *as 1-cells;*
  - *multinatural transformations between such multifunctors as 2-cells;*
  - *vertical composition as composition; and*
  - *identity multinatural transformations as identity 2-cells.*
- *The identity 1-cell* 1_{M} *is the identity multifunctor* 1_{M}.
- *Horizontal composition of 1-cells is the composition of multifunctors.*
- *Horizontal composition of 2-cells is that of multinatural transformations.*

Definition 5.1.21. The *initial operad* I consists of a single object $*$ and a single unary operation given by the unit 1_* on its single object. $\diamond$

Explanation 5.1.22. For $n \neq 1$, the n-ary operation sets I_n are empty. Since multifunctors preserve identity operations, I is indeed initial among operads. More generally, for a multicategory M, a multifunctor $\mathsf{I} \longrightarrow \mathsf{M}$ consists of a choice of object in M. Multinatural transformations between multifunctors

$$F\colon \mathsf{I} \longrightarrow \mathsf{M} \quad\text{and}\quad G\colon \mathsf{I} \longrightarrow \mathsf{M}$$

consist of unary operations $F* \longrightarrow G*$. Thus, for small M, $\mathsf{Multicat}(\mathsf{I},\mathsf{M})$ is isomorphic to the underlying category of objects and unary operations in M. ◇

Definition 5.1.23. We will use the notation $\mathsf{Multicat}$ both for the 2-category of small multicategories and for its underlying 1-category. ◇

Recall from Definition I.6.3.9 the notions of internal adjunction and equivalence in a bicategory or 2-category.

Definition 5.1.24. An *equivalence*, respectively *adjoint equivalence*, of multicategories is an internal equivalence, respectively adjoint equivalence, in the 2-category of multicategories. ◇

Explanation 5.1.25. Suppose M and N are multicategories. A multifunctor

$$F\colon \mathsf{M} \longrightarrow \mathsf{N}$$

is an equivalence if there is a multifunctor $G\colon \mathsf{N} \longrightarrow \mathsf{M}$ together with isomorphisms of multifunctors

$$\eta\colon 1_{\mathsf{M}} \longrightarrow GF \quad\text{and}\quad \varepsilon\colon FG \longrightarrow 1_{\mathsf{N}}.$$

These data are an adjoint equivalence if they satisfy the addional triangle axioms of an internal adjunction in $\mathsf{Multicat}$. ◇

5.2. The Cartesian Structure on Multicategories

Now we discuss the Cartesian product for multicategories.

Definition 5.2.1. The *terminal multicategory* T consists of a single object and a single n-ary operation ι_n for each $n \geq 0$. ◇

Explanation 5.2.2 (The Commutative Operad)**.** The terminal multicategory T is also denoted Com and is known as the *commutative operad.* For a permutative category C, a multifunctor $P\colon \mathsf{T} \longrightarrow \mathsf{End}(\mathsf{C})$ is precisely a commutative monoid structure on the object $P*$. Its unit is given by the nullary operation $P\iota_0$, its multiplication is given by the binary operation $P\iota_2$, and the axioms (associativity and symmetry) follow from the composition and symmetric group actions on $\mathsf{T} = \mathsf{Com}$ being trivial.

For a general multicategory M, a multifunctor $P\colon \mathsf{T} \longrightarrow \mathsf{M}$ is determined by a choice of object $P*$ and a choice of n-ary operation

$$P\iota_n \in \mathsf{M}\big(\underbrace{(P*,\ldots,P*)}_{n \text{ terms}}; P*\big)$$

for each $n \geq 0$, subject to the following compatibility conditions.

Unity: The equality $P\iota_1 = 1_{P*}$ holds.
Composition: The set of operations $P\iota_n$ is closed under composition in M.
Symmetry: Each $P\iota_n$ is fixed by the right symmetric group action. ◇

Definition 5.2.3. Suppose M and N are multicategories. Their *Cartesian product* is the multicategory $\mathsf{M} \times \mathsf{N}$ whose objects are given by pairs (c,d) with $c \in \mathsf{M}$ and

$d \in \mathsf{N}$ and whose n-ary operations are given by the Cartesian product of sets

$$(\mathsf{M}\times\mathsf{N})\big(\langle c,d\rangle\,;\,(c',d')\big) = \mathsf{M}\big(\langle c\rangle\,;\,c'\big)\times\mathsf{N}\big(\langle d\rangle\,;\,d'\big).$$

for $\langle c,d\rangle = ((c_1,d_1),\ldots,(c_n,d_n)) \in \mathsf{Prof}(\mathsf{Ob}\,\mathsf{M}\times\mathsf{Ob}\,\mathsf{N})$. The other data of $\mathsf{M}\times\mathsf{N}$ is specified as follows.

- The right action of a permutation σ is given by the diagonal
$$\mathsf{M}\big(\langle c\rangle\,;\,c'\big)\times\mathsf{N}\big(\langle d\rangle\,;\,d'\big) \xrightarrow{(\sigma,\sigma)} \mathsf{M}\big(\langle c\rangle\sigma\,;\,c'\big)\times\mathsf{N}\big(\langle d\rangle\sigma\,;\,d'\big)$$
- The (c,d)-colored unit is the pair $(1_c,1_d)$.
- The composition is given by permuting factors and then applying the product of the compositions in M and N:

$$\Big[\mathsf{M}\big(\langle c'\rangle\,;\,c''\big)\times\mathsf{N}\big(\langle d'\rangle\,;\,d''\big)\Big]\times\prod_{j=1}^{n}\Big[\mathsf{M}\big(\langle c_j\rangle\,;\,c_j'\big)\times\mathsf{N}\big(\langle d_j\rangle\,;\,d_j'\big)\Big]$$
$$\searrow$$
$$\Big[\mathsf{M}\big(\langle c'\rangle\,;\,c''\big)\times\prod_{j=1}^{n}\mathsf{M}\big(\langle c_j\rangle\,;\,c_j'\big)\Big]\times\Big[\mathsf{N}\big(\langle d'\rangle\,;\,d''\big)\times\prod_{j=1}^{n}\mathsf{N}\big(\langle d_j\rangle\,;\,d_j'\big)\Big]$$
$$\searrow$$
$$\mathsf{M}\big(\langle c\rangle\,;\,c''\big)\times\mathsf{N}\big(\langle d\rangle\,;\,d''\big)$$

  where $\langle c\rangle = \oplus_j\langle c_j\rangle$ and $\langle d\rangle = \oplus_j\langle d_j\rangle$ are the concatenations of the $\langle c_j\rangle$ and $\langle d_j\rangle$, respectively.

The multicategory axioms for these data follow directly from the axioms for M and N together with the axioms for $\times$ as a symmetric monoidal product on Set. $\diamond$

Definition 5.2.4. For multifunctors

$$F\colon\mathsf{M}\longrightarrow\mathsf{M}' \quad\text{and}\quad G\colon\mathsf{N}\longrightarrow\mathsf{N}',$$

the Cartesian product $F\times G$ is given by the pairwise assignment on objects and the product of the corresponding morphisms on sets of operations:

$$\mathsf{M}\big(\langle c\rangle\,;\,c'\big)\times\mathsf{N}\big(\langle d\rangle\,;\,d'\big)\longrightarrow\mathsf{M}'\big(F\langle c\rangle\,;\,Fc'\big)\times\mathsf{N}'\big(G\langle d\rangle\,;\,Gd'\big).$$

As above, the multifunctor axioms for $F\times G$ follow from those of F and G together with the axioms for $\times$ as a symmetric monoidal product on Set. $\diamond$

Proposition 5.2.5. *The category of small multicategories is symmetric monoidal with respect to the Cartesian product. The unit for this product is the terminal multicategory* T.

Proof. Since T has a single object, and each set of n-ary operations is a singleton, the unit isomorphisms

$$\mathsf{T}\times\mathsf{M}\cong\mathsf{M}\cong\mathsf{M}\times\mathsf{T}$$

for each small multicategory M follow from those for the singleton in Set. Likewise, the symmetry for the Cartesian product of multicategories is given by the symmetry of sets. The axioms making $\times$ a symmetric monoidal product in $\mathsf{Multicat}$ follow from those in Set. For example, the hexagon axiom (I.1.2.22) follows from commutativity of the following diagram in Set for each triple of small multicategories L, M, and N and each triple of inputs and outputs

$$\begin{aligned}
&(\langle b\rangle\,;b') \in \mathsf{Prof}(\mathsf{L}) \times \mathsf{Ob}\,\mathsf{L},\\
&(\langle c\rangle\,;c') \in \mathsf{Prof}(\mathsf{M}) \times \mathsf{Ob}\,\mathsf{M}, \text{ and}\\
&(\langle d\rangle\,;d') \in \mathsf{Prof}(\mathsf{N}) \times \mathsf{Ob}\,\mathsf{N}.
\end{aligned}$$

$$\begin{array}{ccc}
 & \mathsf{L}(\langle b\rangle;b') \times [\mathsf{M}(\langle c\rangle;c') \times \mathsf{N}(\langle d\rangle;d')] & \\
\overset{\cong}{\nearrow} & & \overset{1\times\xi}{\searrow} \\
[\mathsf{L}(\langle b\rangle;b') \times \mathsf{M}(\langle c\rangle;c')] \times \mathsf{N}(\langle d\rangle;d') & & \mathsf{L}(\langle b\rangle;b') \times [\mathsf{N}(\langle d\rangle;d') \times \mathsf{M}(\langle c\rangle;c')] \\
\xi\downarrow & & \downarrow\cong \\
\mathsf{N}(\langle d\rangle;d') \times [\mathsf{L}(\langle b\rangle;b') \times \mathsf{M}(\langle c\rangle;c')] & & [\mathsf{L}(\langle b\rangle;b') \times \mathsf{N}(\langle d\rangle;d')] \times \mathsf{M}(\langle c\rangle;c') \\
\overset{\cong}{\searrow} & & \overset{\xi\times 1}{\nearrow} \\
 & [\mathsf{N}(\langle d\rangle;d') \times \mathsf{L}(\langle b\rangle;b')] \times \mathsf{M}(\langle c\rangle;c') &
\end{array}$$

The other axioms are proved in the same manner. □

5.3. Permutative Categories as Pointed Multicategories

In this section we describe pointed multicategories and the canonical basepoint for the multicategory $\mathsf{End}(\mathsf{C})$ associated to a permutative category C in Example 5.1.11. Then we go on to discuss 2-functoriality of $\mathsf{End}(-)$. The multifunctors induced by symmetric monoidal functors do not generally preserve the canonical basepoints. However, we show below that *strictly unital* symmetric monoidal functors yield pointed multifunctors.

The following is a special case of the more general Definition 4.1.1.

Definition 5.3.1. A *pointed multicategory* $(\mathsf{M}, *^{\mathsf{M}}, \iota^{\mathsf{M}})$ consists of a multicategory M together with a multifunctor $\mathsf{T} \longrightarrow \mathsf{M}$ determined by an object $*^{\mathsf{M}}$ and operations ι_n^{M} described in Explanation 5.2.2. Typically we will omit the superscripts when M is clear from context.

A *pointed multifunctor*

$$F: (\mathsf{M}, *^{\mathsf{M}}, \iota^{\mathsf{M}}) \longrightarrow (\mathsf{N}, *^{\mathsf{N}}, \iota^{\mathsf{N}})$$

is a multifunctor $F: \mathsf{M} \longrightarrow \mathsf{N}$ that commutes with the multifunctors $\mathsf{T} \longrightarrow \mathsf{M}$ and $\mathsf{T} \longrightarrow \mathsf{N}$. A *pointed multinatural transformation* between pointed multifunctors F and G is a multinatural transformation $\alpha: F \longrightarrow G$ such that the component $\alpha_{*^{\mathsf{M}}}: F(*^{\mathsf{M}}) \longrightarrow G(*^{\mathsf{M}})$ is the colored unit on $*^{\mathsf{N}}$. With these definitions, composites of pointed multifunctors and multinatural transformations are again pointed. ◇

Definition 5.3.2. We let $\mathsf{Multicat}_*$ denote the 2-category of pointed small multicategories, pointed multifunctors, and pointed multinatural transformations. The compositions and identity cells of $\mathsf{Multicat}_*$ are defined as in $\mathsf{Multicat}$ (see Theorem 5.1.20). ◇

Explanation 5.3.3. As noted in the introduction to Chapter 4, we will be interested in a smash product for pointed multicategories, but not the one obtained from the Cartesian product. We will develop a tensor product for multicategories in Section 5.6 below, and its associated smash product will be the one of interest in the applications of Part 2. ◇

Definition 5.3.4. Suppose $(\mathsf{C}, \oplus, e)$ is a permutative category. The *canonical basepoint* of $\mathsf{End}(\mathsf{C})$ is given by the unit object e, with n-ary basepoint operations $\iota_n = 1_e$ for all $n \geq 0$. ◇

Explanation 5.3.5 (Strictly Unital Symmetric Monoidal Functors)**.** Recall from Definitions 1.1.6 and 1.1.23 that a symmetric monoidal functor

$$F\colon \mathsf{C} \longrightarrow \mathsf{D}$$

has structure morphisms

$$F^0\colon \mathbb{1}^{\mathsf{C}} \longrightarrow F\mathbb{1}^{\mathsf{D}} \quad \text{and}$$
$$F^2_{X,Y}\colon FX \oplus FY \longrightarrow F(X \oplus Y)$$

satisfying axioms for associativity (1.1.9), unity (1.1.10), and compatibility with the symmetry (1.1.18). We say that F is *strictly unital* if F^0 is the identity morphism. If C and D are permutative and F is strictly unital, then the unity axioms imply that $F^2_{X,Y}$ is an identity whenever X or Y is the monoidal unit of C. Recall from Definition 1.1.27 that

$$\mathsf{PermCat}^{\mathsf{su}}$$

denotes the 2-category consisting of permutative categories, strictly unital symmetric monoidal functors, and monoidal natural transformations. ◇

Proposition 5.3.6. *The construction* $\mathsf{End}(\mathsf{C})$ *of Example 5.1.11 defines a 2-functor*

$$\mathsf{End}\colon \mathsf{PermCat} \longrightarrow \mathsf{Multicat}$$

from the 2-category of small permutative categories to the 2-category of small multicategories.

Proof. Given a symmetric monoidal functor $F\colon \mathsf{C} \longrightarrow \mathsf{D}$, we apply F componentwise on tuples of objects in $\mathsf{End}(\mathsf{C})$ to define an assignment $\mathsf{Prof}(\mathsf{C}) \longrightarrow \mathsf{Prof}(\mathsf{D})$. Given an operation

$$f \in \mathsf{End}(\mathsf{C})(\langle X \rangle; X') = \mathsf{C}(\oplus_i X_i, X'),$$

we define $(\mathsf{End}(F))f$ by the composite

$$\oplus_i FX_i \longrightarrow F(\oplus_i X_i) \xrightarrow{Ff} FX'.$$

where the unlabeled morphism is given by iterating F^2 if $\langle X \rangle$ is nonempty, and is given by F^0 if $\langle X \rangle = \langle\rangle$. The associativity and unity axoms for F imply that this definition of $\mathsf{End}(F)$ is compatible with the composition of operations in $\mathsf{End}(\mathsf{C})$ and $\mathsf{End}(\mathsf{D})$. The symmetry axiom ensures that $\mathsf{End}(F)$ commutes with the symmetric group actions on $\mathsf{End}(\mathsf{C})$ and $\mathsf{End}(\mathsf{D})$. Likewise, the components of a monoidal natural transformation $\alpha\colon F \longrightarrow F'$ satisfy the naturality axiom for a multinatural transformation from $\mathsf{End}(F)$ to $\mathsf{End}(F')$.

To verify that End is 2-functorial, first note that $\mathsf{End}(1_{\mathsf{C}}) = 1_{\mathsf{End}(\mathsf{C})}$ by checking that these multifunctors agree on objects and operations. Next, for a composable pair of symmetric monoidal functors, G and F, we have

$$\mathsf{End}(GF) = \mathsf{End}(F) \circ \mathsf{End}(G)$$

because $(GF)^2 = G(F^2) \circ G^2$. □

Proposition 5.3.7. *The 2-functor*

$$\mathsf{End}\colon \mathsf{PermCat} \longrightarrow \mathsf{Multicat}$$

is bijective on 1-cells and 2-cells.

Proof. The proof of Proposition 5.3.6 gives the construction of a multifunctor $\mathsf{End}(\mathsf{C}) \longrightarrow \mathsf{End}(\mathsf{D})$ from a symmetric monoidal functor $\mathsf{C} \longrightarrow \mathsf{D}$. Here we discuss an inverse construction.

For each $\langle X \rangle \in \mathsf{Prof}(\mathsf{C})$, we let $\iota_{\langle X \rangle} \in \mathsf{End}(\mathsf{C})\big(\langle X \rangle\, ; \oplus_i X_i\big)$ denote the operation corresponding to the identity $1_{\oplus_i X_i}$. Given a multifunctor

$$G\colon \mathsf{End}(\mathsf{C}) \longrightarrow \mathsf{End}(\mathsf{D})$$

we define the underlying functor of

$$F\colon \mathsf{C} \longrightarrow \mathsf{D}$$

by restricting G to objects and unary operations. Then we define

$$F^2_{X,Y} = G\iota_{(X,Y)} \quad \text{and} \quad F^0 = G\iota_{\langle\rangle}.$$

Naturality of F^2 with respect to morphisms in C, the associativity axiom (I.1.2.14), and the unity axioms (I.1.2.15) follow from the composition axiom (5.1.15) for G together with the following equalities in $\mathsf{End}(\mathsf{C})$ for $f\colon X \longrightarrow X'$ and $g\colon Y \longrightarrow Y'$ in C:

(1) (functoriality of F^2)

$$\iota_{(X',Y')} \circ (f,g) = (f \oplus g) \circ \iota_{(X,Y)};$$

(2) (associativity axiom)

$$\iota_{(X,Y\oplus Z)} \circ \big(1_X, \iota_{(Y,Z)}\big) = \iota_{(X,Y,Z)} = \iota_{(X\oplus Y,Z)} \circ \big(\iota_{(X,Y)}, 1_Z\big); \quad \text{and}$$

(3) (unity axioms)

$$\iota_{(e,X)} \circ \big(\iota_{\langle\rangle}, 1_X\big) = 1_X \quad \text{and} \quad \iota_{(X,e)} \circ \big(1_X, \iota_{\langle\rangle}\big) = 1_X.$$

The symmetry axiom (I.1.2.26) for F holds by the equivariance axiom (5.1.13) for G.

To see that this construction determines a bijection between monoidal functors $\mathsf{C} \longrightarrow \mathsf{D}$ and multifunctors $\mathsf{End}(\mathsf{C}) \longrightarrow \mathsf{End}(\mathsf{D})$, observe first by the associativity of composition that the values of G on $\iota_{\langle X \rangle}$ for profiles $\langle X \rangle$ of length $n > 2$ are determined by its values on $\iota_{\langle\rangle}$ and $\iota_{(X,Y)}$. Next observe that composition with $\iota_{\langle X \rangle}$ induces a bijection on the top row of the following diagram for any $\langle X \rangle \in \mathsf{Prof}(\mathsf{C})$ and any $Y \in \mathsf{C}$.

$$\begin{array}{ccc}
\mathsf{End}(\mathsf{C})\big(\oplus_i X_i\,;Y\big) & \xrightarrow[\cong]{(-)\circ\iota_{\langle X\rangle}} & \mathsf{End}(\mathsf{C})\big(\langle X\rangle\,;Y\big) \\
{\scriptstyle G}\downarrow & & \downarrow{\scriptstyle G} \\
\mathsf{End}(\mathsf{D})\big(G(\oplus_i X_i)\,;GY\big) & \xrightarrow{(-)\circ(G\iota_{\langle X\rangle})} & \mathsf{End}(\mathsf{D})\big(G\langle X\rangle\,;GY\big)
\end{array}$$

This shows that the values of G on arbitrary n-ary operations are determined by its values on unary operations and on the operations $\iota_{\langle X\rangle}$. Thus G is determined by the data of F constructed above.

Now we turn to the correspondence between monoidal natural transformations of symmetric monoidal functors $\mathsf{C} \longrightarrow \mathsf{D}$ and multinatural transformations of multifunctors $\mathsf{End}(\mathsf{C}) \longrightarrow \mathsf{End}(\mathsf{D})$. Each type of transformation is determined by its components, which are unary operations. As noted in the proof of Proposition 5.3.6, the components of a monoidal natural transformation $F \longrightarrow F'$ already satisfy the additional axioms for multinatural transformations

$$\mathsf{End}(F) \longrightarrow \mathsf{End}(F'),$$

and thus End is bijective on 2-cells. □

Now we discuss preservation of the canonical basepoints.

Lemma 5.3.8. *Suppose* C *and* D *are permutative categories.*

- *Suppose* $F\colon \mathsf{C} \longrightarrow \mathsf{D}$ *is a symmetric monoidal functor. The multifunctor*

$$\mathsf{End}(F)\colon \mathsf{End}(\mathsf{C}) \longrightarrow \mathsf{End}(\mathsf{D})$$

  preserves the canonical basepoints if and only if F is strictly unital.
- *Suppose* $\alpha\colon F \longrightarrow F'$ *is a monoidal natural transformation between strictly unital symmetric monoidal functors* $F, F'\colon \mathsf{C} \longrightarrow \mathsf{D}$. *Then* $\mathsf{End}(\alpha)$ *is a pointed multinatural transformation.*

Proof. Given a symmetric monoidal functor $F\colon \mathsf{C} \longrightarrow \mathsf{D}$, let

$$G = \mathsf{End}(F)\colon \mathsf{End}(\mathsf{C}) \longrightarrow \mathsf{End}(\mathsf{D})$$

denote the corresponding multifunctor. Since F and G are the same assignments on objects, each strictly preserves the unit if and only if the other does. Moreover, the proof of Proposition 5.3.7 shows that $F^0 = G\iota_{\langle\rangle}$ and $F^2_{X,Y} = G\iota_{\langle X,Y\rangle}$. Thus we conclude that F is strictly unital if and only if G is a pointed multifunctor.

Given a monoidal natural transformation $\alpha\colon F \longrightarrow F'$ between strictly unital symmetric monoidal functors

$$F, F'\colon \mathsf{C} \longrightarrow \mathsf{D},$$

the unit condition (1.1.13) for α implies that $\alpha_e = 1_e$. Therefore, $\mathsf{End}(\alpha)$ is a pointed multinatural transformation. □

Using Lemma 5.3.8, we have the following corollary of Propositions 5.3.6 and 5.3.7.

Corollary 5.3.9. *The endomorphism construction* End *restricts to a 2-functor*

$$\mathsf{End}\colon \mathsf{PermCat}^{\mathsf{su}} \longrightarrow \mathsf{Multicat}_*$$

that is bijective on 1-cells and 2-cells.

5.4. Limits and Colimits of Monadic Algebras

This section reviews the general theory of limits and colimits for algebras over monads. We will apply this in Section 5.5 to show that Multicat, the category of small multicategories and multifunctors, is complete and cocomplete.

Definition 5.4.1. A *monad* on a category B is a triple (T, μ, η) in which

- $T : \mathsf{B} \longrightarrow \mathsf{B}$ is a functor and
- $\mu : T^2 \longrightarrow T$, called the *multiplication*, and $\eta : 1_{\mathsf{M}} \longrightarrow T$, called the *unit*, are natural transformations

such that the associativity and unity diagrams

$$\begin{array}{ccc} T^3 & \xrightarrow{T\mu} & T^2 \\ {\scriptstyle \mu T}\downarrow & & \downarrow{\scriptstyle \mu} \\ T^2 & \xrightarrow{\mu} & T \end{array} \qquad \begin{array}{ccccc} 1_{\mathsf{M}} \circ T & \xrightarrow{\eta T} & T^2 & \xleftarrow{T\eta} & T \circ 1_{\mathsf{M}} \\ \| & & {\scriptstyle \mu}\downarrow & & \| \\ T & = & T & = & T \end{array}$$

are commutative. We often abbreviate such a monad to T. ◇

Definition 5.4.2. Suppose (T, μ, η) is a monad on a category B.

(1) A *T-algebra* is a pair (X, θ) consisting of
   - an object X in B and
   - a morphism $\theta : TX \longrightarrow X$, called the *structure morphism*,

   such that the associativity and unity diagrams

(5.4.3)
$$\begin{array}{ccc} T^2X & \xrightarrow{T\theta} & TX \\ {\scriptstyle \mu_X}\downarrow & & \downarrow{\scriptstyle \theta} \\ TX & \xrightarrow{\theta} & X \end{array} \qquad \begin{array}{ccc} X & \xrightarrow{\eta_X} & TX \\ & \diagdown\!\!\diagdown & \downarrow{\scriptstyle \theta} \\ & & X \end{array}$$

   are commutative.

(2) A *morphism of T-algebras*

$$f : (X, \theta^X) \longrightarrow (Y, \theta^Y)$$

   is a morphism $f : X \longrightarrow Y$ in B such that the diagram

$$\begin{array}{ccc} TX & \xrightarrow{Tf} & TY \\ {\scriptstyle \theta^X}\downarrow & & \downarrow{\scriptstyle \theta^Y} \\ X & \xrightarrow{f} & Y \end{array}$$

   is commutative. Identities and composition of T-algebra morphisms are defined in B.

(3) The category of T-algebras is denoted by $\mathsf{Alg}(T)$. ◇

Example 5.4.4. Suppose given an adjunction

$$\mathsf{B} \underset{U}{\overset{L}{\rightleftarrows}} \mathsf{A} \quad (L \dashv U)$$

with unit $\eta : 1_{\mathsf{B}} \longrightarrow UL$ and counit $\varepsilon : LU \longrightarrow 1_{\mathsf{A}}$. The composite functor UL is a monad on B, with unit and composition given, respectively, by

$$\eta : 1_{\mathsf{B}} \longrightarrow UL \quad \text{and} \quad U\varepsilon L : ULUL \longrightarrow UL,$$

where $U\varepsilon L = 1_U * \varepsilon * 1_L$. One can verify that the triangle identities (I.1.1.11) for $L \dashv U$ imply that the associativity and unity diagrams of Definition 5.4.1 commute for $T = UL$.

For any $A \in \mathsf{A}$, there is a canonical UL-algebra structure on UA given by $U\varepsilon = 1_U * \varepsilon$:

$$(UL)(UA) = (ULU)A \xrightarrow{U\varepsilon_A} UA.$$

Naturality of the counit implies that $Uf: UA \longrightarrow UA'$ is a morphism of UL-algebras for each $f: A \longrightarrow A'$ in A. Thus U induces a canonical functor

$$\mathsf{A} \longrightarrow \mathsf{Alg}(UL). \tag{5.4.5}$$

◇

Definition 5.4.6. In the context of Example 5.4.4, we say that $L \dashv U$ is a *monadic adjunction* if the canonical functor (5.4.5) induced by U is an equivalence of categories

$$\mathsf{A} \xrightarrow{\simeq} \mathsf{Alg}(UL). \tag{5.4.7}$$

Moreover, we say that the adjunction $L \dashv U$ is *strictly* monadic if (5.4.7) is an isomorphism of categories.

We say that $U: \mathsf{A} \longrightarrow \mathsf{B}$ is a *monadic functor* if it has a left adjoint L and the pair (L, U) is a monadic adjunction. Similarly, we say that U is *strictly monadic* if it has a left adjoint L such that (L, U) is a strictly monadic adjunction. ◇

Explanation 5.4.8 (Limits of Algebras)**.** For any monad T on a category B, the forgetful functor $\mathsf{Alg}(T) \longrightarrow \mathsf{B}$ creates limits in $\mathsf{Alg}(T)$. That is, given a diagram of T-algebras, one proves that, upon forgetting to B, if the limit of the underlying diagram exists then it has a canonical T-algebra structure and is indeed the limit of the given diagram in $\mathsf{Alg}(T)$. ◇

In many applications of interest one can also obtain colimits in $\mathsf{Alg}(T)$ from those in the underlying category, although the construction is more complicated. We recall several standard definitions and results from the literature and apply them below to the category of small multicategories.

To fix terminology, we begin with the following reformulation of colimits from Definition I.1.1.13.

Definition 5.4.9 (Cocones and Colimits)**.** Suppose that $F: \mathsf{D} \longrightarrow \mathsf{C}$ is a functor of categories. A *cocone* under F is an object $W \in \mathsf{C}$ together with a natural transformation $c: F \longrightarrow \Delta_W$, where Δ_W denotes the constant functor at W, given by the composite in the diagram below.

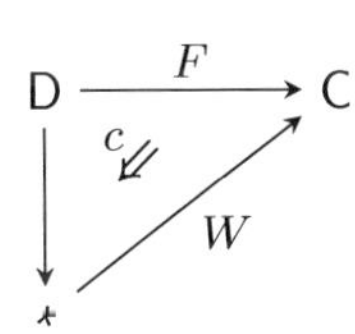

A morphism of cocones $(W, c) \longrightarrow (W', c')$ consists of a morphism $W \longrightarrow W'$ in C whose whiskering with c is equal to c'. A *colimit* for F is an initial object in the category of cocones under F. ◇

Definition 5.4.10. Suppose A is a category and $f, g \in \mathsf{A}(X,Y)$ is a parallel pair of morphisms.

(1) A *coequalizer* of f and g consists of a pair (Z,h)

$$X \overset{f}{\underset{g}{\rightrightarrows}} Y \xrightarrow{\ h\ } Z$$

that is a colimit of the diagram formed by f and g.

(2) A *split coequalizer* of f and g consists of an object $Z \in \mathsf{A}$ together with morphisms h, s, and t

$$X \overset{f}{\underset{g}{\rightrightarrows}} Y \xrightarrow{\ h\ } Z, \qquad t\colon Y \to X, \quad s\colon Z \to Y$$

such that

$$hf = hg, \quad hs = 1_Z, \quad gt = 1_Y, \quad \text{and} \quad ft = sh.$$

We call these the *splitting conditions.*

(3) An *absolute coequalizer* for (f,g) consists of an object and morphism pair (Z,h) such that (FZ, Fh) is a coequalizer of (Ff, Fg) for any functor F with domain A. ◇

Explanation 5.4.11. The splitting conditions in the definition of split coequalizer imply that Z is indeed a coequalizer of f and g. Moreover, since these conditions are preserved by any functor, a split coequalizer is an absolute coequalizer. ◇

Definition 5.4.12. Suppose $U\colon \mathsf{A} \longrightarrow \mathsf{B}$ is a functor. We say that U *strictly creates coequalizers* for a parallel pair of morphisms (f,g) in A if, for each coequalizer (W,k) of (Uf, Ug) in B,

$$UX \overset{Uf}{\underset{Ug}{\rightrightarrows}} UY \xrightarrow{\ k\ } W,$$

there is a unique object and morphism pair (Z,h) in A such that

$$(UZ, Uh) = (W,k)$$

and (Z,h) is a coequalizer of f and g in A. ◇

Theorem 5.4.13 (Beck's Precise Tripleability). *For an adjunction*

$$\mathsf{B} \underset{U}{\overset{L}{\rightleftarrows}} \mathsf{A} \quad (L \dashv U)$$

the following statements are equivalent.

(1) *The adjunction $L \dashv U$ is strictly monadic.*
(2) *U strictly creates coequalizers for parallel pairs (f,g) in A for which (Uf, Ug) has an absolute coequalizer in B.*
(3) *U strictly creates coequalizers for parallel pairs (f,g) in A for which (Uf, Ug) has a split coequalizer in B.*

Proof. We provide a sketch proof, following that of [**ML98**, VI.7 Theorem 1]. We refer the reader there for complete details, with the warning that [**ML98**] omits the term "strictly" from our (2) and (3) but the conditions are identical. See Note 5.8.2 for further explanation of this point.

To show that (1) implies (2), let $T = UL$ and suppose given a parallel pair of morphisms in $\mathsf{A} \cong \mathsf{Alg}(T)$

$$X \overset{f}{\underset{g}{\rightrightarrows}} Y$$

such that (Uf, Ug) has an absolute coequalizer (W, k) in B. We must show two things.

- There is a unique T-algebra structure on W. Letting Z denote the corresponding T-algebra, we have $Z = UW$.
- The pair (Z, k) is a coequalizer of f and g in $\mathsf{A} \cong \mathsf{Alg}(T)$.

Applying T to the coequalizer diagram for W, one has the following diagram in B, where the solid vertical arrows are the T-algebra structure morphisms for UX and UY.

$$(5.4.14)\qquad \begin{array}{ccccc} TUX & \overset{TUf}{\underset{TUg}{\rightrightarrows}} & TUY & \xrightarrow{\ Tk\ } & TW \\ \downarrow & & \downarrow & & \downarrow{\scriptstyle \theta} \\ UX & \overset{Uf}{\underset{Ug}{\rightrightarrows}} & UY & \xrightarrow{\ k\ } & W \end{array}$$

The dashed arrow θ is uniquely determined because, by hypothesis, (W, k) is an absolute coequalizer and therefore (TW, Tk) is a coequalizer of (TUf, TUg).

Next one shows that θ satisfies the associativity and unity diagrams (5.4.3) of a T-algebra structure morphism. The associativity diagram follows from using the absolute coequalizer condition in a second instance to conclude that (T^2W, T^2k) is a coequalizer of (T^2Uf, T^2Ug). This implies that the two composites $T^2W \longrightarrow W$ determined by θ and μ are equal. The unity diagram follows similarly by considering the analogue of (5.4.14) with the units for UX and UY.

Thus the absolute coequalizer condition determines a unique T-algebra structure on $W = UZ$ such that k is a morphism of T-algebras. One uses the absolute coequalizer condition again to argue that Z has the universal property of a coequalizer of f and g in $\mathsf{A} \cong \mathsf{Alg}(T)$. This concludes the argument that (1) implies (2).

It is immediate that (2) implies (3) because each split coequalizer is an example of an absolute coequalizer. To show that (3) implies (1), we begin with the following key observation.

For any monad (T, μ, η) and T-algebra (X, θ), the multiplication and unit structure morphisms make the following diagram a coequalizer in $\mathsf{Alg}(T)$ that becomes a split coequalizer after forgetting down to the underlying category on which T acts.

$$(5.4.15)\qquad T^2X \overset{T\theta}{\underset{\mu_X}{\rightrightarrows}} TX \xrightarrow{\ \theta\ } X, \qquad \eta_{TX} : TX \to T^2X, \quad \eta_X : X \to TX$$

The arrows $T\theta$, μ_X, and θ are morphisms of T-algebras, but the splittings are merely morphisms in the underlying category.

For the argument that (3) implies (1), again let $T = UL$ and note that for any object $A \in \mathsf{A}$ we have a diagram

$$LULUA \underset{LU\varepsilon_A}{\overset{\varepsilon_{LUA}}{\rightrightarrows}} LUA \xrightarrow{\quad \varepsilon_A \quad} A \tag{5.4.16}$$

given by the counit ε of the adjunction $L \dashv U$. Applying U yields a split coequalizer as an instance of (5.4.15) with $X = UA$. The assumption (3) implies therefore that each object, respectively morphism, in A is uniquely determined as a coequalizer, respectively morphism of coequalizers, in (5.4.16). The desired isomorphism in (1) follows by further developing this observation. □

The second major result that we will need from the theory of monad algebras is Theorem 5.4.18 below, concerning filtered colimits.

Definition 5.4.17 (Filtered Colimits)**.** We say that a category J is *filtered* if every finite diagram

$$\mathsf{N} \longrightarrow \mathsf{J}$$

has a cocone.

(1) A *filtered colimit* is a colimit over a filtered diagram category.
(2) We say that a monad $T: \mathsf{B} \longrightarrow \mathsf{B}$ *preserves filitered colimits* if, whenever $Z = \mathsf{colim}_{\mathsf{J}} X$ for a filtered diagram

$$\mathsf{J} \xrightarrow{X} \mathsf{B},$$

then applying T yields a colimit, TZ, of the composite diagram

$$\mathsf{J} \xrightarrow{TX} \mathsf{B}.$$

◇

Theorem 5.4.18. *Suppose T is a monad on a complete and cocomplete category B. If T preserves filtered colimits, then the category of T-algebras, $\mathsf{Alg}(T)$, is complete and cocomplete.*

Explanation 5.4.19. Theorem 5.4.18 appears in many introductory texts. For example, [**Rie16**, Theorem 5.6.12] gives a complete proof following [**Bor94b**, Proposition 4.3.6]. The hypothesis that T preserves filtered colimits is not required for the construction of limits, but is required in order to show that $\mathsf{Alg}(T)$ has colimits. Since neither coequalizers nor coproducts are filtered colimits, the existence of colimits is not an immediate consequence of the hypotheses.

One can construct coproducts of algebras via certain coequalizers of free algebras, so the majority of the work in proving cocompleteness for $\mathsf{Alg}(T)$ is the construction of coequalizers. To construct a coequalizer of algebra morphisms, one constructs a certain sequence of coequalizers in the underlying category, none of which are necessarily T-algebras. However, because sequential colimits are filtered, their colimit is preserved by T and one uses this to define T-algebra structure morphisms. The hypothesis that T preserves filtered colimits is used several more times while verifying the algebra axioms of Definition 5.4.2. ◇

For our application of Theorems 5.4.13 and 5.4.18 below, we will use the following observations about finite products.

Lemma 5.4.20.

(1) *Suppose* C *is a category with products and suppose*

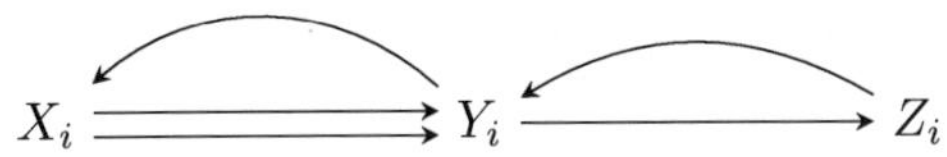

is a split coequalizer in C *for each* i *in an indexing set* I. *Then the product*

is a split coequalizer in C.

(2) *Suppose* J *is a small filtered category,* n *is a natural number, and*

$$F_i \colon \mathsf{J} \longrightarrow \mathsf{Set}$$

is a diagram with colimit Z_i *for each* $i \in \{1, \ldots, n\}$. *Then the product* $\prod_{i=1}^n Z_i$ *is a colimit for* $\prod_{i=1}^n F_i$.

Proof. Statement (1) holds because the splittings are preserved and therefore imply that the resulting diagram is a coequalizer (see Explanation 5.4.11). For statement (2), the key observation is that the colimit for a diagram of sets

$$\mathsf{D} \xrightarrow{X} \mathsf{Set}$$

is given by the quotient of $\coprod_{d \in \mathsf{D}} X(d)$ subject to the relation that, for each $x \in X(d)$ and $x' \in X(d')$ we have $x \sim x'$ if and only if there are morphisms in D

$$h \colon d \longrightarrow t \quad \text{and} \quad h' \colon d' \longrightarrow t$$

such that

$$(Xh)(x) = (Xh')(x')$$

in the set $X(t)$. We call d, respectively d', the *witness* for x, respectively x'. We call (h, h') the *witness* for the equivalence $x \sim x'$.

In the case of $Z_i = \mathsf{colim}_{\mathsf{J}} F_i$, we have the canonical function

$$\mathsf{colim}_{\mathsf{J}} \prod_{i=1}^n F_i \longrightarrow \prod_{i=1}^n Z_i$$

that sends each equivalence class of tuples to the tuple of corresponding equivalence classes. This is well defined because an equivalence of tuples

$$(z_i)_{i=1}^n \sim (z_i')_{i=1}^n$$

implies a componentwise equivalence $z_i \sim z_i'$ for each i.

We show that the canonical function is bijective. To check surjectivity, suppose given $z_i \in Z_i$, each with witness d_i. Because J is filtered, it has a cocone t for the set of objects

$$d_1, \ldots, d_n \in \mathsf{D}.$$

Let

$$h_i \colon d_i \longrightarrow t$$

denote the morphisms of this cocone, and let

$$\overline{z}_i = (F_i h_i)(z_i),$$

so $z_i \sim \overline{z}_i$ for each i, witnessed by $(h_i, 1_j)$. Since each $\overline{z}_i \in F_i(t)$, we have

$$(\overline{z}_i)_{i=1}^n \in \mathsf{colim}_{\mathsf{J}} \prod_{i=1}^n F_i$$

such that the image of this tuple under the canonical function is equivalent to $(z_i)_{i=1}^n$. This shows that the canonical function is surjective.

To verify that the canonical function is injective, suppose given (z_i) and (z_i') such that $z_i \sim z_i'$ for each i, with witnesses

$$h_i \colon d_i \longrightarrow t_i \quad \text{and} \quad h_i' \colon d_i' \longrightarrow t_i.$$

Because J is filtered, there is a cocone s for the finite subdiagram spanned by the h_i and h_i'. Taking the product of composites

$$d_i \longrightarrow t_i \longrightarrow s, \quad \text{respectively} \quad d_i' \longrightarrow t_i \longrightarrow s,$$

over $i \in \{1,\ldots,n\}$ provides a witness for the equivalence of tuples

$$(z_i)_{i=1}^n \sim (z_i')_{i=1}^n \quad \text{in} \quad \mathsf{colim}_{\mathsf{J}} \prod_{i=1}^n F_i.$$

Therefore, the canonical function is injective. □

5.5. Limits and Colimits of Multicategories

We will use Theorems 5.4.13 and 5.4.18 to show that $\mathsf{Multicat}$ is complete and cocomplete. The underlying category A for our application will be the category of small multigraphs, which we now describe.

Definition 5.5.1. A *multigraph* X consists of a class $\mathsf{Vt}X$ of vertices, together with a set $X(\langle c\rangle\,;c')$ for each tuple of vertices $\langle c\rangle = (c_1,\ldots,c_n)$ and c'. We refer to the elements of $X(\langle c\rangle\,;c')$ as *multiedges*, with *source* $\langle c\rangle$ and *target* c'. As with multicategories, we let $\mathsf{Prof}(X)$ denote $\mathsf{Prof}(\mathsf{Vt}X)$.

A morphism of multigraphs

$$f \colon X \longrightarrow Y$$

consists of a function on vertices

$$f \colon \mathsf{Vt}X \longrightarrow \mathsf{Vt}Y$$

together with a function on multiedges

$$f \colon X(\langle x\rangle\,;x') \longrightarrow Y(f\langle x\rangle\,;f(x'))$$

for each $(\langle x\rangle\,;x') \in \mathsf{Prof}(X) \times \mathsf{Vt}X$, with $f\langle x\rangle$ being the tuple whose jth entry is $f(x_j)$. This finishes the definition of a multigraph.

Moreover:

- A multigraph is *small* if its class of vertices is a set.
- The collection of small multigraphs and their morphisms form a category, denoted MGraph.
- The category of multigraphs with a fixed set of vertices C and whose morphisms are the identity on vertices is denoted MGraph^C. ◇

Explanation 5.5.2. Note that MGraph^C is isomorphic to the category of functors and natural transformations

$$\mathsf{Set}^{\mathsf{Prof}(C)\times C},$$

where $\mathsf{Prof}(C) \times C$ is the discrete category with object set $\mathsf{Prof}(C) \times C$. ◇

Definition 5.5.3. Let $\mathsf{Multicat}^C$ denote the subcategory of $\mathsf{Multicat}$ consisting of those multicategories with object set C and those multifunctors that are the identity on objects. Let

$$U^C: \mathsf{Multicat}^C \longrightarrow \mathsf{MGraph}^C$$

denote the functor that takes a multicategory M to its underlying multigraph. The vertices of $U\mathsf{M}$ are the objects of M and the multiedges of $U\mathsf{M}$ are the operations of M. A multifunctor F that is the identity on object set C defines an assignment on sets of operations, and this gives a morphism UF of underlying multigraphs. ◇

Theorem 5.5.4 ([**Yau16**, Theorem 20.3.22])**.** *For a fixed set of objects C, the forgetful functor $U^C: \mathsf{Multicat}^C \longrightarrow \mathsf{MGraph}^C$ has a left adjoint L^C.*

Explanation 5.5.5. We will not need the precise definition of the left adjoint L^C in our discussion below. However, the following information will be useful.

For $X \in \mathsf{MGraph}^C$, the set of operations $L^C X(\langle c\rangle; c')$ is generated under formal composition and permutation by the multiedges of X. To make this precise, [**Yau16**] indexes such composites by certain rooted trees (planar and labeled by C). Then

$$L^C X(\langle c\rangle; c') = \coprod_{[T]} \prod_v X(\langle c_v\rangle; c'_v),$$

where the coproduct is over isomorphism classes of trees, the product is over the internal vertices v of a class representative T, and $(\langle c_v\rangle; c'_v)$ is a labeling of v by objects of C. Thus each element of $L^C X(\langle c\rangle; c')$ is a tuple of composible multiedges of X, whose individual components have profile $(\langle c_v\rangle; c'_v)$ and whose underlying multigraph has the shape of some planar tree T.

For our purposes, the crucial feature of L^C is that the number of internal vertices v in each tree T is finite. Therefore, $L^C X(\langle c\rangle; c')$ is a coproduct of finite products in Set. For further details and proofs, we refer the reader to [**Yau16**]. ◇

Definition 5.5.6. Suppose $f_0: C \longrightarrow D$ is a function between sets. If X is a multigraph with vertex set D, we define $f_0^* X$ to be the multigraph whose vertices are the elements of C and whose multiedges are given by applying f_0:

$$(f_0^* X)(\langle x\rangle; x') = X(f_0\langle x\rangle; f_0(x')). \tag{5.5.7}$$

Then f_0 determines a morphism of multigraphs, using the same notation,

$$f_0^* X \xrightarrow{f_0} X$$

given by f_0 on vertices and by the identity on multiedges. If M is a multicategory with object set C, we define $f_0^* \mathsf{M}$ by applying f_0^* to the underlying multigraph of M. Defining symmetric group actions, units, and composition in $f_0^* \mathsf{M}$ via the equalities (5.5.7), f_0 induces a multifunctor

$$f_0^* \mathsf{M} \xrightarrow{f_0} \mathsf{M}.$$

◇

Explanation 5.5.8. Suppose $f: X \longrightarrow Y$ is a morphism of multigraphs. Using the notation of Definition 5.5.6, f factors uniquely as a composite

$$X \xrightarrow{f_1} f_0^* Y \xrightarrow{f_0} Y$$

where

- $f_0: \mathsf{Vt}X \longrightarrow \mathsf{Vt}Y$ is the function on vertices determined by f and
- $f_1: X \longrightarrow f_0^* Y$ is a multigraph morphism in $\mathsf{MGraph}^{\mathsf{Vt}X}$.

If $f: X \longrightarrow Y$ and $g: Y \longrightarrow Z$ is a composable pair of multigraph morphisms, the factorization of their composite is given by

$$(gf)_0 = g_0 f_0 \quad \text{and} \quad (gf)_1 = (f_0^* g_1) \circ f_1$$

as in the following commutative diagram.

$$\begin{array}{ccccccccc}
 & & & & Y & & & & \\
 & & & \overset{f_0}{\nearrow} & & \overset{g_1}{\searrow} & & & \\
X & \xrightarrow{f_1} & f_0^* Y & & & & g_0^* Z & \xrightarrow{g_0} & Z \\
 & & & \underset{f_0^* g_1}{\searrow} & & \underset{f_0}{\nearrow} & & & \\
 & & & & f_0^* g_0^* Z & & & &
\end{array}$$

Similarly, if $F: \mathsf{M} \longrightarrow \mathsf{N}$ is a multifunctor, F factors uniquely as a composite

$$\mathsf{M} \xrightarrow{F_1} F_0^* \mathsf{N} \xrightarrow{F_0} \mathsf{N}$$

where

- $F_0: \mathsf{Ob}\,\mathsf{M} \longrightarrow \mathsf{Ob}\,\mathsf{N}$ is the function on objects determined by F and
- $F_1: \mathsf{M} \longrightarrow F_0^* \mathsf{N}$ is a multifunctor in $\mathsf{Multicat}^{\mathsf{Ob}\,\mathsf{M}}$. ◇

We will use these factorizations to extend the adjunction of Theorem 5.5.4 as follows.

Proposition 5.5.9. *The forgetful functor* $U: \mathsf{Multicat} \longrightarrow \mathsf{MGraph}$ *has a left adjoint* L *given on each subcategory* MGraph^C *by the adjoint* L^C.

Proof. Each multigraph X has a set of vertices C, so we define $LX = L^C X$. For a multigraph morphism $f: X \longrightarrow Y$ where Y has vertex set $\mathsf{Vt}Y = D$, we factor $f = f_0 f_1$ as in Explanation 5.5.8 and define $L(f_0 f_1)$ to be the composite

$$L^C X \xrightarrow{L^C f_1} L^C (f_0^* Y) \xrightarrow{(f_0)'} L^D Y$$

where $(f_0)'$ is the multifunctor given by f_0 on objects and induced by the identity on sets of multiedges.

The units and counits of each adjunction $L^C \dashv U^C$ from Theorem 5.5.4 provide components of a unit and counit for $L \dashv U$. Naturality with respect to general morphisms in $\mathsf{Multicat}$ and MGraph follows from naturality with respect to each fixed object set C and the uniqueness of the decompositions in Explanation 5.5.8. The triangle identities for $L \dashv U$ follow from the triangle identities for each fixed object set C. □

Explanation 5.5.10 (Products of Multicategories)**.** In Definition 5.2.3 we defined the Cartesian product of multicategories. We can now observe that this Cartesian product is created by taking the product of underlying multigraphs and verifying that the additional structure and axioms of a multicategory hold componentwise. This construction is a special case of the more general construction for limits of monad algebras mentioned in Explanation 5.4.8. ◇

Theorem 5.5.11. *The adjunction*

$$\mathsf{MGraph} \; \underset{U}{\overset{L}{\underset{\longleftarrow}{\overset{\longrightarrow}{\bot}}}} \; \mathsf{Multicat}$$

is strictly monadic.

Proof. We will use Theorem 5.4.13. Therefore, it suffices to show that U strictly creates coequalizers of parallel pairs $F, G\colon \mathsf{Multicat}(\mathsf{M},\mathsf{N})$ such that (UF, UG) has a split coequalizer in MGraph.

Suppose (F, G) is such a pair, and suppose

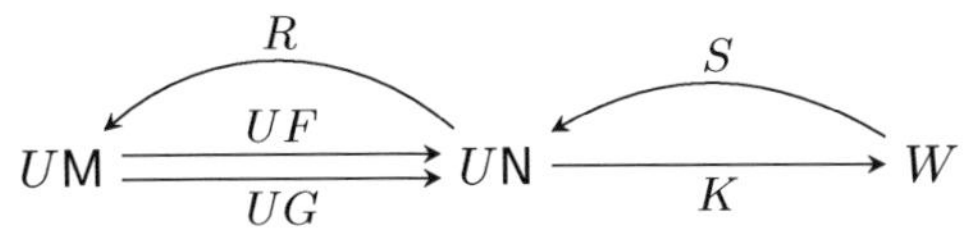
R
S
UF
UM
UN
W
UG
K

is a split coequalizer of multigraphs. Being a split coequalizer implies that each set of multiedges in W with input profile $\langle c\rangle \in \mathsf{Prof}(W)$ and output c' is a split coequalizer

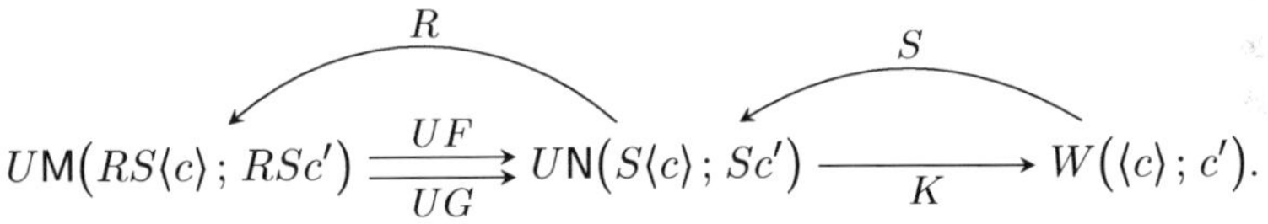
R
S
UF
UM(RS⟨c⟩ ; RSc′)
UN(S⟨c⟩ ; Sc′)
W(⟨c⟩ ; c′).
UG
K

For each $c \in \mathsf{Vt}W$ we define the c-colored unit in W to be the image under K of the edge in $U\mathsf{N}$ corresponding to the (Sc)-colored unit in N.

Now we recall, from Lemma 5.4.20 (1), that split coequalizers commute with finite products. Thus the left- and right-hand columns of the diagram below are split coequalizers, where the labels at left indicate a product of morphisms given by the indicated morphism of multigraphs. The solid horizontal arrows are induced by the composition in M and N, respectively, and therefore commute with the arrows labeled UF and UG (although they do not necessarily commute with the sections S and T). Thus there is a unique induced function between the coequalizers, drawn as the dashed arrow.

(5.5.12)

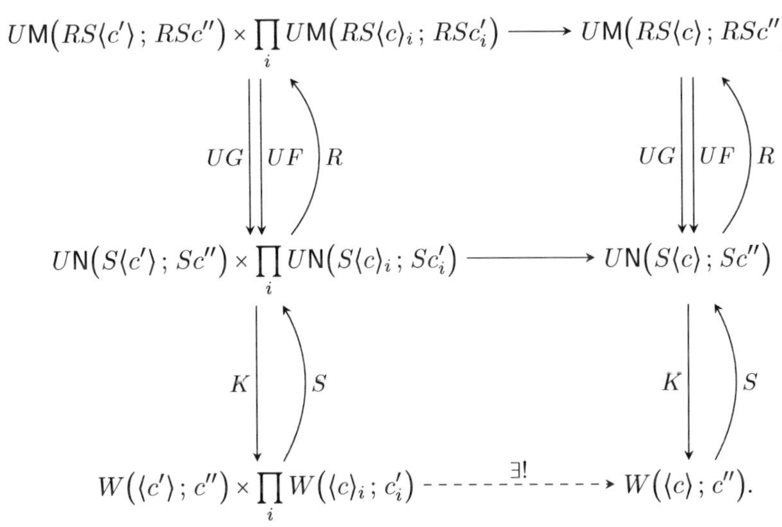
UM(RS⟨c′⟩ ; RSc″) × ∏ UM(RS⟨c⟩ᵢ ; RSc′ᵢ)
UM(RS⟨c⟩ ; RSc″)
UG
UF
R
UG
UF
R
UN(S⟨c′⟩ ; Sc″) × ∏ UN(S⟨c⟩ᵢ ; Sc′ᵢ)
UN(S⟨c⟩ ; Sc″)
K
S
K
S
W(⟨c′⟩ ; c″) × ∏ W(⟨c⟩ᵢ ; c′ᵢ)
∃!
W(⟨c⟩ ; c″).

This defines the composition in W. Similarly, we define the symmetric group actions in W as the unique morphism of coequalizers shown below.

$$\begin{array}{ccc}
U\mathsf{M}\big(RS\langle c\rangle\,;\,RSc'\big) & \xrightarrow{\quad U\sigma \quad} & U\mathsf{M}\big(RS\langle c\rangle\sigma\,;\,RSc'\big) \\
UG \downarrow\downarrow UF \quad \uparrow R & & UG \downarrow\downarrow UF \quad \uparrow R \\
U\mathsf{N}\big(S\langle c\rangle\,;\,Sc'\big) & \xrightarrow{\quad U\sigma \quad} & U\mathsf{N}\big(S\langle c\rangle\sigma\,;\,Sc'\big) \\
K \downarrow \quad \uparrow S & & K \downarrow \quad \uparrow S \\
W\big(\langle c\rangle\,;\,c'\big). & \overset{\exists!}{\dashrightarrow} & W\big(\langle c\rangle\sigma\,;\,c'\big).
\end{array} \tag{5.5.13}$$

Each of the axioms in the definition of a multicategory (Definition 5.1.2) is a commuting diagram whose objects are finite products of sets of operations and whose morphisms are a finite product of the following types:

- compositions,
- permutations,
- identities, and
- morphisms from singleton sets picking out unit operations.

Therefore, the morphisms in each diagram commute with those given by UF, respectively UG, on components.

Again using the fact that split coequalizers commute with finite products, together with functoriality of coequalizers with respect to natural transformations of diagrams, we see that each axiom holds for W and therefore W is a multicategory. For the remainder of this proof we will write UW for the underlying multigraph of the multicategory W.

Now we show that W is a coequalizer in $\mathsf{Multicat}$. Suppose that P is another cocone for F and G. Since UW is the coequalizer of UF and UG, then there is a unique induced morphism of multigraphs $UW \longrightarrow U\mathsf{P}$. A similar argument based on split coequalizers commuting with finite products shows that this morphism satisfies the axioms of a multifunctor $W \longrightarrow \mathsf{P}$. It is uniquely determined by its underlying function of multigraphs, and therefore W is the coequalizer of F and G in $\mathsf{Multicat}$.

The choice of c-colored units in W is the unique one such that K preserves units. Moreover, the uniqueness of the dashed arrow in (5.5.12) shows that the composition in W is uniquely determined by the coequalizer condition. Similarly, the uniqueness of the dashed arrow in (5.5.13) shows that the symmetric group actions in W are uniquely determined. Therefore, our argument above shows that the split coequalizer W has a unique multicategory structure induced by that of M and N via split coequalizers. This verifies that U strictly creates coequalizers for (F, G), and therefore verifies condition (3) of Theorem 5.4.13. $\square$

Theorem 5.5.14. *The category* Multicat *is complete and cocomplete.*

Proof. We have shown that Multicat is monadic over MGraph in Theorem 5.5.11. As noted in Explanation 5.4.8, limits of algebras are created in the underlying category.

To show that Multicat is cocomplete, we apply Theorem 5.4.18. Thus it suffices to verify that the monad $T = UL$ of Theorem 5.5.11 preserves filtered colimits. This follows because, for each multigraph W and each pair $(\langle c\rangle; c') \in \mathsf{Prof}(W) \times \mathsf{Vt}W$, the set of multiedges $ULW(\langle c\rangle; c')$ is given by a coproduct of finite products of sets (see Explanation 5.5.5). Since coproducts commute with arbitrary colimits and we recall, from Lemma 5.4.20 (2), that filtered colimits are preserved by finite products, it follows that T preserves filtered colimits. □

Colimits of multicategories are generally difficult to describe explicitly. However, the following feature is relatively straightforward and will be useful in our further work.

Proposition 5.5.15. *Suppose* D *is a small category and* $X\colon \mathsf{D} \longrightarrow \mathsf{Multicat}$ *is a diagram of multicategories. Then the set of objects of* $\mathsf{colim}\, X$ *is given by the colimit of objects:*

$$\mathsf{Ob}(\mathsf{colim}\, X) \cong \mathsf{colim}(\mathsf{Ob}\, X).$$

Proof. Suppose W is a colimit of X in Multicat and let T be the colimit of $\mathsf{Ob}\, X$ in Set. For each $d \in \mathsf{D}$, let

$$F_d\colon Xd \longrightarrow \mathsf{W} \quad \text{and} \quad g_d\colon \mathsf{Ob}(Xd) \longrightarrow T$$

be the structure morphisms to the respective colimits. We show that the canonical map of sets

$$T \longrightarrow \mathsf{Ob}\,\mathsf{W} \tag{5.5.16}$$

is a bijection.

To see that (5.5.16) is surjective, suppose a is an object of W. Then there is some $d \in \mathsf{D}$ and object $x \in \mathsf{Ob}(Xd)$ such that $F_d(x) = a$. Then $g_d(x)$ is an element of T that maps to a under (5.5.16).

To see that (5.5.16) is injective, suppose given $z, z' \in T$ such that both z and z' map to $a \in \mathsf{Ob}\,\mathsf{W}$ under (5.5.16). There are $d, d' \in \mathsf{D}$ together with

$$x \in \mathsf{Ob}(Xd) \quad \text{and} \quad x' \in \mathsf{Ob}(Xd')$$

such that $g_d(x) = z$ and $g_{d'}(x') = z'$. If z and z' both map to the same object $a \in \mathsf{Ob}(W)$ then there must be some $d'' \in \mathsf{D}$ together with morphisms

$$d \longrightarrow d'' \longleftarrow d'$$

such that the objects $x \in \mathsf{Ob}(Xd)$ and $x' \in \mathsf{Ob}(Xd')$ map to the same object of Xd''. But this implies that $z = z'$ and therefore (5.5.16) is injective. □

5.6. Tensor and Smash Products of Multicategories

In this section we define monoidal products for multicategories (and pointed multicategories) that are part of a symmetric monoidal closed structure.

Definition 5.6.1. Given profiles $\langle c\rangle \in \mathsf{Prof}(C)$ and $\langle d\rangle \in \mathsf{Prof}(D)$ with $m = \operatorname{len}\langle c\rangle$ and $n = \operatorname{len}\langle d\rangle$, we define

$$\begin{aligned} \langle c\rangle \times d_j &= \langle (c_i, d_j)\rangle_i = ((c_1, d_j), (c_2, d_j), \ldots), \\ c_i \times \langle d\rangle &= \langle (c_i, d_j)\rangle_j = ((c_i, d_1), (c_i, d_2), \ldots), \\ \langle c\rangle \otimes \langle d\rangle &= \langle\langle (c_i, d_j)\rangle_i\rangle_j, \quad \text{and} \\ \langle c\rangle \otimes^{\mathrm{t}} \langle d\rangle &= \langle\langle (c_i, d_j)\rangle_j\rangle_i. \end{aligned}$$

Let $\xi^{\otimes} = \xi^{\otimes}_{m,n}$ denote the permutation $\langle c\rangle \otimes \langle d\rangle \leftrightarrow \langle c\rangle \otimes^{\mathrm{t}} \langle d\rangle$ induced by changing order of indexing. ◇

Explanation 5.6.2. In the context of Definition 5.6.1, we note the following relationships to earlier material.

- Viewing the tuples $\langle c\rangle$ and $\langle d\rangle$ as elements of the distortion category (Definition I.4.2.1), the product $\langle c\rangle \otimes \langle d\rangle$ is that of (I.4.2.15) with $(-,-)$ in place of $+$.
- Taking $m = \operatorname{len}\langle c\rangle$ and $n = \operatorname{len}\langle d\rangle$ as objects of the finite ordinal category Σ (Definition I.2.4.1), the permutation $\xi^{\otimes}_{m,n}$ is induced by the multiplicative symmetry (I.2.4.5) between mn and nm. Explanation I.2.4.7 gives a geometric description via the transpose of an $n \times m$ matrix. ◇

Definition 5.6.3. For multigraphs X and Y with vertex classes C and D, respectively, we define a multigraph $X \,\&\, Y$ with vertex class $C \times D$ as follows. Given

$$\langle c, d\rangle = ((c_1, d_1), \ldots, (c_n, d_n)) \in \mathsf{Prof}(C \times D) \quad \text{and} \quad (c', d') \in C \times D,$$

the set of multiedges with source $\langle c, d\rangle$ and target (c', d') is given by the following coproduct over pairs $\langle c''\rangle$, $\langle d''\rangle$ such that $\langle c''\rangle \otimes \langle d''\rangle = \langle c, d\rangle$:

$$(X \,\&\, Y)\big(\langle c, d\rangle\,;\,(c', d')\big) = \coprod_{\langle c''\rangle \otimes \langle d''\rangle = \langle c, d\rangle} X\big(\langle c''\rangle\,;\,c'\big) \times Y\big(\langle d''\rangle\,;\,d'\big). \tag{5.6.4}$$ ◇

Explanation 5.6.5. If the indexing set of the coproduct (5.6.4) is empty, then

$$(X \,\&\, Y)\big(\langle c, d\rangle\,;\,(c', d')\big)$$

is empty. For an example where the indexing set has more than a single element, suppose given $c \in C$ and $d \in D$. Then the pair $((c, d)\,,\,(c, d))$ is equal to both

$$(c) \otimes (d, d) \quad \text{and} \quad (c, c) \otimes (d).$$ ◇

Explanation 5.6.6 (Definition via Kan Extensions)**.** Recall the concept of Kan extensions from Definition I.1.1.18 and Explanation I.1.1.19. Viewing multigraphs as functors

$$X\colon \mathsf{Prof}(X) \times \mathsf{Vt}X \longrightarrow \mathsf{Set},$$

the product $X \,\&\, Y$ is the left Kan extension along the composite

$$\begin{array}{l} \mathsf{Prof}(X) \times \mathsf{Vt}X \times \mathsf{Prof}(Y) \times \mathsf{Vt}Y \\ \qquad \xrightarrow{\ \cong\ } (\mathsf{Prof}(X) \times \mathsf{Prof}(Y)) \times (\mathsf{Vt}X \times \mathsf{Vt}Y) \\ \qquad\qquad \xrightarrow{\ \otimes \times 1 \times 1\ } \mathsf{Prof}(X \times Y) \times (\mathsf{Vt}X \times \mathsf{Vt}Y) \end{array} \tag{5.6.7}$$

that swaps $\mathsf{Vt}X$ with $\mathsf{Prof}(Y)$ and then takes the tensor product of profiles. That is, letting $X \boxtimes Y$ denote the external product

$$(\langle c\rangle, c', \langle d\rangle, d') \longmapsto X(\langle c\rangle; c') \times Y(\langle d\rangle; d'),$$

one verifies that $X \,\&\, Y$ is initial among left extensions as in the diagram below, where the unlabeled arrow is the composite (5.6.7) and the double arrow denotes a natural transformation of functors.

$$\begin{array}{ccc}
\mathsf{Prof}(X) \times \mathsf{Vt}X \times \mathsf{Prof}(Y) \times \mathsf{Vt}Y & \xrightarrow{\;X \boxtimes Y\;} & \mathsf{Set} \\
& \Downarrow & \uparrow{\scriptstyle X \,\&\, Y} \\
\searrow & & \mathsf{Prof}(X \times Y) \times \mathsf{Vt}X \times \mathsf{Vt}Y
\end{array}$$

Since $\mathsf{Prof}(X) \times \mathsf{Vt}X \times \mathsf{Prof}(Y) \times \mathsf{Vt}Y$ is discrete, the coend formula for left Kan extension (Explanation I.1.1.19) reduces to the coproduct (5.6.4). ◇

Definition 5.6.8. Suppose M and N are small multicategories. We define the *sharp product* $\mathsf{M} \# \mathsf{N}$ as the following pushout along morphisms induced by the inclusions

$$\mathsf{Ob}\,\mathsf{M} \hookrightarrow \mathsf{M} \quad \text{and} \quad \mathsf{Ob}\,\mathsf{N} \hookrightarrow \mathsf{N}.$$

$$\begin{array}{ccc}
\mathsf{Ob}\,\mathsf{M} \times \mathsf{Ob}\,\mathsf{N} & \longrightarrow & \coprod_{d \in \mathsf{Ob}\,\mathsf{N}} \mathsf{M} \\
\downarrow & & \downarrow \\
\coprod_{c \in \mathsf{Ob}\,\mathsf{M}} \mathsf{N} & \longrightarrow & \mathsf{M} \# \mathsf{N}
\end{array}$$

◇

Explanation 5.6.9 (Unpacking the Sharp Product)**.** Restricting Definition 5.6.8 to objects and recalling Proposition 5.5.15, we see that

$$\mathsf{Ob}(\mathsf{M} \# \mathsf{N}) \cong \mathsf{Ob}\,\mathsf{M} \times \mathsf{Ob}\,\mathsf{N}.$$

The operations of $\mathsf{M} \# \mathsf{N}$ are generated by operations of the form

$$\phi \times d \in \mathsf{M} \times \{d\} \quad \text{and} \quad c \times \psi \in \{c\} \times \mathsf{N}$$

subject to the following symmetry and compatibility axioms determined by the pushout.

(1) For $(c,d) \in \mathsf{M} \# \mathsf{N}$, we have

$$1_c \times d = 1_{(c,d)} = c \times 1_d.$$

(2) For operations $\phi, \phi_1, \ldots, \phi_n$ in M such that the composite below is defined, we have

$$(\phi \times d) \circ (\phi_1 \times d, \ldots, \phi_n \times d) = (\phi \circ (\phi_1, \ldots, \phi_n)) \times d.$$

(3) For $\sigma \in \Sigma_n$, we have

$$(\phi \times d) \cdot \sigma = (\phi \cdot \sigma) \times d.$$

(4) For operations $\psi, \psi_1, \ldots, \psi_m$ in N such that the composite below is defined, we have

$$(c \times \psi) \circ (c \times \psi_1, \ldots, c \times \psi_m) = c \times (\psi \circ (\psi_1, \ldots, \psi_m)).$$

(5) For $\sigma \in \Sigma_m$, we have

$$(c \times \psi) \cdot \sigma = c \times (\psi \cdot \sigma).$$

The first three conditions imply that, for any $d \in \mathsf{N}$, the assignment $\mathsf{M} \longrightarrow \mathsf{M} \# \mathsf{N}$ given by $\phi \longmapsto \phi{\times}d$ is a multifunctor. Likewise, the first, fourth, and fifth conditions imply that $\psi \longmapsto c{\times}\psi$ is a multifunctor $\mathsf{N} \longrightarrow \mathsf{M}\#\mathsf{N}$ for any $c \in \mathsf{M}$. Taken together, the conditions are equivalent to the requirement that a multifunctor

$$F\colon \mathsf{M} \# \mathsf{N} \longrightarrow \mathsf{P}$$

consists of an assignment on objects,

$$F(c,d) \in \operatorname{Ob}\mathsf{P}$$

for $(c,d) \in \operatorname{Ob}\mathsf{M} \times \operatorname{Ob}\mathsf{N}$ such that each

$$F(c,-)\colon \mathsf{N} \longrightarrow \mathsf{P} \quad \text{and} \quad F(-,d)\colon \mathsf{M} \longrightarrow \mathsf{P}$$

is a multifunctor. ◇

Definition 5.6.10. Suppose given small multicategories M and N along with operations

$$\phi \in \mathsf{M}\big(\langle c\rangle\,;\,c'\big) \quad \text{and} \quad \psi \in \mathsf{N}\big(\langle d\rangle\,;\,d'\big).$$

We define

$$\begin{aligned}
\phi \times \langle d\rangle &= \langle \phi \times d_j\rangle_j \in \prod_j \mathsf{M}\big(\langle c\rangle\,;\,c'\big) \times \{d_j\} \\
\langle c\rangle \times \psi &= \langle c_i \times \psi\rangle_i \in \prod_i \{c_i\} \times \mathsf{N}\big(\langle d\rangle\,;\,d'\big) \\
\phi \otimes \psi &= (c' \times \psi) \circ (\phi \times \langle d\rangle) \in (\mathsf{M} \# \mathsf{N})\big(\langle c\rangle \otimes \langle d\rangle\,;\,(c',d')\big) \\
\phi \otimes^{\mathsf{t}} \psi &= (\phi \times d') \circ (\langle c\rangle \times \psi) \in (\mathsf{M} \# \mathsf{N})\big(\langle c\rangle \otimes^{\mathsf{t}} \langle d\rangle\,;\,(c',d')\big).
\end{aligned}$$

Let $\xi^{\otimes}$ denote the bijection

$$(\mathsf{M} \# \mathsf{N})\big(\langle c\rangle \otimes^{\mathsf{t}} \langle d\rangle\,;\,(c',d')\big) \xrightarrow{\cong} (\mathsf{M} \# \mathsf{N})\big(\langle c\rangle \otimes \langle d\rangle\,;\,(c',d')\big)$$

induced by the permutation $\xi^{\otimes}$ that interchanges order of indexing. ◇

Definition 5.6.11 (Boardman-Vogt Tensor Product of Multicategories)**.** For small multicategories M and N, the tensor products of Definition 5.6.10 give two canonical morphisms of multigraphs

$$(U\mathsf{M}) \mathbin{\&} (U\mathsf{N}) \underset{\xi^{\otimes} \circ \otimes^{\mathsf{t}}}{\overset{\otimes}{\rightrightarrows}} U(\mathsf{M} \# \mathsf{N}). \tag{5.6.12}$$

Taking adjoints, we have two morphisms in $\mathsf{Multicat}$, and we define $\mathsf{M} \otimes \mathsf{N}$ to be their coequalizer.

$$L((U\mathsf{M}) \mathbin{\&} (U\mathsf{N})) \rightrightarrows \mathsf{M} \# \mathsf{N} \dashrightarrow \mathsf{M} \otimes \mathsf{N}$$

For an object $(c,d) \in \mathsf{M} \# \mathsf{N}$, we let $c \otimes d$ denote its image in $\mathsf{M} \otimes \mathsf{N}$. ◇

Explanation 5.6.13 (Definition via Kan Extensions)**.** Continuing from Explanation 5.6.6, the tensor products of Definition 5.6.10 give two different left extensions shown below.

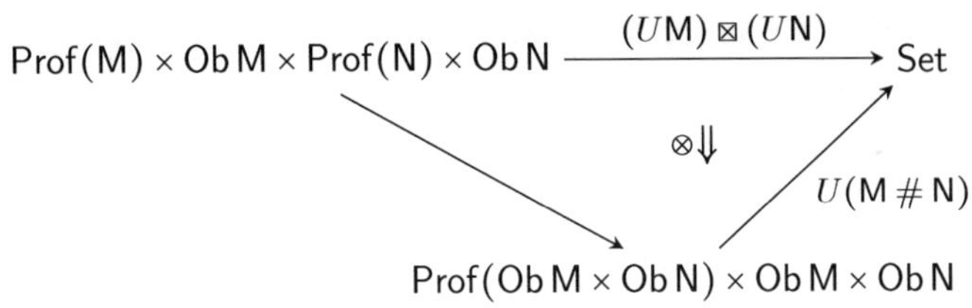

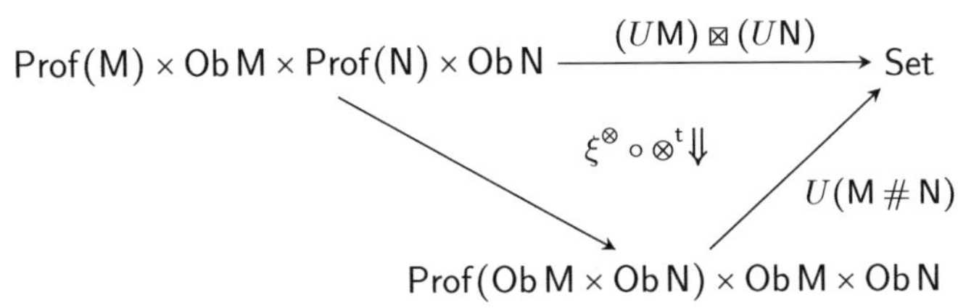

The canonical morphisms of (5.6.12) are the corresponding universal morphisms out of the left Kan extension $(U\mathsf{M}) \,\&\, (U\mathsf{N})$. ◇

Explanation 5.6.14 (Unpacking the Tensor Product)**.** Restricting Definition 5.6.11 to objects, we see that

$$\mathrm{Ob}(\mathsf{M} \otimes \mathsf{N}) \cong \mathrm{Ob}(\mathsf{M} \# \mathsf{N}) \cong \mathrm{Ob}\,\mathsf{M} \times \mathrm{Ob}\,\mathsf{N}.$$

Continuing from our unpacking of the sharp product in Explanation 5.6.9, the operations of $\mathsf{M} \otimes \mathsf{N}$ are generated by

$$\phi \otimes d \in \mathsf{M}\big(\langle c\rangle ; c'\big) \times \{d\} \quad \text{and} \quad c \otimes \psi \in \{c\} \times \mathsf{N}\big(\langle d\rangle ; d'\big)$$

subject to the relations of $\mathsf{M} \# \mathsf{N}$ along with one additional *interchange relation*

$$\phi \otimes \psi = (\phi \otimes^{\mathrm{t}} \psi) \cdot \xi^{\otimes}. \tag{5.6.15}$$

If we draw an operation as an arrow from its input profile to its output object, the interchange relation means that the two composites

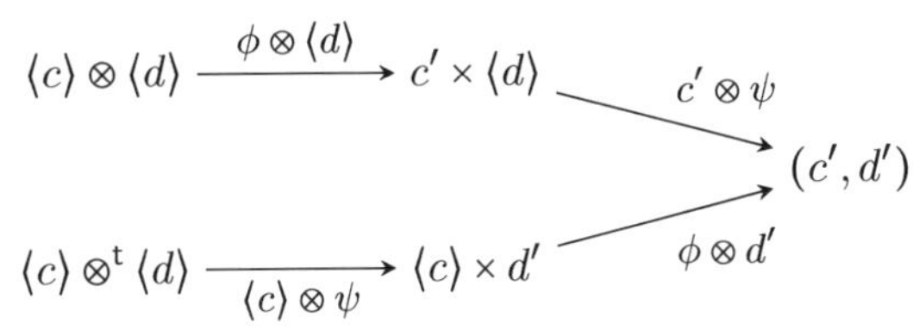

correspond under the bijection

$$\xi^{\otimes} \colon (\mathsf{M} \otimes \mathsf{N})\big(\langle c\rangle \otimes^{\mathrm{t}} \langle d\rangle ; c' \otimes d'\big) \xrightarrow{\cong} (\mathsf{M} \otimes \mathsf{N})\big(\langle c\rangle \otimes \langle d\rangle ; c' \otimes d'\big).$$

A multifunctor

$$F : \mathsf{M} \otimes \mathsf{N} \longrightarrow \mathsf{P}$$

consists of an assignment on objects $F(c,d) \in \mathrm{Ob}\,\mathsf{P}$ for $(c,d) \in \mathrm{Ob}\,\mathsf{M} \times \mathrm{Ob}\,\mathsf{N}$ such that each $F(c,-)$ and $F(-,d)$ is a multifunctor and such that we have

$$F(\phi \otimes \psi) = F(\phi \otimes^{\mathrm{t}} \psi) \cdot \xi^{\otimes} \tag{5.6.16}$$

for each $\phi \in \mathsf{M}\big(\langle c\rangle ; c'\big)$ and $\psi \in \mathsf{N}\big(\langle d\rangle ; d'\big)$. ◇

Definition 5.6.17. For small multicategories M and N, let

$$\beta \colon \mathsf{M} \# \mathsf{N} \longrightarrow \mathsf{N} \# \mathsf{M}$$

denote the multifunctor given on objects by $\beta(c,d) = (d,c)$ and on generating operations by

$$\beta(\phi \times d) = d \times \phi \quad \text{and} \quad \beta(c \times \psi) = \psi \times c.$$

Let

$$\beta \colon \mathsf{M} \otimes \mathsf{N} \longrightarrow \mathsf{N} \otimes \mathsf{M}$$

denote the induced multifunctor on tensor products. ◇

Recall from Definition 5.1.21 the initial operad I has a single object $*$ and single operation 1_*.

Theorem 5.6.18. *The tensor product*

$$\otimes \colon \mathsf{Multicat} \times \mathsf{Multicat} \longrightarrow \mathsf{Multicat}$$

is a symmetric monoidal product on the category of small multicategories, with unit given by the initial operad I *and symmetry given by* β*.*

Proof. Observe first that $(\#, \mathsf{I}, \beta)$ is a symmetric monoidal product on $\mathsf{Multicat}$:

- Functoriality of $\#$ follows from functoriality of the pushout in Definition 5.6.8.
- The associativity and unit isomorphisms follow from Explanation 5.6.9. Multifunctors with domain I are simply a choice of object, and therefore the unit isomorphisms

$$\mathsf{M} \# \mathsf{I} \xrightarrow{\cong} \mathsf{M} \xleftarrow{\cong} \mathsf{I} \# \mathsf{M}$$

  are determined by the identities on M. Likewise we have

$$(\mathsf{L} \# \mathsf{M}) \# \mathsf{N} \xrightarrow{\cong} \mathsf{L} \# (\mathsf{M} \# \mathsf{N})$$

  determined by the identity multifunctors in each variable.
- The universal property of pushouts implies that these satisfy the axioms of a symmetric monoidal category listed in Definitions 1.1.1 and 1.1.23.

Similarly, functoriality of the tensor product follows from functoriality of the coequalizer in Definition 5.6.11. The associativity, unit, and symmetry isomorphisms for the tensor product are induced by those for the sharp product, and satisfy the corresponding axioms by the universal product of coequalizers. □

Motivation 5.6.19 (2-functoriality of the Tensor Product)**.** We have not yet discussed 2-functoriality of the tensor product with respect to multinatural transformations. We will describe this in Section 6.4 and show that it provides a symmetric Cat-monoidal structure as in Definition 1.5.1. ◇

Recall from Definition 5.3.1 a pointed multicategory consists of a multicategory M together with a multifunctor $\mathsf{T} \longrightarrow \mathsf{M}$ determined by an object $*^{\mathsf{M}}$ and operations ι^{M}. The following are special cases of the smash and wedge products defined in Section 4.1.

Definition 5.6.20. Given small pointed multicategories $(\mathsf{M}, *^{\mathsf{M}}, \iota^{\mathsf{M}})$ and $(\mathsf{N}, *^{\mathsf{N}}, \iota^{\mathsf{N}})$, we define the *smash product* as the following pushout in $\mathsf{Multicat}$.

$$\begin{array}{ccc} (\mathsf{M} \otimes \mathsf{T}) \coprod (\mathsf{T} \otimes \mathsf{N}) & \longrightarrow & \mathsf{M} \otimes \mathsf{N} \\ \downarrow & & \downarrow \\ \mathsf{T} & \longrightarrow & \mathsf{M} \wedge \mathsf{N} \end{array} \tag{5.6.21}$$

The *smash unit* multicategory is

$$\mathsf{S} = \mathsf{I}_+ = \mathsf{I} \coprod \mathsf{T}. \tag{5.6.22}$$

◇

Definition 5.6.23. Given small pointed multicategories $(\mathsf{M}, *^{\mathsf{M}}, \iota^{\mathsf{M}})$ and $(\mathsf{N}, *^{\mathsf{N}}, \iota^{\mathsf{N}})$, we define the *wedge product* as the coequalizer in $\mathsf{Multicat}$ of the two structure morphisms

$$\mathsf{T} \underset{*^{\mathsf{N}}}{\overset{*^{\mathsf{M}}}{\rightrightarrows}} \mathsf{M} \coprod \mathsf{N} \dashrightarrow \mathsf{M} \vee \mathsf{N}$$

The wedge product is sometimes also called the *wedge sum*. ◇

In Theorem 5.7.22 below we will show, as a special case of Theorem 4.2.3, that the smash product is part of a symmetric monoidal closed structure on $\mathsf{Multicat}_*$.

5.7. The Internal Hom for Multicategories

In this section we describe internal hom objects for $\mathsf{Multicat}$ and $\mathsf{Multicat}_*$. We will use the following notation for a tuple of multifunctors $\langle F \rangle$.

Definition 5.7.1. Suppose given multicategories M and N together with a tuple of multifunctors $\langle F \rangle = (F_1, \ldots, F_m)$ where each

$$F_i \colon \mathsf{M} \longrightarrow \mathsf{N} \quad \text{for} \quad 1 \le i \le m.$$

Then we use the following notation.

- For $c \in \mathsf{Ob}\,\mathsf{M}$, let $\langle F \rangle c = \langle F_i c \rangle_i$.
- For $\langle c \rangle = (c_1, \ldots, c_n) \in \mathsf{Prof}(\mathsf{Ob}\,\mathsf{M})$, let

$$\langle Fc \rangle = \langle \langle F_i c_j \rangle_i \rangle_j \quad \text{and} \quad \langle Fc \rangle^{\mathsf{t}} = \langle \langle F_i c_j \rangle_j \rangle_i.$$

- For $\phi \in \mathsf{M}(\langle c \rangle; c')$, let

$$\langle F \rangle \phi = \langle F_i \phi \rangle_i \in \prod_i \mathsf{N}(F_i \langle c \rangle; F_i c').$$

◇

Definition 5.7.2. Suppose M and N are small multicategories. The *internal hom multicategory* $\mathsf{Hom}(\mathsf{M}, \mathsf{N})$ is defined as follows. The objects of $\mathsf{Hom}(\mathsf{M}, \mathsf{N})$ are multifunctors $\mathsf{M} \longrightarrow \mathsf{N}$. The operations

$$\alpha \colon \langle F \rangle \longrightarrow G$$

in $\mathsf{Hom}(\mathsf{M}, \mathsf{N})$ are called *transformations* and are given by components

$$\alpha_c \in \mathsf{N}(\langle F \rangle c; Gc)$$

for each $c \in \mathsf{Ob}\,\mathsf{M}$ such that, for each operation $\phi \in \mathsf{M}(\langle c\rangle; c')$, the following *naturality condition* holds, with $\alpha_{\langle c\rangle} = \langle \alpha_{c_j}\rangle_j$, $\operatorname{len}\langle F\rangle = m$, and $\operatorname{len}\langle c\rangle = n$:

$$(G\phi) \circ \alpha_{\langle c\rangle} = (\alpha_{c'} \circ \langle F\rangle\phi) \cdot \xi^{\otimes}_{m,n}. \tag{5.7.3}$$

The unit operation $1_F: F \longrightarrow F$ is given by the identity multinatural transformation whose component at c is 1_{Fc}. The composition and symmetric group actions on operations of $\mathsf{Hom}(\mathsf{M},\mathsf{N})$ are given componentwise by those in N.

We describe (5.7.3) further in Explanation 5.7.4. In Lemma 5.7.6 we show that (5.7.3) is closed under composition. Using this, we complete the proof that $\mathsf{Hom}(\mathsf{M},\mathsf{N})$ is a multicategory in Proposition 5.7.11. ◇

Explanation 5.7.4 (Operations in $\mathsf{Hom}(\mathsf{M},\mathsf{N})$). If we draw an operation as an arrow from its input profile to its output object, the equality (5.7.3) means that the two composites

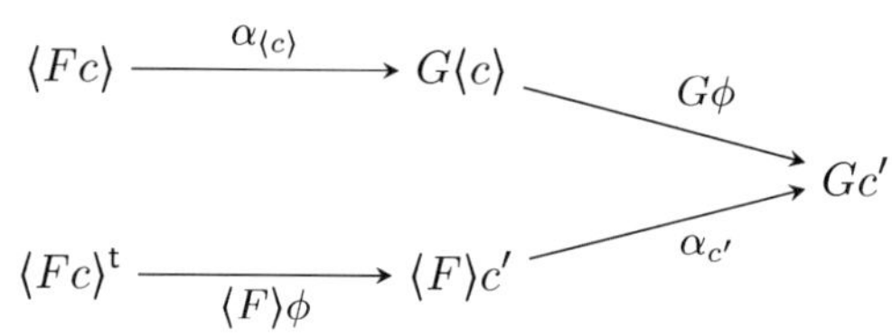

correspond under the bijection

$$\xi^{\otimes}: \mathsf{N}(\langle Fc\rangle^{\mathsf{t}}; Gc') \xrightarrow{\cong} \mathsf{N}(\langle Fc\rangle; Gc'). \qquad ◇$$

Explanation 5.7.5 (Category of Unary Operations in $\mathsf{Hom}(\mathsf{M},\mathsf{N})$). In the case that $\langle F\rangle$ is a singleton, then $\xi^{\otimes}$ is an identity and the equality (5.7.3) reduces to the naturality condition of Definition 5.1.17. Therefore, the set of unary operations $\mathsf{Hom}(\mathsf{M},\mathsf{N})(F,G)$ is precisely the set of multinatural transformations from F to G. Thus the underlying category of objects and unary operations of $\mathsf{Hom}(\mathsf{M},\mathsf{N})$ is $\mathsf{Multicat}(\mathsf{M},\mathsf{N})$. ◇

Lemma 5.7.6. *In the context of Definition 5.7.2, the naturality condition* (5.7.3) *is closed under composition.*

Proof. Suppose given multifunctors

$$E_{i,k}, F_i, G: \mathsf{M} \longrightarrow \mathsf{N} \quad \text{for} \quad 1 \le i \le m \quad \text{and} \quad 1 \le k \le p_i.$$

Let

$$\langle F\rangle = \langle F_i\rangle_i, \qquad \langle E_i\rangle = \langle E_{i,k}\rangle_k, \qquad \langle E\rangle = \oplus_i \langle E_i\rangle = \langle\langle E_{i,k}\rangle_k\rangle_i,$$
$$\text{and} \quad p = p_1 + \cdots + p_m.$$

So $\langle F\rangle$ has length m, each $\langle E_i\rangle$ has length p_i, and $\langle E\rangle$ has length $p = \sum_i p_i$. Then suppose given transformations

$$\alpha \in \mathsf{Hom}(\mathsf{M},\mathsf{N})(\langle F\rangle; G) \quad \text{and} \quad \omega_i \in \mathsf{Hom}(\mathsf{M},\mathsf{N})(\langle E_i\rangle; F_i).$$

We must show that the composite $\alpha \circ \langle\omega\rangle$ satisfies the naturality condition (5.7.3). That is, we must show

$$(G\phi) \circ \big(\alpha \circ \langle\omega\rangle\big)_{\langle c\rangle} = \big((\alpha \circ \langle\omega\rangle)_{c'} \circ \langle E\rangle\phi\big) \cdot \xi^{\otimes}_{p,n} \tag{5.7.7}$$

for each $\langle c\rangle = (c_1, \ldots, c_n) \in \mathsf{Prof}(\mathsf{M})$ and each operation $\phi \in \mathsf{M}\big(\langle c\rangle\,; c'\big)$. For this purpose, we introduce the following notation.

- As in Definition 5.7.1, we let

$$\langle F\rangle\phi = \langle F_i\phi\rangle_i \in \prod_{i=1}^{m} \mathsf{N}\big(F_i\langle c\rangle\,; F_i c'\big)$$

  for $\phi \in \mathsf{M}\big(\langle c\rangle\,; c'\big)$.
- For $\langle c\rangle = (c_1, \ldots, c_n) \in \mathsf{Prof}(\mathsf{M})$ we let

$$\langle \omega_c\rangle = \langle\langle(\omega_i)_{c_j}\rangle_i\rangle_j \quad \text{and} \quad \langle\omega_c\rangle^{\mathsf{t}} = \langle\langle(\omega_i)_{c_j}\rangle_j\rangle_i$$

- As described in Explanation 5.6.2, we use subscripts $\xi^{\otimes}_{m,n}$ to indicate the component of $\xi^{\otimes}$ (I.2.4.5) at mn. So, for example, we have

$$\langle\omega_c\rangle^{\mathsf{t}} \cdot \xi^{\otimes}_{m,n} = \langle\omega_c\rangle. \tag{5.7.8}$$

- We write $\langle p\rangle n$ for the tuple of indices

$$\oplus_{j=1}^{n}\langle p\rangle = (\underbrace{p_1, \ldots, p_m}_{1}, \ldots, \underbrace{p_1, \ldots, p_m}_{n}).$$

  So $\langle p\rangle n$ has length mn.
- We use the notation $\xi^{\otimes}_{m,n}\langle\langle p\rangle n\rangle$ for the block permutation (II.1.1.19) induced by $\xi^{\otimes}_{m,n}$ on mn subintervals whose lengths are given by $\langle p\rangle n$. So $\xi^{\otimes}_{m,n}\langle\langle p\rangle n\rangle$ permutes the subintervals as $\xi^{\otimes}_{m,n}$ permutes the elements of mn, but leaves the relative order within each block unchanged.

The desired equality (5.7.7) is given by the following computation. In this computation we use the naturality conditions (5.7.3) for α and ω, together with the multicategory axioms for N and the following equality relating (block) transposition permutations:

$$\Big(\oplus_{i=1}^{m} \xi^{\otimes}_{p_i,n}\Big) \cdot \xi^{\otimes}_{m,n}\langle\langle p\rangle n\rangle = \xi^{\otimes}_{(p_1+\cdots+p_m),n} = \xi^{\otimes}_{p,n} \tag{5.7.9}$$

We explain the equality (5.7.9) in Explanation 5.7.10 below. Now we give the computation showing (5.7.7).

$$\begin{aligned}
\big(G\phi\big)\circ\big(\alpha\circ\langle\omega\rangle\big)_{\langle c\rangle} &= \big(G\phi\big)\circ\big\langle\alpha_{c_j}\circ\langle\omega\rangle_{c_j}\big\rangle_j \\
&\qquad \text{by definition,} \\
&= \big((G\phi)\circ\alpha_{\langle c\rangle}\big)\circ\langle\omega_c\rangle \\
&\qquad \text{by associativity,} \\
&= \big((\alpha_{c'}\circ\langle F\rangle\phi)\cdot\xi^{\otimes}_{m,n}\big)\circ\langle\omega_c\rangle \\
&\qquad \text{by naturality (5.7.3) for } \alpha, \\
&= \big((\alpha_{c'}\circ\langle F\rangle\phi)\circ\langle\omega_c\rangle^{\mathrm{t}}\big)\cdot\xi^{\otimes}_{m,n}\langle\langle p\rangle n\rangle \\
&\qquad \text{by (5.7.8) and top equivariance,} \\
&= \big(\alpha_{c'}\circ\big\langle F_i\phi\circ(\omega_i)_{\langle c\rangle}\big\rangle_i\big)\cdot\xi^{\otimes}_{m,n}\langle\langle p\rangle n\rangle \\
&\qquad \text{by associativity,} \\
&= \big(\alpha_{c'}\circ\big\langle((\omega_i)_{c'}\circ\langle E_i\rangle\phi)\cdot\xi^{\otimes}_{p_i,n}\big\rangle_i\big)\cdot\xi^{\otimes}_{m,n}\langle\langle p\rangle n\rangle \\
&\qquad \text{by naturality (5.7.3) for } \omega, \\
&= \big(\alpha_{c'}\circ\big\langle(\omega_i)_{c'}\circ\langle E_i\rangle\phi\big\rangle_i\big)\cdot\Big(\big(\oplus_{i=1}^{m}\xi^{\otimes}_{p_i,n}\big)\cdot\xi^{\otimes}_{m,n}\langle\langle p\rangle n\rangle\Big) \\
&\qquad \text{by bottom equivariance and group action,} \\
&= \big(\alpha_{c'}\circ\big\langle(\omega_i)_{c'}\circ\langle E_i\rangle\phi\big\rangle_i\big)\cdot\xi^{\otimes}_{p,n} \\
&\qquad \text{by the hexagon equality (5.7.9),} \\
&= \big((\alpha\circ\langle\omega\rangle)_{c'}\circ\big\langle\langle E_i\rangle\phi\big\rangle_i\big)\cdot\xi^{\otimes}_{p,n} \\
&\qquad \text{by associativity,} \\
&= \big((\alpha\circ\langle\omega\rangle)_{c'}\circ\langle E\rangle\phi\big)\cdot\xi^{\otimes}_{p,n} \\
&\qquad \text{by definition.}
\end{aligned}$$

Therefore, $\alpha\circ\langle\omega\rangle$ is a transformation. □

Explanation 5.7.10 (Block Transposition Relation)**.** Here we explain the equality (5.7.9),

$$\Big(\oplus_{i=1}^{m}\xi^{\otimes}_{p_i,n}\Big)\cdot\xi^{\otimes}_{m,n}\langle\langle p\rangle n\rangle = \xi^{\otimes}_{(p_1+\cdots+p_m),n} = \xi^{\otimes}_{p,n},$$

used in the proof of Lemma 5.7.6 above. We first give explicit formulas showing the two sides are equal. Then we give a second, more geometric argument, using the interpretation of $\xi^{\otimes}$ as matrix transposition.

Using the formulas of Definition I.2.4.1 and (II.1.1.19), both sides of (5.7.9) send

$$A = (j-1)p + p_1 + \cdots + p_{i-1} + k$$

to

$$B = n(p_1 + \cdots + p_{i-1} + k - 1) + j$$

for

$$i \in \{1,\ldots,m\}, \qquad j \in \{1,\ldots,n\}, \quad \text{and} \quad k \in \{1,\ldots,p_i\}.$$

The block permutation $\xi^{\otimes}_{m,n}\langle\langle p\rangle n\rangle$ sends A to

$$A' = n(p_1 + \cdots + p_{i-1}) + (j-1)p_i + k.$$

The block sum

$$\oplus_{i=1}^{m}\xi^{\otimes}_{p_i,n}$$

then sends A' to B.

Alternatively, we also have the following geometric interpretation of (5.7.9) in terms of matrix transposition. A special case of this argument, where all of the p_i are equal, is used in Explanation I.2.4.14 to explain the hexagon axiom for Σ. Suppose

$$P_i = [\bullet,\ldots,\bullet] \quad \text{for} \quad 1 \le j \le n \quad \text{and} \quad 1 \le i \le m$$

is a $1\times p_i$ matrix with p_i objects, and consider the $n \times p$ matrix

$$M = \begin{bmatrix} P_{11} & \cdots & P_{1m} \\ \vdots & \ddots & \vdots \\ P_{n1} & \cdots & P_{nm} \end{bmatrix}$$

with each P_{ji} a copy of P_i for $1 \le j \le n$. Then the left hand side of (5.7.9) is the composite of permutations along the top and right of the following diagram, while the right hand side is the other permutation, with $(-)^T$ denoting transpose of matrices.

$$\begin{array}{ccc} \begin{bmatrix} P_{11} & \cdots & P_{1m} \\ \vdots & \ddots & \vdots \\ P_{n1} & \cdots & P_{nm} \end{bmatrix} & \xrightarrow{\xi^{\otimes}_{m,n}\langle\langle p\rangle n\rangle} & \begin{bmatrix} P_{11} & \cdots & P_{n1} \\ \vdots & \ddots & \vdots \\ P_{1m} & \cdots & P_{nm} \end{bmatrix} = \begin{bmatrix} p_1 \otimes n \\ \vdots \\ p_m \otimes n \end{bmatrix} \\ & \searrow^{\xi^{\otimes}_{p,n}} & \Big\downarrow \oplus_{i=1}^{m}\xi^{\otimes}_{p_i,n} \\ & & \begin{bmatrix} P^T_{11} & \cdots & P^T_{n1} \\ \vdots & \ddots & \vdots \\ P^T_{1m} & \cdots & P^T_{nm} \end{bmatrix} = \begin{bmatrix} n \otimes p_1 \\ \vdots \\ n \otimes p_m \end{bmatrix} \end{array}$$

The two equalities in the above diagram are instances of the convention for products in Σ, as described in Explanation I.2.4.7. For example, each $p_i \otimes n$ is viewed either as a row of $p_i n$ objects arranged into n intervals, each with p_i objects, or as an $n \times p_i$ matrix. ◇

Proposition 5.7.11. *In the context of Definition 5.7.2,* $\mathsf{Hom}(\mathsf{M},\mathsf{N})$ *is a multicategory.*

Proof. Lemma 5.7.6 shows that the naturality condition (5.7.3) is closed under composition. Closure of the naturality condition under permutations follows from the top and bottom equivariance axioms for composition in N.

The four multicategory axioms for $\mathsf{Hom}(\mathsf{M},\mathsf{N})$ follow from those of N. For example, let $\mathsf{H} = \mathsf{Hom}(\mathsf{M},\mathsf{N})$ and suppose given

- $F''' \in \mathsf{H}$,
- $\langle F''\rangle \in \mathsf{Prof}(\mathsf{H})$ of length n,
- $\langle F'_j\rangle \in \mathsf{Prof}(\mathsf{H})$ of length k_j for each $j \in \{1,\ldots,n\}$, and
- $\langle F_{j,i}\rangle \in \mathsf{Prof}(\mathsf{H})$ of length $\ell_{j,i}$ for each $i \in \{1,\ldots,k_j\}$.

Then, using the notation conventions for concatenation from (5.1.4), the associativity diagram

$$
\begin{array}{ccc}
 & & \mathsf{H}\big(\langle F'\rangle\,;\,F'''\big)\times\prod_{j=1}^{n}\Big[\prod_{i=1}^{k_j}\mathsf{H}\big(\langle F_{j,i}\rangle\,;\,F'_{j,i}\big)\Big] \\
 & \nearrow^{(\gamma,1)} & \downarrow{\scriptstyle\gamma} \\
\mathsf{H}\big(\langle F''\rangle\,;\,F'''\big)\times\Big[\prod_{j=1}^{n}\mathsf{H}\big(\langle F'_j\rangle\,;\,F''_j\big)\Big]\times\prod_{j=1}^{n}\Big[\prod_{i=1}^{k_j}\mathsf{H}\big(\langle F_{j,i}\rangle\,;\,F'_{j,i}\big)\Big] & & \\
\text{permute}\,\Big\downarrow\,\cong & & \mathsf{H}\big(\langle F\rangle\,;\,F'''\big) \\
\mathsf{H}\big(\langle F''\rangle\,;\,F'''\big)\times\prod_{j=1}^{n}\Big[\mathsf{H}\big(\langle F'_j\rangle\,;\,F''_j\big)\times\prod_{i=1}^{k_j}\mathsf{H}\big(\langle F_{j,i}\rangle\,;\,F'_{j,i}\big)\Big] & & \uparrow{\scriptstyle\gamma} \\
 & \searrow_{(1,\prod_j\gamma)} & \\
 & & \mathsf{H}\big(\langle F''\rangle\,;\,F'''\big)\times\prod_{j=1}^{n}\mathsf{H}\big(\langle F_j\rangle\,;\,F''_j\big)
\end{array}
$$

commutes because, for

- $\alpha''\in\mathsf{H}\big(\langle F''\rangle\,;\,F'''\big)$,
- $\alpha'_j\in\mathsf{H}\big(\langle F'_j\rangle\,;\,F''_j\big)$ for $j\in\{1,\ldots,n\}$, and
- $\alpha_{j,i}\in\mathsf{H}\big(\langle F_{j,i}\rangle\,;\,F'_{j,i}\big)$ for $j\in\{1,\ldots,n\}$ and $i\in\{1,\ldots,k_j\}$,

the two composites around the diagram, which are

$$(\alpha''\circ\langle\alpha'_j\rangle)\circ\langle\langle\alpha_{j,i}\rangle_i\rangle_j \quad\text{and}\quad \alpha''\circ\langle\alpha'_j\circ\langle\alpha_{j,i}\rangle_i\rangle_j,$$

are given componentwise by the corresponding two composites around the following diagram for each $c\in\mathsf{M}$, and are therefore equal.

$$
\begin{array}{ccc}
 & & \mathsf{N}\big(\langle F'\rangle c\,;\,F'''c\big)\times\prod_{j=1}^{n}\Big[\prod_{i=1}^{k_j}\mathsf{N}\big(\langle F_{j,i}\rangle c\,;\,F'_{j,i}c\big)\Big] \\
 & \nearrow^{(\gamma,1)} & \downarrow{\scriptstyle\gamma} \\
\mathsf{N}\big(\langle F''\rangle c\,;\,F'''c\big)\times\Big[\prod_{j=1}^{n}\mathsf{N}\big(\langle F'_j\rangle c\,;\,F''_jc\big)\Big]\times\prod_{j=1}^{n}\Big[\prod_{i=1}^{k_j}\mathsf{N}\big(\langle F_{j,i}\rangle c\,;\,F'_{j,i}c\big)\Big] & & \\
\text{permute}\,\Big\downarrow\,\cong & & \mathsf{N}\big(\langle F\rangle c\,;\,F'''c\big) \\
\mathsf{N}\big(\langle F''\rangle c\,;\,F'''c\big)\times\prod_{j=1}^{n}\Big[\mathsf{N}\big(\langle F'_j\rangle c\,;\,F''_jc\big)\times\prod_{i=1}^{k_j}\mathsf{N}\big(\langle F_{j,i}\rangle c\,;\,F'_{j,i}c\big)\Big] & & \uparrow{\scriptstyle\gamma} \\
 & \searrow_{(1,\prod_j\gamma)} & \\
 & & \mathsf{N}\big(\langle F''\rangle c\,;\,F'''c\big)\times\prod_{j=1}^{n}\mathsf{N}\big(\langle F_j\rangle c\,;\,F''_jc\big)
\end{array}
$$

The other axioms of Definition 5.1.2 are verified componentwise in the same way. □

In the special case $\mathsf{N} = \mathsf{End}(\mathsf{C})$ for a permutative category C, the following result gives a description of $\mathsf{Hom}(\mathsf{M},\mathsf{N})$ arising from the monoidal product on C.

Lemma 5.7.12. *Suppose* M *is a small multicategory and* $(\mathsf{C},\oplus,e,\xi)$ *is a small permutative category.*

(1) *Taking the pointwise monoidal product of multifunctors induces a permutative structure on* $\mathsf{Multicat}(\mathsf{M},\mathsf{End}(\mathsf{C}))$.
(2) *There is an isomorphism of multicategories*

$$\mathsf{End}\big(\mathsf{Multicat}(\mathsf{M},\mathsf{End}(\mathsf{C}))\big) \cong \mathsf{Hom}(\mathsf{M},\mathsf{End}(\mathsf{C})).$$

Proof. First we explain the pointwise monoidal product. Suppose given multifunctors

$$F, F' \in \mathsf{Multicat}(\mathsf{M},\mathsf{End}(\mathsf{C})).$$

For each $c \in \mathsf{M}$ and $\phi \in \mathsf{M}(\langle c\rangle\,;c')$, let $\xi_\oplus$ denote the shuffle isomorphism

$$\bigoplus_{c_i\in\langle c\rangle}(Fc_i \oplus F'c_i) \xrightarrow{\ \xi_\oplus\ } \Big(\bigoplus_{c_i\in\langle c\rangle} Fc_i\Big) \oplus \Big(\bigoplus_{c_i\in\langle c\rangle} F'c_i\Big)$$

induced by the permutation $\xi^{\otimes}_{2,\mathrm{len}\langle c\rangle}$ (I.2.4.5) in the finite ordinal category. Then we define

$$(F\oplus F')c = (Fc)\oplus(F'c) \quad\text{and}\quad (F\oplus F')\phi = ((F\phi)\oplus(F'\phi))\circ\xi_\oplus.$$

For multinatural transformations

$$\alpha\colon F \longrightarrow G \quad\text{and}\quad \alpha'\colon F' \longrightarrow G'$$

we define

$$(\alpha\oplus\alpha')_c = \alpha_c \oplus \alpha'_c.$$

The monoidal unit is the constant multifunctor at $e \in \mathsf{C}$, and the permutative structure for C implies that this is a strict symmetric monoidal product for

$$\mathsf{Multicat}(\mathsf{M},\mathsf{End}(\mathsf{C})).$$

Second we verify the asserted isomorphism. To begin, observe that the objects and unary operations on both sides are the same by definition. For operations

$$\eta \in \mathsf{Hom}(\mathsf{M},\mathsf{End}(\mathsf{C}))(\langle F\rangle\,;F'),$$

we have, by definition,

$$\eta_c\colon \langle F\rangle c = \bigoplus_{F_i\in\langle F\rangle} F_i c \longrightarrow F'c$$

and in this context the equality (5.7.3) is equivalent to commutativity of the following diagram in C for each $\phi \in \mathsf{M}\big(\langle c\rangle\,;\,c'\big)$.

$$\begin{array}{ccc}
\bigoplus_i \bigoplus_j F_i c_j & \xrightarrow{\ \langle F\rangle\phi\ } & \bigoplus_i F_i c' \\
\xi \downarrow \cong & & \\
\bigoplus_j \bigoplus_i F_i c_j & & \downarrow \eta_{c'} \\
\eta_{\langle c\rangle} \downarrow & & \\
\bigoplus_j F' c_j & \xrightarrow{\ \ F'\phi\ \ } & F'c'
\end{array}$$

Inverting the shuffle isomorphism, this is precisely the condition in Definition 5.1.17 for η to define a multinatural transformation

$$\oplus_i F_i \longrightarrow F.$$

Thus the data of an operation in $\mathsf{Hom}(\mathsf{M}, \mathsf{End}(\mathsf{C}))$ is precisely the same as that of an operation in

$$\mathsf{End}\big(\mathsf{Multicat}(\mathsf{M}, \mathsf{End}(\mathsf{C}))\big).$$

This correspondence preserves compositions and symmetric group actions because these are induced by the permutative structure of C. □

Proposition 5.7.13. *Given small multicategories* M, N, *and* P, *there are isomorphisms of categories*

$$\mathsf{Multicat}(\mathsf{M}, \mathsf{Hom}(\mathsf{N}, \mathsf{P})) \cong \mathsf{Multicat}(\mathsf{M} \otimes \mathsf{N}, \mathsf{P}) \cong \mathsf{Multicat}(\mathsf{N}, \mathsf{Hom}(\mathsf{M}, \mathsf{P}))$$

that are natural in M, N, *and* P.

Proof. We discuss the first isomorphism; the second follows from the first by applying the symmetry β. We define a functor

$$\Theta\colon \mathsf{Multicat}(\mathsf{M}, \mathsf{Hom}(\mathsf{N}, \mathsf{P})) \longrightarrow \mathsf{Multicat}(\mathsf{M} \otimes \mathsf{N}, \mathsf{P})$$

as follows. Given

$$H\colon \mathsf{M} \longrightarrow \mathsf{Hom}(\mathsf{N}, \mathsf{P})$$

we first define an assignment

$$\Theta H\colon \mathsf{M} \otimes \mathsf{N} \longrightarrow \mathsf{P}$$

beginning with $\mathsf{M} \times \mathsf{N}$ and then descending to $\mathsf{M} \# \mathsf{N}$ and $\mathsf{M} \otimes \mathsf{N}$.

- For $(c, d) \in \mathsf{M} \times \mathsf{N}$, let

$$(\Theta H)(c, d) = (Hc)d.$$

- For $\phi \in \mathsf{M}\big(\langle c\rangle\,;\,c'\big)$ and $\psi \in \mathsf{N}\big(\langle d\rangle\,;\,d'\big)$, let

$$(\Theta H)(\phi \times d) = (H\phi)_d \quad \text{and} \quad (\Theta H)(c \times \psi) = (Hc)\psi.$$

Now by definition

$$(\Theta H)(c,-) = (Hc)(-)$$

is a multifunctor from N to P for each object c in M. Next we observe that

$$(\Theta H)(-,d)\colon \mathsf{M} \longrightarrow \mathsf{P}$$

is a multifunctor for each $d \in \mathsf{N}$. The basic reason for this is that the composition of operations in $\mathsf{Hom}(\mathsf{N},\mathsf{P})$ is defined componentwise. For example, given composable operations ϕ' and $\langle\phi\rangle$ in M, we have and equality of operations in $\mathsf{Hom}(\mathsf{N},\mathsf{P})$,

$$H(\phi') \circ H\langle\phi\rangle = H(\phi' \circ \langle\phi\rangle),$$

because H is a multifunctor. Now because composition of operations in $\mathsf{Hom}(\mathsf{N},\mathsf{P})$ is defined componentwise, we have the following for each $d \in \mathsf{N}$:

$$\begin{aligned}(\Theta H)(\phi' \times d) \circ \langle(\Theta H)(\phi_i \times d)\rangle_i &= H(\phi')_d \circ (H\langle\phi\rangle)_d\\ &= (H(\phi' \circ \langle\phi\rangle))_d\\ &= (\Theta H)((\phi' \circ \langle\phi\rangle) \times d).\end{aligned}$$

The rest of the multifunctor axioms for $(\Theta H)(-,d)$ are verified similarly.

Therefore, recalling Explanation 5.6.9, ΘH gives a well-defined multifunctor on $\mathsf{M} \# \mathsf{N}$. To verify that this assignment descends to a unique multifunctor on the tensor product $\mathsf{M} \otimes \mathsf{N}$, we recall from Explanation 5.6.14 that it suffices to show

$$(\Theta H)(\phi \otimes \psi) = (\Theta H)((\phi \otimes^{\mathsf{t}} \psi) \cdot \xi^{\otimes})$$

for all operations

$$\phi \in \mathsf{M}(\langle c\rangle\,; c') \quad \text{and} \quad \psi \in \mathsf{N}(\langle d\rangle\,; d').$$

Using the decompositions of Definition 5.6.10 and the definition of ΘH, we must verify

$$((Hc')\psi) \circ \langle(H\phi)_{d_j}\rangle_j = \big((H\phi)_{d'} \circ \langle(Hc_i)\rangle_i\psi\big) \cdot \xi^{\otimes}.$$

This equality follows directly from the defining equality (5.7.3) at the operation

$$\psi \in \mathsf{N}(\langle d\rangle\,; d')$$

with

$$\alpha = (H\phi), \quad F_i = Hc_i, \quad \text{and} \quad G = Hc'.$$

Therefore

$$\Theta H\colon \mathsf{M} \otimes \mathsf{N} \longrightarrow \mathsf{P}$$

is a well-defined multifunctor for each

$$H \in \mathsf{Multicat}(\mathsf{M}, \mathsf{Hom}(\mathsf{N},\mathsf{P})).$$

For a multinatural transformation

$$\eta \in \mathsf{Multicat}(\mathsf{M}, \mathsf{Hom}(\mathsf{N},\mathsf{P}))(H,H'),$$

we define

$$\Theta\eta \in \mathsf{Multicat}(\mathsf{M} \otimes \mathsf{N},\mathsf{P})(\Theta H, \Theta H')$$

via components

$$(\Theta\eta)_{c\otimes d} = (\eta_c)_d.$$

One verifies the analogue of equality (5.7.3) for $\Theta\eta$ on generating operations $\phi \otimes d$ and $c\otimes\psi$ using, respectively, componentwise composition of operations in $\mathsf{Hom}(\mathsf{N},\mathsf{P})$

at the object d and the analogue of (5.7.3) for each η_c at the operation ψ. This shows that Θ defines a function

$$\mathsf{Multicat}(\mathsf{M},\mathsf{Hom}(\mathsf{N},\mathsf{P}))(H,H') \xrightarrow{\Theta} \mathsf{Multicat}(\mathsf{M}\otimes\mathsf{N},\mathsf{P})(\Theta H,\Theta H')$$

for each pair

$$H,H' \in \mathsf{Multicat}(\mathsf{M},\mathsf{Hom}(\mathsf{N},\mathsf{P})).$$

Finally, one verifies that Θ preserves identities and composition because these are determined componentwise:

$$(\Theta 1_H)_{c\otimes d} = ((1_H)_c)_d \quad \text{and}$$

$$(\Theta(\eta'\eta))_{c\otimes d} = ((\eta'\eta)_c)_d = (\eta'_c)_d \circ (\eta_c)_d) = ((\Theta\eta')\circ(\Theta\eta))_{c\otimes d}$$

for composable multinatural transformations η' and η and for each $c\otimes d \in \mathsf{M}\otimes\mathsf{N}$. This finishes the definition of a functor

$$\Theta\colon \mathsf{Multicat}(\mathsf{M},\mathsf{Hom}(\mathsf{N},\mathsf{P})) \longrightarrow \mathsf{Multicat}(\mathsf{M}\otimes\mathsf{N},\mathsf{P}).$$

Next we define a functor

$$\Psi\colon \mathsf{Multicat}(\mathsf{M}\otimes\mathsf{N},\mathsf{P}) \longrightarrow \mathsf{Multicat}(\mathsf{M},\mathsf{Hom}(\mathsf{N},\mathsf{P}))$$

as follows. Given a multifunctor $K\colon \mathsf{M}\otimes\mathsf{N} \longrightarrow \mathsf{P}$ we first define

$$(\Psi K)c = K(c\otimes(-))\colon \mathsf{N} \longrightarrow \mathsf{P}$$

and

$$(\Psi K)\phi = K(\phi\otimes -) \in \mathsf{Hom}(\mathsf{N},\mathsf{P})\big((\Psi K)\langle c\rangle\,;\,(\Psi K)c'\big)$$

for

$$\big(\langle c\rangle\,;\,c'\big) \in \mathsf{Prof}(\mathsf{Ob}\,\mathsf{M})\times\mathsf{Ob}\,\mathsf{M} \quad \text{and} \quad \phi\in\mathsf{M}\big(\langle c\rangle\,;\,c'\big).$$

Given a multinatural transformation $\omega \in \mathsf{Multicat}(\mathsf{M}\otimes\mathsf{N},\mathsf{P})(K,K')$, we define

$$(\Psi\omega)_c = \omega_{c\otimes -}\colon K(c\otimes -) \longrightarrow K'(c\otimes -).$$

Checking the definitions, one verifies the following.

(1) $(\Psi K)c$ is a multifunctor $\mathsf{N} \longrightarrow \mathsf{P}$ for each $c \in \mathsf{M}$ and $(\Psi K)\phi$ is an operation of $\mathsf{Hom}(\mathsf{N},\mathsf{P})$ for each $\phi \in \mathsf{M}\big(\langle c\rangle\,;\,c'\big)$. Therefore, ΨK defines a function on objects and operations of M taking values in $\mathsf{Hom}(\mathsf{N},\mathsf{P})$.

(2) ΨK satisfies the axioms of a multifunctor

$$\Psi K\colon \mathsf{M} \longrightarrow \mathsf{Hom}(\mathsf{N},\mathsf{P})$$

for each $K \in \mathsf{Multicat}(\mathsf{M}\otimes\mathsf{N},\mathsf{P})$.

(3) $\Psi\omega$ satisfies the axioms of an operation $\Psi K \longrightarrow \Psi K'$ for each

$$\omega \in \mathsf{Multicat}(\mathsf{M}\otimes\mathsf{N},\mathsf{P})(K,K').$$

Therefore, Ψ defines a function on the objects and morphisms

$$\mathsf{Multicat}(\mathsf{M}\otimes\mathsf{N},\mathsf{P}) \longrightarrow \mathsf{Multicat}(\mathsf{M},\mathsf{Hom}(\mathsf{N},\mathsf{P})).$$

(4) Ψ is functorial.

Lastly, one verifies that Θ and Ψ define an inverse bijection on objects and morphisms and, therefore, define an isomorphism of categories. Naturality of the isomorphism with respect to multifunctors of M, N, and P follows from the definitions of Θ and Ψ. □

The isomorphism of Proposition 5.7.13 shows that the internal Hom provides a closed structure for $(\mathsf{Multicat}, \otimes, \beta)$. Combining Proposition 5.7.13 with Theorems 5.5.14 and 5.6.18 we have the following.

Theorem 5.7.14. *The category of small multicategories* $\mathsf{Multicat}$ *is complete, cocomplete, symmetric monoidal, and closed. The monoidal product is given by the tensor product of Definition 5.6.11 and the closed structure is given by the internal hom of Definition 5.7.2.*

Recalling Example 3.9.2, the symmetric monoidal closed structure for $\mathsf{Multicat}$ also makes it tensored and cotensored over itself. Thus we record the following internal $\otimes$-Hom adjunction.

Lemma 5.7.15. *In the context of Proposition 5.7.13, there are isomorphisms of multicategories*

$$\mathsf{Hom}(\mathsf{M}, \mathsf{Hom}(\mathsf{N}, \mathsf{P})) \cong \mathsf{Hom}(\mathsf{M} \otimes \mathsf{N}, \mathsf{P}) \cong \mathsf{Hom}(\mathsf{N}, \mathsf{Hom}(\mathsf{M}, \mathsf{P})) \tag{5.7.16}$$

that are natural in M, N, *and* P.

Restricting to pointed multicategories, we now define a pointed internal hom. This is a special case of Definition 4.2.1.

Definition 5.7.17. Suppose M and N are small pointed multicategories. We define the *pointed hom* multicategory as the following pullback in $\mathsf{Multicat}$.

$$\begin{array}{ccc} \mathsf{Hom}_*(\mathsf{M},\mathsf{N}) & \longrightarrow & \mathsf{T} \\ \big\downarrow & & \big\downarrow \\ \mathsf{Hom}(\mathsf{M},\mathsf{N}) & \longrightarrow & \mathsf{Hom}(\mathsf{T},\mathsf{N}) \end{array} \tag{5.7.18}$$

The composite

$$\mathsf{T} \cong \mathsf{Hom}(\mathsf{M},\mathsf{T}) \longrightarrow \mathsf{Hom}(\mathsf{M},\mathsf{N}) \longrightarrow \mathsf{Hom}(\mathsf{T},\mathsf{N})$$

induced by the structure morphisms for M and N is equal to the vertical morphism in the diagram, and therefore induces a canonical structure morphism $\mathsf{T} \longrightarrow \mathsf{Hom}_*(\mathsf{M},\mathsf{N})$ making $\mathsf{Hom}_*(\mathsf{M},\mathsf{N})$ a pointed multicategory. ◇

Explanation 5.7.19. The objects and operations of $\mathsf{Hom}_*(\mathsf{M},\mathsf{N})$ can be identified as follows. The objects are those multifunctors $\mathsf{M} \longrightarrow \mathsf{N}$ that commute with the structure morphisms from T. The n-ary operations are those operations η in $\mathsf{Hom}(\mathsf{M},\mathsf{N})$, between pointed multifunctors, such that

$$\eta_{*^{\mathsf{M}}} = \iota_n^{\mathsf{N}},$$

with $*^{\mathsf{M}}$ the basepoint of M. Specializing Explanation 5.7.5, we see that the underlying category of objects and unary operations of $\mathsf{Hom}_*(\mathsf{M},\mathsf{N})$ is $\mathsf{Multicat}_*(\mathsf{M},\mathsf{N})$. That is, we have an isomorphism of categories

$$\mathsf{Multicat}(\mathsf{I}, \mathsf{Hom}_*(\mathsf{M},\mathsf{N})) \cong \mathsf{Multicat}_*(\mathsf{M},\mathsf{N}). \tag{5.7.20}$$

◇

The following pointed version of Lemma 5.7.12 will be used in the proof of Proposition 10.6.7.

Lemma 5.7.21. *Suppose* M *is a small pointed multicategory and* $(\mathsf{C}, \oplus, e, \xi)$ *is a small permutative category.*

(1) *The pointwise monoidal product on* $\mathsf{Multicat}(\mathsf{M}, \mathsf{End}(\mathsf{C}))$ *restricts to make*
$$\mathsf{Multicat}_*(\mathsf{M}, \mathsf{End}(\mathsf{C}))$$
a permutative category.

(2) *There is an isomorphism of multicategories*
$$\mathsf{End}\big(\mathsf{Multicat}_*(\mathsf{M}, \mathsf{End}(\mathsf{C}))\big) \cong \mathsf{Hom}_*(\mathsf{M}, \mathsf{End}(\mathsf{C})).$$

Proof. First, if
$$F, F' \in \mathsf{Multicat}_*(\mathsf{M}, \mathsf{End}(\mathsf{C}))$$
then so is $F \oplus F'$, because C is permutative. Likewise, if α and α' are pointed multinatural transformations between pointed multifunctors, then so is $\alpha \oplus \alpha'$. The pointwise monoidal unit, which is the constant multifunctor at the unit $e \in \mathsf{C}$, is a pointed multifunctor. The symmetry isomorphism of
$$\mathsf{Multicat}(\mathsf{M}, \mathsf{End}(\mathsf{C})),$$
induced by that of C, is pointed because C is permutative.

Second, we observe that the correspondence of operations established in the proof of Lemma 5.7.12 restricts to a correspondence between pointed operations. Using Explanation 5.7.19, we see that n-ary operations
$$\eta \in \mathsf{Hom}_*(\mathsf{M}, \mathsf{End}(\mathsf{C}))\big(\langle F \rangle ; F\big)$$
have components
$$\eta_c \colon \langle F \rangle c = \bigoplus_{F_i \in \langle F \rangle} F_i c \longrightarrow F' c$$
such that
$$\eta_{*\mathsf{M}} = \iota_n = 1_e \colon \bigoplus_{F_i \in \langle F \rangle} e \longrightarrow e.$$
Since this is precisely the condition for n-ary operations of
$$\mathsf{End}(\mathsf{Multicat}_*(\mathsf{M}, \mathsf{End}(\mathsf{C}))),$$
the result follows. □

Now applying Theorem 4.2.3 to the complete and cocomplete symmetric monoidal closed category
$$(\mathsf{Multicat}, \otimes, \mathsf{I}, \mathsf{Hom})$$
with terminal object T, we have the following.

Theorem 5.7.22. *The category of pointed small multicategories* $\mathsf{Multicat}_*$ *is complete, cocomplete, symmetric monoidal, and closed. The monoidal product is given by the smash product of Definition 5.6.20 with monoidal unit*
$$\mathsf{S} = \mathsf{I} \amalg \mathsf{T}.$$
The closed structure is given by the pointed hom of Definition 5.7.17.

Recall from Corollary 5.3.9 that taking endomorphism multicategories provides a 2-functor
$$\mathsf{PermCat}^{\mathsf{su}} \xrightarrow{\mathsf{End}} \mathsf{Multicat}_*$$
from permutative categories and strictly unital functors to pointed multicategories and pointed multifunctors. The smash product of multicategories does not restrict along this 2-functor, as we now show.

Proposition 5.7.23. *The symmetric monoidal structure on* $\mathsf{Multicat}_*$ *in Theorem 5.7.22 does not restrict to a symmetric monoidal structure on* $\mathsf{PermCat}^{\mathsf{su}}$ *along the 2-functor* End *in Corollary 5.3.9.*

Proof. The monoidal unit in $\mathsf{Multicat}_*$ is $\mathsf{S} = \mathsf{I} \coprod \mathsf{T}$ with object set $\{1,0\}$, where 1 denotes the unique object of I and 0 denotes the unique object of T, the basepoint of S. The entries of this 2-object multicategory are given by

$$\mathsf{S}(\varepsilon_1,\ldots,\varepsilon_n\,;\varepsilon) = \begin{cases} \mathsf{T}(0,\ldots,0\,;0) = * & \text{if } \varepsilon_1 = \cdots = \varepsilon_n = \varepsilon = 0, \\ \mathsf{I}(1\,;1) = * & \text{if } \varepsilon_1 = \varepsilon = 1 \text{ with } n = 1, \text{ and} \\ \varnothing & \text{otherwise.} \end{cases}$$

If the symmetric monoidal structure on $\mathsf{Multicat}_*$ restricts to one on $\mathsf{PermCat}^{\mathsf{su}}$ along End, then

$$\mathsf{S} = \mathsf{End}(\mathsf{J})$$

for some permutative category $(\mathsf{J},\oplus,0)$ with object set $\{0,1\}$. But then there are equalities

$$\begin{aligned} \varnothing = \mathsf{S}(0,1\,;1) &= \mathsf{End}(\mathsf{J})(0,1\,;1) \\ &= \mathsf{J}(0\oplus 1;1) = \mathsf{J}(1;1) \neq \varnothing, \end{aligned}$$

which cannot happen. □

5.8. Notes

5.8.1 (Multicategories). What we call a multicategory is also called a *symmetric multicategory*, with the plain term *multicategory* reserved for the non-symmetric definition. The terms *operad*, a *symmetric operad*, and a *colored operad* are also common. The book [**Yau16**] is a gentle introduction to multicategories.

Historically, multicategories without symmetric group actions were introduced by Lambek [**Lam69**]. May [**May72**] introduced the term *operad* for a one-object multicategory. The tensor product was introduced by Boardman and Vogt [**BV73**] and proved to give a symmetric monoidal closed structure on the category of small multicategories and multifunctors by Moerdijk and Toen [**MT10**]. Homotopy theory of multicategories is discussed in [**MT10, WY18**]. Applications of multicategories outside of pure mathematics can be found in [**Yau18, Yau20**]. ◇

5.8.2 (Creating Coequalizers). In the context of Definition 5.4.12 we caution that some authors, notably [**ML98**], write that *U creates coequalizers* for what we have called *strictly* creating coequalizers in Definition 5.4.12. We have included the extra adjective because a more general condition, without the adjective "strictly", is often used in the literature and is equivalent to the comparison (5.4.7)

$$\mathsf{A} \longrightarrow \mathsf{Alg}(UL)$$

being a mere equivalence of categories. See, for example, [**Rie16**, Section 5] for further development. ◇

5.8.3 (Monadicity Theorems). The original reference for tripleability/monadicity theorems is the thesis of Beck [**Bec67**]. Our proof of Theorem 5.4.13 sketches that of [**ML98**, VI.7 Theorem 1], which follows [**Bec67**]. Variations and generalizations of Theorem 5.4.13 are given in [**Bec67**] and the later literature. See [**ML98**, Section VI.7] and [**Rie16**, Section 5.5] for further discussion and references. ◇

5.8.4 (Finite Products and Filtered Colimits)**.** Our proof of Lemma 5.4.20 (2), that finite products commute with filtered colimits in Set, is an adaptation of a more general argument in [**Rie16**, Theorem 3.8.9] showing that general finite limits commute with filtered colimits in Set. The argument is special to the category Set, as it depends on the construction of colimits there. ◇

5.8.5 (Symmetric Monoidal Closed Structure)**.** Our presentation of the tensor product in Definition 5.6.11 follows that of Elmendorf-Mandell [**EM09**] and is equivalent to the tensor product introduced by Boardman-Vogt [**BV73**]. We refer the reader to [**MT10**] for additional development.

Our presentations of the internal hom and its pointed variant also follow [**EM09**]. A more general properad analogue of Proposition 5.7.13 and Theorem 5.7.14 is given by Hackney-Robertson-Yau in [**HRY15**, Theorem 4.30 and Corollary 4.31]. ◇

5.8.6 (Closed Symmetric Multicategories)**.** The category Multicat of small multicategories is closed symmetric monoidal by Theorem 5.7.14. This implies that Multicat is a *closed symmetric multicategory* in the sense of [**Zak18**, 1.2]. The internal hom objects are the small multicategories

$$\mathsf{Hom}(\mathsf{A}_1 \otimes \cdots \otimes \mathsf{A}_n; \mathsf{B})$$

for $n \geq 0$ and small multicategories $\mathsf{A}_1, \ldots, \mathsf{A}_n, \mathsf{B}$. The evaluation morphism is the second isomorphism in Proposition 5.7.13 with

$$\begin{aligned} \mathsf{M} &= \mathsf{A}_1 \otimes \cdots \otimes \mathsf{A}_n \\ \mathsf{N} &= \mathsf{Hom}(\mathsf{A}_1 \otimes \cdots \otimes \mathsf{A}_n; \mathsf{B}) \\ \mathsf{P} &= \mathsf{B}. \end{aligned}$$

This notion of a closed symmetric multicategory is the symmetric analogue of a *closed multicategory* in the sense of [**Lam69**, p. 106] and [**Man12**, 3.6], which considered multicategories without symmetric group actions. The same remarks also apply to the category of pointed small multicategories $\mathsf{Multicat}_*$, which is closed symmetric monoidal by Theorem 5.7.22 with the pointed hom in Definition 5.7.17. ◇

CHAPTER 6

Enriched Multicategories

In this chapter we define multicategories enriched over a symmetric monoidal category $\mathsf{V} = (\mathsf{V}, \otimes, \xi)$. To simplify the presentation, we give the definitions under the assumption that V is permutative. These provide the general definitions via strictification, and we describe that process in Explanation 6.1.19 below. This will be important because in our two applications of interest V is symmetric monoidal but not permutative. The first is $\mathsf{V} = \mathsf{Cat}$, and will be discussed below. The second uses $\mathsf{V} = \mathsf{Cat}_*$, the category of small pointed categories, and $\mathsf{V} = \mathsf{sSet}_*$, the category of pointed simplicial sets. The latter will be discussed in Chapter 7.

In Section 6.1 we define the 2-category of V-enriched multicategories, multifunctors, and multinatural transformations. In Section 6.2 we discuss change of enrichment along a symmetric monoidal functor. In Section 6.3 we consider a symmetric monoidal V-category K, as defined in Section 1.4, and show that $\mathsf{End}(\mathsf{K})$ is V-enriched as a multicategory.

In Sections 6.4 through 6.6 we apply the preceding general theory to the case $\mathsf{V} = \mathsf{Cat}$ and $\mathsf{K} = \mathsf{Multicat}_*$. This begins in Section 6.4 where we extend Theorem 5.7.22 to show that $\mathsf{Multicat}_*$ is a symmetric Cat-monoidal 2-category and therefore a Cat-enriched multicategory.

Our main reason for considering the Cat-enriched multicategory structure on $\mathsf{Multicat}_*$, rather than the Cat-monoidal structure from which it arises, is the following. As noted in Proposition 5.7.23, the symmetric monoidal structure of $\mathsf{Multicat}_*$ does not restrict along the 2-functor

$$\mathsf{PermCat}^{\mathsf{su}} \xrightarrow{\mathsf{End}} \mathsf{Multicat}_*$$

of Section 5.3. In contrast, the multicategory structure *does* restrict along inclusions such as End.

We describe this induced multicategory structure in Section 6.5. The objects in this multicategory are small permutative categories. Each category

$$\mathsf{PermCat}^{\mathsf{su}}\big(\langle \mathsf{C} \rangle \,;\, \mathsf{D}\big)$$

has multilinear functors as objects (Definition 6.5.4) and multilinear transformations as morphisms (Definition 6.5.11).

The Cat-enriched multicategory structure on $\mathsf{PermCat}^{\mathsf{su}}$ is of fundamental importance for our applications in Part 2 because it encodes the algebraic structures of interest for our K-theory multifunctors. It is the domain of the Elmendorf-Mandell K theory multifunctor defined in Chapter 10. Therefore, in Section 6.6 we give a second, self-contained and detailed proof that $\mathsf{PermCat}^{\mathsf{su}}$ is a Cat-enriched multicategory. As we explain there, the multicategory structure on $\mathsf{PermCat}^{\mathsf{su}}$ is a generalization of the 2-category of small symmetric monoidal categories, symmetric monoidal functors, and monoidal natural transformations. Keeping this connection in mind, the proofs that (i) the various parts of $\mathsf{PermCat}^{\mathsf{su}}$ are well defined and (ii) the Cat-enriched multicategory axioms are satisfied, are not conceptually difficult. However, they require a nontrivial amount of notation to keep track of the many lists of objects. See also Note 6.7.2. By reversing the linearity constraints, the detailed proofs in Section 6.6 can also be used for colax multilinear functors and colax multilinear transformations; see Definition 10.7.24 for $\mathsf{PermCat}^{\mathsf{su}}_{\mathsf{co}}$.

6.1. Enriched Multicategories

Throughout this section we suppose that $\mathsf{V} = (\mathsf{V}, \otimes, \xi)$ is a permutative category. Recall from Definition 1.1.23 this means that the associativity and unit isomorphisms of V are identities. However in some of the axioms it will be useful to explicitly write the units

$$\lambda: \mathbb{1} \otimes X \longrightarrow X \quad \text{and} \quad \rho: X \otimes \mathbb{1} \longrightarrow X$$

even though they are identities.

Definition 6.1.1. Suppose V is a permutative category. A V-*enriched multicategory* $(\mathsf{M}, \gamma, 1)$ consists of the following data.

- M is equipped with a class $\mathsf{Ob}\,\mathsf{M}$ of *objects*. We write $\mathsf{Prof}(\mathsf{M})$ for $\mathsf{Prof}(\mathsf{Ob}\,\mathsf{M})$.
- For $c' \in \mathsf{Ob}\,\mathsf{M}$ and $\langle c \rangle = (c_1, \ldots, c_n) \in \mathsf{Prof}(\mathsf{M})$, M is equipped with an object of V
$$\mathsf{M}\big(\langle c \rangle\,; c'\big) = \mathsf{M}\big(c_1, \ldots, c_n\,; c'\big) \in \mathsf{V}$$
called the *n-ary operation object* with *input profile* $\langle c \rangle$ and *output* c'.
- For $\big(\langle c \rangle\,; c'\big) \in \mathsf{Prof}(\mathsf{M}) \times \mathsf{Ob}\,\mathsf{M}$ as above and a permutation $\sigma \in \Sigma_n$, M is equipped with an isomorphism in V
$$\mathsf{M}\big(\langle c \rangle\,; c'\big) \xrightarrow[\cong]{\sigma} \mathsf{M}\big(\langle c \rangle\sigma\,; c'\big),$$
called the *right action* or the *symmetric group action*, in which
$$\langle c \rangle\sigma = \big(c_{\sigma(1)}, \ldots, c_{\sigma(n)}\big)$$
is the right permutation of $\langle c \rangle$ by σ.
- For $c \in \mathsf{Ob}\,\mathsf{M}$, M is equipped with a morphism
$$1_c: \mathbb{1} \longrightarrow \mathsf{M}\big(c\,; c\big),$$
called the *c-colored unit*.
- For $c'' \in \mathsf{Ob}\,\mathsf{M}$, $\langle c' \rangle = (c'_1, \ldots, c'_n) \in \mathsf{Prof}(\mathsf{M})$, and $\langle c_j \rangle = (c_{j,1}, \ldots, c_{j,k_j}) \in \mathsf{Prof}(\mathsf{M})$ for each $j \in \{1, \ldots, n\}$, let $\langle c \rangle = \oplus_j \langle c_j \rangle \in \mathsf{Prof}(\mathsf{M})$ be the concatenation of the $\langle c_j \rangle$. Then M is equipped with a morphism in V
$$\mathsf{M}\big(\langle c' \rangle\,; c''\big) \otimes \bigotimes_{j=1}^{n} \mathsf{M}\big(\langle c_j \rangle\,; c'_j\big) \xrightarrow{\gamma} \mathsf{M}\big(\langle c \rangle\,; c''\big) \tag{6.1.2}$$
called the *composition*.

These data are required to satisfy the following axioms.

Symmetric Group Action: For $(\langle c\rangle; c') \in \mathsf{Prof}(\mathsf{M}) \times \mathsf{Ob}\,\mathsf{M}$ with $n = \operatorname{len}\langle c\rangle$ and $\sigma, \tau \in \Sigma_n$, the following diagram in V commutes.

$$\begin{array}{ccc}
\mathsf{M}(\langle c\rangle; c') & \xrightarrow{\ \sigma\ } & \mathsf{M}(\langle c\rangle\sigma; c') \\
 & \searrow^{\sigma\tau} & \downarrow{\scriptstyle \tau} \\
 & & \mathsf{M}(\langle c\rangle\sigma\tau; c')
\end{array}$$

Moreover, the identity permutation in Σ_n acts as the identity morphism of $\mathsf{M}(\langle c\rangle; c')$.

Associativity: Suppose given

- $c''' \in \mathsf{Ob}\,\mathsf{M}$,
- $\langle c''\rangle = (c''_1, \ldots, c''_n) \in \mathsf{Prof}(\mathsf{M})$,
- $\langle c'_j\rangle = (c'_{j,1}, \ldots, c'_{j,k_j}) \in \mathsf{Prof}(\mathsf{M})$ for each $j \in \{1, \ldots, n\}$, and
- $\langle c_{j,i}\rangle = (c_{j,i,1}, \ldots, c_{j,i,\ell_{j,i}}) \in \mathsf{Prof}(\mathsf{M})$ for each $j \in \{1, \ldots, n\}$ and each $i \in \{1, \ldots, k_j\}$,

such that $k_j = \operatorname{len}\langle c'_j\rangle > 0$ for at least one j. For each j, let $\langle c_j\rangle = \oplus_{i=1}^{k_j}\langle c_{j,i}\rangle$ denote the concatenation of the $\langle c_{j,i}\rangle$. Let $\langle c\rangle = \oplus_{j=1}^{n}\langle c_j\rangle$ denote the concatenation of the $\langle c_j\rangle$. Let $\langle c'\rangle = \oplus_{j=1}^{n}\langle c'_j\rangle$ denote the concatenation of the $\langle c'_j\rangle$.

Then the *associativity diagram* below commutes.

(6.1.3)

$$\begin{array}{ccc}
 & \mathsf{M}(\langle c'\rangle; c''') \otimes \bigotimes_{j=1}^{n}\Big[\bigotimes_{i=1}^{k_j} \mathsf{M}(\langle c_{j,i}\rangle; c'_{j,i})\Big] & \\
 {\scriptstyle (\gamma,1)}\nearrow & & \searrow{\scriptstyle \gamma} \\
\mathsf{M}(\langle c''\rangle; c''') \otimes \Big[\bigotimes_{j=1}^{n} \mathsf{M}(\langle c'_j\rangle; c''_j)\Big] \otimes \bigotimes_{j=1}^{n}\Big[\bigotimes_{i=1}^{k_j} \mathsf{M}(\langle c_{j,i}\rangle; c'_{j,i})\Big] & & \\
 \text{permute}\ \downarrow \cong & & \mathsf{M}(\langle c\rangle; c''') \\
\mathsf{M}(\langle c''\rangle; c''') \otimes \bigotimes_{j=1}^{n}\Big[\mathsf{M}(\langle c'_j\rangle; c''_j) \otimes \bigotimes_{i=1}^{k_j} \mathsf{M}(\langle c_{j,i}\rangle; c'_{j,i})\Big] & & \\
 {\scriptstyle (1,\otimes_j \gamma)}\searrow & & \nearrow{\scriptstyle \gamma} \\
 & \mathsf{M}(\langle c''\rangle; c''') \otimes \bigotimes_{j=1}^{n} \mathsf{M}(\langle c_j\rangle; c''_j) &
\end{array}$$

Unity: Suppose $c' \in \mathsf{Ob}\,\mathsf{M}$.

(1) If $\langle c\rangle = (c_1, \ldots, c_n) \in \mathsf{Prof}(\mathsf{M})$ has length $n \geq 1$, then the following *right unity diagram* is commutative. Here $\mathbb{1}^n$ is the n-fold monoidal product of $\mathbb{1}$ with itself and the unlabeled morphism is given by a composite of (strict) units in V.

$$
(6.1.4)\qquad
\begin{array}{ccc}
\mathsf{M}\big(\langle c\rangle\,;c'\big)\otimes \mathbb{1}^n & \longrightarrow & \mathsf{M}\big(\langle c\rangle\,;c'\big)\\
{\scriptstyle 1\otimes(\otimes_j 1_{c_j})}\big\downarrow & & \big\downarrow{\scriptstyle 1}\\
\mathsf{M}\big(\langle c\rangle\,;c'\big)\otimes \bigotimes_{j=1}^{n}\mathsf{M}\big(c_j\,;c_j\big) & \xrightarrow{\gamma} & \mathsf{M}\big(\langle c\rangle\,;c'\big)
\end{array}
$$

(2) For any $\langle c\rangle \in \mathsf{Prof}(\mathsf{M})$, the *left unity diagram* below is commutative, where the unlabeled morphism is given by a (strict) unit in V.

$$
(6.1.5)\qquad
\begin{array}{ccc}
\mathbb{1}\otimes\mathsf{M}\big(\langle c\rangle\,;c'\big) & \longrightarrow & \mathsf{M}\big(\langle c\rangle\,;c'\big)\\
{\scriptstyle 1_{c'}\otimes 1}\big\downarrow & & \big\downarrow{\scriptstyle 1}\\
\mathsf{M}\big(c'\,;c'\big)\otimes\mathsf{M}\big(\langle c\rangle\,;c'\big) & \xrightarrow{\gamma} & \mathsf{M}\big(\langle c\rangle\,;c'\big)
\end{array}
$$

Equivariance: Suppose that in the definition of γ (6.1.2), $\operatorname{len}\langle c_j\rangle = k_j \geq 0$.

(1) For each $\sigma \in \Sigma_n$, the following *top equivariance diagram* is commutative.

(6.1.6)

$$
\begin{array}{ccc}
\mathsf{M}\big(\langle c'\rangle\,;c''\big)\otimes \bigotimes_{j=1}^{n}\mathsf{M}\big(\langle c_j\rangle\,;c_j'\big) & \xrightarrow{(\sigma,\sigma^{-1})} & \mathsf{M}\big(\langle c'\rangle\sigma\,;c''\big)\otimes \bigotimes_{j=1}^{n}\mathsf{M}\big(\langle c_{\sigma(j)}\rangle\,;c_{\sigma(j)}'\big)\\
{\scriptstyle\gamma}\big\downarrow & & \big\downarrow{\scriptstyle\gamma}\\
\mathsf{M}\big(\langle c_1\rangle,\ldots,\langle c_n\rangle\,;c''\big) & \xrightarrow{\sigma\langle k_{\sigma(1)},\ldots,k_{\sigma(n)}\rangle} & \mathsf{M}\big(\langle c_{\sigma(1)}\rangle,\ldots,\langle c_{\sigma(n)}\rangle\,;c''\big)
\end{array}
$$

Here $\sigma\langle k_{\sigma(1)},\ldots,k_{\sigma(n)}\rangle \in \Sigma_{k_1+\cdots+k_n}$ is right action of the block permutation (II.1.1.19) that permutes the n consecutive blocks of lengths $k_{\sigma(1)}$, ..., $k_{\sigma(n)}$ as σ permutes $\{1,\ldots,n\}$, leaving the relative order within each block unchanged.

(2) Given permutations $\tau_j \in \Sigma_{k_j}$ for $1 \leq j \leq n$, the following *bottom equivariance diagram* is commutative.

$$
(6.1.7)\qquad
\begin{array}{ccc}
\mathsf{M}\big(\langle c'\rangle\,;c''\big)\otimes \bigotimes_{j=1}^{n}\mathsf{M}\big(\langle c_j\rangle\,;c_j'\big) & \xrightarrow{(1,\otimes_j\tau_j)} & \mathsf{M}\big(\langle c'\rangle\,;c''\big)\otimes \bigotimes_{j=1}^{n}\mathsf{M}\big(\langle c_j\rangle\tau_j\,;c_j'\big)\\
{\scriptstyle\gamma}\big\downarrow & & \big\downarrow{\scriptstyle\gamma}\\
\mathsf{M}\big(\langle c_1\rangle,\ldots,\langle c_n\rangle\,;c''\big) & \xrightarrow{\tau_1\times\cdots\times\tau_n} & \mathsf{M}\big(\langle c_1\rangle\tau_1,\ldots,\langle c_n\rangle\tau_n\,;c''\big)
\end{array}
$$

Here the block sum $\tau_1\times\cdots\times\tau_n \in \Sigma_{k_1+\cdots+k_n}$ (II.1.1.8) is the image of $(\tau_1,\ldots,\tau_n)$ under the canonical inclusion

$$\Sigma_{k_1}\times\cdots\times\Sigma_{k_n} \longrightarrow \Sigma_{k_1+\cdots+k_n}.$$

This finishes the definition of a V-enriched multicategory. A V-enriched multicategory is *small* if its class of objects is a set. ◇

Definition 6.1.8. A V-enriched multicategory with only one object is called a V-*enriched operad.* If M is a V-enriched operad, then its object of n-ary operations is denoted by $\mathsf{M}_n \in \mathsf{V}$. ◇

Example 6.1.9. Suppose M is a V-enriched multicategory and c is an object of M. Then $\mathsf{End}(c)$ is the V-enriched operad consisting of the single object c and n-ary operation object

$$\mathsf{End}(c)_n = \mathsf{M}\big(\langle c\rangle\,;c\big) \in \mathsf{V},$$

where $\langle c \rangle$ denotes the constant n-tuple at c. The symmetric group action, unit, and composition of $\mathsf{End}(c)$ are given by those of M. ◇

Definition 6.1.10. A V-*enriched multifunctor* $F : \mathsf{M} \longrightarrow \mathsf{N}$ between V-enriched multicategories M and N consists of the following data:

- an assignment
$$F : \operatorname{Ob}\mathsf{M} \longrightarrow \operatorname{Ob}\mathsf{N},$$
where $\operatorname{Ob}\mathsf{M}$ and $\operatorname{Ob}\mathsf{N}$ are the classes of objects of M and N, respectively, and
- for each $(\langle c \rangle; c') \in \mathsf{Prof}(\mathsf{M}) \times \operatorname{Ob}\mathsf{M}$ with $\langle c \rangle = (c_1, \ldots, c_n)$, a morphism in V
$$F : \mathsf{M}(\langle c \rangle; c') \longrightarrow \mathsf{N}(F\langle c \rangle; Fc'),$$
where $F\langle c \rangle = (Fc_1, \ldots, Fc_n)$.

These data are required to preserve the symmetric group action, the colored units, and the composition in the following sense.

Symmetric Group Action: For each $(\langle c \rangle; c')$ as above and each permutation $\sigma \in \Sigma_n$, the following diagram in V is commutative.

$$\begin{array}{ccc} \mathsf{M}(\langle c \rangle; c') & \xrightarrow{F} & \mathsf{N}(F\langle c \rangle; Fc') \\ {\scriptstyle\sigma}\downarrow{\scriptstyle\cong} & & {\scriptstyle\sigma}\downarrow{\scriptstyle\cong} \\ \mathsf{M}(\langle c \rangle\sigma; c') & \xrightarrow{F} & \mathsf{N}(F\langle c \rangle\sigma; c') \end{array} \tag{6.1.11}$$

Units: For each $c \in \operatorname{Ob}\mathsf{M}$, the following diagram in V is commutative.

$$\begin{array}{ccc} & & \mathsf{M}(c; c) \\ & \nearrow^{1_c} & \downarrow{\scriptstyle F} \\ \mathbb{1} & \searrow_{1_{Fc}} & \mathsf{N}(Fc; Fc) \end{array} \tag{6.1.12}$$

Composition: For c'', $\langle c'_j \rangle$, and $\langle c \rangle = \oplus_j \langle c_j \rangle$ as in the definition of γ (5.1.3), the following diagram in V is commutative.

$$\begin{array}{ccc} \mathsf{M}(\langle c' \rangle; c'') \otimes \bigotimes_{j=1}^{n} \mathsf{M}(\langle c_j \rangle; c'_j) & \xrightarrow{(F, \otimes_j F)} & \mathsf{N}(F\langle c' \rangle; Fc'') \otimes \bigotimes_{j=1}^{n} \mathsf{N}(F\langle c_j \rangle; Fc'_j) \\ {\scriptstyle\gamma}\downarrow & & \downarrow{\scriptstyle\gamma} \\ \mathsf{M}(\langle c \rangle; c'') & \xrightarrow{\quad F \quad} & \mathsf{N}(F\langle c \rangle; Fc'') \end{array} \tag{6.1.13}$$

This finishes the definition of a V-enriched multifunctor.

Moreover:

(1) For another V-enriched multifunctor $G : \mathsf{N} \longrightarrow \mathsf{P}$ between V-enriched multicategories, where P has object class $\operatorname{Ob}\mathsf{P}$, the *composition* $GF : \mathsf{M} \longrightarrow \mathsf{P}$ is the V-enriched multifunctor defined by composing the assignments on objects
$$\operatorname{Ob}\mathsf{M} \xrightarrow{F} \operatorname{Ob}\mathsf{N} \xrightarrow{G} \operatorname{Ob}\mathsf{P}$$
and the morphisms on n-ary operations
$$\mathsf{M}(\langle c \rangle; c') \xrightarrow{F} \mathsf{N}(F\langle c \rangle; Fc') \xrightarrow{G} \mathsf{P}(GF\langle c \rangle; GFc').$$

(2) The *identity* V*-enriched multifunctor* $1_{\mathsf{M}} : \mathsf{M} \longrightarrow \mathsf{M}$ is defined by the identity assignment on objects and the identity morphism on n-ary operations.
(3) A V*-enriched operad morphism* is a V-enriched multifunctor between two V-enriched multicategories with one object.

The proof of Lemma 5.1.16 generalizes to the V-enriched case, with $\otimes$ in place of $\times$, and shows that composition of V-enriched multifunctors is well defined, associative, and unital with respect to identity multifunctors. ◇

Definition 6.1.14. Suppose P is a V-enriched operad and M is a V-enriched multicategory. A P*-algebra* in M is a pair

$$(c, \theta)$$

consisting of an object c in M and a V-enriched multifunctor

$$\theta \colon \mathsf{P} \longrightarrow \mathsf{M}$$

that sends the single object of P to c. Equivalently, θ is a V-enriched operad morphism

$$\theta \colon \mathsf{P} \longrightarrow \mathsf{End}(c).$$ ◇

Definition 6.1.15. Suppose $F, G : \mathsf{M} \longrightarrow \mathsf{N}$ are V-enriched multifunctors as in Definition 6.1.10. A V*-enriched multinatural transformation* $\alpha : F \longrightarrow G$ consists of morphisms in V

$$\alpha_c \colon \mathbb{1} \longrightarrow \mathsf{N}\big(Fc; Gc\big) \quad \text{for} \quad c \in \mathsf{Ob}\,\mathsf{M}$$

such that the following V*-naturality diagram* in V commutes for each $\big(\langle c\rangle; c'\big) \in \mathsf{Prof}(\mathsf{M}) \times \mathsf{Ob}\,\mathsf{M}$ with $\langle c\rangle = (c_1, \ldots, c_n)$.

(6.1.16)
$$\begin{array}{ccc}
\mathbb{1} \otimes \mathsf{M}\big(\langle c\rangle; c'\big) & \xrightarrow{\alpha_{c'} \otimes F} & \mathsf{N}\big(Fc'; Gc'\big) \otimes \mathsf{N}\big(F\langle c\rangle; Fc'\big) \\
{\scriptstyle \lambda^{-1}} \nearrow {\scriptstyle 1} & & \searrow {\scriptstyle \gamma} \\
\mathsf{M}\big(\langle c\rangle; c'\big) & & \mathsf{N}\big(F\langle c\rangle; Gc'\big) \\
\searrow {\scriptstyle 1} & & \nearrow {\scriptstyle \gamma} \\
\mathsf{M}\big(\langle c\rangle; c'\big) \otimes \bigotimes_{j=1}^{n} \mathbb{1} & \xrightarrow{G \otimes \bigotimes_{j=1}^{n} \alpha_{c_j}} & \mathsf{N}\big(G\langle c\rangle; Gc'\big) \otimes \bigotimes_{j=1}^{n} \mathsf{N}\big(Fc_j; Gc_j\big)
\end{array}$$

This finishes the definition of a V-enriched multinatural transformation.

- Each α_c is called a *component* of α.
- The *identity* V*-enriched multinatural transformation* $1_F : F \longrightarrow F$ has components

$$(1_F)_c = 1_{Fc} \quad \text{for} \quad c \in \mathsf{Ob}\,\mathsf{M}.$$ ◇

Definition 6.1.17. Suppose $\alpha : F \longrightarrow G$ is a V-enriched multinatural transformation between V-enriched multifunctors as in Definition 6.1.15.

(1) Suppose $\beta : G \longrightarrow H$ is a V-enriched multinatural transformation for a V-enriched multifunctor $H : \mathsf{M} \longrightarrow \mathsf{N}$. The *vertical composition*

$$\beta\alpha : F \longrightarrow H$$

is the V-enriched multinatural transformation with components at $c \in \mathsf{Ob}\,\mathsf{M}$ given by the following composites in V.

$$\begin{array}{ccc} \mathbb{1} & \xrightarrow{(\beta\alpha)_c} & \mathsf{N}(Fc;Hc) \\ {\scriptstyle \lambda^{-1}}\big\downarrow{\scriptstyle 1} & & \big\uparrow{\scriptstyle \gamma} \\ \mathbb{1}\otimes\mathbb{1} & \xrightarrow{\beta_c\otimes\alpha_c} & \mathsf{N}(Gc;Hc)\otimes\mathsf{N}(Fc;Gc) \end{array}$$

(2) Suppose $\alpha' : F' \longrightarrow G'$ is a V-enriched multinatural transformation for V-enriched multifunctors $F', G' : \mathsf{N} \longrightarrow \mathsf{P}$. The *horizontal composition*

$$\alpha' * \alpha : F'F \longrightarrow G'G$$

is the V-enriched multinatural transformation with components at $c \in \mathsf{Ob}\,\mathsf{M}$ given by the following composites in V.

$$\begin{array}{ccc} \mathbb{1} & \xrightarrow{(\alpha'*\alpha)_c} & \mathsf{P}(F'Fc;G'Gc) \\ & & \big\uparrow{\scriptstyle \gamma} \\ {\scriptstyle \lambda^{-1}}\big\downarrow{\scriptstyle 1} & & \mathsf{P}(F'Gc;G'Gc)\otimes\mathsf{P}(F'Fc;F'Gc) \\ & & \big\uparrow{\scriptstyle 1\otimes F'} \\ \mathbb{1}\otimes\mathbb{1} & \xrightarrow{\alpha'_{Gc}\otimes\alpha_c} & \mathsf{P}(F'Gc;G'Gc)\otimes\mathsf{N}(Fc;Gc) \end{array}$$

◇

Theorem 6.1.18. *There is a 2-category* $\mathsf{V}\text{-}\mathsf{Multicat}$ *consisting of the following data.*

- *Its objects are small* V*-enriched multicategories.*
- *For small* V*-enriched multicategories* M *and* N*, the hom category* $\mathsf{V}\text{-}\mathsf{Multicat}(\mathsf{M},\mathsf{N})$ *has:*
  - V*-enriched multifunctors* $\mathsf{M} \longrightarrow \mathsf{N}$ *as 1-cells;*
  - V*-enriched multinatural transformations as 2-cells;*
  - *vertical composition as composition; and*
  - *identity* V*-enriched multinatural transformations as identity 2-cells.*
- *The identity 1-cell* 1_{M} *is the identity* V*-enriched multifunctor* 1_{M}*.*
- *Horizontal composition of 1-cells is the composition of* V*-enriched multifunctors.*
- *Horizontal composition of 2-cells is that of* V*-enriched multinatural transformations.*

Explanation 6.1.19 (Enrichment over Non-Strict V). We extend the definitions in this and subsequent sections to general symmetric monoidal V by using the same data and then, in each axiom, choosing an association of iterated monoidal products and inserting associativity isomorphisms so that the new diagrams correspond to the given ones under the symmetric strictification

$$\mathsf{V} \longrightarrow \mathsf{V}_{\mathsf{st}}$$

of Theorem 1.1.42. Because strictification is an equivalence, the strict diagrams commute if and only if their preimages in V commute. ◇

6.2. Change of Enriching Categories

In this section we discuss change of enrichment along a symmetric monoidal functor

$$U: \mathsf{V} \longrightarrow \mathsf{W}.$$

We continue the assumption that V and W are permutative categories. The definitions and results of this section are extended to general V and W via strictification as in Explanation 6.1.19, replacing U with the composite

$$\mathsf{V}_{\mathsf{st}} \xrightarrow{R} \mathsf{V} \xrightarrow{U} \mathsf{W} \xrightarrow{L} \mathsf{W}_{\mathsf{st}}.$$

Definition 6.2.1. Suppose $(\mathsf{V}, \otimes, \mathbb{1})$ and $(\mathsf{W}, \otimes, \mathbb{1})$ are permutative categories and

$$U: \mathsf{V} \longrightarrow \mathsf{W}$$

is a symmetric monoidal functor. For a V-enriched multicategory M we define a W-enriched multicategory M_U with the following data.

- The objects of M_U are those of M.
- The n-ary operation objects are given by

$$\mathsf{M}_U\big(\langle c\rangle\,;c'\big) = U\mathsf{M}\big(\langle c\rangle\,;c'\big)$$

  for $c' \in \operatorname{Ob}\mathsf{M}$ and $\langle c\rangle \in \operatorname{Prof}(\mathsf{M})$.
- The symmetric group action is given by

$$U\mathsf{M}\big(\langle c\rangle\,;c'\big) \xrightarrow{U\sigma} U\mathsf{M}\big(\langle c\rangle\sigma\,;c'\big)$$

  for $\sigma \in \Sigma_n$ where n is the length of $\langle c\rangle$.
- The c-colored unit for $c \in \operatorname{Ob}\mathsf{M}$ is given by

$$\mathbb{1} \xrightarrow{U^0} U\mathbb{1} \xrightarrow{U1_c} U\mathsf{M}\big(c;c\big).$$

- The composition for M_U is given as follows for $c'' \in \operatorname{Ob}\mathsf{M}$, $\langle c'\rangle \in \operatorname{Prof}(\mathsf{M})$, and $\langle c_j\rangle \in \operatorname{Prof}(\mathsf{M})$ for each $j \in \{1,\ldots,n\}$. Let $\langle c\rangle = \oplus_j\langle c_j\rangle \in \operatorname{Prof}(\mathsf{M})$ be the concatenation of the $\langle c_j\rangle$. Then the composition of M_U is given by

(6.2.2)
$$\begin{array}{c} U\mathsf{M}\big(\langle c'\rangle\,;c''\big) \otimes \bigotimes_{j=1}^{n} U\mathsf{M}\big(\langle c_j\rangle\,;c'_j\big) \\ \big\downarrow \\ U\Big(\mathsf{M}\big(\langle c'\rangle\,;c''\big) \otimes \bigotimes_{j=1}^{n} \mathsf{M}\big(\langle c_j\rangle\,;c'_j\big)\Big) \xrightarrow{U\gamma} U\mathsf{M}\big(\langle c\rangle\,;c''\big) \end{array}$$

  where the vertical morphism is the U-coherent map given by iterates of U^2 and is unique by Epstien's Coherence Theorem 1.1.44. ◇

Proposition 6.2.3. *In the context of Definition 6.2.1, M_U is a W-enriched multicategory.*

Proof. Each of the axioms of Definition 6.1.1 for M_U follows by applying U to the corresponding axiom for M and then using naturality of U^2 and symmetric monoidal axioms of U. □

Definition 6.2.4. Suppose $(\mathsf{V},\otimes,\mathbb{1})$ and $(\mathsf{W},\otimes,\mathbb{1})$ are permutative categories and

$$U\colon \mathsf{V} \longrightarrow \mathsf{W}$$

is a symmetric monoidal functor. Suppose M and N are V-enriched multicategories and

$$F\colon \mathsf{M} \longrightarrow \mathsf{N}$$

is a V-enriched multifunctor. We define a W-enriched multifunctor

$$F_U\colon \mathsf{M}_U \longrightarrow \mathsf{N}_U$$

with the following data.

- The assignment on objects is given by that of F.
- The morphism on operation objects is given by

$$UF\colon U\mathsf{M}\big(\langle c\rangle\,;c'\big) \longrightarrow U\mathsf{N}\big(F\langle c\rangle\,;Fc'\big)$$

for $c \in \operatorname{Ob}\mathsf{M}$ and $\langle c\rangle \in \mathsf{Prof}(\mathsf{M})$. ◇

Proposition 6.2.5. *In the context of Definition 6.2.4, F_U is a W-enriched multifunctor.*

Proof. The symmetric group action diagram (6.1.11) and unit diagram (6.1.12) for F_U both commute by functoriality of U and the corresponding diagrams for F. For the composition diagram (6.1.13), first note that U^2, and hence also the U-coherent maps defining composition in M_U and N_U (6.2.2), are natural with respect to morphisms in V. Then the composition diagram for F_U commutes by applying U to the corresponding diagram for F, using functoriality of U and naturality of the U-coherent maps in (6.2.2). □

Definition 6.2.6. Suppose $(\mathsf{V},\otimes,\mathbb{1})$ and $(\mathsf{W},\otimes,\mathbb{1})$ are permutative categories and

$$U\colon \mathsf{V} \longrightarrow \mathsf{W}$$

is a symmetric monoidal functor. Suppose M and N are V-enriched multicategories and

$$\alpha\colon F \longrightarrow G\colon \mathsf{M} \longrightarrow \mathsf{N}$$

are V-enriched multifunctors and a V-enriched multinatural transformation, respectively. We define a W-enriched multinatural transformation

$$\alpha_U\colon F_U \longrightarrow G_U$$

with components

$$\mathbb{1} \xrightarrow{U^0} U\mathbb{1} \xrightarrow{U\alpha} U\mathsf{N}\big(Fc;Gc\big).$$ ◇

Proposition 6.2.7. *In the context of Definition 6.2.6, α_U is a W-enriched multinatural transformation.*

Proof. Verification that the W-naturality diagram (6.1.16) holds for α_U is similar to the proof of Proposition 6.2.5 for F_U. We apply U to the V-naturality diagram for α and then use functoriality of U and naturality of U^0 and of the U-coherent maps appearing in (6.2.2) for M_U and N_U. □

Definition 6.2.8. Suppose V and W are permutative categories and

$$U : \mathsf{V} \longrightarrow \mathsf{W}$$

is a symmetric monoidal functor. Then there is a *change of enrichment* 2-functor

$$(-)_U : \mathsf{V}\text{-}\mathsf{Multicat} \longrightarrow \mathsf{W}\text{-}\mathsf{Multicat}$$

given by Definitions 6.2.1, 6.2.4, and 6.2.6. ◇

Proposition 6.2.9. *In the context of Definition 6.2.8,*

$$(-)_U : \mathsf{V}\text{-}\mathsf{Multicat} \longrightarrow \mathsf{W}\text{-}\mathsf{Multicat}$$

is a 2-functor.

Proof. Change of enrichment preserves identities and composition of multifunctors by functoriality of U. Preservation of identity multinatural transformations holds by the definition of identity operations via composition with U^0. Horizontal and vertical composition of multinatural transformations are preserved by functoriality of U and naturality of U^0 and U^2. □

6.3. Enriched Endomorphism Multicategories

The two main types of enriched multicategories we will consider are enriched operads and those arising from enriched symmetric monoidal categories. We describe the latter in this section.

Again throughout this section we suppose that $\mathsf{V} = (\mathsf{V}, \otimes, \xi)$ is a permutative category. However, as outlined in Explanation 6.1.19, we will implicitly extend the definitions here to general symmetric monoidal V via strictification.

Convention 6.3.1 (Left Normalized Iterated Products)**.** Suppose K is a monoidal V-category and $\langle X \rangle = (X_1, \ldots, X_n)$ is a tuple of objects of K. Recall from Definition 2.5.3 the *left normalized product* is defined to be

$$\boxtimes \langle X \rangle = (\cdots((X_1 \boxtimes X_2) \boxtimes X_3)\cdots) \boxtimes X_n.$$

For tuples of objects $\langle X_j \rangle = (X_{j,1}, \ldots, X_{j,k_j})$ with $j \in \{1, \ldots, n\}$, let $\langle X \rangle$ denote their concatenation. By the coherence for monoidal V-categories, Theorem 2.5.6, there is a unique canonical V-map giving an isomorphism

$$\boxtimes \langle X \rangle \xrightarrow{\cong} \mathop{\boxtimes}_{j=1}^{n} \Big(\mathop{\boxtimes}_{i=1}^{k_j} X_{j,i} \Big).$$

We call this the *normalization map* and write

$$\mathsf{K}\Big(\mathop{\boxtimes}_{j=1}^{n} \Big(\mathop{\boxtimes}_{i=1}^{k_j} X_{j,i} \Big), Y \Big) \xrightarrow[\mathsf{norm}]{\cong} \mathsf{K}(\boxtimes \langle X \rangle, Y)$$

for the induced morphism on hom objects. ◇

Example 6.3.2. In the context of Convention 6.3.1, if $n = 3$ and $(k_1, k_2, k_3) = (2, 2, 1)$, then the normalization map at left below is given by an (inverse) associator. If $n = 3$ and $(k_1, k_2, k_3) = (2, 0, 1)$, then the normalization map at right below is given by a right unitor.

$$\begin{array}{ccc} ((X_{1,1} \boxtimes X_{1,2}) \boxtimes (X_{2,1} \boxtimes X_{2,2})) \boxtimes X_{3,1} & & ((X_{1,1} \boxtimes X_{1,2}) \boxtimes \mathrm{I}) \boxtimes X_{3,1} \\ \mathsf{norm} \downarrow \cong & & \mathsf{norm} \downarrow \cong \\ (((X_{1,1} \boxtimes X_{1,2}) \boxtimes X_{2,1}) \boxtimes X_{2,2}) \boxtimes X_{3,1} & & (X_{1,1} \boxtimes X_{1,2}) \boxtimes X_{3,1} \end{array}$$

◇

Definition 6.3.3 (Enriched Endomorphism Multicategory)**.** Suppose that V is a permutative category and K is a symmetric monoidal V-category. The V-*enriched endomorphism multicategory* of K, denoted $\mathsf{End}(\mathsf{K})$, is a V-enriched multicategory defined as follows.

- The objects are those of K.
- For an object $X' \in \mathsf{K}$ and a tuple $\langle X \rangle \in \mathsf{Prof}(\mathsf{K})$, we define the V-object of operations

$$\mathsf{End}(\mathsf{K})\big(\langle X \rangle\,; X'\big) = \mathsf{K}(\boxtimes\langle X \rangle, X'),$$

  where $\boxtimes\langle X \rangle$ denotes the left normalized product.
- For $\langle X \rangle$ and X' as above and a permutation $\sigma \in \Sigma_n$, the right action of σ is defined as the following composite.

$$(6.3.4)\qquad \begin{array}{ccc} \mathsf{K}(\boxtimes\langle X \rangle, X') & \xrightarrow{\quad\sigma\quad} & \mathsf{K}(\boxtimes\langle X \rangle\sigma, X') \\ \rho^{-1} \downarrow & & \uparrow m \\ \mathsf{K}(\boxtimes\langle X \rangle, X') \otimes \mathbb{1} & \xrightarrow{1 \otimes \beta_\sigma} & \mathsf{K}(\boxtimes\langle X \rangle, X') \otimes \mathsf{K}(\boxtimes\langle X \rangle\sigma, \boxtimes\langle X \rangle) \end{array}$$

  In the above diagram, β_σ denotes the V-natural isomorphism that permutes coordinates according to the permutation σ. The existence and uniqueness of such a V-natural isomorphism follows from the symmetric form of coherence, Theorem 2.5.6.
- For each $X \in \mathsf{K}$ the unit of $\mathsf{End}(\mathsf{K})$ is given by that of K,

$$1_X \colon \mathbb{1} \longrightarrow \mathsf{K}(X, X) = \mathsf{End}(\mathsf{K})\big(X\,; X\big).$$

- The composition γ in $\mathsf{End}(\mathsf{K})$ is defined as the following composite for tuples of objects X'', $\langle X' \rangle = (X'_1, \ldots, X'_n)$, and $\langle X_j \rangle = (X_{j,1}, \ldots, X_{j,k_j})$ with $j \in \{1, \ldots, n\}$ and with $\langle X \rangle$ being the concatenation of the $\langle X_j \rangle$.

$$
\begin{array}{ccc}
\mathsf{K}(\boxtimes\langle X'\rangle, X'') \otimes \bigotimes_{j=1}^{n} \mathsf{K}(\boxtimes_{i=1}^{k_j} X_{j,i}, X'_j) & \xrightarrow{\ \gamma\ } & \mathsf{K}(\boxtimes\langle X\rangle, X'') \\
{\scriptstyle 1 \otimes \boxtimes_{j=1}^{n-1}} \Big\downarrow & & \Big\uparrow {\scriptstyle \mathsf{norm}\ \cong} \\
\mathsf{K}(\boxtimes\langle X'\rangle, X'') \otimes \mathsf{K}(\boxtimes_{j=1}^{n} \boxtimes_{i=1}^{k_j} X_{j,i}, \boxtimes\langle X'\rangle) & \xrightarrow{\ m\ } & \mathsf{K}(\boxtimes_{j=1}^{n} \boxtimes_{i=1}^{k_j} X_{j,i}, X'')
\end{array}
\tag{6.3.5}
$$

◇

Proposition 6.3.6. *Suppose that* V *is a permutative category and* K *is a symmetric monoidal* V*-category. Then* $\mathsf{End}(\mathsf{K})$ *is a* V*-enriched multicategory.*

Proof. We will verify commutativity of the associativity diagram (6.1.3) for $\mathsf{End}(\mathsf{K})$. The other axioms are similar. We use the following notation for W, Z_j, Y_{ji}, and X_{jik} objects of K with $j \in \{1, \ldots, n\}$, $i \in 1, \ldots, m_j$, and $k \in \{1, \ldots, \ell_{ji}\}$.

$$
\begin{aligned}
\mathsf{E}^{ZW} &= \mathsf{K}(\boxtimes_{j=1}^{n} Z_j, W) & \mathsf{E}^{YZ} &= \mathsf{K}(\boxtimes_{j=1}^{n} \boxtimes_{i=1}^{m_j} Y_{ji}, \boxtimes_{j=1}^{n} Z_j) \\
\mathsf{E}_j^{YZ} &= \mathsf{K}(\boxtimes_{i=1}^{m_j} Y_{ji}, Z_j) & \mathsf{E}^{YW} &= \mathsf{K}(\boxtimes_{j=1}^{n} \boxtimes_{i=1}^{m_j} Y_{ji}, W) \\
\mathsf{E}_{ji}^{XY} &= \mathsf{K}(\boxtimes_{k=1}^{\ell_{ji}} X_{jik}, Y_{ji}) & \mathsf{E}^{XY} &= \mathsf{K}(\boxtimes_{j=1}^{n} \boxtimes_{i=1}^{m_j} \boxtimes_{k=1}^{\ell_{ji}} X_{jik}, \boxtimes_{j=1}^{n} \boxtimes_{i=1}^{m_j} Y_{ji}) \\
\mathsf{E}_j^{XY} &= \mathsf{K}(\boxtimes_{i=1}^{m_j} \boxtimes_{k=1}^{\ell_{ji}} X_{jik}, \boxtimes_{i=1}^{m_j} Y_{ji}) & \mathsf{E}_j^{XZ} &= \mathsf{K}(\boxtimes_{i=1}^{m_j} \boxtimes_{k=1}^{\ell_{ji}} X_{jik}, Z_j) \\
\mathsf{E}^{XZ} &= \mathsf{K}(\boxtimes_{j=1}^{n} \boxtimes_{i=1}^{m_j} \boxtimes_{k=1}^{\ell_{ji}} X_{jik}, \boxtimes_{j=1}^{n} Z_j) & \mathsf{E}^{XW} &= \mathsf{K}(\boxtimes_{j=1}^{n} \boxtimes_{i=1}^{m_j} \boxtimes_{k=1}^{\ell_{ji}} X_{jik}, W)
\end{aligned}
$$

Thus, for example, we have

$$\mathsf{End}(\mathsf{K})\big(\langle Z\rangle ; W\big) = \mathsf{K}(\boxtimes\langle Z\rangle, W) = \mathsf{E}^{ZW}.$$

Moreover, we use the following notation.

- The concatenation of the $\langle Y_j\rangle$ is denoted $\langle Y\rangle$ and

$$\mathsf{E}^{YW}_{\mathsf{norm}} = \mathsf{End}(\mathsf{K})(\boxtimes\langle Y\rangle, W).$$

- For each j, the concatenation of the $\langle X_{ji}\rangle$ is denoted $\langle X_j\rangle$ and

$$\mathsf{E}^{XZ}_{j\,\mathsf{norm}} = \mathsf{End}(\mathsf{K})(\boxtimes\langle X_j\rangle, Z_j).$$

- The concatenation of the $\langle X_j\rangle$ is denoted $\langle X\rangle$ and

$$\mathsf{E}^{XW}_{\mathsf{norm}} = \mathsf{End}(\mathsf{K})(\boxtimes\langle X\rangle, W).$$

The associativity diagram for $\mathsf{End}(\mathsf{K})$ is the outer diagram below, where the left and right vertical composites are, by definition, $\gamma \otimes 1$ and $1 \otimes \bigotimes_j \gamma$, respectively.

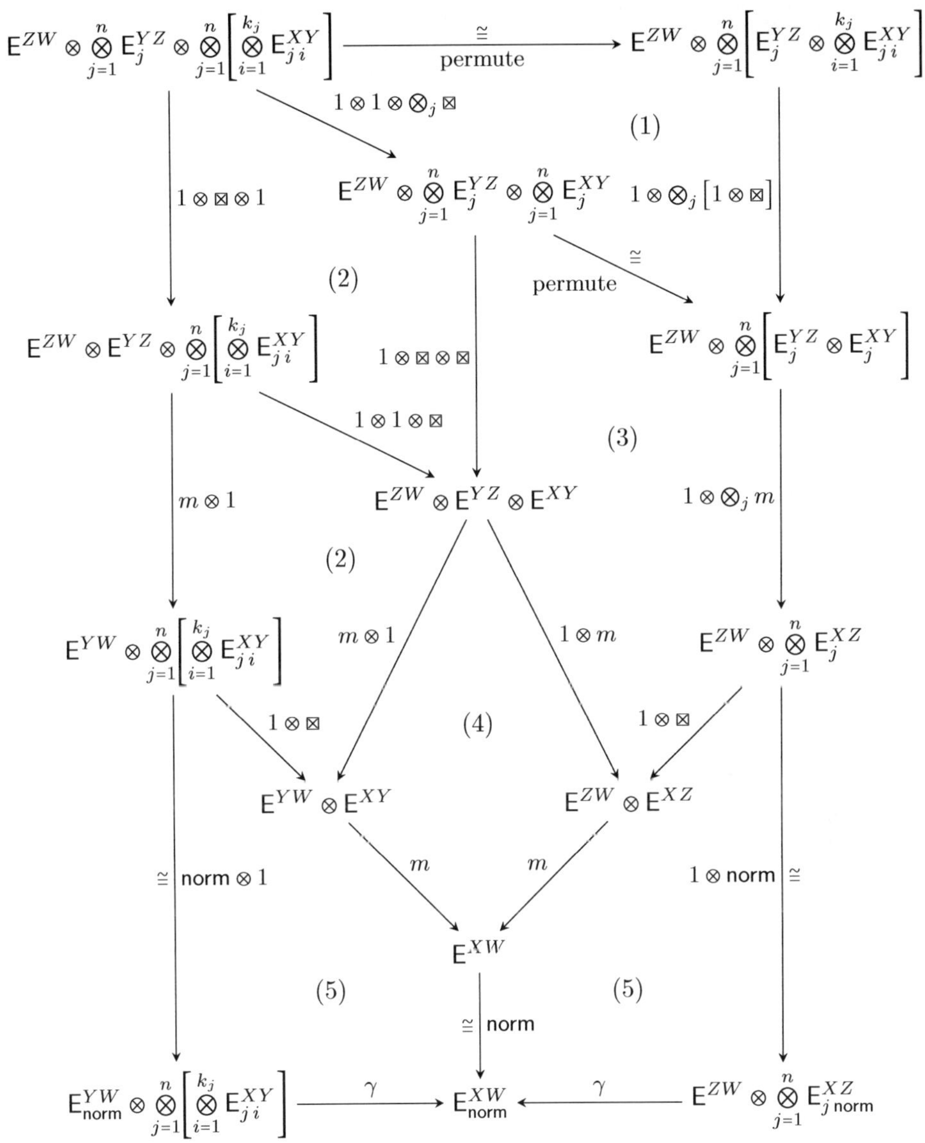

Commutativity of each of the inner regions is verified as follows.

(1) Use naturality of the symmetry, ξ, in V.
(2) Use functoriality of the monoidal product, $\otimes$.
(3) Use enriched functoriality of $\boxtimes$ (interchange; see Explanation 1.4.7).
(4) Use associativity of the composition m in K (1.2.2).
(5) Use functoriality of $\otimes$ and naturality of m with respect to morphisms in V. □

The following is a special case of Example 6.1.9.

Definition 6.3.7 (Enriched Endomorphism Operad)**.** Suppose V is a permutative category and K is a symmetric monoidal V-category. For each object $X \in \mathsf{K}$, the *endomorphism operad* of X is the V-enriched multicategory with a single object X and with n-ary operations

$$\mathsf{End}(X)_n = \mathsf{End}(\mathsf{K})\big(\langle X\rangle\,;\,X\big)$$

where $\langle X\rangle$ is the length-n tuple whose entries are all the single object X. Symmetric group actions, unit, and compositions are defined as in $\mathsf{End}(\mathsf{K})$. ◇

Definition 6.3.8. Suppose that V is a permutative category, K and L are symmetric monoidal V-categories, and

$$F\colon \mathsf{K} \longrightarrow \mathsf{L}$$

is a symmetric monoidal V-functor. We define a V-enriched multifunctor

$$\mathsf{End}(F)\colon \mathsf{End}(\mathsf{K}) \longrightarrow \mathsf{End}(\mathsf{L})$$

with assignment on objects given by F and morphisms on operation objects

$$\mathsf{End}(F)\colon \mathsf{End}(\mathsf{K})\big(\langle X\rangle\,;\,X'\big) \longrightarrow \mathsf{End}(\mathsf{L})\big(F\langle X\rangle\,;\,X'\big) \tag{6.3.9}$$

defined as follows. For each $n \geq 1$, there is an F-coherent map with components

$$F^n\colon \mathbb{1} \longrightarrow \mathsf{L}\big(\,\boxtimes\,(F\langle X\rangle)\,,\,F(\boxtimes\langle X\rangle)\,\big)$$

provided by iterated application of F^2, where $\langle X\rangle$ has length n. This is unique by Enriched Epstein's Coherence Theorem 2.5.8. Then (6.3.9) is given by applying F to mapping objects

$$F\colon \mathsf{K}(\boxtimes\langle X\rangle, X') \longrightarrow \mathsf{L}(F(\boxtimes\langle X\rangle), FX')$$

and then precomposing with the components of F^n. ◇

Proposition 6.3.10. *In the context of Definition 6.3.8,*

$$\mathsf{End}(F)\colon \mathsf{End}(\mathsf{K}) \longrightarrow \mathsf{End}(\mathsf{L})$$

is a V*-enriched multifunctor.*

Proof. The unit condition (6.1.12) is given by the identity axiom (1.2.6) for F. We apply the coherence Theorem 2.5.8 to show that the diagrams for preservation of the symmetric group action (6.1.11) and composition (6.1.13) commute.

By definition, the morphisms given by $\mathsf{End}(F)$ on operation objects are F-coherent maps. The permutation action (6.3.4) and composition (6.3.5) of $\mathsf{End}(\mathsf{K})$ and $\mathsf{End}(\mathsf{L})$ are given by composites of the monoidal product $\boxtimes$, symmetry β, and the normalization maps of Convention 6.3.1. All of these are (permuted) canonical V-maps and hence also F-coherent maps. This shows that (6.1.11) and (6.1.13) for $\mathsf{End}(F)$ consist entirely of F-coherent maps. Therefore their respective composites are equal by Theorem 2.5.8. □

6.4. The Multicategory of Small Multicategories

Now we use the theory above to show that the 2-categories $\mathsf{Multicat}$ and $\mathsf{Multicat}_*$ have the structure of Cat-enriched multicategories. Although Cat is not a permutative category, we understand the definition of Cat-enriched multicategory via strictification, as outlined in Explanation 6.1.19.

We first show that each of $(\mathsf{Multicat},\otimes)$ and $(\mathsf{Multicat}_*,\wedge)$ is symmetric Cat-monoidal in the sense of Definition 1.5.1. Then we apply Definition 6.3.7 and Proposition 6.3.6 to obtain Cat-enriched multicategory structures.

To show that $(\mathsf{Multicat}, \otimes)$ is symmetric Cat-monoidal, we must first extend the Boardman-Vogt tensor product of Definition 5.6.11 to multinatural transformations.

Definition 6.4.1. Suppose that

$$F, F' \colon \mathsf{M} \longrightarrow \mathsf{M}' \quad \text{and} \quad G, G' \colon \mathsf{N} \longrightarrow \mathsf{N}'$$

are multifunctors and suppose moreover that

$$\theta \colon F \longrightarrow F' \quad \text{and} \quad \omega \colon G \longrightarrow G'$$

are multinatural transformations. Define a multinatural transformation

$$\theta \otimes \omega \colon F \otimes G \longrightarrow F' \otimes G'$$

via components

$$(\theta \otimes \omega)_{(c,d)} = \theta_c \times \omega_d$$

for $c \in \mathsf{M}$ and $d \in \mathsf{N}$. ◇

Proposition 6.4.2. *In the context of Definition 6.4.1, $\theta \otimes \omega$ is a multinatural transformation $F \otimes G \longrightarrow F' \otimes G'$.*

Proof. We verify the naturality condition of Definition 5.1.17 for generating operations $\phi \otimes d$ and $c \otimes \psi$ with

$$\phi \in \mathsf{M}(\langle c \rangle ; c') \quad \text{and} \quad \psi \in \mathsf{N}(\langle d \rangle ; d').$$

Let $(\theta \otimes \omega)_{\langle c \rangle \otimes d}$ denote the tuple of unary operations $\theta_{c_i} \times \omega_d$ indexed by the objects of $\langle c \rangle \otimes d$. Similarly let $(\theta \otimes \omega)_{c \otimes \langle d \rangle}$ denote the tuple indexed by $c \otimes \langle d \rangle$. The equalities

$$\begin{aligned} \big(F'\phi \otimes G'd\big) \circ \big((\theta \otimes \omega)_{\langle c \rangle \otimes d}\big) &= \big(\theta_{c'} \times \omega_d\big) \circ \big(F\phi \otimes Gd\big) \quad \text{and} \\ \big(F'c \otimes G'\psi\big) \circ \big((\theta \otimes \omega)_{c \otimes \langle d \rangle}\big) &= \big(\theta_c \times \omega_{d'}\big) \circ \big(Fc \otimes G\psi\big) \end{aligned}$$

follow from the naturality conditions for θ and ω, respectively, together with the unit conditions

$$1_{G'd}\omega_d = \omega_d 1_{Gd} \quad \text{and} \quad 1_{F'c}\theta_c = \theta_c 1_{Fc}.$$ □

Theorem 6.4.3. *The tensor product of small multicategories is a 2-functor*

$$\mathsf{Multicat} \times \mathsf{Multicat} \xrightarrow{\otimes} \mathsf{Multicat}$$

and $(\mathsf{Multicat}, \otimes)$ *is a symmetric* Cat*-monoidal 2-category.*

Proof. The 2-functoriality of $\otimes$ follows because the units and composition of multinatural transformations are given componentwise Recalling Definition 6.4.1, the components of $\theta \otimes \omega$ are given pairwise by those of θ and ω. Therefore, $\otimes$ is 2-functorial.

To show that $\mathsf{Multicat}$ is symmetric Cat-monoidal, we will apply Theorem 2.5.1. So we need to continue showing that the data of the symmetric monoidal structure on $\mathsf{Multicat}$ extend to 2-functors and 2-natural transformations.

To that end, we first note that the unit

$$* \xrightarrow{\mathsf{I}} \mathsf{Multicat}$$

extends to a 2-functor by sending the unique 2-cell of the terminal 2-category to the identity multinatural transformation on the identity multifunctor 1_{I}. Next we show that the symmetry β is 2-natural. Suppose given multifunctors

$$F, F' \colon \mathsf{M} \longrightarrow \mathsf{N} \quad \text{and} \quad G, G' \colon \mathsf{M}' \longrightarrow \mathsf{N}'$$

together with multinatural transformations

$$\theta \colon F \longrightarrow F' \quad \text{and} \quad \omega \colon G \longrightarrow G'.$$

The whiskering of $\omega \otimes \theta$ with $\beta_{\mathsf{M},\mathsf{M}'}$,

$$\mathsf{M} \otimes \mathsf{M}' \xrightarrow{\beta} \mathsf{M}' \otimes \mathsf{M} \underset{G' \otimes F'}{\overset{G \otimes F}{\rightrightarrows}} \mathsf{N}' \otimes \mathsf{N}, \qquad \Downarrow \omega \otimes \theta$$

has component at an object $c \otimes c' \in \mathsf{M} \otimes \mathsf{M}'$ given by

$$(\omega \otimes \theta)_{\beta(c \otimes c')} = \omega_{c'} \otimes \theta_c.$$

On the other hand, the whiskering of $\beta_{\mathsf{N},\mathsf{N}'}$ with $\theta \otimes \omega$,

$$\mathsf{M} \otimes \mathsf{M}' \underset{F' \otimes G'}{\overset{F \otimes G}{\rightrightarrows}} \mathsf{N} \otimes \mathsf{N}' \xrightarrow{\beta} \mathsf{N}' \otimes \mathsf{N}, \qquad \Downarrow \theta \otimes \omega$$

has component at an object $c \otimes c' \in \mathsf{M} \otimes \mathsf{M}'$ given by

$$\beta(\theta_c \otimes \omega_{c'}) = \omega_{c'} \otimes \theta_c.$$

Therefore, β is 2-natural.

Similar arguments for the unit and associativity isomorphisms show that these are 2-natural as well. Ultimately their 2-naturality follows from the 2-naturality of the corresponding data for the Cartesian product, pushouts, and coequalizers. Since the data of the symmetric monoidal structure on $(\mathsf{Multicat}, \otimes)$ are the underlying functors and natural transformations of Cat-enriched data, Theorem 2.5.1 shows that $(\mathsf{Multicat}, \otimes)$ is symmetric Cat-monoidal. □

The proof of Theorem 6.4.3 descends to the smash product of pointed multicategories and gives the following.

Theorem 6.4.4. *The smash of small multicategories is a 2-functor*

$$\mathsf{Multicat}_* \times \mathsf{Multicat}_* \xrightarrow{\wedge} \mathsf{Multicat}_*$$

and $(\mathsf{Multicat}_*, \wedge)$ *is a symmetric* Cat*-monoidal 2-category.*

By applying Proposition 6.3.6 to $(\mathsf{Multicat}, \otimes)$ and $(\mathsf{Multicat}_*, \wedge)$, we have the following.

Corollary 6.4.5. *Each of* $\mathsf{Multicat}$ *and* $\mathsf{Multicat}_*$ *has the structure of* Cat*-enriched multicategory induced by the products* $\otimes$ *and* $\wedge$*, respectively.*

Convention 6.4.6 (Multicategories $\mathsf{Multicat}$ and $\mathsf{Multicat}_*$). Instead of the overly-cumbersome $\mathsf{End}(\mathsf{Multicat})$ and $\mathsf{End}(\mathsf{Multicat}_*)$, we will continue to use the notation $\mathsf{Multicat}$ and $\mathsf{Multicat}_*$ even when regarding these as Cat-enriched multicategories. ◇

6.5. Permutative Categories and Multilinearity

Recall from Definition 1.1.27 that $\mathsf{PermCat}^{\mathsf{su}}$ denotes the 2-category of small permutative categories, strictly unital symmetric monoidal functors, and monoidal natural transformations. Moreover, recall from Corollary 5.3.9 that taking endomorphism multicategories provides a 2-functor

$$\mathsf{End}\colon \mathsf{PermCat}^{\mathsf{su}} \longrightarrow \mathsf{Multicat}_*$$

that is bijective on 1- and 2-cells.

Therefore, we have a multicategory structure on $\mathsf{PermCat}^{\mathsf{su}}$ given by restricting that of $\mathsf{Multicat}_*$. Likewise, the Cat-enriched multicategory structure on $\mathsf{Multicat}_*$ restricts to such a structure for $\mathsf{PermCat}^{\mathsf{su}}$. This is in contrast to Proposition 5.7.23, which shows that the symmetric monoidal structure of $\mathsf{Multicat}_*$ does not restrict along End.

Definition 6.5.1. We define a Cat-enriched multicategory structure on $\mathsf{PermCat}^{\mathsf{su}}$ as the sub-multicategory of $\mathsf{Multicat}_*$ given by the image of End. Recalling Corollary 5.3.9, this is the full sub-multicategory whose objects are given by $\mathsf{End}(\mathsf{C})$ for permutative categories C. As an extension of Convention 6.4.6, we continue to use the notation $\mathsf{PermCat}^{\mathsf{su}}$ for the Cat-enriched multicategory. ◇

The Cat-enriched multicategory structure of $\mathsf{PermCat}^{\mathsf{su}}$ is fundamental to our K-theory applications in Part 2 because it is the domain of the Elmendorf-Mandell K-theory multifunctor defined in Chapter 10. Therefore, in Section 6.6 we give a second, more direct proof that $\mathsf{PermCat}^{\mathsf{su}}$ is a Cat-enriched multicategory.

In the remainder of this section we give more detailed descriptions of the categories of operations

$$\mathsf{PermCat}^{\mathsf{su}}\big(\langle \mathsf{C}\rangle\,;\,\mathsf{D}\big)$$

for permutative categories $\mathsf{C}_1, \ldots, \mathsf{C}_n$, and D. Such a category is, by definition, the category of multifunctors

$$F\colon \wedge_{i=1}^{n} \mathsf{End}(\mathsf{C}_i) \longrightarrow \mathsf{End}(\mathsf{D}) \tag{6.5.2}$$

and multinatural transformations between them.

Definition 6.5.3. Suppose $\langle X\rangle \in \prod_j \mathsf{C}_j$ is a tuple of objects.

- For an object $X_i' \in \mathsf{C}_i$ with $i \in \{1, \ldots, n\}$, we let $\langle X \circ_i X_i'\rangle$ denote the tuple whose jth entry is X_j for $j \neq i$, and whose ith entry is X_i'.
- Similarly, for objects $X_i' \in \mathsf{C}_i$ and $X_k' \in \mathsf{C}_k$ with $i, k \in \{1, \ldots, n\}$ and $i \neq k$, we let $\langle X \circ_i X_i' \circ_k X_k'\rangle$ denote the tuple whose jth entry is that of $\langle X \circ_i X_i'\rangle$ for $j \neq k$ and whose kth entry is X_k'. ◇

Definition 6.5.4 (Multilinear Functors)**.** Suppose $\mathsf{C}_1, \ldots, \mathsf{C}_n$, and D are permutative categories. An *n-linear functor* from $\langle \mathsf{C}\rangle$ to D is a functor

$$F\colon \mathsf{C}_1 \times \cdots \times \mathsf{C}_n \longrightarrow \mathsf{D}$$

together with, for each $i \in \{1, \ldots, n\}$, a natural transformation F_i^2 called the *ith linearity constraint* with components

$$F_i^2\colon F\langle X \circ_i X_i\rangle \oplus F\langle X \circ_i X_i'\rangle \longrightarrow F\langle X \circ_i (X_i \oplus X_i')\rangle$$

for $\langle X\rangle \in \prod_j \mathsf{C}_j$ and $X_i' \in \mathsf{C}_i$. These data satisfy the following axioms.

Unity: If any $X_j = e$, the unit of C_j, then $F\langle X\rangle = e$, the unit in D. Moreover, $F\langle f \circ_j 1_e\rangle = 1_e$ for any morphisms $f_i \in \mathsf{C}_i$ for $i \neq j$.

Constraint Unity: If any $X_j = e$ or if $X_i' = e$, then F_i^2 is an identity morphism.

Constraint Associativity: The following diagram commutes for each $i \in \{1,\ldots,n\}$ and $\langle X\rangle \in \prod_j \mathsf{C}_j$, with $X_i', X_i'' \in \mathsf{C}_i$.

(6.5.5)

$$\begin{array}{ccc}
F\langle X \circ_i X_i\rangle \oplus F\langle X \circ_i X_i'\rangle \oplus F\langle X \circ_i X_i''\rangle & \xrightarrow{1 \oplus F_i^2} & F\langle X \circ_i X_i\rangle \oplus F\langle X \circ_i (X_i' \oplus X_i'')\rangle \\
{\scriptstyle F_i^2 \oplus 1}\downarrow & & \downarrow{\scriptstyle F_i^2} \\
F\langle X \circ_i (X_i \oplus X_i')\rangle \oplus F\langle X \circ_i X_i''\rangle & \xrightarrow{F_i^2} & F\langle X \circ_i (X_i \oplus X_i' \oplus X_i'')\rangle
\end{array}$$

Constraint Symmetry: The following diagram commutes for each $i \in \{1,\ldots,n\}$ and $\langle X\rangle \in \prod_j \mathsf{C}_j$, with $X_i' \in \mathsf{C}_i$.

(6.5.6)

$$\begin{array}{ccc}
F\langle X \circ_i X_i\rangle \oplus F\langle X \circ_i X_i'\rangle & \xrightarrow{F_i^2} & F\langle X \circ_i (X_i \oplus X_i')\rangle \\
{\scriptstyle \xi}\downarrow & & \downarrow{\scriptstyle F\langle 1 \circ_i \xi\rangle} \\
F\langle X \circ_i X_i'\rangle \oplus F\langle X \circ_i X_i\rangle & \xrightarrow{F_i^2} & F\langle X \circ_i (X_i' \oplus X_i)\rangle
\end{array}$$

Constraint 2-By-2: The following diagram commutes for each

$$i \neq k \in \{1,\ldots,n\}, \quad \langle X\rangle \in \textstyle\prod_j \mathsf{C}_j, \quad X_i' \in \mathsf{C}_i, \quad \text{and} \quad X_k' \in \mathsf{C}_k.$$

(6.5.7)

$$\begin{array}{ccc}
 & F\langle X \circ_i (X_i \oplus X_i') \circ_k X_k\rangle \oplus F\langle X \circ_i (X_i \oplus X_i') \circ_k X_k'\rangle & \\
{\scriptstyle F_i^2 \oplus F_i^2}\nearrow & & \searrow{\scriptstyle F_k^2} \\
\begin{array}{c} F\langle X \circ_i X_i \circ_k X_k\rangle \oplus F\langle X \circ_i X_i' \circ_k X_k\rangle \\ \oplus F\langle X \circ_i X_i \circ_k X_k'\rangle \oplus F\langle X \circ_i X_i' \circ_k X_k'\rangle \end{array} & & \\
{\scriptstyle 1 \oplus \xi \oplus 1}\downarrow & & F\langle X \circ_i (X_i \oplus X_i') \circ_k (X_k \oplus X_k')\rangle \\
\begin{array}{c} F\langle X \circ_i X_i \circ_k X_k\rangle \oplus F\langle X \circ_i X_i \circ_k X_k'\rangle \\ \oplus F\langle X \circ_i X_i' \circ_k X_k\rangle \oplus F\langle X \circ_i X_i' \circ_k X_k'\rangle \end{array} & & \\
{\scriptstyle F_k^2 \oplus F_k^2}\searrow & & \nearrow{\scriptstyle F_i^2} \\
 & F\langle X \circ_i X_i \circ_k (X_k \oplus X_k')\rangle \oplus F\langle X \circ_i X_i' \circ_k (X_k \oplus X_k')\rangle &
\end{array}$$

A 0-*linear functor* is a choice of object in D, regarded as a functor

$$F : \mathbb{1} \longrightarrow \mathsf{D}$$

from the empty product. We say that F is a *multilinear functor* if it is n-linear for some $n \geq 0$. In Proposition 6.5.10 below we will show that the operations of $\mathsf{PermCat^{su}}$ are precisely multilinear functors. ◇

Example 6.5.8. In the context of Definition 6.5.4 with $n = 1$, comparing the axioms above with those of Definition 1.1.6 shows that a 1-linear functor is precisely a strictly unital symmetric monoidal functor. ◇

Example 6.5.9 (Ring Category Product)**.** Recall from Definition II.9.1.2 the definition of ring category. If

$$\mathsf{C} = \big(\mathsf{C}, (\oplus, \mathbb{0}, \xi^{\oplus}), (\otimes, \mathbb{1}), (\partial^l, \partial^r)\big)$$

is a ring category, then

$$F = \otimes \colon \mathsf{C} \times \mathsf{C} \longrightarrow \mathsf{C}$$

is a 2-linear functor with $F^2_1 = \partial^l$ and $F^2_2 = \partial^r$. The following table lists the correspondence between axioms of Definition 6.5.4 for $n = 2$ and those of Definition II.9.1.2.

2-Linear Functor	**Ring Category Product**
Unity	Multiplicative Zero
Constraint Unity	Zero Factorization
Constraint Associativity	Internal Factorization
Constraint Symmetry	Symmetry Factorization
Constraint 2-By-2	2-By-2 Factorization

We explain further details of this correspondence when, in the proof of Theorem 11.2.16, we use it along with 1- and 3-linearity to show that ring category structure on C is encoded by action of the Cat-enriched associative operad.

Theorem 11.2.16 is, in turn, used in Theorems 11.5.5, 12.4.5, and 13.4.12 to establish similar correspondences between certain operad actions and, respectively, bipermutative, braided ring, and E_n-monoidal structures on C. In the proof of Theorem 13.4.12, 4-linearity is used to describe the exchanges $\eta^{k,l}$ (Definition 13.4.1). ◇

Proposition 6.5.10. *Suppose $\mathsf{C}_1, \ldots, \mathsf{C}_n$ and D are small permutative categories. The objects of*

$$\mathsf{PermCat^{su}}\big(\langle \mathsf{C}\rangle\,;\,\mathsf{D}\big) = \mathsf{Multicat}_*\big(\textstyle\bigwedge_j \mathsf{End}(\mathsf{C}_j), \mathsf{End}(\mathsf{D})\big)$$

are precisely the multilinear functors $\langle \mathsf{C}\rangle \longrightarrow \mathsf{D}$.

Proof. Recall from the proofs of Proposition 5.3.7 and Lemma 5.3.8 that, in the correspondence between strictly unital symmetric monoidal functors and pointed multifunctors

$$F\colon \mathsf{C} \longrightarrow \mathsf{D} \quad \text{and} \quad G\colon \mathsf{End}(\mathsf{C}) \longrightarrow \mathsf{End}(\mathsf{D}),$$

respectively, the strictly unital aspect of F corresponds to preservation of basepoints by G, the monoidal constraints $F^2_{X,Y}$ correspond to the values of G on $\iota_{(X,Y)}$, and the symmetric monoidal axioms for F correspond to the multifunctoriality axioms for G. Applying this to Definition 6.5.4, we can describe a multilinear functor

$$F\colon \langle \mathsf{C}\rangle \longrightarrow \mathsf{D}$$

as a functor on underlying objects and unary operations of $\wedge_j \mathsf{End}(\mathsf{C}_j)$ such that

- F is strictly unital in each variable;
- F is a symmetric monoidal functor when restricted to each C_i; and
- F satisfies a 2-by-2 interchange condition for the monoidal structures in separate variables.

Recalling our discussion of multifunctors out of a tensor product in Explanation 5.6.14, F corresponds to an assignment on objects

$$\widetilde{F}\colon \prod_i \mathsf{Ob}\,\mathsf{End}(\mathsf{C}_i) \longrightarrow \mathsf{Ob}\,\mathsf{End}(\mathsf{D})$$

such that

- $\widetilde{F}$ is a pointed multifunctor in each variable separately, because in each variable F is strictly unital symmetric monoidal;
- $\widetilde{F}$ satisfies the interchange relation (5.6.15), by the constraint 2-by-2 axiom for F; and
- $\widetilde{F}$ descends to the smash product because it is pointed in each variable.

This proves that multilinear functors $\langle \mathsf{C}\rangle \longrightarrow \mathsf{D}$ correspond to multifunctors

$$\bigwedge_j \mathsf{End}(\mathsf{C}_j) \longrightarrow \mathsf{End}(\mathsf{D}). \qquad \square$$

One can also give a description similar to that of Definition 6.5.4 for the morphisms of

$$\mathsf{PermCat}^{\mathsf{su}}\big(\langle \mathsf{C}\rangle\,;\,\mathsf{D}\big).$$

Definition 6.5.11 (Multilinear Transformations)**.** Suppose $\mathsf{C}_1,\ldots,\mathsf{C}_n$, and D are permutative categories. Suppose

$$F, F'\colon \langle \mathsf{C}\rangle \longrightarrow \mathsf{D}$$

are n-linear functors. An *n-linear transformation* is a natural transformation of underlying functors

$$\alpha\colon F \longrightarrow F'$$

that satisfies the following two *multilinearity conditions*.

(1) The diagram

$$(6.5.12)\qquad \begin{array}{ccc} F\langle X \circ_i X_i\rangle \oplus F\langle X \circ_i X_i'\rangle & \xrightarrow{\;F_i^2\;} & F\langle X \circ_i (X_i \oplus X_i')\rangle \\ {\scriptstyle \alpha\oplus\alpha}\big\downarrow & & \big\downarrow{\scriptstyle \alpha} \\ F'\langle X \circ_i X_i\rangle \oplus F'\langle X \circ_i X_i'\rangle & \xrightarrow{\;(F')_i^2\;} & F'\langle X \circ_i (X_i \oplus X_i')\rangle \end{array}$$

commutes for each $i \in \{1,\ldots,n\}$ and $\langle X\rangle \in \prod_j \mathsf{C}_j$ with $X_i' \in \mathsf{C}_i$.

(2) The component of α at a tuple $\langle X\rangle$ is an identity if any $X_i = e$.

As we explain in the proof of Proposition 6.5.13 below, the two multilinearity conditions make α a monoidal natural transformation in each variable separately. We say that α is a *multilinear transformation* if it is n-linear for some $n \geq 0$. $\diamond$

Proposition 6.5.13. *Suppose $\mathsf{C}_1,\ldots,\mathsf{C}_n$ and D are small permutative categories. Suppose that F and F' are multilinear functors, regarded as objects of*

$$\mathsf{PermCat}^{\mathsf{su}}\big(\langle \mathsf{C}\rangle\,;\,\mathsf{D}\big) = \mathsf{Multicat}_*\big(\bigwedge_j \mathsf{End}(\mathsf{C}_j), \mathsf{End}(\mathsf{D})\big).$$

The morphisms between F and F' in $\mathsf{PermCat}^{\mathsf{su}}\big(\langle \mathsf{C}\rangle\,;\mathsf{D}\big)$ *are precisely the multilinear transformations* $F \longrightarrow F'$.

Proof. By definition, the morphisms in $\mathsf{PermCat}^{\mathsf{su}}\big(\langle \mathsf{C}\rangle\,;\mathsf{D}\big)$ are the pointed multinatural transformations between multifunctors

$$\bigwedge_j \mathsf{End}(\mathsf{C}_j) \longrightarrow \mathsf{End}(\mathsf{D}).$$

Recalling the proofs of Proposition 5.3.7 and Lemma 5.3.8, such a multinatural transformation corresponds to a natural transformation of multilinear functors,

$$\alpha\colon F \longrightarrow F',$$

that is monoidal in each variable separately. Since F and F' are strictly unital in each variable, the two conditions of Definition 1.1.12 for monoidal naturality are, for each variable separately, precisely the two multinaturality conditions of Definition 6.5.11. □

6.6. The Multicategory of Small Permutative Categories

In this section, we provide a direct proof that $\mathsf{PermCat}^{\mathsf{su}}$ is a Cat-enriched multicategory. The fact that $\mathsf{PermCat}^{\mathsf{su}}$ is a Cat-enriched multicategory is crucial in Corollaries 11.3.16, 11.6.12, 12.5.3, and 13.5.2, where $\mathsf{PermCat}^{\mathsf{su}}$ appears as the domain of the Elmendorf-Mandell K-theory multifunctor. See also Note 6.7.2.

Definitions. The objects in $\mathsf{PermCat}^{\mathsf{su}}$ are small permutative categories. For small permutative categories $\mathsf{C}_1, \ldots, \mathsf{C}_n$, and D, the category

$$\mathsf{PermCat}^{\mathsf{su}}\big(\langle \mathsf{C}\rangle\,;\mathsf{D}\big) = \mathsf{PermCat}^{\mathsf{su}}\big(\langle \mathsf{C}_1, \ldots, \mathsf{C}_n\rangle\,;\mathsf{D}\big)$$

is defined as follows.

- Its objects are n-linear functors (Definition 6.5.4)
$$\mathsf{C}_1 \times \cdots \times \mathsf{C}_n \longrightarrow \mathsf{D}.$$
- Its morphisms are multilinear transformations (Definition 6.5.11).
- Identity morphisms are identity natural transformations.
- Categorical composition is the vertical composition of natural transformations (Definition I.1.1.8).

Units. For each small permutative category $(\mathsf{C}, \oplus, \mathbb{0}, \xi^{\oplus})$, the C-colored unit in the category $\mathsf{PermCat}^{\mathsf{su}}\big(\mathsf{C}\,;\mathsf{C}\big)$ is the identity functor $1_{\mathsf{C}}\colon \mathsf{C} \longrightarrow \mathsf{C}$ with the identity linearity constraint

$$1_{A\oplus B}\colon A \oplus B \longrightarrow A \oplus B \quad \text{for} \quad A, B \in \mathsf{C}. \tag{6.6.1}$$

Equivariance. With $\langle \mathsf{C}\rangle = (\mathsf{C}_1, \ldots, \mathsf{C}_n)$ and $\sigma \in \Sigma_n$, the symmetric group action

$$\mathsf{PermCat}^{\mathsf{su}}\big(\langle \mathsf{C}\rangle\,;\mathsf{D}\big) \xrightarrow[\cong]{\sigma} \mathsf{PermCat}^{\mathsf{su}}\big(\langle \mathsf{C}\rangle\sigma\,;\mathsf{D}\big) \tag{6.6.2}$$

sends an n-linear functor

$$\big(F, \{F^2_j\}_{1\le j\le n}\big)\colon \mathsf{C}_1 \times \cdots \times \mathsf{C}_n \longrightarrow \mathsf{D}$$

to the composite functor

$$\begin{array}{ccc} \mathsf{C}_{\sigma(1)} \times \cdots \times \mathsf{C}_{\sigma(n)} & \xrightarrow{F^\sigma} & \mathsf{D} \\ & {\scriptstyle \sigma}\searrow^{\cong} \quad \nearrow{\scriptstyle F} & \\ & \mathsf{C}_1 \times \cdots \times \mathsf{C}_n & \end{array}$$

with σ permuting the entries from the left. For $1 \le j \le n$, the jth linearity constraint of F^σ is $F^2_{\sigma(j)}$ applied to appropriately permuted sequences of objects. In other words, for objects

- $\langle A \rangle = (A_1, \ldots, A_n) \in \prod_{i=1}^n \mathsf{C}_{\sigma(i)}$ and
- $A'_j \in \mathsf{C}_{\sigma(j)}$,

with the $\circ_j$ notation in Definition 6.5.3, the jth linearity constraint $(F^\sigma)^2_j$ is the following composite in D.

$$(6.6.3)\qquad \begin{array}{ccc} F^\sigma\langle A \rangle \oplus F^\sigma\langle A \circ_j A'_j \rangle & \xrightarrow{(F^\sigma)^2_j} & F^\sigma\langle A \circ_j (A_j \oplus A'_j)\rangle \\ \| & & \| \\ F\big(\sigma\langle A\rangle\big) \oplus F\big(\sigma\langle A \rangle \circ_{\sigma(j)} A'_j\big) & \xrightarrow{F^2_{\sigma(j)}} & F\big(\sigma\langle A \rangle \circ_{\sigma(j)} (A_j \oplus A'_j)\big) \end{array}$$

The symmetric group action sends a multilinear transformation $\alpha\colon F \longrightarrow G$ between n-linear functors $F, G \in \mathsf{PermCat}^{\mathsf{su}}\big(\langle \mathsf{C} \rangle ; \mathsf{D}\big)$ to the horizontal composite (Definition I.1.1.8)

$$(6.6.4)\qquad \alpha^\sigma = \alpha * 1_\sigma \colon F^\sigma \longrightarrow G^\sigma.$$

Composition of multilinear functors. For the multicategorical composition γ, suppose

- $\langle \mathsf{B}_j \rangle = (\mathsf{B}_{j,1}, \ldots, \mathsf{B}_{j,k_j})$ is a k_j-tuple of small permutative categories for each $1 \le j \le n$ and
- $\langle \mathsf{B} \rangle = \oplus_{j=1}^n \langle \mathsf{B}_j \rangle$ is their concatenation.

For multilinear functors

- $\big(F, \{F^2_j\}_{1 \le j \le n}\big) \in \mathsf{PermCat}^{\mathsf{su}}\big(\langle \mathsf{C} \rangle ; \mathsf{D}\big)$ and
- $\big(H_j, \{H^2_{j,i}\}_{1 \le i \le k_j}\big) \in \mathsf{PermCat}^{\mathsf{su}}\big(\langle \mathsf{B}_j \rangle ; \mathsf{C}_j\big)$ for $1 \le j \le n$,

the composition

$$(6.6.5)\qquad \mathsf{PermCat}^{\mathsf{su}}\big(\langle \mathsf{C} \rangle ; \mathsf{D}\big) \times \prod_{j=1}^n \mathsf{PermCat}^{\mathsf{su}}\big(\langle \mathsf{B}_j \rangle ; \mathsf{C}_j\big) \xrightarrow{\gamma} \mathsf{PermCat}^{\mathsf{su}}\big(\langle \mathsf{B} \rangle ; \mathsf{D}\big)$$

is defined by

$$(6.6.6)\qquad \begin{aligned} &\gamma\Big(\big(F, \{F^2_j\}_{1 \le j \le n}\big), \Big(\big(H_j, \{H^2_{j,i}\}_{1 \le i \le k_j}\big)\Big)_{1 \le j \le n}\Big) \\ &= \Big(F \circ \textstyle\prod_j H_j, \Big\{\big(F \circ \prod_j H_j\big)^2_l\Big\}_{1 \le l \le k_1 + \cdots + k_n}\Big). \end{aligned}$$

Below we will abbreviate the object in (6.6.6) to $\gamma(F, (H_j))$.

In (6.6.6), $F \circ \prod_j H_j$ is the composite functor

$$\prod_{j=1}^n \prod_{i=1}^{k_j} \mathsf{B}_{j,i} \xrightarrow{\prod_j H_j} \prod_{j=1}^n \mathsf{C}_j \xrightarrow{F} \mathsf{D}.$$

To explain its linearity constraints, consider

- objects $W_{j,i} \in \mathsf{B}_{j,i}$ for $1 \le j \le n$ and $1 \le i \le k_j$,
- an object $W'_{j,i} \in \mathsf{B}_{j,i}$ for a particular choice of (j,i) with $l = k_1 + \cdots + k_{j-1} + i$,
- $\langle W_j \rangle = (W_{j,1}, \ldots, W_{j,k_j}) \in \prod_{i=1}^{k_j} \mathsf{B}_{j,i}$,
- $\langle W \rangle = \oplus_{j=1}^{n} \langle W_j \rangle \in \prod_{j=1}^{n} \prod_{i=1}^{k_j} \mathsf{B}_{j,i}$ their concatenation, and
- $\langle HW \rangle = \big(H_1 \langle W_1 \rangle, \ldots, H_n \langle W_n \rangle\big) \in \prod_{j=1}^{n} \mathsf{C}_j$.

Note that

$$(6.6.7)\qquad \begin{aligned} \langle W_j \circ_i W'_{j,i} \rangle &= \big(\overbrace{W_{j,1}, \ldots, W_{j,i-1}}^{\text{empty if } i=1}, W'_{j,i}, \overbrace{W_{j,i+1}, \ldots, W_{j,k_j}}^{\text{empty if } i=k_j}\big) \\ \langle W_j \circ_i (W_{j,i} \oplus W'_{j,i}) \rangle &= \big(\underbrace{W_{j,1}, \ldots, W_{j,i-1}}_{\text{empty if } i=1}, W_{j,i} \oplus W'_{j,i}, \underbrace{W_{j,i+1}, \ldots, W_{j,k_j}}_{\text{empty if } i=k_j}\big). \end{aligned}$$

The lth linearity constraint $\big(F \circ \prod_j H_j\big)_l^2$ in (6.6.6) is defined as the following composite in D.

(6.6.8)

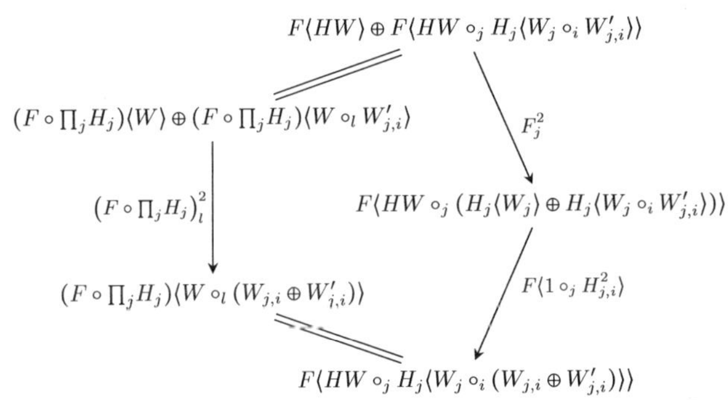

Note that if $n = 1 = k_1$, then $\big(FH, (FH)^2\big)$ in (6.6.6) is precisely the composite of strictly unital symmetric monoidal functors in Definition II.1.3.12.

Composition of multilinear transformations. For multilinear transformations

- $\alpha \colon F \longrightarrow F'$ in $\mathsf{PermCat^{su}}\big(\langle \mathsf{C} \rangle ; \mathsf{D}\big)$ and
- $\beta_j \colon H_j \longrightarrow H'_j$ in $\mathsf{PermCat^{su}}\big(\langle \mathsf{B}_j \rangle ; \mathsf{C}_j\big)$ for each $1 \le j \le n$, with $\langle \beta \rangle = (\beta_1, \ldots, \beta_n)$,

the multicategorical composition is defined as the horizontal composite

$$(6.6.9)\qquad \gamma\big(\alpha, \langle \beta \rangle\big) = \alpha * \textstyle\prod_{j=1}^{n} \beta_j \colon F \circ \prod_j H_j \longrightarrow F' \circ \prod_j H'_j$$

of natural transformations (Definition I.1.1.8).

Proofs. By Theorem 6.4.4, Convention 6.4.6, and Definition 6.5.1, the definitions above define a Cat-enriched multicategory $\mathsf{PermCat^{su}}$. The rest of this section contains an alternative direct proof of this fact in several steps.

Lemma 6.6.10. *$\gamma(F, (H_j))$ in (6.6.6) satisfies the constraint 2-by-2 axiom (6.5.7).*

Proof. There are two cases. For the first case, suppose $p \ne q \in \{1, \ldots, n\}$, $1 \le a \le k_p$, and $1 \le b \le k_q$. Consider objects

- $\langle W_j \rangle = (W_{j,1}, \ldots, W_{j,k_j}) \in \prod_{i=1}^{k_j} \mathsf{B}_{j,i}$ for $1 \le j \le n$,
- $W'_{p,a} \in \mathsf{B}_{p,a}$, $\langle W'_p \rangle = \langle W_p \circ_a W'_{p,a} \rangle \in \prod_{i=1}^{k_p} \mathsf{B}_{p,i}$,
- $W'_{q,b} \in \mathsf{B}_{q,b}$, $\langle W'_q \rangle = \langle W_q \circ_b W'_{q,b} \rangle \in \prod_{i=1}^{k_q} \mathsf{B}_{q,i}$,

- $\langle W\rangle = \oplus_{j=1}^n \langle W_j\rangle \in \prod_{j=1}^n \prod_{i=1}^{k_j} \mathsf{B}_{j,i}$,
- $X_j = H_j\langle W_j\rangle \in \mathsf{C}_j$ for $1 \le j \le n$,
- $X'_p = H_p\langle W'_p\rangle$, $X'_q = H_q\langle W'_q\rangle$,
- $\langle X\rangle = (X_1, \ldots, X_n) \in \prod_{j=1}^n \mathsf{C}_j$,
- $Y_p = H_p\langle W_p \circ_a (W_{p,a} \oplus W'_{p,a})\rangle \in \mathsf{C}_p$,
- $Y_q = H_q\langle W_q \circ_b (W_{q,b} \oplus W'_{q,b})\rangle \in \mathsf{C}_q$, and
- $U = F\langle X \circ_p (X_p \oplus X'_p) \circ_q (X_q \oplus X'_q)\rangle \in \mathsf{D}$.

By the definition (6.6.8) of the linearity constraints $(F \circ \prod_j H_j)^2_l$, the constraint 2-by-2 diagram (6.5.7) for $\gamma(F, (H_j))$ for the above objects is the outer diagram below.

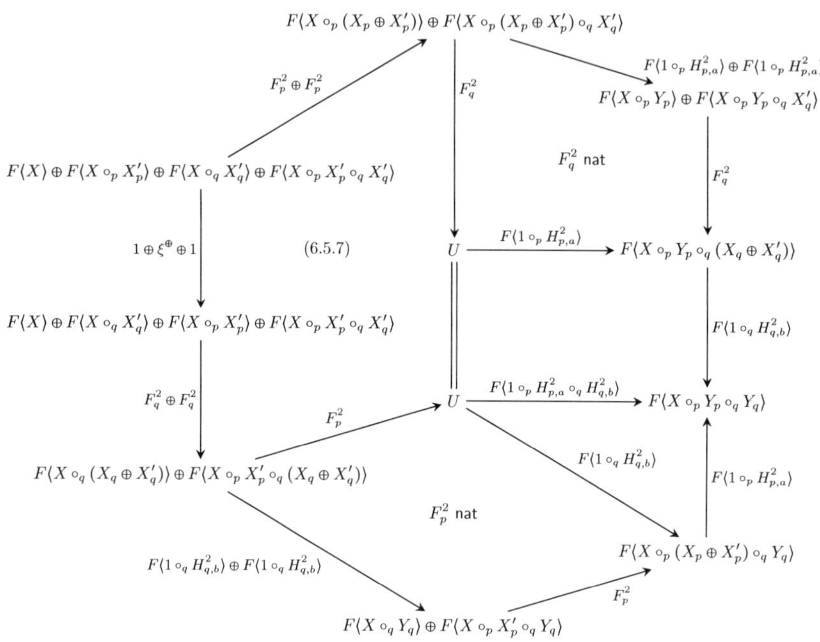

- The upper left subdiagram is commutative by the constraint 2-by-2 axiom (6.5.7) for F.
- The top right and the bottom subdiagrams are commutative by the naturality of, respectively, the linearity constraints F^2_q and F^2_p.
- The two unlabeled subdiagrams are commutative by the functoriality of F.

This proves the first case of the constraint 2-by-2 axiom.

For the other case of the constraint 2-by-2 axiom (6.5.7) for $\gamma(F, (H_j))$ in (6.6.6), suppose $p = q \in \{1, \ldots, n\}$ and $a \ne b \in \{1, \ldots, k_p\}$. Consider objects

- $\langle W_j\rangle \in \prod_i \mathsf{B}_{j,i}$ and $\langle W\rangle = \oplus_{j=1}^n \langle W_j\rangle \in \prod_j \prod_i \mathsf{B}_{j,i}$ as above,
- $W'_{p,a} \in \mathsf{B}_{p,a}$, $\langle W^a_p\rangle = \langle W_p \circ_a W'_{p,a}\rangle$,
- $W'_{p,b} \in \mathsf{B}_{p,b}$, $\langle W^b_p\rangle = \langle W_p \circ_b W'_{p,b}\rangle$,
- $\langle W^{a,b}_p\rangle = \langle W_p \circ_a W'_{p,a} \circ_b W'_{p,b}\rangle$,
- $X_j = H_j\langle W_j\rangle \in \mathsf{C}_j$, $\langle X\rangle = (X_1, \ldots, X_n) \in \prod_j \mathsf{C}_j$,
- $X^a_p = H_p\langle W^a_p\rangle$, $X^b_p = H_p\langle W^b_p\rangle$, $X^{a,b}_p = H_p\langle W^{a,b}_p\rangle$,
- $Y_a = H_p\langle W_p \circ_a (W_{p,a} \oplus W'_{p,a})\rangle$,

- $Y_b = H_p\langle W_p \circ_b (W_{p,b} \oplus W'_{p,b})\rangle$,
- $Z_a = H_p\langle W_p \circ_a (W_{p,a} \oplus W'_{p,a}) \circ_b W'_{p,b}\rangle$,
- $Z_b = H_p\langle W_p \circ_a W'_{p,a} \circ_b (W_{p,b} \oplus W'_{p,b})\rangle$,
- $Z = H_p\langle W_p \circ_a (W_{p,a} \oplus W'_{p,a}) \circ_b (W_{p,b} \oplus W'_{p,b})\rangle$,
- $U_1 = F\langle X \circ_p (X_p \oplus X_p^a \oplus X_p^b \oplus X_p^{a,b})\rangle$,
- $U_2 = F\langle X\rangle \oplus F\langle X \circ_p (X_p^a \oplus X_p^b \oplus X_p^{a,b})\rangle$,
- $U_3 = F\langle X\rangle \oplus F\langle X \circ_p (X_p^a \oplus X_p^b)\rangle \oplus F\langle X \circ_p X_p^{a,b}\rangle$,
- $U_4 = F\langle X\rangle \oplus F\langle X \circ_p (X_p^b \oplus X_p^a)\rangle \oplus F\langle X \circ_p X_p^{a,b}\rangle$,
- $U_5 = F\langle X\rangle \oplus F\langle X \circ_p (X_p^b \oplus X_p^a \oplus X_p^{a,b})\rangle$, and
- $U_6 = F\langle X \circ_p (X_p \oplus X_p^b \oplus X_p^a \oplus X_p^{a,b})\rangle$.

The objects X_p^a, X_p^b, $X_p^{a,b}$, Y_a, Y_b, Z_a, Z_b, and Z are in C_p, and each $U_?$ is in D. The constraint 2-by-2 diagram (6.5.7) for $\gamma(F,(H_j))$ for the above objects is the outer diagram below.

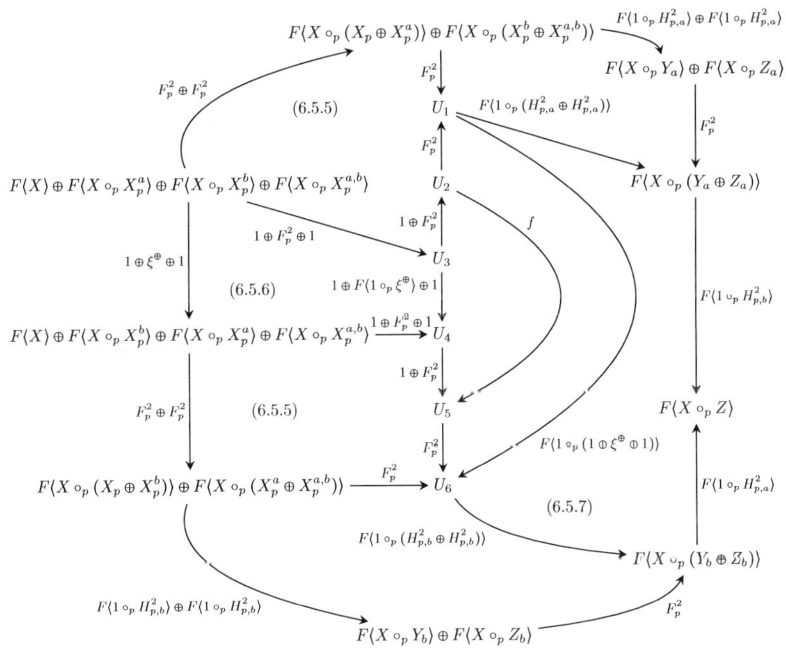

- The morphism $f: U_2 \longrightarrow U_5$ is
$$1 \oplus F\langle 1 \circ_p (\xi^{\oplus} \oplus 1)\rangle.$$
- The four unlabeled subdiagrams are commutative by the naturality of F_p^2.
- Each of the two subdiagrams labeled by (6.5.5) is commutative by the pth linearity constraint associativity axiom (6.5.5) for F twice, as in (II.9.2.17) with F_p^2 in place of F^2.
- The middle left trapezoid is commutative by the pth linearity constraint symmetry axiom (6.5.6) for F.
- The remaining subdiagram is obtained from the constraint 2-by-2 diagram (6.5.7) for H_p by applying $F\langle 1 \circ_p -\rangle$, so it is commutative.

This proves the second case of the constraint 2-by-2 axiom (6.5.7) for $\gamma(F,(H_j))$ in (6.6.6). □

Lemma 6.6.11. *$\gamma(F,(H_j))$ in (6.6.6) is a k-linear functor in* $\mathsf{PermCat}^{\mathsf{su}}(\langle \mathsf{B}\rangle ; \mathsf{D})$ *with* $k = k_1 + \cdots + k_n$.

Proof. The naturality of the lth linearity constraint $\left(F \circ \prod_j H_j\right)_l^2$ in (6.6.8) follows from the functoriality of F and the naturality of F_j^2 and $H_{j,i}^2$. Next we check that $\gamma(F,(H_j))$ satisfies the axioms in Definition 6.5.4 for a k-linear functor from $\langle \mathsf{B}\rangle$ to D. Its constraint 2-by-2 axiom (6.5.7) is verified in Lemma 6.6.10. Its unity axiom and constraint unity axiom follow from those of F and the H_j's.

For the constraint associativity axiom (6.5.5), consider the objects in (6.6.7) and

- $W''_{j,i} \in \mathsf{B}_{j,i}$,
- $\langle W_j^1\rangle = \langle W_j \circ_i W'_{j,i}\rangle$,
- $\langle W_j^2\rangle = \langle W_j \circ_i W''_{j,i}\rangle$,
- $\langle W_j^{01}\rangle = \langle W_j \circ_i (W_{j,i} \oplus W'_{j,i})\rangle$,
- $\langle W_j^{12}\rangle = \langle W_j \circ_i (W'_{j,i} \oplus W''_{j,i})\rangle$,
- $\langle W_j^{012}\rangle = \langle W_j \circ_i (W_{j,i} \oplus W'_{j,i} \oplus W''_{j,i})\rangle$,
- $\langle HW_j\rangle = H_j\langle W_j\rangle$,
- $\langle HW_j^?\rangle = H_j\langle W_j^?\rangle \in \mathsf{C}_j$ for $? \in \{1,2,01,12,012\}$,
- $\langle HW^?\rangle = \langle HW \circ_j \langle HW_j^?\rangle\rangle \in \mathsf{D}$,
- $\langle HW_j\rangle^{0,1} = \langle HW_j\rangle \oplus \langle HW_j^1\rangle$,
- $\langle HW_j\rangle^{1,2} = \langle HW_j^1\rangle \oplus \langle HW_j^2\rangle$,
- $\langle HW_j\rangle^{01,2} = \langle HW_j^{01}\rangle \oplus \langle HW_j^2\rangle$,
- $\langle HW_j\rangle^{0,12} = \langle HW_j\rangle \oplus \langle HW_j^{12}\rangle$, and
- $\langle HW_j\rangle^{0,1,2} = \langle HW_j\rangle \oplus \langle HW_j^1\rangle \oplus \langle HW_j^2\rangle$.

The constraint associativity axiom (6.5.5) for $\gamma(F,(H_j))$ in (6.6.6) for the objects above is the outer diagram in D below.

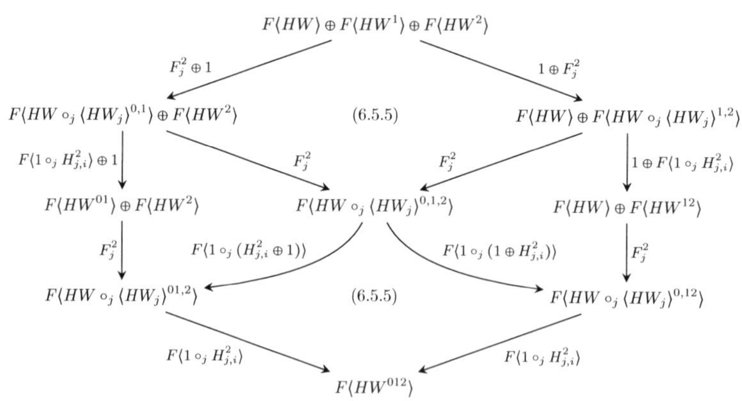

- The top diamond is commutative by the jth linearity constraint associativity axiom (6.5.5) for F.
- The bottom diamond is obtained from the ith linearity constraint associativity diagram (6.5.5) for H_j by applying $F\langle 1 \circ_j -\rangle$, so it is commutative.
- The left and right triangles are commutative by the naturality of F_j^2.

This proves the constraint associativity axiom (6.5.5) for $\gamma(F,(H_j))$.

The constraint symmetry axiom (6.5.6) for $\gamma(F,(H_j))$ is the outer diagram in D below.

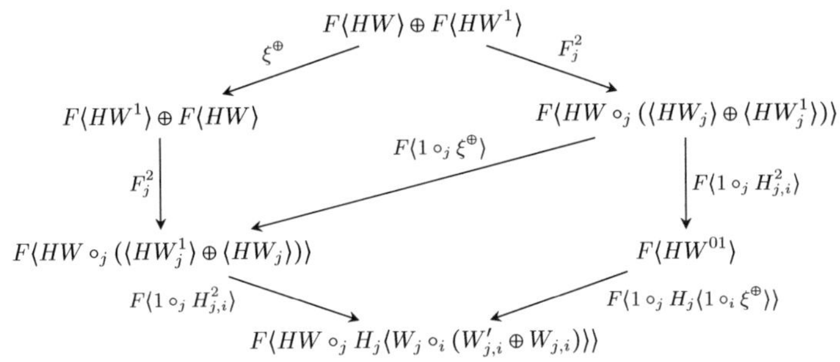

The top subdiagram is commutative by the jth linearity constraint symmetry axiom (6.5.6) for F. The bottom subdiagram is obtained from the ith linearity constraint symmetry diagram (6.5.6) for H_j by applying $F\langle 1 \circ_j -\rangle$, so it is commutative. This finishes the proof that $\gamma(F,(H_j))$ in (6.6.6) is a k-linear functor. □

Lemma 6.6.12. *γ in (6.6.5) is a functor.*

Proof. Lemma 6.6.11 shows that γ is well defined on multilinear functors.

To see that the horizontal composite natural transformation

$$\gamma(\alpha,\langle\beta\rangle) = \alpha * \textstyle\prod_j \beta_j \colon F \circ \prod_j H_j \longrightarrow F' \circ \prod_j H'_j$$

in (6.6.9) is a multilinear transformation, we check the two conditions in Definition 6.5.11. The second condition, which says that the component of $\gamma(\alpha,\langle\beta\rangle)$ at a tuple with at least one entry $\mathbb{0}$ is the identity morphism $1_{\mathbb{0}}$, follows from the corresponding property for α and the β_j's.

The other condition is the commutativity of the diagram (6.5.12) for $\gamma(\alpha,\langle\beta\rangle)$. To prove this, consider the objects in (6.6.7) and in the proof of Lemma 6.6.11. We extend those notations to H' and β. For example, we have the morphisms

- $\beta_{j,\langle W_j\rangle} \colon \langle HW_j\rangle = H_j\langle W_j\rangle \longrightarrow H'_j\langle W_j\rangle = \langle H'W_j\rangle$,
- $\beta = \beta_{\langle W\rangle} = \big(\beta_{1,\langle W_1\rangle},\ldots,\beta_{n,\langle W_n\rangle}\big) \colon \langle HW\rangle \longrightarrow \langle H'W\rangle$,
- $\beta^1 = \beta_{\langle W^1\rangle} = \langle \beta_{\langle W\rangle} \circ_j \beta_{j,\langle W_j^1\rangle}\rangle \colon \langle HW^1\rangle \longrightarrow \langle H'W^1\rangle$, and
- $\beta_{\langle W^{01}\rangle} = \langle \beta_{\langle W\rangle} \circ_j \beta_{j,\langle W_j^{01}\rangle}\rangle \colon \langle HW^{01}\rangle \longrightarrow \langle H'W^{01}\rangle$.

The diagram (6.5.12) for $\gamma(\alpha, \langle\beta\rangle)$ is the outer diagram in D below.

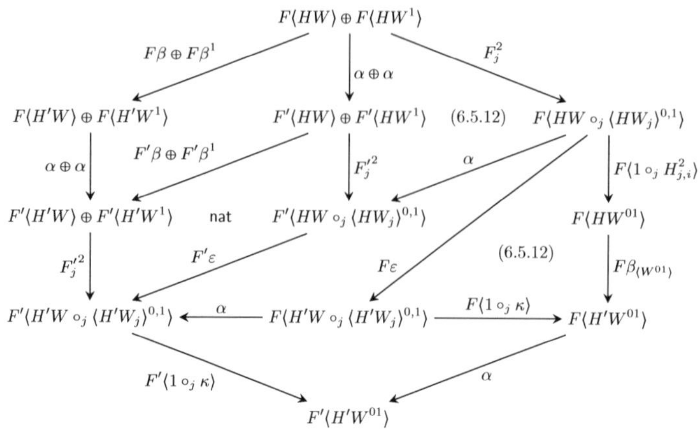

- The morphism ε is
$$\langle \beta_{\langle W\rangle} \circ_j (\beta_{j,\langle W_j\rangle} \oplus \beta_{j,\langle W_j^1\rangle})\rangle.$$
- The morphism
$$\kappa \colon \langle H'W_j\rangle^{0,1} \longrightarrow \langle H'W_j^{01}\rangle$$
is the ith linearity constraint of H'_j.
- The three unlabeled subdiagrams are commutative by the naturality of α.
- The subdiagram labeled by nat is commutative by the naturality of F'^2_j.
- The upper right subdiagram is commutative by (6.5.12) for α.
- The remaining subdiagram is obtained from the diagram (6.5.12) for β by applying $F\langle \beta_{\langle W\rangle} \circ_j -\rangle$, so it is commutative.

This shows that γ is well defined on multilinear transformations.

Moreover, γ preserves identity multilinear transformations because horizontal composition preserves identity natural transformations. Finally, γ preserves the categorical composition in its (co)domain by the following facts:

- The categorical composition in each category $\mathsf{PermCat}^{\mathsf{su}}\big(\langle \mathsf{C}\rangle ; \mathsf{D}\big)$ is vertical composition of natural transformations.
- On multilinear transformations, γ is defined by horizontal composition of natural transformations.
- Natural transformations satisfy the middle-four exchange property
$$(\theta'\theta) * (\phi'\phi) = (\theta' * \phi')(\theta * \phi)$$
with respect to vertical composition and horizontal composition, which are denoted by, respectively, concatenation and $*$.

This finishes the proof that γ is a functor. □

Theorem 6.6.13. $\mathsf{PermCat}^{\mathsf{su}}$ *is a* Cat*-enriched multicategory.*

Proof. We check the axioms in Definition 6.1.1.

- The symmetric group action (6.6.2) is a well-defined right Σ_n-action because it is given by pre-composition with $\sigma \in \Sigma_n$ or 1_σ.

- The equivariance axioms (6.1.6) and (6.1.7) also follow from the fact that the symmetric group action (6.6.2) in $\mathsf{PermCat}^{\mathsf{su}}$ is defined by precomposition with a permutation.
- The unity axioms (6.1.4) and (6.1.5) follow from the fact that, for each small permutative category C, the C-colored unit in $\mathsf{PermCat}^{\mathsf{su}}$ is the identity functor 1_{C} with the identity linearity constraint (6.6.1).

For the associativity axiom (6.1.3) on objects, consider multilinear functors

- $\big(F,\{F^2_j\}_{1\le j\le n}\big) \in \mathsf{PermCat}^{\mathsf{su}}\big(\langle \mathsf{C}\rangle\,;\,\mathsf{D}\big)$,
- $\big(H_j,\{H^2_{j,i}\}_{1\le i\le k_j}\big) \in \mathsf{PermCat}^{\mathsf{su}}\big(\langle \mathsf{B}_j\rangle\,;\,\mathsf{C}_j\big)$ for $1\le j\le n$, and
- $\big(G_{j,i},\{G^2_{j,i,h}\}_{1\le h\le l_{j,i}}\big) \in \mathsf{PermCat}^{\mathsf{su}}\big(\langle \mathsf{A}_{j,i}\rangle\,;\,\mathsf{B}_{j,i}\big)$ with

$$\langle \mathsf{A}_{j,i}\rangle = \big(\mathsf{A}_{j,i,1},\ldots,\mathsf{A}_{j,i,l_{j,i}}\big)$$

an $l_{j,i}$-tuple of small permutative categories for $1\le j\le n$ and $1\le i\le k_j$.

Then each composite in the associativity diagram (6.1.3) sends the tuple

$$\begin{gathered}\Big(\big(F,\{F^2_j\}\big),\big((H_j,\{H^2_{j,i}\})\big)_{1\le j\le n},\Big(\big((G_{j,i},\{G^2_{j,i,h}\})\big)_{1\le i\le k_j}\Big)_{1\le j\le n}\Big)\\ \in \mathsf{PermCat}^{\mathsf{su}}\big(\langle \mathsf{C}\rangle\,;\,\mathsf{D}\big)\times\prod_{j=1}^{n}\mathsf{PermCat}^{\mathsf{su}}\big(\langle \mathsf{B}_j\rangle\,;\,\mathsf{C}_j\big)\times\prod_{j=1}^{n}\prod_{i=1}^{k_j}\mathsf{PermCat}^{\mathsf{su}}\big(\langle \mathsf{A}_{j,i}\rangle\,;\,\mathsf{B}_{j,i}\big)\end{gathered}\tag{6.6.14}$$

to the composite functor

$$\prod_{j=1}^{n}\prod_{i=1}^{k_j}\prod_{h=1}^{l_{j,i}}\mathsf{A}_{j,i,h}\xrightarrow{\prod_j\prod_i G_{j,i}}\prod_{j=1}^{n}\prod_{i=1}^{k_j}\mathsf{B}_{j,i}\xrightarrow{\prod_j H_j}\prod_{j=1}^{n}\mathsf{C}_j\xrightarrow{F}\mathsf{D}.$$

To see that their corresponding linearity constraints are equal, for $1\le j\le n$, $1\le i\le k_j$, and $1\le h\le l_{j,i}$, consider

- $l_j = l_{j,1}+\cdots+l_{j,k_j}$,
- objects $V_{j,i,h}\in \mathsf{A}_{j,i,h}$,
- for a particular choice of (j,i,h), an object $V'_{j,i,h}\in\mathsf{A}_{j,i,h}$ with $p=l_1+\cdots+l_{j-1}$ and $q=l_{j,1}+\cdots+l_{j,i-1}$,
- $\langle V_{j,i}\rangle = (V_{j,i,1},\ldots,V_{j,i,l_{j,i}})\in\prod_{h=1}^{l_{j,i}}\mathsf{A}_{j,i,h}$,
- $\langle V'_{j,i}\rangle = \langle V_{j,i}\circ_h V'_{j,i,h}\rangle$,
- $\langle V''_{j,i}\rangle = \langle V_{j,i}\circ_h (V_{j,i,h}\oplus V'_{j,i,h})\rangle$,
- $\langle V_j\rangle = \oplus_{i=1}^{k_j}\langle V_{j,i}\rangle\in\prod_{i=1}^{k_j}\prod_{h=1}^{l_{j,i}}\mathsf{A}_{j,i,h}$,
- $\langle V'_j\rangle = \langle V_j\circ_{q+h} V'_{j,i,h}\rangle$,
- $\langle V\rangle = \oplus_{j=1}^{n}\langle V_j\rangle\in\prod_{j=1}^{n}\prod_{i=1}^{k_j}\prod_{h=1}^{l_{j,i}}\mathsf{A}_{j,i,h}$,
- $\langle V'\rangle = \langle V\circ_{p+q+h} V'_{j,i,h}\rangle$,
- $\langle G_jV_j\rangle = \big(G_{j,1}\langle V_{j,1}\rangle,\ldots,G_{j,k_j}\langle V_{j,k_j}\rangle\big)\in\prod_{i=1}^{k_j}\mathsf{B}_{j,i}$,
- $\langle G_jV'_j\rangle = \langle G_jV_j\circ_i G_{j,i}\langle V'_{j,i}\rangle\rangle$,
- $\langle HGV\rangle = \big(H_1\langle G_1V_1\rangle,\ldots,H_n\langle G_nV_n\rangle\big)\in\prod_{j=1}^{n}\mathsf{C}_j$, and
- $\langle HGV'\rangle = \langle HGV\circ_j H_j\langle G_jV'_j\rangle\rangle$.

By the definition (6.6.8) and the functoriality of F, each composite in the associativity diagram (6.1.3), when applied to the multilinear functors in (6.6.14) and the objects above, has $(p+q+h)$th linearity constraint the following composite in D.

$$\begin{array}{c}
F\langle HGV\rangle \oplus F\langle HGV'\rangle \\
\Big\downarrow {\scriptstyle F_j^2} \\
F\langle HGV \circ_j (H_j\langle G_jV_j\rangle \oplus H_j\langle G_jV_j'\rangle)\rangle \\
\Big\downarrow {\scriptstyle F\langle 1\circ_j H_{j,i}^2\rangle} \\
F\langle HGV \circ_j H_j\langle G_jV_j \circ_i (G_{j,i}\langle V_{j,i}\rangle \oplus G_{j,i}\langle V_{j,i}'\rangle)\rangle\rangle \\
\Big\downarrow {\scriptstyle F\langle 1\circ_j H_j\langle 1\circ_i G_{j,i,h}^2\rangle\rangle} \\
F\langle HGV \circ_j H_j\langle G_jV_j \circ_i G_{j,i}\langle V_{j,i}''\rangle\rangle\rangle
\end{array}$$

This proves that γ in (6.6.5) satisfies the associativity axiom (6.1.3) on multilinear functors.

Since multilinear transformations are natural transformations with additional properties, but not additional data, the associativity of γ on multilinear transformations follows from the associativity of horizontal composition of natural transformations. □

6.7. Notes

6.7.1 (Enriched Multicategories). For further development of the theory of enriched multicategories, we refer the reader to [**YJ15, Yau20**]. ◇

6.7.2 (Small Permutative Categories). In Section 6.6, we proved in detail that $\mathsf{PermCat}^{\mathsf{su}}$ is a Cat-enriched multicategory. This fact was stated in [**EM06**, 1.1], but a detailed proof was not given there. Our proof in Section 6.6 that $\mathsf{PermCat}^{\mathsf{su}}$ is a Cat-enriched multicategory is a generalization of the proof that small symmetric monoidal categories, symmetric monoidal functors, and monoidal natural transformations form a 2-category. There are two main differences between the cases of $\mathsf{PermCat}^{\mathsf{su}}$ and small symmetric monoidal categories:

- The constraint 2-by-2 axiom (6.5.7) only happens in multilinear functors. So Lemma 6.6.10, which proves the constraint 2-by-2 axiom for $\gamma(F,(H_j))$, does not have an analogue for symmetric monoidal functors.
- The notation for $\mathsf{PermCat}^{\mathsf{su}}$ is much more complicated because it involves lists of objects instead of one object. ◇

Part 2

Algebraic K-Theory

CHAPTER 7

Homotopy Theory Background

In Part 2 we develop the E_n-monoidal K-theory spectra associated to small E_n-monoidal categories. In this chapter we review the parts of homotopy theory that will be necessary for the constructions and their applications.

Convention 7.0.1 (K-theory and J-theory)**.** Throughout Part 2, and throughout this entire work, we use the umbrella term *K-theory* for any construction taking values in a category of spectra. This is the standard usage in the literature and is motivated by Quillen's definition of the higher algebraic K-groups of a commutative ring R as the homotopy groups of an associated spectrum KR [**Qui73, Gra76**]. See Note 10.8.10 for further discussion of this point.

The K-theory constructions we develop will all take values in the category of symmetric spectra and will all be either:

- functors, such as K^{Se} (Definition 8.5.1),
- (enriched) symmetric monoidal functors, such as $\mathsf{K}^{\mathcal{F}}$ (Definition 8.2.5) and $\mathsf{K}^{\mathcal{G}}$ (Definition 9.3.14), or
- (enriched) multifunctors, such as K^{EM} (Definition 10.3.32).

We follow [**EM06, EM09**] in using the umbrella term *J-theory* for constructions that take values in certain pointed diagrams of small categories. The two cases of interest for us will be Γ-categories (Definition 8.1.17) and $\mathcal{G}_*$-categories (Definition 9.2.1). The motivation for this terminology is that construction of the respective J-theory is a key step in both the Segal and Elmendorf-Mandell K-theory constructions. See also Note 8.6.1. $\diamond$

Now we describe the material in this chapter.

Organization. In Section 7.1 we review simplicial objects in a general category, focusing on the special case of simplicial sets. In Section 7.2 we describe the nerve functor from small categories to simplicial sets. This functor provides the fundamental connection between categorical algebra and homotopy theory through which all of our applications factor.

Section 7.3 introduces a certain diagram category of pointed simplicial sets known as symmetric sequences. These are sequences of pointed simplicial sets with compatible symmetric group actions. Symmetric spectra are defined in Section 7.4 as left modules over the symmetric sequence formed by the simplicial spheres, S^n. The Segal and Elmendorf-Mandell K-theory functors defined in Chapters 8, 9, and 10 will take values in the category of symmetric spectra.

In Section 7.5 we show that the category of symmetric spectra is complete and cocomplete. In Section 7.6 we describe its smash product and internal hom. The operad actions that define E_n-symmetric spectra for $1 \le n \le \infty$ in Chapters 11, 12, and 13 are defined in this symmetric monoidal closed structure.

In Section 7.7 we review the definitions of Quillen model categories, which provide a general framework for homotopy theory. In Section 7.8 we describe several important examples, including model structures for pointed simplicial sets and for symmetric spectra.

7.1. Simplicial Objects

Simplicial sets, which we define below, give a combinatorial approach to working with topological spaces. We will also use simplicial categories, and therefore begin with a presentation of simplicial objects in general categories.

Definition 7.1.1. For natural numbers $n \geq 0$, the totally ordered set $\{0 < 1 < \cdots < n\}$ is denoted $\underline{n}$. We let Δ denote the category whose objects are $\underline{n}$ for all $n \geq 0$ and whose morphisms are order-preserving functions.

The morphisms of Δ are generated by

$$\begin{aligned} d^i &: \underline{n-1} \longrightarrow \underline{n}, \quad 0 \leq i \leq n, \quad \text{and} \\ s^i &: \underline{n+1} \longrightarrow \underline{n}, \quad 0 \leq i \leq n, \end{aligned}$$

known as the *coface* and *codegeneracy* morphisms, respectively. The coface d^i is the unique order-preserving injection whose image does not contain $i \in \underline{n}$. The codegeneracy s^i is the unique order-preserving surjection such that the preimage of $i \in \underline{n}$ is a two-element set.

The coface and codegeneracy morphisms are subject to the following *cosimplicial identities.*

$$\begin{aligned} d^j d^i &= d^i d^{j-1} && \text{if } i < j \\ s^j d^i &= d^i s^{j-1} && \text{if } i < j \\ s^j d^j &= 1 = s^j d^{j+1} \\ s^j d^i &= d^{i-1} s^j && \text{if } i > j+1 \\ s^j s^i &= s^i s^{j+1} && \text{if } i \leq j \end{aligned}$$

◇

The cofaces, codegeneracies, and cosimplicial identities are a generating set of morphisms and relations for Δ.

Explanation 7.1.2. In the context of Definition 7.1.1, the coface and codegeneracy maps are given by the following formulas. For $0 \leq i \leq n$,

$$d^i(k) = \begin{cases} k, & \text{if } k < i \\ k+1, & \text{if } k \geq i \end{cases} \quad \text{and} \quad s^i(k) = \begin{cases} k, & \text{if } k \leq i \\ k-1, & \text{if } k > i. \end{cases}$$

◇

Explanation 7.1.3. It will sometimes be useful to regard the totally ordered set $\underline{n}$ as a category with a unique morphism $i \longrightarrow j$ for each $i \leq j$ in $\underline{n}$. Because the morphisms in Δ are order-preserving, they yield functors between the corresponding categories. ◇

Definition 7.1.4. Suppose C is a category. A *simplicial* C*-object* is a functor

$$X : \Delta^{\mathsf{op}} \longrightarrow \mathsf{C}.$$

We let $X_n = X\underline{n}$ and follow the usual convention of letting d_i, respectively s_i, denote Xd^i, respectively Xs^i. These are called *face* and *degeneracy* morphisms. When necessary for clarity, we write d_i^X and s_i^X to indicate the functor X. Morphisms

$f: X \longrightarrow Y$ of simplicial objects are given by natural transformations, and we let f_n denote the compoent of f at $\underline{n}$. The category of simplicial C-objects is denoted sC. ◇

Our two most frequent uses of this notion will be the cases C = Set or C = Cat, yielding the category of simplicial sets, sSet, and the category of simplicial small categories, sCat, respectively.

Example 7.1.5 (One-Point Simplicial Set)**.** The *one-point simplicial set* is denoted $*$ and is defined by the constant functor at the one-point set $*$. ◇

Explanation 7.1.6. The face and degeneracy morphisms are subject to the following *simplicial identities.*

$$\begin{aligned} d_i d_j &= d_{j-1} d_i \quad \text{if} \quad i < j \\ d_i s_j &= s_{j-1} d_i \quad \text{if} \quad i < j \\ d_j s_j &= 1 = d_{j+1} s_j \\ d_i s_j &= s_j d_{i-1} \quad \text{if} \quad i > j+1 \\ s_i s_j &= s_{j+1} s_i \quad \text{if} \quad i \le j \end{aligned}$$

◇

Explanation 7.1.7 (Bisimplicial Sets)**.** In the case C = sSet, we call simplicial sSet-objects *bisimplicial sets*, and we implicitly identify s(sSet) with the category of functors and natural transformations $\Delta^{\mathsf{op}} \times \Delta^{\mathsf{op}} \longrightarrow \mathsf{Set}$. Composing with the diagonal on Δ^{op} gives a functor from bisimplicial sets to simplicial sets. ◇

Definition 7.1.8. For a simplicial set X, the elements of X_n are called *n-simplices.* Those which are in the image of a degeneracy s_i are called *degenerate* n-simplices. Often we focus only on the nondegenerate simplices. ◇

The following definitions provide important basic examples of simplicial sets.

Definition 7.1.9. The *standard n-simplex*, $\Delta^n \in \mathsf{sSet}$, is the simplicial set defined by the represented functor $\Delta(-,\underline{n})$. The *fundamental simplex* $\iota_n = 1_{\underline{n}}$ is the identity morphism in Δ^n_n. The *boundary*, $\partial\Delta^n$, is the smallest simplicial subset of Δ^n containing $d_i\iota_n$ for all $0 \le i \le n$.

The *k-horn* $\Lambda^n_k \subset \Delta^n$ is the simplicial subset generated by all the faces $d_i\iota_n$ except for the kth face, $d_k\iota_n$. ◇

Explanation 7.1.10. The standard n-simplex may be visualized as follows. Its 0-simplices are the elements of $\underline{n}$, which may be thought of as vertices. Its nondegenerate 1-simplices are pairs $i < j$ in $\underline{n}$, which may be thought of as edges from vertex i to vertex j. Its nondegenerate 2-simplices are triples $i < j < k$ in $\underline{n}$, which may be thought of as oriented triangles. Diagrams for $n = 2$ and $n = 3$ are drawn below.

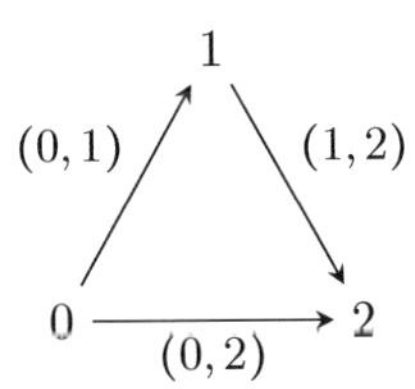

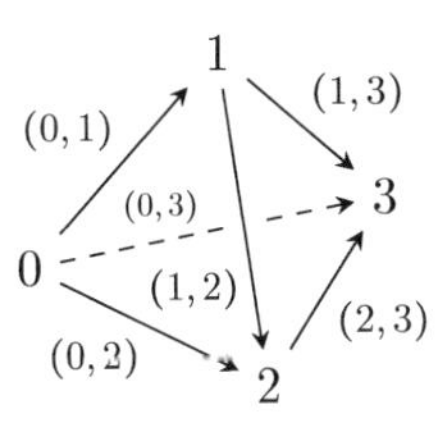

There are no nondegenerate simplices of Δ^n above dimension n. Each Δ^n has a *long spine* consisting of n 1-simplices $(i, i+1)$ and a *short spine* consisting of the 1-simplex $(0, n)$.

The kth horn Λ^n_k consists of all faces that include the kth 0-simplex. For example, Λ^2_1 consists of the 1-simplices

$$0 \xrightarrow{(0,1)} 1 \xrightarrow{(1,2)} 2$$

and Λ^3_1 consists of three triangular faces, omitting the face $(0,2,3)$. ◇

Definition 7.1.11. Since $\Delta^n = \Delta(-,\underline{n})$ is covariant in $\underline{n}$, we have a functor

$$\Delta^?\colon \Delta \longrightarrow \mathsf{sSet}$$

whose value at $\underline{n} \in \Delta$ is Δ^n. ◇

Definition 7.1.12. The *simplicial circle* is denoted S^1 and is defined to be the quotient $\Delta^1/\partial\Delta^1$. For $0 \le j \le n+1$, let $h_j \in \Delta(\underline{n},\underline{1})$ denote the function

$$h_j(i) = \begin{cases} 0, & \text{if } i < j \\ 1, & \text{if } i \ge j. \end{cases}$$

Then the n-simplices of S^1 form an $(n+1)$-element set given by

$$S^1_n = \Delta(\underline{n},\underline{1})/(h_{n+1} = h_0).$$

Identifying S^1_n with $\underline{n}$ via $h_j \leftrightarrow j$ for $0 \le j \le n$, then the face $d_i\colon S^1_n \longrightarrow S^1_{n-1}$ is the unique order-preserving surjection such that $d_i(i+1) = i$ and $s_i\colon S^1_n \longrightarrow S^1_{n+1}$ is the unique order-preserving injection whose image skips $i+1$. Thus all of the faces and degeneracies preserve the elements $0 \leftrightarrow h_0$, and we take this as a simplicial basepoint of S^1. The only nondegenerate simplex is $1 \leftrightarrow h_1 \in S^1_1$. ◇

Definition 7.1.13. We let Top denote the category of compactly generated weak Hausdorff spaces and continuous functions. For each $n \ge 0$ we let $\triangle^n$ denote the *topological n-simplex*

$$\triangle^n = \left\{(t_0,\ldots,t_n) \mid 0 \le t_i \le 1,\ \textstyle\sum_i t_i = 1\right\} \subset \mathbb{R}^{(n+1)}.$$ ◇

Definition 7.1.14. Given a simplicial set X, the *geometric realization* $|X|$ is the topological space defined as the coequalizer

$$\operatorname{coeq}\Bigg(\coprod_{\underline{n} \to \underline{m} \in \Delta} X_m \times \triangle^n \rightrightarrows \coprod_{\underline{n}\in\Delta} X_n \times \triangle^n \Bigg).$$

Geometric realization defines a functor

$$|-|\colon \mathsf{sSet} \longrightarrow \mathsf{Top}$$

and has a right adjoint given by the *total singular complex*

$$\mathsf{Sing}\colon \mathsf{Top} \longrightarrow \mathsf{sSet}.$$ ◇

Explanation 7.1.15. The coequalizer in the definition of geometric realization is the coend

$$\int^{\underline{n}\in\Delta} \mathsf{sSet}(\Delta^n, X) \times \triangle^n \cong \int^{\underline{n}\in\Delta} X_n \times \triangle^n.$$ ◇

Example 7.1.16 (Simplices and Circles)**.** The geometric realization of the standard n-simplex Δ^n is the topological n-simplex, $\triangle^n$. The geometric realization of the simplicial circle is a topological circle. ◇

As a diagram category, sC inherits limits and colimits from C. We record several of the key consequences here, recalling from the following.

- Tensors and cotensors are discussed in Section 3.9.
- Mapping objects in diagram categories are defined in Definition 3.5.19

Proposition 7.1.17. *Suppose that* C *has all limits and colimits.*

(1) *The category* C *is tensored and cotensored over* Set, *with*

$$X \otimes S = \coprod_S X \quad \text{and} \quad X^S = \prod_S X$$

for $X \in \mathsf{C}$ *and* $S \in \mathsf{Set}$.

(2) *Limits and colimits in* sC *exist and are computed objectwise.*

(3) *If* C *is monoidal (resp. braided monoidal, resp. symmetric monoidal), then so is* sC, *with the monoidal product computed objectwise. The monoidal unit of* sC *is the constant simplicial object at the monoidal unit of* C.

(4) *If* $\mathsf{C} = (\mathsf{C}, \otimes, \mathbb{1}, [-,-])$ *is symmetric monoidal closed, then* sC *is enriched, tensored, and cotensored over* C *with*

$$\mathsf{Map}(X,Y) = \int_{n \in \Delta^{\mathsf{op}}} [X_n, Y_n]$$

$$(X \otimes A)_n = X_n \otimes A$$

$$(X^A)_n = [A, X_n]$$

where $n \geq 0$, X *and* Y *are objects of* sC, *and* A *is an object of* C.

Proof.

(1) The tensor and cotensor over Set are provided by the isomorphisms

$$\mathsf{C}(\textstyle\coprod_S X, Y) \cong \prod_S \mathsf{C}(X,Y) \cong \mathsf{Set}(S, \mathsf{C}(X,Y))$$

and

$$\mathsf{C}(Y, \textstyle\prod_S X) \cong \prod_S \mathsf{C}(X,Y) \cong \mathsf{Set}(S, \mathsf{C}(X,Y))$$

for a set S and objects X, Y of C.

(2) For a small category $\mathcal{D}$ and a functor

$$F : \mathcal{D} \longrightarrow \mathsf{sC},$$

the definitions

$$(\mathsf{lim}_{\mathcal{D}} F)_n = \mathsf{lim}_{\mathcal{D}}(F_n) \quad \text{and} \quad (\mathsf{colim}_{\mathcal{D}} F)_n = \mathsf{colim}_{\mathcal{D}}(F_n)$$

define simplicial objects in C by universality of the limits and colimits.

(3) If C has monoidal product $\otimes$, the monoidal product of simplicial objects X and Y is given by

$$(X \otimes Y)_n = X_n \otimes Y_n.$$

This simplicial object is the composite of

$$\Delta^{\mathsf{op}} \xrightarrow{\text{diag}} \Delta^{\mathsf{op}} \times \Delta^{\mathsf{op}} \xrightarrow{X \times Y} \mathsf{C} \times \mathsf{C} \xrightarrow{\otimes} \mathsf{C}$$

where diag denotes the diagonal functor. The monoidal, respectively braided monoidal, respectively symmetric monoidal structure and axioms follow from those of $(\mathsf{C}, \otimes)$.

(4) When C is symmetric monoidal closed, then sC is a special case of the mapping diagram category denoted (Δ^{op})-C in Definition 3.5.19. This provides a C-enrichment by Lemma 3.5.24. The tensor and cotensor isomorphisms

$$\mathsf{Map}(X \otimes A, Y) \cong [A, \mathsf{Map}(X,Y)]$$

and

$$\mathsf{Map}(X, Y^A) \cong [A, \mathsf{Map}(X,Y)]$$

are given by commuting ends with $[-,-]$ and by the closed structure of C, as in Explanation 3.9.16. □

Explanation 7.1.18. When $(\mathsf{C}, \otimes) = (\mathsf{Set}, \times)$, the equalizer formula (3.5.7) shows that

$$\mathsf{Map}(X,Y) = \mathsf{sSet}(X,Y)$$

is the set of natural transformations, that is, simplicial set morphisms, from X to Y. See Example 3.8.14 for a discussion of the case $(\mathsf{C}, \otimes) = (\mathsf{Cat}, \times)$, where $\mathsf{Map}(X,Y)$ is the category whose objects are 2-natural transformations and whose morphisms are modifications. ◇

Proposition 7.1.19. *The category of simplicial sets is symmetric monoidal closed with product given by the Cartesian product*

$$(X \times Y)_n = X_n \times Y_n$$

and internal hom given by

$$\mathsf{Hom}(X,Y)_n = \mathsf{Map}(X \times \Delta^n, Y). \tag{7.1.20}$$

Proof. The symmetric monoidal structure is that of Proposition 7.1.17 (3). Writing $[-,-]$ for $\mathsf{Set}(-,-)$ and recalling

$$\Delta^n_m = \Delta(\underline{m}, \underline{n}) = \Delta^{\mathsf{op}}(\underline{n}, \underline{m})$$

from Definition 7.1.9, the closed monoidal adjunction is given as follows:

$$\begin{aligned}
\mathsf{Map}(X, \mathsf{Hom}(Y,Z)) &= \int_{\underline{n}\in\Delta^{\mathsf{op}}} \big[X_n, \mathsf{Hom}(Y,Z)_n\big] = \int_{\underline{n}\in\Delta^{\mathsf{op}}} \big[X_n, \mathsf{Map}(Y \times \Delta^n, Z)\big] \\
&= \int_{\underline{n}\in\Delta^{\mathsf{op}}} \Big[X_n, \int_{\underline{m}\in\Delta^{\mathsf{op}}} [Y_m \times \Delta^n_m, Z_m]\Big] \\
&\cong \int_{\underline{m}\in\Delta^{\mathsf{op}}} \int_{\underline{n}\in\Delta^{\mathsf{op}}} \big[\Delta^n_m \times X_n, [Y_m, Z_m]\big] \\
&\cong \int_{\underline{m}\in\Delta^{\mathsf{op}}} \Big[\int^{\underline{n}\in\Delta^{\mathsf{op}}} \Delta^{\mathsf{op}}(\underline{n}, \underline{m}) \times X_n, [Y_m, Z_m]\Big] \\
(\star) \qquad &\cong \int_{\underline{m}\in\Delta^{\mathsf{op}}} \big[X_m, [Y_m, Z_m]\big] \\
&\cong \mathsf{Map}(X \times Y, Z).
\end{aligned}$$

In the above computation, the isomorphism labeled ☆ is an application of the V-Yoneda Density Theorem 3.7.8 with $\mathsf{V} = \mathsf{Set}$. The other isomorphisms follow from definitions, moving co/ends in or out of hom objects (Explanation 3.5.9), commuting ends, and the closed structure for Set. □

In Chapter 4 we discussed the general theory of pointed objects, smash products, and pointed homs.

Definition 7.1.21. If C has a terminal object, $*$, then the constant simplicial C-object at $*$ is terminal in sC. Applying Definition 4.1.1 to sC, we have the category of pointed simplicial objects in C, denoted sC_*. ◇

Explanation 7.1.22. Since the basepoint of sC is constant at $*$, and morphisms of simplicial objects are defined levelwise, the category of pointed simplicial objects in C is naturally isomorphic to the category of simplicial objects in C_*. Thus the notation sC_* can be read as either $\mathsf{s}(\mathsf{C}_*)$ or $(\mathsf{sC})_*$. ◇

In the important special case $\mathsf{C} = \mathsf{Set}$, we have the following.

Definition 7.1.23 (Pointed Simplicial Sets)**.** The category of *pointed simplicial sets*, sSet_*, is the category of functors

$$\Delta^{\mathsf{op}} \longrightarrow \mathsf{Set}_*$$

and pointed natural transformations between them. It is symmetric monoidal closed with smash product $X \wedge Y$ given levelwise

$$(X \wedge Y)_n = X_n \wedge Y_n$$

for $X, Y \in \mathsf{sSet}_*$ and with internal hom object defined levelwise by

$$\mathsf{Hom}_{\mathsf{sSet}_*}(X, Y)_n = \mathsf{sSet}_*(X \wedge \Delta^n_+, Y).$$

The closed monoidal adjunction is given by a computation similar to that of Proposition 7.1.19 but with $(\mathsf{V}, \otimes) = (\mathsf{Set}_*, \wedge)$. ◇

Explanation 7.1.24. In Definitions 4.1.6 and 4.2.1 we give general constructions of smash product and pointed hom in a complete and cocomplete symmetric monoidal category $(\mathsf{C}, \otimes, \mathbb{1}, T)$ with terminal object T. Using functoriality of pushouts and pullbacks, and commuting ends with pullbacks, the smash product and internal hom defined in Definition 7.1.23 are equivalent descriptions of those given by Definitions 4.1.6 and 4.2.1 for $(\mathsf{sSet}, \times, *, *)$. ◇

7.2. Simplicial Homotopy and Nerve

In this section we define simplicial homotopy and the nerve functor from small categories to simplicial sets.

Definition 7.2.1 (Simplicial Homotopy)**.** Suppose

$$f, g\colon X \longrightarrow Y$$

are morphisms of simplicial sets X and Y. A *simplicial homotopy* from f to g is a morphism of simplicial sets, H, such that the following diagram commutes.

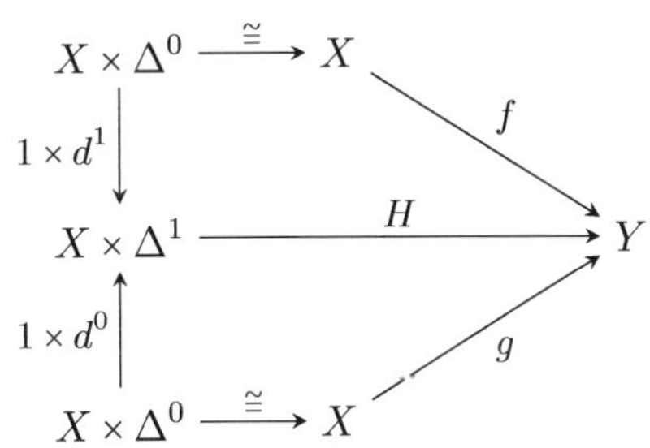

In the above diagram, the unlabeled isomorphisms are the projection isomorphism

$$X \times \Delta^0 \cong X.$$

This finishes the definition of a simplicial homotopy. We say that a morphism

$$f: X \longrightarrow Y$$

is a *simplicial homotopy equivalence* if there is a morphism

$$f': Y \longrightarrow X$$

together with simplicial homotopies from 1_X to $f'f$ and from ff' to 1_Y. ◇

Explanation 7.2.2. We note that simplicial homotopy does not provide an equivalence relation without additional fibrancy conditions on X. We refer the reader to [**GJ09**, Section 1.6] for further development of the homotopy theory of simplicial sets. ◇

Definition 7.2.3. Suppose that C is a small category. For each totally ordered set $\underline{p}$, regarded as a category, we let $N_p\mathsf{C}$ denote the set of functors $\underline{p} \longrightarrow \mathsf{C}$. This is the set of objects of the functor category $\mathsf{C}^{\underline{p}}$. Then

$$N_{(-)}\mathsf{C} = \mathsf{Ob}\left(\mathsf{C}^{(-)}\right): \Delta^{\mathsf{op}} \longrightarrow \mathsf{Set}$$

is a simplicial set called the *nerve* of C. Thus the nerve defines a functor

$$N: \mathsf{Cat} \longrightarrow \mathsf{sSet}.$$

The nerve has a left adjoint

$$h: \mathsf{sSet} \longrightarrow \mathsf{Cat}$$

that takes a simplicial set X to a category hX defined as follows. The objects of hX are the elements of X_0. The morphisms of hX are generated by the 1-simplices $\sigma \in X_1$, where each σ is regarded as a morphism

$$\sigma: d_1\sigma \longrightarrow d_0\sigma.$$

These morphisms are subject to the relation

$$(d_0\sigma) \circ (d_2\sigma) \sim d_1\sigma$$

for all 2-simplices σ. ◇

Explanation 7.2.4. The set of 0-simplices $N_0\mathsf{C}$ is given by the set of objects of C. For a 1-simplex $f: u \longrightarrow v$ in $N_1\mathsf{C}$,

$$d_0(f) = v \quad \text{and} \quad d_1(f) = u.$$

For $p > 1$, the set of p-simplices $N_p\mathsf{C}$ is given by p-tuples $(f_1, \ldots, f_p)$ of composable morphisms

$$\bullet \xrightarrow{f_1} \bullet \xrightarrow{f_2} \bullet \cdots \bullet \xrightarrow{f_p} \bullet.$$

The faces d_i are given by composing f_i with f_{i+1} for $0 < i < p$. For $i = 0$ or $i = p$, respectively, d_i is given by deleting f_1 or f_p, respectively. The morphisms f_i can be thought of as the long spine of the simplex $(f_1, \ldots, f_p)$, and their composites form its other 1-dimensional faces. The degeneracy s_i is given by inserting an identity morphism at position i. The unity and associativity axioms for composition (Definition I.1.1.1) correspond to the simplicial identities $d_i d_{i+1} = d_i d_i$ and $d_j s_j = 1 = d_{j+1} s_j$. The other simplicial identities listed in Explanation 7.1.6 are a matter of reindexing. ◇

Proposition 7.2.5. *Suppose* C *and* D *are small categories.*

(1) *A natural transformation between functors*

$$F, G \colon \mathsf{C} \longrightarrow \mathsf{D}$$

induces a simplicial homotopy from NF *to* NG*.*

(2) *An adjunction*

$$\mathsf{C} \underset{R}{\overset{L}{\rightleftarrows}} \mathsf{D}$$

with $L \dashv R$ *induces a simplicial homotopy equivalence on nerves:*

$$NL \colon N\mathsf{C} \longrightarrow N\mathsf{D}.$$

Proof. The second assertion follows from the first. To prove the first assertion, let $\mathbb{I}$ denote a category with two objects a single non-identity morphism. Let 0, respectively 1, denote the source, respectively target, of the nonidentity morphism. Let

$$i_0, i_1 \colon \mathbb{1} \longrightarrow \mathbb{I}$$

denote the two functors from the terminal category to $\mathbb{I}$, sending the unique object to 0 and 1, respectively. Then a natural transformation

$$\alpha \colon F \longrightarrow G$$

provides a functor, also denoted α, such that the following diagram commutes in Cat, where the unlabeled isomorphisms are given by the projection away from $\mathbb{1}$.

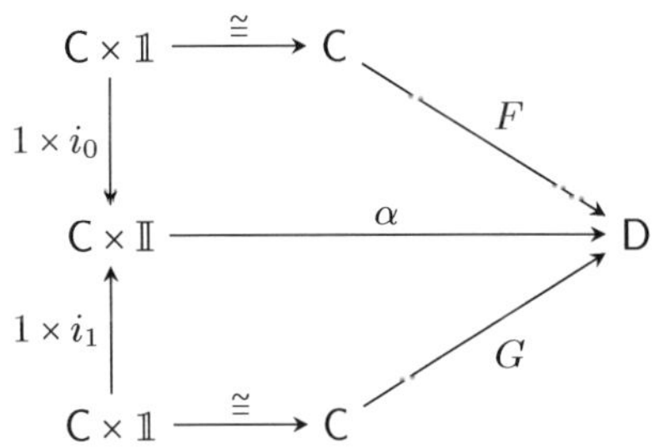

In the above diagram, the functor α sends a pair of morphisms

$$(a \xrightarrow{f} b\,,\, 0 \longrightarrow 1) \quad \text{in} \quad \mathsf{C} \times \mathbb{I}$$

to either of the two composites in the following naturality diagram for α.

$$\begin{array}{ccc} Fa & \xrightarrow{Ff} & Fb \\ {\scriptstyle \alpha_a}\downarrow & & \downarrow{\scriptstyle \alpha_b} \\ Ga & \xrightarrow{Gf} & Gb \end{array}$$

Since the nerve functor N is a right adjoint, it commutes with the Cartesian product. Taking nerve of the diagram above therefore provides a simplicial homotopy from NF to NG. $\square$

Example 7.2.6 (Bar Construction)**.** Suppose that G is a group, and let ΣG denote the 1-object category with morphisms given by G. Then the nerve $N(\Sigma G)$ is the *simplicial bar construction* on G. It has a single 0-simplex, and the set of p-simplices for $p > 0$ is given by p-tuples of elements of G. The faces d_i for $0 < i < p$ are given by multiplying group elements, while d_0 and d_p are given by deleting the first and last element of the tuple, respectively. The degeneracies are given by inserting identities in the appropriate positions. ◇

Example 7.2.7 (Nerve of a Simplicial Category)**.** Suppose A is a simplicial category. Composing with the nerve, we have a simplicial object in sSet:

$$\Delta^{\mathsf{op}} \xrightarrow{A} \mathsf{Cat} \xrightarrow{N} \mathsf{sSet}.$$

This yields a bisimplicial set, and restriction to the diagonal of $\Delta^{\mathsf{op}} \times \Delta^{\mathsf{op}}$ yields a simplicial set

$$\overline{N}A \colon \Delta^{\mathsf{op}} \longrightarrow \mathsf{Set}$$

given by $(\overline{N}A)_p = N_p A_p$ for $\underline{p} \in \Delta^{\mathsf{op}}$. ◇

Example 7.2.8 (Classifying Spaces)**.** For a small category C, the geometric realization of the nerve $N\mathsf{C}$ is called the *classifying space* of C and denoted $B\mathsf{C} = |N\mathsf{C}|$. If $\mathsf{C} = \Sigma G$ is a 1-object groupoid, then $B\Sigma G$ is an Eilenberg-Mac Lane space of type $(G, 1)$. It is usually denoted BG and called the classifying space of G (instead of ΣG). ◇

Proposition 7.2.9. *The nerve is a strong symmetric monoidal functor*

$$N \colon (\mathsf{Cat}, \times, \mathbb{1}) \longrightarrow (\mathsf{sSet}, \times, *).$$

Proof. Recall from Definition 7.2.3 that N is a right adjoint. Therefore, N commutes with small limits and, in particular, preserves terminal objects and commutes with Cartesian products. □

7.3. Symmetric Sequences of Pointed Simplicial Sets

This section describes symmetric sequences in sSet_*. These are used to describe symmetric spectra in Section 7.4. Recall from Definition I.2.4.1 the finite ordinal category Σ whose objects are natural numbers and morphisms are permutations.

Definition 7.3.1. A *symmetric sequence* $X = \{X_k\}_{k \geq 0}$ is a functor

$$X \colon \Sigma \longrightarrow \mathsf{sSet}_*.$$

Thus each $X_k = Xk$ is an object of sSet_* and is equipped with an action of the symmetric group

$$\Sigma_k \longrightarrow \mathsf{sSet}_*(X_k, X_k)_0.$$

The category of symmetric sequences is the category of functors $\Sigma \longrightarrow \mathsf{sSet}_*$ and natural transformations between them, denoted sSet_*^{Σ}. ◇

Explanation 7.3.2. A symmetric sequence consists of pointed simplicial sets X_k for $k \geq 0$ together with a basepoint-preserving action of the symmetric group Σ_k on each X_k. A morphism $X \longrightarrow Y$ of symmetric sequences consists of Σ_k-equivariant and basepoint-preserving morphisms $X_k \longrightarrow Y_k$ for each $k \geq 0$. ◇

We will use the additive structure of Σ below along with the unitary enrichment Definition 3.8.9 over sSet_*. Recall the Day convolution and hom diagram from Definition 3.7.3. The following is a special case of the convolution product (3.7.4) for $\mathsf{V} = \mathsf{sSet}_*$ and $\mathcal{D} = \Sigma$.

Definition 7.3.3. Suppose that X and Y are two symmetric sequences. The *Day convolution*, $X \,\Box\, Y$, is a symmetric sequence given by

$$(X \,\Box\, Y)_k = \bigvee_{p+q=k} \Sigma_k \times_{\Sigma_p \times \Sigma_q} X_p \wedge Y_q.$$

The fibre product is taken with respect to the inclusion

$$i_{p,q}\colon \Sigma_p \times \Sigma_q \hookrightarrow \Sigma_{(p+q)}$$

where Σ_p permutes the first p elements and Σ_q permutes the last q elements.

The unit object I is the symmetric sequence given by $I_0 = S^0$ and $I_k = *$ for $k > 0$. The unit isomorphisms

$$\lambda\colon I \,\Box\, X \longrightarrow X \quad \text{and} \quad \rho\colon X \,\Box\, I \longrightarrow X$$

are induced by the isomorphisms

$$\Sigma_{(p+q)} \times_{\Sigma_p \times \Sigma_q} (* \wedge X_q) \cong * \quad \text{and} \quad \Sigma_{(p+q)} \times_{\Sigma_p \times \Sigma_q} (X_p \wedge *) \cong *$$

along with the unit isomorphisms for S^0 in sSet_*. The associativity isomorphism

$$\alpha\colon (X \,\Box\, Y) \,\Box\, Z \xrightarrow{\cong} X \,\Box\, (Y \,\Box\, Z)$$

is given on summands by the composite isomorphisms shown below, where the first and third are the universal isomorphisms for iterated pullbacks and the second is induced by the associativity in sSet_*.

$$\begin{array}{c}
\Sigma_{(p+q+r)} \times_{\Sigma_{(p+q)} \times \Sigma_r} ((\Sigma_{p+q} \times_{\Sigma_p \times \Sigma_q} X_p \wedge Y_q) \wedge Z_r) \\
\downarrow \cong \\
\Sigma_{(p+q+r)} \times_{\Sigma_p \times \Sigma_q \times \Sigma_r} ((X_p \wedge Y_q) \wedge Z_r) \\
\downarrow \cong \\
\Sigma_{(p+q+r)} \times_{\Sigma_p \times \Sigma_q \times \Sigma_r} (X_p \wedge (Y_q \wedge Z_r)) \\
\downarrow \cong \\
\Sigma_{(p+q+r)} \times_{\Sigma_p \times \Sigma_{(q+r)}} (X_p \wedge (\Sigma_{q+r} \times_{\Sigma_q \times \Sigma_r} Y_q \wedge Z_r))
\end{array}$$

There is a symmetry isomorphism

$$\xi\colon X \,\Box\, Y \longrightarrow Y \,\Box\, X$$

given on summands by the isomorphism

$$\xi_{p,q}\colon \Sigma_{(p+q)} \times_{\Sigma_p \times \Sigma_q} X_p \wedge Y_q \xrightarrow{\cong} \Sigma_{(q+p)} \times_{\Sigma_q \times \Sigma_p} Y_q \wedge X_p$$

induced by the symmetry on sSet_* and the block permutation $\chi_{p,q} \in \Sigma_{(p+q)}$ that swaps the first p and last q elements. Naturality of the symmetry, associativity, and unit isomorphisms follows by naturality of the canonical morphisms together with naturality of the corresponding data in sSet_*. ◇

Definition 7.3.4. For each $n \in \Sigma$, let $\Sigma[n]$ denote the symmetric sequence given by the corepresented functor

$$\Sigma[n] = \Sigma(\underline{n}, -)\colon \Sigma \longrightarrow \mathsf{sSet}_*,$$

with $\Sigma(\underline{n}, \underline{m})$ being the one point simplicial set for $m \neq n$ and the constant simplicial set Σ_n with basepoint the identity permutation for $m = n$. ◇

Definition 7.3.5. Suppose X and Y are symmetric sequences and A is a pointed simplicial set.

(1) The *simplicial tensor* is the symmetric sequence $X \wedge A$ defined for each $k \geq 0$ as

$$(X \wedge A)_k = X_k \wedge A$$

with the diagonal action of Σ_k acting trivially on A. As a functor out of Σ, it is the composite

$$\Sigma \xrightarrow{X} \mathsf{sSet}_* \xrightarrow{(-)\wedge A} \mathsf{sSet}_*.$$

(2) The *simplicial cotensor* is the symmetric sequence X^A defined for each $k \geq 0$ as

$$(X^A)_k = (X_k)^A$$

with the pointwise action of Σ_k. As a functor out of Σ, it is the composite

$$\Sigma \xrightarrow{X} \mathsf{sSet}_* \xrightarrow{(-)^A} \mathsf{sSet}_*.$$

(3) The *symmetric mapping object* is the pointed simplicial set

$$\mathsf{Map}_\Sigma(X,Y) = \mathsf{sSet}_*^\Sigma(X \wedge \Delta_+^?, Y),$$

where $\Delta_+^?$ is given by adjoining a disjoint basepoint to $\Delta^?$ (see Definition 7.1.11).

(4) The *symmetric hom object* is the symmetric sequence

$$\mathsf{Hom}_\Sigma(X,Y) = \mathsf{Map}_\Sigma(X \,\square\, \Sigma[-]_+, Y)$$

where $\Sigma[-]_+$ is given by adjoining a disjoint basepoint to $\Sigma[-]$.

For the closed monoidal adjunction, natural isomorphisms

$$\mathsf{Map}_\Sigma(X, \mathsf{Hom}_\Sigma(Y,Z)) \cong \mathsf{Map}_\Sigma(X \,\square\, Y, Z)$$

are given similarly to those of Proposition 7.1.19, using the V-Yoneda Density Theorem 3.7.8 with $\mathsf{V} = \mathsf{sSet}_*$. ◇

Definition 7.3.6. For each $n \geq 0$

$$i_n \colon \mathsf{sSet}_* \longrightarrow \mathsf{sSet}_*^\Sigma$$

be the functor that sends a pointed simplicial set X to the symmetric sequence whose nth term is $X \wedge (\Sigma_n)_+$ and whose other terms are the terminal simplicial set $*$. A morphism of pointed simplicial sets $f \colon A \longrightarrow B$ induces a morphism of symmetric sequences $i_n A \longrightarrow i_n B$ given at level n by $f \wedge \mathrm{id}$.

Let

$$\mathsf{ev}_n \colon \mathsf{sSet}_*^\Sigma \longrightarrow \mathsf{sSet}_*$$

be the functor that sends a symmetric sequence X to the pointed simplicial set X_n and that sends a morphism of symmetric sequences to its component at n. For each $n \geq 0$ these are an adjoint pair of functors

$$\mathsf{sSet}_* \underset{\mathsf{ev}_n}{\overset{i_n}{\rightleftarrows}} \mathsf{sSet}_*^\Sigma. \quad (\perp)$$

It follows from the definition of the Day convolution $\square$ that i_0 is strong symmetric monoidal and ev_0 is symmetric monoidal. ◇

The following result is a special case of Theorem 3.7.22 combined with Theorem 3.9.8. Additional discussion of this application appears in [**HSS00, Hov01, Ric20**].

Theorem 7.3.7. *Equipped with the Day convolution product and symmetric hom, the category of symmetric sequences,* sSet_*^Σ*, is a symmetric monoidal closed category with*

- *monoidal product given by Day convolution of Definition 7.3.3 and*
- *internal hom given by the symmetric hom objects of Definition 7.3.5.*

Moreover, the adjunction (i_0, ev_0) *makes* sSet_*^Σ *enriched, tensored, and cotensored over* sSet_**. The* sSet_* *enrichment is given by the symmetric mapping objects of Definition 7.3.5.*

Definition 7.3.8. For $X \in \mathsf{sSet}_*$ and $k > 0$ we let $X^{\wedge k}$ denote the k-fold smash product defined iteratively by $X^{\wedge k+1} = X \wedge X^{\wedge k}$. We call this the *right normalized* convention for smash powers. The empty smash product $X^{\wedge 0}$ is the unit, S^0.

The symmetric group action of Σ_p on $X^{\wedge p}$ is given by the unique coherence isomorphism in sSet_* (from Theorem 1.1.41) that permutes the p copies of X. We call this the *action by permuting factors.* ◇

Recall from Definition I.1.2.8 the concept of a monoid in a monoidal category.

Proposition 7.3.9. *Suppose* $X \in \mathsf{sSet}_*$ *and let* $\mathsf{Sym}(X)$ *be the symmetric sequence given by* $\mathsf{Sym}(X)_k = X^{\wedge k}$ *with* Σ_k *acting by permuting factors. Then* $\mathsf{Sym}(X)$ *is a commutative monoid in the category of symmetric sequences.*

Proof. The multiplication morphism

$$\mu\colon \mathsf{Sym}(X) \,\square\, \mathsf{Sym}(X) \longrightarrow \mathsf{Sym}(X)$$

is defined on summands

$$\mu^\Sigma_{p,q}\colon \Sigma_{(p+q)} \times_{\Sigma_p \times \Sigma_q} X^{\wedge p} \wedge X^{\wedge q} \longrightarrow X^{\wedge(p+q)}$$

by the associativity in sSet_* and the action of Σ_{p+q} permuting factors. The monoid associativity and unity axioms follow from the Symmetric Coherence Theorem Theorem 1.1.41 for sSet_*.

To see that $\mathsf{Sym}(X)$ is a commutative monoid, we have the following commuting diagram for each summand of $(\mathsf{Sym}(X) \,\square\, \mathsf{Sym}(X))_{(p+q)}$.

$$\begin{array}{ccc}
\Sigma_{(p+q)} \times_{\Sigma_p \times \Sigma_q} X^{\wedge p} \wedge X^{\wedge q} & \xrightarrow{\;\xi_{p,q}\;} & \Sigma_{(q+p)} \times_{\Sigma_q \times \Sigma_p} X^{\wedge q} \wedge X^{\wedge p} \\
& {\scriptstyle \mu^\Sigma_{p,q}} \searrow \quad \swarrow {\scriptstyle \mu^\Sigma_{q,p}} & \\
& X^{\wedge(p+q)} &
\end{array}$$

□

7.4. Symmetric Spectra

Our model for the stable homotopy category will be the category of symmetric spectra. It has a symmetric monoidal smash product induced by the convolution product on symmetric sequences.

We begin with the following important special case of Definition 7.3.8.

Definition 7.4.1. For $p \geq 0$, the *simplicial p-sphere* is the p-fold smash product

$$S^p = (S^1)^{\wedge p},$$

using the right normalized convention of Definition 7.3.8. The *symmetric sphere* is the symmetric sequence $S = \mathsf{Sym}(S^1)$. ◇

Definition 7.4.2. Suppose $(\mathsf{C}, \otimes, \mathbb{1}, \alpha, \lambda, \rho)$ is a monoidal category, and (A, μ, η) is a monoid in C. A *left A-module*, also called a left module over A, is a pair

$$(M, \theta)$$

consisting of

- an object $M \in \mathsf{C}$ and
- a *structure morphism* $\theta \colon A \otimes M \longrightarrow M$

such that the following associativity and unity diagrams commute.

$$\begin{array}{ccc} (A \otimes A) \otimes M & \xrightarrow{\alpha} & A \otimes (A \otimes M) \\ & & \big\downarrow 1\otimes\theta \\ \big\downarrow \mu\otimes 1 & & A \otimes M \\ & & \big\downarrow \theta \\ A \otimes M & \xrightarrow{\theta} & M \end{array} \qquad \begin{array}{ccc} \mathbb{1} \otimes M & \xrightarrow{\lambda} & M \\ {\scriptstyle \eta\otimes 1}\big\downarrow & & \big\| \\ A \otimes M & \xrightarrow{\theta} & M \end{array}$$

A *morphism*

$$f \colon (M, \theta^M) \longrightarrow (N, \theta^N)$$

of left A-modules is a morphism $f \colon M \longrightarrow N \in \mathsf{C}$ such that the diagram

$$\begin{array}{ccc} A \otimes M & \xrightarrow{\theta^M} & M \\ {\scriptstyle 1\otimes f}\big\downarrow & & \big\downarrow f \\ A \otimes N & \xrightarrow{\theta^N} & N \end{array}$$

commutes. The identity morphism of a left A-module (M, θ) is $1_M \in \mathsf{C}$, and composition is defined in C. The above data define the category of left A-modules. ◇

Example 7.4.3. In the context of Definition 7.4.2, (A, μ) is a left A-module. ◇

Explanation 7.4.4 (Right Modules)**.** In the context of Definition 7.4.2, there is a similar definition of *right A-module* N, with structure morphism

$$N \otimes A \longrightarrow N.$$

If $(\mathsf{C}, \otimes)$ is symmetric monoidal and A is a commutative monoid in C, then the category of right A-modules is isomorphic to the category of left A-modules via the symmetry of C. ◇

Definition 7.4.5. The category of *symmetric spectra*, SymSp, is the category consisting of left modules over the symmetric sphere and left module morphisms. ◇

Proposition 7.5.5 below shows that SymSp is complete and cocomplete. Theorem 7.6.15 below shows that SymSp is symmetric monoidal closed.

Explanation 7.4.6. We have the following explicit description of symmetric spectra. An action

$$\rho \colon S \,\Box\, X \longrightarrow X$$

has summands for each $k = p + q \geq 0$

$$\rho^{\Sigma}_{p,q} \colon \Sigma_{(p+q)} \times_{\Sigma_p \times \Sigma_q} S^p \wedge X_q \longrightarrow X_{p+q}.$$

The action ρ is determined by morphisms

$$\rho_{p,q}\colon S^p \wedge X_q \longrightarrow X_{p+q}$$

for $p, q \geq 0$ with the following properties.

Unity: The morphism

$$\rho_{0,q}\colon S^0 \wedge X_q \longrightarrow X_q \tag{7.4.7}$$

is the unit isomorphism for the smash product.

Associativity: The following diagram commutes for each r, p, and q, where the unlabeled vertical morphism is the smash product of the associativity isomorphisms $S^r \wedge S^p \cong S^{r+p}$ with the identity on X_q.

$$\begin{array}{ccc} (S^r \wedge S^p) \wedge X_q & \xrightarrow{a} & S^r \wedge (S^p \wedge X_q) \\ \downarrow & & \downarrow 1 \wedge \rho_{p,q} \\ S^{r+p} \wedge X_q & & S^r \wedge X_{p+q} \\ & \searrow^{\rho_{r+p,q}} \quad X_{r+p+q} \quad {}^{\rho_{r,p+q}}\swarrow & \end{array} \tag{7.4.8}$$

Equivariance: Each $\rho_{p,q}$ is equivariant with respect to the $(\Sigma_p \times \Sigma_q)$-action on the source and the inclusion

$$i_{p,q}\colon \Sigma_p \times \Sigma_q \hookrightarrow \Sigma_{(p+q)}$$

where Σ_p permutes the first p elements and Σ_q permutes the last q elements.

An S-module morphism $X \longrightarrow X'$ consists of a sequence of morphisms

$$f_k\colon X_k \longrightarrow X'_k$$

that are equivariant with respect to the action of Σ_k on X_k and X'_k and that commute with the structure morphisms $\rho_{p,q}$. ◇

Basic Examples of Symmetric Spectra.

Example 7.4.9 (The Sphere Spectrum)**.** By Example 7.4.3, the symmetric sphere S with its monoid multiplication is a symmetric spectrum. ◇

Example 7.4.10 (Suspension Spectra)**.** Suppose K is a pointed simplicial set. The *suspension spectrum* $\Sigma^\infty K$ is the symmetric spectrum with

$$(\Sigma^\infty K)_n = S^n \wedge K \in \mathsf{sSet}_* \quad \text{for} \quad n \geq 0.$$

The Σ_n-action on $S^n \wedge K$ is given by permuting the n smash factors in

$$S^n = (S^1)^{\wedge n}.$$

For $p, q \geq 0$, the left S-module structure morphism $\rho_{p,q}$ is the composite

$$\begin{array}{ccc} S^p \wedge (S^q \wedge K) & \xrightarrow{\rho_{p,q}} & S^{p+q} \wedge K \\ & \cong\searrow \qquad \nearrow\cong & \\ & (S^p \wedge S^q) \wedge K & \end{array}$$

of associativity isomorphisms if $p, q > 0$. It is

- the left unit isomorphism $\lambda_{S^q \wedge K}$ if $p = 0$ and
- $1_{S^p} \wedge \lambda_K$ if $q = 0$.

The fact that $\rho_{p,q}$ is well defined and the unity, associativity, and equivariance conditions in Explanation 7.4.6 follow from the Symmetric Coherence Theorem 1.1.41 for sSet_*. ◇

Example 7.4.11 (Eilenberg-Mac Lane Spectra). Suppose R is a ring. The *Eilenberg-Mac Lane spectrum* HR is the symmetric spectrum with

$$(HR)_n = R \otimes S^n \in \mathsf{sSet}_* \quad \text{for} \quad n \geq 0.$$

For $k \geq 0$, its set of k-simplices is the free left R-module

$$(R \otimes S^n)_k = \bigoplus_{(\underline{k}^n)^\flat} R$$

with

$$\underline{k}^n \cong \underline{k}^{\wedge n} = (S^n)_k$$

the set of k-simplices in the n-sphere S^n. The basepoint of $(R \otimes S^n)_k$ is the element 0. The simplicial structure morphisms in $(HR)_n$ are induced by those in S^n.

To describe the symmetric sequence structure on HR, we represent a typical direct sum generator in $(HR)_n$ in the form

$$r_{i_1,\ldots,i_n} \qquad \text{with } r \in R \text{ and each } i_j \in \underline{k}^\flat.$$

The Σ_n-action on $(HR)_n$ is given on these generators by

$$\sigma \cdot r_{i_1,\ldots,i_n} = r_{i_{\sigma^{-1}(1)},\ldots,i_{\sigma^{-1}(n)}} \quad \text{for} \quad \sigma \in \Sigma_n.$$

The left S-module structure morphisms on HR,

$$S^p \wedge (R \otimes S^q) \xrightarrow{\rho_{p,q}} R \otimes S^{p+q} \quad \text{for} \quad p, q \geq 0,$$

are defined on k-simplices by

$$\begin{array}{ccc} \underline{k}^{\wedge p} \wedge \Big(\bigoplus_{(\underline{k}^q)^\flat} R \Big) & \xrightarrow{(\rho_{p,q})_k} & \bigoplus_{(\underline{k}^{p+q})^\flat} R \\ \big((h_1,\ldots,h_p), r_{i_1,\ldots,i_q}\big) & \longmapsto & r_{h_1,\ldots,h_p,i_1,\ldots,i_q} \end{array}$$

for $k \geq 0$ and $h_1,\ldots,h_p,i_1,\ldots,i_q \in \underline{k}^\flat$. Moreover, since $(\rho_{p,q})_k$ preserves the basepoint, we have

$$(\rho_{p,q})_k(0,r) = 0.$$

The unity, associativity, and equivariance conditions in Explanation 7.4.6 hold because they do for direct sum generators. ◇

Chapters 11, 12, and 13 have further examples of symmetric spectra with extra structure arising from categorical data. We will revisit Examples 7.4.9 through 7.4.11 in

- Examples 11.3.13 through 11.3.15 in the context of strict ring symmetric spectra and
- Examples 11.6.7, 11.6.9, and 11.6.10 in the context of E_∞-symmetric spectra.

7.5. Limits and Colimits of Symmetric Spectra

The category of symmetric spectra is defined as the category of left S-modules, in the sense of Definition 7.4.2, over the symmetric sphere S. Recall the concepts of a monad and an algebra over a monad in Definitions 5.4.1 and 5.4.2. Using Proposition 7.5.1 below, we can also regard SymSp as the category of algebras over the monad $S \mathbin{\square} -$ in the category sSet_*^{Σ} of symmetric sequences.

Modules as Monadic Algebras.

Proposition 7.5.1. *Suppose* $(\mathsf{C}, \otimes, \mathbb{1}, \alpha, \lambda, \rho)$ *is a monoidal category, and* (A, μ, η) *is a monoid in* C.

(1) *There is a monad*

$$(A \otimes -, \pi, \nu)$$

on C *with multiplication* π *and unit* ν *defined as the following composites for each object* $X \in \mathsf{C}$.

$$\begin{array}{ccc} A \otimes (A \otimes X) & \xrightarrow{\pi_X} & A \otimes X \\ & {\scriptstyle \alpha^{-1}} \searrow \quad \nearrow {\scriptstyle \mu \otimes 1} & \\ & (A \otimes A) \otimes X & \end{array} \qquad \begin{array}{ccc} X & \xrightarrow{\nu_X} & A \otimes X \\ & {\scriptstyle \lambda^{-1}} \searrow \quad \nearrow {\scriptstyle \eta \otimes 1} & \\ & \mathbb{1} \otimes X & \end{array}$$

(2) *There is a canonical isomorphism between*
- *the category of left* A*-modules in Definition 7.4.2 and*
- *the category of algebras over the monad* $(A \otimes -, \pi, \nu)$.

Proof. For assertion (1), we check the commutativity of the associativity and unity diagrams in Definition 5.4.1 for $(A \otimes -, \pi, \nu)$. Since each diagram asserts the equality of parallel natural transformations, it suffices to check each diagram at an object $X \in \mathsf{C}$. The monad associativity diagram for $(A \otimes -, \pi, \nu)$ at an object X is the outer diagram in C below, where all the tensor symbols are omitted to save space.

$$\begin{array}{ccccc} A(A(AX)) & \xrightarrow{1\alpha^{-1}} & A((AA)X) & \xrightarrow{1(\mu 1)} & A(AX) \\ & & \downarrow{\scriptstyle \alpha^{-1}} & & \downarrow{\scriptstyle \alpha^{-1}} \\ \downarrow{\scriptstyle \alpha^{-1}} & & (A(AA))X & \xrightarrow{(1\mu)1} & (AA)X \\ & & \downarrow{\scriptstyle \alpha^{-1}1} & & \\ (AA)(AX) & \xrightarrow{\alpha^{-1}} & ((AA)A)X & & \downarrow{\scriptstyle \mu 1} \\ {\scriptstyle \mu 1}\downarrow & & \downarrow{\scriptstyle (\mu 1)1} & & \\ A(AX) & \xrightarrow{\alpha^{-1}} & (AA)X & \xrightarrow{\mu 1} & AX \end{array}$$

- The upper left rectangle is commutative by the pentagon axiom (1.1.3) in C.
- The lower right rectangle is commutative by the associativity axiom of the monoid (A, μ, η).
- The lower left and upper right rectangles are commutative by the naturality of α.

The monad unity diagram for $(A \otimes -, \pi, \nu)$ at an object X is the outer diagram in C below.

$$
\begin{array}{ccccccccc}
A(AX) & \xleftarrow{\eta 1} & \mathbb{1}(AX) & \xleftarrow{\lambda^{-1}} & AX & \xrightarrow{1\lambda^{-1}} & A(\mathbb{1}X) & \xrightarrow{1(\eta 1)} & A(AX) \\
& & \downarrow{\scriptstyle \alpha^{-1}} & & \| & & \downarrow{\scriptstyle \alpha^{-1}} & & \\
\downarrow{\scriptstyle \alpha^{-1}} & \overset{(\eta 1)1}{\swarrow} & (\mathbb{1}A)X & \overset{\lambda 1}{\searrow} & \| & \overset{\rho 1}{\swarrow} & (A\mathbb{1})X & \overset{(1\eta)1}{\searrow} & \downarrow{\scriptstyle \alpha^{-1}} \\
(AA)X & & \xrightarrow{\mu 1} & & AX & & \xleftarrow{\mu 1} & & (AA)X
\end{array}
$$

In the left half of the diagram above,

- the left and right trapezoids are commutative by, respectively, the naturality of α and the left unity property (1.1.5) in C, and
- the bottom triangle is commutative by the left unity axiom in the monoid (A, μ, η).

Similarly, the right half of the previous diagram is commutative by the naturality of α, the unity axiom (1.1.2) in C, and the right unity axiom in the monoid (A, μ, η). This proves assertion (1).

For assertion (2), the associativity and unity diagrams for a left A-module in Definition 7.4.2 are precisely those for an $(A \otimes -)$-algebra in (5.4.3), after inverting the natural isomorphisms α and λ. Morphisms, identity morphisms, and composition of left A-modules also correspond to those of $(A \otimes -)$-algebras. $\square$

Proposition 7.5.1 applies to the symmetric sphere in Definition 7.4.1 and yields the following description of the category SymSp of symmetric spectra.

Corollary 7.5.2. *There is a canonical isomorphism of categories*

$$\mathsf{SymSp} \cong \mathsf{Alg}(S \,\square\, -),$$

with S the symmetric sphere in the symmetric monoidal category sSet_*^{Σ}.

Limits and Colimits of Modules. The next observation is a variant of Theorem 5.4.18 that applies to left modules, with different assumptions and a much simpler proof. We will use it with the symmetric sphere.

Proposition 7.5.3. *Suppose (A, μ, η) is a monoid in a complete and cocomplete monoidal category $(\mathsf{C}, \otimes, \mathbb{1}, \alpha, \lambda, \rho)$ such that*

$$A \otimes - \colon \mathsf{C} \longrightarrow \mathsf{C}$$

preserves small colimits. Then for the monad $A\otimes-$ in Proposition 7.5.1, its category of algebras, $\mathsf{Alg}(A \otimes -)$, has all small limits and colimits, which are, furthermore, preserved by the forgetful functor

$$U \colon \mathsf{Alg}(A \otimes -) \longrightarrow \mathsf{C}.$$

Proof. For a functor F as in

$$\mathsf{D} \xrightarrow{F} \mathsf{Alg}(A \otimes -) \xrightarrow{U} \mathsf{C}$$

with D a small category, the composite $UF \colon \mathsf{D} \longrightarrow \mathsf{C}$ has a colimit in C. The natural morphism

$$\operatorname{colim}(A \otimes UF) \xrightarrow[\cong]{\omega} A \otimes (\operatorname{colim} UF)$$

is an isomorphism by assumption. We equip the object $\mathsf{colim}\, UF \in \mathsf{C}$ with the structure morphism ϕ below.

$$\begin{array}{ccc} A \otimes (\mathsf{colim}\, UF) & \xrightarrow{\phi} & \mathsf{colim}\, UF \\ & \cong \searrow \quad \nearrow \mathsf{colim}_{d \in \mathsf{D}}\, \theta_d & \\ & \mathsf{colim}\,(A \otimes UF) & \end{array}$$

For each $d \in \mathsf{D}$,

$$A \otimes (UFd) \xrightarrow{\theta_d} UFd$$

is the $(A \otimes -)$-algebra structure morphism of Fd. By

- the universal property of colimits and
- the assumption that $A \otimes -$ preserves small colimits,

the associativity and unity axioms (5.4.3) for the pair

$$\big(\mathsf{colim}\, UF, \phi\big)$$

to be an $(A \otimes -)$-algebra reduce to those for each Fd. The fact that it is a colimit of F follows from the definition of the isomorphism ω.

Similarly, a limit of F in $\mathsf{Alg}(A \otimes -)$ consists of the object $\mathsf{lim}\, UF \in \mathsf{C}$ and structure morphism φ defined by the commutative diagrams

$$\begin{array}{ccc} A \otimes (\mathsf{lim}\, UF) & \xrightarrow{\varphi} & \mathsf{lim}\, UF \\ {\scriptstyle 1 \otimes p_d} \downarrow & & \downarrow {\scriptstyle p_d} \\ A \otimes (UFd) & \xrightarrow{\theta_d} & UFd \end{array}$$

in C for $d \in \mathsf{D}$. Here

$$\mathsf{lim}\, UF \xrightarrow{p_d} UFd$$

is a structure morphism of the limit. □

Explanation 7.5.4. In Proposition 7.5.3, the condition that $A \otimes -$ preserves small colimits is true if C is symmetric monoidal closed. For example, this is the case when

- A is the symmetric sphere S in Definition 7.4.1 and
- C is the category of symmetric sequences sSet_*^{Σ} in Definition 7.3.1. ◇

Now we apply the results above to SymSp.

Proposition 7.5.5. *The category of symmetric spectra,* SymSp, *is complete and cocomplete.*

Proof. By Proposition 7.1.17, sSet_*^{Σ}, on which the monad $S \square -$ acts, is complete, cocomplete, and symmetric monoidal closed. We finish the proof using Corollary 7.5.2, Proposition 7.5.3, and Explanation 7.5.4. □

7.6. Smash Products, Internal Hom, and (Co)tensored Structure of Symmetric Spectra

Proposition 7.5.5 shows that the category of symmetric spectra is complete and cocomplete. Therefore we have the following constructions of monoidal product and internal hom. Since S is a commutative monoid in sSet_*^{Σ}, these provide a symmetric monoidal closed structure for SymSp.

Smash Product.

Definition 7.6.1. The *smash product* of symmetric spectra X and Y is denoted

$$X \wedge Y = X \square_S Y$$

and is given by the coequalizer in sSet_*^Σ

$$(X \square S) \square Y \underset{\overline{\rho}^X \square 1}{\overset{(1 \square \rho^Y) \circ a}{\rightrightarrows}} X \square Y \longrightarrow X \square_S Y$$

where

$$\rho^Y : S \square Y \longrightarrow Y$$

is the left action of S on Y and $\overline{\rho}^X$ is the right action of S on X defined as the composite

$$\overline{\rho}^X = \rho^X \circ \xi : X \square S \longrightarrow S \square X \longrightarrow X$$

of the symmetry ξ with ρ^X, the left action of S on X. The left S-module structure of $X \square_S Y$ is given by the left action of S on X. ◇

The following observation is useful for constructing morphisms out of a smash product of symmetric spectra. We will use it in Proposition 11.3.2 to describe strict ring symmetric spectra.

Proposition 7.6.2. *Suppose (X, ρ^X), (Y, ρ^Y), and (Z, ρ^Z) are symmetric spectra. Then a morphism*

$$X \square_S Y \xrightarrow{\overline{f}} Z \in \mathsf{SymSp}$$

determines and is uniquely determined by a morphism

$$X \square Y \xrightarrow{f} Z \in \mathsf{sSet}_*^\Sigma$$

such that the diagram

$$\begin{array}{ccccc}
(X \square S) \square Y & \xrightarrow[\cong]{\xi \square 1} & (S \square X) \square Y & \xrightarrow[\cong]{a} & S \square (X \square Y) \\
{\scriptstyle a} \downarrow {\scriptstyle \cong} & & \downarrow {\scriptstyle \rho^X \square 1} & & \\
X \square (S \square Y) & & X \square Y & & \downarrow {\scriptstyle 1 \square f} \\
{\scriptstyle 1 \square \rho^Y} \downarrow & & \downarrow {\scriptstyle f} & & \\
X \square Y & \xrightarrow{f} & Z & \xleftarrow{\rho^Z} & S \square Z
\end{array} \tag{7.6.3}$$

in sSet_*^Σ commutes.

Proof. Given a morphism $\overline{f} : X \square_S Y \longrightarrow Z$ of symmetric spectra, the morphism f is defined as the composite

$$X \square Y \longrightarrow X \square_S Y \xrightarrow{\overline{f}} Z \in \mathsf{sSet}_*^\Sigma.$$

- The left half of the diagram (7.6.3) follows from the coequalizer that defines the smash product $X \square_S Y$ in Definition 7.6.1.
- The right half of (7.6.3) expresses the compatibility of $\overline{f}$ with the left S-action.

This argument can also be used in reverse. □

Explanation 7.6.4. Using Definition 7.3.3 of the Day convolution and Explanation 7.4.6, we can further unpack the morphism $f\colon X \mathbin{\square} Y \longrightarrow Z$ in Proposition 7.6.2 as the family of $(\Sigma_p \times \Sigma_q)$-equivariant morphisms

$$X_p \wedge Y_q \xrightarrow{f_{p,q}} Z_{p+q} \in \mathsf{sSet}_* \quad \text{for} \quad p, q \geq 0.$$

The right half of the diagram (7.6.3) is equivalent to the diagram

$$\begin{array}{ccc} (S^n \wedge X_p) \wedge Y_q & \xrightarrow[\cong]{a} & S^n \wedge (X_p \wedge Y_q) \\ {\scriptstyle \rho^X_{n,p} \wedge 1} \downarrow & & \\ X_{n+p} \wedge Y_q & & \downarrow {\scriptstyle 1 \wedge f_{p,q}} \\ {\scriptstyle f_{n+p,q}} \downarrow & & \\ Z_{n+p+q} & \xleftarrow{\rho^Z_{n,p+q}} & S^n \wedge Z_{p+q} \end{array} \tag{7.6.5}$$

in sSet_* for $n, p, q \geq 0$. With the abbreviation

$$\Sigma_n \times \Sigma_p = \Sigma_{n,p}, \tag{7.6.6}$$

the left half of the diagram (7.6.3) is equivalent to the diagram

$$\begin{array}{ccc} \Sigma_{(p+n+q)} \times_{\Sigma_{p+n,q}} \Big(\Sigma_{(p+n)} \times_{\Sigma_{p,n}} X_p \wedge S^n\Big) \wedge Y_q & \xrightarrow{\xi_{p,n} \wedge 1} & \Sigma_{(p+n+q)} \times_{\Sigma_{n+p,q}} \Big(\Sigma_{(n+p)} \times_{\Sigma_{n,p}} S^n \wedge X_p\Big) \wedge Y_q \\ {\scriptstyle a}\downarrow{\scriptstyle \cong} & & \downarrow {\scriptstyle \rho^X_{n,p} \wedge 1} \\ \Sigma_{(p+n+q)} \times_{\Sigma_{p,n+q}} X_p \wedge \Big(\Sigma_{(n+q)} \times_{\Sigma_{n,q}} S^n \wedge Y_q\Big) & & \Sigma_{(p+n+q)} \times_{\Sigma_{n+p,q}} X_{n+p} \wedge Y_q \\ {\scriptstyle 1 \wedge \rho^Y_{n,q}} \downarrow & & \downarrow {\scriptstyle f_{n+p,q}} \\ \Sigma_{(p+n+q)} \times_{\Sigma_{p,n+q}} X_p \wedge Y_{n+q} & \xrightarrow{f_{p,n+q}} & Z_{n+p+q} \end{array} \tag{7.6.7}$$

in sSet_* for $n, p, q \geq 0$. ◇

Internal Hom.

Definition 7.6.8. The *internal hom* for symmetric spectra X and Y is given by the equalizer in sSet_*^{Σ}

$$\mathsf{Hom}_S(X, Y) \longrightarrow \mathsf{Hom}_\Sigma(X, Y) \rightrightarrows \mathsf{Hom}_\Sigma(S \mathbin{\square} X, Y) \tag{7.6.9}$$

where the two parallel arrows are adjoint to the two composites in the diagram below, determined by the S-module structures of X and Y denoted ρ^X and ρ^Y, respectively.

$$\begin{array}{ccc} \mathsf{Hom}_\Sigma(X, Y) \mathbin{\square} (S \mathbin{\square} X) & \longrightarrow & S \mathbin{\square} Y \\ {\scriptstyle 1 \mathbin{\square} \rho^X} \downarrow & & \downarrow {\scriptstyle \rho^Y} \\ \mathsf{Hom}_\Sigma(X, Y) \mathbin{\square} X & \longrightarrow & Y \end{array}$$

The upper horizontal morphism is the composite of associativity and symmetry

$$\mathsf{Hom}_{\Sigma}(X,Y)\,\square\,(S\,\square\,X) \longrightarrow S\,\square\,(\mathsf{Hom}_{\Sigma}(X,Y)\,\square\,X)$$

together with evaluation. The lower horizontal morphism is evaluation. The left S-module structure of $\mathsf{Hom}_S(X,Y)$ is induced by that of Y. ◇

Proposition 7.6.10. *Suppose (A,ρ^A), (X,ρ^X), and (Y,ρ^Y) are symmetric spectra. Then a morphism*

$$A \xrightarrow{\overline{f}} \mathsf{Hom}_S(X,Y) \in \mathsf{SymSp}$$

determines and is uniquely determined by a morphism

$$A\,\square\,X \xrightarrow{f} Y \in \mathsf{sSet}_*^{\Sigma}$$

such that the diagram

$$\begin{array}{ccccc}
A\,\square\,(S\,\square\,X) & \xrightarrow{\cong} & S\,\square\,(A\,\square\,X) & \xrightarrow{a^{-1}} & (S\,\square\,A)\,\square\,X \\
 & & \downarrow{\scriptstyle 1\square f} & & \\
\downarrow{\scriptstyle 1\square\rho^X} & & S\,\square\,Y & & \downarrow{\scriptstyle \rho^A\square 1} \\
 & & \downarrow{\scriptstyle \rho^Y} & & \\
A\,\square\,X & \xrightarrow{f} & Y & \xleftarrow{f} & A\,\square\,X
\end{array} \tag{7.6.11}$$

commutes, with $\cong$ the unique coherence isomorphism that swaps A and S.

Proof. Given a morphism $\overline{f} \in \mathsf{SymSp}$ as stated above, the morphism f is the adjoint of the composite

$$A \xrightarrow{\overline{f}} \mathsf{Hom}_S(X,Y) \longrightarrow \mathsf{Hom}_{\Sigma}(X,Y) \in \mathsf{sSet}_*^{\Sigma}$$

under the $(\square,\mathsf{Hom}_{\Sigma})$-adjunction in Theorem 7.3.7.

- The left half of the diagram (7.6.11) follows from the equalizer that defines the internal hom Hom_S in (7.6.9).
- The right half of (7.6.11) expresses the compatibility of $\overline{f} \in \mathsf{sSet}_*^{\Sigma}$ with the left S-action.

This argument can also be used in reverse. □

Explanation 7.6.12. Similar to Explanation 7.6.4, the morphism $f\colon A\,\square\,X \longrightarrow Y$ in Proposition 7.6.10 is equivalent to the family of $(\Sigma_p\times\Sigma_q)$-equivariant morphisms

$$A_p \wedge X_q \xrightarrow{f_{p,q}} Y_{p+q} \in \mathsf{sSet}_* \quad \text{for} \quad p,q \geq 0.$$

The right half of the diagram (7.6.11) is equivalent to the diagram

$$\begin{array}{ccc}
S^p \wedge (A_n \wedge X_q) & \underset{\cong}{\xrightarrow{a^{-1}}} & (S^p \wedge A_n) \wedge X_q \\
\downarrow{\scriptstyle 1\wedge f_{n,q}} & & \\
S^p \wedge Y_{n+q} & & \downarrow{\scriptstyle \rho^A_{p,n}\wedge 1} \\
\downarrow{\scriptstyle \rho^Y_{p,n+q}} & & \\
Y_{p+n+q} & \xleftarrow{f_{p+n,q}} & A_{p+n} \wedge X_q
\end{array}$$

in sSet_* for $n,p,q \geq 0$. With the abbreviation in (7.6.6), the left half of the diagram (7.6.11) is equivalent to the diagram

$$\begin{array}{ccc}
 & & \Sigma_{(p+n+q)} \times_{\Sigma_{p,n+q}} S^p \wedge \left(\Sigma_{n+q} \times_{\Sigma_{n,q}} A_n \wedge X_q\right) \\
 & \nearrow^{a(\xi_{n,p}\wedge 1)a^{-1}} & \downarrow{\scriptstyle 1\wedge f_{n,q}} \\
\Sigma_{(n+p+q)} \times_{\Sigma_{n,p+q}} A_n \wedge \left(\Sigma_{(p+q)} \times_{\Sigma_{p,q}} S^p \wedge X_q\right) & & \Sigma_{(p+n+q)} \times_{\Sigma_{p,n+q}} S^p \wedge Y_{n+q} \\
\downarrow{\scriptstyle 1\wedge\rho^X_{p,q}} & & \downarrow{\scriptstyle \rho^Y_{p,n+q}} \\
\Sigma_{(n+p+q)} \times_{\Sigma_{n,p+q}} A_n \wedge X_{p+q} & \xrightarrow{f_{n,p+q}} & Y_{n+p+q}
\end{array}$$

in sSet_* for $n,p,q \geq 0$. ◇

Simplicial Tensored and Cotensored Structure. Next we note, for each $n \geq 0$, there is a level-n adjunction with pointed simplicial sets.

Definition 7.6.13. For each $n \geq 0$ we compose the adjunction (i_n, ev_n) of Definition 7.3.6 with the free-forgetful adjunction for left S-modules to obtain an adjunction

$$\mathsf{sSet}_* \underset{\mathsf{ev}_n}{\overset{F_n}{\rightleftarrows}} \mathsf{SymSp}$$

where $\mathsf{ev}_n X = X_n$ for a symmetric spectrum X and $F_n A = S \,\square\, i_n A$ for a pointed simplicial set A. With respect to the smash product $\wedge = \square_S$, the left adjoint F_0 is strong symmetric monoidal and the right adjoint ev_0 is symmetric monoidal. ◇

Explanation 7.6.14. Suppose $K \in \mathsf{sSet}_*$ and $n,k \geq 0$. Then $F_n K$ is the symmetric spectrum with

$$\begin{aligned}
(F_n K)_k &= (S \,\square\, i_n K)_k \\
&= \bigvee_{p+q=k} \Sigma_k \times_{\Sigma_p \times \Sigma_q} S^p \wedge (i_n K)_q \\
&= \begin{cases} * & \text{if } k < n \text{ and} \\ \Sigma_k \times_{\Sigma_{k-n}\times\Sigma_n} S^{k-n} \wedge \left(K \wedge (\Sigma_n)_+\right) & \text{if } k \geq n. \end{cases}
\end{aligned}$$

The left S-action

$$S^p \wedge (F_n K)_q \xrightarrow{\rho_{p,q}} (F_n K)_{p+q} \quad \text{for} \quad p,q \geq 0$$

is the basepoint inclusion if $q < n$. If $q \geq n$, then $\rho_{p,q}$ is induced by the canonical isomorphism

$$S^p \wedge S^{q-n} \xrightarrow[\cong]{\mu^S_{p,q-n}} S^{p+q-n}.$$

In particular, we have

$$F_0 K = \Sigma^\infty K,$$

the suspension spectrum of K in Example 7.4.10. ◇

An adjunction between $\wedge = \square_S$ and Hom_S follows formally from their constructions as coequalizers and equalizers, respectively. Thus we have the following result, with the second half following from Theorem 3.9.8.

Theorem 7.6.15. *The category of symmetric spectra,* SymSp, *is a symmetric monoidal closed category with*

- *monoidal product given by* $\wedge = \square_S$,
- *internal hom given by* Hom_S, *and*
- *monoidal unit the symmetric sphere* S.

Moreover, the adjunction (F_0, ev_0) *makes* SymSp *enriched, tensored, and cotensored over* sSet_*.

Applying Theorems 2.4.10 and 3.3.2 with the change of enrichment by ev_0, we have the following.

Corollary 7.6.16. *The symmetric monoidal structure* $(\mathsf{SymSp}, \wedge, S)$ *is* sSet_*-*enriched.*

Explanation 7.6.17. Suppose $X \in \mathsf{SymSp}$ and $K \in \mathsf{sSet}_*$. Then the tensored and cotensored structures in Theorem 7.6.15 are given by

$$X \wedge K = X \square_S \Sigma^\infty K$$
$$X^K = \mathsf{Hom}_S(\Sigma^\infty K, X),$$

where we used the fact $F_0 = \Sigma^\infty$ in Explanation 7.6.14. ◇

Definition 7.6.18. Because the equalizer (7.6.9) is computed levelwise, we define Map_S as the following special case for $X, Y \in \mathsf{SymSp}$:

$$\mathsf{Map}_S(X,Y) = \mathsf{Hom}_S(X,Y)_0$$

is the equalizer

$$\mathsf{Map}_S(X,Y) \longrightarrow \mathsf{Map}_\Sigma(X,Y) \rightrightarrows \mathsf{Map}_\Sigma(S \square X, Y).$$

Therefore, from Definition 7.3.5 (3) we have

$$\mathsf{Map}_S(X,Y) \cong \mathsf{SymSp}(X \wedge \Delta^?_+, Y) \tag{7.6.19}$$

as pointed simplicial sets. ◇

Explanation 7.6.20 (Underlying Symmetric Sequence of Hom Object)**.** Recall from Definition 7.6.13 the adjunction (F_n, ev_n) and from Definition 7.3.4 the symmetric sequence $\Sigma[n] = \Sigma(\underline{n}, -)$ for each $n \geq 0$. We have $F_n S^0 = S \square \Sigma[n]_+$ and therefore, for each $n \geq 0$ and each $W \in \mathsf{SymSp}$, a composite of natural isomorphisms of pointed simplicial sets

$$\mathsf{Map}_S(S \square \Sigma[n]_+, W) \cong \mathsf{Map}_{\mathsf{sSet}_*}(S^0, \mathsf{ev}_n W) \cong W_n.$$

Since $\Sigma[-]_+$ is a functor from Σ^{op} to sSet_*^Σ, these isomorphisms are equivariant with respect to the induced action of Σ_n and we have an isomorphism of symmetric sequences

$$\mathsf{Map}_S(S \square \Sigma[-]_+, W) \cong W.$$

Applying this to $W = \mathsf{Hom}_S(X,Y)$ for symmetric spectra X and Y, the $\wedge$-Hom_S adjunction gives an isomorphism of symmetric sequences

$$\mathsf{Hom}_S(X,Y) \cong \mathsf{Map}_S(S \square \Sigma[-]_+, \mathsf{Hom}_S(X,Y)) \cong \mathsf{Map}_S(X \wedge (S \square \Sigma[-]_+), Y).$$

Combining this with (7.6.19) we have

$$\mathsf{Hom}_S(X,Y) \cong \mathsf{SymSp}((X \wedge (S \square \Sigma[-]_+)) \wedge \Delta^?_+, Y). \tag{7.6.21}$$

◇

Recall from Definition 6.3.7 the enriched endomorphism operad of an object X in a symmetric monoidal V-category with V permutative. The category of pointed simplicial sets is symmetric monoidal closed, but is not strictly associative or unital. Nevertheless, as noted in Explanation 6.1.19, we extend Definitions 6.1.1 and 6.3.3 to the sSet_*-enriched case via the Symmetric Strictification Theorem 1.1.42. Then we have the following.

Definition 7.6.22. Suppose X is a symmetric spectrum. The *endomorphism simplicial operad* of X is the sSet_*-enriched operad $\mathsf{End}(X)$ whose nth term for $n \geq 0$ is

$$\mathsf{End}(X)_n = \mathsf{Map}_S(X^{\wedge n}, X) = \mathsf{SymSp}(X^{\wedge n} \wedge \Delta^?_+, X). \qquad \diamond$$

7.7. Quillen Model Categories

Definition 7.7.1. Suppose M is a category.

(1) For morphisms $f : A \longrightarrow B$ and $g : C \longrightarrow D$ in M, we write $f \boxslash g$ if for each solid-arrow commutative diagram

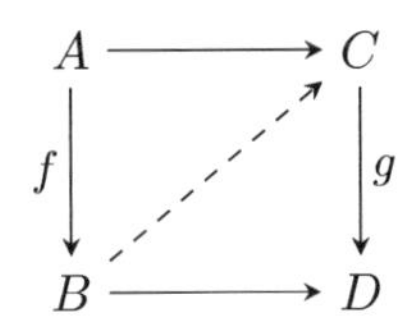

in M, there is a morphism $B \longrightarrow C$ that makes the entire diagram commutative.

(2) For a class $\mathcal{A}$ of morphisms in M, we define the classes of morphisms

$$\begin{aligned} {}^{\boxslash}\mathcal{A} &= \{f \in \mathsf{M} \mid f \boxslash a \text{ for all } a \in \mathcal{A}\}, \\ \mathcal{A}^{\boxslash} &= \{g \in \mathsf{M} \mid a \boxslash g \text{ for all } a \in \mathcal{A}\}. \end{aligned}$$

(3) We say that a pair $(\mathcal{L}, \mathcal{R})$ of classes of morphisms in M *functorially factors* M if each morphism h in M has a functorial factorization $h = gf$ such that $f \in \mathcal{L}$ and $g \in \mathcal{R}$.

(4) A *weak factorization system* in M is a pair $(\mathcal{L}, \mathcal{R})$ of classes of morphisms in M such that
   (i) $(\mathcal{L}, \mathcal{R})$ functorially factors M,
   (ii) $\mathcal{L} = {}^{\boxslash}\mathcal{R}$, and
   (iii) $\mathcal{R} = \mathcal{L}^{\boxslash}$. $\diamond$

Explanation 7.7.2. The lifting properties imply a number of closure properties. Among them are the following (see, e.g., [**MP12**, Proposition 14.1.8]):

- Each of $\mathcal{L}$ and $\mathcal{R}$ contains all isomorphisms.
- Each of $\mathcal{L}$ and $\mathcal{R}$ is closed under composition and retracts.
- The left class, $\mathcal{L}$, is closed under pushouts and coproducts.
- The right class, $\mathcal{R}$, is closed under pullbacks and products. $\diamond$

Definition 7.7.3. A *model category* is a complete and cocomplete category M equipped with three classes of morphisms $(\mathcal{W}, \mathcal{C}, \mathcal{F})$, called *weak equivalences*, *cofibrations*, and *fibrations*, that satisfy the following two axioms:

2-out-of-3: For any morphisms f and g in M such that the composition gf is defined, if any two of the three morphisms f, g, and gf are in $\mathcal{W}$, then so is the third.

WFS: $(\mathcal{C}, \mathcal{F} \cap \mathcal{W})$ and $(\mathcal{C} \cap \mathcal{W}, \mathcal{F})$ are weak factorization systems.

This finishes the definition of a model category. We have the following additional concepts in a model category M.

(1) An *acyclic (co)fibration* is a morphism that is both a (co)fibration and a weak equivalence.
(2) An object $A \in \mathsf{M}$ is *cofibrant* if the unique morphism $\varnothing \longrightarrow A$ from the initial object is a cofibration.
(3) An object $A \in \mathsf{M}$ if *fibrant* if the unique morphism $A \longrightarrow *$ to the terminal object is a fibration.
(4) A *cofibrant replacement* for $A \in \mathsf{M}$ is an object A^c obtained by the factorization of $\varnothing \longrightarrow A$ as a cofibration followed by an acyclic fibration:

$$\varnothing \longrightarrow A^c \xrightarrow{\sim} A.$$

(5) A *fibrant replacement* for $A \in \mathsf{M}$ is an object A^f obtained by the factorization of $A \longrightarrow *$ as an acyclic cofibration followed by a fibration:

$$A \xrightarrow{\sim} A^f \longrightarrow *.$$

(6) A *cylinder object* for $A \in \mathsf{M}$ is an object $\mathsf{Cyl}(A) \in \mathsf{M}$ obtained by factoring the fold morphism $A \coprod A \longrightarrow A$ as a cofibration followed by an acyclic fibration:

$$A \coprod A \longrightarrow \mathsf{Cyl}(A) \xrightarrow{\sim} A.$$

(7) A *path object* for $A \in \mathsf{M}$ is an object $\mathsf{Path}(A) \in \mathsf{M}$ obtained by factoring the diagonal $A \longrightarrow A \times A$ into an acyclic cofibration followed by a fibration:

$$A \xrightarrow{\sim} \mathsf{Path}(A) \longrightarrow A \times A.$$

(8) Functoriality of the weak factorization system $(\mathcal{C}, \mathcal{F} \cap \mathcal{W})$, respectively $(\mathcal{C} \cap \mathcal{W}, \mathcal{F})$, implies that the cofibrant replacements and cylinder objects, respectively fibrant replacements and path objects, are functorial.
(9) If A is fibrant, then A^c is also fibrant. Likewise, if B is cofibrant then B^f is also cofibrant. We let $(-)^{cf} = ((-)^c)^f$ and $(-)^{fc} = ((-)^f)^c$ denote the composites. The lifting properties imply there is a canonical weak equivalence $A^{cf} \xrightarrow{\sim} A^{fc}$ for each $A \in \mathsf{M}$. ◇

Definition 7.7.4. Suppose that $f, g : A \longrightarrow B$ are morphisms in a model category M.

- A *left homotopy* from f to g is a morphism $h : \mathsf{Cyl}(A) \longrightarrow B$ such that the following diagram in M commutes.

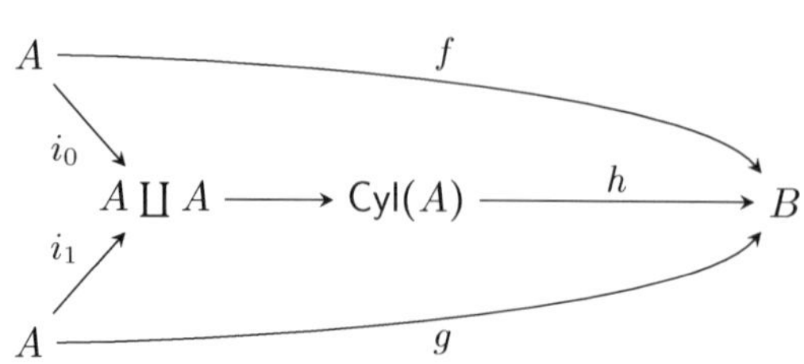

- A *right homotopy* from f to g is a morphism $h: A \longrightarrow \mathsf{Path}(B)$ such that the following diagram in M commutes.

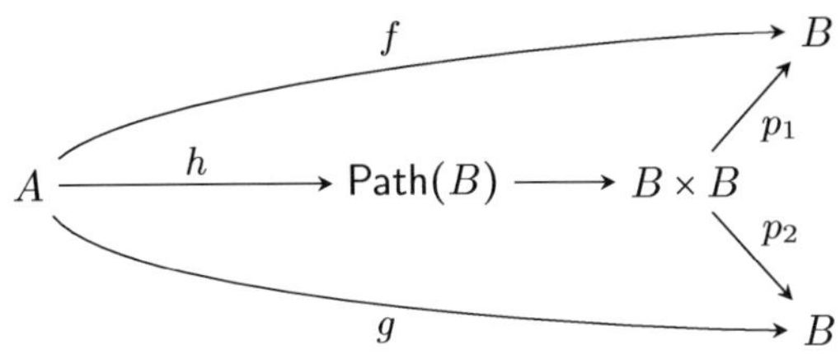

◇

Proposition 7.7.5 ([**MP12**, Proposition 14.3.11])**.** *Suppose $f, g: A \longrightarrow B$ are morphisms in a model category M.*

- *If A is cofibrant and f is left homotopic to g, then f is right homotopic to g.*
- *If B is fibrant and f is right homotopic to g, then f is left homotopic to g.*

Definition 7.7.6. Suppose $f, g: A \longrightarrow B$ are morphisms in a model category M, and suppose that A is cofibrant and B is fibrant. We say that f is *homotopic* to g if f is left, or equivalently right, homotopic to g. When each of A and B is both fibrant and cofibrant, we say that f is a *homotopy equivalence* if there is a morphism $f': B \longrightarrow A$ such that the composites $f'f$ and ff' are homotopic to the respective identities. ◇

Definition 7.7.7. Let M_{cf} denote the full subcategory of M whose objects are both fibrant and cofibrant. For objects $A, B \in \mathsf{M}_{cf}$, let $[A, B]$ denote the set of homotopy equivalence classes of morphisms $A \longrightarrow B$. ◇

Theorem 7.7.8 ([**MP12**, Theorem 14.3.15])**.** *Suppose that $f: A \longrightarrow B$ is a morphism in a model category M, and suppose that each of A and B is both fibrant and cofibrant. Then f is a homotopy equivalence if and only if it is a weak equivalence.*

Theorem 7.7.8 is the key result that justifies the following definition.

Definition 7.7.9. Suppose that M is a model category. The *homotopy category*, $\mathsf{Ho}\mathsf{M}$ is defined to have the same objects as M and morphism sets

$$\mathsf{HoM}(A, B) = [A^{cf}, B^{cf}].$$

By functoriality of factorizations, there is a canonical functor

$$\gamma: \mathsf{M} \longrightarrow \mathsf{HoM}$$

that is the identity on objects and sends a morphism $f: A \longrightarrow B$ to the homotopy class of f^{cf}. ◇

Explanation 7.7.10. One can show that there is a natural homotopy equivalence $(A^c)^f \xrightarrow{\sim} (A^f)^c$, and thus HoM does not depend on which order one takes fibrant and cofibrant replacements. See [**MP12**, Section 14.4]. ◇

Definition 7.7.11. Suppose that C is a category and $\mathcal{W}$ is a class of morphisms in C. We say that a functor $\lambda: \mathsf{C} \longrightarrow \mathsf{C}'$ is a *localization* of C at $\mathcal{W}$ if the following two conditions hold.

(1) For each morphism $f \in \mathsf{W}$, the morphism λf is an isomorphism.

(2) If $\kappa: \mathsf{C} \longrightarrow \mathsf{D}$ is a functor as in the solid-arrow diagram below such that κf is an isomorphism for each $f \in \mathcal{W}$, then there is a unique functor $\overline{\kappa}$ such that $\overline{\kappa}\lambda = \kappa$.

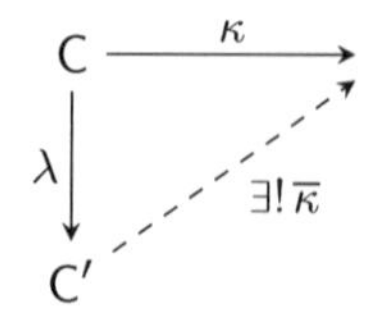

When such a localization exists, it is denoted $\mathsf{C}[\mathcal{W}^{-1}]$. ◇

Theorem 7.7.12 ([**MP12**, Theorem 14.4.7]). *Suppose* M *is a model category with weak equivalences* $\mathcal{W}$*. The functor* $\gamma: \mathsf{M} \longrightarrow \mathit{Ho}\mathsf{M}$ *is a localization of* M *at* $\mathcal{W}$*.*

Explanation 7.7.13. In the context of Definition 7.7.11, one can always define morphisms in $\mathsf{C}[\mathcal{W}^{-1}]$ as a quotient of a free graph. However, in general this produces has a proper class of morphisms between two objects instead of a set. The machinery of model categories guarantees that the morphisms in $\mathit{Ho}\mathsf{M}$ between two objects do form a set, and thus $\mathit{Ho}\mathsf{M}$ is a category in the same Grothendieck universe as M. ◇

Now we turn to functors between model categories, and the conditions under which we have an induced functor on homotopy categories. Certainly functors that preserve weak equivalences will induce functors on homotopy categories. However one has similar results under more general hypotheses, and these are needed in most applications of interest.

Definition 7.7.14. Suppose that $F: \mathsf{M} \longrightarrow \mathsf{N}$ is a functor of model categories. A *left functor* of F is a pair (L, λ) consisting of a functor

$$L: \mathit{Ho}\mathsf{M} \longrightarrow \mathit{Ho}\mathsf{N}$$

and a natural transformation λ as in the diagram below, where γ_{M} and γ_{N} are the canonical localizations.

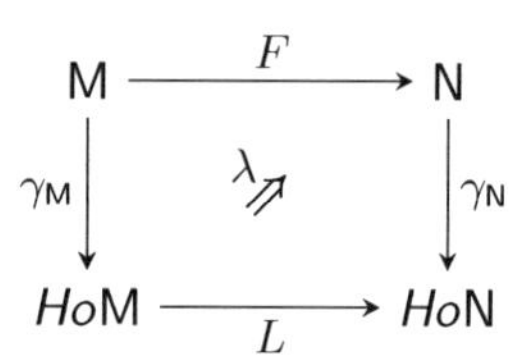

A morphism of left functors $(L, \lambda) \longrightarrow (L, \lambda')$ is a natural transformation $\eta: L \longrightarrow L'$ such that the following equality of pasting diagrams holds.

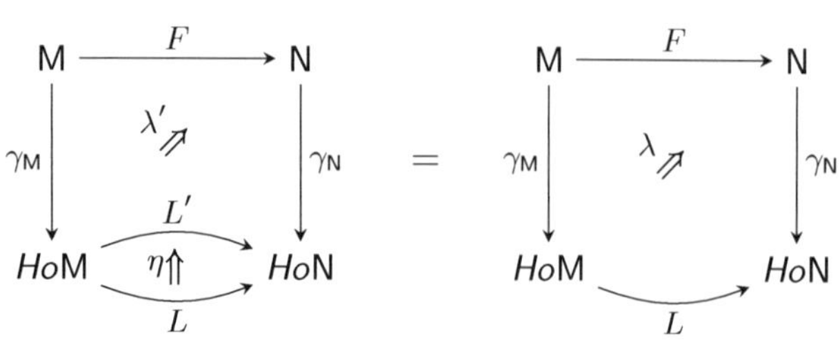

A *left derived functor* for F is a terminal object $(\mathbb{L}F, \tilde{\lambda})$ among left functors of F. ◇

Definition 7.7.15. Suppose that $F\colon \mathsf{M} \longrightarrow \mathsf{N}$ is a functor of model categories. A *right functor* of F is a pair (R, ρ) consisting of a functor

$$R\colon Ho\mathsf{M} \longrightarrow Ho\mathsf{N}$$

and a natural transformation ρ as in the diagram below, where γ_M and γ_N are the canonical localizations.

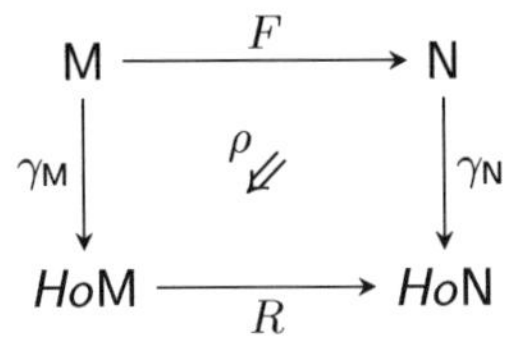

A morphism of right functors $(R, \rho) \longrightarrow (R, \rho')$ is a natural transformation $\eta\colon R \longrightarrow R'$ such that the following equality of pasting diagrams holds.

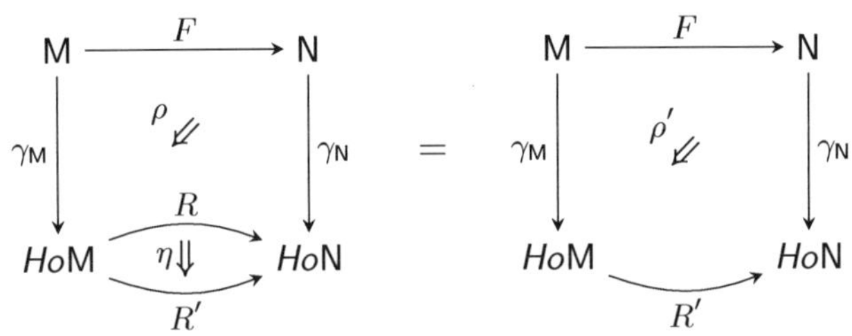

A *right derived functor* for F is an initial object $(\mathbb{R}F, \tilde{\rho})$ among right functors of F. ◇

Explanation 7.7.16. Since initial and terminal objects are unique up to unique isomorphism, it is usual to refer to *the* left and right derived functors of F. ◇

The following result gives a set of conditions that often occur in applictions and under which one has left and right derived functors.

Proposition 7.7.17 (**[MP12**, Propositions 16.1.3 and 16.1.4]). *Suppose that $F\colon \mathsf{M} \longrightarrow \mathsf{N}$ is a functor of model categories.*

(1) *If F takes acyclic cofibrations between cofibrant objects to weak equivalences, then $F \circ (-)^c$ carries weak equivalences in M to isomorphisms in $Ho\mathsf{N}$. The induced functor*

$$\mathbb{L}F\colon Ho\mathsf{M} \longrightarrow Ho\mathsf{N}$$

is the left derived functor of F.

(2) *If F takes acyclic fibrations between fibrant objects to weak equivalences, then $F \circ (-)^f$ carries weak equivalences in M to isomorphisms in $Ho\mathsf{N}$. The induced functor*

$$\mathbb{R}F\colon Ho\mathsf{M} \longrightarrow Ho\mathsf{N}$$

is the right derived functor of F.

The next result records a useful set of equivalent conditions, and follows from an application of adjoints and the lifting properties of weak factorization systems.

Lemma 7.7.18. *Suppose that M and N are model categories and suppose that (F, U) is an adjunction*

$$F\colon \mathsf{M} \rightleftarrows \mathsf{N}\colon U.$$

Then the following conditions are equivalent.

(1) F *preserves cofibrations and* U *preserves fibrations.*
(2) F *preserves cofibrations and acyclic cofibrations.*
(3) U *preserves fibrations and acyclic fibrations.*
(4) F *preserves acyclic cofibrations and* U *preserves acyclic fibrations.*

Definition 7.7.19. We say that an adjoint pair of functors (F,U) is a *Quillen adjunction* if the equivalent conditions in Lemma 7.7.18 are satisfied. We say that (F,U) is a *Quillen equivalence* if it is a Quillen adjunction with the following property. For any cofibrant $X \in \mathsf{M}$ and fibrant $Y \in \mathsf{N}$, a morphism $FX \longrightarrow Y$ is a weak equivalence in N if and only if its adjoint $X \longrightarrow UY$ is a weak equivalence in M. In this case, we denote the adjunction with $\simeq_Q$, as in the following.

$$F\colon \mathsf{M} \underset{}{\overset{\simeq_Q}{\rightleftarrows}} \mathsf{N}\colon U.$$

◇

The definitions of Quillen adjunction and Quillen equivalence are designed to induce the corresponding structures on homotopy categories, as the following result explains.

Proposition 7.7.20 ([**MP12**, Proposition 16.2.2])**.** *Suppose that* (F,U) *is a Quillen adjunction. Then the derived functors* $\mathbb{L}F$ *and* $\mathbb{R}U$ *exist and form an adjoint pair of functors between* HoM *and* HoN*. If* (F,U) *is a Quillen equivalence, then* $(\mathbb{L}F,\mathbb{R}U)$ *is an adjoint equivalence.*

Now we discuss cofibrantly generated model categories. The concept of a small object A in the following definition essentially means that a morphism from A to the codomain of a sufficiently long composition must factor through some stage. We regard an ordinal as a category with a unique morphism $i \longrightarrow j$ if and only if $i \le j$. See [**Pin14, SV02**] for further discussion of cardinals and ordinals.

Definition 7.7.21. Suppose C is a cocomplete category, and $\mathcal{I}$ is a collection of morphisms in C.

(1) For an ordinal α, an *α-sequence* in C is a colimit-preserving functor

$$X : \alpha \longrightarrow \mathsf{C}.$$

The induced morphism

$$X_0 \longrightarrow \mathsf{colim}_{\beta<\alpha} X_\beta$$

is called a *transfinite composition.*

(2) For an ordinal α and an α-sequence X, if each morphism $X_\beta \longrightarrow X_{\beta+1}$ belongs to $\mathcal{I}$ for each ordinal β satisfying $\beta+1<\alpha$, then we call the above morphism a *transfinite composition of morphisms in* $\mathcal{I}$.
(3) A *relative $\mathcal{I}$-cell complex* is a transfinite composition of pushouts of morphisms in $\mathcal{I}$. The collection of relative $\mathcal{I}$-cell complexes is denoted by $\mathsf{Cell}(\mathcal{I})$.
(4) For an object A in C and a cardinal κ, we say that A is *κ-small relative to $\mathcal{I}$* if for
   - each regular cardinal $\alpha \ge \kappa$ and

- each α-sequence X in C with each morphism $X_\beta \longrightarrow X_{\beta+1}$ in $\mathcal{I}$ for each ordinal β satisfying $\beta + 1 < \alpha$,

the induced morphism of sets

$$\mathsf{colim}_{\beta<\alpha}\mathsf{C}(A, X_\beta) \longrightarrow \mathsf{C}\big(A, \mathsf{colim}_{\beta<\alpha}X_\beta\big)$$

is a bijection.

(5) We say that A is *small relative to* $\mathcal{I}$ if it is κ-small relative to $\mathcal{I}$ for some cardinal κ.

(6) We say that $\mathcal{I}$ *permits the small object argument* if the domain of each morphism in $\mathcal{I}$ is small relative to $\mathsf{Cell}(\mathcal{I})$.

◇

Definition 7.7.22. A model category $(\mathsf{M}, \mathcal{W}, \mathcal{C}, \mathcal{F})$ is *cofibrantly generated* if it is equipped with two sets $\mathcal{I}$ and $\mathcal{J}$ of morphisms such that the following three statements hold:

(i) Both $\mathcal{I}$ and $\mathcal{J}$ permit the small object argument.
(ii) $\mathcal{F} = \mathcal{J}^{\boxslash}$.
(iii) $\mathcal{F} \cap \mathcal{W} = \mathcal{I}^{\boxslash}$.

In this case, the morphisms in $\mathcal{I}$ are called *generating cofibrations* and the morphisms in $\mathcal{J}$ are called *generating acyclic cofibrations.* ◇

In particular, in a cofibrantly generated model category, fibrations are detected by the set $\mathcal{J}$, and acyclic fibrations are detected by the set $\mathcal{I}$. The cofibrations and acyclic cofibrations are defined by lifting against acyclic fibrations and fibrations, respectively. These include the morphisms of $\mathcal{I}$, respectively $\mathcal{J}$, but generally others as well.

Definition 7.7.23. Suppose $(\mathsf{M}, \wedge, S^0)$ is a symmetric monoidal category. The *pushout product* of morphisms $f: A \longrightarrow B$ and $g: C \longrightarrow D$ is the morphism $f \,\square\, g$ induced by the universal property of the pushout in the following diagram.

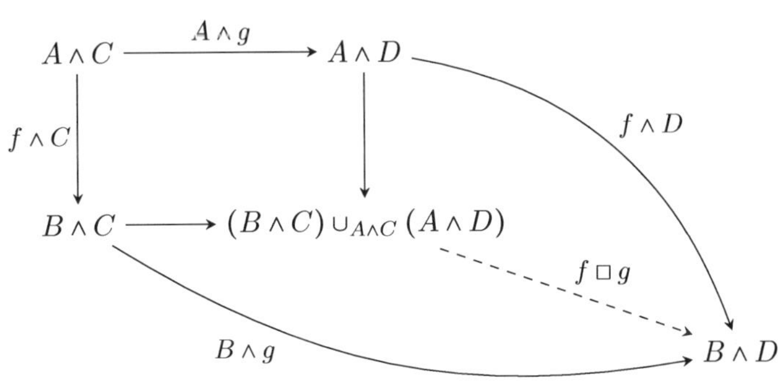

◇

Definition 7.7.24. Suppose $(\mathsf{M}, \wedge, S^0)$ is a symmetric monoidal closed category and that M has a model structure. We say that $(\mathsf{M}, \wedge, S^0)$ is a *monoidal model category* if the following axiom holds.

Pushout product axiom: If f and g are cofibrations, then the pushout product $f \,\square\, g$ is a cofibration and is acyclic if either f or g is acyclic. ◇

7.8. Examples of Quillen Model Categories

Example 7.8.1 (Small Categories [**JT91, Rez∞**]). The category Cat of small categories and functors has a *standard model structure* given as follows.

Fibrations: A functor $F: \mathsf{C} \longrightarrow \mathsf{D}$ is a *fibration* if for each
- object $X \in \mathsf{C}$ and
- isomorphism $f: FX \xrightarrow{\cong} Y \in \mathsf{D}$,

there exists an isomorphism $g: X \xrightarrow{\cong} Z \in \mathsf{C}$ such that

$$FZ = Y \quad \text{and} \quad Fg = f.$$

Cofibrations: A functor is a *cofibration* if it is injective on objects.
Weak equivalences: A functor is a *weak equivalence* if it is an equivalence of categories.

This is a monoidal model category with respect to the Cartesian product. It is cofibrantly generated with the following data.

Generating acyclic cofibration: $\mathcal{J}$ contains the single functor

$$\{0\} \xrightarrow{\quad t \quad} J$$

that sends the object 0 in the domain to 0 in the codomain $J = \{0 \xrightarrow{\cong} 1\}$, which has two objects and a single isomorphism between them.
Generating cofibrations: The set

$$\mathcal{I} = \{u, v, w\}$$

contains the following three functors. The functor

$$\varnothing \xrightarrow{\quad u \quad} \{0\}$$

is the unique functor. Suppose
- $V_0 = \{0, 1\}$ is a discrete category with two objects, and
- $V = \{0 \longrightarrow 1\}$ contains two objects and one non-identity morphism between them.

The functor

$$V_0 \xrightarrow{\quad v \quad} V$$

is the identity function on objects. Suppose $W = \{0 \rightrightarrows 1\}$ contains two objects and a parallel pair of non-identity morphisms. The functor

$$W \xrightarrow{\quad w \quad} V$$

is the identity on objects and sends both non-identity morphisms in W to the unique non-identity morphism in V.

A functor is a fibration if and only if it has the right lifting property with respect to $\mathcal{J} = \{t\}$. A functor is an acyclic fibration if and only if it has the right lifting property with respect to $\mathcal{I} = \{u, v, w\}$. ◇

Example 7.8.2 (Chain Complexes [**Hov99**, Section 2.3]). Suppose that R is a commutative ring, and $\mathsf{Ch}(R)$ the category of chain complexes over R. The *projective model structure* on $\mathsf{Ch}(R)$ consists of the following.

Weak equivalences: $\mathcal{W}$ is the set of homology isomorphisms.
Fibrations: $\mathcal{F}$ is the set of levelwise surjections.

The cofibrations are defined by lifting against the acyclic fibrations. This model structure is also known as the *standard model structure*

There is a second model structure on $\mathsf{Ch}(R)$ that has the same weak equivalences (homology isomorphisms), but whose cofibrations are levelwise injections and whose fibrations are defined by lifting against acyclic cofibrations. Although these model structures are distinct, the identity functor is a Quillen equivalence between them. ◇

Example 7.8.3 (Simplicial Sets [**MP12**, Section 17.5])**.** The standard model structure on sSet is a cofibrantly generated monoidal model structure, given as follows.

Generating cofibrations: $\mathcal{C}$ is the set of boundary inclusions

$$\partial\Delta^n \longrightarrow \Delta^n.$$

Generating acyclic cofibrations: $\mathcal{J}$ is the set of horn inclusions

$$\Lambda^n_k \longrightarrow \Delta^n.$$

Fibrations: Morphisms in $\mathcal{F} = \mathcal{J}^\boxslash$ are called *Kan fibrations*, and the fibrant objects are called *Kan complexes*.

Weak equivalences: $\mathcal{W}$ is the set of morphsisms $f\colon X \longrightarrow Y$ such that

$$f^*\colon [Y,Z] \longrightarrow [X,Z]$$

is a bijection for all Kan complexes Z, where $[X,Z]$ denotes left homotopy equivalence classes of morphisms with respect to the cylinder object $\mathsf{Cyl}(X) = X \times \Delta^1$. ◇

Explanation 7.8.4. In the context of Example 7.8.3, the cofibrations in sSet are levelwise inclusions. In particular, every simplicial set is cofibrant. A simplicial homotopy equivalence (Definition 7.2.1) is a left homotopy in the sense of Definition 7.7.4. Therefore, every simplicial homotopy equivalence is a weak equivalence in sSet. By Theorem 7.7.8, the weak equivalences between Kan complexes are precisely the simplicial homotopy equivalences. ◇

Example 7.8.5 (Topological Spaces [**MP12**, Section 17.2])**.** The *Quillen model structure* on Top is a cofibrantly generated monoidal model structure defined as follows.

Generating Cofibrations: $\mathcal{I}$ is the set of inclusions

$$S^{n-1} \longrightarrow D^n$$

of the topological $(n-1)$-sphere as the boundary of the n-disk for $n \geq 0$, where S^{-1} denotes the empty set.

Generating Acyclic Cofibrations: $\mathcal{J}$ is the set of inclusions

$$D^n \cong D^n \times \{0\} \longrightarrow D^n \times I$$

where I is the unit interval. ◇

Theorem 7.8.6. *With the model structures defined in Examples 7.8.3 and 7.8.5, the geometric realization and total singular complex functors define a Quillen equivalence.*

$$|-|\colon \mathsf{Top} \underset{\longleftarrow}{\overset{\longrightarrow}{\simeq_Q}} \mathsf{sSet}\colon \mathsf{Sing}$$

The category of symmetric spectra, SymSp, has a model structure called the *stable model structure* that we now explain.

Definition 7.8.7. A morphism $f = \{f_k\}_{k\geq 0}$ of symmetric spectra is a *level equivalence* if each f_k is a weak equivalence of pointed simplicial sets. It is a *level fibration* if each f_k is a Kan fibration. ◇

Explanation 7.8.8. The weak equivalences of symmetric spectra are known as *stable equivalences* and are detected in cohomology. They include the level equivalences, but are more general. ◇

Theorem 7.8.9 ([**HSS00**, Theorem 3.4.4, Corollary 5.3.8]). *The category of symmetric spectra is a monoidal model category with weak equivalences given by the stable equivalences and fibrations given by the level fibrations.*

Explanation 7.8.10. The model structure on symmetric spectra is Quillen equivalent to other modern models of stable homotopy theory, including orthogonal spectra and S-modules. See [**MMSS01**] for further details. ◇

7.9. Notes

7.9.1 (Simplicial Objects). See [**Fri12**] for an introduction to simplicial sets. For further details and development of simplicial objects, see [**Cur71, GZ67, GJ09, May92, Rie14, Ric20**]. Discussion of limits and colimits in diagram categories can be found in [**Rie16**], and the resulting enrichment properties are in [**Kel05**]. ◇

7.9.2 (Nerve and Geometric Realization). The nerve and geometric realization functors described in Definitions 7.1.14 and 7.2.3, respectively, together with their adjoints, arise from the theory of Kan extensions and are discussed in [**Rie14**]. Additional references include [**Ric20, MP12**]. ◇

7.9.3 (Compactly Generated Topological Spaces). With the usual product topology, the category of all topological spaces fails to be symmetric monoidal closed with respect to the Cartesian product. Moreover, the induced smash product fails to be associative. Restricting to compactly generated weak Hausdorff spaces, as we do in Definition 7.1.13, avoids the point-set pathologies of the larger category. Familiar classes of spaces, including CW complexes, compact spaces, locally compact spaces, topological manifolds, and metrizable spaces are all weak Hausdorff and compactly generated. See [**Rie14**, Section 6.1] for definitions and a detailed overview. For further reference see [**Bor94b**, Section 7.1], [**May99**, Section 5], and [**MS06**, Section 1.7]. Further references include [**Bro64, GZ67, Kel55, May72, Ste67**]. In the literature, these spaces are sometimes called *k-spaces* and *Kelley spaces*. ◇

7.9.4 (Symmetric Spectra). The category of symmetric spectra was developed in Hovey-Shipley-Smith [**HSS00**]. See [**HSS00**, Proposition 1.2.10] for completeness and cocompleteness as in Proposition 7.5.5. See [**HSS00**, Theorem 2.2.10] for the monoidal product and internal hom as in Theorem 7.6.15. For further development, we also refer the reader to [**MMSS01, Ric20**] and [**Hov01**]. The latter of those discusses symmetric spectra in a general pointed bicomplete category C.

Symmetric spectra are one of several prominent models for stable homotopy theory with a symmetric monoidal product. Others include the S-modules of Elmendorf-Kriz-Mandell-May [**EKMM97**] and the orthogonal spectra of Mandell-May-Schwede-Shipley [**MMSS01**]. All of these, and others, are defined and compared in [**MMSS01**]. ◇

7.9.5 (Sequential Spectra). The earliest and simplest model for stable homotopy theory is the category of *sequential spectra*, $\mathsf{Sp}^{\mathbb{N}}$ introduced by Lima [**Lim59**]. The objects are sequences of pointed simplicial sets together with structure morphisms

$$X_n \in \mathsf{sSet}_* \quad \text{and} \quad \sigma_n: S^1 \wedge X_n \longrightarrow X_{n+1}, \quad \text{for} \quad n \geq 0.$$

The morphisms of sequential spectra are sequences of morphisms in sSet_* commuting with the structure morphisms σ. One motivating interest in sequential spectra is that they represent generalized cohomology theories of spaces (see, for example, [**May99**]).

One can describe sequential spectra as modules over the sequential sphere spectrum $\{S^n\}$. However, the construction of a smash product as in Section 7.4 fails because the sequential sphere spectrum is not a commutative monoid. See [**HSS00**, Section 2.3] for further details on this point. The more sophisticated categories of spectra, including symmetric spectra, were motivated by interest in symmetric monoidal constructions in stable homotopy theory.

The homotopy groups of a sequential spectrum X are given by a limit of the simplicial homotopy groups

$$\pi_n X = \lim_{k \to \infty} \pi_{n+k} X_k$$

that, for each n, is guaranteed to stabilize by the Freudenthal Suspension Theorem. Weak equivalences of sequential spectra are morphisms that induce isomorphisms on π_n for all n, and these are part of a Quillen model structure described in [**BF78**]. The essential requirement for any category of spectra is that its homotopy category be equivalent to the homotopy category of sequential spectra.

There is a forgetful functor

$$U: \mathsf{SymSp} \longrightarrow \mathsf{Sp}^{\mathbb{N}}$$

given by forgetting the symmetric group actions. If f is a weak equivalence of symmetric spectra (i.e. a stable equivalence), then Uf is a weak equivalence (i.e. a π_*-isomorphism) of sequential spectra. However the converse generally does not hold. Nevertheless, [**HSS00**] shows that there is a Quillen equivalence between the (model) category of symmetric spectra and that of sequential spectra. We refer the reader there for detailed definitions of the model structures and proof of the Quillen equivalence. ◇

7.9.6 (Model Categories). The concept of a model category is originally due to Quillen [**Qui67**]. Our presentation generally follows [**MP12**]. Many additional examples of model categories can be found in [**WY18, WY20**]. The first of these, [**WY18**], gives a general approach to (semi-)model structures on the category of algebras over a colored operad in a monoidal model category. For additional references, see [**DS95, Hov99, Hir03, Rie11**]. ◇

CHAPTER 8

Segal K-Theory of Permutative Categories

In this chapter we discuss the Segal K-theory spectra associated to small permutative categories. This construction, an extension of that given by Segal [**Seg74**], takes values in symmetric spectra. The construction has three components.

(1) There is a functor

$$\mathsf{K}^{\mathcal{F}} \colon \Gamma\text{-}\mathsf{sSet} \longrightarrow \mathsf{SymSp}$$

where Γ-sSet is a certain category of (pointed) diagrams of simplicial sets. We define Γ-sSet in Section 8.1 and $\mathsf{K}^{\mathcal{F}}$ in Section 8.2.

(2) Composition with the nerve defines a functor of (pointed) diagram categories

$$N_* \colon \Gamma\text{-}\mathsf{Cat} \longrightarrow \Gamma\text{-}\mathsf{sSet}$$

given in Definition 8.1.18.

(3) There is a functor

$$\mathsf{J}^{\mathsf{Se}} \colon \mathsf{PermCat}^{\mathsf{su}} \longrightarrow \Gamma\text{-}\mathsf{Cat}$$

that constructs Γ-categories from small permutative categories. The definition of J^{Se} is in Section 8.5.

The Segal K-theory functor, K^{Se}, is the composite

$$\mathsf{PermCat}^{\mathsf{su}} \xrightarrow{\mathsf{J}^{\mathsf{Se}}} \Gamma\text{-}\mathsf{Cat} \xrightarrow{N_*} \Gamma\text{-}\mathsf{sSet} \xrightarrow{\mathsf{K}^{\mathcal{F}}} \mathsf{SymSp}.$$

In the treatment below, we give two different but equivalent constructions of the functor J^{Se}. The first, denoted $\mathsf{C}^{\mathcal{F}}$ for a small permutative category C, is in Section 8.3. It is more straightforward and consists of systems of objects indexed by certain pairs of disjoint finite sets. This is the construction originally given by Segal. There are three variants of this construction, each with respective benefits, and all give level-equivalent symmetric spectra.

The second construction is the one we denote J^{Se}. Its construction uses certain diagram multicategories that we call *partition multicategories*. These package the combinatorial structure given by partitions of finite sets and make the categorical properties of J^{Se} more apparent. For example, we describe the 2-functoriality of J^{Se} in these terms. Moreover, the approach via partition multicategories is more readily comparable to our constructions of Elmendorf-Mandell K-theory, denoted K^{EM}, in Chapter 10. We describe the partition multicategories in Section 8.4 and then define J^{Se} and K^{Se} in Section 8.5.

The following table summarizes Segal K-theory and the main constructions in this chapter. The upper portion concerns the definition of Segal K-theory and the constructions $(-)^{\mathcal{F}}$ due to Segal. The lower portion concerns details related to the construction of J^{Se} via partition multicategories.

Segal K-theory	$\mathsf{K}^{\mathsf{Se}} = \mathsf{K}^{\mathcal{F}} N_* \mathsf{J}^{\mathsf{Se}}$	(8.5.1)
indexing category	$\mathcal{F}$	(8.1.1)
K-theory	$\mathsf{K}^{\mathcal{F}}\colon \Gamma\text{-}\mathsf{sSet} \longrightarrow \mathsf{SymSp}$	(8.2.5)
nerve	$N_*\colon \Gamma\text{-}\mathsf{Cat} \longrightarrow \Gamma\text{-}\mathsf{sSet}$	(8.1.18)
Segal Γ-categories	$(-)^{\mathcal{F}}, (-)^{\mathcal{F}}_{\mathsf{lax}}, (-)^{\mathcal{F}}_{\mathsf{co}}$	(8.3.14)
comparison $(-)^{\mathcal{F}}, (-)^{\mathcal{F}}_{\mathsf{lax}}, (-)^{\mathcal{F}}_{\mathsf{co}}$	8.3.21, 8.5.2	
Segal J-theory	$\mathsf{J}^{\mathsf{Se}}\colon \mathsf{PermCat}^{\mathsf{su}} \longrightarrow \Gamma\text{-}\mathsf{Cat}$	
definition	$\mathsf{J}^{\mathsf{Se}} = \mathsf{J}^{\mathcal{M}} \circ \mathsf{End}$	(8.5.1)
endomorphism multicategories	$\mathsf{End}\colon \mathsf{PermCat}^{\mathsf{su}} \longrightarrow \mathsf{Multicat}_*$	(5.1.11, 5.3.9)
partition multicategories	$\mathcal{M}\colon \mathcal{F}^{\mathsf{op}} \longrightarrow \mathsf{Multicat}_*$	(8.4.1, 8.4.7)
partition J-theory	$\mathsf{J}^{\mathcal{M}}\colon \mathsf{Multicat}_* \longrightarrow \Gamma\text{-}\mathsf{Cat}$	(8.4.10)
Segal J-theory and Γ-categories	$\mathsf{J}^{\mathsf{Se}} \cong (-)^{\mathcal{F}}_{\mathsf{lax}}$	(8.4.8)
level equivalence	$\mathsf{K}^{\mathsf{Se}} \xrightarrow{\sim} \mathsf{K}^{\mathsf{EM}}$	(10.6.10)

Reading Guide.

(1) For the essential definitions, read Definitions 8.1.1, 8.1.8, and 8.1.16 through 8.1.18, together with the statement of Proposition 8.2.6.
(2) For Segal's construction alone, it suffices to read the material from (1) and then Definitions 8.3.1, 8.3.9, 8.3.12, and 8.3.14 followed by the statement of Proposition 8.3.13.
(3) For our second, equivalent construction, the material in Section 8.3 may be skipped. After (1), read Definition 8.4.1 and the statement of Proposition 8.4.7. Then read Definitions 8.4.10 and 8.5.1. After that, the items in (2) may be read as a concrete simplification of this material.
(4) For the material on partition multicategories in Section 8.4, the reader will want to be familiar with the basic definitions of multicategories and multifunctors from Section 5.1. However, the additional material in Chapter 5 will not play a major role in this chapter.
(5) Go back and read the rest of this chapter.

8.1. Categories of Γ-Objects

In this section we introduce Γ-objects and their morphisms. The two important cases of interest for us will be Γ-simplicial sets and, later, Γ-categories.

Definition 8.1.1. Let FinSet_* denote the category whose objects are pointed finite sets and whose morphisms are basepoint-preserving functions. Let $\mathcal{F}$ denote the full subcategory of FinSet_* whose objects are pointed finite sets $\underline{n} = \{0, \ldots, n\}$ for natural numbers $n \geq 0$. The pointed finite set $\underline{0}$ is an initial and terminal object for both FinSet_* and $\mathcal{F}$, providing a basepoint such that the inclusion

$$\mathcal{F} \longrightarrow \mathsf{FinSet}_*$$

is a pointed functor. $\diamond$

Explanation 8.1.2. Note that FinSet_* is not small, just as the category of all (not-necessarily-pointed) finite sets is not small. However, $\mathcal{F}$ is a small skeleton for FinSet_*. The inclusion

$$\mathcal{F} \longrightarrow \mathsf{FinSet}_*$$

is fully faithful by definition and is essentially surjective because each pointed finite a is (non-uniquely) isomorphic to $\underline{n-1}$, where $n = |a|$ is the cardinality of a. Therefore the inclusion is an equivalence of categories. ◇

Definition 8.1.3. We make the following notational conventions for pointed finite sets $a \in \mathsf{FinSet}_*$.

- Unless otherwise specified, the basepoint of a general pointed finite set a is denoted $*$.
- The basepoint of $\underline{n}$ is the element 0.
- We let $a^\flat$ denote the *punctured* finite set $a\backslash\{*\}$, obtained by removing the basepoint.
- We let $|a|$ denote the cardinality of a. ◇

Explanation 8.1.4. For a natural number n, we have

$$\underline{n}^\flat = \{1, \ldots, n\}$$

and $|\underline{n}| = n + 1$. The notation $(-)^\flat$ was introduced in Definition 4.3.8, and $a^\flat$ may be read as *a-flat* or *a-punctured.* ◇

Definition 8.1.5. Suppose $\underline{m}$ and $\underline{n}$ are objects of $\mathcal{F}$. The *lexicographic order* for $\underline{m} \wedge \underline{n}$ is the bijection

$$L \colon \underline{m} \wedge \underline{n} \cong \underline{mn}$$

given by

$$L(x,y) = \begin{cases} 0 & \text{if } x = 0 \quad \text{or} \quad y = 0, \\ n(x-1) + y & \text{if } x > 0 \quad \text{and} \quad y > 0. \end{cases}$$

We will also use L to denote the composite

$$\bigvee_{\underline{m}^\flat} \underline{n} \cong \underline{m} \wedge \underline{n} \cong \underline{mn},$$

where the first bijection identifies the wedge summand indexed by $i \in \underline{m}^\flat$ with the elements $(i, y) \in \underline{m} \wedge \underline{n}$. ◇

Explanation 8.1.6. Using the lexicographic order, the smash product of pointed finite sets induces a monoidal product on $\mathcal{F}$ that we also denote $\wedge$. Elementary algebra with the formula for L shows that this product is strictly associative and unital with strict monoidal unit $\underline{1}$, so $(\mathcal{F}, \wedge, \underline{1})$ is a permutative category. Note that its symmetry is generally not trivial, and is induced by that of FinSet_*

$$\underline{m} \wedge \underline{n} \cong \underline{n} \wedge \underline{m}.$$

When we take the smash products of objects in $\mathcal{F}$ below, we will use the lexicographic order unless otherwise stated. ◇

Explanation 8.1.7. One may visualize the lexicographical order

$$L \colon \underline{m} \wedge \underline{n} \cong \underline{mn}$$

as an $(m+1) \times (n+1)$ matrix, with $\underline{m}$ and $\underline{n}$ indexing, respectively, the rows and the columns. Each entry in the first row and the first column yields $0 \in \underline{mn}$ under

L. For the other entries, one counts from left to right in each row, from top to bottom. With this matrix interpretation, the symmetry isomorphism

$$\underline{m} \wedge \underline{n} \cong \underline{n} \wedge \underline{m}$$

corresponds to taking the transpose.

The associativity of $\wedge$ corresponds to a 3-dimensional matrix. In terms of algebra, the associativity of $\wedge$ means the commutativity of the following diagram for all m, n, and p.

$$\begin{array}{ccc} (\underline{m} \wedge \underline{n}) \wedge \underline{p} & \xrightarrow{L \wedge 1} & \underline{mn} \wedge \underline{p} \\ \cong \downarrow & & \\ \underline{m} \wedge (\underline{n} \wedge \underline{p}) & & \downarrow L \\ 1 \wedge L \downarrow & & \\ \underline{m} \wedge \underline{np} & \xrightarrow{L} & \underline{mnp} \end{array}$$

To check its commutativity, consider elements $x \in \underline{m}$, $y \in \underline{n}$, and $z \in \underline{p}$. If one of x, y, or z is 0, then each of the two composites in the previous diagram sends (x, y, z) to 0. If $x, y, z > 0$, then both composites send (x, y, z) to

$$np(x-1) + p(y-1) + z = p\big(n(x-1) + y - 1\big) + z$$

in $\underline{mnp}$. ◇

Definition 8.1.8. Suppose $(\mathsf{C}, *)$ is a pointed category with $*$ terminal in C. A Γ-*object* in C is a pointed functor

$$X \colon (\mathcal{F}, \underline{0}) \longrightarrow (\mathsf{C}, *).$$

The category of Γ-objects in C, denoted Γ-C, is

$$\mathsf{Cat}_*\big((\mathcal{F}, \underline{0}), (\mathsf{C}, *)\big),$$

the category of pointed functors from $(\mathcal{F}, \underline{0})$ to $(\mathsf{C}, *)$ and pointed natural transformations. ◇

Explanation 8.1.9 (Notation Γ)**.** The category Γ-C would be denoted $\mathcal{F}_*$-C in the notation of Definition 4.3.32. The usage of Γ is a historical artifact of Segal's construction [**Seg74**], which defined a category Γ that is isomorphic to $\mathcal{F}^{\mathsf{op}}$ and then defined Γ-objects as pointed functors out of Γ^{op}. The usage of Γ in this context is a fixture of the literature. ◇

Explanation 8.1.10 (Categories of Γ-objects)**.** Some authors work with diagrams on FinSet instead of $\mathcal{F}$, as doing so avoids choices of specific finite sets $\underline{n}$. We define Γ-objects as functors out of $\mathcal{F}$, the skeletal replacement of FinSet, so that the collection of Γ-objects forms a category. Moreover, we implicitly use the canonical ordering on each $\underline{n}$ in constructions that are indexed over the elements of $\underline{n}$. ◇

Explanation 8.1.11 (Canonial Basepoints for Γ-Objects)**.** In the context of Definition 8.1.8, recall from Definition 4.1.1 that C_* denotes the category of objects and morphisms under $*$ in C. Because $\underline{0}$ is also initial in $\mathcal{F}$, there is a canonical isomorphism of categories

$$\Gamma\text{-}\mathsf{C} \cong \Gamma\text{-}(\mathsf{C}_*). \tag{8.1.12}$$

For each Γ-object X, the unique morphisms

$$\underline{0} \longrightarrow \underline{n}$$

induce canonical basepoints

$$* = X\underline{0} \longrightarrow X\underline{n}.$$

By functoriality, each Γ-object in C thereby determines a unique Γ-object in C_*, and conversely. Morphisms of Γ-objects (pointed natural transformations between pointed functors) necessarily preserve the canonical basepoints, and thus we have a bijection between the morphisms of Γ-C and those of Γ-(C_*). ◇

Recall the notion of model category from Definition 7.7.3.

Definition 8.1.13. In the context of Definition 8.1.8, suppose furthermore that C is a model category. A *levelwise weak equivalence* of Γ-objects in C is a morphism f such that each $f_{\underline{n}}$ is a weak equivalence in the model structure for C. ◇

Definition 8.1.14. Suppose X is a Γ-object in C and suppose C has products. For $n > 0$, the nth *Segal map* is the morphism

$$p_n \colon X\underline{n} \longrightarrow \prod_{i \in \underline{n}^\flat} X\underline{1}$$

induced by the maps $\delta_i \colon \underline{n} \longrightarrow \underline{1}$ with $\delta_i(j) = 0$ for $j \neq i$ and $\delta_i(i) = 1$. The 0th Segal map, p_0, is the identity on the terminal object $*$.

Suppose, moreover, that C is a model category. We say that X is *special* if the Segal maps p_n are weak equivalences for all natural numbers $n \geq 0$. ◇

Explanation 8.1.15. In the context of Definition 8.1.14, the Segal map p_0 is an identity by definition and the Segal map p_1 is an identity by functoriality of X. So the first nontrivial Segal map is

$$p_2 \colon X\underline{2} \longrightarrow X\underline{1} \times X\underline{1}.$$ ◇

The two types of Γ-objects we will see most frequently below are the following. Because we will make frequent use of the canonical basepoints for Γ-objects, we take these as part of the definitions.

Definition 8.1.16 (Γ-Simplicial Sets)**.** A Γ-*simplicial set* is a Γ-object in sSet_*. We let

$$\Gamma\text{-}\mathsf{sSet}$$

denote the category of Γ-objects and morphisms in sSet_*. By (8.1.12) this is canonically isomorphic to the category of Γ-objects and morphisms in sSet. ◇

Recall from Definition I.6.3.1 the notion of modification between transformations.

Definition 8.1.17 (Γ-Categories)**.** A Γ-*category* is a Γ-object in Cat_*. We let

$$\Gamma\text{-}\mathsf{Cat}$$

denote the category of Γ-objects and morphisms in Cat_*. By (8.1.12) this is canonically isomorphic to the category of Γ-objects and morphisms in Cat.

We extend Γ-Cat to a 2-category by taking *pointed modifications*, that is, modifications θ between pointed natural transformations such that the component at $\underline{0}$ is the identity natural transformation on the identity functor of $\mathbb{1}$. We will use the notation Γ-Cat for both this 2-category and its underlying 1-category. ◇

Recall the nerve functor N from Definition 7.2.3.

Definition 8.1.18. We let

$$N_*: \Gamma\text{-}\mathsf{Cat} \longrightarrow \Gamma\text{-}\mathsf{sSet}$$

denote the functor induced by composition with the nerve functor N. For a Γ-category X, the Γ-simplicial set N_*X is the composite

$$\mathcal{F} \xrightarrow{X} \mathsf{Cat} \xrightarrow{N} \mathsf{sSet}.$$

For a morphism of Γ-categories f, the morphism N_*f is the whiskering of f (as a natural transformation) with N. ◇

8.2. Symmetric Spectra from Γ-Simplicial Sets

Now we define the K-theory of Γ-simplicial sets as a functor

$$\mathsf{K}^{\mathcal{F}}: \Gamma\text{-}\mathsf{sSet} \longrightarrow \mathsf{SymSp}.$$

Recall S^p denotes the p-fold smash product of the simplicial circle, S^1 (Definition 7.4.1). In Definition 7.1.12 the set of n-simplices $(S^1)_n$ is identified with $\underline{n}$. We will also need the following identification of the n-simplices $(S^p)_n$ for each p and n.

Definition 8.2.1 ($\mathcal{F}$-Sphere)**.** Taking the lexicographic ordering (Definition 8.1.5) defines a bijection

$$(S^p)_n = \underline{n}^{\wedge p} \cong \underline{n^p}$$

for $p > 0$. Let $\overline{S}^0 = \underline{1}$, the constant simplicial set, and for $p > 0$ let $\overline{S}^p$ be the pointed simplicial set whose n-simplicies are $\underline{n^p}$ and such that $S^p \cong \overline{S}^p$ is an isomorphism of simplicial sets. Let $\overline{S}$ be the symmetric sequence determined by these isomorphisms, so that $S \cong \overline{S}$ is an isomorphism of symmetric spectra. We refer to $\overline{S}$ as the *$\mathcal{F}$-sphere.* ◇

The structure morphisms for the symmetric spectra we construct will be defined using the following morphisms.

Definition 8.2.2. Suppose X is a Γ-simplicial set. For each $\underline{m}$ and $\underline{n}$ in $\mathcal{F}$, and for each $i \in \underline{m}^\flat$, let

$$h_i: \underline{n} \longrightarrow \bigvee_{\underline{m}^\flat} \underline{n} \cong \underline{mn}.$$

be the composite of the structure morphism indexed by the inclusion $\{i\} \subset \underline{m}^\flat$ and the isomorphism given by lexicographic ordering (Definition 8.1.5). Applying X, we define

$$\eta_{m,n,i} = Xh_i: X\underline{n} \longrightarrow X\underline{mn}.$$

Taking the wedge sum, we have

$$\eta_{m,n}: \underline{m} \wedge X\underline{n} \cong \bigvee_{i \in \underline{m}^\flat} X\underline{n} \longrightarrow X\underline{mn}. \tag{8.2.3}$$

These morphisms are natural with respect to morphisms of $\underline{m}$ and $\underline{n}$ in $\mathcal{F}$. ◇

Explanation 8.2.4. The map $h_i: \underline{n} \longrightarrow \underline{mn}$ in Definition 8.2.2 is given by

$$h_i(j) = \begin{cases} 0 & \text{if } j = 0 \text{ and} \\ n(i-1) + j & \text{if } j \in \underline{n}^\flat. \end{cases}$$

Using the matrix interpretation of $\underline{m} \wedge \underline{n}$ in Explanation 8.1.7, h_i corresponds to the inclusion of the row indexed by $i \in \underline{m}^\flat$. ◇

In the following definition we will construct pointed bisimplicial sets as functors

$$\Delta^{\mathsf{op}} \longrightarrow \mathsf{sSet}_*$$

and take the diagonal as discussed in Explanation 7.1.7 to obtain a pointed simplicial set.

Definition 8.2.5. Suppose X is a Γ-simplicial set. The *K-theory* of X is a symmetric spectrum $\mathsf{K}^{\mathcal{F}}X$ defined as follows.

For each natural number $k \geq 0$, let $(\mathsf{K}^{\mathcal{F}}X)k$ be the pointed simplicial set obtained by taking the diagonal of the pointed bisimplicial set $X \circ \overline{S}^k$:

$$\Delta^{\mathsf{op}} \xrightarrow{\overline{S}^k} \mathcal{F} \xrightarrow{X} \mathsf{sSet}_*.$$

The components $\eta_{m,n}$ defined in (8.2.3) assemble to give a natural transformation in the diagram below.

$$\begin{array}{ccccc} & & \mathcal{F}\times\mathcal{F} & \xrightarrow{1\times X} & \mathcal{F}\times\mathsf{sSet}_* \\ & \nearrow^{\overline{S}^p\times\overline{S}^q} \;\Delta^{\mathsf{op}}\times\Delta^{\mathsf{op}} & \downarrow{\scriptstyle\wedge} & \Swarrow\eta & \downarrow{\scriptstyle\wedge} \\ \Delta^{\mathsf{op}} \xrightarrow{\text{diag}} & & & & \\ \Delta^{\mathsf{op}} & \xrightarrow[\overline{S}^{p+q}]{} & \mathcal{F} & \xrightarrow[X]{} & \mathsf{sSet}_* \end{array}$$

Taking the diagonal of these pointed bisimplicial sets and using the isomorphsm $S \cong \overline{S}$, we obtain morphisms in sSet_*

$$\rho_{p,q}\colon S^p \wedge (\mathsf{K}^{\mathcal{F}}X)q \xrightarrow{\cong} \overline{S}^p \wedge (\mathsf{K}^{\mathcal{F}}X)q \longrightarrow (\mathsf{K}^{\mathcal{F}}X)(p+q).$$

In Proposition 8.2.6 we show that these data define a symmetric spectrum $\mathsf{K}^{\mathcal{F}}X$ for each Γ-simplicial set X and moreover that $\mathsf{K}^{\mathcal{F}}$ determines a functor

$$\mathsf{K}^{\mathcal{F}}\colon \Gamma\text{-}\mathsf{sSet} \longrightarrow \mathsf{SymSp}.$$

In Explanation 8.2.7 we give an alternate description of each $(\mathsf{K}^{\mathcal{F}}X)k$ in terms of its n-simplices. ◇

Proposition 8.2.6. *The data* $\{(\mathsf{K}^{\mathcal{F}}X)k,\ \rho_{p,q} \mid k,p,q \geq 0\}$ *of Definition 8.2.5 define a functor*

$$\mathsf{K}^{\mathcal{F}}\colon \Gamma\text{-}\mathsf{sSet} \longrightarrow \mathsf{SymSp}.$$

Proof. We will use the description in Explanation 7.4.6 to verify that each pair $(\mathsf{K}^{\mathcal{F}}X, \rho)$ is a symmetric spectrum. The unity condition (7.4.7) follows because $\overline{S}^0$ is the constant simplicial set $\underline{1}$ and each $\eta_{1,n}$ is the identity. The associativity condition (7.4.8) holds because associativity of the lexicographic smash product L (Definition 8.1.5) implies that the following diagram for components of η commutes for each $\underline{\ell}$, $\underline{m}$, and $\underline{n}$ in $\mathcal{F}$.

$$\begin{array}{ccc} \underline{\ell}\wedge\underline{m}\wedge X\underline{n} & \xrightarrow{1\wedge\eta_{m,n}} & \underline{\ell}\wedge X(\underline{mn}) \\ {\scriptstyle L\wedge 1}\downarrow & & \downarrow{\scriptstyle \eta_{\ell,mn}} \\ \underline{\ell m}\wedge X\underline{n} & \xrightarrow[\eta_{\ell m,n}]{} & X(\underline{\ell mn}) \end{array}$$

The action of Σ_k on $(\mathsf{K}^{\mathcal{F}}X)k$ is given by permuting the factors of $\overline{S}^k$. The equivariance of $\rho_{p,q}$ with respect to the standard inclusion $i_{p,q}\colon \Sigma_p \times \Sigma_q \hookrightarrow \Sigma_{p+q}$ then follows from the definition of the Σ_p action on the smash powers $S^p \cong \overline{S}^p$ (see Definitions 7.3.8 and 7.4.1).

Since the definition of η is natural with respect to morphisms of Γ-simplicial sets (i.e., natural transformations of functors), so are the structure morphisms $\rho_{p,q}$. This shows that $\mathsf{K}^{\mathcal{F}}$ takes values in the category of symmetric spectra and morphisms thereof. Functoriality follows because each step in the construction of $(\mathsf{K}^{\mathcal{F}}X)k$ is functorial. $\square$

Explanation 8.2.7 (Simplices of $\mathsf{K}^{\mathcal{F}}X$). Here we give a more explicit definition of the simplices of $(\mathsf{K}^{\mathcal{F}}X)k$ from Definition 8.2.5. This is the description often given in the literature. For each $\underline{n}$, evaluating X at $\overline{S}^k_n \in \mathcal{F}$ gives a simplicial set

$$\left(\underline{m} \longmapsto (X(\overline{S}^k_n))_m\right).$$

The pointed simplicial set $(\mathsf{K}^{\mathcal{F}}X)k$ is given by its diagonal:

$$(\mathsf{K}^{\mathcal{F}}X)k = X(\overline{S}^k) = \left(\underline{n} \longmapsto (X(\overline{S}^k_n))_n\right).$$

The structure morphisms

$$\Sigma X(\overline{S}^{k-1}) \longrightarrow X(\overline{S}^k)$$

are given, for each $\underline{n}$, by

$$\overline{S}^1_n \wedge (X(\overline{S}^{k-1}_n))_n \cong \bigvee_{i \in (\overline{S}^1_n)^\flat} (X(\overline{S}^{k-1}_n))_n \longrightarrow (X(\overline{S}^k_n))_n$$

induced by the inclusions

$$h_i \colon \overline{S}^{k-1}_n \cong \{0,i\} \wedge \overline{S}^{k-1}_n \longrightarrow \overline{S}^1_n \wedge \overline{S}^{k-1}_n \cong \overline{S}^k_n$$

for $i \in (\overline{S}^1_n)^\flat$. $\diamond$

8.3. Γ-Categories from Permutative Categories

In this section we define Segal's construction of Γ-categories from small permutative categories. There are three variants of this construction, and they all yield equivalent K-theory spectra by Theorems 8.3.21 and 8.5.2.

Definition 8.3.1. Suppose $(\mathsf{C}, \oplus, e)$ is a small permutative category and $\underline{n}$ is an object of $\mathcal{F}$. An *$\underline{n}$-system* in C is a pair

$$(C, \rho) = \{C_s, \rho_{s,t}\}$$

consisting of

- a system of objects, with $C_s \in \mathsf{C}$ for each basepoint-free subset $s \subset \underline{n}^\flat$, and
- a system of morphisms, with

  $$\rho_{s,t} \colon C_s \oplus C_t \longrightarrow C_{s \cup t}$$

  for each pair of disjoint basepoint-free subsets:

  $$s, t \subset \underline{n}^\flat \quad \text{and} \quad s \cap t = \varnothing.$$

We call $\rho_{s,t}$ the *(s,t)-gluing morphism*. These data are subject to the following axioms.

Object Unity: For the empty subset we have

$$C_\varnothing = e. \tag{8.3.2}$$

Gluing Unity: The $(\varnothing, t)$- and $(s, \varnothing)$-gluing morphisms are identities:

$$\rho_{\varnothing,t} = 1_{C_t} \quad \text{and} \quad \rho_{s,\varnothing} = 1_{C_s}. \tag{8.3.3}$$

Gluing Symmetry: The following diagram commutes for each pair of disjoint basepoint-free subsets s and t, where ξ denotes the symmetry isomorphism of C.

$$\begin{array}{ccc} C_s \oplus C_t & \xrightarrow{\rho_{s,t}} & C_{s \cup t} \\ {\scriptstyle \xi}\downarrow & & \downarrow{\scriptstyle 1} \\ C_t \oplus C_s & \xrightarrow{\rho_{t,s}} & C_{t \cup s} \end{array} \tag{8.3.4}$$

Gluing Associativity: The following diagram commutes for each triple of pairwise disjoint basepoint-free subsets s, t, and u.

$$\begin{array}{ccc} C_s \oplus C_t \oplus C_u & \xrightarrow{\rho_{s,t} \oplus 1} & C_{s \cup t} \oplus C_u \\ {\scriptstyle 1 \oplus \rho_{t,u}}\downarrow & & \downarrow{\scriptstyle \rho_{s \cup t,u}} \\ C_s \oplus C_{t \cup u} & \xrightarrow{\rho_{s,t \cup u}} & C_{s \cup t \cup u} \end{array} \tag{8.3.5}$$

This finishes the definition of an $\underline{n}$-system in C. We also have the following variant definitions and terms.

- When helpful for clarity, we also call an $\underline{n}$-system a *lax $\underline{n}$-system.*
- A *strong $\underline{n}$-system* is an $\underline{n}$-system for which all of the gluing morphisms $\rho_{s,t}$ are isomorphisms.
- A *colax $\underline{n}$-system* in C is an $\underline{n}$-system in C^{op}. That is, the direction of each (s, t)-gluing morphism is reversed but the same axioms (with reversed arrows $\rho_{s,t}$) are satisfied. ◇

Definition 8.3.6. Suppose $(\mathsf{C}, \oplus, e)$ is a small permutative category and suppose given $\underline{n}$-systems

$$\{C_s, \rho_{s,t}\} \quad \text{and} \quad \{C'_s, \rho'_{s,t}\}$$

for $\underline{n}$ in $\mathcal{F}$. A *morphism of $\underline{n}$-systems*, denoted

$$\{\alpha_s\} \colon \{C_s, \rho_{s,t}\} \longrightarrow \{C'_s, \rho'_{s,t}\},$$

consists of component morphisms

$$\alpha_s \colon C_s \longrightarrow C'_s,$$

for each basepoint-free subset $s \subset \underline{n}^{\flat}$. These components are subject to the following axioms.

Unitary: For the empty subset we have

$$\alpha_{\varnothing} = 1_e. \tag{8.3.7}$$

Gluing Compatibility: The following diagram commutes for each pair of disjoint basepoint-free subsets s and t.

$$\begin{array}{ccc} C_s \oplus C_t & \xrightarrow{\rho_{s,t}} & C_{s \cup t} \\ {\scriptstyle \alpha_s \oplus \alpha_t}\downarrow & & \downarrow{\scriptstyle \alpha_{s \cup t}} \\ C'_s \oplus C'_t & \xrightarrow{\rho'_{s,t}} & C'_{s \cup t} \end{array} \tag{8.3.8}$$

This finishes the definition of a morphism of $\underline{n}$-systems. We also have the following variant definitions.

- A morphism of strong $\underline{n}$-systems is defined as above, with $\{C_s, \rho_{s,t}\}$ and $\{C'_s, \rho'_{s,t}\}$ strong $\underline{n}$-systems.
- A morphism of colax $\underline{n}$-systems is defined as above, with $\{C_s, \rho_{s,t}\}$ and $\{C'_s, \rho'_{s,t}\}$ colax $\underline{n}$-systems, but the gluing compatibility axiom (8.3.8) has the variance of ρ reversed as in the following diagram.

$$\begin{array}{ccc} C_s \oplus C_t & \xleftarrow{\rho_{s,t}} & C_{s\cup t} \\ {\scriptstyle \alpha_s \oplus \alpha_t}\downarrow & & \downarrow{\scriptstyle \alpha_{s\cup t}} \\ C'_s \oplus C'_t & \xleftarrow{\rho'_{s,t}} & C'_{s\cup t} \end{array}$$

The identity morphism for a lax, strong, or colax $\underline{n}$-system $\{C_s, \rho_{s,t}\}$ consists of identities

$$\alpha_s = 1_{C_s} \quad \text{for} \quad s \subset \underline{n}^\flat.$$

Composition of morphisms is defined componentwise. This composition is associative and unital because the composition in C is so. ◇

Definition 8.3.9. Suppose $(\mathsf{C}, \oplus, e)$ is a small permutative category and suppose given $\underline{n}$ in $\mathcal{F}$. Define a pointed category $\mathsf{C}^{\mathcal{F}}\underline{n}$ as follows.

- The objects of $\mathsf{C}^{\mathcal{F}}\underline{n}$ are the strong $\underline{n}$-systems in C.
- The morphisms of $\mathsf{C}^{\mathcal{F}}\underline{n}$ are the morphisms of strong $\underline{n}$-systems.
- The basepoint of $\mathsf{C}^{\mathcal{F}}\underline{n}$ is the constant $\underline{n}$-system with

  $$C_s = e \quad \text{and} \quad \rho_{s,t} = 1_e$$

  for all disjoint basepoint-free subsets:

  $$s, t \subset \underline{n}^\flat \quad \text{and} \quad s \cap t = \varnothing.$$

This finishes the definition of $\mathsf{C}^{\mathcal{F}}\underline{n}$. We also have the following variant definitions and notation.

- To emphasize that the objects of $\mathsf{C}^{\mathcal{F}}\underline{n}$ are the strong $\underline{n}$-systems, we also use the notation

  $$\mathsf{C}^{\mathcal{F}}_{\cong}\underline{n} = \mathsf{C}^{\mathcal{F}}\underline{n}.$$

- We let

  $$\mathsf{C}^{\mathcal{F}}_{\mathrm{lax}}\underline{n}$$

  denote the category of lax $\underline{n}$-systems and morphisms of such. The basepoint is that of $\mathsf{C}^{\mathcal{F}}_{\cong}\underline{n}$.
- We let

  $$\mathsf{C}^{\mathcal{F}}_{\mathrm{co}}\underline{n}$$

  denote the category of colax $\underline{n}$-systems and morphisms of such. The basepoint is that of $\mathsf{C}^{\mathcal{F}}_{\cong}\underline{n}$. ◇

In the following examples, we describe explicitly the categories $\mathsf{C}^{\mathcal{F}}\underline{n}$ for $n = 0, 1, 2$.

Example 8.3.10. There are canonical isomorphisms

$$\mathsf{C}^{\mathcal{F}}\underline{0} \cong \mathbb{1}$$
$$\mathsf{C}^{\mathcal{F}}\underline{1} \cong \mathsf{C}$$

with $\mathbb{1}$ the terminal category. ◇

Example 8.3.11. Since

$$2^{\underline{2}^{\flat}} = \{\varnothing, \{1\}, \{2\}, \{1,2\}\},$$

up to a canonical isomorphism, $\mathsf{C}^{\mathcal{F}}\underline{2}$ may be described as follows. An object in $\mathsf{C}^{\mathcal{F}}\underline{2}$ is a quadruple

$$\left(X_1, X_2, X_{12}, \rho^X : X_1 \oplus X_2 \xrightarrow{\cong} X_{12}\right)$$

consisting of

- three objects $X_1, X_2, X_{12} \in \mathsf{C}$ and
- an isomorphism $\rho^X \in \mathsf{C}$ as indicated.

A morphism

$$\left(X_1, X_2, X_{12}, \rho^X\right) \xrightarrow{\alpha = (\alpha_1, \alpha_2, \alpha_{12})} \left(Y_1, Y_2, Y_{12}, \rho^Y\right) \in \mathsf{C}^{\mathcal{F}}\underline{2}$$

consists of morphisms

$$X_a \xrightarrow{\alpha_a} Y_a \in \mathsf{C} \quad \text{for} \quad a \in \{1, 2, 12\}$$

such that the diagram

$$\begin{array}{ccc} X_1 \oplus X_2 & \xrightarrow{\rho^X} & X_{12} \\ {\scriptstyle \alpha_1 \oplus \alpha_2} \downarrow & & \downarrow {\scriptstyle \alpha_{12}} \\ Y_1 \oplus Y_2 & \xrightarrow{\rho^Y} & Y_{12} \end{array}$$

commutes.

- The identity morphism is given by
$$1_{(X_1, X_2, X_{12}, \rho^X)} = \left(1_{X_1}, 1_{X_2}, 1_{X_{12}}\right).$$
- Composition is defined entrywise in C.

This finishes the description of $\mathsf{C}^{\mathcal{F}}\underline{2}$. ◇

Now we define $\mathsf{C}^{\mathcal{F}}$ on the morphisms of $\mathcal{F}$.

Definition 8.3.12. Suppose $(\mathsf{C}, \oplus, e)$ is a small permutative category and suppose given a morphism $\underline{n}$ in $\mathcal{F}$

$$\psi : \underline{n} \longrightarrow \underline{m}.$$

Define a pointed functor

$$\mathsf{C}^{\mathcal{F}}\psi : \mathsf{C}^{\mathcal{F}}\underline{n} \longrightarrow \mathsf{C}^{\mathcal{F}}\underline{m}$$

as follows.

- Given an $\underline{n}$-system $\{C_s, \rho_{s,t}\}$, define an $\underline{m}$-system $\{C_r^{\psi}, \rho_{r,q}^{\psi}\}$ with
$$C_r^{\psi} = C_{\psi^{-1}(r)} \quad \text{and}$$
$$\rho_{r,q}^{\psi} = \rho_{\psi^{-1}(r), \psi^{-1}(q)}$$
for disjoint basepoint-free subsets:
$$r, q \subset \underline{m}^{\flat} \quad \text{and} \quad r \cap q = \varnothing.$$

- Given a morphism of $\underline{n}$-systems $\{\alpha_s\}$, define a morphism of $\underline{m}$-systems $\{\alpha_r^\psi\}$ with
$$\alpha_r^\psi = \alpha_{\psi^{-1}(r)} \quad \text{for} \quad r \subset \underline{m}^\flat.$$

These definitions give well-defined $\underline{m}$-systems and morphisms thereof because taking inverse images of a basepoint-preserving function preserves the properties of being basepoint-free and disjoint. If C_s, respectively $\rho_{s,t}$, is constant at e, respectively 1_e, then so is C_r^ψ, respectively $\rho_{r,q}^\psi$.

Define the pointed functor $\mathsf{C}^{\mathcal{F}}\psi$ by the following assignment on objects and morphisms:

$$\{C_s, \rho_{s,t}\} \longmapsto \{C_r^\psi, \rho_{r,q}^\psi\} \quad \text{and}$$
$$\{\alpha_s\} \longmapsto \{\alpha_r^\psi\}.$$

Functoriality of $\mathsf{C}^{\mathcal{F}}\psi$ follows because composition and identities in $\mathsf{C}^{\mathcal{F}}\underline{n}$ are defined componentwise.

This finishes the definition of $\mathsf{C}^{\mathcal{F}}\psi$. We also have the following variant definitions and notation extending that of Definition 8.3.9.

- For emphasis or clarity we sometimes also use the notation
$$\mathsf{C}^{\mathcal{F}}_{\cong}\psi = \mathsf{C}^{\mathcal{F}}\psi.$$
- We let
$$\mathsf{C}^{\mathcal{F}}_{\mathrm{lax}}\psi \colon \mathsf{C}^{\mathcal{F}}_{\mathrm{lax}}\underline{n} \longrightarrow \mathsf{C}^{\mathcal{F}}_{\mathrm{lax}}\underline{m}$$
and
$$\mathsf{C}^{\mathcal{F}}_{\mathrm{co}}\psi \colon \mathsf{C}^{\mathcal{F}}_{\mathrm{co}}\underline{n} \longrightarrow \mathsf{C}^{\mathcal{F}}_{\mathrm{co}}\underline{m}$$
be the pointed functors given by the same definitions as above for lax, respectively colax, $\underline{n}$-systems and morphisms. ◇

Taken together, Definitions 8.3.1, 8.3.6, 8.3.9, and 8.3.12 give assignments on objects and morphisms

$$\mathsf{C}^{\mathcal{F}} = \mathsf{C}^{\mathcal{F}}_{\cong} \colon \mathcal{F} \longrightarrow \mathsf{Cat}_*,$$
$$\mathsf{C}^{\mathcal{F}}_{\mathrm{lax}} \colon \mathcal{F} \longrightarrow \mathsf{Cat}_*, \quad \text{and}$$
$$\mathsf{C}^{\mathcal{F}}_{\mathrm{co}} \colon \mathcal{F} \longrightarrow \mathsf{Cat}_*.$$

Now we show that these are Γ-categories.

Proposition 8.3.13. *Suppose C is a small permutative category. Each of the assignments*
$$\mathsf{C}^{\mathcal{F}} = \mathsf{C}^{\mathcal{F}}_{\cong}, \quad \mathsf{C}^{\mathcal{F}}_{\mathrm{lax}}, \quad \textit{and} \quad \mathsf{C}^{\mathcal{F}}_{\mathrm{co}}$$
defines a Γ-category.

Proof. First note that if $\psi = 1_{\underline{n}}$ then the definition of $\mathsf{C}^{\mathcal{F}}\psi$ will be the identity functor. Functoriality with respect to a composable pair ψ and ϕ follows because $(\psi\phi)^{-1} = \phi^{-1}\psi^{-1}$. The same argument applies to show that $\mathsf{C}^{\mathcal{F}}_{\mathrm{lax}}$ and $\mathsf{C}^{\mathcal{F}}_{\mathrm{co}}$ are functorial. □

Definition 8.3.14. Suppose C is a small permutative category. We call

- $\mathsf{C}^{\mathcal{F}} = \mathsf{C}^{\mathcal{F}}_{\cong}$ the *strong Segal Γ-category* of C,
- $\mathsf{C}^{\mathcal{F}}_{\mathrm{lax}}$ the *lax Segal Γ-category* of C, and
- $\mathsf{C}^{\mathcal{F}}_{\mathrm{co}}$ the *colax Segal Γ-category* of C. ◇

Now we relate the three Segal Γ-category constructions.

Definition 8.3.15. Suppose $(\mathsf{C}, \oplus, e, \xi)$ is a small permutative category and

$$\nu \in \{\cong, \mathrm{lax}, \mathrm{co}\}$$

is one of the variant subscripts for $\mathsf{C}^{\mathcal{F}}$. For each $\underline{n}$ in $\mathcal{F}$, let $(e, \ldots, e)$ be the basepoint of C^n and define pointed functors

$$P_\nu \colon \mathsf{C}^{\mathcal{F}}_\nu \underline{n} \longrightarrow \mathsf{C}^n \quad \text{and} \quad Q_\nu \colon \mathsf{C}^n \longrightarrow \mathsf{C}^{\mathcal{F}}_\nu \underline{n}$$

as follows. For objects $\{C_s, \rho_{s,t}\}$ and morphisms $\{\alpha_s\}$ in $\mathsf{C}^{\mathcal{F}}_\nu \underline{n}$, define

$$\begin{aligned} P_\nu\{C_s, \rho_{s,t}\} &= (C_{\{1\}}, \ldots, C_{\{n\}}) \quad \text{and} \\ P_\nu\{\alpha_s\} &= (\alpha_{\{1\}}, \ldots, \alpha_{\{n\}}). \end{aligned}$$

For objects $(C_1, \ldots, C_n)$ and morphisms $(\alpha_1, \ldots, \alpha_n)$ in C^n, define

$$Q_\nu(C_1, \ldots, C_n) = \begin{cases} \{\oplus_{i \in s} C_i \,,\, \xi_{s,t}\}, & \text{if } \nu \in \{\cong, \mathrm{lax}\} \\ \{\oplus_{i \in s} C_i \,,\, \xi_{s,t}^{-1}\}, & \text{if } \nu = \mathrm{co}, \end{cases}$$

and

$$Q_\nu(\alpha_1, \ldots, \alpha_n) = \{\oplus_{i \in s} \alpha_i\}$$

where

$$\xi_{s,t} \colon \Big(\bigoplus_{i \in s} C_i\Big) \oplus \Big(\bigoplus_{i \in t} C_i\Big) \longrightarrow \bigoplus_{i \in (s \cup t)} C_i$$

is the unique morphism given by permuting summands with the symmetry of C. Functoriality of P_ν and Q_ν follows because composition and identities in each of the categories $\mathsf{C}^{\mathcal{F}}_\nu \underline{n}$ and C^n are given componentwise. Each of P_ν and Q_ν is a pointed functor by definition of the basepoints in $\mathsf{C}^{\mathcal{F}}_\nu \underline{n}$ and the assumption that e is a strict unit for C. ◇

Proposition 8.3.16. *In the context of Definition 8.3.15,*

- *$(P_\cong, Q_\cong)$ is an adjoint equivalence,*
- *P_{lax} is a right adjoint, with left adjoint Q_{lax}, and*
- *P_{co} is a left adjoint, with right adjoint Q_{co}.*

Proof. For each variant

$$\nu \in \{\cong, \mathrm{lax}, \mathrm{co}\},$$

the composite

$$\mathsf{C}^n \xrightarrow{Q_\nu} \mathsf{C}^{\mathcal{F}}_\nu \underline{n} \xrightarrow{P_\nu} \mathsf{C}^n$$

is the identity functor. The other composite

$$\mathsf{C}^{\mathcal{F}}_\nu \underline{n} \xrightarrow{P_\nu} \mathsf{C}^n \xrightarrow{Q_\nu} \mathsf{C}^{\mathcal{F}}_\nu \underline{n}$$

is given on objects and morphisms of $\mathsf{C}^{\mathcal{F}}_\nu \underline{n}$ by the following assignments:

$$\{C_s, \rho_{s,t}\} \longmapsto \begin{cases} \{\oplus_{i \in s} C_{\{i\}} \,,\, \xi_{s,t}\} & \text{for } \nu = \cong,\ \nu - \mathrm{lax} \\ \{\oplus_{i \in s} C_{\{i\}} \,,\, \xi_{s,t}^{-1}\} & \text{for } \nu = \mathrm{co} \end{cases}$$

and

$$\{\alpha_s\} \longmapsto \{\oplus_{i \in s} \alpha_{\{i\}}\}.$$

If the variant ν is either $\cong$ or lax, then the gluing morphisms of each $\underline{n}$-system provide morphisms

$$\widehat{\rho}_s\colon \bigoplus_{i\in s} C_{\{i\}} \longrightarrow C_s \tag{8.3.17}$$

for each basepoint-free subset $s \subset \underline{n}^\flat$. These are uniquely determined by the gluing associativity condition (8.3.5). Using gluing associativity again together with gluing symmetry (8.3.4), the morphisms $\widehat{\rho}_s$ satisfy the gluing compatibility condition (8.3.8) with the morphisms $\xi_{s,t}$ and, therefore, define a morphism

$$\{\widehat{\rho}_s\}\colon \{\oplus_{i\in s} C_{\{i\}}\ ,\ \xi_{s,t}\} \longrightarrow \{C_s, \rho_{s,t}\}. \tag{8.3.18}$$

If the variant ν is co, the colax gluing morphisms of a colax $\underline{n}$-system provide morphisms $\widehat{\rho}_s^{\,\mathrm{co}}$ that go in the opposite direction of (8.3.17). Then we have a morphism of colax $\underline{n}$-systems

$$\{\widehat{\rho}_s^{\,\mathrm{co}}\}\colon \{C_s, \rho_{s,t}\} \longrightarrow \{\oplus_{i\in s} C_{\{i\}}, \xi_{s,t}^{-1}\}.$$

In any of the three cases for ν, the morphisms (8.3.18) are natural with respect to morphisms $\{\alpha_s\}$ by the gluing compatibility (8.3.8) for $\{\alpha_s\}$ with the morphisms $\rho_{s,t}$. Thus we have a natural transformation γ with components

$$\gamma_{\{C_s, \rho_{s,t}\}} = \{\widehat{\rho}_s\}$$

that provides either a counit (if ν is either $\cong$ or lax) or a unit (if ν is co). Since the gluing morphisms for the basepoint of $\mathsf{C}^{\mathcal{F}}_\nu \underline{n}$ are identities, the component of γ at the basepoint is its identity.

In any of the three cases for ν, one verifies the triangle identities with the following computations.

- For $\{C_s, \rho_{s,t}\}$ in $\mathsf{C}^{\mathcal{F}}_\nu \underline{n}$,
$$P_\nu \gamma_{\{C_s, \rho_{s,t}\}} = P_\nu \{\widehat{\rho}_s\} = (\widehat{\rho}_{\{1\}}, \ldots, \widehat{\rho}_{\{n\}}) = 1_{(C_{\{1\}}, \ldots, C_{\{n\}})}.$$
- For $(C_1, \ldots, C_n)$ in C^n,
$$\begin{aligned}\gamma_{Q_\nu(C_1,\ldots,C_n)} &= \begin{cases} \{\widehat{\xi}_s\} & \text{if } \ \nu = \cong,\ \nu = \mathrm{lax}, \\ \{\widehat{\xi^{-1}}_s^{\,\mathrm{co}}\} & \text{if } \ \nu = \mathrm{co}, \end{cases} \\ &= \left\{1_{\left(\oplus_{i\in s} C_{\{i\}}\right)}\right\}.\end{aligned}$$

The assertion that $(P_\cong, Q_\cong)$ is an adjoint equivalence follows because the components of γ are given by the gluing morphisms $\rho_{s,t}$. □

Observe that each strong $\underline{n}$ system is lax and, by taking inverses of the gluing morphisms, also colax. Thus we have forgetful functors, given by the identities on morphisms

$$\mathsf{C}^{\mathcal{F}}_{\mathrm{co}} \underline{n} \longleftarrow \mathsf{C}^{\mathcal{F}}_{\cong} \underline{n} \longrightarrow \mathsf{C}^{\mathcal{F}}_{\mathrm{lax}} \underline{n} \tag{8.3.19}$$

for each $\underline{n}$ in $\mathcal{F}$. These are natural with respect to morphisms in $\mathcal{F}$ and therefore we have the following.

Definition 8.3.20. Suppose C is a small permutative category. The *levelwise inclusions*

$$\mathsf{C}^{\mathcal{F}}_{\mathrm{co}} \longleftarrow \mathsf{C}^{\mathcal{F}}_{\cong} \longrightarrow \mathsf{C}^{\mathcal{F}}_{\mathrm{lax}}$$

are the morphisms of Γ-categories induced by the forgetful functors (8.3.19). ◇

Recall from Definition 8.1.18 that

$$N_*: \Gamma\text{-}\mathsf{Cat} \longrightarrow \Gamma\text{-}\mathsf{sSet}$$

denotes the functor induced by composition with the nerve functor N.

Theorem 8.3.21. *Suppose* C *is a small permutative category. Each of the* Γ-*simplicial sets*

$$N_*\mathsf{C}^{\mathcal{F}}_{\cong}, \quad N_*\mathsf{C}^{\mathcal{F}}_{\mathrm{lax}}, \quad \text{and} \quad N_*\mathsf{C}^{\mathcal{F}}_{\mathrm{co}}$$

is special. Moreover, the levelwise inclusions

$$\mathsf{C}^{\mathcal{F}}_{\mathrm{co}} \longleftarrow \mathsf{C}^{\mathcal{F}}_{\cong} \longrightarrow \mathsf{C}^{\mathcal{F}}_{\mathrm{lax}}$$

induce levelwise weak equivalences of Γ-*simplicial sets.*

Proof. For each variant

$$\nu \in \{\cong, \mathrm{lax}, \mathrm{co}\}$$

we have an isomorphism of categories

$$\mathsf{C}^{\mathcal{F}}_{\nu}\underline{1} \cong \mathsf{C}.$$

Since the nerve functor is a right adjoint, we have an isomorphism of simplicial sets

$$N(\mathsf{C}^n) \cong (N\mathsf{C})^n$$

for each $n \geq 0$. Recalling Proposition 7.2.5, an adjunction of categories induces a simplicial homotopy equivalence on nerves. Therefore the adjunctions of Proposition 8.3.16 provide simplicial homotopy equivalences in the composite below

$$N\mathsf{C}^{\mathcal{F}}_{\nu}\underline{n} \xrightarrow{\simeq} N(\mathsf{C}^n) \cong (N\mathsf{C})^n \cong (N\mathsf{C}^{\mathcal{F}}_{\nu}\underline{1})^n$$

for each $\underline{n}$ in $\mathcal{F}$. We recall from Explanation 7.8.4 that each simplicial homotopy equivalence is a weak equivalence. This completes the proof that each $N\mathsf{C}^{\mathcal{F}}_{\nu}$ is special.

For the second assertion we note, moreover, that the levelwise inclusions commute with the functors

$$Q_{\nu}: \mathsf{C}^n \longrightarrow \mathsf{C}^{\mathcal{F}}_{\nu}\underline{n}$$

for each variant ν and each $\underline{n}$ in $\mathcal{F}$. Therefore the levelwise inclusions are weak equivalences by the 2-out-of-3 property of Definition 7.7.3. □

8.4. Partition Multicategories

For each small permutative category C and each object $\underline{n}$ of $\mathcal{F}$, the objects of the category $\mathsf{C}^{\mathcal{F}}\underline{n}$ are systems indexed by disjoint pairs of basepoint-free subsets of $\underline{n}$; the morphisms are similarly indexed. The partition multicategories we introduce in this section provide an abstract approach to such indexed systems. In Proposition 8.4.8 we show that the Γ-categories constructed via partition multicategories are isomorphic to those constructed via $(-)^{\mathcal{F}}_{\mathrm{lax}}$.

Our approach via partition multicategories will make the categorical properties of the Segal J-theory and K-theory constructions in Section 8.5 more transparent. Moreover, our development of Elmendorf-Mandell K-theory in Chapter 10 depends on a generalization of the material here.

Definition 8.4.1. Suppose a is a pointed finite set. We define the *partition multicategory*, $\mathcal{M}a$, as follows. The object set of $\mathcal{M}a$ is 2^{a^b}, the set of basepoint-free subsets of a. For an n-tuple of subsets

$$\langle s\rangle = (s_1,\ldots,s_n) \in \mathsf{Prof}(2^{a^b})$$

and a subset $t \in 2^{a^b}$, we define the set of operations, $(\mathcal{M}a)\big(\langle s\rangle\,;t\big)$, to be a 1-element set if $\langle s\rangle$ is a partition of t, and empty otherwise. If $\langle s\rangle$ is a partition of t, we let $\iota_{\langle s\rangle}$ denote the single operation in $(\mathcal{M}a)\big(\langle s\rangle\,;t\big)$.

Thus

$$(\mathcal{M}a)\big(\langle s\rangle\,;t\big) = \begin{cases} \{\iota_{\langle s\rangle}\} & \text{if } s_i \cap s_j = \varnothing \text{ for all } i \neq j \text{ and } t = \coprod_i s_i, \\ \varnothing & \text{otherwise.} \end{cases}$$

The empty set $\varnothing \in 2^{a^b}$ provides a basepoint for $\mathcal{M}a$ with the unique operations in

$$(\mathcal{M}a)\big(\underbrace{(\varnothing,\ldots,\varnothing)}_{n \text{ terms}}\,;\varnothing\big) \quad \text{for} \quad n \geq 0.$$

This defines the objects, operations, and basepoint of $\mathcal{M}a$. In Proposition 8.4.2 we show that these data satisfy the axioms of a pointed multicategory. ◇

Proposition 8.4.2. *In the context of Definition 8.4.1, $\mathcal{M}a$ is a pointed multicategory.*

Proof. For unit operations we have

$$1_t = \iota_t \in (\mathcal{M}a)\big(t\,;t\big).$$

For symmetric group actions, suppose $\langle s\rangle \in \mathsf{Prof}(2^{a^b})$ has length n. If $(\mathcal{M}a)\big(\langle s\rangle\,;t\big)$ is nonempty (and therefore a singleton) then so is

$$(\mathcal{M}a)\big(\langle s\rangle\sigma\,;t\big) \quad \text{for} \quad \sigma \in \Sigma_n.$$

Therefore, the symmetric group actions

$$(\mathcal{M}a)\big(\langle s\rangle\,;t\big) \xrightarrow[\cong]{\sigma} (\mathcal{M}a)\big(\langle s\rangle\sigma\,;t\big) \quad \text{for} \quad \sigma \in \Sigma_n$$

are uniquely determined. For composition, suppose $\langle s\rangle \in \mathsf{Prof}(2^{a^b})$ has length n and suppose given

$$\langle r_i\rangle \in \mathsf{Prof}(2^{a^b}) \quad \text{for} \quad i \in \{1,\ldots,n\}.$$

Let $\langle r\rangle$ denote the concatenation of the $\langle r_i\rangle$. If

$$(\mathcal{M}a)\big(\langle s\rangle\,;t\big) \times \prod_{i=1}^{n} (\mathcal{M}a)\big(\langle r_i\rangle\,;s_i\big)$$

is nonempty, then so is

$$(\mathcal{M}a)\big(\langle r\rangle\,;t\big).$$

Therefore, the composition

$$(\mathcal{M}a)\big(\langle s\rangle\,;t\big) \times \prod_{i=1}^{n} (\mathcal{M}a)\big(\langle r_i\rangle\,;s_i\big) \longrightarrow (\mathcal{M}a)\big(\langle r\rangle\,;t\big)$$

is uniquely determined. Each of the axioms in Definition 5.1.2 consists of diagrams where either the common domain of the two composites is empty and there is nothing to verify, or the common domain is a singleton and therefore so is each of the other operation sets in the diagram. □

Explanation 8.4.3. For a pointed finite set a, let C denote the discrete permutative category whose objects are those of the commutative monoid

$$(2^{a^\flat}, \cup, \varnothing).$$

Then the objects of $\mathcal{M}a$ are those of $\mathsf{End}(\mathsf{C})$ but the operations are only those whose input profiles consist of pairwise disjoint subsets. $\diamond$

In the following examples, we describe explicitly the partition multicategories $\mathcal{M}\underline{n}$ for $n = 0, 1, 2$.

Example 8.4.4. The partition multicategory $\mathcal{M}\underline{0}$ has object set

$$2^{\underline{0}^\flat} = \{\varnothing\}.$$

Each set of operations has a single element,

$$\mathcal{M}\underline{0}(\langle\varnothing\rangle; \varnothing) = \{\iota^n\},$$

where $\langle\varnothing\rangle = (\varnothing, \ldots, \varnothing)$ contains n copies of the empty set. Its multicategory structure is given as follows.

- The $\varnothing$-colored unit is $\iota^1 \in \mathcal{M}\underline{0}(\varnothing; \varnothing)$.
- Each ι^n is fixed by the right Σ_n-action.
- The composition is given by

$$\iota^n \circ (\iota^{k_1}, \ldots, \iota^{k_n}) = \iota^{k_1 + \cdots + k_n}$$

  for $n \geq 1$ and $k_1, \ldots, k_n \geq 0$.

Therefore, there is a canonical isomorphism

$$\mathcal{M}\underline{0} \cong \mathsf{T}$$

with the terminal multicategory T (Definition 5.2.1). $\diamond$

Example 8.4.5. The partition multicategory $\mathcal{M}\underline{1}$ has object set

$$2^{\underline{1}^\flat} = \{\varnothing, \{1\}\}.$$

Its nonempty sets of operations are

$$\mathcal{M}\underline{1}(\langle\varnothing\rangle; \varnothing) = \{\iota^n\}$$
$$\mathcal{M}\underline{1}((\varnothing, \ldots, \{1\}, \ldots, \varnothing); \{1\}) = \{\pi_j^n\}$$

for $n \geq 0$ and $1 \leq j \leq n$. In the definition of π_j^n above,

$$(\varnothing, \ldots, \{1\}, \ldots, \varnothing)$$

has length n with $\{1\}$ in the jth entry and $\varnothing$ in other entries. The multicategory structure involving only ι^n for $n \geq 0$ is as in $\mathcal{M}\underline{0}$ in Example 8.4.4.

The rest of the multicategory structure in $\mathcal{M}\underline{1}$ is given as follows.

- The $\{1\}$-colored unit is

$$\pi_1^1 \in \mathcal{M}\underline{1}(\{1\}; \{1\}).$$

- The right Σ_n-action is given by

$$\pi_j^n \cdot \sigma = \pi_{\sigma^{-1}(j)}^n \quad \text{for} \quad \sigma \in \Sigma_n.$$

The composition involving π_j^n is given by

$$\pi_j^n \circ \left(\iota^{k_1}, \ldots, \iota^{k_{j-1}}, \pi_p^{k_j}, \iota^{k_{j+1}}, \ldots, \iota^{k_n}\right) = \pi_{k_1 + \cdots + k_{j-1} + p}^{k_1 + \cdots + k_n}$$

for $1 \leq p \leq k_j$. This finishes the description of $\mathcal{M}\underline{1}$. $\diamond$

Example 8.4.6. The partition multicategory $\mathcal{M}\underline{2}$ has object set

$$2^{\underline{2}^\flat} = \{\varnothing, \{1\}, \{2\}, \{1,2\}\}.$$

Its nonempty sets of operations are

$$\begin{aligned}\mathcal{M}\underline{2}(\langle\varnothing\rangle\,;\varnothing) &= \{\iota^n\}\\ \mathcal{M}\underline{2}((\varnothing,\ldots,\{1\},\ldots,\varnothing)\,;\{1\}) &= \{\pi^n_j\}\\ \mathcal{M}\underline{2}((\varnothing,\ldots,\{2\},\ldots,\varnothing)\,;\{2\}) &= \{\tau^n_j\}\\ \mathcal{M}\underline{2}((\varnothing,\ldots,\{1\},\ldots,\{2\},\ldots,\varnothing)\,;\{1,2\}) &= \{\theta^n_{i,j}\}\end{aligned}$$

for $n \geq 0$, $1 \leq j \leq n$ in the second and third lines, and $1 \leq i \neq j \leq n$ in the last line.

- ι^n and π^n_j are as in $\mathcal{M}\underline{1}$ in Example 8.4.5.
- In the definition of τ^n_j,

$$(\varnothing,\ldots,\{2\},\ldots,\varnothing)$$

  has length n with $\{2\}$ in the jth entry and $\varnothing$ in other entries.
- In the definition of $\theta^n_{i,j}$, $\{1\}$ and $\{2\}$ are in, respectively, the ith and the jth entries in

$$(\varnothing,\ldots,\{1\},\ldots,\{2\},\ldots,\varnothing),$$

  with both $i < j$ and $i > j$ possible. All other entries are $\varnothing$.

The multicategory structure in $\mathcal{M}\underline{2}$ involving only ι^n and π^n_j is as in $\mathcal{M}\underline{1}$ in Example 8.4.5. The multicategory structure involving only ι^n and τ^n_j is also as in $\mathcal{M}\underline{1}$, with τ^n_j playing the role of π^n_j.

The rest of the multicategory structure in $\mathcal{M}\underline{2}$ is given as follows.

- The $\{1,2\}$-colored unit is

$$\theta^2_{1,2} \in \mathcal{M}\underline{2}((\{1\},\{2\})\,;\{1,2\}).$$

- The right Σ_n-action is given by

$$\theta^n_{i,j}\cdot\sigma = \theta^n_{\sigma^{-1}(i),\sigma^{-1}(j)} \quad\text{for}\quad \sigma\in\Sigma_n.$$

The composition involving $\theta^n_{i,j}$ is given by

$$\theta^n_{i,j}\circ\left(\iota^{k_1},\ldots,\pi^{k_i}_p,\ldots,\tau^{k_j}_q,\ldots,\iota^{k_n}\right) = \theta^{k_1+\cdots+k_n}_{k_1+\cdots+k_{i-1}+p,\,k_1+\cdots+k_{j-1}+q}$$

for $1 \leq i \neq j \leq n$, $1 \leq p \leq k_i$, and $1 \leq q \leq k_j$. For $1 \leq r \leq n$, the rth entry in

$$\left(\iota^{k_1},\ldots,\pi^{k_i}_p,\ldots,\tau^{k_j}_q,\ldots,\iota^{k_n}\right)$$

is

$$\begin{cases}\iota^{k_r} & \text{if } r \neq i,j,\\ \pi^{k_i}_p & \text{if } r = i, \text{ and}\\ \tau^{k_j}_q & \text{if } r = j.\end{cases}$$

This finishes the description of $\mathcal{M}\underline{2}$. ◇

Proposition 8.4.7. *The assignment*

$$\underline{n} \longmapsto \mathcal{M}\underline{n}$$

is the assignment on objects of a pointed functor

$$\mathcal{M}\colon (\mathcal{F}^{\mathsf{op}}, \underline{0}) \longrightarrow (\mathsf{Multicat}_*, \mathsf{T})$$

taking values in the category of small pointed multicategories.

Proof. Suppose $\psi\colon \underline{n} \longrightarrow \underline{m}$ is a morphism of pointed finite sets. Since ψ preserves the basepoint, taking inverse images provides a function from subsets of $\underline{m}^\flat$ to subsets of $\underline{n}^\flat$:

$$2^{\underline{m}^\flat} \longrightarrow 2^{\underline{n}^\flat}, \quad u \longmapsto \psi^{-1}(u).$$

We denote this function $\tilde{\psi}$. Since taking inverse images preserves disjunction, $\tilde{\psi}$ defines a pointed multifunctor

$$\tilde{\psi}\colon \mathcal{M}\underline{m} \longrightarrow \mathcal{M}\underline{n}$$

that is an identity if ψ is an identity. For composable functions ϕ and ψ in $\mathcal{F}$, we have $(\phi\psi)^{-1} = \psi^{-1}\phi^{-1}$ and thus $\mathcal{M}$ is (contravariantly) functorial. $\square$

We use the partition multicategories to give a reformulation of the lax Segal Γ-categories associated to a small permutative category. This reformulation will be used in our comparison with the more general Elmendorf-Mandell K-theory in Chapter 10. More immediately, we will use this reformulation to verify functoriality of the lax Segal Γ-categories as the input permutative category C varies.

Proposition 8.4.8. *Suppose C is a small permutative category. For each $\underline{n}$ in $\mathcal{F}$ there is an isomorphism of pointed categories*

$$\mathsf{Multicat}_*(\mathcal{M}\underline{n}, \mathsf{End}(\mathsf{C})) \cong \mathsf{C}^{\mathcal{F}}_{\mathrm{lax}}\underline{n}.$$

These isomorphisms are natural with respect to morphisms in $\mathcal{F}$.

Proof. The fundamental reason for this result is that the operations $\iota_{\langle c\rangle}$ of length $n > 2$ can be decomposed as composites of binary operations $\iota_{(s,t)}$. This implies that pointed multifunctors from $\mathcal{M}\underline{n}$ to $\mathsf{End}(\mathsf{C})$ are determined by their values on binary operations, in the following two specific ways.

Suppose given a pointed multifunctor

$$F\colon \mathcal{M}\underline{n} \longrightarrow \mathsf{End}(\mathsf{C}).$$

(1) For a tuple $\langle c\rangle$ of length $n > 2$, the composition axiom (5.1.15) implies that $F\iota_{\langle c\rangle}$ is determined by its value on the operations $\iota_{(s,t)}$ for disjoint pairs $s, t \in 2^{\underline{n}^\flat}$. In particular, for pairwise disjoint subsets $s, t, u \in 2^{\underline{n}^\flat}$ the ternary operation $\iota_{(s,t,u)}$ is equal to both of the following composites:

$$\iota_{(s\cup t,u)} \circ \big(\iota_{(s,t)}, \iota_u\big) = \iota_{(s,t,u)} = \iota_{(s,t\cup u)} \circ \big(\iota_s, \iota_{(t,u)}\big).$$

Therefore the images of these composites under F are equal.

(2) Preservation of the symmetric group action (5.1.13) holds if and only if each $F\iota_{(s,t)}$ commutes with the transposition of s and t as in the following diagram.

$$\begin{array}{ccc} Ft \oplus Fs & \xrightarrow{\ \xi\ } & Fs \oplus Ft \\ & {\scriptstyle F\iota_{(t,s)}}\searrow \quad \swarrow{\scriptstyle F\iota_{(s,t)}} & \\ & F(s \cup t) & \end{array}$$

Moreover, the composite

$$(\mathcal{M}\underline{n})\big((\varnothing, t)\,;t\big) \times \big((\mathcal{M}\underline{n})(\langle\rangle\,;\varnothing) \times (\mathcal{M}\underline{n})(t\,;t)\big) \longrightarrow (\mathcal{M}\underline{n})(t\,;t)$$

takes the unique triple of operations in the source to the unique operation in the target, that is the identity on t. Therefore, after applying F, we have the following commuting diagram in C.

$$\begin{array}{ccccc} & e \oplus Ft & \xrightarrow{F\iota_{()} \oplus 1_{Ft}} & e \oplus Ft & \\ {}^{1}\nearrow & & & & \searrow^{F\iota_{(\varnothing,t)}} \\ Ft & & \xrightarrow{\quad 1_{Ft} \quad} & & Ft \end{array}$$

If F is a pointed multifunctor then $F\iota_{()} = 1_e$ and therefore $F\iota_{(\varnothing,t)} = 1_{Ft}$.

These observations show that the data of an object in $\mathsf{C}^{\mathcal{F}}_{\mathrm{lax}}\underline{n}$ is equivalent to the data of a pointed multifunctor $\mathcal{M}\underline{n} \longrightarrow \mathsf{End}(\mathsf{C})$. The object unity axiom (8.3.2) and gluing unity axiom (8.3.3) correspond to the requirement that F be pointed. The gluing symmetry axiom (8.3.4) corresponds to the requirement that F preserves symmetry as in (2) above. The gluing associativity axiom (8.3.5) corresponds to the composition axiom for F as in (1) above. Therefore, the correspondence

$$F \leftrightarrow \{(Fs, F\iota_{(s,t)})\}$$

establishes a bijection on objects between $\mathsf{Multicat}_*(\mathcal{M}\underline{n}, \mathsf{End}(\mathsf{C}))$ and $\mathsf{C}^{\mathcal{F}}_{\mathrm{lax}}\underline{n}$.

We have two similar observations for pointed multinatural transformations $\theta\colon F \longrightarrow F'$ between pointed multifunctors

$$F, F'\colon \mathcal{M}\underline{n} \longrightarrow \mathsf{End}(\mathsf{C}).$$

(1) The naturality condition of Definition 5.1.17 holds if and only if θ is natural with respect to the pairwise operations $\iota_{(s,t)}$, as in the following diagram.

$$\begin{array}{ccc} Fs \oplus Ft & \xrightarrow{F\iota_{(s,t)}} & F(s \cup t) \\ {\scriptstyle \theta_s \oplus \theta_t}\downarrow & & \downarrow{\scriptstyle \theta_{s \cup t}} \\ F's \oplus F't & \xrightarrow{F'\iota_{(s,t)}} & F'(s \cup t) \end{array}$$

(2) Because F, F', and θ are pointed, the basepoint component $\theta_{\varnothing}$ is the identity 1_e.

Thus the correspondence

$$\theta \leftrightarrow \{\theta_s\}$$

establishes a bijection between the morphisms of $\mathsf{Multicat}_*(\mathcal{M}\underline{n}, \mathsf{End}(\mathsf{C}))$ and those of $\mathsf{C}^{\mathcal{F}}_{\mathrm{lax}}\underline{n}$. The pointed condition for θ corresponds to the unity condition (8.3.7) and the naturality condition for θ corresponds to the gluing compatibility axiom (8.3.8).

To verify naturality of these isomorphisms with respect to a morphism of pointed finite sets

$$\psi\colon \underline{n} \longrightarrow \underline{m} \quad \text{in} \quad \mathcal{F},$$

recall from Proposition 8.4.7 that $\mathcal{M}\psi = \tilde{\psi}$ is given by the function

$$2^{\underline{m}^\flat} \longrightarrow 2^{\underline{n}^\flat}$$

that takes inverse images of subsets. Then the functor

$$\mathsf{Multicat}_*(\mathcal{M}\underline{n}, \mathsf{End}(\mathsf{C})) \longrightarrow \mathsf{Multicat}_*(\mathcal{M}\underline{m}, \mathsf{End}(\mathsf{C}))$$

induced by $\mathcal{M}\psi$ is given on multifunctors F and multinatural transformations θ by

$$(\tilde{\psi}^* F)s = F(\psi^{-1}(s)) \quad \text{and} \quad (\tilde{\psi}^*\theta)_s = \theta_{\psi^{-1}(s)}.$$

Restricted to binary operations, this is precisely Definition 8.3.12 for $\mathsf{C}^{\mathcal{F}}_{\text{lax}}\psi$. □

Explanation 8.4.9. Variants of Proposition 8.4.8 can be given for the strong and colax Segal Γ-category constructions. However we focus on the lax case because it is the one we will use to compare with the Elmendorf-Mandell construction in Chapter 10. ◇

Definition 8.4.10. Let $\mathsf{J}^{\mathcal{M}}$ denote the 2-functor

$$\mathsf{J}^{\mathcal{M}} = \mathsf{Multicat}_*(\mathcal{M}, -) \colon \mathsf{Multicat}_* \longrightarrow \Gamma\text{-}\mathsf{Cat}.$$

For each small pointed multicategory P, the Γ-category $\mathsf{J}^{\mathcal{M}}\mathsf{P}$ is

$$\mathsf{J}^{\mathcal{M}}\mathsf{P} = \mathsf{Multicat}_*(\mathcal{M}(-), \mathsf{P}) \colon \mathcal{F} \longrightarrow \mathsf{Cat}.$$

We call this the *partition J-theory* of multicategories, or the *$\mathcal{M}$-partition J-theory* to emphasize the dependence on $\mathcal{M}$. ◇

Explanation 8.4.11. In the context of Definition 8.4.10, suppose given small pointed multicategories, pointed multifunctors and a pointed multinatural transformation:

$$H, H' \colon \mathsf{P} \longrightarrow \mathsf{P}' \quad \text{and} \quad \theta \colon H \longrightarrow H' \quad \text{in} \quad \mathsf{Multicat}_*.$$

Then

$$(\mathsf{J}^{\mathcal{M}}H)_{\underline{n}}, (\mathsf{J}^{\mathcal{M}}H')_{\underline{n}} \colon (\mathsf{J}^{\mathcal{M}}\mathsf{P})\underline{n} \longrightarrow (\mathsf{J}^{\mathcal{M}}\mathsf{P}')\underline{n}$$

are given by composition with H and H', denoted H_* and H'_*, respectively. The component of $\mathsf{J}^{\mathcal{M}}\theta$ at $\underline{n}$ is given by whiskering with θ:

$$\mathsf{Multicat}_*(\mathcal{M}\underline{n}, \mathsf{P}) \underset{H'_*}{\overset{H_*}{\rightrightarrows}} \mathsf{Multicat}_*(\mathcal{M}\underline{n}, \mathsf{P}') \qquad \Downarrow \theta * (1_-)$$

◇

8.5. Segal J-Theory and K-Theory

Recall from Example 5.1.11 the endomorphism multicategory $\mathsf{End}(\mathsf{C})$ for a permutative category C. Recall from Corollary 5.3.9 that this construction provides a 2-functor

$$\mathsf{End} \colon \mathsf{PermCat}^{\mathsf{su}} \longrightarrow \mathsf{Multicat}_*.$$

From Definition 8.4.10 we have the 2-functor

$$\mathsf{J}^{\mathcal{M}} = \mathsf{Multicat}_*(\mathcal{M}, -) \colon \mathsf{Multicat}_* \longrightarrow \Gamma\text{-}\mathsf{Cat}.$$

Definition 8.5.1 (Segal J-theory and K-theory)**.** Suppose C is a small permutative category. The *Segal J-theory* of C is the Γ-category

$$\mathsf{J}^{\mathsf{Se}}\mathsf{C} = \mathsf{Multicat}_*(\mathcal{M}(-), \mathsf{End}(\mathsf{C})) \colon \mathcal{F} \longrightarrow \mathsf{Cat}_*.$$

As a 2-functor, J^{Se} is the composite

$$\mathsf{PermCat}^{\mathsf{su}} \xrightarrow{\mathsf{End}} \mathsf{Multicat}_* \xrightarrow{\mathsf{J}^{\mathcal{M}}} \Gamma\text{-}\mathsf{Cat}.$$

Taking levelwise nerves, $N_*\mathsf{J}^{\mathsf{Se}}\mathsf{C}$ is a Γ-simplicial set. We define the *Segal K-theory* of C to be the symmetric spectrum

$$\mathsf{K}^{\mathsf{Se}}\mathsf{C} = \mathsf{K}^{\mathcal{F}}(N_*\mathsf{J}^{\mathsf{Se}}\mathsf{C})$$

given by applying the K-theory construction of Definition 8.2.5 to $N_*\mathsf{J}^{\mathsf{Se}}\mathsf{C}$. As a functor, K^{Se} is the composite

$$\mathsf{PermCat}^{\mathsf{su}} \xrightarrow{\mathsf{End}} \mathsf{Multicat}_* \xrightarrow{\mathsf{J}^{\mathcal{M}}} \Gamma\text{-}\mathsf{Cat} \xrightarrow{N_*} \Gamma\text{-}\mathsf{sSet} \xrightarrow{\mathsf{K}^{\mathcal{F}}} \mathsf{SymSp}. \qquad \diamond$$

Recall from Section 8.3 the three Γ-simplicial sets $\mathsf{C}^{\mathcal{F}}_{\nu}$ associated to a small permutative category C, where ν is one of the variants

$$\nu \in \{\cong, \mathrm{lax}, \mathrm{co}\}.$$

The three Γ-simplicial sets are levelwise weakly equivalent by Theorem 8.3.21. Therefore, by Proposition 8.4.8 we have the following.

Theorem 8.5.2. *For each small permutative category C and each variant*

$$\nu \in \{\cong, \mathrm{lax}, \mathrm{co}\},$$

there is a level equivalence of symmetric spectra

$$\mathsf{K}^{\mathcal{F}}(N_*\mathsf{C}^{\mathcal{F}}_{\nu}) \simeq \mathsf{K}^{\mathsf{Se}}\mathsf{C}.$$

Recall from Lemma 5.7.21 that for any small pointed multicategory M, the category

$$\mathsf{Multicat}_*(\mathsf{M}, \mathsf{End}(\mathsf{C}))$$

has a permutative structure given by the pointwise monoidal product. We will compare this with the following.

Definition 8.5.3. Suppose a is a pointed finite set, regarded as a discrete category, and $(\mathsf{C}, \oplus, e)$ is a small permutative category. The *pointwise monoidal product* on

$$\mathsf{Cat}_*(a, \mathsf{C}),$$

the category of pointed functors and natural transformations, is given as follows. For pointed functors

$$F, F' \colon a \longrightarrow \mathsf{C},$$

define a pointed functor $F \oplus F'$ by its assignment on elements:

$$(F \oplus F')x = (Fx) \oplus (F'x) \quad \text{for} \quad x \in a.$$

For pointed natural transformations

$$\theta \colon F \longrightarrow G \quad \text{and} \quad \theta' \colon F' \longrightarrow G',$$

define a pointed natural transformation $\theta \oplus \theta'$ with components

$$(\theta \oplus \theta')_x = \theta_x \oplus \theta'_x \quad \text{for} \quad x \in a.$$

The monoidal unit is the constant functor at $e \in \mathsf{C}$. The associativity and unit isomorphisms are identities because C is permutative. The symmetry is given elementwise by that of C and the hexagon axiom (1.1.26) follows from that of C. $\diamond$

The next result is a reformulation of the adjunction Proposition 8.3.16 for $\mathsf{C}^{\mathcal{F}}_{\mathrm{lax}}$. We repeat the proof with additional details because this result will be used in Proposition 10.6.7 for our comparison with the J-theory of Elmendorf-Mandell.

Proposition 8.5.4. *Suppose* $(\mathsf{C}, \oplus, e)$ *is a small permutative category, equipped with basepoint* e. *Suppose* a *is a pointed finite set. There are strictly unital strong symmetric monoidal functors*

$$\mathsf{Cat}_*(a, \mathsf{C}) \underset{R}{\overset{L}{\rightleftarrows}} \mathsf{Multicat}_*(\mathcal{M}a, \mathsf{End}(\mathsf{C}))$$

and a strong symmetric monoidal adjunction with the following defined below:

- *left adjoint* $L = (-)^{\oplus}$,
- *right adjoint* $R = \overline{(-)}$,
- *unit* $\eta\colon \mathrm{Id} \longrightarrow RL$ *the identity, and*
- *counit* $\varepsilon\colon LR \longrightarrow \mathrm{Id}$.

If, moreover, C *is a groupoid, then* ε *is a natural isomorphism and* (L, R) *is a strong symmetric monoidal equivalence.*

Proof. Suppose F is a pointed multifunctor

$$F\colon \mathcal{M}a \longrightarrow \mathsf{End}(\mathsf{C}).$$

Let $*$ denote the basepoint of a. Restricting to singleton subsets of $a^\flat$ yields an assignment

$$RF = \overline{F}\colon a \longrightarrow \mathrm{Ob}\,\mathsf{C}$$

$$(x \in a) \longmapsto \begin{cases} e & \text{if } \ x = * \\ F\{x\} & \text{otherwise.} \end{cases}$$

If F is constant at $e \in \mathsf{C}$, then so is $\overline{F}$.

A pointed multinatural transformation $\theta\colon F \longrightarrow F'$ has, in particular, unary operations

$$\theta_x\colon F\{x\} \longrightarrow F'\{x\}$$

for each $x \in a^\flat$. These give a natural transformation

$$R\theta = \overline{\theta}\colon \overline{F} \longrightarrow \overline{F}'$$

with components

$$\overline{\theta}_x = \begin{cases} 1_e & \text{if } \ x = * \\ \theta_x & \text{otherwise.} \end{cases}$$

If $\eta\colon G \longrightarrow G'$ is another pointed multinatural transformation between pointed multifunctors

$$G, G'\colon \mathcal{M}a \longrightarrow \mathsf{End}(\mathsf{C}),$$

we have

$$\overline{F \oplus G} = \overline{F} \oplus \overline{G} \quad \text{and} \quad \overline{\theta \oplus \eta} = \overline{\theta} \oplus \overline{\eta}.$$

This defines a strict symmetric monoidal functor

$$R = \overline{(-)}\colon \mathsf{Multicat}_*(\mathcal{M}a, \mathsf{End}(\mathsf{C})) \longrightarrow \mathsf{Cat}_*(a, \mathsf{C}).$$

For a reverse construction, we first fix a total order on $a^\flat$ for the remainder of this proof. Suppose P is an assignment

$$P\colon a \longrightarrow \mathsf{C}$$

with $P* = e$. We define a pointed multifunctor

$$LP = P^{\oplus}\colon \mathcal{M}a \longrightarrow \mathsf{End}(\mathsf{C})$$

by sending a subset $s \in 2^{a^\flat}$ to the sum

$$P^{\oplus}s = \bigoplus_{x \in s} Px$$

in C, where the order of summation is given by the chosen total ordering on a. If $\langle c \rangle$ is a partition of s, define $P^{\oplus}$ on the single element $\iota_{\langle c \rangle}$ of $\mathcal{M}a(\langle c \rangle; s)$ to be the unique coherence isomorphism

$$\bigoplus_{c_i} \bigoplus_{x \in c_i} Px \longrightarrow \bigoplus_{x \in s} Px$$

in C that permutes the summands Px, given by the Symmetric Coherence Theorem 1.1.41. If any c_i is empty, we implicitly compose with the (strict) unit isomorphisms of C, as noted in Examples 5.1.10 and 5.1.11. The multifunctor axioms for $P^{\oplus}$ follow from Theorem 1.1.41 applied to C. Note that $P^{\oplus}\varnothing = e$ because the empty sum in C is the unit object. Similarly, $P^{\oplus}\iota_{\langle\rangle}$ is the identity on the empty sum and therefore $P^{\oplus}$ is a pointed multifunctor.

A pointed natural transformation

$$\omega \colon P \longrightarrow P', \quad \text{for} \quad P, P' \in \mathsf{Cat}_*(a, C),$$

has components

$$\omega_x \colon Px \longrightarrow P'x, \quad \text{for} \quad x \in a,$$

and these assemble to give

$$(L\omega)_s = \omega^{\oplus}_s = \bigoplus_{x \in s} \omega_x, \quad \text{for} \quad s \subset a^\flat.$$

The naturality axiom for $\omega^{\oplus}_s$ follows from the naturality of the associativity and unit isomorphisms in C.

Now we turn to the data and axioms making $L = (-)^{\oplus}$ a symmetric monoidal functor. Given another assignment $Q \colon a \longrightarrow \mathsf{C}$, we define

$$P^{\oplus} \oplus Q^{\oplus} \xrightarrow{\cong} (P \oplus Q)^{\oplus}$$

to have component at $s \subset a^\flat$ given by the shuffle permutation

$$\Big(\bigoplus_{x \in s} Px\Big) \oplus \Big(\bigoplus_{x \in s} Qx\Big) \xrightarrow{\cong} \bigoplus_{x \in s}(Px \oplus Qx). \tag{8.5.5}$$

To see that these components give a well-defined isomorphism in

$$\mathsf{Multicat}_*(\mathcal{M}a, \mathsf{End}(\mathsf{C})),$$

note that the values of $P^{\oplus}$ and $Q^{\oplus}$ on the operations $\iota_{\langle c \rangle}$ are also given by permutations of summands and, therefore, the components (8.5.5) satisfy the naturality condition of Definition 5.1.17 by the Symmetric Coherence Theorem 1.1.41. Naturality of the symmetry in C implies that the structure isomorphism for $(-)^{\oplus}$ is natural with respect to morphisms in $\mathsf{Cat}_*(\mathcal{M}a, \mathsf{C})$.

In both $\mathsf{Multicat}(\mathcal{M}a, \mathsf{End}(\mathsf{C}))$ and $\mathsf{Cat}_*(a, \mathsf{C})$, the unit for the pointwise monoidal product is the constant functor at the unit $e \in \mathsf{C}$. Thus $(-)^{\oplus}$ strictly preserves the unit. The symmetric monoidal functor axioms for $(-)^{\oplus}$ also follow from Theorem 1.1.41. This completes the definition of strictly unital symmetric monoidal functors

$$R = \overline{(-)} : \mathsf{Multicat}_*(\mathcal{M}a, \mathsf{End}(\mathsf{C})) \longrightarrow \mathsf{Cat}_*(a, \mathsf{C}) \quad \text{and}$$
$$L = (-)^{\oplus} \colon \mathsf{Cat}_*(a, \mathsf{C}) \longrightarrow \mathsf{Multicat}_*(\mathcal{M}a, \mathsf{End}(\mathsf{C})).$$

Now we discuss the unit η and counit ε of this adjunction. First, we have $\overline{(-)^{\oplus}} = 1$ because restricting to singletons gives equalities

$$\overline{P^{\oplus}} = P \quad \text{and} \quad \overline{\omega^{\oplus}} = \omega$$

for each $P \in \mathsf{Cat}_*(a, \mathsf{C})$ and each $\omega \in \mathsf{Cat}_*(a, \mathsf{C})(P, P')$. Therefore we take η to be the identity. For the other composite, we define a multinatural transformation with components

$$\varepsilon_F \colon \overline{F}^{\oplus} \longrightarrow F$$

as follows. For each subset $s \in 2^{a^\flat}$, let $\langle s^\bullet \rangle$ denote the tuple formed by all singleton subsets of s, ordered by the chosen ordering on a, and let $F\langle s^\bullet \rangle$ denote the corresponding tuple given by applying F to each of the singletons $\{x\} \in \langle s^\bullet \rangle$. If s is empty then the tuple of singletons is the empty tuple. The unary operations in $\mathsf{End}(\mathsf{C})$ are morphisms in C, and we let $\varepsilon_{F;s}$ be the morphism

$$\varepsilon_{F;s} \colon \overline{F}^{\oplus} s = \bigoplus_{x \in s} F\{x\} \longrightarrow Fs$$

given by applying F to $\iota_{\langle s^\bullet \rangle}$, the single element of $\mathcal{M}a\big(\langle s^\bullet \rangle; s\big)$. If $s = \varnothing$, then

$$\varepsilon_{F;s} = F\iota_{\langle\rangle} = 1_e$$

because F is pointed.

The naturality condition of Definition 5.1.17 for these components is the following equality:

$$(F\iota_{\langle c \rangle}) \circ \Big(\bigoplus_{c_i \in \langle c \rangle} \varepsilon_{F;c_i} \Big) = \varepsilon_{F;s} \circ \big(\overline{F}^{\oplus} \iota_{\langle c \rangle}\big). \tag{8.5.6}$$

Since

$$\varepsilon_{F;s} = F\iota_{\langle s^\bullet \rangle} \quad \text{and} \quad \varepsilon_{F;c_i} = F\iota_{\langle c_i^\bullet \rangle} \quad \text{for} \quad s \in 2^{a^\flat}, c_i \in \langle c \rangle,$$

the naturality condition is equivalent to commutativity in C of the outer rectangle (8.5.7) below. In that diagram, each $\langle c_i^\bullet \rangle$ denotes the tuple of singletons in c_i, with the order inherited from a; the tuple $\langle c^\bullet \rangle$ is the concatenation of the $\langle c_i^\bullet \rangle$, in the order determined by $\langle c \rangle$; and s is the union of the subsets c_i, with the order determined as a subset of $a^\flat$.

$$\begin{array}{ccc} \displaystyle\bigoplus_{c_i \in \langle c \rangle} \bigoplus_{x \in c_i} F\{x\} & \xrightarrow{\bigoplus_{c_i \in \langle c \rangle} F\iota_{\langle c_i^\bullet \rangle}} & \displaystyle\bigoplus_{c_i \in \langle c \rangle} Fc_i \\ {\scriptstyle \overline{F}^{\oplus}\iota_{\langle c \rangle}} \downarrow & \searrow {\scriptstyle F\iota_{\langle c^\bullet \rangle}} & \downarrow {\scriptstyle F\iota_{\langle c \rangle}} \\ \displaystyle\bigoplus_{x \in s} F\{x\} & \xrightarrow[F\iota_{\langle s^\bullet \rangle}]{} & Fs \end{array} \tag{8.5.7}$$

The upper-right triangle commutes because the composition in $\mathcal{M}a$ gives

$$\iota_{\langle c \rangle} \circ \big(\iota_{\langle c^\bullet{}_1 \rangle}, \ldots, \iota_{\langle c^\bullet{}_n \rangle} \big) = \iota_{\langle c^\bullet \rangle}$$

and F preserves composition. The morphism $\overline{F}^{\oplus}\iota_{\langle c\rangle}$ is, by definition, given by permuting terms. Therefore the equivariance condition (5.1.13) for F implies that the lower-left triangle commutes. This concludes the definition of the component

$$\varepsilon_F\colon \overline{F}^{\oplus} \longrightarrow F$$

and the proof that it is a multinatural transformation.

For a multinatural transformation $\theta\colon F \longrightarrow F'$, the naturality of θ at the operation $\iota_{\langle s^\bullet\rangle}$ ensures that the following diagram commutes for each $s \subset a^\flat$.

$$\begin{array}{ccc}
\bigoplus\limits_{x\in s} F\{x\} & \xrightarrow{\ \overline{\theta}^{\oplus}_s\ } & \bigoplus\limits_{x\in s} F'\{x\} \\
{\scriptstyle \varepsilon_{F;s}}\big\downarrow & & \big\downarrow{\scriptstyle \varepsilon_{F';s}} \\
Fs & \xrightarrow{\ \theta_s\ } & F's
\end{array}$$

Therefore, the components of ε at F and F' commute with θ and $\overline{\theta}^{\oplus}$ because the relevant composite multinatural transformations have the same components at s for each $s \in \mathcal{M}a$. Hence

$$\varepsilon\colon LR = \overline{(\ )}^{\oplus} \longrightarrow \mathrm{Id}$$

is a natural transformation of functors.

Since η is the identity, the triangle identities for this adjunction reduce to the equalities

$$\varepsilon_{P^{\oplus}} = 1_{P^{\oplus}} \quad \text{and} \quad \overline{\varepsilon_F} = 1_{\overline{F}}$$

for P and F as above. Both of these equalities follow directly from the definitions.

Verifying that ε is monoidal natural requires commutativity of the following rectangle for pointed multifunctors F and G and subsets $s \subset a^\flat$.

$$\begin{array}{ccc}
\Big(\bigoplus\limits_{x\in s} F\{x\}\Big) \oplus \Big(\bigoplus\limits_{x\in s} G\{x\}\Big) & \xrightarrow{\ F\iota_{\langle s^\bullet\rangle} \oplus G\iota_{\langle s^\bullet\rangle}\ } & Fs \oplus Gs \\
{\scriptstyle \cong}\big\downarrow & & \big\downarrow{\scriptstyle 1} \\
\Big(\bigoplus\limits_{x\in s} (F \oplus G)\{x\}\Big) & \xrightarrow{\ (F\oplus G)\iota_{\langle s^\bullet\rangle}\ } & (F \oplus G)s
\end{array}$$

Commutativity of this diagram follows by definition of $(F \oplus G)$ on operations, as given in the proof of Lemma 5.7.12. Therefore, the functors $R = \overline{(\ )}$ and $L = (\ )^{\oplus}$ determine a symmetric monoidal adjunction. Since the components of ε are given by morphisms in C, this adjunction is an equivalence if C is a groupoid. □

8.6. Notes

8.6.1 (Segal K-Theory). The K-theory functor for small permutative categories first developed by Segal [**Seg74**] uses the Γ-simplicial set $\mathsf{C}^{\mathcal{F}}$ of Section 8.3 instead of the approach via the partition multicategories $\mathcal{M}\underline{n}$ given in Section 8.4. The K-theory construction via $\mathsf{C}^{\mathcal{F}}$ is the standard one. See, for example, [**May78, EM06, Man10**].

Our definitions of J^{Se}, and K^{Se} in Section 10.3, in terms of the partition multicategories $\mathcal{M}a$ are equivalent to those of [**EM09**] using the *terminal parameter* *multicategory for modules*, denoted E there. Our justifications for attaching Segal's name to these constructions are the isomorphisms of Proposition 8.4.8 and the level equivalences of Theorem 8.5.2. In the literature, including [**GJO17b, Man10**], the name *Segal K-theory* is sometimes used for J^{Se}. ◇

8.6.2 (Variants of Segal K-theory)**.** Our discussion of the lax and colax variants for $\mathsf{C}^{\mathcal{F}}$, together with the proof that they result in levelwise equivalent Γ-simplicial sets (Theorem 8.3.21) follows the sketch given in [**Man10**, Section 3]. See [**GJO17a**] for a general approach to equivalences of homotopy theories between categories of strict, strong, and lax algebra morphisms for a monad. Examples given there include categories of small symmetric monoidal categories, Γ-categories, and n-fold monoidal categories. ◇

8.6.3 (2-Functoriality of $\mathsf{C}^{\mathcal{F}}_{\nu}$)**.** One can show the 2-functoriality of $\mathsf{C}^{\mathcal{F}}_{\mathrm{lax}}$ directly, without explicit use of the isomorphism in Proposition 8.4.8. There are similar 2-functoriality results for $\mathsf{C}^{\mathcal{F}}_{\cong}$ and $\mathsf{C}^{\mathcal{F}}_{\mathrm{co}}$. However, the natural domain of definition for $\mathsf{C}^{\mathcal{F}}_{\cong}$ is

$$\mathsf{PermCat}^{\mathsf{sus}},$$

whose morphisms are given by strictly unital *strong* monoidal functors (Definition 1.1.27). Likewise, the natural domain of definition for $\mathsf{C}^{\mathcal{F}}_{\mathrm{co}}$ consists of small permutative categories and *colax symmetric monoidal* functors, also known as *oplax symmetric monoidal* functors or *lax symmetric comonoidal* functors. These are defined with the same axioms as symmetric monoidal functors but their monoidal and unit constraints, (1.1.7) and (1.1.8) respectively, have the opposite variance. Strictly unital colax symmetric monoidal functors are the colax 1-linear functors in Definition 10.7.24. These are used in $\mathsf{PermCat}^{\mathsf{su}}_{\mathsf{co}}$ and Proposition 10.7.27. ◇

8.6.4 (Γ-Simplicial Sets and Connective Spectra)**.** In Note 7.9.5 we recall the earlier and simpler notion of sequential spectra. A sequential spectrum is called *connective* if its homotopy groups π_n are trivial for $n < 0$. We let $\mathsf{Sp}^{\mathbb{N}}_{\geq 0}$ denote the category of connective sequential spectra. A sequential spectrum X is a *weak Ω-spectrum* if the adjoint structure morphisms

$$X_n \longrightarrow \Omega X_{n+1}$$

are weak equivalences of pointed simplicial sets, where Ω denotes the pointed simplicial loop functor $\mathsf{Hom}_{\mathsf{sSet}_*}(S^1, -)$.

When introducing Γ-simplicial sets in [**Seg74**], Segal shows that the spectrum associated to a special Γ-simplicial set is a weak Ω-spectrum. Therefore by Theorem 8.3.21 each spectrum $\mathsf{K}^{\mathsf{Se}}\mathsf{C}$, with C a small permutative category, is a weak Ω-spectrum.

Segal [**Seg74**] also develops homotopy theory for Γ-simplicial sets and its relation to that of connective spectra. Bousfield-Friedlander [**BF78**] extend this to model structures for both Γ-sSet and $\mathsf{Sp}^{\mathbb{N}}_{\geq 0}$, with the weak equivalences of Γ-sSet being created by Segal's K-theory functor $\mathsf{K}^{\mathcal{F}}$ and with the fibrant objects being the weak Ω-spectra. They go on to show that there is a Quillen equivalence of model categories

$$\Gamma\text{-}\mathsf{sSet} \simeq_Q \mathsf{Sp}^{\mathbb{N}}_{\geq 0}\,. \tag{8.6.5}$$

◇

8.6.6 (Permutative Categories and Connective Spectra). Continuing from the previous discussion in Note 8.6.4, one can define weak equivalences of small permutative categories to be the strictly unital symmetric monoidal functors that induce weak equivalences of Γ-simplicial sets (that is, those functors that induce π_*-isomorphisms of associated sequential spectra). It follows from work of Thomason [**Tho95**], for the more general category of small symmetric monoidal categories and all symmetric monoidal functors, that when one inverts the respective weak equivalences there is an equivalence of homotopy categories

$$\mathit{Ho}(\mathsf{PermCat}^{\mathsf{su}}) \simeq \mathit{Ho}(\Gamma\text{-}\mathsf{sSet}). \tag{8.6.7}$$

Combining (8.6.7) with (8.6.5) means that, up to weak equivalence, all connective spectra can be obtained from the Segal K-theory of small permutative categories. In [**Man10**], Mandell gives a second proof of (8.6.7) by defining a functor

$$P = \mathcal{P} \circ \mathcal{S} \colon \Gamma\text{-}\mathsf{sSet} \longrightarrow \mathsf{PermCat},$$

via a certain Grothendieck construction, and showing it is inverse to $N_* \mathsf{J}^{\mathsf{Se}}$. This work is extended to 2-dimensional category theory, for symmetric monoidal 2-categories and Γ-2-categories, in [**GJO17b**]. $\diamond$

CHAPTER 9

Categories of $\mathcal{G}_*$-Objects

This is the first of two chapters generalizing the Segal K-theory functor

$$\mathsf{K}^{\mathsf{Se}}\colon \mathsf{PermCat}^{\mathsf{su}} \longrightarrow \mathsf{SymSp},$$

from Chapter 8, to a simplicially-enriched multifunctor

$$\mathsf{K}^{\mathsf{EM}}\colon \mathsf{PermCat}^{\mathsf{su}} \longrightarrow \mathsf{SymSp},$$

due to Elmendorf-Mandell [**EM06, EM09**]. The material in this chapter concerns the replacement of $\mathcal{F}$ and its associated K-theory

$$\Gamma\text{-}\mathsf{Cat} \xrightarrow{N_*} \Gamma\text{-}\mathsf{sSet} \xrightarrow{\mathsf{K}^{\mathcal{F}}} \mathsf{SymSp}$$

with another diagram category $\mathcal{G}$ and an associated K-theory functor

$$\mathcal{G}_*\text{-}\mathsf{Cat} \xrightarrow{N_*} \mathcal{G}_*\text{-}\mathsf{sSet} \xrightarrow{\mathsf{K}^{\mathcal{G}}} \mathsf{SymSp}.$$

The corresponding J-theory, J^{EM}, is given in Chapter 10.

Beyond simply replacing $\mathcal{F}$ and $\mathsf{K}^{\mathcal{F}}$, this chapter also describes the relevant enriched monoidal structures. These are essential for our applications to K-theory of E_n-monoidal categories.

(1) Theorem 9.2.15 shows that $\mathcal{G}_*$-Cat and $\mathcal{G}_*$-sSet are symmetric monoidal closed categories and are enriched, as symmetric monoidal categories, over Cat_* and sSet_*, respectively.

(2) Theorem 9.2.19 shows that the nerve functor gives $\mathcal{G}_*$-Cat an enrichment over sSet_* and that

$$N_*\colon \mathcal{G}_*\text{-}\mathsf{Cat} \longrightarrow \mathcal{G}_*\text{-}\mathsf{sSet}$$

is symmetric monoidal as a sSet_*-enriched functor.

(3) Theorem 9.4.9 shows that

$$\mathsf{K}^{\mathcal{G}}\colon \mathcal{G}_*\text{-}\mathsf{sSet} \longrightarrow \mathsf{SymSp}$$

is symmetric monoidal as a sSet_*-enriched functor.

Organization. In Section 9.1 we define the category $\mathcal{G}$, which replaces $\mathcal{F}$. The objects of $\mathcal{G}$ consist of certain tuples of objects from $\mathcal{F}$, and $\mathcal{G}$ has a permutative structure given by concatenation of tuples. Taking the smash product of pointed finite sets, with lexicographic ordering, provides a functor

$$\wedge\colon \mathcal{G} \longrightarrow \mathcal{F} \tag{9.0.1}$$

described in Definition 9.1.15.

In Section 9.2 we show that $\mathcal{G}_*$-objects in C form a symmetric monoidal closed category, where C is a complete and cocomplete symmetric monoidal closed category (Theorem 9.2.15). Our two examples of interest are $\mathsf{C} = \mathsf{sSet}_*$ and $\mathsf{C} = \mathsf{Cat}_*$.

- Our method of proof uses the general theory for (pointed) diagram categories developed in Section 4.3.
- We also apply the general theory of enriched monoidal categories and functors from Chapters 2 and 3 to show that N_* is enriched symmetric monoidal (Theorem 9.2.19).
- Of potentially independent interest, our method of proof also shows that Γ-C is a symmetric monoidal closed category (Explanation 9.2.18), with product induced by the smash product of pointed finite sets in $\mathcal{F}$. See Note 9.5.5 for further generalization.

In Section 9.3 we define the associated K-theory functor

$$\mathsf{K}^{\mathcal{G}} \colon \mathcal{G}_*\text{-}\mathsf{sSet} \longrightarrow \mathsf{SymSp},$$

which replaces $\mathsf{K}^{\mathcal{F}}$. We also show, as Proposition 9.3.16, that $\mathsf{K}^{\mathcal{G}}$ agrees with $\mathsf{K}^{\mathcal{F}}$ along the functor

$$\Gamma\text{-}\mathsf{sSet} \longrightarrow \mathcal{G}_*\text{-}\mathsf{sSet}$$

induced by the smash product of pointed finite sets (9.0.1).

In Section 9.4 we show that $\mathsf{K}^{\mathcal{G}}$ is a sSet_*-enriched symmetric monoidal functor (Theorem 9.4.9). We show, as a result of the compatibility with $\mathsf{K}^{\mathcal{F}}$, that $\mathsf{K}^{\mathcal{F}}$ is also symmetric monoidal as a sSet_*-enriched functor (Theorem 9.4.18).

9.1. The Category $\mathcal{G}$

Recall from Definition 8.1.1 that $\mathcal{F}$ denotes the category whose objects are pointed finite sets $\underline{n} = \{0, \ldots, n\}$ for $n \geq 0$ and whose morphisms are basepoint-preserving functions. In this section we define another category, $\mathcal{G}$, whose objects are tuples of objects from $\mathcal{F}$, subject to certain identifications.

Motivation 9.1.1 (Smash Products of Pointed Finite Sets). Recall, as described in Explanation 8.1.6, that $(\mathcal{F}, \wedge, \underline{1})$ is a permutative category. Here, $\wedge$ denotes the smash product of pointed finite sets implicitly composed with the lexicographic order isomorphism

$$\underline{m} \wedge \underline{n} \cong \underline{mn}.$$

The germ of the idea for a K-theory functor that preserves multiplicative structure is that it ought to be encoded via the smash product in $\mathcal{F}$. Working this out with $\mathcal{F}$ alone, one encounters a crucial obstruction that we clarify in Note 10.8.6.

The diagram category $\mathcal{G}$, to be described in Definition 9.1.7 below, resolves the difficulty. Although presented differently, the germ of the idea remains the same. With that in mind, one might conceive of the objects and morphisms of $\mathcal{G}$ as something like "pre-smash-products" of pointed finite sets and their morphisms. We take tuples of objects of $\mathcal{F}$, but with certain technical identifications corresponding to the following properties of the smash product.

- The 1-point set $\underline{0}$ is a null object for $(\mathcal{F}, \wedge, \underline{1})$, in the sense of Definition 4.3.3: $\underline{0}$ is both initial and terminal, and smash product of any object with $\underline{0}$ results in $\underline{0}$. Thus, for a q-tuple of objects $(\underline{n}_1, \ldots, \underline{n}_q)$, we will have

  $$\bigwedge_j \underline{n}_j = \underline{0}$$

  if any $n_j = 0$.

- The unit for $\wedge$ is $\underline{1}$. Thus, for a q-tuple and an r-tuple,
$$(\underline{n}_1, \ldots, \underline{n}_q) \quad \text{and} \quad (\underline{m}_1, \ldots, \underline{m}_r),$$
we will have
$$\bigwedge_j \underline{n}_j = \bigwedge_k \underline{m}_k$$
whenever the tuples differ by insertion or deletion of $\underline{1}$ in various positions. There will of course be other circumstances where the two are equal, depending on the prime factors of the numbers involved, but here we focus on equalities that are a result of extraneous multiplication by 1.
- The product $\wedge$ is symmetric monoidal, but not strictly so. Thus we will have isomorphisms, but generally not equalities, given by permuting the terms of a smash product.
- For each $q > 0$ there is at least one q-tuple whose smash product is $\underline{0}$. But there is no such tuple for $q = 0$; the smash product over the empty tuple is $\underline{1}$. We will need an extra object $*$ to account for this, and it will be identified with all q-tuples for $q > 0$ whose smash products are $\underline{0}$. $\diamond$

With the observations of Motivation 9.1.1 in mind, we give the definition of $\mathcal{G}$ in terms of the following preliminary definitions. These will account for insertions terms $\underline{1}$, permutation of terms, and identifications of tuples whose smash products are $\underline{0}$ in $\mathcal{F}$.

Definition 9.1.2. Let Inj denote the category whose objects are *unpointed finite sets*
$$\overline{p} = \underline{p}^\flat = \begin{cases} \{1, \ldots, p\} & \text{if } p > 0, \\ \varnothing & \text{if } p = 0, \end{cases}$$
for each natural number $p \geq 0$, and whose morphisms are injections
$$f : \overline{q} \hookrightarrow \overline{p}. \qquad \diamond$$

Definition 9.1.3. For each unpointed finite set $\overline{q}$ with $q > 0$, define
$$\mathcal{F}^{(q)} = \mathcal{F}^{\wedge q},$$
the q-fold smash power of pointed categories, where $\mathcal{F}$ has basepoint $\underline{0}$. We denote the objects of $\mathcal{F}^{(q)}$ as
$$\langle \underline{n} \rangle = (\underline{n}_1, \ldots, \underline{n}_q) \quad \text{for} \quad \underline{n}_i \in \mathcal{F}$$
where $\langle \underline{n} \rangle$ is identified with the basepoint of $\mathcal{F}^{(q)}$ if any $\underline{n}_i = \underline{0}$. We denote the morphisms of $\mathcal{F}^{(q)}$ as
$$\langle \psi \rangle = (\psi_1, \ldots, \psi_q)$$
for morphisms ψ_i in $\mathcal{F}$. A tuple $\langle \psi \rangle$ is a zero morphism if any ψ_i factors through $\underline{0}$ in $\mathcal{F}$.

For $q = 0$, define
$$\mathsf{Ob}\, \mathcal{F}^{(0)} = \{*, \langle\rangle\}$$
where $*$ is the basepoint object and $\langle\rangle$ is the empty tuple. Define the morphisms of $\mathcal{F}^{(0)}$ such that $*$ is both initial and terminal and the only nonzero morphisms are identities. That is,

- there are unique morphisms
$$\langle\rangle \longrightarrow * \longrightarrow \langle\rangle,$$

- the only endomorphism of $*$ is its identity, and
- the endomorphisms of $\langle\rangle$ consist of its identity and a zero morphism given by the unique composite through $*$.

We note that $\mathcal{F}^{(0)}$ is distinct from the smash product over the empty indexing set in the category of small pointed categories. We give an alternate description and further details in Note 9.5.2, but they will not be necessary for our further work. $\diamond$

Definition 9.1.4. Suppose

$$f\colon \overline{q} \hookrightarrow \overline{p} \quad \text{in} \quad \mathsf{Inj}$$

is an injection of unpointed finite sets. We define a functor

$$f_*\colon \mathcal{F}^q \longrightarrow \mathcal{F}^p$$

called the *reindexing injection* as follows. For each q-tuple of pointed finite sets

$$\langle \underline{n} \rangle = (\underline{n}_1, \ldots, \underline{n}_q) \in \mathcal{F}^q,$$

let $\underline{n}_\varnothing$ denote the pointed finite set $\underline{1}$ and define $f_*\langle \underline{n} \rangle$ to be the p-tuple whose jth entry is $\underline{n}_{f^{-1}(j)}$. For each q-tuple of pointed functions

$$\langle \psi \rangle = (\psi_1, \ldots, \psi_q) \in \mathcal{F}^q,$$

let $\psi_\varnothing$ denote the identity $1_{\underline{1}}$ and define $f_*\langle \psi \rangle$ to be the p-tuple whose jth entry is $\psi_{f^{-1}(j)}$.

If any $\underline{n}_i = \underline{0}$, then the $f(i)$th coordinate of $f_*\langle n \rangle$ is $\underline{0}$. Therefore, f_* descends to give pointed functors

$$f_*\colon \mathcal{F}^{(q)} \longrightarrow \mathcal{F}^{(p)}.$$

For $q = 0$ and f the unique injection

$$\varnothing \hookrightarrow \overline{p},$$

we have

$$f_*\colon \mathcal{F}^{(0)} \longrightarrow \mathcal{F}^{(p)} \quad \text{with} \quad f_*\langle\rangle = (\underbrace{\underline{1}, \ldots, \underline{1}}_{p \text{ terms}}). \qquad \diamond$$

Explanation 9.1.5. For an injection $f\colon \overline{q} \longrightarrow \overline{p}$, the entries of $f_*\langle \underline{n} \rangle$ are reindexed by f, with $\underline{1}$ inserted for each index not in the image of f. Equivalently, f_* takes the ith coordinate of $\mathcal{F}^q$ to the $f(i)$th coordinate of $\mathcal{F}^p$, and has the value $\underline{1}$ in each coordinate of $\mathcal{F}^p$ not in the image of f. $\diamond$

Example 9.1.6. Consider the injection

$$f\colon \overline{2} \longrightarrow \overline{3}, \qquad (1 \longmapsto 3;\ 2 \longmapsto 1)$$

Then

$$f_*(\underline{n}_1, \underline{n}_2) = (\underline{n}_2, \underline{1}, \underline{n}_1).$$

$\diamond$

Definition 9.1.7. We define a small pointed category $\mathcal{G}$ as follows.

Objects: The set of objects is the wedge of pointed sets

$$\mathsf{Ob}\,\mathcal{G} = \bigvee_{q \ge 0} \mathsf{Ob}\big(\mathcal{F}^{(q)}\big).$$

The distinct objects of $\mathcal{G}$ consist of a basepoint object $*$ together with q-tuples of pointed finite sets

$$\langle \underline{n} \rangle = (\underline{n}_1, \ldots, \underline{n}_q)$$

where $q \geq 0$ and each $\underline{n}_i \in \mathcal{F}$ with $n_i > 0$. Each q-tuple $\langle \underline{n} \rangle$ for which some $n_i = 0$ is identified with the basepoint object of $\mathcal{F}^{(q)}$ and thus also with the basepoint $*$ of the wedge.

Morphisms: The basepoint $*$ is both initial and terminal in $\mathcal{G}$. The set of morphisms from a q-tuple $\langle \underline{n} \rangle$ to a p-tuple $\langle \underline{m} \rangle$ is

$$\begin{aligned} \mathcal{G}(\langle \underline{n} \rangle, \langle \underline{m} \rangle) &= \bigvee_{f \in \mathsf{Inj}(\overline{q}, \overline{p})} \Big(\mathcal{F}^{(p)}\big(f_* \langle \underline{n} \rangle, \langle \underline{m} \rangle\big) \Big) \\ &= \bigvee_{f \in \mathsf{Inj}(\overline{q}, \overline{p})} \Big(\bigwedge_{j=1}^{p} \mathcal{F}\big(\underline{n}_{f^{-1}(j)}, \underline{m}_j\big) \Big) \end{aligned} \tag{9.1.8}$$

where $\underline{n}_{\varnothing} = \underline{1}$ as in Definition 9.1.4. For $p = 0$, the empty smash product of pointed sets is the unit for the smash product—a two-element set. In this case we must also have $q = 0$, and

$$\mathcal{G}(\langle \rangle, \langle \rangle) = \mathcal{F}^{(0)}(\langle \rangle, \langle \rangle)$$

is a two-element set consisting of a zero morphism and an identity morphism.

In (9.1.8) for $p > 0$ we denote a morphism by a pair $(f, \langle \psi \rangle)$, where

$$f : \overline{q} \hookrightarrow \overline{p} \quad \text{in} \quad \mathsf{Inj}$$

and

$$\langle \psi \rangle : f_* \langle \underline{n} \rangle \longrightarrow \langle \underline{m} \rangle$$

is a morphism in $\mathcal{F}^{(p)}$. Thus each

$$\psi_j : \underline{n}_{f^{-1}(j)} \longrightarrow \underline{m}_j$$

is a morphism in $\mathcal{F}$. A morphism $(f, \langle \psi \rangle)$ is identified with the zero morphism

$$\langle \underline{n} \rangle \longrightarrow * \longrightarrow \langle \underline{m} \rangle$$

if there exists a component morphism

$$\psi_j : \underline{n}_{f^{-1}(j)} \longrightarrow \underline{m}_j$$

that is a zero morphism in $\mathcal{F}$, factoring through $\underline{0}$.

Identities: The identity on a q-tuple $\langle \underline{n} \rangle$ is given by the pair $(1_{\overline{q}}, 1_{\langle \underline{n} \rangle})$.

Composition: The composite of morphisms

$$\langle \underline{n} \rangle \xrightarrow{(f, \langle \psi \rangle)} \langle \underline{m} \rangle \xrightarrow{(g, \langle \phi \rangle)} \langle \underline{\ell} \rangle$$

is given by the pair $(gf, \langle \phi \rangle \circ g_* \langle \psi \rangle)$.

This finishes the definition of $\mathcal{G}$. The composition in $\mathcal{G}$ is associative and unital since $(gf)_* = g_* \circ f_*$ for composable injections f and g. Moreover, we note the following.

- We call q the *length* of a q-tuple $\langle \underline{n} \rangle$.
- The *zero morphism* from $\langle \underline{n} \rangle$ to $\langle \underline{m} \rangle$ is the unique morphism factoring through the basepoint $*$.
- Each morphism set $\mathcal{G}(\langle \underline{n} \rangle, \langle \underline{m} \rangle)$ is a pointed set with basepoint the zero morphism.
- For readability, we sometimes use a semicolon and write $\mathcal{G}\big(\langle \underline{n} \rangle ; \langle \underline{m} \rangle\big)$ for the set of morphisms in $\mathcal{G}$ from $\langle \underline{n} \rangle$ to $\langle \underline{m} \rangle$.

In Note 9.5.3 we give an alternate description of $\mathcal{G}$ as a Grothendieck construction, but that description will not be necessary for our further work. ◇

Explanation 9.1.9. We discuss some special cases of Definition 9.1.7.

(1) Each of the following objects in $\mathcal{F}^{(1)}$ and $\mathcal{F}^{(2)}$ are identified with the basepoint in $\mathcal{G}$, for any $\underline{n}$ in $\mathcal{F}$.

$$* = (\underline{0}) = (\underline{0}, \underline{n}) = (\underline{n}, \underline{0}).$$

(2) Let d^2 and s^0 denote the following morphisms in $\mathcal{F}$, respectively:

$$\begin{aligned} d^2&: \underline{1} \longrightarrow \underline{2} && (1 \longmapsto 1)\\ s^0&: \underline{2} \longrightarrow \underline{1} && (1 \longmapsto 0; 2 \longmapsto 1). \end{aligned}$$

Then neither d^2 nor s^0 is a zero morphism, but the composite

$$\underline{1} \xrightarrow{d^2} \underline{2} \xrightarrow{s^0} \underline{1}$$

is a zero morphism in $\mathcal{F}$. The composite in $\mathcal{G}$

$$(\underline{1}, \underline{n}, \underline{m}) \xrightarrow{(1_{\overline{3}}, (d^2, 1_{\underline{n}}, 1_{\underline{m}}))} (\underline{2}, \underline{n}, \underline{m}) \xrightarrow{(1_{\overline{3}}, (s^0, 1_{\underline{n}}, 1_{\underline{m}}))} (\underline{1}, \underline{n}, \underline{m})$$

is a zero morphism for any $\underline{n}$ and $\underline{m}$ in $\mathcal{F}$.

(3) Suppose $q = 2$, $p = 3$, and $\langle \underline{\ell} \rangle$ has length 4. Suppose f and g are the following injections, respectively:

$$\begin{aligned} f&: \overline{2} \longrightarrow \overline{3}, && (1 \longmapsto 3;\ 2 \longmapsto 1)\\ g&: \overline{3} \longrightarrow \overline{4}, && (1 \longmapsto 4;\ 2 \longmapsto 1;\ 3 \longmapsto 2). \end{aligned}$$

Then the second coordinate of the composite

$$\langle \underline{n} \rangle \xrightarrow{(f, \langle \psi \rangle)} \langle \underline{m} \rangle \xrightarrow{(g, \langle \phi \rangle)} \langle \underline{\ell} \rangle$$

is the following morphism in $\mathcal{F}^{(4)}$:

$$(\underline{1}, \underline{n}_1, \underline{1}, \underline{n}_2) \xrightarrow{(\psi_2, \psi_3, 1_{\underline{1}}, \psi_1)} (\underline{m}_2, \underline{m}_3, \underline{1}, \underline{m}_1) \xrightarrow{(\phi_1, \phi_2, \phi_3, \phi_4)} (\underline{\ell}_1, \underline{\ell}_2, \underline{\ell}_3, \underline{\ell}_4).$$

(4) In the case $q = 0$, we have the empty tuple $\langle\rangle$ of length zero. The only nonzero morphism from $\langle\rangle$ to $\langle\rangle$ is the identity. If $\langle \underline{m} \rangle$ has length $p > 0$, then a morphism

$$\langle\rangle \longrightarrow \langle \underline{m} \rangle$$

is determined by $(i_p, \langle \psi \rangle)$ where

$$i_p: \overline{0} = \varnothing \hookrightarrow \overline{p}$$

is the unique inclusion and

$$\langle \psi \rangle: \langle \underline{1} \rangle \longrightarrow \langle \underline{m} \rangle \quad \text{in} \quad \mathcal{F}^{(p)}$$

where $\langle \underline{1} \rangle$ is the constant p-tuple at $\underline{1}$. Such a morphism factors uniquely as the composite

$$\langle\rangle \xrightarrow{(i_p, 1_{\langle \underline{1} \rangle})} \langle \underline{1} \rangle \xrightarrow{(1, \langle \psi \rangle)} \langle \underline{m} \rangle. \tag{9.1.10}$$

◇

Explanation 9.1.11 ($\mathcal{G}$ as a Small Skeleton)**.** Recall, from Explanation 8.1.2, that $\mathcal{F}$ is a small skeleton for FinSet_*. The category $\mathcal{G}$ is a small skeleton for a similarly-constructed category whose objects are finite tuples of pointed finite sets and whose morphisms are given by tuples of morphisms along with injections of indexing sets. We work with diagrams on $\mathcal{G}$ for the same reasons mentioned in Explanation 8.1.10: The collection of $\mathcal{G}_*$-objects in a pointed category C will form a category because $\mathcal{G}$ is small. Moreover each $\underline{n}_i$ has a canonical total ordering that will be used in the constructions below. ◇

We now discuss a permutative structure on $\mathcal{G}$ given by concatenation of tuples. Recall from Definition 4.3.3 an object T of a symmetric monoidal category is called a null object if it is null for the monoidal product and is initial and terminal.

Definition 9.1.12. The *concatenation product*

$$\mathcal{G} \times \mathcal{G} \xrightarrow{\oplus} \mathcal{G}$$

is defined as follows. The concatenation isomorphism of tuples

$$\mathcal{F}^q \times \mathcal{F}^{q'} \cong \mathcal{F}^{q+q'}$$

descends to

$$\mathcal{F}^{(q)} \wedge \mathcal{F}^{(q')} \xrightarrow{\oplus} \mathcal{F}^{(q+q')}$$

for each $q, q' \geq 0$. This defines

$$\langle \underline{n} \rangle \oplus \langle \underline{n}' \rangle \quad \text{and} \quad \langle \psi \rangle \oplus \langle \psi' \rangle$$

for tuples of pointed finite sets and pointed functions, respectively. Since $\oplus$ is a pointed functor, the basepoint $*$ is a null object for this product in the sense of Definition 4.3.3.

Since $*$ is null, the product of any morphism with a morphism from, respectively to, the basepoint is uniquely determined. Given morphisms

$$(f, \langle \psi \rangle) \in \mathcal{G}(\langle \underline{n} \rangle, \langle \underline{m} \rangle) \quad \text{and} \quad (f', \langle \psi' \rangle) \in \mathcal{G}(\langle \underline{n}' \rangle, \langle \underline{m}' \rangle),$$

we first define the concatenation of injections

$$f : \overline{q} \hookrightarrow \overline{p} \quad \text{and} \quad f' : \overline{q'} \hookrightarrow \overline{p'}$$

to be the injection

$$(f \oplus f')(i) = \begin{cases} f(i) & \text{for} \quad i \in \{1, \ldots, q\} \\ p + f'(i - q) & \text{for} \quad i \in \{q+1, \ldots, q+q'\}. \end{cases}$$

Then the concatenation product of $(f, \langle \psi \rangle)$ and $(f', \langle \psi' \rangle)$ is given by the pair

$$(f, \langle \psi \rangle) \oplus (f', \langle \psi' \rangle) = (f \oplus f', \langle \psi \rangle \oplus \langle \psi' \rangle).$$

For checking functoriality, the essential observation is that for morphisms

$$\begin{aligned} (f, \langle \psi \rangle) &\in \mathcal{G}(\langle \underline{n} \rangle, \langle \underline{m} \rangle), \\ (f', \langle \psi' \rangle) &\in \mathcal{G}(\langle \underline{n}' \rangle, \langle \underline{m}' \rangle), \\ (g, \langle \phi \rangle) &\in \mathcal{G}(\langle \underline{m} \rangle, \langle \underline{\ell} \rangle), \quad \text{and} \\ (g', \langle \phi' \rangle) &\in \mathcal{G}(\langle \underline{m}' \rangle, \langle \underline{\ell}' \rangle), \end{aligned}$$

we have

$$(g \oplus g')_*(\langle \psi \rangle \oplus \langle \psi' \rangle) = (g_* \langle \psi \rangle) \oplus (g'_* \langle \psi' \rangle).$$

The concatentation product is strictly associative, and the empty tuple $\langle\rangle$ provides a strict unit. This concludes the definition and verification that $(\mathcal{G}, \oplus, \langle\rangle)$ is a strict monoidal category.

The symmetry components

$$\xi_{\langle \underline{n}\rangle,\langle \underline{n'}\rangle} \colon \langle \underline{n}\rangle \oplus \langle \underline{n'}\rangle \longrightarrow \langle \underline{n'}\rangle \oplus \langle \underline{n}\rangle \tag{9.1.13}$$

are given by pairs $(\gamma_{q,q'}, 1)$ where the first entry

$$\gamma_{q,q'} \colon \overline{q+q'} \longrightarrow \overline{q'+q}$$

is the block permutation swapping the first q elements with the last q' elements. This is the permutation denoted $\tau\langle q, q'\rangle$ in (II.1.2.2). The second entry is the identity on

$$(\gamma_{q,q'})_*\big(\langle \underline{n}\rangle \oplus \langle \underline{n'}\rangle\big) = \langle \underline{n'}\rangle \oplus \langle \underline{n}\rangle.$$

We take the components of ξ involving the basepoint $*$ to be the identity on $*$. The components of ξ involving the empty tuple $\langle\rangle$ are identities because the block permutations $\gamma_{0,q'}$ and $\gamma_{q,0}$ are identities.

We verify that the components $\xi_{\langle \underline{n}\rangle,\langle \underline{n'}\rangle}$ are natural and that $(\mathcal{G}, \oplus, \langle\rangle, \xi)$ is a permutative category in Proposition 9.1.14. ◇

Proposition 9.1.14. *The concatenation product*

$$\mathcal{G} \times \mathcal{G} \xrightarrow{\ \oplus\ } \mathcal{G}$$

makes

$$(\mathcal{G}, \oplus, \langle\rangle, \xi, *)$$

a permutative category with null object $*$.

Proof. The descriptions in Definition 9.1.12 show that $(\mathcal{G}, \oplus, \langle\rangle)$ is a strict monoidal category with null object $*$. The components ξ are defined there, and now we verify that they are natural with respect to morphisms in $\mathcal{G}$. There is nothing to check for monoidal products involving $*$ because all such products are equal to $*$ and the relevant components are identities. Next, for each pair of morphisms

$$\begin{aligned} (f, \langle\psi\rangle) &\in \mathcal{G}(\langle \underline{n}\rangle, \langle \underline{m}\rangle) \quad \text{and} \\ (f', \langle\psi'\rangle) &\in \mathcal{G}(\langle \underline{n'}\rangle, \langle \underline{m'}\rangle) \end{aligned}$$

we verify that the following diagram commutes in $\mathcal{G}$.

$$\begin{array}{ccc} \langle \underline{n}\rangle \oplus \langle \underline{n'}\rangle & \xrightarrow{\ \xi_{\langle \underline{n}\rangle,\langle \underline{n'}\rangle}\ } & \langle \underline{n'}\rangle \oplus \langle \underline{n}\rangle \\ {\scriptstyle (f,\langle\psi\rangle)\oplus(f',\langle\psi'\rangle)}\downarrow & & \downarrow{\scriptstyle (f',\langle\psi'\rangle)\oplus(f,\langle\psi\rangle)} \\ \langle \underline{m}\rangle \oplus \langle \underline{m'}\rangle & \xrightarrow{\ \xi_{\langle \underline{m}\rangle,\langle \underline{m'}\rangle}\ } & \langle \underline{m'}\rangle \oplus \langle \underline{m'}\rangle \end{array}$$

Commutativity for the first components follows because the block permutations γ are natural with respect to concatenation of injections f and f'. Commutativity for the second components follows because the second component of ξ is an identity and $\gamma_*(\langle\psi\rangle \oplus \langle\psi'\rangle) = \langle\psi'\rangle \oplus \langle\psi\rangle$. The symmetry axiom (1.1.24) follows from the equality of permutations

$$\gamma_{q',q}\gamma_{q,q'} = 1_{\overline{q+q'}}.$$

The unit axiom (1.1.25) for ξ holds by definition. The hexagon axiom (1.1.26) follows from the equality of block permutations shown in the following commutative diagram.

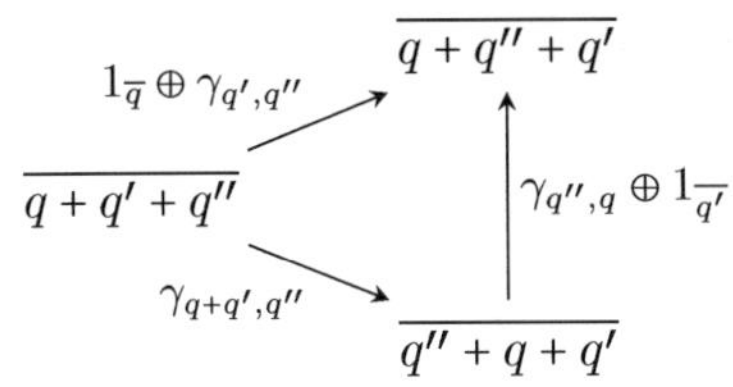

This completes the proof that $(\mathcal{G}, \oplus, \langle\rangle, \xi)$ is a permutative category. □

Now we compare the category $\mathcal{G}$ and its concatenation product to $\mathcal{F}$ and its smash product.

Definition 9.1.15. The smash product of pointed finite sets defines a strict symmetric monoidal pointed functor

$$\wedge : (\mathcal{G}, \oplus, \langle\rangle, *) \longrightarrow (\mathcal{F}, \wedge, \underline{1}, \underline{0})$$

whose value at a tuple of pointed finite sets $\langle \underline{n} \rangle$ is

$$\wedge\langle \underline{n} \rangle = \underline{n_1 \cdots n_q}$$

where $\langle \underline{n} \rangle$ has length $q > 0$. We define

$$\wedge\langle\rangle = \underline{1} \quad \text{and} \quad \wedge * = \underline{0}.$$

To define $\wedge$ on morphisms in $\mathcal{G}$, suppose given

$$(f, \langle\psi\rangle) : \langle \underline{n} \rangle \longrightarrow \langle \underline{m} \rangle$$

with $\langle \underline{n} \rangle \in \mathcal{F}^{(q)}$ and $\langle \underline{m} \rangle \in \mathcal{F}^{(p)}$. Recall from Definition 8.1.5 the lexicographic ordering bijections on FinSet_*. The value of $\wedge$ on $(f, \langle\psi\rangle)$ is given by composing the lexicographic ordering bijections

$$\bigwedge_{k=1}^{q} \underline{n}_k \cong \underline{\textstyle\coprod_k n_k} \quad \text{and} \quad \bigwedge_{j=1}^{p} \underline{m}_j \cong \underline{\textstyle\coprod_j m_j}$$

with the composite

$$\bigwedge_{k=1}^{q} \underline{n}_k \xrightarrow[\text{permute}]{\cong} \bigwedge_{\substack{j \in \overline{p}, \\ f^{-1}(j) \neq \varnothing}} \underline{n}_{f^{-1}(j)} \xrightarrow{\cong} \bigwedge_{j=1}^{p} \underline{n}_{f^{-1}(j)} \xrightarrow{\wedge_j \psi_j} \bigwedge_{j=1}^{p} \underline{m}_j \tag{9.1.16}$$

consisting of:

(1) the permutation induced by f, followed by
(2) unit isomorphisms for the entries of $f_*\langle \underline{n} \rangle$ not indexed by f, and then by
(3) the smash product of morphisms $\psi_j \in \langle\psi\rangle$.

If any $\underline{n}_i = \underline{0}$, then (9.1.16) is the zero morphism to $\wedge_j \underline{m}_j$.

If $(f, \langle\psi\rangle) = (1, \langle 1 \rangle)$ is an identity morphism in $\mathcal{G}$, then the morphism (9.1.16) is the composite of identity morphisms. That $\wedge$ preserves composition follows from functoriality of the smash product on $\mathcal{F}$ along with naturality of its unit and symmetry isomorphisms. The monoidal constraint of $\wedge$ is the identity because $\wedge$ for $\mathcal{F}$ implicitly composes with the lexicographic isomorphisms. ◇

Example 9.1.17. Consider the injection

$$f\colon \overline{2} \longrightarrow \overline{3}, \qquad (1 \longmapsto 3;\ 2 \longmapsto 1)$$

and suppose that

$$(f, \langle\psi\rangle)\colon (\underline{n}_1\,, \underline{n}_2) \longrightarrow (\underline{m}_1\,, \underline{m}_2\,, \underline{m}_3)$$

is a morphism in $\mathcal{G}$. Then $\wedge(f, \langle\psi\rangle)$ is the composite

$$\underline{n_1 n_2} \cong \underline{n}_1 \wedge \underline{n}_2 \xrightarrow{\cong} \underline{n}_2 \wedge \underline{n}_1 \xrightarrow{\cong} \underline{n}_2 \wedge \underline{1} \wedge \underline{n}_1 \xrightarrow{\wedge_j \psi_j} \underline{m}_1 \wedge \underline{m}_2 \wedge \underline{m}_3 \cong \underline{m_1 m_2 m_3}.$$

◇

9.2. Symmetric Monoidal Closed Structure for $\mathcal{G}_*$-Objects

In this section we describe $\mathcal{G}_*$-objects and symmetric monoidal closed structures for categories of such. The material here is an application of the general theory of enriched (pointed) diagram categories in Section 4.3, which itself depends on the preceding material in Part 1. However, for the applications here, the reader need only be familiar with the definitions and statements of results that we reference.

Definition 9.2.1. A *$\mathcal{G}_*$-object* in a pointed category $(\mathsf{C}, *)$ is a pointed functor

$$X\colon (\mathcal{G}, *) \longrightarrow (\mathsf{C}, *).$$

The category of $\mathcal{G}_*$-objects in C, denoted

$$\mathcal{G}_*\text{-}\mathsf{C} = \mathsf{Cat}_*\big((\mathcal{G}, *), (\mathsf{C}, *)\big) \tag{9.2.2}$$

is the category of pointed functors and natural transformations. ◇

Explanation 9.2.3 (Canonical Basepoints for $\mathcal{G}_*$-Objects)**.** In the context of Definition 9.2.1, there is a canonical isomorphism of categories

$$\mathcal{G}_*\text{-}\mathsf{C} \cong \mathcal{G}_*\text{-}(\mathsf{C}_*). \tag{9.2.4}$$

For each $\mathcal{G}_*$-object X, the unique morphisms

$$* \longrightarrow \langle \underline{n} \rangle$$

induce canonical basepoints

$$* = X* \longrightarrow X\langle \underline{n} \rangle$$

for each $\langle \underline{n} \rangle$ in $\mathcal{G}$. By functoriality, each $\mathcal{G}_*$-object in C thereby determines a unique $\mathcal{G}_*$-object in C_*, and conversely. Morphisms of $\mathcal{G}_*$-objects (pointed natural transformations between pointed functors) necessarily preserve these basepoints, and thus we have a bijection between the morphisms of $\mathcal{G}_*$-C and those of $\mathcal{G}_*$-(C_*).

The theory of pointed diagrams from Section 4.3 is formulated for $\mathcal{G}_*$-(C_*), and so we will use (9.2.4) implicitly to apply it below. We've given the definition of $\mathcal{G}_*$-C as in (9.2.2) so that it is not necessary to explicitly note the canonical basepoints in our further work below. ◇

Recall from Definition 4.3.23 and Corollary 4.3.30 the Day convolution, hom diagram, and monoidal unit for pointed diagrams out of a small symmetric monoidal category with a null object as its basepoint. Applying this theory to $\mathcal{G}$, we have the following preliminary definitions.

Definition 9.2.5. Suppose $(\mathsf{C}, \otimes, \mathbb{1}, *)$ is a symmetric monoidal closed category that is complete and cocomplete with chosen terminal object $*$. In the notation of Section 4.3, taking

$$(\mathcal{D}, \boxdot, e, \mathrm{T}^{\mathcal{D}}) = (\mathcal{G}, \oplus, \langle\rangle, *),$$

we have the following.

(1) For $\langle \underline{a} \rangle$ and $\langle \underline{b} \rangle$ in $\mathcal{G}$, we use the notation

$$\mathcal{G}^{\flat}(\langle \underline{a} \rangle ; \langle \underline{b} \rangle) = \big(\mathcal{G}(\langle \underline{a} \rangle, \langle \underline{b} \rangle)\big)^{\flat} \tag{9.2.6}$$

for the subset of nonzero morphisms.

(2) From Definition 4.3.11, we let $\widehat{\mathcal{G}}$ denote $\mathcal{G}$ equipped with the pointed unitary enrichment over C_* defined by

$$\widehat{\mathcal{G}}(\langle \underline{a} \rangle, \langle \underline{b} \rangle) = \bigvee_{\mathcal{G}^{\flat}(\langle \underline{a} \rangle ; \langle \underline{b} \rangle)} E \tag{9.2.7}$$

where $E = \mathbb{1} \coprod *$ is the monoidal unit of $(\mathsf{C}_*, \wedge)$. The empty wedge is the terminal object $*$ of C.

(3) The monoidal product of C_*-categories, called the tensor product in Definition 1.3.3, is denoted $\wedge$ and, in particular, $\widehat{\mathcal{G}} \wedge \widehat{\mathcal{G}}$ has hom objects given by smash products. $\diamond$

The following specializes Definition 4.3.23 and the adjunction (4.3.36) to $\mathcal{G}_*$-C.

Definition 9.2.8. Suppose $(\mathsf{C}, \otimes, \mathbb{1}, *)$ is a symmetric monoidal closed category that is complete and cocomplete with chosen terminal object $*$. Suppose given $\mathcal{G}_*$-objects X and Y. We define a mapping object $\mathsf{Map}_*(X, Y)$ in C_* and $\mathcal{G}_*$-objects

$$X \wedge Y, \quad \mathsf{Hom}_*(X, Y), \quad \text{and} \quad J$$

as follows. Each of $X \wedge Y$, $\mathsf{Hom}_*(X, Y)$, and J are pointed diagrams, so they send the basepoint $*$ of $\mathcal{G}$ to that of C.

- The Day convolution product of X and Y is

$$X \wedge Y = \int^{(\langle \underline{a} \rangle, \langle \underline{b} \rangle) \in \widehat{\mathcal{G}} \wedge \widehat{\mathcal{G}}} \bigvee_{\mathcal{G}^{\flat}(\langle \underline{a} \rangle \oplus \langle \underline{b} \rangle ; -)} \big(X\langle a \rangle \wedge Y\langle b \rangle\big). \tag{9.2.9}$$

- The hom diagram for X and Y is

$$\mathsf{Hom}_*(X, Y) = \int_{(\langle \underline{b} \rangle, \langle \underline{c} \rangle) \in \widehat{\mathcal{G}} \wedge \widehat{\mathcal{G}}} \prod_{\mathcal{G}^{\flat}(- \oplus \langle \underline{b} \rangle ; \langle \underline{c} \rangle)} [X\langle \underline{b} \rangle, Y\langle \underline{c} \rangle]_* \tag{9.2.10}$$

$$\cong \int_{\langle \underline{b} \rangle \in \widehat{\mathcal{G}}} [X\langle \underline{b} \rangle, Y(- \oplus \langle \underline{b} \rangle)]_* \tag{9.2.11}$$

where $[-,-]_*$ denotes the pointed internal hom of $(\mathsf{C}_*, \wedge)$.

- The mapping object for X and Y is

$$\mathsf{Map}_*(X, Y) = \int_{\langle \underline{b} \rangle \in \widehat{\mathcal{G}}} [X\langle \underline{b} \rangle, Y\langle \underline{b} \rangle]_* = \big(\mathsf{Hom}_*(X, Y)\big)\langle\rangle, \tag{9.2.12}$$

where the second equality holds because $\langle\rangle$ is a strict unit in $\mathcal{G}$.

- The monoidal unit diagram is

$$J = \widehat{\mathcal{G}}(\langle\rangle, -) = \bigvee_{\mathcal{G}^{\flat}(\langle\rangle ; -)} E. \tag{9.2.13}$$

These define a symmetric monoidal closed structure for $\mathcal{G}_*$-C by Corollary 4.3.30 and the isomorphism (9.2.4).

Moreover, evaluation at the empty tuple $\langle\rangle$ defines a symmetric monoidal functor $\mathsf{ev}_{\langle\rangle}$ that has a strong symmetric monoidal left adjoint $L_{\langle\rangle}$:

$$L_{\langle\rangle}\colon \mathsf{C}_* \rightleftarrows \mathcal{G}_*\text{-}\mathsf{C} \colon \mathsf{ev}_{\langle\rangle}. \tag{9.2.14}$$

The left adjoint $L_{\langle\rangle}$ is defined for $A \in \mathsf{C}_*$ by composition

$$A \wedge J\colon \widehat{\mathcal{G}} \xrightarrow{J} \mathsf{C}_* \xrightarrow{A\wedge -} \mathsf{C}_*. \qquad \diamond$$

Applying Theorem 4.3.37 in this case, we have the following.

Theorem 9.2.15. *Suppose* $(\mathsf{C}, \otimes, \mathbb{1}, *)$ *is a symmetric monoidal closed category that is complete and cocomplete with chosen terminal object* $*$*. Then* $\mathcal{G}_*$-C *is a complete and cocomplete symmetric monoidal closed category with*

- *monoidal product given by the Day convolution* $\wedge$ (9.2.9),
- *internal hom given by* Hom_* (9.2.10), *and*
- *monoidal unit given by* J (9.2.13).

Moreover, the adjunction $(L_{\langle\rangle}, \mathsf{ev}_{\langle\rangle})$ *makes* $\mathcal{G}_*$-C *enriched, tensored, and cotensored over* C_* *with mapping objects given by* Map_*.

Our applications of Theorem 9.2.15 will have $\mathsf{C} = \mathsf{sSet}$, so $\mathsf{C}_* = \mathsf{sSet}_*$ and $\mathsf{C} = \mathsf{Cat}$, so $\mathsf{C}_* = \mathsf{Cat}_*$.

Explanation 9.2.16 (Mapping Objects for Diagrams in Cat_*). Example 3.8.14 describes the following. Suppose $\mathcal{D}$ is a small symmetric monoidal category with unitary enrichment $\widehat{\mathcal{D}}$ over Cat. If X and Y are diagrams from $\widehat{\mathcal{D}}$ to Cat then, by the equalizer formula (3.5.7), the mapping object

$$\mathsf{Map}(X,Y) = \int_{b\in\widehat{\mathcal{D}}} \mathsf{Cat}(X_b, Y_b)$$

is the category of 2-natural transformations and modifications between X and Y regarded as 2-functors.

Similarly, now suppose $\widehat{\mathcal{D}}$ is the pointed unitary enrichment (Definition 4.3.11) over $(\mathsf{Cat}_*, \wedge)$. If X and Y are diagrams from $\widehat{\mathcal{D}}$ to Cat_*, then the objects of the pointed category

$$\mathsf{Map}_*(X,Y) = \int_{b\in\widehat{\mathcal{D}}} \mathsf{Cat}_*(X_b, Y_b)$$

are 2-natural transformations α such that each component

$$\alpha_b\colon X_b \longrightarrow Y_b$$

is a pointed functor. The basepoint of $\mathsf{Map}_*(X,Y)$ has each component at $b \in \mathcal{D}$ given by the constant functor to the basepoint of Y_b. The morphisms of $\mathsf{Map}_*(X,Y)$ from α to α' are given by modifications Θ such that each component

$$\Theta_b\colon \alpha_b \longrightarrow \alpha'_b$$

is a pointed natural transformation. That is, for each $b \in \mathcal{D}$ the component of Θ at the basepoint $\top$ of X_b is the identity morphism of $\alpha_b(\top) = \alpha'_b(\top)$. $\diamond$

Explanation 9.2.17 (Mapping Objects for Diagrams in sSet_*). Here we give a similar explanation to that of Explanation 9.2.16. Suppose that $\mathcal{D}$ is a small symmetric monoidal category and suppose that $\widehat{\mathcal{D}}$ is its strictly unitary enrichment over $(\mathsf{sSet}_*, \wedge)$. Recall from Proposition 7.1.19 that the internal hom for pointed simplicial sets has set of n-simplices given by

$$\mathsf{Hom}^{\mathsf{sSet}_*}(A,B)_n = \mathsf{sSet}_*(A \wedge \Delta^n_+, B) = \int_{\underline{m}\in\Delta^{\mathsf{op}}} \mathsf{Set}_*(A_m \wedge \Delta^n_m, B_m).$$

for pointed simplicial sets A and B.

Now if X and Y are diagrams from $\widehat{\mathcal{D}}$ to sSet_*, we write $X_{b,n}$ for the n-simplices of X_b with $b \in \mathcal{D}$. A similar application of the equalizer formula (3.5.7) shows that n-simplices of the mapping object $\mathsf{Map}_*(X,Y)$

$$\begin{aligned} \mathsf{Map}_*(X,Y)_n &= \int_{b\in\widehat{\mathcal{D}}} \mathsf{sSet}_*(X_{b,-} \wedge \Delta^n_+, Y_{b,-}) \\ &= \int_{b\in\widehat{\mathcal{D}}} \int_{\underline{m}\in\Delta^{\mathsf{op}}} \mathsf{Set}_*(X_{b,m} \wedge \Delta^n_m, Y_{b,m}) \end{aligned}$$

are given by natural transformations between functors

$$X \wedge \Delta^n, Y\colon \mathcal{D} \times \Delta^{\mathsf{op}} \longrightarrow \mathsf{Set}_*.$$

Equivalently, $\mathsf{Map}_*(X,Y)$ is isomorphic to the internal hom for simplicial objects of $\mathcal{D}_*$-Set. ◇

Explanation 9.2.18 (Symmetric Monoidal Closed Structure for Γ-Objects)**.** One can also apply the theory of Section 4.3 to $\mathcal{F}$, taking

$$(\mathcal{D}, \boxdot, e, \mathrm{T}^{\mathcal{D}}) = (\mathcal{F}, \wedge, \underline{1}, \underline{0}).$$

With C as in Definition 9.2.8, and with $(\mathcal{G}, \oplus, \langle\rangle)$ replaced by $(\mathcal{F}, \wedge, \underline{1})$, the Day convolution, hom diagram, and monoidal unit for

$$\mathcal{F}_*\text{-}\mathsf{C} = \Gamma\text{-}\mathsf{C}$$

are defined via formulas similar to those of Definition 9.2.8. By Corollary 4.3.30 this defines a symmetric monoidal closed structure for Γ-C. In Theorem 9.4.18 below we will see that the functor

$$\mathsf{K}^{\mathcal{F}}\colon \Gamma\text{-}\mathsf{sSet} \longrightarrow \mathsf{SymSp}$$

of Proposition 8.2.6 is symmetric monoidal as a sSet_*-functor. However one encounters difficulties when trying to develop symmetric monoidal structure for

$$\mathsf{J}^{\mathsf{Se}}\colon \mathsf{PermCat}^{\mathsf{su}} \longrightarrow \Gamma\text{-}\mathsf{Cat}.$$

See Note 10.8.6 for additional remarks on this point. ◇

As with Γ-categories and Γ-simplicial sets, we will apply the nerve functor N (Definition 7.2.3) to obtain $\mathcal{G}_*$-simplicial sets from $\mathcal{G}_*$-categories. Now we show that N is compatible with the symmetric monoidal structures discussed above. We will use several of the general results from Chapters 2 and 3 related to change of enrichments, self-enrichments, and enriched diagram categories.

Recall from Proposition 7.2.9 that the nerve is a strong symmetric monoidal functor

$$N\colon (\mathsf{Cat}, \times, \mathbb{1}) \longrightarrow (\mathsf{sSet}_*, \times, *).$$

Theorem 9.2.19. *The nerve functor N induces the following.*

(1) *There is a sSet_*-enrichment of $\mathcal{G}_*$-Cat with hom objects given by*

$$N\big(\mathsf{Map}_*(X,Y)\big)$$

for $\mathcal{G}_*$-categories X and Y.

(2) *Nerve induces a symmetric monoidal* sSet_**-functor*

$$N_* \colon \mathcal{G}_*\text{-}\mathsf{Cat} \longrightarrow \mathcal{G}_*\text{-}\mathsf{sSet}$$

with the following data.

- *For a* $\mathcal{G}_*$*-category* X, N_*X *is the* $\mathcal{G}_*$*-simplicial set given by the composite*

$$\mathcal{G} \xrightarrow{X} \mathsf{Cat}_* \xrightarrow{N} \mathsf{sSet}_*.$$

- *For* $\mathcal{G}_*$*-categories* X *and* Y, *the morphism on hom objects*

$$N_* \colon N\big(\mathsf{Map}_*(X,Y)\big) \longrightarrow \mathsf{Map}_*(N_*X, N_*Y)$$

sends a 0-simplex $\alpha \colon X \longrightarrow Y$ *in the source to the morphism of simplicial sets* $N_*\alpha$ *given by the whiskering of* α *with* N

$$\mathcal{G} \underset{Y}{\overset{X}{\overset{\Downarrow \alpha}{\rightrightarrows}}} \mathsf{Cat}_* \xrightarrow{N} \mathsf{sSet}_*. \tag{9.2.20}$$

Proof. Recall from Definition 9.2.5 that $\widehat{\mathcal{G}}$ denotes the pointed unitary enrichment (9.2.7) of $\mathcal{G}$ over $(\mathsf{C}_*, \wedge, E)$, for a complete and cocomplete symmetric monoidal closed category $(\mathsf{C}, \otimes, \mathbb{1}, *)$ with terminal object $*$ and

$$E = \mathbb{1} \coprod *.$$

The symmetric monoidal closed structure on $\mathcal{G}_*$-C, described in Theorem 9.2.15, is given via the equivalence (4.3.35)

$$\mathcal{G}_*\text{-}\mathsf{C} \simeq \widehat{\mathcal{G}}\text{-}(\mathsf{C}_*) = \mathsf{C}_*\mathsf{Cat}(\widehat{\mathcal{G}}, \underline{\mathsf{C}_*}).$$

The right hand side above denotes the category of C_*-enriched functors and natural transformations from $\widehat{\mathcal{G}}$ to $\underline{\mathsf{C}_*}$, where the latter is the canonical self-enrichment of C_* (Definition 3.1.5).

Our two cases of interest are

$$(\mathsf{C}, \otimes, \mathbb{1}, *) = (\mathsf{Cat}, \times, \mathbb{1}, \mathbb{1})$$
$$(\mathsf{C}, \otimes, \mathbb{1}, *) = (\mathsf{sSet}, \times, *, *).$$

We take $N\mathbb{1} = *$ so that N preserves the monoidal unit and terminal object strictly. Therefore, with $\widehat{\mathcal{G}}$ denoting the pointed unitary enrichment for Cat_*, the change of enrichment $\widehat{\mathcal{G}}_N$ is the pointed unitary enrichment of $\mathcal{G}$ for sSet_*.

Since N is strong symmetric monoidal (Proposition 7.2.9), change of enrichment and composition with N defines a symmetric monoidal functor

$$N_* \colon \widehat{\mathcal{G}}\text{-}(\mathsf{Cat}_*) \longrightarrow (\widehat{\mathcal{G}}_N)\text{-}(\mathsf{sSet}_*) = \widehat{\mathcal{G}}\text{-}(\mathsf{sSet}_*).$$

by Proposition 3.8.4. On the right hand side we have used the observation above that $\widehat{\mathcal{G}}_N$ is the pointed unitary enrichment of $\mathcal{G}$ for sSet_*.

Next, we take the canonical self-enrichments of both $\widehat{\mathcal{G}}$-(Cat_*) and $\widehat{\mathcal{G}}$-(sSet_*), using the same notation for readability. The standard enrichment of $U = N_*$ (Definition 3.3.3) is a symmetric monoidal $\underline{\mathsf{W}}$-functor

$$U = N_* \colon (\widehat{\mathcal{G}}\text{-}(\mathsf{Cat}_*))_U \longrightarrow \widehat{\mathcal{G}}\text{-}(\mathsf{sSet}_*) \tag{9.2.21}$$

with $\mathsf{W} = \widehat{\mathcal{G}}\text{-}(\mathsf{sSet}_*)$. For a pair of $\mathcal{G}_*$-categories X and Y, the morphism on hom objects is

$$N_*(\mathsf{Hom}_*(X,Y)) \longrightarrow \mathsf{Hom}_*(N_*X, N_*Y) \tag{9.2.22}$$

given by adjunction, the monoidal structure of N_*, and evaluation (see Definition 3.3.3).

Now we use the evaluation at the monoidal unit $\langle\rangle$ in $\mathcal{G}$. By Lemma 3.8.1, $\mathsf{ev}_{\langle\rangle}$ is a symmetric monoidal functor

$$\widehat{\mathcal{G}}\text{-}\mathsf{sSet} \longrightarrow \mathsf{sSet}.$$

Changing enrichment of $U = N_*$ along $\mathsf{ev}_{\langle\rangle}$, as in Definition 2.1.1, makes

$$(U)_{\mathsf{ev}_{\langle\rangle}} = (N_*)_{\mathsf{ev}_{\langle\rangle}} \colon \big((\widehat{\mathcal{G}}\text{-}\mathsf{Cat}_*)_U\big)_{\mathsf{ev}_{\langle\rangle}} \longrightarrow \big(\widehat{\mathcal{G}}\text{-}\mathsf{sSet}_*\big)_{\mathsf{ev}_{\langle\rangle}} \tag{9.2.23}$$

a symmetric monoidal sSet_*-functor.

For ease of notation we usually write N_* for the change of enrichment $(N_*)_{\mathsf{ev}\langle\rangle}$. The domain of (9.2.23) has hom objects given as follows for $\mathcal{G}_*$-categories X and Y:

$$\mathsf{ev}_{\langle\rangle}\Big(N_*\big(\mathsf{Hom}_*(X,Y)\big)\Big) = N\big(\mathsf{Hom}_*(X,Y)_{\langle\rangle}\big) \cong N\big(\mathsf{Map}_*(X,Y)\big).$$

This gives the sSet_*-enrichment of $\mathcal{G}_*$-Cat. By definition (Theorem 9.2.15) the codomain of (9.2.23) is the sSet_*-enrichment of $\mathcal{G}_*$-sSet, with hom objects also denoted Map_*. This finishes the proof that N_* defines a symmetric monoidal sSet_*-functor.

For $\mathcal{G}_*$-categories X and Y, the morphism of simplicial sets

$$N\big(\mathsf{Map}_*(X,Y)\big) \longrightarrow \mathsf{Map}_*(N_*X, N_*Y) \tag{9.2.24}$$

is given by evaluating (9.2.22) at the empty tuple, $\langle\rangle$. Checking components at objects of the categories $X\langle \underline{m}\rangle$, for $\langle \underline{m}\rangle$ in $\mathcal{G}$, confirms that (9.2.24) is given on a 0-simplex α by the whiskering (9.2.20) as asserted. ☐

9.3. Symmetric Spectra from $\mathcal{G}_*$-Simplicial Sets

In this section we construct a symmetric spectrum $\mathsf{K}^{\mathcal{G}}X$ associated to a $\mathcal{G}_*$-simplicial set X. We begin with necessary preliminary definitions and then give the construction in Definition 9.3.14. We prove functoriality of $\mathsf{K}^{\mathcal{G}}$ and compare it with $\mathsf{K}^{\mathcal{F}}$ in Proposition 9.3.16.

Recall from Definition 9.1.15 the smash product of pointed finite sets defines a functor

$$\wedge \colon \mathcal{G} \longrightarrow \mathcal{F}.$$

Definition 9.3.1. For each natural number p, we have the *inclusion of length-p tuples*

$$i_p \colon \mathcal{F}^{(p)} \longrightarrow \mathcal{G} \tag{9.3.2}$$

given by the identity on objects and by

$$\langle\psi\rangle \longmapsto (1_{\overline{p}}, \langle\psi\rangle).$$

on morphisms. ◇

Explanation 9.3.3. For $p > 0$, the composite

$$\mathcal{F}^{(p)} \xrightarrow{\ i_p\ } \mathcal{G} \xrightarrow{\ \wedge\ } \mathcal{F}$$

is the p-fold smash product of pointed finite sets and pointed morphisms. For $p = 0$, the composite sends $*$ to $\underline{0}$ and $\langle\rangle$ to $\underline{1}$. ◇

Definition 9.3.4. For each $p > 0$ we define a functor from the p-fold Cartesian product

$$c_p\colon \mathcal{F}^p \longrightarrow \mathcal{F}^{(p)} \tag{9.3.5}$$

by iterating the canonical projection

$$c\colon \mathsf{A} \times \mathsf{B} \longrightarrow \mathsf{A} \wedge \mathsf{B} \tag{9.3.6}$$

in the definition of the smash product (4.1.7) for pointed categories A and B with $\mathsf{C} = \mathsf{Cat}$. For $p = 0$ we define

$$c_0\colon \mathcal{F}^0 = \mathbb{1} \longrightarrow \mathcal{F}^{(0)} \tag{9.3.7}$$

by sending the unique object of $\mathbb{1}$ to the empty tuple $\langle\rangle$. We note that c_0 is *not* a pointed functor. ◇

Now we define a functor for $\mathcal{G}_*$-simplicial sets that restricts to that of Definition 8.2.5 along the functor

$$\Gamma\text{-}\mathsf{sSet} \xrightarrow{\wedge^*} \mathcal{G}_*\text{-}\mathsf{sSet} \tag{9.3.8}$$

induced by the smash product $\wedge\colon \mathcal{G} \longrightarrow \mathcal{F}$ of Definition 9.1.15. In the following discussion, recall from Definition 8.2.1 the $\mathcal{F}$-sphere $\overline{S}$ is a symmetric spectrum with n-simplices of the pth term given by

$$\overline{S}^p_n = \underline{n}^p.$$

Under the composite

$$\mathcal{F}^{(p)} \xrightarrow{i_p} \mathcal{G} \xrightarrow{\wedge} \mathcal{F}$$

we have

$$\underbrace{(\overline{S}^1_n, \ldots, \overline{S}^1_n)}_{p} \longmapsto \overline{S}^p_n.$$

In particular, for $p = 0$, the empty tuple $\langle\rangle$ is sent to $\overline{S}^0_n$.

The structure morphisms for the symmetric spectra we construct will be defined using the following morphisms.

Definition 9.3.9. Suppose X is a $\mathcal{G}_*$-simplicial set. For each $\underline{m} \in \mathcal{F}$ and each q-tuple $\langle\underline{n}\rangle \in \mathcal{G}$ with $q \geq 0$ we combine the inclusion i_1 with the concatenation $\oplus$ in $\mathcal{G}$ to form the $(q+1)$-tuple

$$(i_1\underline{m}) \oplus \langle\underline{n}\rangle = (\underline{m}, \underline{n}_1, \ldots, \underline{n}_q).$$

For each $x \in \underline{m}^\flat$, let

$$h_x\colon \underline{1} \longrightarrow \underline{m}$$

be the pointed function that sends 1 to x. Let

$$f_1\colon \overline{q} \longrightarrow \overline{q+1}$$

be given by $f_1(i) = i + 1$. Then for each $x \in \underline{m}^\flat$ we have a morphism in $\mathcal{G}$

$$(f_1, (i_1 h_x) \oplus 1_{\langle\underline{n}\rangle})\colon \langle\underline{n}\rangle \longrightarrow (i_1\underline{m}) \oplus \langle\underline{n}\rangle.$$

Applying X to this morphism, we have a morphism in sSet_*

$$\eta_{\underline{m},\langle\underline{n}\rangle,x} = X(f_1, (i_1 h_x) \oplus 1_{\langle\underline{n}\rangle})\colon X\langle\underline{n}\rangle \longrightarrow X((i_1\underline{m}) \oplus \langle\underline{n}\rangle).$$

Taking the wedge, we have

$$\eta_{\underline{m},\langle\underline{n}\rangle}\colon \underline{m} \wedge X\langle\underline{n}\rangle \cong \bigvee_{x\in\underline{m}^\flat} X\langle\underline{n}\rangle \longrightarrow X((i_1\underline{m}) \oplus \langle\underline{n}\rangle). \tag{9.3.10}$$

If $m = 0$, then $(i_1\underline{0}) \oplus \langle\underline{n}\rangle$ has first entry $\underline{0}$, so it is the null object $*$ in $\mathcal{G}$. Since $X(*) = *$, $\eta_{\underline{0},\langle\underline{n}\rangle}$ is the identity morphism 1_*.

If $\langle\underline{m}\rangle \in \mathcal{F}^{(p)}$ is a p-tuple of pointed finite sets with $p > 1$, we define

$$\eta_{\langle\underline{m}\rangle,\langle\underline{n}\rangle}\colon \underline{m}_1 \wedge (\underline{m}_2 \wedge (\cdots(\underline{m}_p \wedge X\langle\underline{n}\rangle)\cdots)) \longrightarrow X((i_p\langle\underline{m}\rangle) \oplus \langle\underline{n}\rangle) \tag{9.3.11}$$

inductively as the composite

$$\begin{array}{ccc} & \underline{m}_1 \wedge X((i_{p-1}\langle\underline{m}'\rangle) \oplus \langle\underline{n}\rangle) & \\ {\scriptstyle 1_{\underline{m}_1} \wedge \eta_{\langle\underline{m}'\rangle,\langle\underline{n}\rangle}} \nearrow & & \searrow {\scriptstyle \eta_{\underline{m}_1,\langle\underline{n}'\rangle}} \\ \underline{m}_1 \wedge (\underline{m}_2 \wedge (\cdots(\underline{m}_p \wedge X\langle\underline{n}\rangle)\cdots)) & & X((i_p\langle\underline{m}\rangle) \oplus \langle\underline{n}\rangle) \end{array}$$

where

$$\langle\underline{m}'\rangle = (\underline{m}_2, \ldots, \underline{m}_p) \quad \text{and} \quad \langle\underline{n}'\rangle = (i_{p-1}\langle\underline{m}'\rangle) \oplus \langle\underline{n}\rangle.$$

Similarly to (9.3.10), if any $m_i = 0$, then

$$(i_p\langle\underline{m}\rangle) \oplus \langle\underline{n}\rangle$$

has an entry $\underline{0}$, so it is $*$ in $\mathcal{G}$. Since $X(*) = *$, (9.3.11) is 1_*.

These morphisms are natural with respect to $\langle\underline{m}\rangle \in \mathcal{F}^{(p)}$ and $\langle\underline{n}\rangle \in \mathcal{G}$. For $p = 0$, we define

$$\eta_{\langle\rangle,\langle\underline{n}\rangle} = 1\colon X\langle\underline{n}\rangle \longrightarrow X\langle\underline{n}\rangle \tag{9.3.12}$$

to be the identity of $X\langle\underline{n}\rangle$. For the basepoint $* \in \mathcal{G}$, we recall that $*$ is a null object with respect to the concatenation in $\mathcal{G}$ and, for p-tuples $\langle\underline{m}\rangle$ with $p \geq 0$, define

$$\eta_{\langle\underline{m}\rangle,*} = 1_*, \tag{9.3.13}$$

the identity on the initial and terminal pointed simplicial set. ◇

Generalizing the constructions of Definition 8.2.5, we will construct pointed bisimplicial sets as functors

$$\Delta^{\mathsf{op}} \longrightarrow \mathsf{sSet}_*$$

and take the diagonal as discussed in Explanation 7.1.7 to obtain a pointed simplicial set.

Definition 9.3.14. Suppose X is a $\mathcal{G}_*$-simplicial set. We define a pointed simplicial set $(\mathsf{K}^{\mathcal{G}}X)k$ for each natural number $k \geq 0$ and structure morphisms

$$\rho_{p,q}\colon S^p \wedge (\mathsf{K}^{\mathcal{G}}X)q \longrightarrow (\mathsf{K}^{\mathcal{G}}X)(p+q)$$

for natural numbers $p, q \geq 0$. In Proposition 9.3.16 below we will show that these data define a functor

$$\mathsf{K}^{\mathcal{G}}\colon \mathcal{G}_*\text{-}\mathsf{sSet} \longrightarrow \mathsf{SymSp}.$$

For each natural number $k \geq 0$, let $(\mathsf{K}^{\mathcal{G}}X)k$ be the pointed simplicial set obtained by taking the diagonal of the pointed bisimplicial set given by the following composite:

$$\Delta^{\mathsf{op}} \xrightarrow{\text{diag}} (\Delta^{\mathsf{op}})^k \xrightarrow{(\overline{S}^1)^k} \mathcal{F}^k \xrightarrow{c_k} \mathcal{F}^{(k)} \xrightarrow{i_k} \mathcal{G} \xrightarrow{X} \mathsf{sSet}_*.$$

In this composite, the first functor is the diagonal, the second is the k-fold Cartesian product of $\overline{S}^1$, the third is the canonical functor to the smash product (9.3.5), and the fourth is the inclusion of length k-tuples (9.3.2).

The components of η defined in (9.3.11), (9.3.12), and (9.3.13) assemble to give a natural transformation in the following diagram, whose terms we explain below.
(9.3.15)

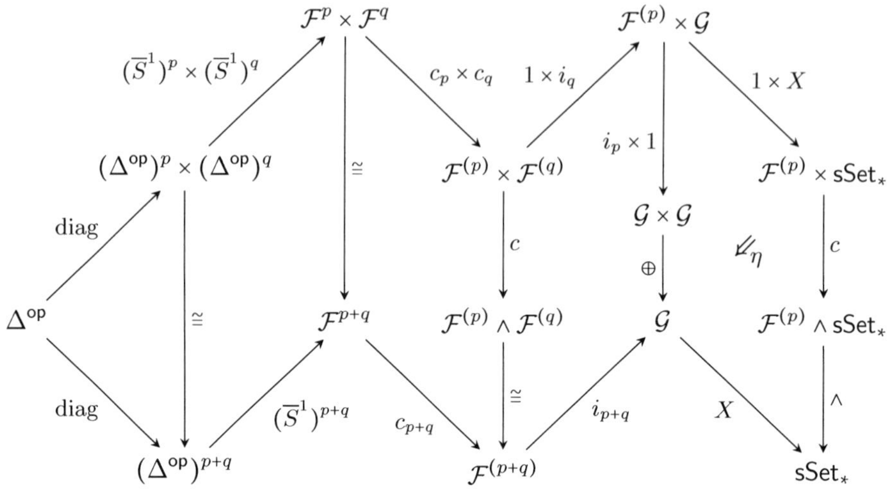

In the diagram (9.3.15) above:

- The unnamed vertical isomorphisms are given by associativity isomorphisms (for $p, q > 0$) or unit isomorphisms (for $p = 0$ or $q = 0$) of the Cartesian or smash products.
- The vertical morphisms c are given by the canonical functors to the smash product (9.3.6).
- The vertical morphism labeled $\wedge$ is the iterated levelwise smash product of pointed finite sets with simplicial sets:

$$\mathcal{F}^{(p)} \wedge \mathsf{sSet}_* \xrightarrow{(1_{\mathcal{F}^{(p-1)}}, \wedge)} \mathcal{F}^{(p-1)} \wedge \mathsf{sSet}_* \cdots \longrightarrow \mathcal{F} \wedge \mathsf{sSet}_* \xrightarrow{\wedge} \mathsf{sSet}_*.$$

- The natural transformation η has components $\eta_{\langle \underline{m} \rangle, \langle \underline{n} \rangle}$ given in (9.3.11).
- The unlabeled parallelogram and triangular regions commute.

The upper, respectively lower, composite around the boundary of (9.3.15) is the pointed bisimplicial set

$$\overline{S}^p \wedge X(\underbrace{\overline{S}^1, \ldots, \overline{S}^1}_{q}), \quad \text{respectively} \quad X(\underbrace{\overline{S}^1, \ldots, \overline{S}^1}_{p+q}),$$

where we have made use of the equality $\overline{S}^p = (\overline{S}^1)^{\wedge p}$ in the diagram category $\mathsf{Cat}(\Delta^{\mathsf{op}}, \mathcal{F})$. Taking diagonals and using the isomorphism $S \cong \overline{S}$, we have morphisms in sSet_*

$$\rho_{p,q} \colon S^p \wedge (\mathsf{K}^{\mathcal{G}} X)q \xrightarrow{\cong} \overline{S}^p \wedge (\mathsf{K}^{\mathcal{G}} X)q \longrightarrow (\mathsf{K}^{\mathcal{G}} X)(p+q).$$

This completes the definition of $\rho_{p,q}$. ◇

As an example of the construction $\mathsf{K}^{\mathcal{G}}$, Lemma 9.4.3 below shows that $\mathsf{K}^{\mathcal{G}}J$ is the symmetric sphere spectrum, where J is the monoidal unit of $\mathcal{G}_*$-sSet from (9.2.13).

Proposition 9.3.16. *The data* $\{(\mathsf{K}^{\mathcal{G}}X)k,\ \rho_{p,q} \mid k,p,q \geq 0\}$ *of Definition 9.3.14 define a functor*

$$\mathsf{K}^{\mathcal{G}}\colon \mathcal{G}_*\text{-sSet} \longrightarrow \mathsf{SymSp}$$

such that the following diagram commutes.

(9.3.17)
$$\begin{array}{ccc} \Gamma\text{-sSet} & & \\ & \searrow^{\mathsf{K}^{\mathcal{F}}} & \\ \wedge^* \big\downarrow & & \mathsf{SymSp} \\ & \nearrow_{\mathsf{K}^{\mathcal{G}}} & \\ \mathcal{G}_*\text{-sSet} & & \end{array}$$

Proof. We first show that the construction $(\mathsf{K}^{\mathcal{G}}X,\rho)$ of Definition 9.3.14 is a symmetric spectrum. For this, we use the description of symmetric spectra from Explanation 7.4.6. The unity condition (7.4.7) holds because each $\eta_{\langle\rangle,\langle \underline{n}\rangle}$ is an identity. The associativity condition (7.4.8) for ρ follows from the inductive definition of η. The equivariance condition follows because the action of Σ_k on $(\mathsf{K}^{\mathcal{G}}X)k$ is given by permuting the coordinates of $(\overline{S}^1)^k$.

The components of η defined in (9.3.11), (9.3.12), and (9.3.13) are natural with respect to morphisms of $\mathcal{G}_*$-simplicial sets (that is, natural transformations of functors) and hence the structure morphisms $\rho_{p,q}$ are too. This shows that $\mathsf{K}^{\mathcal{G}}$ takes values in the category of symmetric spectra and morphisms thereof. Functoriality of $\mathsf{K}^{\mathcal{G}}$ with respect to morphisms of $\mathcal{G}_*$-simplicial sets follows because each step in the construction of $(\mathsf{K}^{\mathcal{G}}X)k$ is functorial.

Commutativity of the diagram (9.3.17) follows by checking the definitions of $(\mathsf{K}^{\mathcal{F}}X)k$ and $\rho_{p,q}$. The key fact is that our definitions of

$$\wedge\colon \mathcal{G} \longrightarrow \mathcal{F} \quad \text{and} \quad \overline{S}^k \in \mathsf{Cat}(\Delta^{\mathrm{op}},\mathcal{F})$$

ensure that we have, for each n, an equality

$$\wedge(\underbrace{\overline{S}^1_n,\ldots,\overline{S}^1_n}_{k}) = \overline{S}^k_n. \qquad \square$$

9.4. $\mathsf{K}^{\mathcal{G}}$ is Symmetric Monoidal

Recall from Corollary 7.6.16 that the category of symmetric spectra is symmetric monoidal as a sSet_*-category. Applying Theorem 9.2.15 in the case

$$(\mathsf{C},\otimes,\mathbb{1},*) = (\mathsf{sSet},\times,\mathbb{1},\mathbb{1})$$

shows that $\mathcal{G}_*$-sSet is enriched over sSet_* and is symmetric monoidal as a sSet_*-category. In this section, we show that the functor $\mathsf{K}^{\mathcal{G}}$ in Proposition 9.3.16 is a symmetric monoidal sSet_*-functor (Definition 1.4.18).

- Lemma 9.4.3 shows that $\mathsf{K}^{\mathcal{G}}$ sends the monoidal unit in $\mathcal{G}_*$-sSet to the symmetric sphere, which is the monoidal unit in SymSp.
- Lemma 9.4.6 computes $\mathsf{K}^{\mathcal{G}}$ of a triple product in $\mathcal{G}_*$-sSet.
- Theorem 9.4.9 is the main result of this section, which uses both Lemmas 9.4.3 and 9.4.6.

- Theorem 9.4.18 shows that the other two functors in (9.3.17), $\mathsf{K}^{\mathcal{F}}$ and $\wedge^*$, are also symmetric monoidal sSet_*-functors. See Note 9.5.5 for further generalization.

Monoidal Units. First recall the following structure.

- $(\mathsf{sSet}_*, \wedge, S^0)$ is the symmetric monoidal closed category of pointed simplicial sets (Definition 7.1.23).
- The category $\mathcal{G}$ is a permutative category (Proposition 9.1.14) with
  - the concatenation product $\oplus$ and
  - the empty tuple $\langle\rangle$ as the monoidal unit.
- From (9.2.6) we use

$$\mathcal{G}^\flat(\langle \underline{a}\rangle ; \langle \underline{b}\rangle) = \big(\mathcal{G}(\langle \underline{a}\rangle, \langle \underline{b}\rangle)\big)^\flat$$

  to denote the set of nonzero morphisms from $\langle \underline{a}\rangle$ to $\langle \underline{b}\rangle$ in $\mathcal{G}$.
- In the symmetric monoidal closed category $\mathcal{G}_*$-sSet, the monoidal unit (9.2.13) is defined by

(9.4.1)
$$J\langle \underline{m}\rangle = \bigvee_{\mathcal{G}^\flat(\langle\rangle ; \langle \underline{m}\rangle)} S^0 \quad\text{for}\quad \langle \underline{m}\rangle \in \mathcal{G} \quad\text{and}\quad J* = *.$$

  The rest of the symmetric monoidal structure in $\mathcal{G}_*$-sSet is induced by the Day convolution and hom diagrams as described in Definition 9.2.8.
- For $X \in \mathcal{G}_*$-sSet and $m, p \geq 0$, we write

(9.4.2)
$$(\mathsf{K}^{\mathcal{G}} X)_{p,m} = X\big(\underline{m}^{\oplus p}\big)_m$$

  for the pointed set of m-simplices in the pointed simplicial set $(\mathsf{K}^{\mathcal{G}} X)_p$ (Definition 9.3.14).

The following preliminary observation says that the symmetric sphere (Definitions 7.4.1 and 8.2.1) is the K-theory of the monoidal unit $J \in \mathcal{G}_*$-sSet. It is both an example of the construction $\mathsf{K}^{\mathcal{G}}$ and an important part of Theorem 9.4.9 below, which asserts that $\mathsf{K}^{\mathcal{G}}$ is a symmetric monoidal sSet_*-functor.

Lemma 9.4.3. *There is a canonical isomorphism*

(9.4.4)
$$S \cong \mathsf{K}^{\mathcal{G}} J$$

of symmetric spectra with

- *S the symmetric sphere,*
- *$\mathsf{K}^{\mathcal{G}} \colon \mathcal{G}_*$-$\mathsf{sSet} \longrightarrow \mathsf{SymSp}$ the functor in Proposition 9.3.16, and*
- *$J \in \mathcal{G}_*$-sSet the monoidal unit.*

Proof. We compute the symmetric spectrum $\mathsf{K}^{\mathcal{G}} J$ and show that it is canonically isomorphic to S. By Explanation 9.1.9 (4), a morphism

$$\langle\rangle \longrightarrow \langle \underline{m}\rangle = \big(\underline{m_1}, \ldots, \underline{m_p}\big) \in \mathcal{G}$$

consists of the unique injection

$$\varnothing = \overline{0} \longrightarrow \overline{p} = \{1, \ldots, p\}$$

together with pointed functions

$$\underline{1} \xrightarrow{\ \psi_i\ } \underline{m_i} \quad \text{for} \quad 1 \leq i \leq p.$$

Applied to (9.4.1), this implies that

$$J\langle \underline{m} \rangle = \bigvee_{\mathcal{G}^{\flat}\left(\langle\rangle ; \langle \underline{m} \rangle\right)} S^0 \cong \bigwedge_{i=1}^{p} \underline{m_i} \in \mathsf{sSet}_*. \tag{9.4.5}$$

For $p, n \geq 0$, the pointed simplicial set $(\mathsf{K}^g J)_p$ has, as its set of n-simplices, the pointed set

$$\begin{aligned}(\mathsf{K}^g J)_{p,n} &= J\left(\underline{n}^{\oplus p}\right)_n \\ &\cong \underline{n}^{\wedge p} \\ &= (S^p)_n.\end{aligned}$$

- The first equality above is from (9.4.2).
- The isomorphism is by (9.4.5).
- The last equality is from Definitions 7.1.12 and 7.1.23.

Fixing p and varying n, the pointed simplicial set structure of $(\mathsf{K}^g J)_p$ comes from the $\underline{n}$ variable as in Definition 7.1.12. So the above isomorphisms yield an isomorphism of pointed simplicial sets

$$(\mathsf{K}^g J)_p \cong S^p.$$

The Σ_p-action on $(\mathsf{K}^g J)_{p,n}$ comes from permuting the p copies of $\underline{n}$ in $\underline{n}^{\oplus p}$. So the Σ_p-action on $(\mathsf{K}^g J)_p$ corresponds to that on S^p under the above isomorphism.

By Definition 9.3.9, for $q \geq 0$, the symmetric spectrum structure morphism

$$S^1 \wedge (\mathsf{K}^g J)_q \longrightarrow (\mathsf{K}^g J)_{1+q}$$

is induced by the following composite for $\langle \underline{n} \rangle = \left(\underline{n_1}, \ldots, \underline{n_q}\right)$, $m \geq 0$, $x \in \underline{m}^{\flat}$, and $h_x \colon \underline{1} \longrightarrow \underline{m}$ the pointed function with $h_x(1) = x$.

$$\begin{array}{ccc}
J\langle \underline{n} \rangle & \xrightarrow{\quad \eta_{\underline{m}, \langle \underline{n} \rangle, x} \quad} & J\left(i_1 \underline{m} \oplus \langle \underline{n} \rangle\right) \\
\| & & \| \\
\underline{n_1} \wedge \cdots \wedge \underline{n_q} & & \underline{m} \wedge \underline{n_1} \wedge \cdots \wedge \underline{n_q} \\
& \searrow^{\cong} \qquad \nearrow^{h_x \wedge 1} & \\
& \{0,1\} \wedge \underline{n_1} \wedge \cdots \wedge \underline{n_q} &
\end{array}$$

So the S^1-action on $\mathsf{K}^g J$ corresponds to the structure morphism

$$S^1 \wedge S^q \xrightarrow{\ \cong\ } S^{1+q}$$

on the symmetric sphere. By the associativity condition (7.4.8) in Explanation 7.4.6, for $p, q \geq 0$, the symmetric spectrum structure morphism

$$S^p \wedge (\mathsf{K}^g J)_q \longrightarrow (\mathsf{K}^g J)_{p+q}$$

also corresponds to that for the symmetric sphere. □

K-Theory of Triple Products. To facilitate the proof of Theorem 9.4.9, the next result contains some preliminary computation. Recall from Theorem 9.2.15 with

$$(\mathsf{C}, \otimes, \mathbb{1}, *) = (\mathsf{sSet}, \times, \mathbb{1}, \mathbb{1})$$

that $\mathcal{G}_*$-sSet is a symmetric monoidal sSet_*-category with

- monoidal product $\wedge$ in (9.2.9) induced by Day convolution and
- J in (9.4.1) as the monoidal unit.

We use the notation (9.4.2) below. Moreover, for indexing of coends we write

$$\widehat{\mathcal{G}}^2 = \widehat{\mathcal{G}} \wedge \widehat{\mathcal{G}}$$

and

$$\widehat{\mathcal{G}}^3 = \widehat{\mathcal{G}} \wedge (\widehat{\mathcal{G}} \wedge \widehat{\mathcal{G}}) \quad \text{or} \quad (\widehat{\mathcal{G}} \wedge \widehat{\mathcal{G}}) \wedge \widehat{\mathcal{G}}$$

with context determining the latter.

Lemma 9.4.6. *For $X, Y, Z \in \mathcal{G}_*$-sSet and $m, p \geq 0$, there are isomorphisms of pointed sets*

$$(9.4.7)\qquad \begin{aligned} &\mathsf{K}^{\mathcal{G}}\big(X \wedge (Y \wedge Z)\big)_{p,m} \\ &\quad\cong \int^{(\langle \underline{a}\rangle, \langle \underline{b}\rangle, \langle \underline{c}\rangle) \in \widehat{\mathcal{G}}^3} \bigvee_{\mathcal{G}^\flat(\langle \underline{a}\rangle \oplus \langle \underline{b}\rangle \oplus \langle \underline{c}\rangle;\, \underline{m}^{\oplus p})} X\langle \underline{a}\rangle_m \wedge \big(Y\langle \underline{b}\rangle_m \wedge Z\langle \underline{c}\rangle_m\big) \\ &\mathsf{K}^{\mathcal{G}}\big((X \wedge Y) \wedge Z\big)_{p,m} \\ &\quad\cong \int^{(\langle \underline{a}\rangle, \langle \underline{b}\rangle, \langle \underline{c}\rangle) \in \widehat{\mathcal{G}}^3} \bigvee_{\mathcal{G}^\flat(\langle \underline{a}\rangle \oplus \langle \underline{b}\rangle \oplus \langle \underline{c}\rangle;\, \underline{m}^{\oplus p})} \big(X\langle \underline{a}\rangle_m \wedge Y\langle \underline{b}\rangle_m\big) \wedge Z\langle \underline{c}\rangle_m \end{aligned}$$

that are natural in X, Y, and Z.

Proof. Applying (9.4.2) to $X \wedge Y$, there are pointed sets as follows.

$$(9.4.8)\qquad \begin{aligned} \mathsf{K}^{\mathcal{G}}(X \wedge Y)_{p,m} &= (X \wedge Y)\big(\underline{m}^{\oplus p}\big)_m \\ &= \int^{(\langle \underline{a}\rangle, \langle \underline{b}\rangle) \in \widehat{\mathcal{G}}^2} \bigvee_{\mathcal{G}^\flat(\langle \underline{a}\rangle \oplus \langle \underline{b}\rangle;\, \underline{m}^{\oplus p})} X\langle \underline{a}\rangle_m \wedge Y\langle \underline{b}\rangle_m \end{aligned}$$

Using (9.4.8) twice, the first isomorphism in (9.4.7) follows from the following computation.

$$\begin{aligned} &\mathsf{K}^{\mathcal{G}}\big(X \wedge (Y \wedge Z)\big)_{p,m} \\ &= \int^{\langle \underline{a}\rangle, \langle \underline{d}\rangle} \bigvee_{\mathcal{G}^\flat(\langle \underline{a}\rangle \oplus \langle \underline{d}\rangle;\, \underline{m}^{\oplus p})} X\langle \underline{a}\rangle_m \wedge (Y \wedge Z)\langle \underline{d}\rangle_m \\ &= \int^{\langle \underline{a}\rangle, \langle \underline{d}\rangle} \bigvee_{\mathcal{G}^\flat(\langle \underline{a}\rangle \oplus \langle \underline{d}\rangle;\, \underline{m}^{\oplus p})} X\langle \underline{a}\rangle_m \wedge \Bigg[\int^{\langle \underline{b}\rangle, \langle \underline{c}\rangle} \bigvee_{\mathcal{G}^\flat(\langle \underline{b}\rangle \oplus \langle \underline{c}\rangle;\, \langle \underline{d}\rangle)} Y\langle \underline{b}\rangle_m \wedge Z\langle \underline{c}\rangle_m\Bigg] \\ &\cong \int^{\langle \underline{a}\rangle, \langle \underline{b}\rangle, \langle \underline{c}\rangle} \Bigg[\bigvee_{\int^{\langle \underline{d}\rangle} \mathcal{G}^\flat(\langle \underline{a}\rangle \oplus \langle \underline{d}\rangle;\, \underline{m}^{\oplus p}) \times \mathcal{G}^\flat(\langle \underline{b}\rangle \oplus \langle \underline{c}\rangle;\, \langle \underline{d}\rangle)} X\langle \underline{a}\rangle_m \wedge \big(Y\langle \underline{b}\rangle_m \wedge Z\langle \underline{c}\rangle_m\big)\Bigg] \\ &\cong \int^{\langle \underline{a}\rangle, \langle \underline{b}\rangle, \langle \underline{c}\rangle} \bigvee_{\mathcal{G}^\flat(\langle \underline{a}\rangle \oplus \langle \underline{b}\rangle \oplus \langle \underline{c}\rangle;\, \underline{m}^{\oplus p})} X\langle \underline{a}\rangle_m \wedge \big(Y\langle \underline{b}\rangle_m \wedge Z\langle \underline{c}\rangle_m\big) \end{aligned}$$

- The first isomorphism above uses the fact that $X\langle \underline{a}\rangle_m \wedge -$ is a left adjoint, which commutes with small colimits. Then it switches the outer wedge and the coend $\int^{\langle\underline{b}\rangle,\langle\underline{c}\rangle}$, which is possible because they are both colimits. Next it switches the order of the coends from

$$\int^{\langle\underline{a}\rangle,\langle\underline{d}\rangle}\int^{\langle\underline{b}\rangle,\langle\underline{c}\rangle} \quad \text{to} \quad \int^{\langle\underline{a}\rangle,\langle\underline{b}\rangle,\langle\underline{c}\rangle}\int^{\langle\underline{d}\rangle}.$$

Then it combines the two wedges and brings $\int^{\langle\underline{d}\rangle}$ down to the index of the combined wedge, which is the only place where $\langle\underline{d}\rangle$ appears.

- The last isomorphism uses the isomorphism (3.7.16) on the wedge index.

Each step in each of the isomorphisms above is natural in X, Y, and Z. This proves the first isomorphism in (9.4.7). The second isomorphism in (9.4.7) is proved similarly using (i) the commutativity of $- \wedge Z\langle\underline{c}\rangle_m$ with small colimits and (ii) the isomorphism (3.7.15). □

Main Result. For the following result, first recall from

- Definitions 1.4.17 and 1.4.18 a symmetric monoidal V-functor, which is, furthermore, *unital* if the unit constraint is invertible, and
- Theorem 7.6.15 that the category SymSp is a symmetric monoidal sSet_*-category with
  - the smash product $\square_S$ (Definition 7.6.1) as the monoidal product and
  - the symmetric sphere S as the monoidal unit.

Theorem 9.4.9. *The functor*

$$\mathsf{K}^g\colon \mathcal{G}_*\text{-}\mathsf{sSet} \longrightarrow \mathsf{SymSp}$$

in Proposition 9.3.16 extends to a unital symmetric monoidal sSet_**-functor.*

Proof. Recall that the monoidal product in $\mathcal{G}_*$-sSet is defined by that of the enriched diagram category $(\widehat{\mathcal{G}})$-(sSet_*), where $\widehat{\mathcal{G}}$ has the pointed unitary sSet_* enrichment (9.2.7). Therefore, by Theorem 2.5.1 (2), it suffices to extend K^g to a unital symmetric monoidal functor. This consists of the following steps:

(1) We first define the unit and monoidal constraints for K^g.
(2) Then we check that they are well defined.
(3) Then we show that they satisfy the axioms of a symmetric monoidal functor.

The unit constraint. The canonical isomorphism

$$S \xrightarrow[\cong]{(\mathsf{K}^g)^0} \mathsf{K}^g J$$

in (9.4.4) is the unit constraint for K^g.

The monoidal constraint. By Proposition 7.6.2 and Explanation 7.6.4, for $X, Y \in \mathcal{G}_*$-sSet, the (X,Y) component of the monoidal constraint

$$\mathsf{K}^g X \,\square_S\, \mathsf{K}^g Y \xrightarrow{(\mathsf{K}^g)^2_{X,Y}} \mathsf{K}^g(X \wedge Y) \in \mathsf{SymSp} \tag{9.4.10}$$

is uniquely determined by a family of $(\Sigma_p \times \Sigma_q)$-equivariant morphisms

$$(\mathsf{K}^g X)_p \wedge (\mathsf{K}^g Y)_q \xrightarrow{(\mathsf{K}^g)^2_{X,Y;p,q}} \mathsf{K}^g(X \wedge Y)_{p+q} \in \mathsf{sSet}_* \tag{9.4.11}$$

for $p, q \geq 0$ such that the diagrams (7.6.5) and (7.6.7) are commutative. Using (9.4.8), at level $m \geq 0$, the morphism $(\mathsf{K}^{\mathcal{G}})^2_{X,Y;p,q,m}$ in (9.4.11) is defined as the following composite of pointed functions.

(9.4.12)

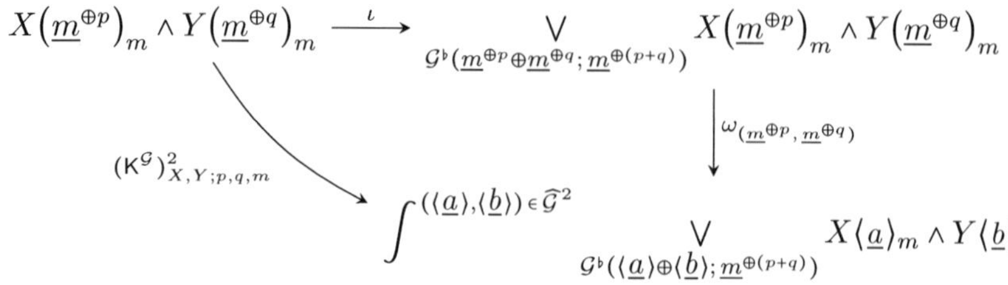

- ι is the wedge summand inclusion for the identity morphism of
$$\underline{m}^{\oplus p} \oplus \underline{m}^{\oplus q} = \underline{m}^{\oplus(p+q)} \in \mathcal{G}.$$
- $\omega_{(\underline{m}^{\oplus p},\, \underline{m}^{\oplus q})}$ is the coend structure morphism (Definition 3.5.1) for
$$\langle \underline{a} \rangle = \underline{m}^{\oplus p} \quad\text{and}\quad \langle \underline{b} \rangle = \underline{m}^{\oplus q}.$$

As m varies, the collection of morphisms in (9.4.12) is a morphism $(\mathsf{K}^{\mathcal{G}})^2_{X,Y;p,q}$ of pointed simplicial sets. Its naturality with respect to X and Y follows from the functoriality of $\wedge$ and the naturality of coends. Next we check equivariance and the commutativity of the diagrams (7.6.5) and (7.6.7).

Equivariance. To show that the morphism $(\mathsf{K}^{\mathcal{G}})^2_{X,Y;p,q}$ is $(\Sigma_p \times \Sigma_q)$-equivariant, it suffices to show that, for permutations $(\sigma, \tau) \in \Sigma_p \times \Sigma_q$, the following diagram of pointed functions is commutative, where most instances of $\oplus$ are omitted to save space.

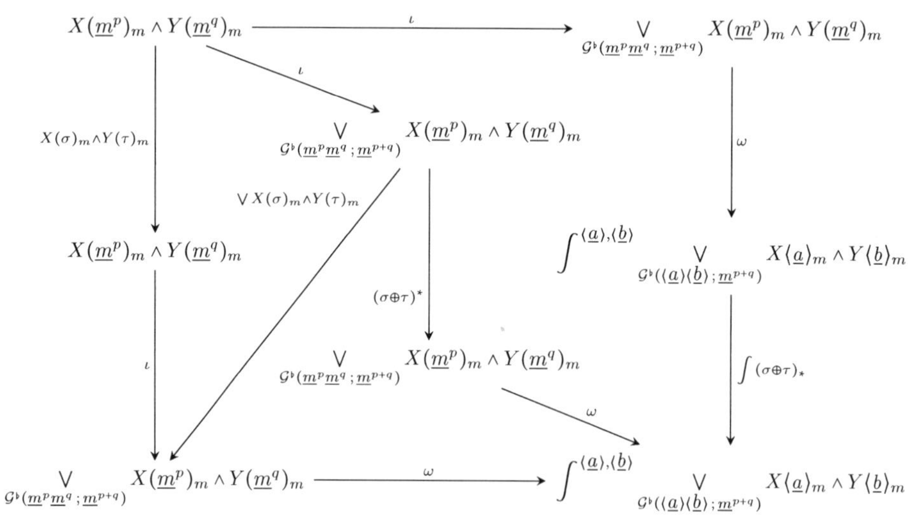

- In the above diagram, ι and ω are as in (9.4.12).
- In the morphism $X(\sigma)_m \wedge Y(\tau)_m$,
$$\sigma = (\sigma, \{1\}) \colon \underline{m}^{\oplus p} \longrightarrow \underline{m}^{\oplus p} \in \mathcal{G}$$
consists of the permutation
$$\sigma \colon \overline{p} \longrightarrow \overline{p}$$

and the entrywise identity function

$$\{1\}: \sigma_* \underline{m}^{\oplus p} \longrightarrow \underline{m}^{\oplus p}.$$

The morphism

$$\tau = (\tau, \{1\}): \underline{m}^{\oplus q} \longrightarrow \underline{m}^{\oplus q} \in \mathcal{G}$$

is defined similarly.

- The morphism $(\sigma \oplus \tau)^*$ is the precomposition with

$$\sigma \oplus \tau: \underline{m}^{\oplus p} \oplus \underline{m}^{\oplus q} \longrightarrow \underline{m}^{\oplus p} \oplus \underline{m}^{\oplus q} \in \mathcal{G}$$

in the wedge index. The morphism $(\sigma \oplus \tau)_*$ is the post-composition with $\sigma \oplus \tau$ in the wedge index.
- The two subdiagrams containing the left vertical composite and the right vertical composite are commutative by definition. The quadrilateral containing the bottom horizontal arrow is commutative by the definition of the coend in the lower right corner.

The commutativity of the above diagram proves that the morphism $(\mathsf{K}^g)^2_{X,Y;p,q}$ in (9.4.11) is $(\Sigma_p \times \Sigma_q)$-equivariant.

Compatibility with S-action. To prove the commutativity of the diagram (7.6.5) for $(\mathsf{K}^g)^2_{X,Y;p,q}$, by the associativity condition in Explanation 7.4.6 of the structure morphisms in symmetric spectra, it suffices to consider the case $n = 1$. At level $m \geq 0$ with $n = 1$, (7.6.5) is the following diagram of pointed functions, with α an associativity isomorphism and ρ the S^1-action.

$$\begin{array}{ccc}
\big((S^1)_m \wedge (\mathsf{K}^g X)_{p,m}\big) \wedge (\mathsf{K}^g Y)_{q,m} & \xrightarrow{\alpha} & (S^1)_m \wedge \big((\mathsf{K}^g X)_{p,m} \wedge (\mathsf{K}^g Y)_{q,m}\big) \\
\downarrow{\scriptstyle \rho\wedge 1} & & \\
(\mathsf{K}^g X)_{1+p,m} \wedge (\mathsf{K}^g Y)_{q,m} & & \downarrow{\scriptstyle 1\wedge(\mathsf{K}^g)^2} \\
\downarrow{\scriptstyle (\mathsf{K}^g)^2} & & \\
\mathsf{K}^g(X \wedge Y)_{1+p+q,m} & \xleftarrow{\rho} & (S^1)_m \wedge \mathsf{K}^g(X \wedge Y)_{p+q,m}
\end{array} \tag{9.4.13}$$

To show that this diagram is commutative, first recall the following from Definition 9.3.9.

- For each $x \in \underline{m}^\flat$, $h_x: \underline{1} \longrightarrow \underline{m}$ is the pointed function with

$$h_x(1) = x.$$

- For $p \geq 0$, $f_1: \overline{p} \longrightarrow \overline{1+p}$ is the pointed function given by

$$f_1(j) = j+1 \quad \text{for} \quad j \in \overline{p} = \{1, \ldots, p\}.$$

- We write

$$i_1^x = \big(f_1, (i_1 h_x) \oplus 1_{\underline{m}^{\oplus p}}\big): \underline{m}^{\oplus p} \longrightarrow \underline{m}^{\oplus(1+p)} \in \mathcal{G},$$

so there is an equality

$$X(i_1^x) = \eta_{\underline{m}, \underline{m}^{\oplus p}, x}: X\big(\underline{m}^{\oplus p}\big) \longrightarrow X\big(\underline{m}^{\oplus(1+p)}\big) \in \mathsf{sSet}_*.$$

The S^1-action (9.3.10) on $\mathsf{K}^g X$ is defined by these morphisms $X(i_1^x)$ for $x \in \underline{m}^\flat$.

Using (9.4.8), to show that (9.4.13) is commutative, it suffices to show that the outer diagram below is commutative for $x \in \underline{m}^\flat$, with the symbol $\oplus$ for concatenation omitted to save space.

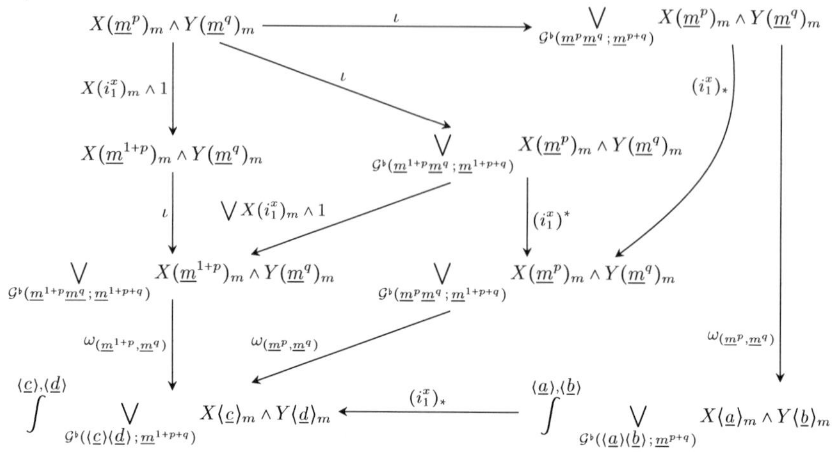

- In the above diagram, each ι is a wedge summand inclusion corresponding to an identity morphism in the wedge index. The upper left triangle and the upper right subdiagram are commutative by definition.
- The lower left parallelogram is commutative by the definition of the coend in the lower left corner.
- The lower right subdiagram is commutative by the naturality of coends.

This proves the commutativity of the diagram (9.4.13), which, in turn, implies the commutativity of (7.6.5) for $(\mathsf{K}^g)^2_{X,Y;p,q}$. A similar argument shows that (7.6.7) is commutative. This finishes the construction of the monoidal constraint $(\mathsf{K}^g)^2$.

Next we check the symmetric monoidal functor axioms in Definitions 1.1.6 and 1.1.17 for $(\mathsf{K}^g, (\mathsf{K}^g)^2, (\mathsf{K}^g)^0)$.

Unity. The left unity axiom in (1.1.10) asserts the commutativity of the following diagram in SymSp.

$$\begin{array}{ccc} S \,\square_S\, \mathsf{K}^g X & \xrightarrow{\ \lambda\ } & \mathsf{K}^g X \\ {\scriptstyle (\mathsf{K}^g)^0 \square_S 1}\downarrow & & \uparrow{\scriptstyle \mathsf{K}^g \lambda} \\ \mathsf{K}^g J \,\square_S\, \mathsf{K}^g X & \xrightarrow{(\mathsf{K}^g)^2} & \mathsf{K}^g(J \wedge X) \end{array} \tag{9.4.14}$$

To show that (9.4.14) is commutative, it suffices to show that the following diagram of pointed functions is commutative for $p, m \geq 0$.

$$\begin{array}{ccc} (S^1)_m \wedge (\mathsf{K}^g X)_{p,m} & \xrightarrow{\ \lambda\ } & (\mathsf{K}^g X)_{1+p,m} \\ {\scriptstyle (\mathsf{K}^g)^0 \wedge 1}\downarrow & & \uparrow{\scriptstyle \mathsf{K}^g \lambda} \\ (\mathsf{K}^g J)_{1,m} \wedge (\mathsf{K}^g X)_{p,m} & \xrightarrow{(\mathsf{K}^g)^2_{J,X;1,p,m}} & \mathsf{K}^g(J \wedge X)_{1+p,m} \end{array} \tag{9.4.15}$$

It suffices to show that the diagram (9.4.15) is commutative when restricted to each $x \in \underline{m}^\flat \subset (S^1)_m$. By the definitions

- of the canonical isomorphism (9.4.4)

$$(\mathsf{K}^g)^0 \colon S \xrightarrow{\cong} \mathsf{K}^g J,$$

- of $(\mathsf{K}^g)^2$ in (9.4.12), and
- of the left unit isomorphism

$$\lambda\colon J \wedge X \xrightarrow{\cong} X$$

induced by that of the Day convolution product in Definition 3.7.17,

the diagram (9.4.15) restricted to $x \in \underline{m}^\flat$ is the outer diagram below, where the symbol $\oplus$ is omitted to save space.

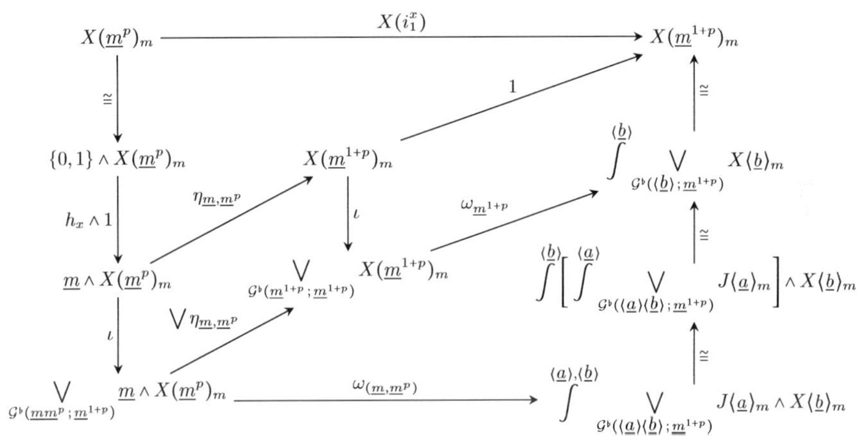

- Each ι is a wedge summand inclusion corresponding to an identity morphism in the wedge index. Each ω is a coend structure morphism.
- The morphism $\eta_{\underline{m},\underline{m}^p}$ is the one in (9.3.10) at level m. The upper left subdiagram is commutative by the definition of $\eta_{\underline{m},\underline{m}^p}$.
- The lower left parallelogram is commutative by the naturality of wedges.
- The upper right parallelogram is commutative by the definition of the upper right vertical isomorphism.
- The lower right subdiagram is commutative by the definitions of the two vertical isomorphisms.

This proves the left unity axiom in (1.1.10). The right unity axiom is proved by a similar argument.

Associativity. The associativity axiom (1.1.9) for K^g asserts the commutativity of the following diagram in SymSp, where the bottom α is the associativity isomorphism for the Day convolution product in Definition 3.7.17.

(9.4.16)
$$\begin{array}{ccc}
(\mathsf{K}^g X \square_S \mathsf{K}^g Y) \square_S \mathsf{K}^g Z & \xrightarrow{\alpha} & \mathsf{K}^g X \square_S (\mathsf{K}^g Y \square_S \mathsf{K}^g Z) \\
\downarrow{\scriptstyle (\mathsf{K}^g)^2 \square_S 1} & & \downarrow{\scriptstyle 1 \square_S (\mathsf{K}^g)^2} \\
\mathsf{K}^g(X \wedge Y) \square_S \mathsf{K}^g Z & & \mathsf{K}^g X \square_S \mathsf{K}^g(Y \wedge Z) \\
\downarrow{\scriptstyle (\mathsf{K}^g)^2} & & \downarrow{\scriptstyle (\mathsf{K}^g)^2} \\
\mathsf{K}^g\big((X \wedge Y) \wedge Z\big) & \xrightarrow{\mathsf{K}^g \alpha} & \mathsf{K}^g\big(X \wedge (Y \wedge Z)\big)
\end{array}$$

By (9.4.7) and the definition (9.4.12) of $(\mathsf{K}^g)^2_{X,Y;p,q,m}$, to show that (9.4.16) is commutative, it suffices to show that, for $m,p,q,r \geq 0$, the following diagram of pointed functions is commutative.

$$\begin{array}{ccc}
\big(X(\underline{m}^{\oplus p})_m \wedge Y(\underline{m}^{\oplus q})_m\big) \wedge Z(\underline{m}^{\oplus r})_m & \xrightarrow{\alpha} & X(\underline{m}^{\oplus p})_m \wedge \big(Y(\underline{m}^{\oplus q})_m \wedge Z(\underline{m}^{\oplus r})_m\big) \\
\downarrow{\scriptstyle \iota} & & \downarrow{\scriptstyle \iota} \\
\bigvee \big(X(\underline{m}^{\oplus p})_m \wedge Y(\underline{m}^{\oplus q})_m\big) \wedge Z(\underline{m}^{\oplus r})_m & & \bigvee X(\underline{m}^{\oplus p})_m \wedge \big(Y(\underline{m}^{\oplus q})_m \wedge Z(\underline{m}^{\oplus r})_m\big) \\
\downarrow{\scriptstyle \omega} & & \downarrow{\scriptstyle \omega} \\
\int \bigvee \big(X\langle \underline{a}\rangle_m \wedge Y\langle \underline{b}\rangle_m\big) \wedge Z\langle \underline{c}\rangle_m & \xrightarrow{\int \bigvee \alpha} & \int \bigvee X\langle \underline{a}\rangle_m \wedge \big(Y\langle \underline{b}\rangle_m \wedge Z\langle \underline{c}\rangle_m\big)
\end{array}$$

- In the above diagram, α is an associativity isomorphism.
- In the middle level, each wedge is indexed by the pointed set

$$\mathcal{G}^\flat\big(\underline{m}^{\oplus p} \oplus \underline{m}^{\oplus q} \oplus \underline{m}^{\oplus r}\,;\, \underline{m}^{\oplus (p+q+r)}\big).$$

  Each ι is the wedge summand inclusion corresponding to the identity morphism of $\underline{m}^{\oplus (p+q+r)}$.
- In the bottom level, each wedge is indexed by the pointed set

$$\mathcal{G}^\flat\big(\langle \underline{a}\rangle \oplus \langle \underline{b}\rangle \oplus \langle \underline{c}\rangle\,;\, \underline{m}^{\oplus (p+q+r)}\big).$$

  Each coend is indexed by

$$\big(\langle \underline{a}\rangle, \langle \underline{b}\rangle, \langle \underline{c}\rangle\big) \in \widehat{\mathcal{G}}^3.$$

  Each ω is the coend structure morphism for

$$\langle \underline{a}\rangle = \underline{m}^{\oplus p}, \quad \langle \underline{b}\rangle = \underline{m}^{\oplus q}, \quad \text{and} \quad \langle \underline{c}\rangle = \underline{m}^{\oplus r}.$$

The above diagram is commutative by the definition of the bottom horizontal morphism $\int \bigvee \alpha$.

Symmetry. The compatibility axiom (1.1.18) for K^g asserts the commutativity of the following diagram in SymSp, where the bottom ξ is the symmetry isomorphism for the Day convolution product in Definition 3.7.17.

$$\begin{array}{ccc}
\mathsf{K}^g X \,\square_S\, \mathsf{K}^g Y & \xrightarrow{\xi} & \mathsf{K}^g Y \,\square_S\, \mathsf{K}^g X \\
\downarrow{\scriptstyle (\mathsf{K}^g)^2} & & \downarrow{\scriptstyle (\mathsf{K}^g)^2} \\
\mathsf{K}^g(X \wedge Y) & \xrightarrow{\mathsf{K}^g \xi} & \mathsf{K}^g(Y \wedge X)
\end{array} \tag{9.4.17}$$

Similar to the proof of the associativity axiom (9.4.16) above, the commutativity of (9.4.17) follows from (9.4.8), the definition (9.4.12) of $(\mathsf{K}^g)^2_{X,Y;p,q,m}$, and the definition of $(\mathsf{K}^g\xi)_{p,m}$. □

Theorem 9.4.18. *The functors*

$$\begin{array}{ccc}
\Gamma\text{-}\mathsf{sSet} & \xrightarrow{\mathsf{K}^{\mathcal{F}}} & \mathsf{SymSp} \\
\downarrow{\scriptstyle \wedge^*} & & \\
\mathcal{G}_*\text{-}\mathsf{sSet} & &
\end{array}$$

in (9.3.17) *are symmetric monoidal* sSet_**-functors.*

Proof. Since

$$\mathsf{K}^{\mathcal{F}} = \mathsf{K}^{\mathcal{G}} \circ \wedge^*,$$

by Theorem 9.4.9 it suffices to show that $\wedge^*$ is a symmetric monoidal sSet_*-functor. Since each of $\mathcal{F}$ and $\mathcal{G}$ is given the pointed unitary enrichment over sSet_* (Definition 4.3.11), as in the first paragraph of the proof of Theorem 9.4.9, it suffices to show that $\wedge^*$ is a symmetric monoidal functor. By Theorem 3.7.28, it suffices to show that

$$\widehat{\mathcal{G}} \xrightarrow{\wedge} \widehat{\mathcal{F}}$$

is a symmetric monoidal sSet_*-functor. Once again, by the pointed unitary enrichment, it remains to observe that

$$\mathcal{G} \xrightarrow{\wedge} \mathcal{F}$$

in Definition 9.1.15 is a strict symmetric monoidal functor that preserves the null objects. □

9.5. Notes

9.5.1 (References)**.** The category $\mathcal{G}$ in Definition 9.1.7 and the associated K-theory functor $\mathsf{K}^{\mathcal{G}}$ in Definition 9.3.14 were first developed in [**EM06, EM09**]. Theorem 9.2.15, which follows from the more general Theorem 4.3.37, is stated as [**EM09**, Theorem 5.6]. A similar construction of symmetric spectra from certain diagram categories is given in [**BO20**] (see Note 9.5.3 below). ◇

9.5.2 ($\mathcal{F}^{(0)}$)**.** The category of small pointed categories and pointed functors, Cat_*, has a symmetric monoidal smash product whose unit E is a discrete category with two objects. So the category $\mathcal{F}^{(0)}$ of Definition 9.1.3 is not the monoidal unit of $(\mathsf{Cat}_*, \wedge)$ and, therefore, not the smash product over the empty indexing set in $(\mathsf{Cat}_*, \wedge)$. However, $\mathcal{F}^{(0)}$ is the monoidal unit for the smash product in Cat_0, the category of pointed small categories whose basepoints are initial and terminal, together with pointed functors. All of the other smash powers $\mathcal{F}^{(q)}$ for $q > 0$ are the same whether taken in Cat_* or Cat_0. Further discussion of Cat_0 as a module category is given in [**EM09**, Section 5], and this perspective is used in the definition of $\mathcal{G}$ as a Grothendieck construction in Note 9.5.3. ◇

9.5.3 (Alternate Description of $\mathcal{G}$)**.** For those readers familiar with Grothendieck constructions (see e.g., [**BW12**, Chapter 12] or [**JY21**, Chapter 10]), the category $\mathcal{G}$ of Definition 9.1.7 can be described similarly.

Recall from Definition 9.1.2 that Inj denotes the category whose objects are unpointed finite sets $\overline{q} = \{1, \ldots, q\}$ for $q \geq 0$ and whose morphisms are injections. As in Note 9.5.2, let Cat_0 denote the category of small pointed categories whose basepoints are both initial and terminal, together with pointed functors. Then, as described further in Note 9.5.2,

$$\mathcal{F}^{(-)} \colon \mathsf{Inj} \longrightarrow \mathsf{Cat}_0$$

defines a functor that sends an object $\overline{p}$ to the p-fold smash power in Cat_0 and sends an injection $f \colon \overline{p} \hookrightarrow \overline{q}$ to the functor f_* defined in Definition 9.1.7. Then $\mathcal{G}$ is the Grothendieck construction

$$\int_{\mathsf{Inj}} \mathcal{F}^{(-)}$$

formed in Cat_0. This construction is similar to the usual Grothendieck construction, but uses the coproduct (wedge) and smash product in Cat_0 instead of the disjoint union and Cartesian product in Cat. This is the definition given in [**EM09**, Section 5], denoted $\mathcal{G}_*$ there.

Alternatively, one can make a Grothendieck construction in $(\mathsf{Cat}, \times)$

$$\mathcal{H} = \int_{\mathsf{Inj}} \mathcal{F}^{\times(-)},$$

where $\mathcal{F}^{\times(q)}$ denotes the q-fold Cartesian product. Then $\mathcal{G}$ can be defined as a quotient of $\mathcal{H}$, identifying all tuples $\langle \underline{n} \rangle$ having an entry $\underline{0}$ to a basepoint $*$ and all morphisms factoring through such tuples to the zero morphism in $\mathcal{G}_*$. This description is given in [**EM06**]; see Note 9.5.4 for an explanation of the notation change. Bohmann and Osorno, in [**BO20**, Section 5], use a similar Grothendieck construction, denoted $\mathcal{E}$, over Cartesian products of Δ^{op} in Cat. They then define $\mathcal{E}_*$-objects as diagrams that factor through a quotient similar to that defining $\mathcal{G}$ from $\mathcal{H}$. ◇

9.5.4 (Notation Warning for $\mathcal{G}_*$)**.** We warn the reader comparing with the work of Elmendorf-Mandell [**EM06, EM09**] that the category we denote $\mathcal{G}$ in Definition 9.1.7 is denoted $\mathcal{G}_*$ in those references. Moreover, those references use $\mathcal{G}$ to denote what we call $\mathcal{H}$ in Note 9.5.3. We have made this change of notation because the category we call $\mathcal{G}$ is the one that parallels most closely the skeleton of finite sets, $\mathcal{F}$. Moreover, then we can reserve the notation $\mathcal{G}_*$ for the pointed diagram categories $\mathcal{G}_*$-Cat and $\mathcal{G}_*$-sSet. See Note 10.8.5 for further comparison of notation between [**EM06, EM09**] and this work. ◇

9.5.5 (Generality of $\mathsf{K}^{\mathcal{G}}$)**.** In Theorem 9.4.9, we showed that the K-theory construction

$$\mathsf{K}^{\mathcal{G}} \colon \mathcal{G}_*\text{-}\mathsf{sSet} \longrightarrow \mathsf{SymSp}$$

is unital symmetric monoidal as a sSet_*-functor. This theorem holds more generally with sSet_* replaced by the category sC_* of pointed simplicial objects in a complete, cocomplete, and Cartesian closed category C.

- The relevant structure on $\mathcal{G}_*$-C, the category of $\mathcal{G}_*$-objects in C, is still given by Theorem 9.2.15.
- Instead of symmetric spectra based on pointed simplicial sets (Explanation 7.4.6), one considers symmetric spectra based on sC_* as defined in, for example, [**EM06**, 7.1].

The constructions and proof of Theorem 9.4.9 translate to this more general setting essentially verbatim. This more general form of the theorem implies [**EM06**, 7.4], which says that $\mathsf{K}^{\mathcal{G}}$, denoted by I there, is an enriched multifunctor. ◇

CHAPTER 10

Elmendorf-Mandell K-Theory of Permutative Categories

This is the second of two chapters generalizing the Segal K-theory functor

$$\mathsf{K}^{\mathsf{Se}} \colon \mathsf{PermCat}^{\mathsf{su}} \longrightarrow \mathsf{SymSp},$$

from Chapter 8, to a simplicially-enriched multifunctor

$$\mathsf{K}^{\mathsf{EM}} \colon \mathsf{PermCat}^{\mathsf{su}} \longrightarrow \mathsf{SymSp},$$

due to Elmendorf-Mandell [**EM06, EM09**]. Chapter 9 describes the sSet_*-enriched symmetric monoidal categories and functors

$$\mathcal{G}_*\text{-}\mathsf{Cat} \xrightarrow{N_*} \mathcal{G}_*\text{-}\mathsf{sSet} \xrightarrow{\mathsf{K}^{\mathcal{G}}} \mathsf{SymSp}.$$

In this chapter we describe $\mathcal{G}_*$-diagrammatic replacements of the partition multicategory $\mathcal{M}$ and its associated J-theory,

$$\mathsf{PermCat}^{\mathsf{su}} \xrightarrow{\mathsf{End}} \mathsf{Multicat}_* \xrightarrow{\mathsf{J}^{\mathcal{M}}} \Gamma\text{-}\mathsf{Cat} \qquad (\text{composite } \mathsf{J}^{\mathsf{Se}})$$

with a partition multicategory for tuples, $\mathcal{T}$, and an associated J-theory

$$\mathsf{PermCat}^{\mathsf{su}} \xrightarrow{\mathsf{End}} \mathsf{Mod}^{\mathcal{M}\underline{1}} \xrightarrow{\mathsf{J}^{\mathcal{T}}} \mathcal{G}_*\text{-}\mathsf{Cat}. \qquad (\text{composite } \mathsf{J}^{\mathsf{EM}})$$

It is important to note the change in the codomain of End (and the domain of $\mathsf{J}^{\mathcal{T}}$) from $\mathsf{Multicat}_*$ to $\mathsf{Mod}^{\mathcal{M}\underline{1}}$—the category of left modules over $\mathcal{M}\underline{1}$ (Example 8.4.5). We discuss the reasons for this change in the Literature subsection below. For now we only need to note that End factors through $\mathsf{Mod}^{\mathcal{M}\underline{1}}$ (Lemma 10.2.14).

Theorem 10.6.10 shows that the two constructions

$$\mathsf{K}^{\mathsf{Se}} = \mathsf{K}^{\mathcal{F}} N_* \mathsf{J}^{\mathsf{Se}} \quad \text{and} \quad \mathsf{K}^{\mathsf{EM}} = \mathsf{K}^{\mathcal{G}} N_* \mathsf{J}^{\mathsf{EM}}$$

are naturally level-equivalent as functors from $\mathsf{PermCat}^{\mathsf{su}}$ to SymSp. But, as with $\mathcal{G}$ and $\mathsf{K}^{\mathcal{G}}$ in Chapter 9, we do more than simply replacing $\mathsf{J}^{\mathcal{M}}$ with $\mathsf{J}^{\mathcal{T}}$. We also describe the relevant enriched monoidal structures.

(1) The smash product of small pointed multicategories restricts to a symmetric monoidal product for $\mathsf{Mod}^{\mathcal{M}\underline{1}}$, with monoidal unit $\mathcal{M}\underline{1}$. With respect to this structure, $\mathsf{Mod}^{\mathcal{M}\underline{1}}$ is a Cat_*-enriched symmetric monoidal category (Definition 10.1.36).

(2) Theorem 10.3.17 shows that

$$\mathsf{J}^{\mathcal{T}} \colon \mathsf{Mod}^{\mathcal{M}\underline{1}} \longrightarrow \mathcal{G}_*\text{-}\mathsf{Cat}$$

is symmetric monoidal as a Cat_*-enriched functor.

(3) Lemma 10.2.14 shows that

$$\mathsf{End}\colon \mathsf{PermCat}^{\mathsf{su}} \longrightarrow \mathsf{Mod}^{\mathcal{M}\underline{1}}$$

is a Cat-enriched multifunctor.

(4) Changing enrichments to sSet via the nerve, the Elmendorf-Mandell K-theory, K^{EM}, is the composite sSet-enriched multifunctor

$$\overbrace{\mathsf{PermCat}^{\mathsf{su}} \xrightarrow{\mathsf{End}} \mathsf{Mod}^{\mathcal{M}\underline{1}} \xrightarrow{\mathsf{J}^{\mathcal{T}}} \mathcal{G}_*\text{-}\mathsf{Cat}}^{\mathsf{J}^{\mathsf{EM}}} \xrightarrow{N_*} \mathcal{G}_*\text{-}\mathsf{sSet} \xrightarrow{\mathsf{K}^{\mathcal{G}}} \mathsf{SymSp}.$$

As a consequence of the enriched multifunctoriality, K^{EM} preserves operad actions. We state this result as Theorem 10.3.33 and apply it in Chapters 11, 12, and 13.

The following table summarizes the Segal and Elmendorf-Mandell K-theory constructions, extending the table given in the introduction to Chapter 8. Each N_* is induced by composition with the nerve functor

$$N\colon \mathsf{Cat} \longrightarrow \mathsf{sSet},$$

from Definition 7.2.3.

	$\mathsf{K}^{\mathsf{Se}} = \mathsf{K}^{\mathcal{F}} N_* \mathsf{J}^{\mathsf{Se}}$	(8.5.1)	$\mathsf{K}^{\mathsf{EM}} = \mathsf{K}^{\mathcal{G}} N_* \mathsf{J}^{\mathsf{EM}}$	(10.3.32)
indexing category	$\mathcal{F}$	(8.1.1)	$\mathcal{G}$	(9.1.7)
K-theory	$\mathsf{K}^{\mathcal{F}}\colon \Gamma\text{-}\mathsf{sSet} \longrightarrow \mathsf{SymSp}$	(8.2.5)	$\mathsf{K}^{\mathcal{G}}\colon \mathcal{G}_*\text{-}\mathsf{sSet} \longrightarrow \mathsf{SymSp}$	(9.3.14)
nerve	$N_*\colon \Gamma\text{-}\mathsf{Cat} \longrightarrow \Gamma\text{-}\mathsf{sSet}$	(8.1.18)	$N_*\colon \mathcal{G}_*\text{-}\mathsf{Cat} \longrightarrow \mathcal{G}_*\text{-}\mathsf{sSet}$	(9.2.19)
$\Gamma/\mathcal{G}_*$-categories	$(-)^{\mathcal{F}}, (-)^{\mathcal{F}}_{\mathsf{lax}}, (-)^{\mathcal{F}}_{\mathsf{co}}$	(8.3.14)	$(-)^{\mathcal{G}}_{\cong}, (-)^{\mathcal{G}}, (-)^{\mathcal{G}}_{\mathsf{co}}$	(10.4.20)
variants equiv.	(8.5.2)		(10.7.16, 10.7.19)	
J-theory	$\mathsf{J}^{\mathsf{Se}}\colon \mathsf{PermCat}^{\mathsf{su}} \longrightarrow \Gamma\text{-}\mathsf{Cat}$		$\mathsf{J}^{\mathsf{EM}}\colon \mathsf{PermCat}^{\mathsf{su}} \longrightarrow \mathcal{G}_*\text{-}\mathsf{Cat}$	
definition	$\mathsf{J}^{\mathsf{Se}} = \mathsf{J}^{\mathcal{M}} \circ \mathsf{End}$	(8.5.1)	$\mathsf{J}^{\mathsf{EM}} = \mathsf{J}^{\mathcal{T}} \circ \mathsf{End}$	(10.3.25)
End	$\mathsf{End}\colon \mathsf{PermCat}^{\mathsf{su}} \longrightarrow \mathsf{Mod}^{\mathcal{M}\underline{1}} \hookrightarrow \mathsf{Multicat}_*$			(5.1.11, 5.3.9, 10.2.14)
partitions	$\mathcal{M}\colon \mathcal{F}^{\mathsf{op}} \longrightarrow \mathsf{Multicat}_*$	(8.4.1)	$\mathcal{T}\colon \mathcal{G}^{\mathsf{op}} \longrightarrow \mathsf{Mod}^{\mathcal{M}\underline{1}}$	(10.3.1)
partition J-theory	$\mathsf{J}^{\mathcal{M}}\colon \mathsf{Multicat}_* \longrightarrow \Gamma\text{-}\mathsf{Cat}$	(8.4.10)	$\mathsf{J}^{\mathcal{T}}\colon \mathsf{Mod}^{\mathcal{M}\underline{1}} \longrightarrow \mathcal{G}_*\text{-}\mathsf{Cat}$	(10.3.9)
equiv. desc. J	$\mathsf{J}^{\mathsf{Se}} \cong (-)^{\mathcal{F}}_{\mathsf{lax}}$	(8.4.8)	$\mathsf{J}^{\mathsf{EM}} \cong (-)^{\mathcal{G}}$	(10.5.1)
level equiv.	$\mathsf{K}^{\mathsf{Se}} \xrightarrow{\sim} \mathsf{K}^{\mathsf{EM}}$			(10.6.10)

Organization. Sections 10.1 and 10.2 discuss smash products of the partition multicategories $\mathcal{M}a$ from Section 8.4. It follows from Proposition 10.1.6 that $\mathcal{M}\underline{1}$ is a commutative monoid with respect to this product. Important properties of $\mathcal{M}\underline{1}$-modules are discussed in the remainder of Section 10.1 and in Section 10.2.

Section 10.3 defines a partition multifunctor for $\mathcal{G}$ via smash products of the partition multicategories $\mathcal{M}\underline{n}$. This is denoted $\mathcal{T}$ and defines a partition J-theory

$$\mathsf{J}^{\mathcal{T}}\colon \mathsf{Mod}^{\mathcal{M}\underline{1}} \longrightarrow \mathcal{G}_*\text{-}\mathsf{Cat}.$$

The main result of this section is Theorem 10.3.17, which shows that $\mathsf{J}^{\mathcal{T}}$ is symmetric monoidal as a Cat_*-enriched functor. Combining this with the previous material from Chapter 9, we define the Elmendorf-Mandell J- and K-theory multifunctors in Definitions 10.3.25 and 10.3.32, respectively.

Section 10.4 defines $\mathcal{G}_*$-categories $\mathsf{C}^{\mathcal{G}}$, including lax, strong, and colax variants (Definition 10.4.20), that generalize the Segal Γ-categories $\mathsf{C}^{\mathcal{F}}$ from Section 8.3. Section 10.5 shows that the constructions $\mathsf{J}^{\mathsf{EM}}\mathsf{C}$ and $\mathsf{C}^{\mathcal{G}} = \mathsf{C}^{\mathcal{G}}_{\mathrm{lax}}$ give naturally isomorphic $\mathcal{G}_*$-categories (Proposition 10.5.1). Definition 10.5.3 uses the strong and colax variants of $(-)^{\mathcal{G}}$ to define strong and colax Elmendorf-Mandell K-theory, denoted $\mathsf{K}^{\mathsf{EM}}_{\cong}$ and $\mathsf{K}^{\mathsf{EM}}_{\mathsf{co}}$.

Section 10.6 is devoted to proving that the Segal and Elmendorf-Mandell K-theory symmetric spectra are naturally level-equivalent constructions. The main result is Theorem 10.6.10, and depends on several preliminary comparisons for partition multicategories.

Section 10.7 compares the K-theory symmetric spectra given by the three variant constructions $\mathsf{C}^{\mathcal{G}} = \mathsf{C}^{\mathcal{G}}_{\mathsf{lax}}$, $\mathsf{C}^{\mathcal{G}}_{\cong}$, and $\mathsf{C}^{\mathcal{G}}_{\mathsf{co}}$. The main results are Theorems 10.7.16 and 10.7.19, which show that the resulting symmetric spectra are level equivalent.

Reading Guide. The material in this chapter depends on the smash product of pointed multicategories from Section 5.6 as well as the material on enriched multicategories from Chapter 6. A familiarity with the general definitions and main results will be useful. Throughout this chapter we give references to particular definitions and results from earlier material as needed.

(1) For $\mathcal{M}\underline{1}$-modules, read Definition 10.1.1 and the statement of Proposition 10.1.6, followed by Explanations 10.1.10 and 10.1.11. Then read Definition 10.1.12 and the important properties of $\mathcal{M}\underline{1}$-modules given in the statement of Proposition 10.1.28. Finally, read Definition 10.1.36 followed by Proposition 10.2.7, Definition 10.2.13, and the statement of Lemma 10.2.14.
(2) For the partition J-theory $\mathsf{J}^{\mathcal{T}}$, read Definition 10.3.1, Explanation 10.3.2, and the statement of Theorem 10.3.17.
(3) For J^{EM} and K^{EM}, read Definition 10.3.25 and the following text up through Definition 10.3.32. Read both the statement and the short proof of Theorem 10.3.33. Then read the statement of Theorem 10.6.10 for the level equivalence between K^{Se} and K^{EM}.
(4) For the reader who is only interested in the further applications to E_n-monoidal structures, in Chapters 11, 12, and 13, the remaining content of this chapter is not necessary. Go back and finish reading Sections 10.1 through 10.3 and 10.6. Then go on to the subsequent chapters.
(5) For the reader who is interested in the $\mathcal{G}_*$-categories $\mathsf{C}^{\mathcal{G}}$ and variants, read all of Section 10.4, the statement of Proposition 10.5.1, and Definition 10.5.3. Then read the statements of Theorems 10.7.16 and 10.7.19.
(6) Go back and read the rest of this chapter.

The detailed proofs and variant definitions in this chapter and the previous one have several additional purposes.

Cat-Enriched Multifunctors. In the definitions of K^{Se} and K^{EM} via the respective composites

$$\mathsf{PermCat}^{\mathsf{su}} \xrightarrow{\mathsf{End}} \mathsf{Multicat}_* \xrightarrow{\mathsf{J}^{\mathcal{M}}} \Gamma\text{-}\mathsf{Cat} \xrightarrow{N_*} \Gamma\text{-}\mathsf{sSet} \xrightarrow{\mathsf{K}^{\mathcal{F}}} \mathsf{SymSp}$$

and

$$\mathsf{PermCat^{su}} \xrightarrow{\mathsf{End}} \mathsf{Mod}^{\mathcal{M}\underline{1}} \xrightarrow{\mathsf{J}^{\mathcal{T}}} \mathcal{G}_*\text{-}\mathsf{Cat} \xrightarrow{N_*} \mathcal{G}_*\text{-}\mathsf{sSet} \xrightarrow{\mathsf{K}^{\mathcal{G}}} \mathsf{SymSp},$$

we see that almost all of the categories and functors are symmetric monoidal. The only exceptions are the following.

- There is not a symmetric monoidal structure on $\mathsf{PermCat^{su}}$ inherited from either $\mathsf{Multicat}_*$ (Proposition 5.7.23), or the proper subcategory $\mathsf{Mod}^{\mathcal{M}\underline{1}}$ (Proposition 10.2.17). It is merely a Cat-enriched multicategory.
- As a consequence of the previous item, End is merely a Cat-enriched multifunctor. This is why $\mathsf{K^{EM}}$ is defined as a multifunctor instead of a symmetric monoidal functor.
- The monoidal constraint for $\mathcal{M}$ is not an isomorphism, and thus $\mathsf{J}^{\mathcal{M}}$ is not a symmetric monoidal functor (see Explanation 10.1.10 and Note 10.8.6). This is why, for our applications to K-theory of E_n-monoidal categories, we use $\mathsf{K^{EM}}$ instead of $\mathsf{K^{Se}}$.

Literature. We alluded above to the importance of $\mathsf{Mod}^{\mathcal{M}\underline{1}}$ as the domain of $\mathsf{J}^{\mathcal{T}}$. This is because there are subtle errors in [**EM09**] and subsequent literature regarding the domain of $\mathsf{J}^{\mathcal{T}}$. In our treatment we note the following.

(1) $\mathsf{Mod}^{\mathcal{M}\underline{1}}$ is a proper subcategory of $\mathsf{Multicat}_*$ (Examples 10.2.15 and 10.2.16). Notably, it does not contain the unit pointed multicategory, S.
(2) $\mathsf{J}^{\mathcal{T}}$ does *not* extend, as a monoidal functor, to all of $\mathsf{Multicat}_*$ (see Question A.5.1).

This final item is contrary to the statement of [**EM09**, Theorem 1.3] and some other assertions there. We give a careful account with corrections in Note 10.8.2. All of the incorrect statements are corrected by restricting $\mathsf{J}^{\mathcal{T}}$ to $\mathsf{Mod}^{\mathcal{M}\underline{1}}$. The essential difficulty is the *unit constraint* for $\mathsf{J}^{\mathcal{T}}$, and we discuss it further in Note 10.8.3 and Question A.5.1.

We point out that the important structure result for $\mathcal{M}\underline{1}$-modules, Proposition 10.1.28, is also given in [**EM09**]. See Note 10.8.1 for further details about that result.

Lax versus Strong K-Theory. The lax construction $\mathsf{K^{EM}} = \mathsf{K^{EM}_{lax}}$ has a significant difference from the strong construction $\mathsf{K^{EM}_{\cong}}$ that is not present for $\mathsf{K^{Se}_{lax}}$ and $\mathsf{K^{Se}_{\cong}}$. First, in similarity with the case of Segal K-theory, Theorem 10.7.16 shows that there is a level equivalence

$$\mathsf{K^{EM}_{\cong}C} \longrightarrow \mathsf{K^{EM}_{lax}C}$$

for each small permutative category C. As noted in Explanation 10.7.23, the domain of definition for $\mathsf{K^{EM}_{\cong}}$ is the subcategory $\mathsf{PermCat^{sus}}$ consisting of strictly unital *strong* symmetric monoidal functors. All of these statements are similarly true for $\mathsf{K^{Se}_{lax}}$ and $\mathsf{K^{Se}_{\cong}}$.

Moreover, one obtains an equivalence of homotopy categories between $\mathsf{PermCat^{su}}$ and $\mathsf{PermCat^{sus}}$ upon inverting those morphisms that induce weak-equivalences on Segal K-theory spectra. These are discussed, for example, in [**Man10**, Theorem 3.9] and [**GJO17a**, Theorem 2.15]. These equivalences are obtained from general (2-)monadic strictification theory, and they imply that, for the purposes of Segal K-theory, there is no homotopy-theoretic difference between $\mathsf{PermCat^{su}}$ and $\mathsf{PermCat^{sus}}$.

However, such strictification functors generally do *not* preserve additional structure such as the multicategory structure present for small permutative categories. General monoidal strictification theory, such as that of [**WY19**, Theorems 4.4 and 4.6], does not apply because $\mathsf{PermCat}^{\mathsf{su}}$ does not have a monoidal structure.

Therefore, for the purposes of E_n-monoidal structures on Elmendorf-Mandell K-theory spectra, the difference between the domains of definition $\mathsf{PermCat}^{\mathsf{su}}$ and $\mathsf{PermCat}^{\mathsf{sus}}$ is a significant one. For example (see Explanation 10.7.23), small ring categories are monoids in $\mathsf{PermCat}^{\mathsf{su}}$. But monoids in $\mathsf{PermCat}^{\mathsf{sus}}$ are small ring categories with *invertible* factorization morphisms. A general small ring category C can be regarded as an object of $\mathsf{PermCat}^{\mathsf{sus}}$ by forgetting structure, but then one cannot detect the monoid structure on $\mathsf{K}^{\mathsf{Se}}_{\cong}\mathsf{C}$. The non-invertible factorization morphisms have been forgotten. Similar issues arise for other E_n-algebras in $\mathsf{PermCat}^{\mathsf{su}}$ and we discuss them in Question A.5.7.

10.1. The Partition Product

Recall the partition multicategory $\mathcal{M}$ of Definition 8.4.1, whose basepoint is the empty set

$$\varnothing \in 2^{a^\flat}.$$

Examples 8.4.4 through 8.4.6 give explicit descriptions of $\mathcal{M}\underline{0}$, $\mathcal{M}\underline{1}$, and $\mathcal{M}\underline{2}$. Recall from Definition 5.6.20 and Theorem 5.7.22 the smash product of pointed multicategories. In this section we record several details related to smash products of partition multicategories.

Definition 10.1.1. For a pair of finite pointed sets a and b, the Cartesian product of subsets induces a multifunctor

$$\prod_{a,b} : \mathcal{M}a \wedge \mathcal{M}b \longrightarrow \mathcal{M}(a \wedge b) \tag{10.1.2}$$

called the *partition product*. The assignment on objects is given by the Cartesian product of subsets, noting that

$$s \times t \subset (a^\flat \times b^\flat) \cong (a \wedge b)^\flat \quad \text{for} \quad s \subset a^\flat \quad \text{and} \quad t \subset b^\flat.$$

Recalling Explanation 5.6.14, the generating operations of

$$\mathcal{M}a \otimes \mathcal{M}b$$

are of the form

$$\iota_{\langle s\rangle} \otimes t \in \mathcal{M}a\big(\langle s\rangle\,;\, s'\big) \times \{t\} \quad \text{and} \quad s \otimes \iota_{\langle t\rangle} \in \{s\} \times \mathcal{M}b\big(\langle t\rangle\,;\, t'\big)$$

where

- $s \subset a^\flat$, $t \subset b^\flat$,
- $\langle s\rangle$ is a partition of s' in $\mathcal{M}a$, and
- $\langle t\rangle$ is a partition of t' in $\mathcal{M}b$.

In this context we define partitions of $s' \times t$ and $s \times t'$ by, respectively,

$$\langle s\rangle \times t = \langle s_i \times t\rangle_i \quad \text{and} \quad s \times \langle t\rangle = \langle s \times t_j\rangle_j.$$

Then we define $\prod_{a,b}$ on generating operations by

$$\iota_{\langle s\rangle} \otimes t \longmapsto \iota_{\langle s\rangle \times t} \quad \text{and} \quad s \otimes \iota_{\langle t\rangle} \longmapsto \iota_{s \times \langle t\rangle}.$$

This assignment preserves the interchange relation of the tensor product as in (5.6.16) because the two composites

$$\begin{array}{l} \langle s\rangle \otimes \langle t\rangle \xrightarrow{\iota_{\langle s\rangle} \otimes \langle t\rangle} s' \times \langle t\rangle \xrightarrow{s' \otimes \iota_{\langle t\rangle}} (s',t') \\ \langle s\rangle \otimes^{\mathsf{t}} \langle t\rangle \xrightarrow[\langle s\rangle \otimes \iota_{\langle t\rangle}]{} \langle s\rangle \times t' \xrightarrow{\iota_{\langle s\rangle} \otimes t'} (s',t') \end{array} \tag{10.1.3}$$

are sent to

$$\iota_{\langle s\rangle \otimes \langle t\rangle} \quad \text{and} \quad \iota_{\langle s\rangle \otimes^{\mathsf{t}} \langle t\rangle},$$

respectively, and these are interchanged by the action of the transposition reindexing denoted $\xi^{\otimes}$ in Definition 5.6.10 and Explanation 5.6.14.

The definition of $\prod_{a,b}$ descends to the smash product of $\mathcal{M}a$ and $\mathcal{M}b$ because the Cartesian product of any set with the empty set is again empty. Therefore, each generating operation of the form

$$\iota_{\langle s\rangle} \times \varnothing \quad \text{or} \quad \varnothing \times \iota_{\langle t\rangle}$$

is sent to a partition of the empty set, the basepoint of $\mathcal{M}(a \wedge b)$.

Each of the axioms in Definition 5.1.12 for $F = \prod_{a,b}$ consists of diagrams where either the common domain of the two composites is empty and there is nothing to verify, or the common domain is a singleton and, therefore, so is each of the other operation sets in the diagram. ◇

Lemma 10.1.4. *The partition product components $\prod_{a,b}$ are natural with respect to morphisms of pointed finite sets.*

Proof. Suppose given pointed finite sets and morphisms

$$\psi : a \longrightarrow c \quad \text{and} \quad \phi : b \longrightarrow d.$$

Recalling Proposition 8.4.7, we have

$$\tilde{\psi} = \psi^{-1} : 2^{c^b} \longrightarrow 2^{a^b} \quad \text{and} \quad \tilde{\phi} = \phi^{-1} : 2^{d^b} \longrightarrow 2^{b^b}.$$

The following diagram of sets commutes, where the horizontal morphisms are given by Cartesian products of subsets.

$$\begin{array}{ccc} 2^{c^b} \times 2^{d^b} & \xrightarrow{- \times -} & 2^{(c\wedge d)^b} \\ {\scriptstyle \tilde{\psi} \times \tilde{\phi}} \big\downarrow & & \big\downarrow {\scriptstyle \widetilde{\psi \wedge \phi}} \\ 2^{a^b} \times 2^{b^b} & \xrightarrow{- \times -} & 2^{(a\wedge b)^b} \end{array}$$

Therefore, the following diagram of multicategories commutes on objects.

$$\begin{array}{ccc} \mathcal{M}c \times \mathcal{M}d & \xrightarrow{\Pi} & \mathcal{M}(c\wedge d) \\ {\scriptstyle \tilde{\psi} \times \tilde{\phi}} \big\downarrow & & \big\downarrow {\scriptstyle \widetilde{\psi \wedge \phi}} \\ \mathcal{M}a \times \mathcal{M}b & \xrightarrow{\Pi} & \mathcal{M}(a\wedge b) \end{array} \tag{10.1.5}$$

Commutativity of the above diagram on generating operations of the form

$$\iota_{\langle s\rangle} \wedge t \quad \text{or} \quad s \wedge \iota_{\langle t\rangle}$$

follows, as for objects, because taking inverse images commutes with taking Cartesian products. This finishes the proof of naturality for $\prod$. □

Proposition 10.1.6. *The partition multicategory $\mathcal{M}$ defines a symmetric monoidal functor*

$$(\mathcal{F}^{\mathsf{op}}, \wedge, \underline{1}) \longrightarrow (\mathsf{Multicat}_*, \wedge, \mathsf{S}).$$

Proof. The unit constraint

$$\mathcal{M}^0 \colon \mathsf{S} \longrightarrow \mathcal{M}\underline{1} \tag{10.1.7}$$

is the pointed multifunctor determined by sending the unique object of I to $\{1\} \subset \underline{1}^\flat$. The monoidal constraint is given by the composite of the partition product $\prod$ with the lexicographic isomorphisms

$$\mathcal{M}^2_{\underline{m},\underline{n}} \colon \mathcal{M}\underline{m} \wedge \mathcal{M}\underline{n} \xrightarrow{\prod_{\underline{m},\underline{n}}} \mathcal{M}(\underline{m} \wedge \underline{n}) \cong \mathcal{M}(\underline{mn}). \tag{10.1.8}$$

Naturality of $\mathcal{M}^2$ is given by that of $\prod$ (Lemma 10.1.4).

The associativity axiom (1.1.9) for $\mathcal{M}^2 = \prod$ is that the following diagram commutes for each triple of objects $\underline{\ell}$, $\underline{m}$, and $\underline{n}$ in $\mathcal{F}$.

$$\begin{array}{ccc}
(\mathcal{M}\underline{\ell} \wedge \mathcal{M}\underline{m}) \wedge \mathcal{M}\underline{n} & \xrightarrow{\alpha} & \mathcal{M}\underline{\ell} \wedge (\mathcal{M}\underline{m} \wedge \mathcal{M}\underline{n}) \\
\Big\downarrow {\scriptstyle \prod \wedge 1} & & \Big\downarrow {\scriptstyle 1 \wedge \prod} \\
\mathcal{M}(\underline{\ell} \wedge \underline{m}) \wedge \mathcal{M}\underline{n} & & \mathcal{M}\underline{\ell} \wedge \mathcal{M}(\underline{m} \wedge \underline{n}) \\
\Big\downarrow {\scriptstyle \prod} & & \Big\downarrow {\scriptstyle \prod} \\
\mathcal{M}((\underline{\ell} \wedge \underline{m}) \wedge \underline{n}) & \xrightarrow{\mathcal{M}(\alpha^{-1})} & \mathcal{M}(\underline{\ell} \wedge (\underline{m} \wedge \underline{n}))
\end{array} \tag{10.1.9}$$

Commutativity of the above diagram uses associativity of the Cartesian product for subsets. Generating operations of the form

$$(\iota_{\langle s\rangle} \otimes t) \otimes v, \quad (s \otimes \iota_{\langle t\rangle}) \otimes v, \quad \text{and} \quad ((s,t)) \otimes \iota_{\langle v\rangle}$$

are sent, via either composite of (10.1.9), to

$$\iota_{\langle s\rangle \times (t \times v)}, \quad \iota_{s \times (\langle t\rangle \times v)}, \quad \text{and} \quad \iota_{s \times (t \times \langle v\rangle)},$$

respectively.

The left unity axiom (1.1.10) follows because

$$\textstyle\prod_{\underline{1},\underline{n}} \colon \mathcal{M}\underline{1} \wedge \mathcal{M}\underline{n} \longrightarrow \mathcal{M}\underline{n}$$

sends the object $(\{1\}, s)$ to $1 \times s \cong s \subset \underline{n}^\flat$. The right unity axiom is similar. The braiding axiom (1.1.18) follows by symmetry of the Cartesian product. □

Explanation 10.1.10. The monoidal constraint $\mathcal{M}^2$ determined by $\prod$ is generally far from an *isomorphism* of pointed multicategories, and thus $\mathcal{M}$ is not *strong* as a monoidal functor. This is the main reason for introducing the category $\mathcal{G}$ and the construction $\mathcal{T}$ in Definition 10.3.1 below. ◇

Explanation 10.1.11 ($\mathcal{M}\underline{1}$-modules)**.** Since $\underline{1}$ is the monoidal unit in $\mathcal{F}$, it follows from Proposition 10.1.6 that $\mathcal{M}\underline{1}$ is a commutative monoid in $\mathsf{Multicat}_*$. The unit morphism

$$\eta = \mathcal{M}^0 \colon \mathsf{S} \longrightarrow \mathcal{M}\underline{1}$$

is the unit constraint (10.1.7). The multiplication morphism is given by

$$\pi = \textstyle\prod_{\underline{1},\underline{1}} \colon \mathcal{M}\underline{1} \wedge \mathcal{M}\underline{1} \longrightarrow \mathcal{M}\underline{1}.$$

For each pointed finite set b, the partition product multifunctors

$$\begin{aligned} \textstyle\prod_{\underline{1},b} \colon \mathcal{M}\underline{1} \wedge \mathcal{M}b &\longrightarrow \mathcal{M}(\underline{1} \wedge b) \cong \mathcal{M}b \\ \textstyle\prod_{b,\underline{1}} \colon \mathcal{M}b \wedge \mathcal{M}\underline{1} &\longrightarrow \mathcal{M}(b \wedge \underline{1}) \cong \mathcal{M}b \end{aligned}$$

make $\mathcal{M}b$ a bimodule over $\mathcal{M}\underline{1}$. ◇

Definition 10.1.12. The 2-category of left $\mathcal{M}\underline{1}$-modules, denoted $\mathsf{Mod}^{\mathcal{M}\underline{1}}$, has objects and 1-cells given by modules and module morphisms as in Proposition 7.5.1. For $\mathcal{M}\underline{1}$-modules N and N' together with $\mathcal{M}\underline{1}$-module morphisms

$$F, F' \colon \mathsf{N} \longrightarrow \mathsf{N}',$$

the set of $\mathcal{M}\underline{1}$-module 2-cells consists of pointed multinatural transformations

$$\theta \colon F \longrightarrow F'$$

such that the whiskerings indicated in the following diagram are equal, where μ denotes the module structure for both N and N'.

(10.1.13)

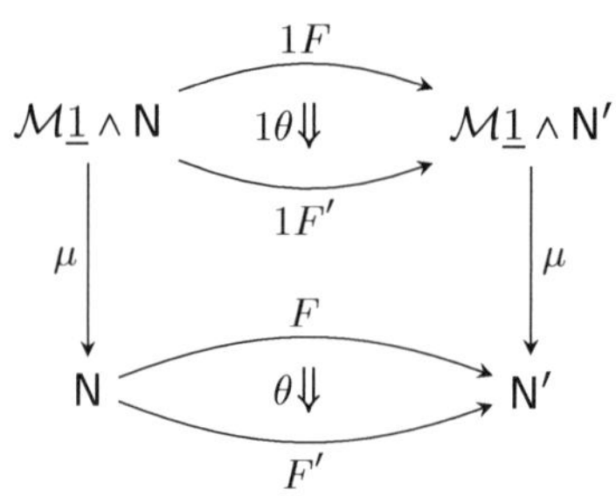

Identities and compositions in $\mathsf{Mod}^{\mathcal{M}\underline{1}}$ are given by those of $\mathsf{Multicat}_*$.

We will use the same notation for the underlying 1-category of $\mathsf{Mod}^{\mathcal{M}\underline{1}}$. In Proposition 10.1.28 below we show that $\mathsf{Mod}^{\mathcal{M}\underline{1}}$ is a full sub-2-category of $\mathsf{Multicat}_*$, with the same monoidal product. This defines a Cat_*-enriched symmetric monoidal structure on $\mathsf{Mod}^{\mathcal{M}\underline{1}}$ in Definition 10.1.36. ◇

Corollary 10.1.14. *The category of $\mathcal{M}\underline{1}$-modules is complete and cocomplete.*

Proof. The result follows from Proposition 7.5.1 (2) together with Proposition 7.5.3 and Explanation 7.5.4. □

We will use the following special cases below.

Lemma 10.1.15. *For each pointed finite set b, the partition products for $\underline{1}$ and b are isomorphisms*

$$\begin{aligned} \textstyle\prod_{\underline{1},b} \colon \mathcal{M}\underline{1} \wedge \mathcal{M}b &\xrightarrow{\cong} \mathcal{M}(\underline{1} \wedge b) \cong \mathcal{M}b \\ \textstyle\prod_{b,\underline{1}} \colon \mathcal{M}b \wedge \mathcal{M}\underline{1} &\xrightarrow{\cong} \mathcal{M}(b \wedge \underline{1}) \cong \mathcal{M}b. \end{aligned} \tag{10.1.16}$$

Proof. We will show that $\prod_{\underline{1},b}$ is an isomorphism. The assertion for $\prod_{b,\underline{1}}$ then follows because $\mathcal{M}$ is symmetric monoidal (Proposition 10.1.6). If $b = \underline{0}$, there is nothing to prove because both sides of (10.1.16) are isomorphic to the terminal multicategory, T. For the remainder of the proof we assume that b has more than one element.

For the remainder of this proof we write $\prod$ for $\prod_{\underline{1},b}$. We recall and extend the notation of Example 8.4.5 as follows for $t \subset b^\flat$.

- Let
$$(t)^n_j = (\varnothing, \ldots, t, \ldots, \varnothing)$$
denote the n-tuple with all entries $\varnothing$ except for t in position j.
- Let $(\varnothing)^n$ denote the n-tuple with all entries $\varnothing$.
- For $b = \underline{1}$, let $(1)^n_j = (\{1\})^n_j$.
- The nonempty sets of operations in $\mathcal{M}\underline{1}$ are singletons
$$\mathcal{M}\underline{1}\big((\varnothing)^n\,;\varnothing\big) = \{\iota^n\}$$
$$\mathcal{M}\underline{1}\big((1)^n_j\,;\{1\}\big) = \{\pi^n_j\}.$$
We will also write $\pi^n_j(1) = \pi^n_j$.
- Define operations ι^n and $\pi^n_j(t)$ in $\mathcal{M}b$ as
$$\mathcal{M}b\big((\varnothing)^n\,;\varnothing\big) = \{\iota^n\}$$
$$\mathcal{M}b\big((t)^n_j\,;t\big) = \{\pi^n_j(t)\}.$$
- We use $\wedge$ instead of $\otimes$ to denote operations in $\mathcal{M}\underline{1} \wedge \mathcal{M}b$.

The definition of $\prod$ via Cartesian products of subsets shows that it is bijective on objects. The nontrival generating operations of $\mathcal{M}1 \wedge \mathcal{M}b$, and their images under $\prod$, are

$$\{1\} \wedge \iota_{\langle u\rangle} \longmapsto \iota_{\langle u\rangle} \quad \text{for} \quad \langle u\rangle \text{ a partition of } u' \subset b^\flat$$
and
$$\pi^n_j(1) \wedge t \longmapsto \pi^n_j(t) \quad \text{for} \quad t \subset b^\flat.$$

This shows that $\prod$ is surjective on operations. To show that $\prod$ is injective on operations, we need to show

$$\pi^n_j(1) \wedge t = \{1\} \wedge \pi^n_j(t) \quad \text{for} \quad 1 \le j \le n \quad \text{and} \quad t \subset b^\flat. \tag{10.1.17}$$

Using the interchange relation (10.1.3) in $\mathcal{M}\underline{1} \otimes \mathcal{M}b$ with

$$\begin{aligned} \langle s\rangle &= (1)^n_j & s' &= \{1\} & \iota_{\langle s\rangle} &= \pi^n_j(1)\\ \langle t\rangle &= (t)^n_j & t' &= t & \iota_{\langle t\rangle} &= \pi^n_j(t), \end{aligned}$$

we see that the composites below are identified under the action of $\xi^\otimes$.

$$\begin{array}{ccccc} (1)^n_j \otimes (t)^n_j & \xrightarrow{\pi^n_j(1)\otimes (t)^n_j} & \{1\} \times (t)^n_j & \xrightarrow{\{1\}\otimes \pi^n_j(t)} & \\ & & & & (\{1\},t) \\ (1)^n_j \otimes^{\mathrm{t}} (t)^n_j & \xrightarrow[(1)^n_j \otimes \pi^n_j(t)]{} & (1)^n_j \times t & \xrightarrow{\pi^n_j(1)\otimes t} & \end{array} \tag{10.1.18}$$

Descending to the smash product $\mathcal{M}\underline{1} \wedge \mathcal{M}b$, each of the leftmost tuples of objects is sent to the n^2-tuple whose only non-basepoint entry is $(\{1\}, t)$ in position

$$n(j-1)+j \in \overline{n \cdot n} \quad \leftrightarrow \quad (j,j) \in \overline{n} \times \overline{n},$$

with the correspondence given by the lexicographic ordering (Definition 8.1.5). The other entries are basepoints, using the equalities

$$(\{1\}, \varnothing) = (\varnothing, \varnothing) = (\varnothing, t)$$

in the smash product. Likewise, when descending to $\mathcal{M}\underline{1} \wedge \mathcal{M}b$, the upper middle and lower middle objects are both sent to the n-tuple whose only non-basepoint entry is $(\{1\}, t)$. Letting

$$(1,t)^{n^2}_{n(j-1)+j} \quad \text{and} \quad (1,t)^n_j$$

denote these tuples, (10.1.18) becomes the following in $\mathcal{M}\underline{1} \wedge \mathcal{M}b$, where we use $\wedge$ in place of $\otimes$ and $\times$.

(10.1.19)
$$\begin{array}{ccccc}
(1)^n_j \wedge (t)^n_j & \xrightarrow{\pi^n_j(1) \wedge (t)^n_j} & \{1\} \wedge (t)^n_j & & \\
\| & & \| & \searrow{\scriptstyle \{1\} \wedge \pi^n_j(t)} & \\
(1,t)^{n^2}_{n(j-1)+j} & & (1,t)^n_j & & (\{1\}, t) \\
\| & & \| & \nearrow{\scriptstyle \pi^n_j(1) \wedge t} & \\
(1)^n_j \wedge^{\mathsf{t}} (t)^n_j & \xrightarrow[(1)^n_j \wedge \pi^n_j(t)]{} & (1)^n_j \wedge t & &
\end{array}$$

The two composites in (10.1.19) are equal by the interchange relation. For (10.1.17), we need to show that the two rightmost operations are equal.

To continue on, we make the following simplifications and notation.

- By equivariance in $\mathcal{M}\underline{1} \wedge \mathcal{M}b$, it suffices to prove (10.1.17) in the case $j = 1$.
- To avoid confusion with some additional notation below, we let $s = \{1\}$, $(s)^n_j = (1)^n_j$, and $\pi^n_j(s) = \pi^n_j(1)$.
- We let $\mathsf{P} = \mathcal{M}\underline{1} \wedge \mathcal{M}b$.
- To denote the basepoint of P, we use

$$* = (\varnothing, \varnothing) = (s, \varnothing) = (\varnothing, t).$$

- We let $(*)^n$ denote the n-tuple in P whose entries are all basepoints.
- The operations $\iota^n \wedge t$, $\iota^n \wedge \varnothing$, $s \wedge \iota^n$, and $\varnothing \wedge \iota^n$ are basepoint operations and hence are all equal in P. We let ι^n_{P} denote this operation, so

(10.1.20)
$$\iota^n_{\mathsf{P}} = \iota^n \wedge t = \iota^n \wedge \varnothing = s \wedge \iota^n = \varnothing \wedge \iota^n \in \mathsf{P}\big((*)^n; *\big).$$

- We let

$$\gamma\big(\phi; \psi_1, \ldots, \psi_m\big) = \phi \circ \langle \psi \rangle$$

denote composition of an m-ary operation ϕ with an m-tuple of operations $\langle \psi \rangle$ in P.
- We use $\gamma^{\mathcal{M}\underline{1}}$ and $\gamma^{\mathcal{M}b}$ similarly to denote composition in the indicated multicategories.

Now we show (10.1.17) in the case $j = 1$ and $n = 2$. We will use each of the following computations.

- In $\mathsf{P}\big((s,t)\,;\,(s,t)\big)$ we have

$$\gamma\big(\pi_1^2(s)\wedge t\,;\,1_{(s,t)}\,,\,\iota^0\wedge t\big)=\gamma^{\mathcal{M}\underline{1}}\big(\pi_1^2(s)\,;\,1_s\,,\,\iota^0\big)\wedge t=1_s\wedge t=1_{(s,t)}\tag{10.1.21}$$

because $-\wedge t$ is a multifunctor from $\mathcal{M}\underline{1}$ to P.

- In $\mathsf{P}\big((\varnothing,\varnothing)\,;\,(s,\varnothing)\big)=\mathsf{P}\big(*\,;\,*\big)$ we have

$$\gamma\big(\pi_1^2(s)\wedge\varnothing\,;\,s\wedge\iota^0\,,\,1_{(\varnothing,\varnothing)}\big)=\gamma\big(\iota_{\mathsf{P}}^2\,;\,\iota_{\mathsf{P}}^0\,,\,\iota_{\mathsf{P}}^1\big)=1_*\tag{10.1.22}$$

because $\pi_1^2(s)\wedge\varnothing$ is equal to ι_{P}^2.

- In $\mathsf{P}\big((s,t)\,;\,(s,t)\big)$ we have

$$\gamma\big(s\wedge\pi_1^2(t)\,;\,1_{(s,t)}\,,\,s\wedge\iota^0\big)=s\wedge\gamma^{\mathcal{M}b}\big(\pi_1^2(t)\,;\,1_t\,,\,\iota^0\big)=s\wedge 1_t=1_{(s,t)}\tag{10.1.23}$$

because $s\wedge-$ is a multifunctor from $\mathcal{M}b$ to P.

- In $\mathsf{P}\big((\varnothing,\varnothing)\,;\,(\varnothing,t)\big)=\mathsf{P}\big(*\,;\,*\big)$ we have

$$\gamma\big(\varnothing\wedge\pi_1^2(t)\,;\,\iota^0\wedge t,1_{(\varnothing,\varnothing)}\big)=\gamma\big(\iota_{\mathsf{P}}^2\,;\,\iota_{\mathsf{P}}^0,\iota_{\mathsf{P}}^1\big)=\iota_{\mathsf{P}}^1=1_*\tag{10.1.24}$$

because $\varnothing\wedge\pi_1^2(t)$ is equal to ι_{P}^2.

Next, we apply the above computations:

$$\begin{aligned}
s\wedge\pi_1^2(t)&=\gamma\big(s\wedge\pi_1^2(t)\,;\,1_{(s,t)}\,,\,1_*\big)\\
&\qquad\text{by right unity (5.1.5) in }\mathsf{P}\\
&=\gamma\big(s\wedge\pi_1^2(t)\,;\,\gamma\big(\pi_1^2(s)\wedge t\,;\,1_{(s,t)}\,,\,\iota^0\wedge t\big),\,\gamma\big(\pi_1^2(s)\wedge\varnothing\,;\,s\wedge\iota^0\,,\,1_{(\varnothing,\varnothing)}\big)\big)\\
&\qquad\text{by (10.1.21) and (10.1.22)}\\
&=\gamma\big(\gamma\big(s\wedge\pi_1^2(t)\,;\,\pi_1^2(s)\wedge t\,,\,\pi_1^2(s)\wedge\varnothing\big)\,;\,1_{(s,t)}\,,\,\iota^0\wedge t\,,\,s\wedge\iota^0\,,\,1_{(\varnothing,\varnothing)}\big)\\
&\qquad\text{by associativity (5.1.4) in }\mathsf{P}\\
&=\gamma\big(\gamma\big(\pi_1^2(s)\wedge t\,;\,s\wedge\pi_1^2(t)\,,\,\varnothing\wedge\pi_1^2(t)\big)\,;\,1_{(s,t)}\,,\,s\wedge\iota^0\,,\,\iota^0\wedge t\,,\,1_{(\varnothing,\varnothing)}\big)\\
&\qquad\text{by interchange (10.1.19) and }\iota_{\mathsf{P}}^0\text{ equalities (10.1.20)}\\
&=\gamma\big(\pi_1^2(s)\wedge t\,;\,\gamma\big(s\wedge\pi_1^2(t)\,;\,1_{(s,t)}\,,\,s\wedge\iota^0\big),\,\gamma\big(\varnothing\wedge\pi_1^2(t)\,;\,\iota^0\wedge t\,,\,1_{(\varnothing,\varnothing)}\big)\big)\\
&\qquad\text{by associativity (5.1.4) in }\mathsf{P}\\
&=\gamma\big(\pi_1^2(s)\wedge t\,;\,1_{(s,t)}\,,\,1_*\big)\\
&\qquad\text{by (10.1.23) and (10.1.24)}\\
&=\pi_1^2(s)\wedge t\\
&\qquad\text{by right unity (5.1.5) in }\mathsf{P}.
\end{aligned}$$

This proves (10.1.17) in the case $j=1$ and $n=2$.

For $j=1$ and $n\geq 3$ we note

$$\pi_1^n(s)=\gamma^{\mathcal{M}\underline{1}}\big(\pi_1^2(s)\,;\,\pi_1^2(s)\,,\,\iota^{n-2}\big)\quad\text{in}\quad\mathcal{M}\underline{1}\big(s,\underbrace{\varnothing,\ldots,\varnothing}_{(n-1)};\,s\big)\tag{10.1.25}$$

and

$$\pi_1^n(t)=\gamma^{\mathcal{M}b}\big(\pi_1^2(t)\,;\,\pi_1^2(t)\,,\,\iota^{n-2}\big)\quad\text{in}\quad\mathcal{M}b\big(t,\underbrace{\varnothing,\ldots,\varnothing}_{(n-1)};\,t\big).\tag{10.1.26}$$

Then, using the composites above and (10.1.17) for $n = 2$ we have

$$\begin{aligned} s \wedge \pi_1^n(t) &= s \wedge \gamma^{\mathcal{M}b}\big(\pi_1^2(t)\,;\, \pi_1^2(t)\,,\, \iota^{n-2}\big) \\ &\qquad \text{by (10.1.26)} \\ &= \gamma\big(s \wedge \pi_1^2(t)\,;\, s \wedge \pi_1^2(t)\,,\, s \wedge \iota^{n-2}\big) \\ &\qquad \text{by multifunctoriality of } s \wedge - \\ &= \gamma\big(\pi_1^2(s) \wedge t\,;\, \pi_1^2(s) \wedge t\,,\, \iota^{n-2} \wedge t\big) \\ &\qquad \text{by the case } n = 2 \text{ and } \iota_{\mathsf{P}}^n \text{ equalities (10.1.20)} \\ &= \gamma^{\mathcal{M}\underline{1}}\big(\pi_1^2(s)\,;\, \pi_1^2(s)\,,\, \iota^{n-2}\big) \wedge t \\ &\qquad \text{by multifunctoriality of } - \wedge t \\ &= \pi_1^n(s) \wedge t \\ &\qquad \text{by (10.1.25).} \end{aligned}$$

This proves (10.1.17) in the case $j = 1$ and $n \geq 3$. The case for general j follows by equivariance. This finishes the proof that $\prod_{\underline{1},b}$ is injective on operations, and thus finishes the proof that $\prod_{\underline{1},b}$ is an isomorphism of pointed multicategories. □

As a special case of Lemma 10.1.15, the multiplication for $\mathcal{M}\underline{1}$ is an isomorphism

$$\mathcal{M}\underline{1} \wedge \mathcal{M}\underline{1} \xrightarrow{\cong} \mathcal{M}\underline{1}. \tag{10.1.27}$$

This implies the following. See Note 10.8.4 for discussion of more general cases.

Proposition 10.1.28.

(1) *Suppose* N *is a left* $\mathcal{M}\underline{1}$*-module in* $\mathsf{Multicat}_*$*. Then the structure morphism*

$$\mu\colon \mathcal{M}\underline{1} \wedge \mathsf{N} \longrightarrow \mathsf{N}$$

is an isomorphism with inverse given by the unit

$$\mathsf{N} \xrightarrow[\cong]{\lambda^{-1}} \mathsf{S} \wedge \mathsf{N} \xrightarrow{\eta \wedge 1} \mathcal{M}\underline{1} \wedge \mathsf{N}$$

where λ *is the left unit isomorphism for* $\wedge$ *and* $\eta = \mathcal{M}^0$ *is the unit for* $\mathcal{M}\underline{1}$*.*

(2) *Similarly, the structure morphism for a right module is an isomorphism with inverse given by* $(1 \wedge \eta) \circ \rho^{-1}$*, where* ρ *is the right unit isomorphism for* $\wedge$*.*

(3) *Each small pointed multicategory* N *admits at most one left* $\mathcal{M}\underline{1}$*-module structure and one right* $\mathcal{M}\underline{1}$*-module structure.*

(4) *The 2-category of left, respectively right,* $\mathcal{M}\underline{1}$*-modules is a full sub-2-category of* $\mathsf{Multicat}_*$*.*

(5) *For a left* $\mathcal{M}\underline{1}$*-module* $(\mathsf{N}, \mu^{\mathsf{N}})$ *and a right* $\mathcal{M}\underline{1}$*-module* $(\mathsf{P}, \mu^{\mathsf{P}})$*, the two morphisms in* $\mathsf{Multicat}_*$

$$(\mathsf{P} \wedge \mathcal{M}\underline{1}) \wedge \mathsf{N} \underset{(1 \wedge \mu^{\mathsf{N}}) \circ \alpha}{\overset{\mu^{\mathsf{P}} \wedge 1}{\rightrightarrows}} \mathsf{P} \wedge \mathsf{N} \tag{10.1.29}$$

are equal and so the canonical morphism to the coequalizer

$$\mathsf{P} \wedge \mathsf{N} \longrightarrow \mathsf{P} \wedge_{\mathcal{M}\underline{1}} \mathsf{N}$$

is an isomorphism in $\mathsf{Multicat}_*$*.*

Proof. Let

$$\eta = \mathcal{M}^0 : \mathsf{S} \longrightarrow \mathcal{M}\underline{1} \quad \text{and} \quad \pi = \textstyle\prod_{\underline{1},\underline{1}} : \mathcal{M}\underline{1} \wedge \mathcal{M}\underline{1} \longrightarrow \mathcal{M}\underline{1}$$

denote the unit and multiplication morphisms, respectively, for $\mathcal{M}\underline{1}$. As stated above, π is an isomorphism by Lemma 10.1.15 with $b = \underline{1}$. The left and right unit diagrams for η and π are the following

$$\begin{tikzcd} \mathsf{S}\wedge\mathcal{M}\underline{1} \ar[r,"\eta\wedge 1"] \ar[dr,"\lambda"'] & \mathcal{M}\underline{1}\wedge\mathcal{M}\underline{1} \ar[d,"\pi","\cong"'] & \mathcal{M}\underline{1}\wedge\mathsf{S} \ar[l,"1\wedge\eta"'] \ar[dl,"\rho"] \\ & \mathcal{M}\underline{1} & \end{tikzcd} \tag{10.1.30}$$

We note, therefore, that $(\eta \wedge 1)$ and $(1 \wedge \eta)$ in the above diagram are both isomorphisms.

To show (1), suppose N is a left $\mathcal{M}\underline{1}$-module. Applying $(- \wedge \mathsf{N})$ to (10.1.30) and using the inverse of $(1 \wedge \eta)$, we obtain the upper region of the following diagram. In this diagram we write

$$\mathsf{A} = \mathcal{M}\underline{1}$$

and α denotes the associativity isomorphism for $\wedge$. Here and below we write $\wedge$ as juxtaposition to save space.

$$\begin{tikzcd} & & \mathsf{AN} \ar[dr,"(\rho 1)^{-1}"] & \\ & (\mathsf{SA})\mathsf{N} \ar[ur,"\lambda 1"] \ar[r,"(\eta 1)1"] \ar[d,"\alpha"] & (\mathsf{AA})\mathsf{N} \ar[r,"((1\eta)1)^{-1}"] \ar[d,"\alpha"] & (\mathsf{AS})\mathsf{N} \ar[d,"\alpha"] \\ \mathsf{AN} \ar[ur,"(\lambda 1)^{-1}"] \ar[r,"(\lambda_{\mathsf{AN}})^{-1}"'] \ar[d,"\mu"'] & \mathsf{S}(\mathsf{AN}) \ar[r,"\eta(11)"] \ar[d,"1\mu"] & \mathsf{A}(\mathsf{AN}) \ar[d,"1\mu"'] & \mathsf{A}(\mathsf{SN}) \ar[l,"(1(\eta 1))"'] \ar[dl,"1\lambda"] \\ \mathsf{N} \ar[r,"\lambda^{-1}"] & \mathsf{SN} \ar[r,"\eta 1"] & \mathsf{AN} & \end{tikzcd} \tag{10.1.31}$$

In the above diagram, the uppermost region commutes by (10.1.30). The triangle and two squares on the middle row commute by (1.1.5) and naturality of α. On the bottom row, the leftmost square commutes by naturality of λ and the remaining square commutes by functoriality of $\wedge$. The remaining triangle commutes by the unit condition for left module structure on N.

Now the composite in (10.1.31) from the leftmost vertex AN around the top and right to the bottom right vertex AN is an identity by invertibility of λ and the middle unity axiom (1.1.2). This shows that the composite

$$\mathsf{AN} \xrightarrow{\mu} \mathsf{N} \xrightarrow{\lambda^{-1}} \mathsf{SN} \xrightarrow{\eta 1} \mathsf{AN}$$

is the identity. The reverse composite

$$\mathsf{N} \xrightarrow{\lambda^{-1}} \mathsf{SN} \xrightarrow{\eta 1} \mathsf{AN} \xrightarrow{\mu} \mathsf{N}$$

is the identity by the left unit condition for N. This completes the proof of (1). The corresponding calculation for right module structure shows (2).

The remaining parts of this proof follow from (1) and (2). For assertion (3), suppose

$$\mu, \mu' \colon \mathcal{M}\underline{1} \wedge \mathsf{N} \longrightarrow \mathsf{N}$$

are the multiplication morphisms for two left module structures on N. Then by (1) both μ and μ' are isomorphisms with inverse given by the composite

$$(\eta \wedge 1) \circ \lambda^{-1} \colon N \longrightarrow \mathcal{M}\underline{1} \wedge N.$$

Therefore $\mu = \mu'$. This proves the uniqueness of left module structures. A similar argument using (2) proves the uniqueness of right module structures.

Now we show that assertion (4) for left modules follows from (1). A similar argument shows that the assertion for right modules follows from (2). By (3) the forgetful 2-functor U from left $\mathcal{M}\underline{1}$-modules to pointed multicategories is injective on objects. The forgetful 2-functor U is faithful on 1- and 2-cells because each left-module multifunctor and multinatural transformation is determined by its underlying data in $\mathsf{Multicat}_*$. Thus the 2-category of left $\mathcal{M}\underline{1}$-modules is identified with a sub-2-category of $\mathsf{Multicat}_*$.

Now suppose that N and N' are left $\mathcal{M}\underline{1}$-modules and suppose that

$$F \colon \mathsf{N} \longrightarrow \mathsf{N}' \quad \text{in} \quad \mathsf{Multicat}_*$$

is a pointed multifunctor. Naturality of λ and η implies that the following outer diagram commutes.

(10.1.32)
$$\begin{tikzcd} \mathsf{AN} \ar[r, "1F"] \ar[d, bend left, "\mu"] & \mathsf{AN}' \ar[d, bend right, "(\eta 1)\circ\lambda^{-1}"'] \\ \mathsf{N} \ar[r, "F"] \ar[u, bend left, "(\eta 1)\circ\lambda^{-1}"] & \mathsf{N}' \ar[u, bend right, "\mu"'] \end{tikzcd}$$

By (1), the two morphisms in each column are inverses and so the innermost diagram also commutes. Thus F is an module morphism for $\mathsf{A} = \mathcal{M}\underline{1}$. The same argument for pointed multinatural transformations

$$\theta \colon F \longrightarrow F'$$

shows that the whiskerings indicated in the diagram below are equal, and hence θ is an module 2-cell for $\mathsf{A} = \mathcal{M}\underline{1}$.

(10.1.33)
$$\begin{tikzcd} \mathsf{AN} \ar[r, bend left, "1F"] \ar[r, bend right, "1F'"'] \ar[r, phantom, "1\theta\Downarrow"] \ar[d, "\mu"'] & \mathsf{AN}' \ar[d, "\mu"] \\ \mathsf{N} \ar[r, bend left, "F"] \ar[r, bend right, "F'"'] \ar[r, phantom, "\theta\Downarrow"] & \mathsf{N}' \end{tikzcd}$$

Now we show that assertion (5) follows from (1) and (2). Again using naturality of α and middle unity, we have

$$\big((1\eta)1\big)\circ(\rho^{-1}1) = \big(\alpha^{-1}\circ\big(1(\eta 1)\big)\circ(1\lambda^{-1})\big).$$

By (1) and (2), this provides a common inverse for the two morphisms in (10.1.29). □

Explanation 10.1.34. Lemma 10.1.15 for $b \neq \underline{1}$ follows from Proposition 10.1.28. However, the case $b = \underline{1}$ is required for the proof of Proposition 10.1.28. ◇

The unit

$$\eta = \mathcal{M}^0 \colon \mathsf{S} \longrightarrow \mathcal{M}\underline{1}$$

is not an isomorphism. However, if N has a left $\mathcal{M}\underline{1}$-module structure, then $(\eta \wedge 1_{\mathsf{N}})$ is an isomorphism by Proposition 10.1.28 (1). Similarly, by Proposition 10.1.28 (2), $(1_{\mathsf{N}} \wedge \eta)$ is an isomorphism if N has a right $\mathcal{M}\underline{1}$-module structure. In Proposition 10.2.1 we show that the converse statements hold.

Explanation 10.1.35 (Sub 2-Category of $\mathcal{M}\underline{1}$-Modules)**.** For a commutative monoid A in a symmetric monoidal category C, the category of left A-modules is generally *not* a subcategory of C because a given object of C may admit several distinct A-module structures. The multiplication and unit are additional data of an A-module. For $\mathsf{A} = \mathcal{M}\underline{1}$, Proposition 10.1.28 shows that

- an $\mathcal{M}\underline{1}$-module structure on N is unique if it exists and
- pointed multifunctors and pointed multinatural transformations are automatically compatible with such module structures.

Thus $\mathsf{Mod}^{\mathcal{M}\underline{1}}$ is identified with a full sub-2-category of $\mathsf{Multicat}_*$.

Part (5) of Proposition 10.1.28 shows that the smash product of $\mathcal{M}\underline{1}$-modules is naturally isomorphic to the underlying smash product of pointed multicategories. In Definition 10.1.36 we use this to define the enriched symmetric monoidal structure of $\mathsf{Mod}^{\mathcal{M}\underline{1}}$. ◇

Recall the descriptions of $(\mathsf{Multicat}, \otimes)$ and $(\mathsf{Multicat}_*, \wedge)$ as symmetric Cat-monoidal 2-categories from Theorems 6.4.3 and 6.4.4, respectively. By Proposition 10.1.28, the same verification of 2-naturality for the symmetric monoidal data applies to the sub-2-category of left $\mathcal{M}\underline{1}$-modules in $\mathsf{Multicat}_*$. We extend Definition 10.1.12 as follows.

Definition 10.1.36. Let

$$\big(\mathsf{Mod}^{\mathcal{M}\underline{1}}, \wedge, \mathcal{M}\underline{1}\big)$$

denote the full Cat_*-enriched subcategory of left $\mathcal{M}\underline{1}$-modules in $\mathsf{Multicat}_*$, with hom categories

$$\mathsf{Mod}^{\mathcal{M}\underline{1}}(\mathsf{N},\mathsf{N}') = \mathsf{Multicat}_*(\mathsf{N},\mathsf{N}').$$

By Proposition 10.1.28, the smash product and symmetry of $(\mathsf{Multicat}_*, \wedge)$ restrict to $\mathsf{Mod}^{\mathcal{M}\underline{1}}$. With $\mathcal{M}\underline{1}$ as monoidal unit and unit isomorphisms given by the module structure and symmetry, these data make $(\mathsf{Mod}^{\mathcal{M}\underline{1}}, \wedge)$ a symmetric monoidal Cat_*-category. Each of the axioms of Definition 1.4.13 for $\mathsf{Mod}^{\mathcal{M}\underline{1}}$ follows from the corresponding axioms for $\mathsf{Multicat}_*$ because the forgetful inclusion

$$\mathsf{Mod}^{\mathcal{M}\underline{1}} \hookrightarrow \mathsf{Multicat}_* \tag{10.1.37}$$

is faithful on 2-cells. With this structure, the inclusion (10.1.37) is a symmetric monoidal Cat_*-functor whose monoidal constraint is the identity and whose unit constraint is given by

$$\eta = \mathcal{M}^0 \colon \mathsf{S} \longrightarrow \mathcal{M}\underline{1}. \qquad \diamond$$

10.2. Characterization of $\mathcal{M}\underline{1}$-Modules

As noted in Explanation 10.1.11, $\mathcal{M}\underline{n}$ is a bimodule over $\mathcal{M}\underline{1}$ for each $\underline{n}$ in $\mathcal{F}$. In Propositions 10.2.1 and 10.2.7 we give characterizations of $\mathcal{M}\underline{1}$-module structure. We use these to discuss further examples.

Proposition 10.2.1. *A pointed multicategory* N *is a left* $\mathcal{M}\underline{1}$*-module if and only if*

$$\mathsf{S} \wedge \mathsf{N} \xrightarrow{\eta \wedge 1} \mathcal{M}\underline{1} \wedge \mathsf{N}$$

is an isomorphism. Similarly, N *is a right* $\mathcal{M}\underline{1}$*-module if and only if* $1 \wedge \eta$ *is an isomorphism.*

Proof. We show that the statement for left modules follows from part (1) of Proposition 10.1.28 and the unit properties of the smash product $\wedge$. The statement for right modules follows likewise from Proposition 10.1.28 (2).

If N is a left $\mathcal{M}\underline{1}$-module, then $\eta \wedge 1$ is an isomorphism by Proposition 10.1.28 (1). For the converse, suppose that $\eta \wedge 1$ is an isomorphism and define a multiplication μ as the composite

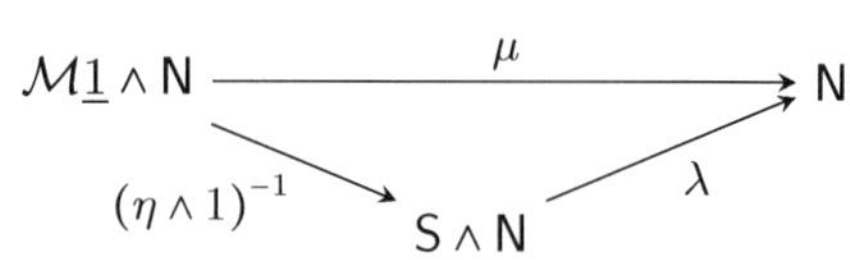

Commutativity of the unity diagram of Definition 7.4.2 for μ is equivalent to that of the above diagram defining μ.

Commutativity of the associativity diagram for μ follows from commutativity of the outer diagram below, where we write $\wedge$ as juxtaposition, $\mathsf{A} = \mathcal{M}\underline{1}$, and $\pi = \Pi_{\underline{1},\underline{1}}$ is the monoid multiplication for $\mathcal{M}\underline{1}$.

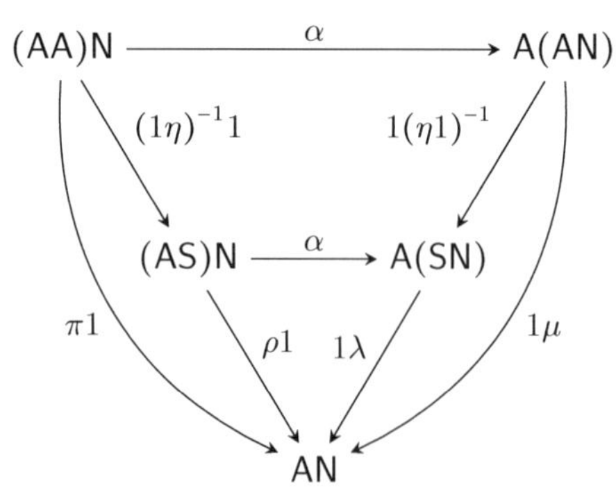

In the above diagram, the upper trapezoid commutes by naturality of the associativity isomorphism α. The inner triangle commutes by the unity axiom (1.1.2) for $\wedge$. The region at left commutes by the right unit axiom for π and the region at right commutes by the definition of μ. Therefore the outer diagram commutes. $\square$

By Proposition 10.2.1, one way to check that an object P in $\mathsf{Multicat}_*$ is *not* a left $\mathcal{M}\underline{1}$-module is to check that $\eta \wedge 1_{\mathsf{P}}$ is not an isomorphism, and likewise for right $\mathcal{M}\underline{1}$-modules.

Example 10.2.2. Consider the monoidal unit (5.6.22)

$$\mathsf{S} = \mathsf{I} \amalg \mathsf{T}.$$

By naturality of the right unitor, $(\eta \wedge 1_{\mathsf{S}})$ is an isomorphism if and only if η is so. Since η is not an isomorphism, then neither is $(\eta \wedge 1_{\mathsf{S}})$. Therefore, by Proposition 10.2.1, S does not have a left $\mathcal{M}\underline{1}$-module structure. Likewise S does not have a right module structure over $\mathcal{M}\underline{1}$. See Example 10.2.16 for a separate argument regarding the non-existence of $\mathcal{M}\underline{1}$-module structure for S. ◇

Uniqueness of the module structures in Proposition 10.1.28 implies the following.

Corollary 10.2.3. *Each pointed multicategory* P *is a left* $\mathcal{M}\underline{1}$*-module if and only if it is a right* $\mathcal{M}\underline{1}$*-module. In this case, the left and right module actions commute with the symmetry of* $\wedge$ *as in the following diagram.*

$$\begin{array}{ccc} \mathcal{M}\underline{1} \wedge \mathsf{P} & \xrightarrow{\ \beta\ } & \mathsf{P} \wedge \mathcal{M}\underline{1} \\ & \searrow \quad \swarrow & \\ & \mathsf{P} & \end{array}$$

Proof. By naturality of the symmetry isomorphism for $\wedge$, we have that $(\eta \wedge 1_{\mathsf{P}})$ is an isomorphism if and only if $(1_{\mathsf{P}} \wedge \eta)$ is so. By Propositions 10.1.28 and 10.2.1 these determine unique left- and right-module structures. □

In the discussion below we use the following notation from Example 8.4.5:

$$\mathcal{M}\underline{1}\big(\langle \varnothing \rangle ; \varnothing\big) = \{\iota^n\}$$
$$\mathcal{M}\underline{1}\big((\varnothing, \ldots, \{1\}, \ldots, \varnothing) ; \{1\}\big) = \{\pi_j^n\}$$

for $n \geq 0$ and $1 \leq j \leq n$. The input of ι^n above is the constant n-tuple whose entries are all the basepoint. The input of π_j^n above consists of a length-n tuple whose only non-basepoint entry is $\{1\}$ in position j. We will also use

$$\iota^n \in \mathsf{T}\big(\langle * \rangle ; *\big)$$

to denote the unique n-ary operation in the terminal multicategory T, and the unique n-ary basepoint operation in any pointed multicategory.

Lemma 10.2.4. *Suppose* P *is a pointed multicategory with composition* γ*. A pointed multifunctor*

$$F \colon \mathcal{M}\underline{1} \longrightarrow \mathsf{P}$$

is uniquely determined by

- *an object* $x = F\{1\}$ *and*
- *an operation* $\pi_1^2(x) = F\pi_1^2$

such that

$$\gamma\big(\pi_1^2(x) ; 1_x, \iota^0\big) = 1_x.$$

Proof. Since F is a pointed multifunctor, we have

$$F\varnothing = * \quad \text{and} \quad F\iota^n = \iota^n \in \mathsf{P}(\langle * \rangle; *)$$

where $*$ denotes the basepoint of P. We let

$$x = F\{1\} \in \mathsf{P}.$$

The nontrivial operations of $\mathcal{M}\underline{1}$ are π_j^n. By equivariance, the value of F on each π_j^n is determined by that of π_1^n. Furthermore, for $n \geq 3$ we have

$$\pi_1^n = \gamma^{\mathcal{M}\underline{1}}\big(\pi_1^2 ; \pi_1^2 , \iota^{n-2}\big) \quad \text{in} \quad \mathcal{M}\underline{1}(\{1\},\underbrace{\varnothing,\ldots,\varnothing}_{(n-1)}; \{1\}).$$

Therefore, by multifunctoriality of F, each $F\pi_1^n$ is determined by $F\pi_1^2$. We define

$$\pi_1^2(x) = F\pi_1^2 \in \mathsf{P}(x, *; x).$$

Lastly, we also have

$$1_{\{1\}} = \gamma^{\mathcal{M}\underline{1}}\big(\pi_1^2 ; 1_{\{1\}} , \iota^0\big) \quad \text{in} \quad \mathcal{M}\underline{1}(\{1\}; \{1\}).$$

Applying F and using multifunctoriality together with preservation of basepoint operations, we have

$$\begin{aligned} 1_x = 1_{F\{1\}} = F(1_{\{1\}}) &= F\big(\gamma^{\mathcal{M}\underline{1}}\big(\pi_1^2 ; 1_{\{1\}} , \iota^0\big)\big) \\ &= \gamma\big(F\pi_1^2 ; F(1_{\{1\}}) , F(\iota^0)\big) \\ &= \gamma\big(\pi_1^2(x) ; 1_x , \iota^0\big). \end{aligned}$$

□

Lemma 10.2.5. *Suppose* $(\mathsf{C}, \oplus, e)$ *is a small permutative category and suppose* $\mathsf{End}(\mathsf{C})$ *is equipped with its canonical basepoint,* e*. There is a natural isomorphism of pointed categories*

$$\mathsf{Multicat}_*(\mathcal{M}\underline{1}, \mathsf{End}(\mathsf{C})) \xrightarrow{\cong} (\mathsf{C}, e) \tag{10.2.6}$$

Proof. With the canonical basepoint e, the n-ary basepoint operations ι^n of $\mathsf{End}(\mathsf{C})$ are given by 1_e viewed as morphisms

$$e = \bigoplus_{i=1}^{n} e \xrightarrow{1_e} e.$$

By Lemma 10.2.4, a pointed multifunctor

$$F : \mathcal{M}\underline{1} \longrightarrow \mathsf{End}(\mathsf{C})$$

is determined by

$$x = F\{1\} \quad \text{and} \quad \pi_1^2(x) = F\pi_1^2 \in \mathsf{End}(\mathsf{C})(x, e; x) = \mathsf{C}(x \oplus e, x)$$

such that the composite of $\pi_1^2(x)$ with

$$\iota^0 = 1_e \in \mathsf{End}(\mathsf{C})(\langle\rangle; e) = \mathsf{C}(e, e) \quad \text{and} \quad 1_x \in \mathsf{End}(\mathsf{C})(x; x) = \mathsf{C}(x, x)$$

is the identity on x. This means that the composite

$$x \oplus e \xrightarrow{1_x \oplus 1_e} x \oplus e \xrightarrow{\pi_1^2(x)} x$$

is equal to 1_x.

Therefore, we must also have $\pi_1^2(x) = 1_x$ and hence F is determined uniquely by $x = F\{1\}$. A pointed multinatural transformation

$$\theta : F \longrightarrow G \quad \text{in} \quad \mathsf{Multicat}_*(\mathcal{M}\underline{1}, \mathsf{End}(\mathsf{C}))$$

is likewise completely determined by the component morphism in C

$$\theta_{\{1\}}: F(\{1\}) \longrightarrow G(\{1\}).$$

Since identity morphisms and composition in $\mathsf{Multicat}_*(\mathcal{M}\underline{1}, \mathsf{End}(\mathsf{C}))$ are defined in C, this gives the isomorphism (10.2.6). Naturality in C follows by construction. □

Proposition 10.2.7. *Suppose* P *is a pointed multicategory with basepoint* $*$. *Then a left* $\mathcal{M}\underline{1}$*-module structure on* P

$$\mu: \mathcal{M}\underline{1} \wedge \mathsf{P} \longrightarrow \mathsf{P}.$$

determines and is uniquely determined by operations

$$\pi_1^2(a) \in \mathsf{P}(a, * \,; a) \quad \textit{for} \quad a \in \mathsf{Ob}\,\mathsf{P}$$

such that the following basepoint, unit, and interchange conditions hold for objects a *in* Ob P *and operations* ϕ *in* $\mathsf{P}(\langle b\rangle ; a)$ *with* $\langle b\rangle = (b_1, \ldots, b_m)$.

$$\pi_1^2(*) = \iota^2 \quad \textit{in} \quad \mathsf{P}(*, * \,; *) \tag{10.2.8}$$

$$\gamma(\pi_1^2(a) ; 1_a, \iota^0) = 1_a \quad \textit{in} \quad \mathsf{P}(a ; a) \tag{10.2.9}$$

$$\gamma(\phi ; \langle \pi_1^2(b_j)\rangle_j) = \gamma(\pi_1^2(a) ; \phi, \iota^m) \cdot \xi_{2,m}^{\otimes} \quad \textit{in} \quad \mathsf{P}(b_1, *, \ldots, b_m, * \,; a) \tag{10.2.10}$$

Proof. Suppose P is a left $\mathcal{M}\underline{1}$-module with multiplication μ. Then μ is

- a pointed multifunctor and
- a unital multiplication.

Therefore, μ must give the following assignments for an object a and an m-ary operation ϕ in P:

$$(\varnothing, a) \longmapsto * \qquad \iota^n \wedge a \longmapsto \iota^n \qquad \varnothing \wedge \phi \longmapsto \iota^m$$

$$(\{1\}, a) \longmapsto a \qquad \pi_j^n \wedge a \longmapsto \pi_j^n(a) = \mu(\pi_j^n \wedge a) \qquad \{1\} \wedge \phi \longmapsto \phi$$

where we take $\mu(\pi_j^n \wedge a)$ as the definition of $\pi_j^n(a)$.

The basepoint condition (10.2.8) follows because μ is a pointed multifunctor. Now each $\mu(- \wedge a)$ is a pointed multifunctor from $\mathcal{M}\underline{1}$ to P and so Lemma 10.2.4 implies the following.

- The operations $\pi_j^n(a)$ are determined via equivariance by $\pi_1^n(a)$.
- The operations $\pi_1^n(a)$ for $n \geq 3$ are determined by
$$\pi_1^2(a) \in \mathsf{P}(a, * \,; a).$$
- The operation $\pi_1^2(a)$ satisfies the unit condition (10.2.9).

For each m-ary operation

$$\phi \in \mathsf{P}(\langle b\rangle ; a)$$

the compatibility of μ with composition means that the two sides of (10.2.10) can be rewritten as

$$\gamma(\phi ; \langle \pi_1^2(b_j)\rangle_j) = \gamma(\mu(\{1\} \wedge \phi) ; \mu(\pi_1^2 \wedge \langle b\rangle)) = \mu(\pi_1^2 \wedge \phi) \tag{10.2.11}$$

and

$$\gamma(\pi_1^2(a) ; \phi, \iota_m) = \gamma(\mu(\pi_1^2 \wedge a) ; \mu((\{1\}, \varnothing) \wedge \phi)) = \mu(\pi_1^2 \wedge^{\mathsf{t}} \phi). \tag{10.2.12}$$

Thus, the interchange relation (5.6.15) in $\mathcal{M}\underline{1} \wedge \mathsf{P}$ implies the interchange condition (10.2.10). This finishes the proof that a left $\mathcal{M}\underline{1}$-module structure determines operations satisfying the indicated conditions.

Conversely, suppose given operations $\pi_1^2(a)$ for each a in $\mathrm{Ob}\,\mathsf{P}$, satisfying the three conditions (10.2.8) through (10.2.10). By Lemma 10.2.4, the unit conditions (10.2.9) imply that each $\pi_1^2(a)$ determines a pointed multifunctor

$$\mu_a: \mathcal{M}\underline{1} \longrightarrow \mathsf{P}.$$

We let

- $\mu(\varnothing,-)$ be the constant pointed multifunctor at $* \in \mathsf{P}$,
- $\mu(\{1\},-)$ be the identity pointed multifunctor on P, and
- $\mu(-,a) = \mu_a$ for each object a of P.

By Explanation 5.6.9 this defines a pointed multifunctor

$$\mu: \mathcal{M}\underline{1} \# \mathsf{P} \longrightarrow \mathsf{P}.$$

Next, because the operations $\mu(\pi_j^n \wedge a)$ in P are determined by multifunctoriality and the operations $\mu(\pi_1^2 \wedge a)$, the interchange conditions (10.2.10) imply, via the equalities (10.2.11) and (10.2.12), that μ satisfies the interchange condition (5.6.16) of Explanation 5.6.14. Thus μ descends to $\mathcal{M}\underline{1} \otimes \mathsf{P}$. Finally, μ descends to $\mathcal{M}\underline{1} \wedge \mathsf{P}$ by the basepoint condition (10.2.8).

We have shown that the three conditions in the statement determine a pointed multifunctor

$$\mu: \mathcal{M}\underline{1} \wedge \mathsf{P} \longrightarrow \mathsf{P}.$$

To see that μ is a left $\mathcal{M}\underline{1}$-module multiplication, we will apply Proposition 10.2.1 together with parts (1) and (3) of Proposition 10.1.28. For this purpose, observe that each of the composites

$$\mathsf{P} \xrightarrow{\lambda^{-1}} \mathsf{S} \wedge \mathsf{P} \xrightarrow{\eta \wedge 1_{\mathsf{P}}} \mathcal{M}\underline{1} \wedge \mathsf{P} \xrightarrow{\mu} \mathsf{P}$$

and

$$\mathcal{M}\underline{1} \wedge \mathsf{P} \xrightarrow{\mu} \mathsf{P} \xrightarrow{\lambda^{-1}} \mathsf{S} \wedge \mathsf{P} \xrightarrow{\eta \wedge 1_{\mathsf{P}}} \mathcal{M}\underline{1} \wedge \mathsf{P}$$

is the identity. The key step, as in the proof of Lemma 10.1.15, is the equality

$$\{1\} \wedge \pi_1^2(a) = \pi_1^2 \wedge a \quad \text{in} \quad \mathcal{M}\underline{1} \wedge \mathsf{P}.$$

The above equality is given by the following computation, generalizing that of Lemma 10.1.15 from the case $\mathsf{P} = \mathcal{M}b$:

$$\begin{aligned}
\{1\} \wedge \pi_1^2(a) &= \gamma\big(\gamma(\{1\} \wedge \pi_1^2(a)\,;\, \pi_1^2 \wedge a\,,\, \pi_1^2 \wedge *)\,;\, 1_{(\{1\},a)}\,,\, \iota^0 \wedge a\,,\, \{1\} \wedge \iota^0\,,\, 1_{(\varnothing,*)}\big)\\
&\qquad \text{by multifunctoriality of } - \wedge a\\
&= \gamma\big(\gamma(\pi_1^2 \wedge a\,;\, \{1\} \wedge \pi_1^2(a)\,,\, \varnothing \wedge \pi_1^2(a))\,;\, 1_{(\{1\},a)}\,,\, \{1\} \wedge \iota^0\,,\, \iota^0 \wedge a\,,\, 1_{(\varnothing,*)}\big)\\
&\qquad \text{by interchange in } \mathcal{M}\underline{1} \wedge \mathsf{P}\\
&= \pi_1^2 \wedge a\\
&\qquad \text{by multifunctoriality of } \{1\} \wedge -.
\end{aligned}$$

For $n \geq 3$ we have

$$\pi_1^n = \gamma\big(\pi_1^2\,;\, \pi_1^2\,,\, \iota^{n-2}\big) \quad \text{in} \quad \mathcal{M}\underline{1}(\{1\}, \underbrace{\varnothing, \ldots, \varnothing}_{n-1}\,;\, \{1\})$$

and, by definition of μ, we have

$$\pi_1^n(a) = \mu(\pi_1^n \wedge a) = \gamma\big(\pi_1^2(a)\,;\, \pi_1^2(a)\,,\, \iota^{n-2}\big) \quad \text{in} \quad \mathsf{P}\big(a, \underbrace{*, \ldots, *}_{n-1}\,;\, a\big).$$

Therefore, again following the proof of Lemma 10.1.15, this implies

$$\pi_j^n \wedge a = \{1\} \wedge \pi_j^n(a)$$

for $n \geq 2$ and $1 \leq j \leq n$.

This shows that μ is an isomorphism with inverse $(\eta \wedge 1) \circ \lambda^{-1}$. Therefore, by Proposition 10.2.1 together with parts (1) and (3) of Proposition 10.1.28, μ is the multiplication for the unique left $\mathcal{M}\underline{1}$-module structure on P. □

Definition 10.2.13. Suppose $(\mathsf{C}, \oplus, e)$ is a small permutative category and $\mathsf{End}(\mathsf{C})$ is equipped with its canonical basepoint, e. Then the *canonical (left) $\mathcal{M}\underline{1}$-module structure* on $\mathsf{End}(\mathsf{C})$ is

$$\mu \colon \mathcal{M}\underline{1} \wedge \mathsf{End}(\mathsf{C}) \longrightarrow \mathsf{End}(\mathsf{C})$$

given as follows for objects a in C and morphisms f in $\mathsf{End}(\mathsf{C})$.

$$\begin{array}{lll} (\varnothing, a) \longmapsto e & \iota^n \wedge a \longmapsto \iota^n = 1_e & \varnothing \wedge f \longmapsto 1_e \\ (\{1\}, a) \longmapsto a & \pi_j^n \wedge a \longmapsto 1_a & \{1\} \wedge f \longmapsto f \end{array}$$

Since e is a strict unit for C, the composite

$$a = a \oplus e \xrightarrow{1_a \oplus 1_e} a \oplus e = a \xrightarrow{1_a} a$$

is equal to the identity on a. The interchange relation (10.2.10) holds for each operation f because both sides are equal to f. By Proposition 10.2.7 this uniquely determines a left $\mathcal{M}\underline{1}$-module structure on $\mathsf{End}(\mathsf{C})$. ◇

Lemma 10.2.14. *Taking the canonical module structures of Definition 10.2.13,* End *factors as a* Cat*-enriched multifunctor through* $\mathsf{Mod}^{\mathcal{M}\underline{1}}$*:*

$$\mathsf{End} \colon \mathsf{PermCat}^{\mathsf{su}} \longrightarrow \mathsf{Mod}^{\mathcal{M}\underline{1}} \hookrightarrow \mathsf{Multicat}_*.$$

Proof. The canonical module structure on $\mathsf{End}(\mathsf{C})$ for each small permutative category C is given by Definition 10.2.13. By Proposition 10.1.28 and Definition 10.1.36 the multifunctors and multinatural transformations of $\mathsf{Mod}^{\mathcal{M}\underline{1}}$ are those of $\mathsf{Multicat}_*$. Moreover, the monoidal product and braiding of $\mathsf{Mod}^{\mathcal{M}\underline{1}}$ are those of $\mathsf{Multicat}_*$. Hence, for $n > 0$, the n-ary operations of $\mathsf{Mod}^{\mathcal{M}\underline{1}}$ and their permutation actions are identified with those of $\mathsf{Multicat}_*$ by Proposition 10.1.28 (4) and Definition 10.1.36.

For $n = 0$, the monoidal unit of $\mathsf{Mod}^{\mathcal{M}\underline{1}}$ is $\mathcal{M}\underline{1}$ and hence a 0-ary operation of $\mathsf{Mod}^{\mathcal{M}\underline{1}}$ is given by a pointed multifunctor out of $\mathcal{M}\underline{1}$. In contrast, each 0-ary operation of $\mathsf{Multicat}_*$ is given by a pointed multifunctor out of S. On 0-ary operations, the inclusion

$$\mathsf{Mod}^{\mathcal{M}\underline{1}} \hookrightarrow \mathsf{Multicat}_*$$

is given by composition along the unit

$$\eta \colon \mathsf{S} \longrightarrow \mathcal{M}\underline{1}.$$

For a small permutative category C, this composition induces a natural isomorphism of pointed categories

$$\mathsf{Mod}^{\mathcal{M}\underline{1}}(\mathcal{M}\underline{1}, \mathsf{End}(\mathsf{C})) = \mathsf{Multicat}_*(\mathcal{M}\underline{1}, \mathsf{End}(\mathsf{C})) \cong \mathsf{Multicat}_*(\mathsf{S}, \mathsf{End}(\mathsf{C}))$$

by Lemma 10.2.5.

These observations show that End factors through $\mathsf{Mod}^{\mathcal{M}\underline{1}}$ as an assignment on objects and n-ary operations for each $n \geq 0$. The Cat_*-enriched multifunctor axioms for End hold in $\mathsf{Mod}^{\mathcal{M}\underline{1}}$ since the inclusion into $\mathsf{Multicat}_*$ is injective on 2-cells. The

Cat-enrichment is obtained by forgetting basepoint objects, that is, multifunctors that factor through T. □

Example 10.2.15 (Non-examples of $\mathcal{M}\underline{1}$-modules)**.** We use Proposition 10.2.7 to give a class of pointed multicategories that are not left $\mathcal{M}\underline{1}$-modules. A similar calculation shows that they also are not right $\mathcal{M}\underline{1}$-modules.

Suppose $(\mathcal{D}, \boxdot, e, \mathrm{T})$ is a small permutative category with null object $\mathrm{T} \neq e$ as basepoint. Recall from Definition 4.3.3 that being null means T is both initial and terminal and, moreover, is fixed by the monoidal product of $\mathcal{D}$. We note two specific cases of interest:

- $(\mathcal{F}, \wedge, \underline{1}, \underline{0})$ (Explanation 8.1.6) and
- $(\mathcal{G}, \oplus, \langle\rangle, *)$ (Proposition 9.1.14).

Let $\mathsf{End}(\mathcal{D})$ be the endomorphism multicategory of $\mathcal{D}$, but equipped with T as its basepoint instead of the monoidal unit e. Thus the nullary basepoint operation

$$\iota^0 \in \mathcal{D}(e, \mathrm{T})$$

is given by the unique morphism to T.

If

$$\mu\colon \mathcal{M}\underline{1} \wedge \mathsf{End}(\mathcal{D}) \longrightarrow \mathsf{End}(\mathcal{D})$$

is a left $\mathcal{M}\underline{1}$-module structure, then by Proposition 10.2.7 we must have, for each object a of $\mathcal{D}$, an operation

$$\pi_1^2(a) \in \mathsf{End}(\mathcal{D})\big(a, \mathrm{T}\,;\, a\big) = \mathcal{D}(a \boxdot \mathrm{T}, a)$$

such that the composite with ι^0 and 1_a is the identity on a, as in (10.2.9). The composition in $\mathsf{End}(\mathcal{D})$ is given by that of $\mathcal{D}$, and because T is null any composite

$$a \boxdot e \xrightarrow{1_a \boxdot \iota^0} a \boxdot \mathrm{T} \xrightarrow{\pi_1^2(a)} a$$

is equal to the zero morphism of a. If a is not the basepoint T, this cannot be the identity morphism of a. Therefore, $\mathsf{End}(\mathcal{D})$ equipped with T as its basepoint is not a left $\mathcal{M}\underline{1}$-module. ◇

Example 10.2.16 (Further Non-examples of $\mathcal{M}\underline{1}$-modules)**.** If Q is a nonempty multicategory and $\mathsf{P} = \mathsf{Q}_+$ is the pointed multicategory obtained by adjoining a disjoint basepoint, then P cannot be a left (or right) $\mathcal{M}\underline{1}$-module because, for any object a in Q, the set of operations

$$\mathsf{P}\big(a, *\,;\, a\big)$$

is empty. Hence there cannot be any operation $\pi_1^2(a)$ as required by Proposition 10.2.7. In particular, the monoidal unit (5.6.22)

$$\mathsf{S} = \mathsf{I} \coprod \mathsf{T}$$

cannot be a left (or right) $\mathcal{M}\underline{1}$-module. ◇

Proposition 5.7.23 shows that the smash product for small pointed multicategories does not restrict, along End, to define a symmetric monoidal structure on the category $\mathsf{PermCat}^{\mathsf{su}}$. The proof there proceeds by showing that there is no permutative category J such that

$$\mathsf{End}(\mathsf{J}) = \mathsf{S},$$

the monoidal unit for $\mathsf{Multicat}_*$. However, Example 10.2.16 shows that S is not a left $\mathcal{M}\underline{1}$-module. Moreover, Lemma 10.2.14 shows that End factors as a Cat-enriched multifunctor

$$\mathsf{PermCat}^{\mathsf{su}} \xrightarrow{\mathsf{End}} \mathsf{Mod}^{\mathcal{M}\underline{1}}.$$

Now we show that this still does not provide a symmetric monoidal structure on $\mathsf{PermCat}^{\mathsf{su}}$.

Proposition 10.2.17. *The symmetric monoidal structure on* $\mathsf{Mod}^{\mathcal{M}\underline{1}}$ *does not restrict to a symmetric monoidal structure on* $\mathsf{PermCat}^{\mathsf{su}}$.

Proof. Suppose to the contrary that the symmetric monoidal structure on $\mathsf{Mod}^{\mathcal{M}\underline{1}}$ restricts to a symmetric monoidal structure on $\mathsf{PermCat}^{\mathsf{su}}$. Then there exists a permutative category C with two objects $\{e, x\}$ such that

$$\mathsf{End}(\mathsf{C}) \cong \mathcal{M}\underline{1},$$

the monoidal unit in $\mathsf{Mod}^{\mathcal{M}\underline{1}}$, as a pointed multicategory, with $\{e, x\}$ corresponding to the objects $\{\varnothing, \{1\}\}$ in $\mathcal{M}\underline{1}$. Since

$$\mathsf{C}(x, x) \neq \varnothing$$

and

$$\begin{aligned} \varnothing &= \mathcal{M}\underline{1}\big(\{1\}, \{1\}\,;\, \{1\}\big) \\ &= \mathsf{End}(\mathsf{C})\big(x, x\,;\, x\big) = \mathsf{C}(x \oplus x, x), \end{aligned}$$

we have that $x \oplus x \neq x$. So $x \oplus x = e$, which implies

$$\begin{aligned} \varnothing \neq \mathsf{C}(x, x) = \mathsf{C}(x \oplus e, x) &= \mathsf{C}(x \oplus x \oplus x, x) \\ &= \mathsf{End}(\mathsf{C})\big(x, x, x\,;\, x\big) \\ &= \mathcal{M}\underline{1}\big(\{1\}, \{1\}, \{1\}\,;\, \{1\}\big). \end{aligned}$$

This contradicts the definition of $\mathcal{M}\underline{1}$. □

The following results describe a closed structure for $\mathsf{Mod}^{\mathcal{M}\underline{1}}$. Recall from Definition 5.7.17 the pointed hom for small pointed multicategories.

Lemma 10.2.18. *For* $\mathsf{M} \in \mathsf{Multicat}_*$ *and* $\mathsf{N} \in \mathsf{Mod}^{\mathcal{M}\underline{1}}$, *the pointed hom multicategory,* $\mathsf{Hom}_*(\mathsf{M}, \mathsf{N})$, *is an* $\mathcal{M}\underline{1}$*-module.*

Proof. The $\mathcal{M}\underline{1}$-module structure morphism for $\mathsf{Hom}_*(\mathsf{M}, \mathsf{N})$,

$$\mathcal{M}\underline{1} \wedge \mathsf{Hom}_*(\mathsf{M}, \mathsf{N}) \longrightarrow \mathsf{Hom}_*(\mathsf{M}, \mathsf{N}),$$

is adjoint to the composite

$$\begin{array}{ccc} (\mathcal{M}\underline{1} \wedge \mathsf{Hom}_*(\mathsf{M}, \mathsf{N})) \wedge \mathsf{M} & & \mathsf{N} \\ {\scriptstyle\cong}\big\downarrow & & \big\uparrow \\ \mathcal{M}\underline{1} \wedge (\mathsf{Hom}_*(\mathsf{M}, \mathsf{N}) \wedge \mathsf{M}) & \xrightarrow{\;1 \wedge \mathsf{ev}\;} & \mathcal{M}\underline{1} \wedge \mathsf{N} \end{array} \tag{10.2.19}$$

where the first isomorphism is the associativity of $\wedge$, the second is given by the evaluation for Hom_*, and the third is the $\mathcal{M}\underline{1}$-module structure of N.

Then the $\mathcal{M}\underline{1}$-module axioms for $\mathsf{Hom}_*(\mathsf{M}, \mathsf{N})$ follow from (10.2.19) and the $\mathcal{M}\underline{1}$-module axioms for N. □

Explanation 10.2.20 (Module Structure of $\mathsf{Hom}_*(\mathsf{M},\mathsf{N})$)**.** The characterization of $\mathcal{M}\underline{1}$-modules in Proposition 10.2.7 gives an alternative explanation of the $\mathcal{M}\underline{1}$-module structure on $\mathsf{Hom}_*(\mathsf{M},\mathsf{N})$ in Lemma 10.2.18. For each pointed multifunctor

$$F \in \mathsf{Hom}_*(\mathsf{M},\mathsf{N}),$$

the requisite operation $\pi_1^2(F)$ is a pointed transformation defined by components

$$\big(\pi_1^2(F)\big)_c = \pi_1^2(Fc) \in \mathsf{N}\big(Fc, *\,; Fc\big), \tag{10.2.21}$$

where the right hand side of the equality is the operation determined by the $\mathcal{M}\underline{1}$-module structure on N. To show that (10.2.21) defines a pointed transformation satisfying the basepoint, unit, and interchange conditions of Proposition 10.2.7, one uses the fact that F is pointed, together with the corresponding conditions componentwise in N. ◇

Proposition 10.2.22. *The category $\mathsf{Mod}^{\mathcal{M}\underline{1}}$ is complete, cocomplete, symmetric monoidal, and closed. The product and internal hom are given by those of $(\mathsf{Multicat}_*, \wedge, \mathsf{Hom}_*)$.*

Proof. Corollary 10.1.14 shows that $\mathsf{Mod}^{\mathcal{M}\underline{1}}$ is complete and cocomplete. Definition 10.1.36 describes the symmetric monoidal structure of $\mathsf{Mod}^{\mathcal{M}\underline{1}}$. For small pointed multicategories M, N, and P, Lemma 10.2.18 shows that the pointed hom multicategory, $\mathsf{Hom}_*(\mathsf{M},\mathsf{N})$, is an $\mathcal{M}\underline{1}$-module. Recall also from Definition 10.1.36 that $\mathsf{Mod}^{\mathcal{M}\underline{1}}$ is full as a subcategory of $\mathsf{Multicat}_*$; this gives the first and final isomorphisms below. Then we have the following, with the second isomorphism from the closed structure of $\mathsf{Multicat}_*$, as in Proposition 5.7.13 and Theorem 5.7.22, and the third isomorphism induced by that of Proposition 10.1.28 (5):

$$\begin{aligned}
\mathsf{Mod}^{\mathcal{M}\underline{1}}\big(\mathsf{M}, \mathsf{Hom}_*(\mathsf{N},\mathsf{P})\big) &\cong \mathsf{Multicat}_*\big(\mathsf{M}, \mathsf{Hom}_*(\mathsf{N},\mathsf{P})\big)\\
&\cong \mathsf{Multicat}_*\big(\mathsf{M}\wedge\mathsf{N}, \mathsf{P}\big)\\
&\cong \mathsf{Multicat}_*\big(\mathsf{M}\wedge_{\mathcal{M}\underline{1}}\mathsf{N}, \mathsf{P}\big)\\
&\cong \mathsf{Mod}^{\mathcal{M}\underline{1}}\big(\mathsf{M}\wedge_{\mathcal{M}\underline{1}}\mathsf{N}, \mathsf{P}\big).
\end{aligned}$$

A similar treatment shows

$$\mathsf{Mod}^{\mathcal{M}\underline{1}}\big(\mathsf{M}\wedge_{\mathcal{M}\underline{1}}\mathsf{N}, \mathsf{P}\big) \cong \mathsf{Mod}^{\mathcal{M}\underline{1}}\big(\mathsf{N}, \mathsf{Hom}_*(\mathsf{M},\mathsf{P})\big).$$

These isomorphisms provide the closed structure for $\mathsf{Mod}^{\mathcal{M}\underline{1}}$. □

Explanation 10.2.23. Recall, from Explanation 5.7.5, that the underlying category of unary operations in each hom object $\mathsf{Hom}(\mathsf{M},\mathsf{N})$ gives the Cat-enrichment of $\mathsf{Multicat}$. The same reasoning shows that the Cat_*-enrichment of $\mathsf{Mod}^{\mathcal{M}\underline{1}}$, in Definition 10.1.36, is given by the underlying categories of the internal hom objects $\mathsf{Hom}_*(\mathsf{M},\mathsf{N})$. ◇

10.3. Elmendorf-Mandell J-Theory and K-Theory

Now we turn to a second K-theory construction, with tuples of finite sets in place of individual finite sets. While moderately more complex, it will yield equivalent symmetric spectra. The additional structure present in $\mathcal{G}_*$-categories will allow us to express the additional multiplicative structure present in ring, bipermutative, braided ring, and E_n-monoidal categories.

Definition 10.3.1. Suppose $\langle a\rangle = (a_1, \ldots, a_q)$ is a tuple of pointed finite sets. Let $\mathcal{T}\langle a\rangle$ denote the smash product of pointed multicategories

$$\mathcal{T}\langle a\rangle = \bigwedge_{k=1}^{q} \mathcal{M}a_k$$

with respect to the basepoints

$$\varnothing \in \mathcal{M}a_k.$$

We let $\varnothing$ also denote the basepoint object of $\mathcal{T}\langle a\rangle$. For the empty tuple we define

$$\mathcal{T}\langle\rangle = \mathcal{M}\underline{1}. \qquad \diamond$$

Explanation 10.3.2. Recalling Proposition 10.1.28 (5), $\mathcal{T}\langle a\rangle$ is equivalently defined as the smash product of $\mathcal{M}a_k$ as left $\mathcal{M}\underline{1}$-modules, taking $\mathcal{M}\underline{1}$ for the empty smash product of such. $\diamond$

Recall from Definition 10.1.36 that we let

$$\left(\mathsf{Mod}^{\mathcal{M}\underline{1}}, \wedge, \mathcal{M}\underline{1}\right)$$

denote the full subcategory of left $\mathcal{M}\underline{1}$-modules in $\mathsf{Multicat}_*$.

Proposition 10.3.3. *The smash product $\mathcal{T}$ of Definition 10.3.1 defines a pointed functor*

$$\mathcal{T}\colon (\mathcal{G}^{\mathsf{op}}, *) \longrightarrow (\mathsf{Mod}^{\mathcal{M}\underline{1}}, \mathsf{T}).$$

Proof. We have $\mathcal{T}$ taking values in $\mathsf{Mod}^{\mathcal{M}\underline{1}}$ by definition. To make $\mathcal{T}$ a pointed functor, we define $\mathcal{T}* = \mathsf{T}$, where $*$ is the basepoint of $\mathcal{G}$. Since T is both initial and terminal in $\mathsf{Multicat}_*$, this defines $\mathcal{T}$ on morphisms in $\mathcal{G}$ either from or to the basepoint $*$.

For tuples $\langle \underline{n}\rangle$ and $\langle \underline{m}\rangle$ of length q and p, respectively, the morphisms from $\langle \underline{n}\rangle$ to $\langle \underline{m}\rangle$ in $\mathcal{G}$ are given by pairs $(f, \langle\psi\rangle)$ as described in Definition 9.1.7. We define

$$\mathcal{T}(f, \langle\psi\rangle)\colon \bigwedge_{\underline{m}_j \in \langle \underline{m}\rangle} \mathcal{M}\underline{m}_j \longrightarrow \bigwedge_{\underline{n}_k \in \langle \underline{n}\rangle} \mathcal{M}\underline{n}_k$$

via the following.

- For each $j \in \overline{p}$, we have
$$\psi_j\colon \underline{n}_{f^{-1}(j)} \longrightarrow \underline{m}_j$$
and so we have an induced multifunctor
$$\tilde{\psi}_j = \mathcal{M}\psi_j\colon \mathcal{M}\underline{m}_j \longrightarrow \mathcal{M}(\underline{n}_{f^{-1}(j)})$$
defined by applying ψ_j^{-1} to subsets, as in the proof of Proposition 8.4.7.
- Recall from Lemma 10.1.15 that the partition product (10.1.2) provides
$$\begin{aligned}\Pi_{\underline{1},\underline{n}}&\colon \mathcal{M}\underline{1} \wedge \mathcal{M}\underline{n} \xrightarrow{\cong} \mathcal{M}(\underline{1} \wedge \underline{n}) \cong \mathcal{M}\underline{n}\\ \Pi_{\underline{n},\underline{1}}&\colon \mathcal{M}\underline{n} \wedge \mathcal{M}\underline{1} \xrightarrow{\cong} \mathcal{M}(\underline{n} \wedge \underline{1}) \cong \mathcal{M}\underline{n}\end{aligned} \tag{10.3.4}$$
for each $\underline{n}$ in $\mathcal{F}$.

We define $\mathcal{T}(f, \langle\psi\rangle)$ to be the composite formed by first taking the smash product of $\tilde{\psi}_j$ over $j \in \overline{p}$, then using (10.3.4) for any factors $\mathcal{M}\underline{n}_{\varnothing}$, and then permuting factors according to f^{-1}:

(10.3.5)

$$\bigwedge_j \mathcal{M}\underline{m}_j \xrightarrow{\wedge_j \tilde{\psi}_j} \bigwedge_j \mathcal{M}(\underline{n}_{f^{-1}(j)}) \xrightarrow{\cong} \bigwedge_{f^{-1}(j) \neq \varnothing} \mathcal{M}(\underline{n}_{f^{-1}(j)}) \xrightarrow[\text{permute}]{\cong} \bigwedge_k \mathcal{M}\underline{n}_k.$$

The associativity (10.1.9) of $\prod$ ensures that iterated applications of (10.3.4) are independent of order.

This definition of $\mathcal{T}$ on morphisms preserves identities because each of the terms in (10.3.5) is an identity when $(f,\langle\psi\rangle) = (1,\langle 1\rangle)$. It preserves composition by naturality (10.1.5) of $\prod$, naturality of the symmetry for $\wedge$ in $\mathsf{Multicat}_*$, and functoriality of the smash product. $\square$

Explanation 10.3.6. We note, in the proof of Proposition 10.3.3 with $\langle\underline{n}\rangle = \langle\rangle$, that

$$\mathcal{T}_{\langle\rangle,\langle\underline{m}\rangle}\colon \mathcal{G}(\langle\rangle,\langle\underline{m}\rangle) \longrightarrow \mathsf{Multicat}_*(\mathcal{T}\langle\underline{m}\rangle,\mathcal{T}\langle\rangle) = \mathsf{Multicat}_*(\mathcal{T}\langle\underline{m}\rangle,\mathcal{M}\underline{1})$$

is given as follows. By Explanation 9.1.9 (4), a morphism from $\langle\rangle$ to $\langle\underline{m}\rangle$ in $\mathcal{G}$ factors uniquely as

$$\langle\rangle \xrightarrow{(i_p,1_{\langle\underline{1}\rangle})} \langle\underline{1}\rangle \xrightarrow{(1,\langle\psi\rangle)} \langle\underline{m}\rangle$$

where $\langle\underline{m}\rangle$ has length p, $\langle\underline{1}\rangle$ denotes the constant p-tuple at $\underline{1}$, and i_p is the unique inclusion from $\varnothing$ to $\overline{p}$. The pointed multifunctor induced by $\mathcal{T}$ is

$$\wedge_j \mathcal{M}\underline{m}_j \xrightarrow{\wedge_j\tilde{\psi}_j} \wedge_j \mathcal{M}\underline{1} \xrightarrow{\cong} \mathcal{M}\underline{1}$$

where the unlabeled isomorphism is the iterate of multiplication isomorphisms for $\mathcal{M}\underline{1}$. Each

$$\psi_j\colon \underline{1} \longrightarrow \underline{m}_j$$

is uniquely determined by an element x of $\underline{m}_j$, namely, the image of $1 \in \underline{1}$. The pointed multifunctor

$$\tilde{\psi}_j = \psi_j^{-1}(-)\colon \mathcal{M}\underline{m}_j \longrightarrow \mathcal{M}\underline{1}$$

sends a subset $s \subset \underline{m}_j^\flat$ to $\{1\}$ if $x \in s$ and to $\varnothing$ if $x \notin s$. $\diamond$

The main feature of $\mathcal{T}$ over the individual factors $\mathcal{M}$ is stated in the following result. Recall from Proposition 10.1.28 that the smash product of left $\mathcal{M}\underline{1}$-modules is the ordinary smash product of pointed multicategories. The important difference is their units.

Proposition 10.3.7. *The functor*

$$\mathcal{T}\colon \mathcal{G}^{\mathsf{op}} \longrightarrow \mathsf{Mod}^{\mathcal{M}\underline{1}}$$

is a strictly unital strong symmetric monoidal functor with respect to the concatenation product in $\mathcal{G}$ and the smash product of pointed multicategories.

Proof. The functor $\mathcal{T}$ has been defined in Proposition 10.3.3. Given objects $\langle\underline{n}\rangle$ and $\langle\underline{n}'\rangle$ in $\mathcal{G}$, the monoidal constraint isomorphisms

$$\mathcal{T}^2\colon \mathcal{T}\langle\underline{n}\rangle \wedge \mathcal{T}\langle\underline{n}'\rangle \xrightarrow{\cong} \mathcal{T}(\langle\underline{n}\rangle \oplus \langle\underline{n}'\rangle) \tag{10.3.8}$$

are defined by the associativity of the smash product or, in the case that $\langle\underline{n}\rangle$ or $\langle\underline{n}'\rangle$ is empty, by the $\mathcal{M}\underline{1}$-module isomorphisms $\prod$ (10.1.16). Naturality of $\mathcal{T}^2$ follows from naturality of the associativity isomorphism and of $\prod$. The unit constraint is the identity because

$$\mathcal{T}\langle\rangle = \mathcal{M}\underline{1}$$

by definition. The associativity axiom (1.1.9) requires an equality of two morphisms

$$\Big(\bigwedge_{\underline{m}_j\in\langle\underline{m}\rangle} \mathcal{M}\underline{m}_j\Big) \wedge \Big(\bigwedge_{\underline{m}'_j\in\langle\underline{m}'\rangle} \mathcal{M}\underline{m}'_j\Big) \wedge \Big(\bigwedge_{\underline{m}''_j\in\langle\underline{m}''\rangle} \mathcal{M}\underline{m}''_j\Big) \longrightarrow \bigwedge_{\underline{r}\in\big(\langle\underline{m}\rangle\oplus\langle\underline{m}'\rangle\oplus\langle\underline{m}''\rangle\big)} \mathcal{M}\underline{r}$$

defined by associativity isomorphisms. Therefore, the associativity axiom follows from Mac Lane's Coherence Theorem 1.1.31. The unity axiom (1.1.10) follows from the definition of the unit constraint being the identity and the strict unit isomorphism in $\mathcal{G}$. The symmetry axiom (1.1.18) follows from the Symmetric Coherence Theorem 1.1.41. □

Definition 10.3.9. Let $\mathsf{J}^{\mathcal{T}}$ denote the Cat_*-functor

$$\mathsf{J}^{\mathcal{T}} = \mathsf{Multicat}_*(\mathcal{T}, -)\colon \mathsf{Mod}^{\mathcal{M}\underline{1}} \longrightarrow \mathcal{G}_*\text{-}\mathsf{Cat}.$$

For each left $\mathcal{M}\underline{1}$-module P, the $\mathcal{G}_*$-category $\mathsf{J}^{\mathcal{T}}\mathsf{P}$ is

$$\mathsf{J}^{\mathcal{T}}\mathsf{P} = \mathsf{Multicat}_*(\mathcal{T}(-), \mathsf{P})\colon \mathcal{G} \longrightarrow \mathsf{Cat}.$$

We call this the *$\mathcal{T}$-partition J-theory* of $\mathcal{M}\underline{1}$-modules, or simply the *partition J-theory* when $\mathcal{T}$ is implied by context. ◇

Explanation 10.3.10 (Comparison and Contrast with $\mathsf{J}^{\mathcal{M}}$)**.** Recall the $\mathcal{M}$-partition J-theory, $\mathsf{J}^{\mathcal{M}}$, from Definition 8.4.10. Explanation 8.4.11 describes the components of $\mathsf{J}^{\mathcal{M}}H$ and $\mathsf{J}^{\mathcal{M}}\theta$ for a pointed multifunctor H and a pointed multinatural transformation θ. One has a similar description for $\mathsf{J}^{\mathcal{T}}$, with $\mathcal{M}$ replaced by $\mathcal{T}$ and $\mathcal{F}$ replaced by $\mathcal{G}$.

The key distinction between $\mathsf{J}^{\mathcal{M}}$ and $\mathsf{J}^{\mathcal{T}}$ is Theorem 10.3.17, which shows that $\mathsf{J}^{\mathcal{T}}$ is *symmetric monoidal* as a Cat_*-functor. See Note 10.8.6 for further discussion of this point. The symmetric monoidal structure of $\mathsf{J}^{\mathcal{T}}$ is the reason that our further applications require $\mathcal{G}$, $\mathcal{T}$, and $\mathsf{J}^{\mathcal{T}}$. This is also the reason that the domain of $\mathsf{J}^{\mathcal{T}}$ must be restricted from $\mathsf{Multicat}_*$ to $\mathsf{Mod}^{\mathcal{M}\underline{1}}$. ◇

Now we define the monoidal constraint and unit constraint for $\mathsf{J}^{\mathcal{T}}$.

Definition 10.3.11. For each pair of left $\mathcal{M}\underline{1}$-modules P and Q we define a pointed functor of categories

$$(\mathsf{J}^{\mathcal{T}})^2_{\mathsf{P},\mathsf{Q}}\colon (\mathsf{J}^{\mathcal{T}}\mathsf{P}) \wedge (\mathsf{J}^{\mathcal{T}}\mathsf{Q}) \longrightarrow \mathsf{J}^{\mathcal{T}}(\mathsf{P} \wedge \mathsf{Q}) \tag{10.3.12}$$

as follows. For each pair of objects $\langle \underline{n} \rangle$ and $\langle \underline{n}' \rangle$ in $\mathcal{G}$, we have

$$\begin{array}{c} \mathsf{Multicat}_*\big(\mathcal{T}\langle \underline{n} \rangle, \mathsf{P}\big) \wedge \mathsf{Multicat}_*\big(\mathcal{T}\langle \underline{n}' \rangle, \mathsf{Q}\big) \\ \downarrow{\scriptstyle \wedge} \\ \mathsf{Multicat}_*\big((\mathcal{T}\langle \underline{n} \rangle) \wedge (\mathcal{T}\langle \underline{n}' \rangle), \mathsf{P} \wedge \mathsf{Q}\big) \\ \downarrow{\scriptstyle (\mathcal{T}^{-2})^*} \\ \mathsf{Multicat}_*\big(\mathcal{T}(\langle \underline{n} \rangle \oplus \langle \underline{n}' \rangle), \mathsf{P} \wedge \mathsf{Q}\big) \end{array} \tag{10.3.13}$$

where $\wedge$ is the smash product of pointed multifunctors and $(\mathcal{T}^{-2})^*$ is precomposition with the inverse of the monoidal constraint for $\mathcal{T}$ (10.3.8).

Then the morphisms (10.3.13) induce a morphism of coends giving the first arrow of the following composite for the Day convolution of $\mathcal{G}_*$-categories, where we write $\mathsf{M}_* = \mathsf{Multicat}_*$ and evaluate at $\langle \underline{m} \rangle$ in $\mathcal{G}$.

(10.3.14)
$$\begin{array}{c}
\big((\mathsf{J}^{\mathcal{T}}\mathsf{P}) \wedge (\mathsf{J}^{\mathcal{T}}\mathsf{Q})\big)_{\langle \underline{m} \rangle} \\
\| \\
\displaystyle\int^{(\langle \underline{n} \rangle, \langle \underline{n}' \rangle) \in \widehat{\mathcal{G}} \wedge \widehat{\mathcal{G}}} \bigvee_{\mathcal{G}^\flat(\langle \underline{n} \rangle \oplus \langle \underline{n}' \rangle, \langle \underline{m} \rangle)} \mathsf{M}_*\big(\mathcal{T}\langle \underline{n} \rangle, \mathsf{P}\big) \wedge \mathsf{M}_*\big(\mathcal{T}\langle \underline{n}' \rangle, \mathsf{Q}\big) \\
\downarrow \\
\displaystyle\int^{(\langle \underline{n} \rangle, \langle \underline{n}' \rangle) \in \widehat{\mathcal{G}} \wedge \widehat{\mathcal{G}}} \bigvee_{\mathcal{G}^\flat(\langle \underline{n} \rangle \oplus \langle \underline{n}' \rangle, \langle \underline{m} \rangle)} \mathsf{M}_*\big(\mathcal{T}(\langle \underline{n} \rangle \oplus \langle \underline{n}' \rangle), \mathsf{P} \wedge \mathsf{Q}\big) \\
\downarrow \\
\mathsf{M}_*\big(\mathcal{T}\langle \underline{m} \rangle, \mathsf{P} \wedge \mathsf{Q}\big)
\end{array}$$

The second arrow in the above diagram is the universal morphism out of the coend that is induced on each summand by

$$\mathsf{M}_*\big(\mathcal{T}(f, \langle \psi \rangle), \mathsf{P} \wedge \mathsf{Q}\big) \quad \text{for} \quad (f, \langle \psi \rangle) \in \mathcal{G}^\flat\big(\langle \underline{n} \rangle \oplus \langle \underline{n}' \rangle, \langle \underline{m} \rangle\big).$$

Therefore, (10.3.14) is induced by the pointed functor that, for each summand

$$(f, \langle \psi \rangle) \in \mathcal{G}^\flat\big(\langle \underline{n} \rangle \oplus \langle \underline{n}' \rangle, \langle \underline{m} \rangle\big),$$

sends pointed multifunctors and pointed multinatural transformations

$$\begin{gathered} \theta \colon F \longrightarrow F' \quad \text{in} \quad \mathsf{M}_*(\mathcal{T}\langle \underline{n} \rangle, \mathsf{P}) \\ \text{and} \\ \omega \colon G \longrightarrow G' \quad \text{in} \quad \mathsf{M}_*(\mathcal{T}\langle \underline{n}' \rangle, \mathsf{Q}) \end{gathered}$$

to the respective composites

(10.3.15)
$$\mathcal{T}\langle \underline{m} \rangle \xrightarrow{\mathcal{T}(f, \langle \psi \rangle)} \mathcal{T}(\langle \underline{n} \rangle \oplus \langle \underline{n}' \rangle) \xrightarrow{\mathcal{T}^{-2}} (\mathcal{T}\langle \underline{n} \rangle) \wedge (\mathcal{T}\langle \underline{n}' \rangle) \underset{F' \wedge G'}{\overset{F \wedge G}{\underset{\Downarrow \theta \wedge \omega}{\rightrightarrows}}} \mathsf{P} \wedge \mathsf{Q}.$$

These definitions on summands satisfy the V-cowedge condition (3.5.2) for $\mathsf{V} = \mathsf{Cat}_*$ by the functoriality of $\mathcal{T}$ and naturality of $\mathcal{T}^2$. We define the components

$$((\mathsf{J}^{\mathcal{T}})^2_{\mathsf{P},\mathsf{Q}})_{\langle \underline{m} \rangle}$$

by (10.3.14).

With respect to morphisms in $\mathcal{G}$, each morphism of (10.3.14) is natural. Thus $(\mathsf{J}^{\mathcal{T}^2})_{\mathsf{P},\mathsf{Q}}$ is a morphism of $\mathcal{G}_*$-categories. By construction, these components are 2-natural with respect to pointed multifunctors and pointed multinatural transformations

$$\mathsf{P} \longrightarrow \mathsf{P}' \quad \text{and} \quad \mathsf{Q} \longrightarrow \mathsf{Q}'.$$

This finishes the definition of $(\mathsf{J}^{\mathcal{T}})^2$. ◇

Now we turn to the unit constraint for $\mathsf{J}^{\mathcal{T}}$. Recall from (9.2.13) that the monoidal unit in $\mathcal{G}_*$-Cat is

$$J = \widehat{\mathcal{G}}(\langle\rangle, -) : \mathcal{G} \longrightarrow \mathsf{Cat}_*.$$

Recall from Explanation 10.3.6 the description of $\mathcal{T}$ on morphisms

$$\langle\rangle \xrightarrow{(i_p, \langle\psi\rangle)} \langle \underline{m} \rangle \quad \text{in} \quad \mathcal{G}.$$

Definition 10.3.16. The unit constraint

$$(\mathsf{J}^{\mathcal{T}})^0 : J \longrightarrow \mathsf{J}^{\mathcal{T}}(\mathcal{M}\underline{1}) = \mathsf{Multicat}_*(\mathcal{T}(-), \mathcal{M}\underline{1})$$

has component at each $\langle \underline{m} \rangle$ given by applying $\mathcal{T}$ to morphisms:

$$\widehat{\mathcal{G}}(\langle\rangle, \langle \underline{m} \rangle) \xrightarrow{(\mathcal{T})_{\langle\rangle, \langle \underline{m}\rangle}} \mathsf{Multicat}_*(\mathcal{T}\langle \underline{m} \rangle, \mathcal{T}\langle\rangle) = \mathsf{Multicat}_*(\mathcal{T}\langle \underline{m} \rangle, \mathcal{M}\underline{1}). \qquad \diamond$$

Theorem 10.3.17. *The functor*

$$\mathsf{J}^{\mathcal{T}} = \mathsf{Multicat}_*(\mathcal{T}, -) : \mathsf{Mod}^{\mathcal{M}\underline{1}} \longrightarrow \mathcal{G}_*\text{-}\mathsf{Cat}$$

is a symmetric monoidal Cat_**-functor.*

Proof. The monoidal constraint $(\mathsf{J}^{\mathcal{T}})^2$ is given in Definition 10.3.11. The unit constraint is given in Definition 10.3.16. Each is a Cat_*-natural transformation by construction. By Theorem 2.5.1 it suffices to verify the symmetric monoidal axioms for the underlying functor $\mathsf{J}^{\mathcal{T}}$ and underlying natural transformations $(\mathsf{J}^{\mathcal{T}})^2$ and $(\mathsf{J}^{\mathcal{T}})^0$.

We first check the unity axioms (1.1.10) for $\mathsf{J}^{\mathcal{T}}$. We will show the left unity axiom and the right is similar. The left unity axiom requires that the following diagram commutes for each left $\mathcal{M}\underline{1}$-module P.

(10.3.18)
$$\begin{array}{ccc}
J \wedge (\mathsf{J}^{\mathcal{T}}\mathsf{P}) & \xrightarrow{\quad \lambda^J \quad} & \mathsf{J}^{\mathcal{T}}\mathsf{P} \\
{\scriptstyle (\mathsf{J}^{\mathcal{T}})^0 \wedge 1}\downarrow & & \uparrow{\scriptstyle \cong} \\
(\mathsf{J}^{\mathcal{T}}\mathcal{M}\underline{1}) \wedge (\mathsf{J}^{\mathcal{T}}\mathsf{P}) & \xrightarrow{\quad (\mathsf{J}^{\mathcal{T}})^2 \quad} & \mathsf{J}^{\mathcal{T}}(\mathcal{M}\underline{1} \wedge \mathsf{P})
\end{array}$$

In the above diagram, the morphism λ^J denotes the left unit isomorphism for the Day convolution (Definition 3.7.17). The unlabeled isomorphism is induced by the left $\mathcal{M}\underline{1}$-module structure of P.

The morphisms in (10.3.18) are morphisms of $\mathcal{G}_*$-categories. We show they are equal by showing their components at each $\langle \underline{m} \rangle$ in $\mathcal{G}$ are equal. At $\langle \underline{m} \rangle$, the $\mathcal{G}_*$-categories on the left side of (10.3.18) are given by the following coends. Here and below we abbreviate $\mathsf{M}_* = \mathsf{Multicat}_*$.

(10.3.19)
$$\begin{aligned}
&(J \wedge (\mathsf{J}^{\mathcal{T}}\mathsf{P}))\langle \underline{m} \rangle = \\
&\qquad \int^{(\langle \underline{n} \rangle, \langle \underline{n}' \rangle) \in \widehat{\mathcal{G}} \wedge \widehat{\mathcal{G}}} \bigvee_{\mathcal{G}^\flat(\langle \underline{n} \rangle \oplus \langle \underline{n}' \rangle, \langle \underline{m} \rangle)} \widehat{\mathcal{G}}(\langle\rangle, \langle \underline{n} \rangle) \wedge \mathsf{M}_*(\mathcal{T}\langle \underline{n}' \rangle, \mathsf{P}) \\
&((\mathsf{J}^{\mathcal{T}}\mathcal{M}\underline{1}) \wedge (\mathsf{J}^{\mathcal{T}}\mathsf{P}))\langle \underline{m} \rangle = \\
&\qquad \int^{(\langle \underline{n} \rangle, \langle \underline{n}' \rangle) \in \widehat{\mathcal{G}} \wedge \widehat{\mathcal{G}}} \bigvee_{\mathcal{G}^\flat(\langle \underline{n} \rangle \oplus \langle \underline{n}' \rangle, \langle \underline{m} \rangle)} \mathsf{M}_*(\mathcal{T}\langle \underline{n}' \rangle, \mathcal{M}\underline{1}) \wedge \mathsf{M}_*(\mathcal{T}\langle \underline{n}' \rangle, \mathsf{P})
\end{aligned}$$

To show that the two composites around (10.3.18) are equal, we show that the morphisms out of the coend (10.3.19) are equal by checking that the morphisms out of each summand are equal. Recall from (9.2.7)

$$\widehat{\mathcal{G}}(\langle\rangle, \langle \underline{m}\rangle) = \bigvee_{\mathcal{G}^\flat(\langle\rangle,\langle \underline{m}\rangle)} \mathbb{1}_+,$$

where $\mathbb{1}$ is the terminal category. For each pair $\langle \underline{n}\rangle$ and $\langle \underline{n}'\rangle$ in $\mathcal{G}$, a summand of

$$\bigvee_{\mathcal{G}^\flat(\langle \underline{n}\rangle\oplus\langle \underline{n}'\rangle,\langle \underline{m}\rangle)} \widehat{\mathcal{G}}(\langle\rangle, \langle \underline{n}\rangle) \wedge \mathsf{M}_*(\mathcal{T}\langle \underline{n}'\rangle, \mathsf{P})$$

together with an object in that summand are determined by

$$\big((f,\langle\psi\rangle),\, (i_q,\langle\phi\rangle),\, F\big) \tag{10.3.20}$$

where q is the length of $\langle \underline{n}\rangle$,

$$\begin{gathered}(f,\langle\psi\rangle) \in \mathcal{G}^\flat(\langle \underline{n}\rangle \oplus \langle \underline{n}'\rangle, \langle \underline{m}\rangle),\\ (i_q,\langle\phi\rangle) \in \mathcal{G}^\flat(\langle\rangle, \langle \underline{n}\rangle), \quad \text{and}\\ F \in \mathsf{M}_*(\mathcal{T}\langle \underline{n}'\rangle, \mathsf{P}).\end{gathered}$$

The left unit λ^J sends the data (10.3.20) to the object of $\mathsf{M}_*(\mathcal{T}\langle \underline{m}\rangle, \mathsf{P})$ (a pointed multifunctor) given by the composite
(10.3.21)

$$\mathcal{T}\langle \underline{m}\rangle \xrightarrow{\mathcal{T}(f,\langle\psi\rangle)} \mathcal{T}(\langle \underline{n}\rangle \oplus \langle \underline{n}'\rangle) \xrightarrow{\mathcal{T}((i_q,\langle\phi\rangle)\oplus 1_{\langle \underline{n}'\rangle})} \mathcal{T}\langle \underline{n}'\rangle \xrightarrow{F} \mathsf{P}.$$

The composite of other morphisms in (10.3.18) sends the data (10.3.20) first to

$$\big((f,\langle\psi\rangle),\, \mathcal{T}(i_q,\langle\phi\rangle),\, F\big),$$

by $(\mathsf{J}^{\mathcal{T}})^0 \wedge 1$, and then to the following composite, where the final isomorphism is the left $\mathcal{M}\underline{1}$-module structure of P.
(10.3.22)

$$\begin{array}{ccccc}\mathcal{T}\langle \underline{m}\rangle & \xrightarrow{\mathcal{T}(f,\langle\psi\rangle)} & \mathcal{T}(\langle \underline{n}\rangle \oplus \langle \underline{n}'\rangle) & & \mathsf{P}\\ & & \Big\downarrow{\scriptstyle \mathcal{T}^{-2}} & & \Big\uparrow{\scriptstyle \cong}\\ & & (\mathcal{T}\langle \underline{n}\rangle) \wedge (\mathcal{T}\langle \underline{n}'\rangle) & \xrightarrow{\mathcal{T}(i_q,\langle\phi\rangle)\wedge F} & \mathcal{M}\underline{1} \wedge \mathsf{P}\end{array}$$

The two composites (10.3.21) and (10.3.22) are equal by commutativity of the following diagram in $\mathsf{M}_* = \mathsf{Multicat}_*$, which we explain below.

$$\begin{array}{ccccc}\mathcal{T}(\langle \underline{n}\rangle \oplus \langle \underline{n}'\rangle) & \xrightarrow{\mathcal{T}((i_q,\langle\phi\rangle)\oplus 1_{\langle \underline{n}'\rangle})} & \mathcal{T}\langle \underline{n}'\rangle & \xrightarrow{F} & \mathsf{P}\\ \Big\downarrow{\scriptstyle \mathcal{T}^{-2}} & & \Big\uparrow{\scriptstyle \mathcal{T}^2_{\langle\rangle,\langle \underline{n}'\rangle}\,\cong} & & \Big\uparrow{\scriptstyle\cong}\\ & \overset{\mathcal{T}(i_q,\langle\phi\rangle)\wedge 1}{\nearrow} & \mathcal{M}\underline{1} \wedge (\mathcal{T}\langle \underline{n}'\rangle) & \overset{1\wedge F}{\searrow} & \\ (\mathcal{T}\langle \underline{n}\rangle) \wedge (\mathcal{T}\langle \underline{n}'\rangle) & & \xrightarrow[\mathcal{T}(i_q,\langle\phi\rangle)\wedge F]{} & & \mathcal{M}\underline{1} \wedge \mathsf{P}\end{array} \tag{10.3.23}$$

In the above diagram, both upward-pointing isomorphisms are given by the left $\mathcal{M}\underline{1}$-module structure. In the case of $\mathcal{T}\langle \underline{n}\rangle$, this module structure is given by the partition product $\prod$ (10.1.2) and is equal to the indicated component of $\mathcal{T}^2$ by definition (10.3.8). Therefore, the trapezoid at left commutes by naturality of $\mathcal{T}^2$ and the trapezoid at right commutes because each pointed multifunctor F is necessarily a left $\mathcal{M}\underline{1}$-module morphism (Proposition 10.1.28 (4)). The lower triangle commutes by functoriality of $\wedge$.

Commutativity of (10.3.23) shows that the data (10.3.20) are sent to the same object of $(\mathsf{J}^{\mathcal{T}}\mathsf{P})\langle\underline{m}\rangle$ by both composites of the unity diagram (10.3.18). A similar computation holds for morphisms, that is, multinatural transformations

$$\theta\colon F \longrightarrow F' \quad \text{in} \quad \mathsf{M}_*(\mathcal{T}\langle\underline{n}'\rangle, \mathsf{P}),$$

by the 2-cell aspect of Proposition 10.1.28 (4). This completes the proof of the left unity axiom (1.1.10) for $\mathsf{J}^{\mathcal{T}}$.

Now we turn to the associativity and braiding axioms for $\mathsf{J}^{\mathcal{T}}$. The associativity axiom (1.1.9) follows, as in the definition of $(\mathsf{J}^{\mathcal{T}})^2$ (10.3.12), by checking on summands as in (10.3.13). For $\langle\underline{n}\rangle$, $\langle\underline{n}'\rangle$, and $\langle\underline{n}''\rangle$ in $\mathcal{G}$ we have the following.

$$\begin{array}{c}
\Big(\mathsf{Multicat}_*(\mathcal{T}\langle\underline{n}\rangle, \mathsf{P}) \wedge \mathsf{Multicat}_*(\mathcal{T}\langle\underline{n}'\rangle, \mathsf{Q})\Big) \wedge \mathsf{Multicat}_*(\mathcal{T}\langle\underline{n}''\rangle, \mathsf{R}) \\
\alpha_* \circ \wedge \circ (\wedge, 1) \Big\downarrow \qquad \Big\downarrow \alpha^* \circ \wedge \circ (1, \wedge) \circ \alpha' \\
\mathsf{Multicat}_*\Big(\big((\mathcal{T}\langle\underline{n}\rangle) \wedge (\mathcal{T}\langle\underline{n}'\rangle)\big) \wedge (\mathcal{T}\langle\underline{n}''\rangle), \mathsf{P} \wedge (\mathsf{Q} \wedge \mathsf{R})\Big) \\
\Big\downarrow \\
\mathsf{Multicat}_*\big(\mathcal{T}(\langle\underline{n}\rangle \oplus \langle\underline{n}'\rangle \oplus \langle\underline{n}''\rangle), \mathsf{P} \wedge (\mathsf{Q} \wedge \mathsf{R})\big)
\end{array}$$

In the above diagram, α' denotes the associator for the smash product of Cat_*-categories (denoted $\alpha^{\otimes}$ in Section 1.4). The morphisms α^* and α_* denote pre- and post-composition with the associativity isomorphism for the smash product in $\mathsf{Multicat}_*$. The upper two morphisms are equal by Cat_*-naturality of

$$\alpha\colon \wedge \circ (\wedge, 1) \longrightarrow \wedge \circ (1, \wedge) \circ \alpha'.$$

The lower morphism is induced by either of the two equal morphisms

$$\mathcal{T}(\langle\underline{n}\rangle \oplus \langle\underline{n}'\rangle \oplus \langle\underline{n}''\rangle) \longrightarrow ((\mathcal{T}\langle\underline{n}\rangle) \wedge (\mathcal{T}\langle\underline{n}'\rangle)) \wedge (\mathcal{T}\langle\underline{n}''\rangle)$$

given by the (inverses of) morphisms in the associativity axiom (1.1.9) for $\mathcal{T}$. The remaining morphisms defining $(\mathsf{J}^{\mathcal{T}})^2$ are given by universal morphisms out of the respective coends, and thus the associativity axiom for $\mathsf{J}^{\mathcal{T}}$ follows.

The braiding axiom (1.1.18) follows from naturality of the symmetry isomorphism for $\wedge$ in $\mathsf{Multicat}_*$ together with Proposition 10.1.28 (1) and (2). □

Recall from Definitions 6.3.3 and 6.3.8 the enriched endomorphism multicategory and multifunctor associated to an enriched symmetric monoidal category and functor, respectively. Then Theorem 10.3.17 implies the following.

Corollary 10.3.24. *Taking* Cat_**-enriched endomorphism multicategories,*

$$\mathsf{J}^{\mathcal{T}}\colon \mathsf{Mod}^{\mathcal{M}\underline{1}} \longrightarrow \mathcal{G}_*\text{-}\mathsf{Cat}$$

is a Cat_**-enriched multifunctor between* Cat_**-enriched multicategories.*

Recall from Definition 10.2.13 the canonical left $\mathcal{M}\underline{1}$-module structure on $\mathsf{End}(\mathsf{C})$ for a small permutative category C. With this choice of module structure, Lemma 10.2.14 shows that End factors, as a Cat-enriched multifunctor, through the category of left $\mathcal{M}\underline{1}$-modules in $\mathsf{Multicat}_*$. Now we consider the composite with the symmetric monoidal Cat_*-functor $\mathsf{J}^{\mathcal{T}}$ from Definition 10.3.9 and Theorem 10.3.17:

$$\mathsf{PermCat}^{\mathsf{su}} \xrightarrow{\mathsf{End}} \mathsf{Mod}^{\mathcal{M}\underline{1}} \xrightarrow{\mathsf{J}^{\mathcal{T}}} \mathcal{G}_*\text{-}\mathsf{Cat}.$$

By Corollary 10.3.24 and forgetting to Cat, this is a composite of Cat-enriched multifunctors.

Definition 10.3.25. We define the *Elmendorf-Mandell J-theory* as the composite Cat-enriched multifunctor

$$\mathsf{J}^{\mathsf{EM}} = \mathsf{J}^{\mathcal{T}} \circ \mathsf{End} \colon \mathsf{PermCat}^{\mathsf{su}} \longrightarrow \mathcal{G}_*\text{-}\mathsf{Cat}. \tag{10.3.26}$$

For a small permutative category C, the associated $\mathcal{G}_*$-category is

$$\mathsf{J}^{\mathsf{EM}}\mathsf{C} = \mathsf{Multicat}_*(\mathcal{T}(-), \mathsf{End}(\mathsf{C})) \colon \mathcal{G} \longrightarrow \mathsf{Cat}_*. \tag{10.3.27}$$

◇

Now we define the Elmendorf-Mandell K-theory multifunctor as a composite of the following. In each case except the first, we have symmetric monoidal sSet-categories and -functors. Recall from Definitions 6.3.3 and 6.3.8 the enriched endomorphism multicategory and multifunctor constructions. Proposition 6.3.10 shows that an enriched symmetric monoidal functor induces an enriched multifunctor on endomorphism multicategories. When applying this result, we will suppress End from the notation.

- By Definition 6.5.1 and Lemma 10.2.14, $\mathsf{PermCat}^{\mathsf{su}}$ is a Cat-enriched multicategory such that

$$\mathsf{End} \colon \mathsf{PermCat}^{\mathsf{su}} \longrightarrow \mathsf{Mod}^{\mathcal{M}\underline{1}} \tag{10.3.28}$$

  is a Cat-enriched multifunctor. Using the same notation, we change enrichment along the nerve functor N as described for enriched multicategories in Section 6.3. Proposition 6.2.9 shows that the change of enrichment makes (10.3.28) a sSet-enriched multifunctor.
- Theorem 10.3.17 shows that

$$\mathsf{J}^{\mathcal{T}} \colon \mathsf{Mod}^{\mathcal{M}\underline{1}} \longrightarrow \mathcal{G}_*\text{-}\mathsf{Cat} \tag{10.3.29}$$

  is a symmetric monoidal Cat_*-functor. Using the same notation, we apply change of enrichment along N, as described for enriched symmetric monoidal categories in Chapter 2. By Theorem 2.4.10 this makes $\mathsf{J}^{\mathcal{T}}$ a symmetric monoidal sSet_*-functor of symmetric monoidal sSet_*-categories. Forgetting basepoints of hom objects (another change of enrichment) and passing to sSet-multicategories via Proposition 6.3.10, we obtain (10.3.29) as a sSet-enriched multifunctor.
- Theorem 9.2.19 shows that taking levelwise nerves induces a symmetric monoidal sSet_*-functor

$$N_* \colon \mathcal{G}_*\text{-}\mathsf{Cat} \longrightarrow \mathcal{G}_*\text{-}\mathsf{sSet}. \tag{10.3.30}$$

  Forgetting basepoints of hom objects and passing to sSet-multicategories via Proposition 6.3.10, we regard (10.3.30) as a sSet-enriched multifunctor.

- Theorem 9.4.9 shows that

$$\mathsf{K}^{\mathcal{G}}\colon \mathcal{G}_*\text{-}\mathsf{sSet} \longrightarrow \mathsf{SymSp} \tag{10.3.31}$$

is a unital symmetric monoidal sSet_*-functor. Forgetting basepoints of hom objects and passing to sSet-multicategories via Proposition 6.3.10, we obtain (10.3.31) as a sSet-enriched multifunctor.

Definition 10.3.32. The *Elmendorf-Mandell K-theory*

$$\mathsf{K}^{\mathsf{EM}}\colon \mathsf{PermCat}^{\mathsf{su}} \longrightarrow \mathsf{SymSp}$$

is the sSet-enriched multifunctor given by the composite

$$\mathsf{PermCat}^{\mathsf{su}} \xrightarrow{\mathsf{End}} \mathsf{Mod}^{\mathcal{M}\underline{1}} \xrightarrow{\mathsf{J}^{\mathcal{T}}} \mathcal{G}_*\text{-}\mathsf{Cat} \xrightarrow{N_*} \mathcal{G}_*\text{-}\mathsf{sSet} \xrightarrow{\mathsf{K}^{\mathcal{G}}} \mathsf{SymSp}$$

of (10.3.28) through (10.3.31). ◇

We will use the sSet-enriched multifunctoriality of K^{EM} as follows. Recall the following notions from Section 6.1 for a symmetric monoidal category V.

(1) A V-enriched operad is a V-enriched multicategory with one object (Definition 6.1.8).
(2) A V-enriched operad morphism is a V-enriched multifunctor between V-enriched operads (Definition 6.1.10).
(3) Suppose that P is a V-enriched operad and M is a V-enriched multicategory. We say that (X,θ) is an algebra over P if X is an object of M and θ is a multifunctor

$$\theta\colon \mathsf{P} \longrightarrow \mathsf{M}$$

that sends the single object of P to X. Equivalently,

$$\theta\colon \mathsf{P} \longrightarrow \mathsf{End}(X)$$

is an operad morphism from P to the endomorphism operad of X (Example 6.1.9 and Definition 6.1.14).

For a Cat-enriched operad P, we let $N\mathsf{P} = \mathsf{P}_N$ denote the sSet-enriched operad given by change of enrichment (Definition 6.2.1) along the nerve functor. The sSet-enriched multifunctoriality of K^{EM} implies the following result.

Theorem 10.3.33. *Suppose P is a Cat-enriched operad and suppose (C,θ) is a P-algebra in $\mathsf{PermCat}^{\mathsf{su}}$. Then $\mathsf{K}^{\mathsf{EM}}\mathsf{C}$ is an $N\mathsf{P}$-algebra in SymSp.*

Proof. Consider the following composite of multifunctors.

$$\begin{array}{ccccc} \mathsf{P} & \xrightarrow{\theta} & \mathsf{PermCat}^{\mathsf{su}} & \xrightarrow{\mathsf{K}^{\mathsf{EM}}} & \mathsf{SymSp} \\ * & \longmapsto & \mathsf{C} & \longmapsto & \mathsf{K}^{\mathsf{EM}}\mathsf{C} \end{array}$$

Then the composite sSet-enriched operad morphism

$$\begin{array}{ccccc} N\mathsf{P} & & & & \mathsf{End}(\mathsf{K}^{\mathsf{EM}}\mathsf{C}) \\ \downarrow{\scriptstyle \theta_N} & & & & \uparrow{\scriptstyle \mathsf{K}^{\mathcal{G}}} \\ N\mathsf{End}(\mathsf{C}) & \xrightarrow{\mathsf{J}^{\mathsf{EM}}_N} & N\mathsf{End}(\mathsf{J}^{\mathsf{EM}}\mathsf{C}) & \xrightarrow{N_*} & \mathsf{End}(N_*\mathsf{J}^{\mathsf{EM}}\mathsf{C}) \end{array}$$

gives $\mathsf{K}^{\mathsf{EM}}\mathsf{C}$ the structure of an $N\mathsf{P}$-algebra. □

In Chapters 11, 12, and 13 we show that the higher monoidal structures on small permutative categories described in Part II.2 are encoded via Cat-enriched operad actions. We apply Theorem 10.3.33 in Corollaries 11.3.16, 12.5.3, and 13.5.2 to show that the associated spectra given by K^{EM} have E_n-monoidal structure for $1 \leq n \leq \infty$.

10.4. Elmendorf-Mandell $\mathcal{G}_*$-categories

In this section we define a $\mathcal{G}_*$-category $\mathsf{C}^{\mathcal{G}} = \mathsf{C}^{\mathcal{G}}_{\mathrm{lax}}$ for each small permutative category C, along with two other variants $\mathsf{C}^{\mathcal{G}}_{\cong}$ and $\mathsf{C}^{\mathcal{G}}_{\mathrm{co}}$. These generalize the three constructions of Γ-categories $\mathsf{C}^{\mathcal{F}}_{\mathrm{lax}}$, $\mathsf{C}^{\mathcal{F}}_{\cong}$, and $\mathsf{C}^{\mathcal{F}}_{\mathrm{co}}$ of Section 8.3. In Proposition 10.5.1 below we show that $\mathsf{J}^{\mathsf{EM}}\mathsf{C}$ and $\mathsf{C}^{\mathcal{G}} = \mathsf{C}^{\mathcal{G}}_{\mathrm{lax}}$ are naturally isomorphic as $\mathcal{G}_*$-categories.

Recall from Definition 6.5.3 that for $\langle s \rangle \in \mathsf{Prof}(S)$ and $t \in S$, we let $\langle s \circ_k t \rangle$ denote the tuple with t replacing s_k. Likewise for $k \neq \ell$ we let $\langle s \circ_k t \circ_l t' \rangle$ denote the tuple with t replacing s_k and t' replacing s_ℓ.

Definition 10.4.1. Suppose $(\mathsf{C}, \oplus, e)$ is a small permutative category and $\langle \underline{n} \rangle$ is an object of $\mathcal{G}$ with length $q > 0$. An *$\langle \underline{n} \rangle$-system* in C is a pair

$$(C, \rho) = \{C_{\langle s \rangle}, \rho_{\langle s \rangle;k,t,u}\}$$

consisting of

- a system of objects, with $C_{\langle s \rangle} \in \mathsf{C}$ for each q-tuple of basepoint-free subsets $s_k \subset \underline{n}_k^\flat$, and
- a system of morphisms, with
$$\rho_{\langle s \rangle;k,t,u} : C_{\langle s \circ_k t \rangle} \oplus C_{\langle s \circ_k u \rangle} \longrightarrow C_{\langle s \rangle}$$
for each $k \in \overline{q}$ and each partition (t, u) of s_k.

We call $\rho_{\langle s \rangle;k,t,u}$ the *(t, u)-gluing morphism of s_k*. These data are subject to the following axioms.

Object Unity: If $s_k = \varnothing$ for any k, then

$$C_{\langle s \rangle} = e. \tag{10.4.2}$$

Gluing Unity: If $s_j = \varnothing$ for any j, or if $t = \varnothing$, or if $u = \varnothing$, then

$$\rho_{\langle s \rangle;k,t,u} = 1. \tag{10.4.3}$$

Gluing Symmetry: For each $k \in \overline{q}$, the following diagram commutes for all partitions (t, u) of s_k.

$$\begin{array}{ccc} C_{\langle s \circ_k t \rangle} \oplus C_{\langle s \circ_k u \rangle} & \xrightarrow{\rho_{\langle s \rangle;k,t,u}} & C_{\langle s \rangle} \\ {\scriptstyle \xi}\big\downarrow & & \big\downarrow{\scriptstyle 1} \\ C_{\langle s \circ_k u \rangle} \oplus C_{\langle s \circ_k t \rangle} & \xrightarrow{\rho_{\langle s \rangle;k,u,t}} & C_{\langle s \rangle} \end{array} \tag{10.4.4}$$

Gluing Associativity: For each $k \in \overline{q}$, the following diagram commutes for all partitions (t, u, v) of s_k.

$$\begin{array}{ccc} C_{\langle s \circ_k t \rangle} \oplus C_{\langle s \circ_k u \rangle} \oplus C_{\langle s \circ_k v \rangle} & \xrightarrow{\rho_{\langle s \circ_k t \cup u \rangle;k,t,u} \oplus 1} & C_{\langle s \circ_k t \cup u \rangle} \oplus C_{\langle s \circ_k v \rangle} \\ {\scriptstyle 1 \oplus \rho_{\langle s \circ_k u \cup v \rangle;k,u,v}}\big\downarrow & & \big\downarrow{\scriptstyle \rho_{\langle s \rangle;k,t \cup u,v}} \\ C_{\langle s \circ_k t \rangle} \oplus C_{\langle s \circ_k u \cup v \rangle} & \xrightarrow{\rho_{\langle s \rangle;k,t,u \cup v}} & C_{\langle s \rangle} \end{array} \tag{10.4.5}$$

Gluing 2-By-2: For each $k \neq \ell$, the following pentagon commutes for all partitions (t,u) of s_k and (v,w) of s_ℓ.

(10.4.6)
$$\begin{array}{ccc}
 & & C_{\langle s \circ_\ell v\rangle} \oplus C_{\langle s \circ_\ell w\rangle} \\
 & \nearrow^{\rho_{\langle s\circ_\ell v\rangle;k,t,u} \oplus \rho_{\langle s\circ_\ell w\rangle;k,t,u}} & \\
C_{\langle s\circ_k t\circ_\ell v\rangle} \oplus C_{\langle s\circ_k u\circ_\ell v\rangle} \oplus C_{\langle s\circ_k t\circ_\ell w\rangle} \oplus C_{\langle s\circ_k u\circ_\ell w\rangle} & & \downarrow \rho_{\langle s\rangle;\ell,v,w} \\
\downarrow 1 \oplus \xi \oplus 1 & & C_{\langle s\rangle} \\
C_{\langle s\circ_k t\circ_\ell v\rangle} \oplus C_{\langle s\circ_k t\circ_\ell w\rangle} \oplus C_{\langle s\circ_k u\circ_\ell v\rangle} \oplus C_{\langle s\circ_k u\circ_\ell w\rangle} & & \uparrow \rho_{\langle s\rangle;k,t,u} \\
 & \searrow_{\rho_{\langle s\circ_k t\rangle;\ell,v,w} \oplus \rho_{\langle s\circ_k u\rangle;\ell,v,w}} & \\
 & & C_{\langle s\circ_k t\rangle} \oplus C_{\langle s\circ_k u\rangle}
\end{array}$$

This finishes the definition of an $\langle \underline{n} \rangle$-system in C. We also have the following variant definitions and terms.

- We will also refer to an $\langle \underline{n} \rangle$-system as a *lax $\langle \underline{n} \rangle$-system*.
- A *strong $\langle \underline{n} \rangle$-system* is an $\langle \underline{n} \rangle$-system for which all of the gluing morphisms $\rho_{\langle s\rangle;k,t,u}$ are isomorphisms.
- A *colax $\langle \underline{n} \rangle$-system* is an $\langle \underline{n} \rangle$-system in C^{op}. That is, the direction of each gluing morphism is reversed but the same axioms (with reversed arrows ρ) are satisfied. ◇

Definition 10.4.7. Suppose $(\mathsf{C}, \oplus, e)$ is a small permutative category and suppose given $\langle \underline{n} \rangle$-systems

$$\{C_{\langle s\rangle}, \rho_{\langle s\rangle;k,t,u}\} \quad \text{and} \quad \{C'_{\langle s\rangle}, \rho'_{\langle s\rangle;k,t,u}\}$$

for $\langle \underline{n} \rangle$ in $\mathcal{G}$ with length $q > 0$. A *morphism of $\langle \underline{n} \rangle$-systems*, denoted

$$\{\alpha_{\langle s\rangle}\} \colon \{C_{\langle s\rangle}, \rho_{\langle s\rangle;k,t,u}\} \longrightarrow \{C'_{\langle s\rangle}, \rho'_{\langle s\rangle;k,t,u}\},$$

consists of component morphisms

$$\alpha_{\langle s\rangle} \colon C_{\langle s\rangle} \longrightarrow C'_{\langle s\rangle}$$

for each q-tuple of basepoint-free subsets $s_k \subset \underline{n}_k^{\flat}$. These components are subject to the following axioms.

Unitary: If $s_k = \varnothing$ for any k, then

(10.4.8)
$$\alpha_{\langle s\rangle} = 1_e.$$

Gluing Compatibility: The following diagram commutes for all k and for each partition (t,u) of s_k.

$$\begin{array}{ccc} C_{\langle s\circ_k t\rangle} \oplus C_{\langle s\circ_k u\rangle} & \xrightarrow{\rho_{\langle s\rangle;k,t,u}} & C_{\langle s\rangle} \\ {\scriptstyle \alpha_{\langle s\circ_k t\rangle}\oplus \alpha_{\langle s\circ_k u\rangle}}\downarrow & & \downarrow{\scriptstyle \alpha_{\langle s\rangle}} \\ C'_{\langle s\circ_k t\rangle} \oplus C'_{\langle s\circ_k u\rangle} & \xrightarrow{\rho'_{\langle s\rangle;k,t,u}} & C'_{\langle s\rangle} \end{array} \tag{10.4.9}$$

This finishes the definition of a morphism of $\underline{n}$-systems. We also have the following variant definitions.

- A morphism of strong $\langle\underline{n}\rangle$-systems is defined as above, with
$$\{C_{\langle s\rangle}, \rho_{\langle s\rangle;k,t,u}\} \quad\text{and}\quad \{C'_{\langle s\rangle}, \rho'_{\langle s\rangle;k,t,u}\}$$
strong $\langle\underline{n}\rangle$-systems.
- A morphism of colax $\langle\underline{n}\rangle$-systems is defined as above, with
$$\{C_{\langle s\rangle}, \rho_{\langle s\rangle;k,t,u}\} \quad\text{and}\quad \{C'_{\langle s\rangle}, \rho'_{\langle s\rangle;k,t,u}\}$$
colax $\langle\underline{n}\rangle$-systems, but the gluing compatibility axiom (10.4.9) has the variance of ρ reversed as in the following diagram.
$$\begin{array}{ccc} C_{\langle s\circ_k t\rangle} \oplus C_{\langle s\circ_k u\rangle} & \xleftarrow{\rho_{\langle s\rangle;k,t,u}} & C_{\langle s\rangle} \\ {\scriptstyle \alpha_{\langle s\circ_k t\rangle}\oplus \alpha_{\langle s\circ_k u\rangle}}\downarrow & & \downarrow{\scriptstyle \alpha_{\langle s\rangle}} \\ C'_{\langle s\circ_k t\rangle} \oplus C'_{\langle s\circ_k u\rangle} & \xleftarrow{\rho'_{\langle s\rangle;k,t,u}} & C'_{\langle s\rangle} \end{array}$$

The identity morphism for a lax, strong, or colax $\langle\underline{n}\rangle$-system $\{C_{\langle s\rangle}, \rho_{\langle s\rangle;k,t,u}\}$ consists of identities

$$\alpha_{\langle s\rangle} = 1_{C_{\langle s\rangle}}$$

for each q-tuple of basepoint-free subsets $s_k \subset \underline{n}_k^\flat$. Composition of morphisms is defined componentwise. This composition is associative and unital because the composition in C is so. ◇

Definition 10.4.10. Suppose $(\mathsf{C},\oplus,e)$ is a small permutative category. If $\langle\underline{n}\rangle$ is an object of $\mathcal{G}$ with length $q>0$, define a pointed category $\mathsf{C}^{\mathcal{G}}\langle\underline{n}\rangle$ as follows.

- The objects of $\mathsf{C}^{\mathcal{G}}\langle\underline{n}\rangle$ are the lax $\langle\underline{n}\rangle$-systems in C.
- The morphisms of $\mathsf{C}^{\mathcal{G}}\langle\underline{n}\rangle$ are the morphisms of lax $\langle\underline{n}\rangle$-systems.
- The basepoint of $\mathsf{C}^{\mathcal{G}}\langle\underline{n}\rangle$ is the constant $\langle\underline{n}\rangle$-system with
$$C_{\langle s\rangle} = e \quad\text{and}\quad \rho_{\langle s\rangle;k,t,u} = 1_e$$
for each $\langle s\rangle$, k, t, and u.

This finishes the definition of $\mathsf{C}^{\mathcal{G}}\langle\underline{n}\rangle$.

For the empty tuple, we define

$$\mathsf{C}^{\mathcal{G}}\langle\rangle = \mathsf{C} \tag{10.4.11}$$

with basepoint given by the unit e. For the basepoint $*$ of $\mathcal{G}$ we define

$$\mathsf{C}^{\mathcal{G}}* = \mathbb{1}, \tag{10.4.12}$$

the terminal category. If $\langle \underline{n} \rangle$ is an object of $\mathcal{G}$ such that some $n_j = 0$, then by the object unity condition (10.4.2) and the gluing unit condition (10.4.3), the only $\langle \underline{n} \rangle$-system is that given by the constant system at $(e, 1_e)$. In this case we define

$$\mathsf{C}^{\mathcal{G}}\langle \underline{n} \rangle = \mathsf{C}^{\mathcal{G}} * = \mathbb{1}.$$

We also have the following variant definitions and notation. In each case we take the value at the empty tuple $\langle \rangle$ to be C as in (10.4.11) and the value at $*$ to be $\mathbb{1}$ as in (10.4.12). This also defines the value at tuples $\langle \underline{n} \rangle$ where some $n_j = 0$.

- To emphasize that the objects of $\mathsf{C}^{\mathcal{G}}\langle \underline{n} \rangle$ are the lax $\langle \underline{n} \rangle$-systems, we also use the notation
$$\mathsf{C}^{\mathcal{G}}_{\text{lax}}\langle \underline{n} \rangle = \mathsf{C}^{\mathcal{G}}\langle \underline{n} \rangle.$$
- For $\langle \underline{n} \rangle$ of positive length, and with each $n_j > 0$, we let
$$\mathsf{C}^{\mathcal{G}}_{\cong}\langle \underline{n} \rangle$$
denote the category of strong $\langle \underline{n} \rangle$-systems and morphisms of such. The basepoint is that of $\mathsf{C}^{\mathcal{G}}\langle \underline{n} \rangle$.
- For $\langle \underline{n} \rangle$ of positive length, and with each $n_j > 0$, we let
$$\mathsf{C}^{\mathcal{G}}_{\text{co}}\langle \underline{n} \rangle$$
denote the category of colax $\langle \underline{n} \rangle$-systems and morphisms of such. The basepoint is that of $\mathsf{C}^{\mathcal{G}}\langle \underline{n} \rangle$. ◇

Explanation 10.4.13. In contrast with Definition 8.3.9 using $\mathsf{C}^{\mathcal{F}}\underline{n}$ for the category of strong $\underline{n}$-systems, we use the unadorned notation $\mathsf{C}^{\mathcal{G}}\langle \underline{n} \rangle$ to denote the category of lax $\langle \underline{n} \rangle$-systems because this will be the one we use most frequently in our applications below. See Explanation 10.7.23 for further discussion of this point. ◇

Example 10.4.14. Suppose $p > 0$ and let $\langle \underline{1} \rangle$ denote the constant p-tuple at $\underline{1}$. Then there is a canonical isomorphism of pointed categories

$$\mathsf{C} \cong \mathsf{C}^{\mathcal{G}}\langle \underline{1} \rangle \tag{10.4.15}$$

where an object C of C corresponds to the $\langle \underline{1} \rangle$-system whose only (possibly) nontrivial object is C and whose gluing morphisms are all identities. A morphism h of C corresponds to the morphism of $\langle \underline{1} \rangle$-systems whose only (possibly) nontrivial component is h. ◇

To define $\mathsf{C}^{\mathcal{G}}$ on morphisms of $\mathcal{G}$, we first make the following preliminary definitions. The complete definition of $\mathsf{C}^{\mathcal{G}}$ on morphisms is Definition 10.4.17 below. Recall from Definition 9.1.4 that for an injection

$$f : \overline{q} \hookrightarrow \overline{p}$$

and a q-tuple of pointed finite sets $\langle \underline{n} \rangle$, the p-tuple of pointed finite sets $f_*\langle \underline{n} \rangle$ has jth entry given by $\underline{n}_{f^{-1}(j)}$.

Definition 10.4.16. Suppose $(\mathsf{C}, \oplus, e)$ is a small permutative category and suppose

$$(f, \langle \psi \rangle) : \langle \underline{n} \rangle \longrightarrow \langle \underline{m} \rangle$$

is a morphism in $\mathcal{G}$ where $\langle \underline{n} \rangle$ has length $q > 0$ and $\langle \underline{m} \rangle$ has length $p > 0$. So

$$f : \overline{q} \hookrightarrow \overline{p}$$

is an injection of sets and

$$\langle \psi \rangle : f_*\langle \underline{n} \rangle \longrightarrow \langle \underline{m} \rangle$$

is a morphism in $\mathcal{F}^p$.

(1) For a p-tuple of basepoint-free subsets $\langle s\rangle$ with

$$s_j \subset (f_*\langle \underline{n}\rangle)_j^\flat = \underline{n}_{f^{-1}(j)}^\flat,$$

let $\tilde{f}_*\langle s\rangle$ be the q-tuple obtained by removing entries not indexed by the image of f and then permuting according to (the inverse of) f. Then for each $\langle \underline{n}\rangle$-system

$$(C,\rho) = \{C_{\langle s\rangle}, \rho_{\langle s\rangle;k,t,u}\} \in \mathsf{C}^g\langle \underline{n}\rangle$$

and each morphism of $\langle \underline{n}\rangle$-systems $\{\alpha_{\langle s\rangle}\}$, we define data for objects and morphisms in $\mathsf{C}^g(f_*\langle \underline{n}\rangle)$ with the following:

$$C^f_{\langle s\rangle} = \begin{cases} C_{\tilde{f}_*\langle s\rangle} & \text{if each } s_j \neq \varnothing,\\ e & \text{otherwise,}\end{cases}$$

$$\rho^f_{\langle s\rangle;k,t,u} = \begin{cases} \rho_{\tilde{f}_*\langle s\rangle;\, f^{-1}(k),\, t,\, u} & \text{if each } s_j \neq \varnothing,\\ 1 & \text{otherwise,}\end{cases}$$

and

$$\alpha^f_{\langle s\rangle} = \begin{cases} \alpha_{\tilde{f}_*\langle s\rangle} & \text{if each } s_j \neq \varnothing,\\ 1 & \text{otherwise.}\end{cases}$$

If, in the definition of $\rho^f_{\langle s\rangle;k,t,u}$, we have $f^{-1}(k) = \varnothing$, then $s_k \subset \overline{1}$ and so at least one of t and u is empty. In this case $\rho^f_{\langle s\rangle;k,t,u}$ is the identity at $C_{\tilde{f}_*\langle s\rangle}$ (if no s_j is empty) or at e (if any s_j is empty). We define a pointed functor

$$\tilde{f}\colon \mathsf{C}^g\langle \underline{n}\rangle \longrightarrow \mathsf{C}^g(f_*\langle \underline{n}\rangle)$$

by $X \longmapsto X^f$ where X is any of

$$C_{\langle s\rangle},\quad \rho_{\langle s\rangle;k,t,u},\quad \text{or}\quad \alpha_{\langle s\rangle}.$$

The definition of $\alpha^f_{\langle s\rangle}$ shows that $\tilde{f}$ is functorial and the definition of $C^f_{\langle s\rangle}$ shows that $\tilde{f}$ is pointed.

(2) For each p-tuple of basepoint-free subsets $\langle s\rangle$ with $s_j \subset \underline{m}_j^\flat$, we let $\langle \psi^{-1}(s)\rangle$ be the tuple whose kth component is $\psi_k^{-1}(s_k)$, a subset of the kth entry in $f_*\langle \underline{n}\rangle$. Then for each $f_*\langle \underline{n}\rangle$-system

$$(C,\rho) = \{C_{\langle s\rangle}, \rho_{\langle s\rangle;k,t,u}\} \in \mathsf{C}^g(f_*\langle \underline{n}\rangle)$$

and each morphism of $f_*\langle \underline{n}\rangle$-systems $\{\alpha_{\langle s\rangle}\}$ we define data for objects and morphisms in $\mathsf{C}^g\langle \underline{m}\rangle$ by letting

$$C^{\langle\psi\rangle}_{\langle s\rangle} = C_{\langle \psi^{-1}(s)\rangle},$$

$$\rho^{\langle\psi\rangle}_{\langle s\rangle;k,t,u} = \rho_{\langle \psi^{-1}(s)\rangle;\, k,\, \psi_k^{-1}(t),\, \psi_k^{-1}(u)}, \quad \text{and}$$

$$\alpha^{\langle\psi\rangle}_{\langle s\rangle} = \alpha_{\langle \psi^{-1} s\rangle}.$$

We define a pointed functor

$$\widetilde{\langle\psi\rangle}\colon \mathsf{C}^g(f_*\langle \underline{n}\rangle) \longrightarrow \mathsf{C}^g\langle \underline{m}\rangle$$

by $X \longmapsto X^{\langle\psi\rangle}$ where X is any of

$$C_{\langle s\rangle},\quad \rho_{\langle s\rangle;k,t,u},\quad \text{or}\quad \alpha_{\langle s\rangle}.$$

The definition of $\alpha^{\langle\psi\rangle}_{\langle s\rangle}$ shows that $\langle\tilde{\psi}\rangle$ is functorial and the definition of $C^{\langle\psi\rangle}_{\langle s\rangle}$ shows that $\langle\tilde{\psi}\rangle$ is pointed.

We use the same definitions and notation above if the systems (C,ρ) are strong or colax. ◇

Now we define $\mathsf{C}^{\mathcal{G}}$, $\mathsf{C}^{\mathcal{G}}_{\cong}$, and $\mathsf{C}^{\mathcal{G}}_{\mathrm{co}}$ on the morphisms of $\mathcal{G}$.

Definition 10.4.17. Suppose $(\mathsf{C},\oplus,e)$ is a small permutative category and suppose given a morphism

$$\Psi\colon N \longrightarrow M$$

in $\mathcal{G}$. If either N or M is the basepoint $*$, then define $\mathsf{C}^{\mathcal{G}}\Psi$ as the corresponding zero morphism in Cat_*.

Otherwise, Ψ is of the form

$$\Psi = (f,\langle\psi\rangle)\colon \langle\underline{n}\rangle \longrightarrow \langle\underline{m}\rangle.$$

If $\langle\underline{m}\rangle$ is the empty tuple, $\langle\rangle$, then so is $\langle\underline{n}\rangle$ and Ψ is either the zero morphism or the identity. In either case $\mathsf{C}^{\mathcal{G}}\Psi$ is uniquely determined.

Now suppose $\langle\underline{m}\rangle$ has length $p>0$. If $\langle\underline{n}\rangle$ is the empty tuple, then Ψ factors uniquely as a composite (9.1.10)

$$\langle\rangle \longrightarrow \langle\underline{1}\rangle \xrightarrow{\langle\psi\rangle} \langle\underline{m}\rangle$$

for $\langle\psi\rangle$ in $\mathcal{F}^p$ (see Explanation 9.1.9 (4)). Then we define $\mathsf{C}^{\mathcal{G}}\Psi$ to be the composite

$$\mathsf{C}^{\mathcal{G}}\langle\rangle = \mathsf{C} \xrightarrow{\cong} \mathsf{C}^{\mathcal{G}}\langle\underline{1}\rangle \xrightarrow{\langle\tilde{\psi}\rangle} \mathsf{C}^{\mathcal{G}}\langle\underline{m}\rangle$$

of the canonical isomorphism (10.4.15) with $\langle\tilde{\psi}\rangle$ as in Definition 10.4.16 above.

Now suppose, moreover, that $\langle\underline{n}\rangle$ has length $q>0$. Then we define $\mathsf{C}^{\mathcal{G}}(f,\langle\psi\rangle)$ to be the composite pointed functor

$$\mathsf{C}^{\mathcal{G}}\langle\underline{n}\rangle \xrightarrow{\tilde{f}} \mathsf{C}^{\mathcal{G}}(f_*\langle\underline{n}\rangle) \xrightarrow{\langle\tilde{\psi}\rangle} \mathsf{C}^{\mathcal{G}}\langle\underline{m}\rangle$$

with $\tilde{f}$ and $\langle\tilde{\psi}\rangle$ as in Definition 10.4.16 above.

This finishes the definition of $\mathsf{C}^{\mathcal{G}}\Psi$, which we sometimes also denote $\mathsf{C}^{\mathcal{G}}_{\mathrm{lax}}\Psi$. We let

$$\mathsf{C}^{\mathcal{G}}_{\cong}\Psi\colon \mathsf{C}^{\mathcal{G}}_{\cong}\langle\underline{n}\rangle \longrightarrow \mathsf{C}^{\mathcal{G}}_{\cong}\langle\underline{m}\rangle$$

and

$$\mathsf{C}^{\mathcal{G}}_{\mathrm{co}}\Psi\colon \mathsf{C}^{\mathcal{G}}_{\mathrm{co}}\langle\underline{n}\rangle \longrightarrow \mathsf{C}^{\mathcal{G}}_{\mathrm{co}}\langle\underline{m}\rangle$$

be the pointed functors given by the same definitions as above for strong, respectively colax, $\langle\underline{n}\rangle$-systems and morphisms. ◇

Taken together, Definitions 10.4.1, 10.4.7, 10.4.10, 10.4.16, and 10.4.17 give assignments on objects and morphisms

$$\begin{aligned} \mathsf{C}^{\mathcal{G}} = \mathsf{C}^{\mathcal{G}}_{\mathrm{lax}}\colon \mathcal{G} &\longrightarrow \mathsf{Cat}_*, \\ \mathsf{C}^{\mathcal{G}}_{\cong}\colon \mathcal{G} &\longrightarrow \mathsf{Cat}_*, \quad \text{and} \\ \mathsf{C}^{\mathcal{G}}_{\mathrm{co}}\colon \mathcal{G} &\longrightarrow \mathsf{Cat}_*. \end{aligned}$$

Now we show these give $\mathcal{G}_*$-categories.

Proposition 10.4.18. *Suppose* C *is a small permutative category. Each of the assignments*

$$\mathsf{C}^{\mathcal{G}} = \mathsf{C}^{\mathcal{G}}_{\mathrm{lax}}, \quad \mathsf{C}^{\mathcal{G}}_{\cong}, \quad \textit{and} \quad \mathsf{C}^{\mathcal{G}}_{\mathrm{co}}$$

defines a $\mathcal{G}_*$*-category.*

Proof. We give the proof for $\mathsf{C}^{\mathcal{G}} = \mathsf{C}^{\mathcal{G}}_{\mathrm{lax}}$; the argument for the other two cases is the same. By definition we have $\mathsf{C}^{\mathcal{G}}* = \mathbb{1}$, so $\mathsf{C}^{\mathcal{G}}$ preserves basepoints. Preservation of identity morphisms $(1, \langle 1\rangle)$ can be checked directly from Definition 10.4.17.

Now we consider functoriality with respect to a composable pair of morphisms

$$\langle \underline{n}\rangle \xrightarrow{(f,\langle\psi\rangle)} \langle \underline{m}\rangle \xrightarrow{(g,\langle\phi\rangle)} \langle \underline{\ell}\rangle. \tag{10.4.19}$$

where $\langle \underline{n}\rangle$, $\langle \underline{m}\rangle$, and $\langle \underline{\ell}\rangle$ are tuples of length q, p, and r, respectively. If $r = 0$ or $p = 0$, then both p and q must be zero and $(f, \langle\psi\rangle)$ is either a zero or identity morphism so functoriality follows. So now suppose both $p > 0$ and $r > 0$. On tuples of basepoint-free and nonempty subsets $\langle s\rangle$, with s_k a subset of the kth entry in $g_*\langle \underline{m}\rangle$, we have

$$\langle \psi^{-1}\tilde{g}_*\langle s\rangle\rangle = \tilde{g}_*\langle (g_*\langle\psi\rangle)^{-1}(s)\rangle.$$

This shows that the following diagram commutes in Cat_*. In the diagram below, if $q = 0$, we let $f_*\langle \underline{n}\rangle$ and $(gf)_*\langle \underline{n}\rangle$ both be the constant tuple $\langle \underline{1}\rangle$ of length p.

$$\begin{array}{ccc}
\mathsf{C}^{\mathcal{G}}(f_*\langle \underline{n}\rangle) & \xrightarrow{\mathsf{C}^{\mathcal{G}}(1,\langle\psi\rangle)} & \mathsf{C}^{\mathcal{G}}\langle \underline{m}\rangle \\
{\scriptstyle \mathsf{C}^{\mathcal{G}}(g,1)}\downarrow & \searrow{\scriptstyle \mathsf{C}^{\mathcal{G}}(g,g_*\langle\psi\rangle)} & \downarrow{\scriptstyle \mathsf{C}^{\mathcal{G}}(g,1)} \\
\mathsf{C}^{\mathcal{G}}((gf)_*\langle \underline{n}\rangle) & \xrightarrow{\mathsf{C}^{\mathcal{G}}(1,g_*\langle\psi\rangle)} & \mathsf{C}^{\mathcal{G}}(g_*\langle \underline{m}\rangle)
\end{array}$$

Then functoriality follows by checking for composites of morphisms $(f, 1)$ with $(g, 1)$ and $(1, \langle\psi'\rangle)$ with $(1, \langle\phi\rangle)$. □

Definition 10.4.20. Suppose C is a small permutative category. We call

- $\mathsf{C}^{\mathcal{G}}_{\mathrm{lax}}$ the *lax Elmendorf-Mandell* $\mathcal{G}_*$*-category* of C,
- $\mathsf{C}^{\mathcal{G}}_{\cong}$ the *strong Elmendorf-Mandell* $\mathcal{G}_*$*-category* of C, and
- $\mathsf{C}^{\mathcal{G}}_{\mathrm{co}}$ the *colax Elmendorf-Mandell* $\mathcal{G}_*$*-category* of C.

When used without an adjective, the phrase *Elmendorf-Mandell* $\mathcal{G}_*$*-category* will refer to the lax case. ◇

In Section 10.7 we give a direct comparison of these three variants of $\mathsf{C}^{\mathcal{G}}$ and show that the respective K-theory symmetric spectra are level equivalent.

10.5. An Equivalent Description of Elmendorf-Mandell J-theory

Now we relate the Elmendorf-Mandell $\mathcal{G}_*$-categories with the J-theory construction J^{EM} via partition multicategories in Definition 10.3.25. This result is a $\mathcal{G}_*$ version of Proposition 8.4.8 for J^{Se} and $\mathsf{C}^{\mathcal{F}}_{\mathrm{lax}}$.

Proposition 10.5.1. *Suppose* C *is a small permutative category. For each* $\langle \underline{n}\rangle \in \mathcal{G}$ *there is an isomorphism of pointed categories*

$$\mathsf{J}^{\mathsf{EM}}\mathsf{C}\langle \underline{n}\rangle = \mathsf{Multicat}_*(\mathcal{T}\langle \underline{n}\rangle, \mathsf{End}(\mathsf{C})) \cong \mathsf{C}^{\mathcal{G}}\langle \underline{n}\rangle.$$

These isomorphisms are the components of a pointed natural isomorphism

$$\mathsf{J}^{\mathsf{EM}}\mathsf{C} \cong \mathsf{C}^{\mathcal{G}}.$$

Proof. The first assertion follows, as in the proof of Proposition 8.4.8, from an explicit description of the left hand side. Recall from Definition 10.3.1 and Explanation 10.3.2 that by taking the smash product in $\mathsf{Mod}^{\mathcal{M}\underline{1}}$ we have

$$\mathcal{T}\langle \underline{n} \rangle = \wedge_k \mathcal{M}\underline{n}_k \tag{10.5.2}$$

for each $\langle \underline{n} \rangle$ in $\mathcal{G}$ with length $q \geq 0$. For $q = 0$, the empty smash product (10.5.2) is $\mathcal{M}\underline{1}$. In this case we have

$$(\mathsf{J}^{\mathsf{EM}}\mathsf{C})\langle\rangle = \mathsf{Multicat}_*(\mathcal{M}\underline{1}, \mathsf{End}(\mathsf{C})) \cong \mathsf{C} = \mathsf{C}^{\mathcal{G}}\langle\rangle$$

by Lemma 10.2.5. Recalling the proof, each pointed multifunctor on the left is determined by where it sends the object $\{1\}$ in $\mathcal{M}\underline{1}$.

For $\langle \underline{n} \rangle$ of length $q > 0$, we introduce the following notation extending that of Definition 6.5.3. Given a partition $\langle s_k \rangle = (s_{k,1}, \ldots, s_{k,q_k})$ of $s_k \subset \underline{n}_k^\flat$ for some k, and subsets $s_\ell \subset \underline{n}_\ell^\flat$ for $\ell \neq k$, let

$$\langle s \circ_k \langle s_k \rangle \rangle \in \mathsf{Prof}(\mathcal{T}\langle \underline{n} \rangle)$$

be the profile of length q_k whose jth entry is $\langle s \circ_k s_{k,j} \rangle$. Let

$$\langle s \circ_k \iota_{\langle s_k \rangle} \rangle \in \mathcal{T}\langle \underline{n} \rangle \big(\langle s \circ_k \langle s_k \rangle \rangle \,;\, \langle s \rangle \big)$$

be the tuple of operations whose ℓth entry is the unit 1_{s_ℓ} for $\ell \neq k$, and whose kth entry is $\iota_{\langle s_k \rangle}$. Recalling Explanation 5.6.14, we note that these operations generate the operations of the tensor product $\otimes_k \mathcal{M}\underline{n}_k$ and hence also of the smash product $\mathcal{T}\langle \underline{n} \rangle$.

Moreover, recalling the argument of Proposition 8.4.8 that the operations of each $\mathcal{M}\underline{n}_k$ are generated by 2-ary operations $\iota_{(t,u)}$, we see that the operations of $\mathcal{T}\langle \underline{n} \rangle$ are generated by $\langle s \circ_k \iota_{(t,u)} \rangle$. Therefore, a pointed multifunctor

$$F \colon \bigwedge_k \mathcal{M}\underline{n}_k \longrightarrow \mathsf{End}(\mathsf{C})$$

determines and is uniquely determined by the following data:

- a family of objects $C_{\langle s \rangle} = F\langle s \rangle \in \mathsf{C}$ for tuples of subsets $s_k \in \underline{n}_k^\flat$, and
- a family of morphisms

$$\rho_{\langle s \rangle; k,t,u} = F\langle s \circ_k \iota_{(t,u)} \rangle \colon C_{\langle s \circ_k t \rangle} \oplus C_{\langle s \circ_k u \rangle} \longrightarrow C_{\langle s \rangle}.$$

These data are subject to relations for symmetry, compatibility, and interchange detailed in Explanations 5.6.9 and 5.6.14, along with relations for the smash product, Definition 5.6.20. One can verify that these correspond to the axioms of Definition 10.4.1 just as was done in the proof of Proposition 8.4.8.

Similarly, a pointed multinatural transformation $\alpha \colon F \longrightarrow F'$ determines and is uniquely determined by a family of morphisms in C,

$$\alpha_{\langle s \rangle} \colon C_{\langle s \rangle} \longrightarrow C'_{\langle s \rangle},$$

satisfying basepoint and naturality axioms corresponding to the axioms for a morphism in $\mathsf{C}^{\mathcal{G}}\langle \underline{n} \rangle$.

Now we verify naturality of the isomorphisms

$$\mathsf{J}^{\mathsf{EM}}\mathsf{C}\langle \underline{n}\rangle \longrightarrow \mathsf{C}^{\mathcal{G}}\langle \underline{n}\rangle$$

with respect to morphisms $(f, \langle\psi\rangle)$ in $\mathcal{G}$. For each $p > 0$, naturality for the morphism

$$(i_p, 1)\colon \langle\rangle \longrightarrow \underline{1}$$

to the constant p-tuple at $\underline{1}$ follows by comparison of Explanation 10.3.6 with Example 10.4.14. Both composites of the naturality square

$$\begin{array}{ccc} \mathsf{Multicat}_*(\mathcal{M}\underline{1}, \mathsf{End}(\mathsf{C})) & \xrightarrow{\quad\cong\quad} & \mathsf{C}^{\mathcal{G}}\langle\rangle \\ {\scriptstyle (\mathcal{T}(i_p,0))^*}\downarrow{\scriptstyle\cong} & & {\scriptstyle\cong}\downarrow{\scriptstyle \mathsf{C}^{\mathcal{G}}(i_p,0)} \\ \mathsf{Multicat}_*(\wedge_{k=1}^{p}\mathcal{M}\underline{1}, \mathsf{End}(\mathsf{C})) & \xrightarrow{\quad\cong\quad} & \mathsf{C}^{\mathcal{G}}\langle\underline{1}\rangle \end{array}$$

are given as follows.

- A pointed multifunctor F is sent to the $\underline{1}$-system whose only (possibly) nontrivial object is $F\{1\}$.
- A pointed multinatural transformation θ is sent to the morphism of $\underline{1}$-systems whose only (possibly) nontrivial component is $\theta_{\{1\}}$.

Naturality with respect to all other morphisms $(f, \langle\psi\rangle)$ follows by comparing the definition of $\mathcal{T}$ on morphisms, written in (10.3.5), with that of $\mathsf{C}^{\mathcal{G}}$, written in Definitions 10.4.16 and 10.4.17. The morphism denoted $\langle\tilde{\psi}\rangle$ corresponds to $\wedge_j \tilde{\psi}_j$ and the morphism denoted $\tilde{f}$ corresponds to the composite given by permutation and the isomorphisms $\prod_{\underline{1},\underline{m}}$ and $\prod_{\underline{m},\underline{1}}$. □

It follows from Definition 10.3.32 and Proposition 10.5.1 that there is a natural isomorphism of symmetric spectra

$$\mathsf{K}^{\mathsf{EM}}\mathsf{C} = \mathsf{K}^{\mathcal{G}} N_* \mathsf{J}^{\mathsf{EM}}\mathsf{C} \cong \mathsf{K}^{\mathcal{G}} N_* \mathsf{C}^{\mathcal{G}}_{\mathrm{lax}}$$

with $\mathsf{C}^{\mathcal{G}}_{\mathrm{lax}} = \mathsf{C}^{\mathcal{G}}$, for small permutative categories C. Using the other two variants of Elmendorf-Mandell $\mathcal{G}_*$-categories in Definition 10.4.20, $\mathsf{C}^{\mathcal{G}}_{\cong}$ and $\mathsf{C}^{\mathcal{G}}_{\mathrm{co}}$, we define the following variants of Elmendorf-Mandell K-theory.

Definition 10.5.3. Suppose C is a small permutative category. We define the symmetric spectra

$$\begin{aligned} \mathsf{K}^{\mathsf{EM}}_{\cong}\mathsf{C} &= \mathsf{K}^{\mathcal{G}} N_* \mathsf{C}^{\mathcal{G}}_{\cong} \quad \text{and} \\ \mathsf{K}^{\mathsf{EM}}_{\mathrm{co}}\mathsf{C} &= \mathsf{K}^{\mathcal{G}} N_* \mathsf{C}^{\mathcal{G}}_{\mathrm{co}}, \end{aligned}$$

which are called, respectively, the *strong Elmendorf-Mandell K-theory* and the *colax Elmendorf-Mandell K-theory* of C. ◇

In Theorems 10.7.16 and 10.7.19, we will show that the three variants of Elmendorf-Mandell K-theory—namely, $\mathsf{K}^{\mathsf{EM}}\mathsf{C}$, $\mathsf{K}^{\mathsf{EM}}_{\cong}\mathsf{C}$, and $\mathsf{K}^{\mathsf{EM}}_{\mathrm{co}}\mathsf{C}$ in Definitions 10.3.32 and 10.5.3—are level equivalent symmetric spectra.

Explanation 10.5.4 (Multifunctoriality of $(-)^{\mathcal{G}}$)**.** By Proposition 10.5.1, the object assignment $\mathsf{C} \longmapsto \mathsf{C}^{\mathcal{G}}$ is part of a Cat-enriched multifunctor

$$\mathsf{PermCat}^{\mathsf{su}} \xrightarrow{\;(-)^{\mathcal{G}}\;} \mathcal{G}_*\text{-}\mathsf{Cat}.$$

Here we describe in detail this Cat-enriched multifunctor. With the exception of the case $n = 0$, the description below is essentially the proof of [**EM06**, Theorem 6.1]. First recall from

- Definitions 6.5.1, 6.5.4, and 6.5.11 that $\mathsf{PermCat}^{\mathsf{su}}$ is the Cat-enriched multicategory with small permutative categories, multilinear functors, and multilinear transformations, and
- Definition 9.2.8 and Theorem 9.2.15 the symmetric monoidal closed category $\mathcal{G}_*$-Cat with the Day convolution product $\wedge$ (9.2.9) and hom diagram (9.2.10), and with the permutative structure on $\mathcal{G}$ given by concatenation of tuples of pointed finite sets.

For small permutative categories $\mathsf{C}_1, \ldots, \mathsf{C}_n$, and D, the functor

$$\mathsf{PermCat}^{\mathsf{su}}\big(\mathsf{C}_1, \ldots, \mathsf{C}_n\,;\,\mathsf{D}\big) \xrightarrow{(-)^{\mathcal{G}}} \mathcal{G}_*\text{-}\mathsf{Cat}\big(\mathsf{C}_1^{\mathcal{G}}, \ldots, \mathsf{C}_n^{\mathcal{G}}\,;\,\mathsf{D}^{\mathcal{G}}\big) \tag{10.5.5}$$

is described as follows.

For $n = 0$, a 0-linear functor

$$\mathbb{1} \xrightarrow{F} \mathsf{D}$$

in $\mathsf{PermCat}^{\mathsf{su}}$ is uniquely determined by a choice of object $x \in \mathsf{D}$. By Lemma 10.2.5, such an object x determines a unique pointed multifunctor

$$U_x\colon \mathcal{M}\underline{1} \longrightarrow \mathsf{End}(\mathsf{D}) \quad \text{with} \quad U_x\{1\} = x. \tag{10.5.6}$$

With J the monoidal unit of $\mathcal{G}_*$-Cat, the image of F in (10.5.5) is given by the composite
(10.5.7)

$$J \xrightarrow{(\mathsf{J}^{\mathcal{T}})^0} \mathsf{J}^{\mathcal{T}}(\mathcal{M}\underline{1}) = \mathsf{Multicat}_*(\mathcal{T}(-), \mathcal{M}\underline{1}) \xrightarrow{(U_x)_*} \mathsf{Multicat}_*(\mathcal{T}(-), \mathsf{End}(\mathsf{D})) \cong \mathsf{D}^{\mathcal{G}}.$$

More explicitly, recall from (9.2.13) that

$$J\langle \underline{m} \rangle = \widehat{\mathcal{G}}(\langle\rangle, \langle \underline{m} \rangle)$$

for each object $\langle \underline{m} \rangle \in \mathcal{G}$. If $\langle \underline{m} \rangle$ has length 0, then the only nonzero morphism $\langle\rangle \longrightarrow \langle\rangle$ in $\mathcal{G}$ is the identity. The composite (10.5.7) sends $1_{\langle\rangle}$ to the object

$$x = F(*) \in \mathsf{D} = \mathsf{D}^{\mathcal{G}}\langle\rangle.$$

If $\langle \underline{m} \rangle$ has length $p > 0$, then, by Explanation 9.1.9 (4), a morphism

$$(i_p, \langle \psi \rangle)\colon \langle\rangle \longrightarrow \langle \underline{m} \rangle \in \mathcal{G}$$

is uniquely determined by the pointed functions

$$\psi_i\colon \underline{1} \longrightarrow \underline{m_i} \quad \text{for} \quad i \in \overline{p}.$$

The composite (10.5.7) sends $(i_p, \langle \psi \rangle)$ to the object $X \in \mathsf{D}^{\mathcal{G}}\langle \underline{m} \rangle$ given as follows. For a p-tuple $\langle s \rangle$ of unpointed finite sets with each $s_i \subset \underline{m_i}^{\flat}$ for $i \in \overline{p}$, the $\langle s \rangle$-component of X is given by

$$X_{\langle s \rangle} = \begin{cases} x & \text{if } \psi_i(1) \in s_i \text{ for each } i \in \overline{p}, \text{ and} \\ e & \text{otherwise.} \end{cases}$$

For the gluing morphisms, suppose $k \in \overline{p}$ and (t, u) is a partition of s_k. Then the (t, u)-gluing morphism

$$X_{\langle s \circ_k t \rangle} \oplus X_{\langle s \circ_k u \rangle} \xrightarrow{\rho_{\langle s \rangle; k, t, u}} X_{\langle s \rangle}$$

is given by

- 1_x if $\psi_i(1) \in s_i$ for each $i \in \overline{p}$, and
- 1_e otherwise.

For $n > 0$, (10.5.5) sends an n-linear functor

$$\mathsf{C}_1 \times \cdots \times \mathsf{C}_n \xrightarrow{(F, \{F_j^2\}_{j=1}^n)} \mathsf{D} \tag{10.5.8}$$

to a $\mathcal{G}_*$-category morphism

$$\bigwedge_{j=1}^{n} \mathsf{C}_j^{\mathcal{G}} = \mathsf{C}_1^{\mathcal{G}} \wedge \cdots \wedge \mathsf{C}_n^{\mathcal{G}} \xrightarrow{F^{\mathcal{G}}} \mathsf{D}^{\mathcal{G}}. \tag{10.5.9}$$

The iterated product on the left-hand side and below are assumed to be left normalized (2.5.4). For $p \geq 0$ and each p-tuple of pointed finite sets

$$\langle \underline{m} \rangle = \big(\underline{m_1}, \ldots, \underline{m_p}\big) \in \mathcal{F}^{(p)},$$

the $\mathcal{G}_*$-category morphism $F^{\mathcal{G}}$ has a component functor

$$\Big(\bigwedge_{j=1}^{n} \mathsf{C}_j^{\mathcal{G}}\Big)\langle \underline{m} \rangle \xrightarrow{F^{\mathcal{G}}\langle \underline{m} \rangle} \mathsf{D}^{\mathcal{G}}\langle \underline{m} \rangle.$$

Since $\bigwedge_{j=1}^{n}$ is an iterated Day convolution product of $\mathcal{G}_*$-categories, the functor $F^{\mathcal{G}}\langle \underline{m} \rangle$ is uniquely determined by component functors

$$\bigwedge_{j=1}^{n} \mathsf{C}_j^{\mathcal{G}}\big(\underline{m_{i_{j-1}+1}}, \ldots, \underline{m_{i_j}}\big) \xrightarrow{F^{\mathcal{G}}\langle \underline{m} \rangle} \mathsf{D}^{\mathcal{G}}\langle \underline{m} \rangle \tag{10.5.10}$$

for sequences $(i_0 = 0, i_1, \ldots, i_n)$ with

$$1 \leq i_1 \leq \cdots \leq i_n = p.$$

The smash product of small categories in (10.5.10) is the one in (4.1.7) for $(\mathsf{Cat}, \times, *)$. Now we address two special cases.

- If $p = 0$, then $\langle \underline{m} \rangle$ is the empty tuple. In (10.5.10), each $\mathsf{C}_j^{\mathcal{G}}\langle\rangle = \mathsf{C}_j$ and $\mathsf{D}^{\mathcal{G}}\langle\rangle = \mathsf{D}$ by definition. The component functor $F^{\mathcal{G}}\langle\rangle$ is given by the given functor F in (10.5.8). It is well defined on the smash product in (10.5.10) by the unity axiom of an n-linear functor.
- If $p > 0$ with some $m_i = 0$, then $\langle \underline{m} \rangle = *$ in $\mathcal{G}$, and $\mathsf{D}^{\mathcal{G}}(*) = \mathbb{1}$, the terminal category. The component functor $F^{\mathcal{G}}(*)$ in (10.5.10) is the unique functor to the terminal category.

Now suppose $p > 0$ and each $m_j > 0$. To describe the functor $F^{\mathcal{G}}\langle \underline{m} \rangle$ in (10.5.10), first note that an object in the domain in (10.5.10) is represented by an n-tuple of objects

$$C = \big((C^1, \rho^1), \ldots, (C^n, \rho^n)\big) \in \prod_{j=1}^{n} \mathsf{C}_j^{\mathcal{G}}\big(\underline{m_{i_{j-1}+1}}, \ldots, \underline{m_{i_j}}\big). \tag{10.5.11}$$

For a p-tuple of pointed finite sets

$$\langle s \rangle = \big(s_1, \ldots, s_p\big) \quad \text{with each} \quad s_i \subset \underline{m_i}^{\flat}, \tag{10.5.12}$$

the object

$$(F^{\mathcal{G}}\langle \underline{m} \rangle)C \in \mathsf{D}^{\mathcal{G}}\langle \underline{m} \rangle$$

has as its $\langle s\rangle$-component the object

$$\big((F^{\mathcal{G}}\langle \underline{m}\rangle)C\big)_{\langle s\rangle} = F\Big\{C^j_{(s_{i_{j-1}+1},\ldots,s_{i_j})}\Big\}_{j=1}^{n} \in \mathsf{D}. \tag{10.5.13}$$

For a partition

$$t \cup u = s_k \quad \text{for some} \quad i_{r-1}+1 \le k \le i_r \quad \text{and} \quad 1 \le r \le n, \tag{10.5.14}$$

the gluing morphism ρ of $(F^{\mathcal{G}}\langle \underline{m}\rangle)C$ is the following composite in D.

$$\begin{array}{ccc}
\big((F^{\mathcal{G}}\langle \underline{m}\rangle)C\big)_{\langle s\circ_k t\rangle} \oplus \big((F^{\mathcal{G}}\langle \underline{m}\rangle)C\big)_{\langle s\circ_k u\rangle} & \xrightarrow{\rho_{\langle s\rangle;k,t,u}} & \big((F^{\mathcal{G}}\langle \underline{m}\rangle)C\big)_{\langle s\rangle} \\
\| & & \| \\
F\Big(\ldots, C^r_{(s_{i_{r-1}+1},\ldots,t,\ldots,s_{i_r})},\ldots\Big) \oplus F\Big(\ldots, C^r_{(s_{i_{r-1}+1},\ldots,u,\ldots,s_{i_r})},\ldots\Big) & & F\Big\{C^j_{(s_{i_{j-1}+1},\ldots,s_{i_j})}\Big\}_{j=1}^{n} \\
\downarrow F^2_r & \nearrow F(1,\ldots,\rho^r,\ldots,1) & \\
F\Big(\ldots, C^r_{(s_{i_{r-1}+1},\ldots,t,\ldots,s_{i_r})} \oplus C^r_{(s_{i_{r-1}+1},\ldots,u,\ldots,s_{i_r})},\ldots\Big) & &
\end{array} \tag{10.5.15}$$

In the above diagram, F^2_r is the rth linearity constraint of the n-linear functor F, and ρ^r is the gluing morphism in the object (C^r,ρ^r) for the indicated sequences of finite sets and partition (10.5.14).

A morphism $\alpha: C \longrightarrow D$ in the domain in (10.5.10) is represented by an n-tuple of morphisms

$$\Big\{(C^j,\rho^j_C) \xrightarrow{\alpha^j} (D^j,\rho^j_D)\Big\}_{j=1}^{n} \in \prod_{j=1}^{n} \mathsf{C}^{\mathcal{G}}_j\big(\underline{m_{i_{j-1}+1}},\ldots,\underline{m_{i_j}}\big).$$

With $\langle s\rangle$ as in (10.5.12), the morphism

$$(F^{\mathcal{G}}\langle \underline{m}\rangle)C \xrightarrow{(F^{\mathcal{G}}\langle \underline{m}\rangle)\alpha} (F^{\mathcal{G}}\langle \underline{m}\rangle)D \in \mathsf{D}^{\mathcal{G}}\langle \underline{m}\rangle$$

has as its $\langle s\rangle$-component the morphism

$$\big((F^{\mathcal{G}}\langle \underline{m}\rangle)\alpha\big)_{\langle s\rangle} = F\Big\{\alpha^j_{(s_{i_{j-1}+1},\ldots,s_{i_j})}\Big\}_{j=1}^{n} \in \mathsf{D}. \tag{10.5.16}$$

This finishes the description of the $\mathcal{G}_*$-category morphism $F^{\mathcal{G}}$ in (10.5.9).

Suppose θ is the following n-linear transformation between n-linear functors.

$$\mathsf{C}_1 \times \cdots \times \mathsf{C}_n \underset{G}{\overset{F}{\rightrightarrows}} \mathsf{D} \qquad \Downarrow\theta \tag{10.5.17}$$

For $n = 0$, θ is uniquely determined by a morphism $f: x \longrightarrow y$, where x and y are the objects determined by F and G. By Lemma 10.2.5, f determines a pointed multinatural transformation

$$U_f: U_x \longrightarrow U_y: \mathcal{M}\underline{1} \longrightarrow \mathsf{End}(\mathsf{D}) \quad \text{with} \quad (U_f)_{\{1\}} = f,$$

where U_x and U_y are as in (10.5.6). This induces a modification

$$(U_f)_*: (U_x)_* \longrightarrow (U_y)_*: \mathsf{J}^{\mathcal{T}}(\mathcal{M}\underline{1}) \longrightarrow \mathsf{Multicat}_*(\mathcal{T}(-),\mathsf{End}(\mathsf{D})) \cong \mathsf{D}^{\mathcal{G}}.$$

Pre-whiskering $(U_f)_*$ with $(\mathsf{J}^{\mathcal{T}})^0$ gives the image of θ in (10.5.5).

Now suppose $n > 0$ in (10.5.17). With C as in (10.5.11) and $\langle s\rangle$ as in (10.5.12), $(\theta^g\langle \underline{m}\rangle)_C$ has as its $\langle s\rangle$-component the morphism

$$\begin{array}{c} ((F^g\langle \underline{m}\rangle)C)_{\langle s\rangle} = F\Big\{C^j_{(s_{i_{j-1}+1},\ldots,s_{i_j})}\Big\}_{j=1}^n \\ \Big\downarrow \theta_{\{C^j_{(s_{i_{j-1}+1},\ldots,s_{i_j})}\}_{j=1}^n} \\ ((G^g\langle \underline{m}\rangle)C)_{\langle s\rangle} = G\Big\{C^j_{(s_{i_{j-1}+1},\ldots,s_{i_j})}\Big\}_{j=1}^n \end{array}$$

in D. This finishes the description of the functor $(-)^g$ in (10.5.5). ◇

10.6. Equivalence Between Segal K-Theory and Elmendorf-Mandell K-Theory

This section is devoted to proving Theorem 10.6.10, which asserts that the symmetric spectra $\mathsf{K}^{\mathsf{Se}}\mathsf{C}$ and $\mathsf{K}^{\mathsf{EM}}\mathsf{C}$ are naturally level equivalent. Our comparison begins with the observation that, for a tuple of pointed finite sets $\langle a\rangle$, we can also form the smash product

$$\wedge_k a_k \cong (\textstyle\prod_k a_k^\flat)_+$$

and apply $\mathcal{M}$. This is the composite of $\mathcal{M}$ with (the opposite of) the functor

$$\wedge : \mathcal{G} \longrightarrow \mathcal{F}$$

discussed in Definition 9.1.15.

The multicategory $\mathcal{T}\langle a\rangle$ relates to $\mathcal{M}(\wedge_k a_k)$ as follows. The following extends the partition product of Definition 10.1.1 to q-fold products for $q > 2$.

Definition 10.6.1. Suppose $\langle a\rangle = (a_1, \ldots, a_q)$ is a q-tuple of pointed finite sets, and $\wedge_k a_k$ is their smash product. For $q > 0$, define

$$\textstyle\prod : \mathcal{T}\langle a\rangle \longrightarrow \mathcal{M}(\wedge_k a_k)$$

by

$$\langle s\rangle \longmapsto \prod_k s_k \subset \prod_k a_k^\flat \subset \wedge_k a_k$$

where $\langle s\rangle$ is a q-tuple of basepoint-free subsets $s_k \subset a_k^\flat$. For $q = 0$, we take $\underline{1}$ as the empty smash product in FinSet_* and define

$$\textstyle\prod = 1_{\mathcal{M}\underline{1}} : \mathcal{M}\underline{1} = \mathcal{T}\langle\rangle \longrightarrow \mathcal{M}\underline{1}.$$

We show that $\prod$ is a well-defined pointed multifunctor in Proposition 10.6.2. ◇

The next result follows by induction on the binary case of Definition 10.1.1, but we also give a detailed direct proof.

Proposition 10.6.2. *In the context of Definition 10.6.1, suppose given a q-tuple of pointed finite sets $\langle a\rangle$ for $q \geq 0$. Then*

$$\textstyle\prod : \mathcal{T}\langle a\rangle \longrightarrow \mathcal{M}(\wedge_k a_k)$$

is a well-defined pointed multifunctor.

Proof. For $q = 0$, $\prod$ is the identity and there is nothing to prove. For $q > 0$, the essential observation that makes $\prod$ well defined is that taking products preserves disjunction. To explain more clearly, recall the following notation from Definition 9.1.4 and the proof of Proposition 10.5.1. Given a partition

$$\langle s_k \rangle = (s_{k,1}, \ldots, s_{k,q_k})$$

of $s_k \subset a_k^\flat$ and subsets $s_\ell \subset a_\ell^\flat$ for $\ell \neq k$,

$$\langle s \circ_k s_{k,j} \rangle \in \mathcal{T}\langle a \rangle$$

denotes the tuple whose ℓth entry is s_ℓ for $\ell \neq k$, and whose kth entry is $s_{k,j}$. Then

$$\langle s \circ_k \langle s_k \rangle \rangle \in \mathsf{Prof}(\mathcal{T}\langle a \rangle)$$

is the profile of length q_k whose jth entry is $\langle s \circ_k s_{k,j} \rangle$. Finally,

$$\langle s \circ_k \iota_{\langle s_k \rangle} \rangle \in \mathcal{T}\big(\langle s \circ_k \langle s_k \rangle \rangle ; \langle s \rangle\big)$$

is the tuple of operations whose ℓth entry is the unit 1_{s_ℓ} for $\ell \neq k$, and whose kth entry is $\iota_{\langle s_k \rangle}$. Recalling Explanation 5.6.14, we note that these operations generate the operations of the smash product $\mathcal{T}\langle a \rangle$.

With this notation, the key observation is the following. Since the $s_{k,j}$ are disjoint for $j \in \{1, \ldots, q_k\}$, the products

$$\textstyle\prod \langle s \circ_k s_{k,j} \rangle = \displaystyle\prod_{u \in \langle s \circ_k s_{k,j} \rangle} u \quad \text{and} \quad \textstyle\prod \langle s \circ_k s_{k,i} \rangle = \displaystyle\prod_{u \in \langle s \circ_k s_{k,i} \rangle} u$$

are disjoint for $j \neq i$. Therefore, $\langle \prod \langle s \circ_k s_{k,j} \rangle \rangle_{j=1}^{q_k}$ is a partition of $\prod \langle s \rangle$ and we define

$$\textstyle\prod \langle s \circ_k \iota_{\langle s_k \rangle} \rangle = \iota_{\langle \prod \langle s \circ_k s_{k,j} \rangle \rangle_{j=1}^{q_k}}.$$

This defines $\prod$ as an assignment on objects and operations. Since there is at most one operation in $\mathcal{M}(\wedge_k a_k)$ with each given input profile and output, commutativity of the multifunctor axioms is automatic. □

Next we have a generalization of the naturality Lemma 10.1.4. Here too the result follows by induction on the binary case, but we also give a detailed direct proof.

Proposition 10.6.3. *The pointed multifunctor* $\prod$ *is natural with respect to the morphisms of* $\mathcal{G}^{\mathsf{op}}$ *and defines a natural transformation of functors*

$$\textstyle\prod : \mathcal{T} \longrightarrow (\mathcal{M} \circ \wedge).$$

Proof. Recalling (9.1.16) and (10.3.5), naturality with respect to a morphism

$$(f, \langle \psi \rangle) : \langle \underline{n} \rangle \longrightarrow \langle \underline{m} \rangle,$$

is verified by commutativity of the outer rectangle in the following diagram. In the upper row we take smash products of left $\mathcal{M}\underline{1}$-modules so that empty smash products are $\mathcal{M}\underline{1}$. In the lower row empty smash products are $\underline{1}$.

$$\begin{array}{ccccccc}
\bigwedge\limits_j \mathcal{M}\underline{m}_j & \xrightarrow{\wedge_j \tilde{\psi}_j} & \bigwedge\limits_j \mathcal{M}(\underline{n}_{f^{-1}(j)}) & \xrightarrow{\cong} & \bigwedge\limits_{f^{-1}(j) \neq \varnothing} \mathcal{M}(\underline{n}_{f^{-1}(j)}) & \xrightarrow[\text{permute}]{\cong} & \bigwedge\limits_k \mathcal{M}\underline{n}_k \\
\Big\downarrow \textstyle\prod & & \Big\downarrow \textstyle\prod & & \Big\downarrow \textstyle\prod & & \Big\downarrow \textstyle\prod \\
\mathcal{M}\big[\bigwedge\limits_j \underline{m}_j\big] & \xrightarrow{\widetilde{\wedge_j \psi_j}} & \mathcal{M}\big[\bigwedge\limits_j (\underline{n}_{f^{-1}(j)})\big] & \xrightarrow{\cong} & \mathcal{M}\Big[\bigwedge\limits_{f^{-1}(j) \neq \varnothing} (\underline{n}_{f^{-1}(j)})\Big] & \xrightarrow[\text{permute}]{\cong} & \mathcal{M}\big[\bigwedge\limits_k \underline{n}_k\big]
\end{array}$$

In the diagram above, the leftmost square commutes because $\wedge_j \psi_j$ is defined componentwise and hence, as a function on tuples of subsets $\langle t \rangle$ with $t_j \subset \underline{m}_j^\flat$, we have

$$\wedge_j(\psi_j^{-1}(t_j)) = (\wedge_j \psi_j)^{-1}(\wedge_j t_j).$$

The two composites in the middle square are equal because both are given by

$$\langle s \rangle \longmapsto \bigwedge_{f^{-1}(j) \neq \varnothing} s_{f^{-1}(j)}$$

for $s_j \subset \underline{n}_j^\flat$. The two composites in the rightmost square are equal because both are given by permuting factors of a smash product according to the action of f^{-1} on its indexing set. □

There are projections

$$\wedge_k a_k \longrightarrow a_k$$

for each k, but these do *not* give well-defined multifunctors

$$\mathcal{M}(\wedge_k a_k) \longrightarrow \mathcal{M}a_k.$$

Disjoint subsets of a product may project to nondisjoint subsets. However, there is an intermediary given as follows. Recall from Example 5.1.9 we let M^u denote the unary multicategory associated to a category M.

Definition 10.6.4. Suppose $\langle a \rangle = (a_1, \ldots, a_q)$ is a tuple of pointed finite sets for $q > 0$. Let

$$\mathcal{P}\langle a \rangle = \Big(\big(\prod_k a_k^\flat \big)^u \Big)_+$$

denote the pointed multicategory obtained by adjoining a disjoint basepoint T to the discrete unary multicategory whose objects are the elements of the product.

Note that the objects of $\mathcal{P}\langle a \rangle$ correspond to the elements of $\wedge_k a_k$, but $\mathcal{P}\langle a \rangle$ differs from the discrete unary multicategory on $\wedge_k a_k$ because a discrete unary multicateory has no non-identity operations and therefore cannot be a pointed multicategory. ◇

Remark 10.6.5. Note that $\mathcal{P}$ is *not* defined on morphisms of $\mathcal{G}$. For a morphism $\psi: \underline{n} \longrightarrow \underline{m}$ in $\mathcal{F}$, the inverse image of an element $x \in \underline{m}$ is generally not a singleton subset of $\underline{n}$. Therefore, the definitions of $\mathcal{M}$ or $\mathcal{T}$ on morphisms of $\mathcal{F}$ or $\mathcal{G}$, respectively, do not specialize to give well-defined morphisms $\mathcal{P}\langle \underline{m} \rangle \longrightarrow \mathcal{P}\langle \underline{n} \rangle$. ◇

Definition 10.6.6. Suppose $\langle a \rangle$ is a q-tuple of pointed finite sets for $q > 0$. We define two pointed multifunctors

$$\mathcal{T}\langle a \rangle \xleftarrow{\ i\ } \mathcal{P}\langle a \rangle \xrightarrow{\ j\ } \mathcal{M}(\wedge_k a_k)$$

by taking elements of $\prod_k a_k^\flat$ to singletons in two different ways. For each tuple of elements $\langle x \rangle = (x_1, \ldots, x_q) \in \prod_k a_k^\flat$, we define

$$\begin{aligned} i\langle x \rangle &= (\{x_1\}, \ldots, \{x_q\}) \quad \text{and} \\ j\langle x \rangle &= \{(x_1, \ldots, x_q)\}. \end{aligned}$$

So i sends $\langle x \rangle$ to the tuple of the singletons and j sends $\langle x \rangle$ to the singleton of the tuple. ◇

Proposition 10.6.7. *Suppose* $\mathsf{C} = (\mathsf{C}, \oplus, e)$ *is a small permutative category. Then for each tuple of pointed finite sets* $\langle a \rangle = (a_1, \ldots, a_q)$ *with* $q > 0$*, the pointed multifunctors* j *and* i *induce adjunctions of pointed categories*

$$\mathsf{Multicat}_*(\mathcal{T}\langle a \rangle, \mathsf{End}(\mathsf{C})) \xrightarrow{i^*} \mathsf{Multicat}_*(\mathcal{P}\langle a \rangle, \mathsf{End}(\mathsf{C}))$$

and

$$\mathsf{Multicat}_*(\mathcal{M}(\wedge_k a_k), \mathsf{End}(\mathsf{C})) \xrightarrow{j^*} \mathsf{Multicat}_*(\mathcal{P}\langle a \rangle, \mathsf{End}(\mathsf{C})).$$

Proof. We begin with the assertion about j^*. A pointed multifunctor from $\mathcal{P}\langle a \rangle$ to $\mathsf{End}(\mathsf{C})$ necessarily sends the non-unary operations of the basepoint T to those of $e \in \mathsf{C}$. Therefore,

$$\mathsf{Multicat}_*(\mathcal{P}\langle a \rangle, \mathsf{End}(\mathsf{C})) = \mathsf{Cat}_*(a, \mathsf{C}),$$

where the basepoint of C is e and

$$a = \Big(\prod_k a_k^\flat\Big)_+ \cong \bigwedge_k a_k. \tag{10.6.8}$$

By applying Proposition 8.5.4 to a and using the isomorphism (10.6.8), we see that j^* is a strictly unital symmetric monoidal right adjoint in $\mathsf{PermCat}^{\mathsf{su}}$.

For the assertion about i^*, we apply Proposition 8.5.4 to each individual pointed finite set a_k to obtain strictly unital symmetric monoidal adjunctions

$$\mathsf{Multicat}_*(\mathcal{M}a_k, \mathsf{End}(\mathsf{C})) \longrightarrow \mathsf{Cat}_*(a_k, \mathsf{C}) = \mathsf{Multicat}_*(\mathcal{P}a_k, \mathsf{End}(\mathsf{C})). \tag{10.6.9}$$

By Corollary 5.3.9, End is 2-functorial with respect to strictly unital symmetric monoidal functors. Applying End to both sides of (10.6.9) and using Lemma 5.7.21 we have, therefore, an internal right adjoint in $\mathsf{Multicat}_*$

$$\mathsf{Hom}_*(\mathcal{M}a_k, \mathsf{End}(\mathsf{C})) \longrightarrow \mathsf{Hom}_*(\mathcal{P}a_k, \mathsf{End}(\mathsf{C}))$$

for each $a_k \in \langle a \rangle$.

Therefore, by the $\wedge$-Hom_* adjunction of Theorem 5.7.22 we have, for example, the following sequence of equivalences and right adjoints in Cat_*:

$$\begin{aligned}
\mathsf{Multicat}_*(\mathcal{M}(a_1) \wedge \mathcal{M}(a_2), \mathsf{End}(\mathsf{C})) &\simeq \mathsf{Multicat}_*(\mathcal{M}(a_1), \mathsf{Hom}_*(\mathcal{M}(a_2), \mathsf{End}(\mathsf{C}))) \\
&\longrightarrow \mathsf{Multicat}_*(\mathcal{M}(a_1), \mathsf{Hom}_*(\mathcal{P}(a_2), \mathsf{End}(\mathsf{C}))) \\
&\simeq \mathsf{Multicat}_*(\mathcal{M}(a_1) \wedge \mathcal{P}(a_2), \mathsf{End}(\mathsf{C})) \\
&\simeq \mathsf{Multicat}_*(\mathcal{P}(a_2), \mathsf{Hom}_*(\mathcal{M}(a_1), \mathsf{End}(\mathsf{C}))) \\
&\longrightarrow \mathsf{Multicat}_*(\mathcal{P}(a_2), \mathsf{Hom}_*(\mathcal{P}(a_1), \mathsf{End}(\mathsf{C}))) \\
&\simeq \mathsf{Multicat}_*(\mathcal{P}(a_1) \wedge \mathcal{P}(a_2), \mathsf{End}(\mathsf{C})).
\end{aligned}$$

Now the assertion for i^* follows from induction and the identification

$$\wedge_k \mathcal{P}(a_k) \cong \mathcal{P}\langle a \rangle. \qquad \square$$

Recall from Definition 9.1.15 the smash product of pointed finite sets defines a pointed functor

$$\wedge : \mathcal{G} \longrightarrow \mathcal{F}.$$

Composition then provides a change-of-diagram functor

$$\Gamma\text{-}\mathsf{Cat} \xrightarrow{- \circ \wedge} \mathcal{G}_*\text{-}\mathsf{Cat}$$

denoted $\wedge^*$ below.

Theorem 10.6.10. *Suppose* C *is a small permutative category. The product*

$$\prod : \mathcal{T} \longrightarrow \mathcal{M}(\wedge(-))$$

of Proposition 10.6.2 induces a level equivalence of symmetric spectra

$$\mathsf{K}^{\mathsf{Se}}\mathsf{C} \longrightarrow \mathsf{K}^{\mathsf{EM}}\mathsf{C}$$

that is natural with respect to strictly unital symmetric monoidal functors.

Proof. Recall from Proposition 10.6.3 that $\prod$ is natural with respect to morphisms in $\mathcal{G}$. Therefore, by functoriality of

$$\mathsf{Multicat}_*(-, \mathsf{End}(\mathsf{C})) : (\mathsf{Multicat}_*)^{\mathsf{op}} \longrightarrow \mathsf{Cat}_*$$
and
$$N : \mathsf{Cat}_* \longrightarrow \mathsf{sSet}_*,$$

we have a morphism of $\mathcal{G}_*$-simplicial sets

$$\begin{aligned} N_*\prod{}^* : N_* \wedge^* \mathsf{J}^{\mathsf{Se}}\mathsf{C} &= N_*\mathsf{Multicat}_*\big(\mathcal{M}(\wedge(-)), \mathsf{End}(\mathsf{C})\big) \\ &\longrightarrow N_*\mathsf{Multicat}_*\big(\mathcal{T}(-), \mathsf{End}(\mathsf{C})\big) = N_*\mathsf{J}^{\mathsf{EM}}\mathsf{C} \end{aligned} \tag{10.6.11}$$

where

- N_* denotes the levelwise composition with N and
- $\wedge^*$ is the functor along the top of the following commutative diagram in Cat, induced by the smash product of pointed finite sets.

$$\begin{array}{ccc} \Gamma\text{-}\mathsf{Cat} & \xrightarrow{\wedge^*} & \mathcal{G}_*\text{-}\mathsf{Cat} \\ {\scriptstyle N_*}\downarrow & & \downarrow{\scriptstyle N_*} \\ \Gamma\text{-}\mathsf{sSet} & \xrightarrow{\wedge^*} & \mathcal{G}_*\text{-}\mathsf{sSet} \end{array}$$

Since the above diagram commutes, and since we have

$$\mathsf{K}^{\mathcal{F}} = \mathsf{K}^{\mathcal{G}} \wedge^*$$

by Proposition 9.3.16, then (10.6.11) becomes

$$\begin{aligned} \mathsf{K}^{\mathsf{Se}}\mathsf{C} &= \mathsf{K}^{\mathcal{F}} N_*\mathsf{J}^{\mathsf{Se}}\mathsf{C} = \mathsf{K}^{\mathcal{G}} \wedge^* N_*\mathsf{J}^{\mathsf{Se}}\mathsf{C} \\ &= \mathsf{K}^{\mathcal{G}} N_* \wedge^* \mathsf{J}^{\mathsf{Se}}\mathsf{C} \xrightarrow{\mathsf{K}^{\mathcal{G}}(N_*\prod^*)} \mathsf{K}^{\mathcal{G}} N_*\mathsf{J}^{\mathsf{EM}}\mathsf{C} = \mathsf{K}^{\mathsf{EM}}\mathsf{C}. \end{aligned} \tag{10.6.12}$$

The morphism of symmetric spectra (10.6.12) is natural with respect to morphisms in $\mathsf{PermCat}^{\mathsf{su}}$ by naturality of $N_*\prod^*$ and functoriality of $\mathsf{K}^{\mathcal{G}}$.

We show (10.6.12) is a level equivalence of symmetric spectra by showing that each component of (10.6.11) at $\langle \underline{n} \rangle$ in $\mathcal{G}$,

$$N\mathsf{Multicat}_*\big(\mathcal{M}(\wedge_k \underline{n}_k), \mathsf{End}(\mathsf{C})\big) \longrightarrow N\mathsf{Multicat}_*\big(\mathcal{T}\langle \underline{n} \rangle, \mathsf{End}(\mathsf{C})\big), \tag{10.6.13}$$

is a simplicial homotopy equivalence.

If $\langle \underline{n} \rangle = \langle \rangle$, then $\prod$ is the identity and there is nothing more to prove. Now suppose $\langle \underline{n} \rangle \in \mathcal{G}$ has length $q > 0$. Then we have the following diagram of pointed multifunctors involving i and j of Definition 10.6.6 along with the component of $\prod$ at $\langle \underline{n} \rangle$.

$$\begin{array}{ccc}
\mathcal{M}(\wedge_k \underline{n}_k) & & \\
{\scriptstyle j}\uparrow & \nwarrow^{\prod} & \\
\mathcal{P}\langle \underline{n} \rangle & \xrightarrow[i]{\hspace{4em}} & \mathcal{T}\langle \underline{n} \rangle
\end{array}$$

This diagram in $\mathsf{Multicat}_*$ commutes because, for a tuple $\langle x \rangle$ of elements $x_k \in \underline{n}_k^\flat$, the product of singleton subsets $\{x_k\}$ is the singleton of the tuple $\langle x \rangle$.

Applying the functor

$$\mathsf{Multicat}_*(-, \mathsf{End}(\mathsf{C})) \colon (\mathsf{Multicat}_*)^{\mathsf{op}} \longrightarrow \mathsf{Cat}_*,$$

we have the following commutative diagram of pointed functors.

$$\begin{array}{ccc}
(\mathsf{J}^{\mathsf{Se}}\mathsf{C})(\wedge_k \underline{n}_k) & & \\
{\scriptstyle j^*}\downarrow & \searrow^{\prod^*} & \\
\mathsf{Multicat}_*(\mathcal{P}\langle \underline{n} \rangle, \mathsf{End}(\mathsf{C})) & \xleftarrow[i^*]{\hspace{4em}} & (\mathsf{J}^{\mathsf{EM}}\mathsf{C})\langle \underline{n} \rangle
\end{array}$$

The vertical and horizontal functors are right adjoints of pointed categories by Proposition 10.6.7 and, therefore, induce simplicial homotopy equivalences on nerves by Proposition 7.2.5. This implies that (10.6.13),

$$(N_* \textstyle\prod^*)_{\langle \underline{n} \rangle} \colon N(\mathsf{J}^{\mathsf{Se}}\mathsf{C}(\wedge_k \underline{n}_k)) \longrightarrow N(\mathsf{J}^{\mathsf{EM}}\mathsf{C}(\langle \underline{n} \rangle)),$$

is a simplicial homotopy equivalence for each $\langle \underline{n} \rangle$ in $\mathcal{G}$. □

10.7. (Co)lax and Strong Elmendorf-Mandell $\mathcal{G}_*$-Categories

Throughout this section, $(\mathsf{C}, \oplus, e, \xi)$ denotes a small permutative category. The main objective of this section is to compare the categories $\mathsf{C}^{\mathcal{G}}_{\mathrm{lax}}\langle \underline{n} \rangle$, $\mathsf{C}^{\mathcal{G}}_{\cong}\langle \underline{n} \rangle$, and $\mathsf{C}^{\mathcal{G}}_{\mathrm{co}}\langle \underline{n} \rangle$ in Definition 10.4.10, which we refer to as, respectively, the *lax*, *strong*, and *colax* versions. Recall that, by definition, if either

- $\langle \underline{n} \rangle = \langle \rangle$ or
- some $n_j = 0$, so $\langle \underline{n} \rangle = *$ in $\mathcal{G}$,

then each of $\mathsf{C}^{\mathcal{G}}_{\mathrm{lax}}\langle \underline{n} \rangle$, $\mathsf{C}^{\mathcal{G}}_{\cong}\langle \underline{n} \rangle$, and $\mathsf{C}^{\mathcal{G}}_{\mathrm{co}}\langle \underline{n} \rangle$ is defined in the same way—either C or $\mathbb{1}$.

After some preliminary definitions, Proposition 10.7.10 shows that $\mathsf{C}^{\mathcal{G}}_{\mathrm{lax}}\langle \underline{n} \rangle$ and $\mathsf{C}^{\mathcal{G}}_{\cong}\langle \underline{n} \rangle$ are related by an adjunction. Theorem 10.7.16 shows that, as $\langle \underline{n} \rangle$ varies in $\mathcal{G}$, the left adjoints assemble into a morphism of $\mathcal{G}_*$-categories, which induces a level equivalence in K-theory symmetric spectra. Proposition 10.7.22 shows that there is, furthermore, a multifunctorial property with respect to the small permutative category C. Theorem 10.7.19 and Proposition 10.7.27 are the corresponding results for the colax and strong versions. Explanation 10.7.23 discusses some advantages of K^{EM} over $\mathsf{K}^{\mathsf{EM}}_{\cong}$.

Levelwise Adjunctions.

Definition 10.7.1. We define an object (10.7.3) and a morphism (10.7.6) here. In Lemma 10.7.8 below, we show that they are well defined. Suppose

$$\langle \underline{n} \rangle = (\underline{n_1}, \ldots, \underline{n_q}) \in \mathcal{F}^{(q)}$$

with $q > 0$ and each $n_j > 0$. Suppose

$$C = (C, \rho) = \{C_{\langle s \rangle}, \rho_{\langle s \rangle;k,t,u}\} \in \mathsf{C}^g\langle \underline{n} \rangle \tag{10.7.2}$$

is an $\langle \underline{n} \rangle$-system as in Definition 10.4.1. Define an object

$$(\widetilde{C}, \widetilde{\rho}) \in \mathsf{C}^g_{\cong}\langle \underline{n} \rangle \tag{10.7.3}$$

as follows. For a tuple

$$\langle s \rangle = (s_1, \ldots, s_q) \quad \text{with each} \quad s_j \subset \underline{n_j}^{\flat}, \tag{10.7.4}$$

define the component object

$$\widetilde{C}_{\langle s \rangle} = \bigoplus_{(x_1,\ldots,x_q) \in \prod_{j=1}^q s_j} C_{(\{x_1\},\ldots,\{x_q\})} \in \mathsf{C},$$

with the sequences

$$(x_1, \ldots, x_q) \in \prod_{j=1}^q s_j \subset \prod_{j=1}^q \underline{n}_j^{\flat} \tag{10.7.5}$$

ordered by the lexicographical order of $\wedge_{j=1}^q \underline{n}_j$ in Definition 8.1.5. For a partition (t, u) of s_k with $1 \leq k \leq q$, the gluing morphism

$$\widetilde{C}_{\langle s \circ_k t \rangle} \oplus \widetilde{C}_{\langle s \circ_k u \rangle} \xrightarrow[\cong]{\widetilde{\rho}_{\langle s \rangle;k,t,u}} \widetilde{C}_{\langle s \rangle} \in \mathsf{C}$$

is

- the identity morphism if t, u, or any s_j is empty, and
- the unique coherence isomorphism in C that permutes factors otherwise.

Moreover, for a morphism

$$\alpha = \{\alpha_{\langle s \rangle} : C_{\langle s \rangle} \longrightarrow D_{\langle s \rangle}\} : C \longrightarrow D \in \mathsf{C}^g\langle \underline{n} \rangle,$$

define a morphism

$$(\widetilde{C}, \widetilde{\rho}_C) \xrightarrow{\widetilde{\alpha}} (\widetilde{D}, \widetilde{\rho}_D) \in \mathsf{C}^g_{\cong}\langle \underline{n} \rangle \tag{10.7.6}$$

with the components

$$\widetilde{\alpha}_{\langle s \rangle} = \bigoplus_{(x_1,\ldots,x_q) \in \prod_{j=1}^q s_j} \alpha_{(\{x_1\},\ldots,\{x_q\})} : \widetilde{C}_{\langle s \rangle} \longrightarrow \widetilde{D}_{\langle s \rangle} \in \mathsf{C}$$

and $\langle s \rangle$ as in (10.7.4). ◇

Explanation 10.7.7. In $\langle s \rangle = (s_1, \ldots, s_q)$ in (10.7.4), suppose that either

- each s_j is a one-element set for $1 \leq j \leq q$, or
- some $s_j = \varnothing$.

Then

$$\widetilde{C}_{\langle s \rangle} = C_{\langle s \rangle} \quad \text{and} \quad \widetilde{\alpha}_{\langle s \rangle} = \alpha_{\langle s \rangle}$$

in Definition 10.7.1. ◇

Lemma 10.7.8. *The constructions* (10.7.3) *and* (10.7.6) *define a functor*

$$\mathsf{C}^{\mathcal{G}}\langle \underline{n} \rangle \xrightarrow{\widetilde{(-)}} \mathsf{C}^{\mathcal{G}}_{\cong}\langle \underline{n} \rangle.$$

Proof. The pair $(\widetilde{C}, \widetilde{\rho})$ in (10.7.3) satisfies the axioms in Definition 10.4.1 for an object in $\mathsf{C}^{\mathcal{G}}_{\cong}\langle \underline{n} \rangle$ by the Symmetric Coherence Theorem 1.1.41 in C, since the gluing isomorphisms $\widetilde{\rho}$ are coherence isomorphisms in C. The collection $\widetilde{\alpha}$ in (10.7.6) is compatible with the gluing isomorphisms in $\widetilde{C}$ and $\widetilde{D}$ by the naturality of the structure isomorphisms in C. So $\widetilde{\alpha}$ is a morphism in $\mathsf{C}^{\mathcal{G}}_{\cong}\langle \underline{n} \rangle$. Identity morphisms and composition in $\mathsf{C}^{\mathcal{G}}\langle \underline{n} \rangle$ are defined componentwise. So the functoriality of $\oplus$ in C implies that $\widetilde{(-)}$ preserves identity morphisms and composition. □

Definition 10.7.9. For each q-tuple of pointed finite sets $\langle \underline{n} \rangle \in \mathcal{F}^{(q)}$ with $q \geq 0$, define two functors

$$\mathsf{C}^{\mathcal{G}}_{\cong}\langle \underline{n} \rangle \underset{R^{\mathsf{C}}_{\langle \underline{n} \rangle}}{\overset{L^{\mathsf{C}}_{\langle \underline{n} \rangle}}{\rightleftarrows}} \mathsf{C}^{\mathcal{G}}\langle \underline{n} \rangle$$

as follows.

- If $\langle \underline{n} \rangle = \langle\rangle$ or if any $n_j = 0$, then $R^{\mathsf{C}}_{\langle \underline{n} \rangle}$ and $L^{\mathsf{C}}_{\langle \underline{n} \rangle}$ are both defined to be identity functors.
- Otherwise:
  - $R^{\mathsf{C}}_{\langle \underline{n} \rangle} = \widetilde{(-)}$ is the functor in Lemma 10.7.8.
  - $L^{\mathsf{C}}_{\langle \underline{n} \rangle}$ is also defined by the assignments (10.7.3) and (10.7.6), applied to the objects and morphisms in $\mathsf{C}^{\mathcal{G}}_{\cong}\langle \underline{n} \rangle$.

The proof of Lemma 10.7.8 applies to $L^{\mathsf{C}}_{\langle \underline{n} \rangle}$ to show that it is a functor. If C and $\langle \underline{n} \rangle$ are understood, then we abbreviate $L^{\mathsf{C}}_{\langle \underline{n} \rangle}$ and $R^{\mathsf{C}}_{\langle \underline{n} \rangle}$ to, respectively, L and R. ◇

Proposition 10.7.10. *In the context of Definition 10.7.9, the following statements hold.*

1. *$L \dashv R$ is an adjunction with unit a natural isomorphism.*
2. *If, in addition, C is a groupoid, then $L \dashv R$ is an adjoint equivalence.*

Proof. There is nothing to check if $\langle \underline{n} \rangle = \langle\rangle$ or if any $n_j = 0$. For the remainder of the proof we assume $q > 0$ and each $n_j > 0$. For the first assertion, we

- define the unit and the counit for (L, R) and
- check the triangle identities (I.1.1.11).

The component of the unit

$$1_{\mathsf{C}^{\mathcal{G}}_{\cong}\langle \underline{n} \rangle} \xrightarrow{\eta} RL$$

at a strong $\langle \underline{n} \rangle$-system

$$C = (C, \rho) = \{C_{\langle s \rangle}, \rho_{\langle s \rangle;k,t,u}\} \in \mathsf{C}^{\mathcal{G}}_{\cong}\langle \underline{n} \rangle \tag{10.7.11}$$

is the morphism

$$C \xrightarrow{\eta_C} RLC \in \mathsf{C}^{\mathcal{G}}_{\cong}\langle \underline{n} \rangle$$

with, for $\langle s \rangle = (s_1, \ldots, s_q)$ as in (10.7.4), components

$$C_{\langle s \rangle} \xrightarrow{(\eta_C)_{\langle s \rangle}} \bigoplus_{(x_1,\ldots,x_q) \in \prod_{j=1}^q s_j} C_{(\{x_1\},\ldots,\{x_q\})} = (RLC)_{\langle s \rangle} \in \mathsf{C}$$

given by

- the identity morphism in each of the two cases in Explanation 10.7.7 and
- composites of inverses of the gluing isomorphisms ρ in C otherwise.

The second case above is well defined by the invertibility of ρ in C and the axioms in Definition 10.4.1. The condition that η_C is a morphism in $\mathsf{C}^g_{\cong}\langle \underline{n} \rangle$ means that the diagram

$$\begin{array}{ccc} C_{\langle s\circ_k t\rangle} \oplus C_{\langle s\circ_k u\rangle} & \xrightarrow{\rho_{\langle s\rangle;k,t,u}} & C_{\langle s\rangle} \\ {\scriptstyle (\eta_C)_{\langle s\circ_k t\rangle}\oplus(\eta_C)_{\langle s\circ_k u\rangle}}\downarrow & & \downarrow{\scriptstyle (\eta_C)_{\langle s\rangle}} \\ (RLC)_{\langle s\circ_k t\rangle} \oplus (RLC)_{\langle s\circ_k u\rangle} & \xrightarrow{\widetilde{\rho}_{\langle s\rangle;k,t,u}} & (RLC)_{\langle s\rangle} \end{array} \tag{10.7.12}$$

is commutative for each partition (t,u) of s_k with $1 \le k \le q$.

- If either each s_j is a one-element set or some $s_j = \varnothing$, then all four arrows in (10.7.12) are identity morphisms.
- In all other cases, each $(\eta_C)_?$ consists of inverses of ρ in C, and $\widetilde{\rho}$ is a coherence isomorphism in C that permutes factors. So (10.7.12) is commutative by the symmetry and associativity axioms (10.4.4) and (10.4.5) for (C,ρ).

This shows that η_C is a morphism in $\mathsf{C}^g_{\cong}\langle \underline{n} \rangle$.

To check the naturality of η, suppose

$$\alpha = \{\alpha_{\langle s\rangle} : C_{\langle s\rangle} \longrightarrow D_{\langle s\rangle}\} : (C,\rho_C) \longrightarrow (D,\rho_C) \in \mathsf{C}^g_{\cong}\langle \underline{n} \rangle$$

is a morphism. It suffices to check the naturality of η at a component $\langle s \rangle$ as in (10.7.4), which asserts the commutativity of the following diagram.

$$\begin{array}{ccc} C_{\langle s\rangle} & \xrightarrow{(\eta_C)_{\langle s\rangle}} & \bigoplus\limits_{(x_1,\ldots,x_q)\in\prod_{j=1}^q s_j} C_{(\{x_1\},\ldots,\{x_q\})} = (RLC)_{\langle s\rangle} \\ {\scriptstyle \alpha_{\langle s\rangle}}\downarrow & & \downarrow{\scriptstyle \oplus\, \alpha_{(\{x_1\},\ldots,\{x_q\})}} \\ D_{\langle s\rangle} & \xrightarrow{(\eta_D)_{\langle s\rangle}} & \bigoplus\limits_{(x_1,\ldots,x_q)\in\prod_{j=1}^q s_j} D_{(\{x_1\},\ldots,\{x_q\})} = (RLD)_{\langle s\rangle} \end{array} \tag{10.7.13}$$

In the two cases in Explanation 10.7.7, the horizontal arrows in (10.7.13) are identity morphisms, and the two vertical arrows are equal. In all other cases, (10.7.13) is commutative by

- the fact that each of $(\eta_C)_{\langle s\rangle}$ and $(\eta_D)_{\langle s\rangle}$ is a composite of inverses of gluing morphisms and
- the compatibility of α with the gluing morphisms in C and D.

We have shown that η is a natural transformation. Moreover, it is a natural isomorphism because each component $(\eta_C)_{\langle s\rangle}$ is either an identity morphism or a composite of isomorphisms.

The component of the counit

$$LR \xrightarrow{\ \varepsilon\ } 1_{\mathsf{C}^g\langle \underline{n}\rangle}$$

at a lax $\langle \underline{n} \rangle$-system $C \in \mathsf{C}^g\langle \underline{n} \rangle$ as in (10.7.2) is the morphism

$$LRC \xrightarrow{\ \varepsilon_C\ } C \in \mathsf{C}^g\langle \underline{n} \rangle$$

with components

$$(LRC)_{\langle s\rangle} = \bigoplus_{(x_1,\ldots,x_q)\in\prod_{j=1}^q s_j} C_{(\{x_1\},\ldots,\{x_q\})} \xrightarrow{(\varepsilon_C)_{\langle s\rangle}} C_{\langle s\rangle} \in \mathsf{C}$$

given by

- the identity morphism in each of the two cases in Explanation 10.7.7 and
- composites of the gluing morphisms ρ in C otherwise.

The second case above is well defined by the axioms in Definition 10.4.1. Reusing the diagrams (10.7.12) and (10.7.13) with each η replaced by ε in the opposite direction, we infer that ε is a natural transformation.

Next we check the triangle identities (I.1.1.11). For the triangle identity that asserts the commutativity of the diagram

$$\begin{array}{ccccc} & & LRL & & \\ & {\scriptstyle L\eta}\nearrow & & \searrow{\scriptstyle \varepsilon L} & \\ L & & \xrightarrow{1_L} & & L \end{array}$$

consider an object $C \in \mathsf{C}^{\mathcal{G}}_{\cong}\langle \underline{n}\rangle$. Then $\big((\varepsilon L)\circ(L\eta)\big)_C$ has components

$$\begin{array}{rcl} (LC)_{\langle s\rangle} = & \displaystyle\bigoplus_{(x_1,\ldots,x_q)\in\prod_{j=1}^q s_j} & C_{(\{x_1\},\ldots,\{x_q\})} \\ & {\scriptstyle (L\eta_C)_{\langle s\rangle} =}\Big\downarrow{\scriptstyle \oplus(\eta_C)_{(\{x_1\},\ldots,\{x_q\})}} & \\ (LRLC)_{\langle s\rangle} = & \displaystyle\bigoplus_{(x_1,\ldots,x_q)\in\prod_{j=1}^q s_j} & C_{(\{x_1\},\ldots,\{x_q\})} \\ & {\scriptstyle (\varepsilon_{LC})_{\langle s\rangle} =}\Big\downarrow{\scriptstyle \widetilde{\rho}} & \\ (LC)_{\langle s\rangle} = & \displaystyle\bigoplus_{(x_1,\ldots,x_q)\in\prod_{j=1}^q s_j} & C_{(\{x_1\},\ldots,\{x_q\})} \end{array}$$

with $\langle s\rangle$ as in (10.7.4).

- In $(L\eta_C)_{\langle s\rangle}$, each factor
$$(\eta_C)_{(\{x_1\},\ldots,\{x_q\})}$$
is an identity morphism by definition, so $(L\eta_C)_{\langle s\rangle}$ is an identity morphism.
- The gluing morphism $\widetilde{\rho}$ in LC is a coherence isomorphism in C that permutes factors. The Symmetric Coherence Theorem 1.1.41 implies that $\widetilde{\rho}$ is the identity morphism in the above diagram.

Therefore, each of $L\eta$ and εL is the identity natural transformation.

The other triangle identity

$$\begin{array}{ccccc} & & RLR & & \\ & {\scriptstyle \eta R}\nearrow & & \searrow{\scriptstyle R\varepsilon} & \\ R & & \xrightarrow{1_R} & & R \end{array}$$

is proved similarly as follows.

- Each component $(\eta_{RC})_{\langle s\rangle}$ is a coherence isomorphism in C that permutes factors. Since its domain and codomain are equal, the uniqueness in Theorem 1.1.41 implies that this is the identity morphism.
- The other natural transformation has components

$$(R\varepsilon_C)_{\langle s\rangle} = \bigoplus_{(x_1,\ldots,x_q)\in\prod_{j=1}^q s_j} (\varepsilon_C)_{(\{x_1\},\ldots,\{x_q\})}.$$

Each displayed component of ε_C is an identity morphism by definition.

Therefore, each of ηR and $R\varepsilon$ is the identity natural transformation. This finishes the proof that (L,R) is an adjunction with unit η and counit ε.

For the second assertion, suppose C is a groupoid. Then each component $(\varepsilon_C)_{\langle s\rangle}$ is either an identity morphism or a composite of isomorphisms. In this case, both the unit η and the counit ε are natural isomorphisms, so (L,R) is an adjoint equivalence. □

Naturality. Next we observe that the left adjoints $L^{\mathsf{C}}_{\langle \underline{n}\rangle}$ are natural with respect to $\langle \underline{n}\rangle \in \mathcal{G}$. To do that, first we provide a simpler description of the left adjoints as follows.

Definition 10.7.14. For each q-tuple of pointed finite sets $\langle \underline{n}\rangle \in \mathcal{F}^{(q)}$, we denote by

$$\mathsf{C}^{\mathcal{G}}_{\cong}\langle \underline{n}\rangle \xrightarrow{I^{\mathsf{C}}_{\langle \underline{n}\rangle}} \mathsf{C}^{\mathcal{G}}\langle \underline{n}\rangle$$

the full subcategory inclusion. If C and $\langle \underline{n}\rangle$ are understood, then we abbreviate $I^{\mathsf{C}}_{\langle \underline{n}\rangle}$ to I. ◇

Proposition 10.7.15. *For each $\langle \underline{n}\rangle \in \mathcal{F}^{(q)}$, the two functors*

$$\mathsf{C}^{\mathcal{G}}_{\cong}\langle \underline{n}\rangle \underset{I}{\overset{L}{\rightrightarrows}} \mathsf{C}^{\mathcal{G}}\langle \underline{n}\rangle$$

in Definitions 10.7.9 and 10.7.14 are naturally isomorphic. Therefore, Proposition 10.7.10 also holds for the adjunction $I \dashv R$.

Proof. If $q = 0$ or any $n_j = 0$ then L and I are both the identity functor. For the remainder of the proof we assume $q > 0$ and each $n_j > 0$. We define a natural isomorphism

$$L \xrightarrow[\cong]{g} I$$

as follows. For each object $C \in \mathsf{C}^{\mathcal{G}}_{\cong}\langle \underline{n}\rangle$ as in (10.7.11) and $\langle s\rangle$ as in (10.7.4), the morphism

$$(LC)_{\langle s\rangle} = \bigoplus_{(x_1,\ldots,x_q)\in\prod_{j=1}^q s_j} C_{(\{x_1\},\ldots,\{x_q\})} \xrightarrow{(g_C)_{\langle s\rangle}} C_{\langle s\rangle} \in \mathsf{C}$$

is given by

- the identity morphism in each of the two cases in Explanation 10.7.7 and
- composites of the gluing morphisms ρ in C otherwise.

The second case above is well defined by the axioms in Definition 10.4.1. Reusing the diagrams (10.7.12) and (10.7.13) with each η replaced by g in the opposite direction, we infer that

- the collection
$$LC \xrightarrow{g_C = \{(g_C)_{\langle s\rangle}\}} C$$
is a morphism in $\mathsf{C}^{\mathcal{G}}\langle \underline{n}\rangle$ and
- $g\colon L \longrightarrow I$ is a natural transformation.

Since the gluing morphisms ρ in C are isomorphisms, each component $(g_C)_{\langle s\rangle}$ is an isomorphism. So g is a natural isomorphism. $\square$

Recall that $\mathsf{C}^{\mathcal{G}}$ and, by restriction, $\mathsf{C}^{\mathcal{G}}_{\cong}$ are $\mathcal{G}_*$-categories by Proposition 10.4.18.

Theorem 10.7.16. *For each small permutative category $(\mathsf{C}, \oplus, e, \xi)$, the functors $I^{\mathsf{C}}_{\langle \underline{n}\rangle}$ in Definition 10.7.14 are the components of a $\mathcal{G}_*$-category morphism*
$$\mathsf{C}^{\mathcal{G}}_{\cong} \xrightarrow{I^{\mathsf{C}}} \mathsf{C}^{\mathcal{G}}.$$
Moreover, the induced morphism of symmetric spectra
$$\mathsf{K}^{\mathsf{EM}}_{\cong}\mathsf{C} = \mathsf{K}^{\mathcal{G}} N_* \mathsf{C}^{\mathcal{G}}_{\cong} \xrightarrow{\mathsf{K}^{\mathcal{G}} N_* I^{\mathsf{C}}} \mathsf{K}^{\mathcal{G}} N_* \mathsf{C}^{\mathcal{G}} \cong \mathsf{K}^{\mathsf{EM}}\mathsf{C}$$
is a level equivalence.

Proof. For each morphism
$$\langle \underline{n}\rangle = (\underline{n}_1, \ldots, \underline{n}_q) \xrightarrow{(f, \langle\psi\rangle)} (\underline{m}_1, \ldots, \underline{m}_p) = \langle \underline{m}\rangle \in \mathcal{G}$$
as in Definition 9.1.7, the diagram of functors

(10.7.17)
$$\begin{array}{ccc} \mathsf{C}^{\mathcal{G}}_{\cong}\langle \underline{n}\rangle & \xrightarrow{I^{\mathsf{C}}_{\langle \underline{n}\rangle}} & \mathsf{C}^{\mathcal{G}}\langle \underline{n}\rangle \\ {\scriptstyle \mathsf{C}^{\mathcal{G}}_{\cong}(f,\langle\psi\rangle)}\downarrow & & \downarrow{\scriptstyle \mathsf{C}^{\mathcal{G}}(f,\langle\psi\rangle)} \\ \mathsf{C}^{\mathcal{G}}_{\cong}\langle \underline{m}\rangle & \xrightarrow{I^{\mathsf{C}}_{\langle \underline{m}\rangle}} & \mathsf{C}^{\mathcal{G}}\langle \underline{m}\rangle \end{array}$$

is commutative because each $I^{\mathsf{C}}_{?}$ is a full subcategory inclusion. So I^{C} is a morphism of $\mathcal{G}_*$-categories.

For the second assertion, by Proposition 10.7.15, I^{C} is levelwise a left adjoint. After taking the nerve, $N_* I^{\mathsf{C}}$ is a levelwise simplicial homotopy equivalence of $\mathcal{G}_*$-simplicial sets. So $\mathsf{K}^{\mathcal{G}} N_* I^{\mathsf{C}}$ is a level equivalence of symmetric spectra. $\square$

Explanation 10.7.18. Using the notation in Definition 10.4.16, we explain the diagram (10.7.17) in more detail. Suppose $q > 0$, $p > 0$, each $n_j > 0$, and each $m_i > 0$. For an object $(C, \rho) \in \mathsf{C}^{\mathcal{G}}_{\cong}\langle \underline{n}\rangle$ as in (10.7.11), denote by
$$(C', \rho') \in \mathsf{C}^{\mathcal{G}}\langle \underline{m}\rangle$$
the image of (C, ρ) under either composite in (10.7.17). For a p-tuple of pointed finite sets
$$\langle s\rangle = (s_1, \ldots, s_p) \subset \mathcal{F}^{(p)} \quad \text{with each} \quad s_i \subset \underline{m}_i^{\flat},$$
the component $C'_{\langle s\rangle}$ is the object
$$C'_{\langle s\rangle} = \begin{cases} C_{\langle \tilde{f}_* \psi^{-1}(s)\rangle} & \text{if each } \psi_i^{-1}(s_i) \neq \varnothing, \\ e & \text{otherwise} \end{cases}$$

with

$$\langle \tilde{f}_* \psi^{-1}(s) \rangle = \left\{ \psi^{-1}_{f(j)} \left(s_{f(j)} \right) \right\}_{1 \le j \le q} \in \mathcal{F}^{(q)}.$$

For a partition

$$t \cup u = s_k \quad \text{with} \quad 1 \le k \le p,$$

the gluing morphism ρ' is

$$\rho'_{\langle s \rangle;k,t,u} = \begin{cases} \rho_{\langle \tilde{f}_* \psi^{-1}(s) \rangle; f^{-1}(k), \psi_k^{-1}(t), \psi_k^{-1}(u)} & \text{if } f^{-1}(k) \ne \varnothing \text{ and each } \psi_i^{-1}(s_i) \ne \varnothing, \\ 1_{C_{\langle \tilde{f}_* \psi^{-1}(s) \rangle}} & \text{if } f^{-1}(k) = \varnothing \text{ and each } \psi_i^{-1}(s_i) \ne \varnothing, \\ 1_e & \text{otherwise.} \end{cases}$$

For a morphism

$$C \xrightarrow{\alpha} D \in \mathsf{C}^{\mathcal{G}}_{\cong}\langle \underline{n} \rangle,$$

its image α' under either composite in (10.7.17) has components

$$\alpha'_{\langle s \rangle} = \begin{cases} \alpha_{\langle \tilde{f}_* \psi^{-1}(s) \rangle} & \text{if each } \psi_i^{-1}(s_i) \ne \varnothing, \\ 1_e & \text{otherwise} \end{cases}$$

with $\langle s \rangle \in \mathcal{F}^{(p)}$ as above. ◇

So far we have compared the lax and the strong variants of the Elmendorf-Mandell $\mathcal{G}_*$-categories of C. The next result compares the colax and the strong variants.

Theorem 10.7.19. *The following statements hold for each small permutative category C and each $\langle \underline{n} \rangle \in \mathcal{F}^{(q)}$.*

(1) *For $q > 0$ and each $n_j > 0$, the constructions* (10.7.3) *and* (10.7.6) *define functors*

$$\mathsf{C}^{\mathcal{G}}_{\mathrm{co}}\langle \underline{n} \rangle \underset{R^{\mathrm{co}}_{\langle \underline{n} \rangle}}{\overset{L^{\mathrm{co}}_{\langle \underline{n} \rangle}}{\rightleftarrows}} \mathsf{C}^{\mathcal{G}}_{\cong}\langle \underline{n} \rangle,$$

where, in

$$R^{\mathrm{co}}_{\langle \underline{n} \rangle}(C, \rho) = (\widetilde{C}, \widetilde{\rho}) \quad \textit{for} \quad (C,\rho) \in \mathsf{C}^{\mathcal{G}}_{\cong}\langle \underline{n} \rangle,$$

the gluing morphisms go in the colax direction, as in

$$\widetilde{C}_{\langle s \rangle} \xrightarrow[\cong]{\widetilde{\rho}_{\langle s \rangle;k,t,u}} \widetilde{C}_{\langle s \circ_k t \rangle} \oplus \widetilde{C}_{\langle s \circ_k u \rangle} \in \mathsf{C}.$$

(2) *The functors $\left(L^{\mathrm{co}}_{\langle \underline{n} \rangle}, R^{\mathrm{co}}_{\langle \underline{n} \rangle}\right)$ form an adjunction with counit a natural isomorphism, where, if $q = 0$ or any $n_j = 0$, we define $L^{\mathrm{co}}_{\langle \underline{n} \rangle}$ and $R^{\mathrm{co}}_{\langle \underline{n} \rangle}$ to both be identity functors. If C is a groupoid, then $\left(L^{\mathrm{co}}_{\langle \underline{n} \rangle}, R^{\mathrm{co}}_{\langle \underline{n} \rangle}\right)$ is, furthermore, an adjoint equivalence.*

(3) *$R^{\mathrm{co}}_{\langle \underline{n} \rangle}$ is naturally isomorphic to the full and faithful functor*

$$\mathsf{C}^{\mathcal{G}}_{\mathrm{co}}\langle \underline{n} \rangle \xleftarrow{I^{\mathrm{co}}_{\langle \underline{n} \rangle}} \mathsf{C}^{\mathcal{G}}_{\cong}\langle \underline{n} \rangle$$

that sends each object

$$(C,\rho) \in \mathsf{C}^{\mathcal{G}}_{\cong}\langle \underline{n} \rangle \quad \textit{to} \quad (C, \rho^{\mathrm{co}}) \in \mathsf{C}^{\mathcal{G}}_{\mathrm{co}}\langle \underline{n} \rangle$$

with each component of ρ^{co} the inverse of the corresponding component of ρ. On morphisms, $I^{\mathrm{co}}_{\langle \underline{n} \rangle}$ is the identity function.

(4) *The functors* $I^{\mathrm{co}}_{\langle \underline{n} \rangle}$ *are the components of a* $\mathcal{G}_*$*-category morphism*

$$\mathsf{C}^{\mathcal{G}}_{\mathrm{co}} \xleftarrow{\;I^{\mathrm{co}}\;} \mathsf{C}^{\mathcal{G}}_{\cong}.$$

Moreover, the induced morphism of symmetric spectra

$$\mathsf{K}^{\mathsf{EM}}_{\mathrm{co}}\mathsf{C} = \mathsf{K}^{\mathcal{G}} N_* \mathsf{C}^{\mathcal{G}}_{\mathrm{co}} \xleftarrow{\;\mathsf{K}^{\mathcal{G}} N_* I^{\mathrm{co}}\;} \mathsf{K}^{\mathcal{G}} N_* \mathsf{C}^{\mathcal{G}}_{\cong} = \mathsf{K}^{\mathsf{EM}}_{\cong}\mathsf{C}$$

is a level equivalence.

Proof. We reuse the proofs of Lemma 10.7.8, Propositions 10.7.10 and 10.7.15, and Theorem 10.7.16 essentially without change. The key point is that, in (10.7.3), $\widetilde{\rho}$ is a coherence isomorphism in C and does not use the gluing morphisms in the given $\langle \underline{n} \rangle$-system (C, ρ). Moreover, for a strong $\langle \underline{n} \rangle$-system (C, ρ), the gluing morphisms ρ are isomorphisms, so they can be inverted to yield colax gluing morphisms. □

Multifunctoriality. Theorem 10.7.16 can be strengthened to include multifunctoriality as follows.

Definition 10.7.20.

- A multilinear functor F (Definition 6.5.4) is *strong* if each linearity constraint F^2_i is a natural isomorphism for $1 \le i \le n$.
- Denote by

$$\mathsf{PermCat}^{\mathsf{sus}}$$

the Cat-enriched full sub-multicategory of $\mathsf{PermCat}^{\mathsf{su}}$ in Definition 6.5.1 consisting of *strong* multilinear functors.

The definitions (6.6.3) and (6.6.8) show that strong multilinear functors are closed under the symmetric group action and the composition. ◇

Explanation 10.7.21 (Multifunctoriality of $(-)^{\mathcal{G}}_{\cong}$)**.** The Cat-enriched multifunctor

$$\mathsf{PermCat}^{\mathsf{su}} \xrightarrow{\;(-)^{\mathcal{G}}\;} \mathcal{G}_*\text{-}\mathsf{Cat}$$

in Explanation 10.5.4 restricts to a Cat-enriched multifunctor

$$\mathsf{PermCat}^{\mathsf{sus}} \xrightarrow{\;(-)^{\mathcal{G}}_{\cong}\;} \mathcal{G}_*\text{-}\mathsf{Cat}.$$

For $(-)^{\mathcal{G}}_{\cong}$, the restriction to *strong* multilinear functors is necessary. In more detail, for an n-linear functor F as in (10.5.8), the $\mathcal{G}_*$-category morphism $F^{\mathcal{G}}_{\cong}$ has the same description as $F^{\mathcal{G}}$ in (10.5.9), but with

- each $\mathsf{C}^{\mathcal{G}}_j$ replaced by $(\mathsf{C}_j)^{\mathcal{G}}_{\cong}$ and
- $\mathsf{D}^{\mathcal{G}}$ replaced by $\mathsf{D}^{\mathcal{G}}_{\cong}$.

In this case, in (10.5.11), the components of the gluing morphisms ρ^j are isomorphisms in C_j. This implies that, in the diagram (10.5.15), the morphism

$$F(1, \ldots, \rho^r, \ldots, 1)$$

is an isomorphism. Therefore, in order for the gluing morphism $\rho_{\langle s \rangle;k,t,u}$ in (10.5.15) to be an isomorphism, we must assume that the linearity constraints $\{F^2_r\}_{r=1}^n$ are natural isomorphisms. In other words, F is a strong n-linear functor. ◇

Proposition 10.7.22. *The $\mathcal{G}_*$-category morphisms I^{C} in Theorem 10.7.16 are the components of a* Cat*-enriched multinatural transformation*

$$\begin{array}{ccc} \mathsf{PermCat}^{\mathsf{sus}} & \xrightarrow{(-)^{\mathcal{G}}_{\cong}} & \mathcal{G}_*\text{-}\mathsf{Cat} \\ & \Downarrow I & \\ \searrow & \mathsf{PermCat}^{\mathsf{su}} & \nearrow_{(-)^{\mathcal{G}}} \end{array}$$

with the unlabeled arrow the sub-multicategory inclusion.

Proof. For I, the naturality condition in Definition 5.1.17 states that, for each strong n-linear functor

$$\mathsf{C}_1 \times \cdots \times \mathsf{C}_n \xrightarrow{F} \mathsf{D} \quad \text{with} \quad n \geq 0,$$

the diagram

$$\begin{array}{ccc} (\mathsf{C}_1)^{\mathcal{G}}_{\cong} \otimes \cdots \otimes (\mathsf{C}_n)^{\mathcal{G}}_{\cong} & \xrightarrow{F^{\mathcal{G}}_{\cong}} & \mathsf{D}^{\mathcal{G}}_{\cong} \\ {\scriptstyle I^{\mathsf{C}_1} \otimes \cdots \otimes I^{\mathsf{C}_n}} \downarrow & & \downarrow {\scriptstyle I^{\mathsf{D}}} \\ \mathsf{C}_1^{\mathcal{G}} \otimes \cdots \otimes \mathsf{C}_n^{\mathcal{G}} & \xrightarrow{F^{\mathcal{G}}} & \mathsf{D}^{\mathcal{G}} \end{array}$$

in $\mathcal{G}_*$-Cat is commutative. This diagram is commutative by the explicit description of $F^{\mathcal{G}}$ and $F^{\mathcal{G}}_{\cong}$ on

- objects in (10.5.13) and (10.5.15) and
- morphisms in (10.5.16)

and the fact that each $I^{\mathsf{C}}_{\langle \underline{m} \rangle}$, for $\mathsf{C} \in \mathsf{PermCat}^{\mathsf{sus}}$ and $\langle \underline{m} \rangle \in \mathcal{G}$, is an inclusion. □

Explanation 10.7.23 (Lax versus Strong K-Theory)**.** The main reason that we consider the lax version

$$\mathsf{K}^{\mathsf{EM}} \cong \mathsf{K}^{\mathcal{G}} N_* (-)^{\mathcal{G}} \colon \mathsf{PermCat}^{\mathsf{su}} \longrightarrow \mathsf{SymSp},$$

instead of the strong version

$$\mathsf{K}^{\mathsf{EM}}_{\cong} = \mathsf{K}^{\mathcal{G}} N_* (-)^{\mathcal{G}}_{\cong} \colon \mathsf{PermCat}^{\mathsf{sus}} \longrightarrow \mathsf{SymSp},$$

as the default Elmendorf-Mandell K-theory is that K^{EM} allows a strictly bigger domain multicategory $\mathsf{PermCat}^{\mathsf{su}}$, as opposed to $\mathsf{PermCat}^{\mathsf{sus}}$. This is a nontrivial difference in practice when further algebraic structures are taken into account. In fact, as we will see in Theorem 11.2.16, monoids in $\mathsf{PermCat}^{\mathsf{su}}$ are small ring categories. On the other hand, the same proof restricted to $\mathsf{PermCat}^{\mathsf{sus}}$ shows that monoids in $\mathsf{PermCat}^{\mathsf{sus}}$ are small ring categories with *invertible* factorization morphisms ∂^l and ∂^r. So general small ring categories, whose factorization morphisms are not necessarily invertible, can only be detected in $\mathsf{PermCat}^{\mathsf{su}}$. Moreover, there are similar remarks for

- Theorem 11.5.5, where a Cat-enriched multifunctor

  $$E\mathsf{As} \longrightarrow \mathsf{PermCat}^{\mathsf{su}}$$

  is a small bipermutative category;

- Theorem 12.4.5, where a Cat-enriched multifunctor

$$\mathsf{Br} \longrightarrow \mathsf{PermCat}^{\mathsf{su}}$$

is a small braided ring category; and
- Theorem 13.4.12, where a Cat-enriched multifunctor

$$\mathsf{Mon}^n \longrightarrow \mathsf{PermCat}^{\mathsf{su}}$$

is a small E_n-monoidal category.

In each of the above cases, replacing the codomain with $\mathsf{PermCat}^{\mathsf{sus}}$ yields the corresponding type of categorical structure with invertible factorization morphisms.

To phrase it in another way, while Theorem 10.7.16 shows that, for each small permutative category C, I^{C} induces a level equivalence

$$\mathsf{K}^{\mathsf{EM}}_{\cong}\mathsf{C} \xrightarrow{\mathsf{K}^{\mathcal{G}} N_* I^{\mathsf{C}}} \mathsf{K}^{\mathsf{EM}}\mathsf{C}$$

of K-theory symmetric spectra, many important algebraic structures preserved by K^{EM}—including those in Theorems 11.2.16, 11.5.5, 12.4.5, and 13.4.12—can only be detected in the bigger multicategory $\mathsf{PermCat}^{\mathsf{su}}$, but not in $\mathsf{PermCat}^{\mathsf{sus}}$. Therefore, when algebraic structures on small permutative categories, as parametrized by operads, are taken into account, the lax version K^{EM} has a significant advantage over the strong version $\mathsf{K}^{\mathsf{EM}}_{\cong}$. ◇

Theorem 10.7.19 can also be strengthened to include multifunctoriality using the definitions below.

Definition 10.7.24 (Colax Multilinear Functors). Suppose $\mathsf{C}_1, \ldots, \mathsf{C}_n$, and D are permutative categories.

- A *colax n-linear functor*

$$F\colon \langle \mathsf{C} \rangle \longrightarrow \mathsf{D}$$

is defined as in Definition 6.5.4 with the direction of each linearity constraint F^2_i reversed, as in

$$F^2_i\colon F\langle X \circ_i (X_i \oplus X'_i)\rangle \longrightarrow F\langle X \circ_i X_i\rangle \oplus F\langle X \circ_i X'_i\rangle,$$

but the same axioms, with the reversed arrows F^2_i, are satisfied.
- A *colax n-linear transformation* between colax n-linear functors

$$F, F'\colon \langle \mathsf{C} \rangle \longrightarrow \mathsf{D}$$

is defined as in Definition 6.5.11 with the reserved arrows F^2_i and $(F')^2_i$ in the diagram (6.5.12).
- Denote by

$$\mathsf{PermCat}^{\mathsf{su}}_{\mathsf{co}}$$

the Cat-enriched multicategory consisting of
  - small permutative categories,
  - colax multilinear functors, and
  - colax multilinear transformations.

The proofs in Section 6.6, with the linearity constraints reversed, show that $\mathsf{PermCat}^{\mathsf{su}}_{\mathsf{co}}$ is a well-defined Cat-enriched multicategory. ◇

Explanation 10.7.25. Each strong n-linear functor $\big(F, \{F_i^2\}_{i=1}^n\big)$ as in Definition 10.7.20 yields a colax n-linear functor by taking the inverse of each linearity constraint F_i^2. This assignment gives a Cat-enriched multifunctor

$$\mathsf{PermCat}^{\mathsf{sus}} \xrightarrow{\ \iota^{\mathrm{co}}\ } \mathsf{PermCat}^{\mathsf{su}}_{\mathsf{co}} \tag{10.7.26}$$

that is the identity assignment on objects and multilinear transformations. Moreover, the object assignment $\mathsf{C} \longmapsto \mathsf{C}^{\mathcal{G}}_{\mathrm{co}}$ is part of a Cat-enriched multifunctor

$$\mathsf{PermCat}^{\mathsf{su}}_{\mathsf{co}} \xrightarrow{\ (-)^{\mathcal{G}}_{\mathrm{co}}\ } \mathcal{G}_*\text{-}\mathsf{Cat}.$$

The assignment on strictly unital colax multilinear functors and colax multilinear transformations are as in Explanation 10.5.4, with

- the linearity constraints F_i^2 reversed and
- $(-)^{\mathcal{G}}_{\mathrm{co}}$ instead of $(-)^{\mathcal{G}}$.

In the current context, each arrow in the diagram (10.5.15) is reversed. ◇

Essentially the same reasoning for Proposition 10.7.22 gives the following result.

Proposition 10.7.27. *The $\mathcal{G}_*$-category morphisms I^{co} in Theorem 10.7.19 (4) are the components of a Cat-enriched multinatural transformation*

$$\begin{array}{ccc} \mathsf{PermCat}^{\mathsf{sus}} & \xrightarrow{\ (-)^{\mathcal{G}}_{\cong}\ } & \mathcal{G}_*\text{-}\mathsf{Cat} \\ & \Downarrow I^{\mathrm{co}} & \\ {\scriptstyle \iota^{\mathrm{co}}}\searrow & & \nearrow {\scriptstyle (-)^{\mathcal{G}}_{\mathrm{co}}} \\ & \mathsf{PermCat}^{\mathsf{su}}_{\mathsf{co}} & \end{array}$$

with ι^{co} as in (10.7.26).

10.8. Notes

10.8.1 (Elmendorf-Mandell K-Theory). Continuing from Note 9.5.1, the constructions $\mathsf{J}^{\mathcal{T}}$, J^{EM}, and K^{EM} were first presented in [**EM06, EM09**]. The first of these uses the $\mathcal{G}_*$-categories $\mathsf{C}^{\mathcal{G}}$ (Section 10.4), and the second uses a description equivalent to that of Section 10.3.

A description of $\mathcal{M}\underline{1}$-modules is also given in [**EM09**] (there called E-modules, where E is the *terminal parameter multicategory for modules*). The important structure result, Proposition 10.1.28, is from [**EM09**, Proposition 5.1] (see Note 10.8.4). However, [**EM09**] contains incorrect statements that $\mathsf{J}^{\mathcal{T}}$ (there called J') extends beyond $\mathsf{Mod}^{\mathcal{M}\underline{1}}$ to all of $\mathsf{Multicat}_*$. The error is related to the monoidal unit, and we explain it further in Question A.5.1 below. We identify the necessary corrections in Note 10.8.2 ◇

10.8.2 (Corrections to [**EM09**]). As mentioned in Note 10.8.1, and discussed further in Question A.5.1 below, the symmetric monoidal functor $\mathsf{J}^{\mathcal{T}}$ does not extend, as a symmetric monoidal functor, to all of $\mathsf{Multicat}_*$. The restriction of its domain to $\mathsf{Mod}^{\mathcal{M}\underline{1}}$ is essential to its satisfying the monoidal unity axioms as in (10.3.18). These, in turn, are essential to the preservation of composition with identity operations in the associated multicategory.

To account for this subtlety, the following statements in [**EM09**] should be corrected:

- In Theorems 1.3 and 5.14, and in Definition 5.15, $\mathbf{Mult}_*$ should be replaced with $\mathsf{Mod}^{\mathcal{M}\underline{1}}$ (there called the category of E-modules).
- On page 2392, lines 1 and 2 (just before the statement of Theorem 1.1), the phrase beginning "with the objects of the larger category being..." should end with "$\mathcal{M}\underline{1}$-modules." Note that $\mathsf{Mod}^{\mathcal{M}\underline{1}}$ is bicomplete and symmetric monoidal closed as desired, by Proposition 10.2.22.
- On page 2432, the second to last sentence of section 5, which reads "The lax structure map for the unit is then given by the unique map to the terminal object," should be removed and replaced with the unit constraint $(\mathsf{J}^{\mathcal{T}})^0$ from Definition 10.3.16.

If the unit constraint $(\mathsf{J}^{\mathcal{T}})^0$ is the unique map to the terminal object in $\mathcal{G}_*$-Cat, then the unit axioms of a monoidal functor would imply that every left and right unit isomorphism factors through the zero morphism. But that is not the case.

In the earlier paper [**EM06**], the construction $\mathsf{J}^{\mathcal{T}}$ (there called J) is only discussed as a multifunctor from $\mathsf{PermCat}^{\mathsf{su}}$. Since

$$\mathsf{End}\colon \mathsf{PermCat}^{\mathsf{su}} \longrightarrow \mathsf{Multicat}_*$$

takes values in $\mathsf{Mod}^{\mathcal{M}\underline{1}}$ (Lemma 10.2.14), no corrections are necessary. In the proof of [**EM06**, Theorem 6.1], the image of a 0-linear functor, as described in Explanation 10.5.4, is left implicit. ◇

10.8.3 (Structure of $\mathsf{J}^{\mathcal{T}}$ on $\mathsf{Multicat}_*$)**.** Question A.5.1 and Note 10.8.2 explain that $\mathsf{J}^{\mathcal{T}}$ cannot extend, as a monoidal functor, to all of $\mathsf{Multicat}_*$. However we do have the underlying 2-functor

$$\mathsf{J}^{\mathcal{T}} = \mathsf{Multicat}_*(\mathcal{T}, -)$$

defined on all of $\mathsf{Multicat}_*$. Moreover, the monoidal constraint $(\mathsf{J}^{\mathcal{T}})^2$ and the verifications of the associativity and symmetry axioms in Theorem 10.3.17 are not restricted to left $\mathcal{M}\underline{1}$-modules. It appears that the unit constraint and unity axioms are the essential reasons that $\mathsf{J}^{\mathcal{T}}$ cannot extend to a symmetric monoidal functor on all of $\mathsf{Multicat}_*$. ◇

10.8.4 (Monoids With Invertible Multiplication)**.** Proposition 10.1.28 shows that the category of $\mathcal{M}\underline{1}$-modules is a full subcategory of $\mathsf{Multicat}_*$. This result holds generally for the subcategory of modules over a monoid M in a monoidal category such that the multiplication map for M is an isomorphism. The result is proved with this generality in [**EM09**, Proposition 5.1]. As noted there, a more general result for algebras over idempotent monads is given in [**Bor94b**, Proposition 4.2.3]. ◇

10.8.5 (Notation Comparison)**.** For comparison with [**EM06, EM09**], we give the following table of corresponding notation.

[**EM06, EM09**]	Our Notation	Reference
$\mathcal{G}_*$	$\mathcal{G}$	Definition 9.1.7
$q: \underline{r} \longrightarrow \underline{s}$	$f: \overline{p} \hookrightarrow \overline{q}$	Definition 9.1.2
$q_*\langle \mathbf{m} \rangle$	$f_*\langle \underline{n} \rangle$	Definition 9.1.4
$(\alpha, q): \langle \mathbf{m} \rangle \longrightarrow \langle \mathbf{n} \rangle$	$(f, \langle \psi \rangle): \langle \underline{n} \rangle \longrightarrow \langle \underline{m} \rangle$	Definition 9.1.7
K^{Seg}, K^{New}	K^{Se}, K^{EM}	Definitions 8.5.1 and 10.3.32
α_*	$\langle \tilde{\psi} \rangle$	Definition 10.4.16 (2)
E	$\mathcal{M}\underline{1}$	Example 8.4.5
$\overline{\mathsf{C}}\langle \mathbf{m} \rangle$	$\mathsf{C}^{\mathcal{G}}\langle \underline{n} \rangle$	Definition 10.4.10
$p_m, q_{\langle m \rangle}$	i^*, j^*	Proposition 10.6.7
J,J'	$\mathsf{J}^{\mathcal{T}}$	Definition 10.3.9

◇

10.8.6 (Segal K-Theory and Symmetric Monoidal Structure)**.** In Explanation 9.2.18 we note that Γ-C carries a symmetric monoidal structure defined similarly (via Day convolution) to that of $\mathcal{G}_*$-C. Theorem 9.4.18 shows that the functor

$$\mathsf{K}^{\mathcal{F}}: \Gamma\text{-}\mathsf{sSet} \longrightarrow \mathsf{SymSp}$$

of Proposition 8.2.6 is symmetric monoidal as a sSet_*-functor. However one encounters difficulties attempting to develop a symmetric monoidal structure for

$$\mathsf{J}^{\mathsf{Se}}: \mathsf{PermCat}^{\mathsf{su}} \longrightarrow \Gamma\text{-}\mathsf{Cat}.$$

First, as noted in Propositions 5.7.23 and 10.2.17, the symmetric monoidal structure for small pointed multicategories does not restrict to a symmetric monoidal structure for $\mathsf{PermCat}^{\mathsf{su}}$. When one extends to use the symmetric monoidal structure on pointed multicategories, as we do for J^{EM} in Section 10.3, one faces a more fundamental difficulty elaborated in [**EM06**, pp. 181–182]. To produce a pairing

$$\mathsf{C}^{\mathcal{F}}\underline{m} \wedge \mathsf{C}^{\mathcal{F}}\underline{n} \longrightarrow \mathsf{C}^{\mathcal{F}}\underline{mn}$$

on objects, one needs to produce an $(\underline{m} \wedge \underline{n})$-system, in the sense of Definition 8.3.1, from an $\underline{m}$-system and an $\underline{n}$-system. The essential problem is that the (basepoint-free) subsets of $\underline{m} \wedge \underline{n}$ cannot be effectively determined by the subsets of $\underline{m}$ and $\underline{n}$ alone. One can take products of subsets, as in Definition 10.1.1 and Proposition 10.6.2, but these will not produce all the necessary subsets of the product. Efforts to resolve that difficulty were the motivation for the development, in [**EM06, EM09**], of $\mathcal{G}$, $\mathsf{K}^{\mathcal{G}}$, J^{EM}, and K^{EM}.

The preceding discussion can be rephrased in terms of the partition multicategory functors

$$\mathcal{M}: \mathcal{F}^{\mathsf{op}} \longrightarrow \mathsf{Multicat}_* \quad \text{and} \quad \mathcal{T}: \mathcal{G}^{\mathsf{op}} \longrightarrow \mathsf{Mod}^{\mathcal{M}\underline{1}}$$

from Definitions 8.4.1 and 10.3.1: both are symmetric monoidal functors, but only $\mathcal{T}$ is strong (see Proposition 10.1.6, Explanation 10.1.10, and Proposition 10.3.7).

The inverse monoidal constraint for $\mathcal{T}$ is used in (10.3.13) and is an essential part of the proof in Theorem 10.3.17 that $\mathsf{J}^{\mathcal{T}}$, and hence also J^{EM}, is a symmetric monoidal functor. ◇

10.8.7 (Multiplicative K-Theory Machines)**.** In addition to the Elmendorf-Mandell K-theory multifunctor, there is also May's construction [**May09a**], which improves the original version in [**May77, May82**]. See also the companion papers [**May09b, May09c**]. A further extension to multiplicative *equivariant* K-theory is given in [**GMMO∞**].

We follow the Elmendorf-Mandell approach in part because it has operadic structure built into it in the sense that K^{EM} is a multifunctor that respects the enrichment. In particular, it allows us to seamlessly include E_n-symmetric spectra for $1 \le n \le \infty$ as the target of one construction K^{EM}. ◇

10.8.8 (Lax versus Strong)**.** In Section 10.7 we compare the lax, colax, and strong variant constructions of Elmendorf-Mandell K-theory, showing that they result in level-equivalent symmetric spectra. A similar result for the variant Segal K-theory constructions is recorded in Theorems 8.3.21 and 8.5.2.

The paper [**BO20**] uses the strong Elmendorf-Mandell K-theory $\mathsf{K}^{\mathsf{EM}}_{\cong}$. The conclusions of Theorem 10.7.16 are mentioned in a footnote in [**BO20**, p. 1215]. For the Cat-enriched multifunctor $(-)^{\mathcal{G}}_{\cong}$, the necessity to restrict to *strong* multilinear functors, which we discuss in Explanation 10.7.21, is mentioned in [**BO20**, Remark 3.11]. ◇

10.8.9 (Waldhausen K-Theory)**.** In addition to the Segal K-theory and the equivalent Elmendorf-Mandell K-theory of a small permutative category, there is also Waldhausen K-theory of a small Waldhausen category [**Wal85**]. When the latter satisfies a split cofibration hypothesis, Waldhausen showed that his K-theory functor and Segal's yield equivalent spectra. The main results in [**BO20**] show that this equivalence of K-theory spectra can be promoted to one between multifunctors, using the multifunctorial Waldhausen K-theory in [**BM11, GH06, Zak18**] and the strong Elmendorf-Mandell K-theory multifunctor $\mathsf{K}^{\mathsf{EM}}_{\cong}$. ◇

10.8.10 (Quillen K-Theory)**.** The K-theory constructions of Waldhausen discussed above in Note 10.8.9, along with those of Segal and Elmendorf-Mandell, are generalizations of Quillen's *higher algebraic K-groups* [**Qui73**]. For a commutative ring R, these higher algebraic K groups are given by the homotopy groups of the space $BGL(R)^+$, where $GL(R)$ is the infinite general linear group of R, its classifying space is $BGL(R)$, and $(-)^+$ is Quillen's *plus construction* that abelianizes π_1 while preserving homology groups of a space.

To relate Segal's work to that of Quillen, the permutative category of interest is $\mathsf{S} = \coprod_n GL_n(R)$, the category whose object set is the natural numbers and whose morphisms are the elements of the general linear groups. Quillen gives an explicit construction, related in [**Gra76**] and also [**Wei13**, Section IV.4], that inverts the monoidal sum in S to yield an infinite loop space equivalent to $\mathbb{Z} \times BGL(R)^+$. This space is the zeroth space of the spectrum $\mathsf{K}^{\mathsf{Se}}(\mathsf{S})$. The Segal and Elmendorf-Mandell constructions thus generalize that of Quillen to arbitrary permutative categories. This is why their constructions are regarded as general forms of K-theory. ◇

CHAPTER 11

K-Theory of Ring and Bipermutative Categories

The Elmendorf-Mandell K-theory multifunctor (Definition 10.3.32)

$$\mathsf{PermCat}^{\mathsf{su}} \xrightarrow{\mathsf{K}^{\mathsf{EM}}} \mathsf{SymSp}$$

sends small permutative categories to symmetric spectra. It provides one of the most important connections between category theory, homotopy theory, and algebraic K-theory. Since SymSp is a symmetric monoidal closed category (Theorem 7.6.15), it makes sense to talk about *structured symmetric spectra*, which are symmetric spectra that are algebras of operads. Among the most important structured symmetric spectra are the *E_n-symmetric spectra* for $1 \le n \le \infty$.

- An E_1-symmetric spectrum is a strict ring symmetric spectrum, that is, a monoid in SymSp. They form the most basic multiplicative structure in symmetric spectra.
- An E_∞-symmetric spectrum is a strict ring symmetric spectrum whose multiplication is commutative up to all higher coherent homotopies. In a model categorical sense, each E_∞-symmetric spectrum is weakly equivalent to a strictly commutative ring symmetric spectrum. For example, the sphere symmetric spectrum in Definition 7.4.1 is an E_∞-symmetric spectrum.
- An E_n-symmetric spectrum for $1 < n < \infty$ sits somewhere between the E_1 and the E_∞ cases. For example, at each prime p, the Brown-Peterson spectrum is an E_4-symmetric spectrum but not an E_∞-symmetric spectrum. See also Question A.4.2.

In this chapter, we apply the Elmendorf-Mandell K-theory multifunctor K^{EM} to ring categories and bipermutative categories (Chapter II.9) to produce, respectively, strict ring symmetric spectra and E_∞-symmetric spectra. These results are due to Elmendorf-Mandell [**EM06**]. See Corollaries 11.3.16 and 11.6.12 and also Note 11.7.4 for the differences between the presentation in [**EM06**] and this chapter. In Chapters 12 and 13, using our braided ring categories and E_n-monoidal categories (Definitions II.9.5.1 and II.10.7.2), the main results in this chapter will be expanded to the E_n cases for $1 < n < \infty$.

Each of Corollaries 11.3.16 and 11.6.12 is a combination of several key facts. First, the Elmendorf-Mandell K-theory multifunctor respects

- the categorical enrichment in the Cat-enriched multicategory $\mathsf{PermCat}^{\mathsf{su}}$ of small permutative categories and
- the simplicial enrichment in the symmetric monoidal closed category SymSp of symmetric spectra.

The fact that K^{EM} is a multifunctor that respects the enrichment implies that a structure in $\mathsf{PermCat}^{\mathsf{su}}$ that is parametrized by a Cat-enriched operad passes along

K^{EM} to symmetric spectra. Therefore, Corollaries 11.3.16 and 11.6.12 will follow once we know the parameter operads in $\mathsf{PermCat}^{\mathsf{su}}$ for, respectively, ring categories and bipermutative categories in Definitions II.9.1.2 and II.9.3.2.

For the E_1 case, the associative operad As parametrizes

- monoids in a general symmetric monoidal category (Proposition 11.1.15) and
- ring categories in the multicategory $\mathsf{PermCat}^{\mathsf{su}}$ (Theorem 11.2.16).

Both of these statements are consequences of the Coherence Theorem 11.1.7 for As. Combining Theorem 11.2.16 with K^{EM} yields Corollary 11.3.16, which says that, for each small ring category C, $\mathsf{K}^{\mathsf{EM}}\mathsf{C}$ is a strict ring symmetric spectrum. Along the same lines, the categorical Barratt-Eccles operad $E\mathsf{As}$ is an E_∞-operad that parametrizes

- permutative categories in Cat (Proposition 11.4.26) and
- bipermutative categories in $\mathsf{PermCat}^{\mathsf{su}}$ (Theorem 11.5.5).

Both of these statements are consequences of the Coherence Theorem 11.4.14 for $E\mathsf{As}$. Combining Theorem 11.5.5 with K^{EM} yields Corollary 11.6.12, which says that, for each small bipermutative category C, $\mathsf{K}^{\mathsf{EM}}\mathsf{C}$ is an E_∞-symmetric spectrum. The following table summaries this discussion.

operad	associative As (11.1.1)	Barratt-Eccles $E\mathsf{As}$ (11.4.10)
$E_?$-operad	E_1 (13.1.23, 13.2.1)	E_∞ (11.6.3)
coherence	11.1.7	11.4.14
in Cat	strict monoidal (11.1.15)	permutative (11.4.26)
in $\mathsf{PermCat}^{\mathsf{su}}$	ring (11.2.16)	bipermutative (11.5.5)
in SymSp	strict ring (11.3.16)	E_∞ (11.6.12)

For open questions related to the Barratt-Eccles operad $E\mathsf{As}$, see Question A.4.3.

Organization. Section 11.1 discusses the associative operad As from scratch. After defining As, Lemma 11.1.4 provides a detailed proof that As is actually an operad. The nth object in As is the nth symmetric group Σ_n. Its operad composition is defined by block sums and block permutations. The Coherence Theorem 11.1.7 describes As in terms of

- two generators, one for the unit and one for the multiplication, and
- two relations, one for unity and one for associativity.

The first application of Theorem 11.1.7 is Proposition 11.1.15, which says that As is the operad for monoids.Section 11.2 proves Theorem 11.2.16, which says that the associative operad As detects ring category structures on small permutative categories. This is another application of the Coherence Theorem 11.1.7 for As. Section 11.3 proves the main result about strict ring symmetric spectra, Corollary 11.3.16, which says that $\mathsf{K}^{\mathsf{EM}}\mathsf{C}$ is a strict ring symmetric spectrum for each small ring category C. This result is a consequence of Theorem 11.2.16 and the Elmendorf-Mandell K-theory multifunctor.

Section 11.4 begins by defining the translation category of a set. Applied to the associative operad, this yields the Cat-enriched Barratt-Eccles operad $E\mathsf{As}$. The Coherence Theorem 11.4.14 describes the Barratt-Eccles operad in terms of

- two generating objects for the unit and the multiplication,
- one generating isomorphism for the symmetry isomorphism, and
- generating relations corresponding to those of a permutative category.

The first application of Theorem 11.4.14 is Proposition 11.4.26, which says that $E\mathsf{As}$ is the Cat-enriched operad for permutative categories.

Section 11.5 proves Theorem 11.5.5, which says that the Barratt-Eccles operad $E\mathsf{As}$ detects bipermutative category structures on small permutative categories. This is another application of the Coherence Theorem 11.4.14 for $E\mathsf{As}$. Section 11.6 proves the main result about E_∞-symmetric spectra, Corollary 11.6.12, which says that $\mathsf{K}^{\mathsf{EM}}\mathsf{C}$ is an E_∞-symmetric spectrum for each small bipermutative category C. This result is a consequence of Theorem 11.5.5 and the Elmendorf-Mandell K-theory multifunctor.

Reading Guide.

(1) Related to the associative operad, read Definition 11.1.1 and the statements of Theorems 11.1.7 and 11.2.16, Proposition 11.1.15, and Corollary 11.3.16.
(2) Related to the Barratt-Eccles operad, read Definitions 11.4.1, 11.4.10, 11.6.1, and 11.6.5 and the statements of Theorems 11.4.14 and 11.5.5, Propositions 11.4.26 and 11.6.3, and Corollary 11.6.12.
(3) Go back and read the rest of this chapter.

11.1. The Associative Operad

Recall from Definition 5.1.2 that an *operad* is a one-object multicategory. In this section, we describe the associative operad explicitly and also in terms of generators and relations (Theorem 11.1.7). As the first application of Theorem 11.1.7, we observe in Proposition 11.1.15 that the associative operad detects monoid structures. Moreover, in Theorem 11.2.16, we use Theorem 11.1.7 to describe ring category structures on small permutative categories in terms of the associative operad. Recall that Σ_n denotes the nth symmetric group, with identity $\mathrm{id}_n \in \Sigma_n$.

Definition 11.1.1. The *associative operad* As is defined by the following data.

Operations: $\mathsf{As}_n = \Sigma_n$ for $n \geq 0$.

Equivariance: The right symmetric group action

$$\mathsf{As}_n \times \Sigma_n \longrightarrow \mathsf{As}_n$$

is the group multiplication in Σ_n.

Unit: The operad unit is $\mathrm{id}_1 \in \mathsf{As}_1 = \Sigma_1$.

Composition: For $n \geq 1$ and $k_j \geq 0$ for $1 \leq j \leq n$, the operad composition

$$\mathsf{As}_n \times \prod_{j=1}^n \mathsf{As}_{k_j} \xrightarrow{\gamma} \mathsf{As}_{k_1+\cdots+k_n}$$

is defined by

$$\gamma\big(\sigma,(\tau_1,\ldots,\tau_n)\big) = \overbrace{\sigma\langle k_1,\ldots,k_n\rangle}^{\text{block permutation}} \cdot \overbrace{(\tau_1 \times \cdots \times \tau_n)}^{\text{block sum}} \tag{11.1.2}$$

for $\sigma \in \Sigma_n$ and $\tau_j \in \Sigma_{k_j}$. On the right-hand side, the block sum and the block permutation are as in, respectively, (II.1.1.8) and (II.1.1.19).

This finishes the definition of As. ◇

Explanation 11.1.3. The permutation in (11.1.2) is given by

$$\begin{aligned}&\big(\sigma\langle k_1,\ldots,k_n\rangle\cdot(\tau_1\times\cdots\times\tau_n)\big)\big(\overbrace{k_1+\cdots+k_{j-1}}^{0 \text{ if } j=1}+i\big)\\&=\underbrace{k_{\sigma^{-1}(1)}+\cdots+k_{\sigma^{-1}(\sigma(j)-1)}}_{0 \text{ if } \sigma(j)=1}+\tau_j(i)\end{aligned}$$

for $1 \le j \le n$ and $1 \le i \le k_j$.

- This permutation first permutes within the jth block via $\tau_j \in \Sigma_{k_j}$ for each $1 \le j \le n$.
- Then it permutes the n resulting consecutive blocks of lengths $k_1, \ldots, k_n$ via $\sigma \in \Sigma_n$.

On the left-hand side, $k_1 + \cdots + k_{j-1} + i$ is the ith element in the jth block. On the right-hand side, its image under the permutation is the $\tau_j(i)$th element in the new $\sigma(j)$th block. ◇

Lemma 11.1.4. As *in Definition 11.1.1 is an operad.*

Proof. We check the operad axioms in Definition 5.1.2 for As.

Unity. The right unity axiom (5.1.5) and the left unity axiom (5.1.6) follow from the following permutation equalities.

$$\begin{aligned}\overbrace{\mathrm{id}_1\times\cdots\times\mathrm{id}_1}^{n}&=\mathrm{id}_n\in\Sigma_n\\\sigma\langle 1,\ldots,1\rangle&=\sigma\\\mathrm{id}_1\langle k_1\rangle&=\mathrm{id}_{k_1}\end{aligned}$$

Equivariance. The top equivariance axiom (5.1.7) holds by the following computation for $\sigma, \sigma' \in \Sigma_n$ with $n \ge 1$ and $\tau_j \in \Sigma_{k_j}$ for $1 \le j \le n$.

$$\begin{aligned}&\gamma\big(\sigma,(\tau_1,\ldots,\tau_n)\big)\cdot\sigma'\langle k_{\sigma'(1)},\ldots,k_{\sigma'(n)}\rangle\\&=\sigma\langle k_1,\ldots,k_n\rangle\cdot(\tau_1\times\cdots\times\tau_n)\cdot\sigma'\langle k_{\sigma'(1)},\ldots,k_{\sigma'(n)}\rangle\\&=\sigma\langle k_1,\ldots,k_n\rangle\cdot\sigma'\langle k_{\sigma'(1)},\ldots,k_{\sigma'(n)}\rangle\cdot\big(\tau_{\sigma'(1)}\times\cdots\times\tau_{\sigma'(n)}\big)\\&=(\sigma\sigma')\langle k_{\sigma'(1)},\ldots,k_{\sigma'(n)}\rangle\cdot\big(\tau_{\sigma'(1)}\times\cdots\times\tau_{\sigma'(n)}\big)\\&=\gamma\Big(\sigma\sigma',\big(\tau_{\sigma'(1)},\ldots,\tau_{\sigma'(n)}\big)\Big)\end{aligned}$$

The bottom equivariance axiom (5.1.8) follows from the fact that block sums preserve products in the sense that

$$(\tau_1\times\cdots\times\tau_n)\cdot(\tau_1'\times\cdots\times\tau_n')=\tau_1\tau_1'\times\cdots\times\tau_n\tau_n' \tag{11.1.5}$$

for $\tau_j' \in \Sigma_{k_j}$ and $1 \le j \le n$.

Associativity. For the associativity axiom (5.1.4), suppose further that $\theta_{j,i} \in \Sigma_{l_{j,i}}$ for $1 \le j \le n$ and $1 \le i \le k_j$. For each $1 \le j \le n$, suppose

$$l_j = l_{j,1} + \cdots + l_{j,k_j}.$$

Then there is an equality of permutations

$$\begin{aligned}(11.1.6)\qquad &\Big(\sigma\langle k_1,\ldots,k_n\rangle\cdot\prod_{j=1}^{n}\tau_j\Big)\langle\overbrace{l_{1,1},\ldots,l_{1,k_1},\ldots,l_{n,1},\ldots,l_{n,k_n}}^{((l_{j,i})_{1\le i\le k_j})_{1\le j\le n}}\rangle\\ &=\sigma\langle l_1,\ldots,l_n\rangle\cdot\prod_{j=1}^{n}\tau_j\langle l_{j,1},\ldots,l_{j,k_j}\rangle\in\Sigma_{l_1+\cdots+l_n}.\end{aligned}$$

Indeed, on each side of the equal sign in (11.1.6), the permutation is given by

$$\Big(\sum_{p=1}^{j-1}l_p\Big)+\Big(\sum_{q=1}^{i-1}l_{j,q}\Big)+h\longmapsto\Big(\sum_{p=1}^{\sigma(j)-1}l_{\sigma^{-1}(p)}\Big)+\Big(\sum_{q=1}^{\tau_j(i)-1}l_{j,\tau_j^{-1}(q)}\Big)+h$$

for $1\le j\le n$, $1\le i\le k_j$, and $1\le h\le l_{j,i}$, with the usual convention that an empty sum means 0.

The associativity axiom (5.1.4) holds in As by the following computation, where the second and the third equalities hold by, respectively, (11.1.6) and (11.1.5).

$$\begin{aligned}&\gamma\Big(\gamma\big(\sigma,(\tau_1,\ldots,\tau_n)\big),\big((\theta_{1,i})_{1\le i\le k_1},\ldots,(\theta_{n,i})_{1\le i\le k_n}\big)\Big)\\ &=\Big(\sigma\langle k_1,\ldots,k_n\rangle\cdot\prod_{j=1}^{n}\tau_j\Big)\langle l_{1,1},\ldots,l_{1,k_1},\ldots,l_{n,1},\ldots,l_{n,k_n}\rangle\cdot\prod_{j=1}^{n}\prod_{i=1}^{k_j}\theta_{j,i}\\ &=\sigma\langle l_1,\ldots,l_n\rangle\cdot\Big[\prod_{j=1}^{n}\tau_j\langle l_{j,1},\ldots,l_{j,k_j}\rangle\Big]\cdot\prod_{j=1}^{n}\prod_{i=1}^{k_j}\theta_{j,i}\\ &=\sigma\langle l_1,\ldots,l_n\rangle\cdot\prod_{j=1}^{n}\Big[\tau_j\langle l_{j,1},\ldots,l_{j,k_j}\rangle\cdot\prod_{i=1}^{k_j}\theta_{j,i}\Big]\\ &=\gamma\Big(\sigma,\big(\gamma(\tau_j,(\theta_{j,1},\ldots,\theta_{j,k_j}))\big)_{1\le j\le n}\Big)\end{aligned}$$

This proves that As is an operad. □

Coherence of the Associative Operad. An *operad morphism* means a multifunctor (Definition 5.1.12) between two one-object multicategories. So an operad morphism $f:\mathsf{P}\longrightarrow\mathsf{Q}$ between two operads P and Q consists of structure morphisms

$$f_n:\mathsf{P}_n\longrightarrow\mathsf{Q}_n\quad\text{for}\quad n\ge 0$$

that preserve the symmetric group action, operad units, and operad composition in the sense of (5.1.13)–(5.1.15). The subscripts in the structure morphisms will often be omitted. The next coherence result characterizes the associative operad in terms of generators and relations.

Theorem 11.1.7. *The associative operad* As *is operadically generated by the permutations*

$$\mathrm{id}_0\in\Sigma_0\quad\textit{and}\quad\mathrm{id}_2\in\Sigma_2$$

and is subject to the following unity and associativity relations.

$$\begin{aligned}(11.1.8)\qquad \gamma\big(\mathrm{id}_2,(\mathrm{id}_0,\mathrm{id}_1)\big)&=\mathrm{id}_1=\gamma\big(\mathrm{id}_2,(\mathrm{id}_1,\mathrm{id}_0)\big)\\ \gamma\big(\mathrm{id}_2,(\mathrm{id}_2,\mathrm{id}_1)\big)&=\mathrm{id}_3=\gamma\big(\mathrm{id}_2,(\mathrm{id}_1,\mathrm{id}_2)\big)\end{aligned}$$

Proof. Suppose $\mathsf{F}(\mu, \mathbb{0})$ is the operad freely generated by the elements

$$\mu \in \mathsf{F}_2 \quad \text{and} \quad \mathbb{0} \in \mathsf{F}_0,$$

and suppose $1 \in \mathsf{F}_1$ is the operad unit. Suppose As' is the quotient of $\mathsf{F}(\mu, \mathbb{0})$ by the relations operadically generated by

$$\begin{cases} \mu(\mathbb{0}, 1) = 1 = \mu(1, \mathbb{0}) \\ \mu(\mu, 1) = \mu(1, \mu) \end{cases} \tag{11.1.9}$$

in which we used the juxtaposition notation for operad composition. An element in $\mathsf{F}(\mu, \mathbb{0})$ and its image in As' will be denoted by the same symbol below. We will show that As' is canonically isomorphic to As.

By (11.1.9) and an induction, each operad composite in As'_n involving only

$$\mu_0 = \mathbb{0} \in \mathsf{As}'_0, \quad \mu_1 = 1 \in \mathsf{As}'_1, \quad \text{and} \quad \mu_2 = \mu \in \mathsf{As}'_2$$

yields the same element, which is denoted by $\mu_n \in \mathsf{As}'_n$. This property will be called the *uniqueness of* μ_n below. There is a unique operad morphism

$$\mathsf{F}(\mu, \mathbb{0}) \xrightarrow{\ \phi\ } \mathsf{As} \quad \text{such that} \quad \begin{cases} \phi(\mu) = \mathrm{id}_2 \in \Sigma_2 \\ \phi(\mathbb{0}) = \mathrm{id}_0 \in \Sigma_0. \end{cases}$$

Since the relations (11.1.8) hold in As, ϕ preserves the relations (11.1.9). So it factors through the quotient operad As' to yield an operad morphism

$$\mathsf{As}' \xrightarrow{\ \phi'\ } \mathsf{As} \quad \text{such that} \quad \phi'(\mu_n) = \mathrm{id}_n \tag{11.1.10}$$

for $n \geq 0$.

In the other direction, there is an operad morphism as follows.

$$\begin{array}{ccc} \mathsf{As} & \xrightarrow{\ \varphi\ } & \mathsf{As}' \\ \Sigma_n \ni \sigma & \longmapsto & \mu_n\sigma \in \mathsf{As}'_n \end{array} \tag{11.1.11}$$

Indeed, φ preserves the operad unit and the symmetric group action by definition. The preservation of the operad composition means the commutativity of the diagram

$$\begin{array}{ccc} \mathsf{As}_n \times \prod_{j=1}^n \mathsf{As}_{k_j} & \xrightarrow{\ \gamma\ } & \mathsf{As}_k \\ {\scriptstyle \varphi \times \prod_j \varphi}\downarrow & & \downarrow{\scriptstyle \varphi} \\ \mathsf{As}'_n \times \prod_{j=1}^n \mathsf{As}'_{k_j} & \xrightarrow{\ \gamma\ } & \mathsf{As}'_k \end{array} \tag{11.1.12}$$

with $k = k_1 + \cdots + k_n$. By

- the equivariance axioms (5.1.7) and (5.1.8) in As and As' and
- the fact that φ preserves the symmetric group action,

it suffices to consider the element

$$\left(\mathrm{id}_n, (\mathrm{id}_{k_j})_{j=1}^n\right) \in \mathsf{As}_n \times \prod_{j=1}^n \mathsf{As}_{k_j}.$$

On this element, the diagram (11.1.12) is commutative by the uniqueness of $\mu_k \in \mathsf{As}'_k$. Finally, observe that the operad morphisms ϕ' in (11.1.10) and φ in (11.1.11) are inverses of each other. □

Monoids as Algebras. An operad, such as As, is regarded as an operad in a permutative category $(\mathsf{C}, \otimes, \mathbb{1})$ with all small coproducts, in which $\otimes$ commutes with small coproducts on each side separately, via the strong symmetric monoidal functor

$$\begin{array}{ccc} \mathsf{Set} & \longrightarrow & \mathsf{C} \\ X & \longmapsto & \coprod_X \mathbb{1}. \end{array} \tag{11.1.13}$$

Recall from Definition I.1.2.28 that a symmetric monoidal category is *closed* if, for each object A, the functor $- \otimes A$ has a right adjoint $[A, -]$, which is called the internal hom. If C is closed, then, by adjunction, $\otimes$ commutes with all small colimits that exist in C on each side separately.

In Proposition 11.1.15 below, we use the C-*enriched endomorphism operad* defined using the internal hom as in Proposition 6.3.6 with $A \in \mathsf{C} = \mathsf{V} = \mathsf{K}$. Its objects are

$$\mathsf{End}(A)_n = [A^{\otimes n}, A] \in \mathsf{C} \quad \text{for} \quad n \geq 0. \tag{11.1.14}$$

Recall from Definition I.1.2.8 that, in a monoidal category $(\mathsf{C}, \otimes, \mathbb{1}, \alpha, \lambda, \rho)$, a *monoid* (A, μ, η) consists of

- an object $A \in \mathsf{C}$,
- a multiplication $\mu: A \otimes A \longrightarrow A$, and
- a unit $\eta: \mathbb{1} \longrightarrow A$

that are associative and unital in the sense that the following diagrams are commutative.

$$\begin{array}{ccc} (A \otimes A) \otimes A & \xrightarrow{\alpha} & A \otimes (A \otimes A) \\ & & \downarrow {\scriptstyle 1_A \otimes \mu} \\ \downarrow {\scriptstyle \mu \otimes 1_A} & & A \otimes A \\ & & \downarrow {\scriptstyle \mu} \\ A \otimes A & \xrightarrow{\mu} & A \end{array} \qquad \begin{array}{ccc} \mathbb{1} \otimes A & \xrightarrow{\lambda_A} & A \\ \downarrow {\scriptstyle \eta \otimes 1_A} & & \| \\ A \otimes A & \xrightarrow{\mu} & A \\ \uparrow {\scriptstyle 1_A \otimes \eta} & & \| \\ A \otimes \mathbb{1} & \xrightarrow{\rho_A} & A \end{array}$$

The next observation is a consequence of the characterization of the associative operad in Theorem 11.1.7.

Proposition 11.1.15. *Suppose* $(\mathsf{C}, \otimes, \mathbb{1})$ *is a permutative closed category with all small coproducts. Then, for each object* $A \in \mathsf{C}$*, a monoid structure on* A *is uniquely determined by an operad morphism* $\mathsf{As} \longrightarrow \mathsf{End}(A)$.

Proof. Given an operad morphism

$$\phi: \mathsf{As} \longrightarrow \mathsf{End}(A),$$

the adjoint of the structure morphism

$$\mathsf{As}_0 = \mathbb{1} \xrightarrow{\phi_0} [\mathbb{1}, A] = \mathsf{End}(A)_0 \tag{11.1.16}$$

is the monoid unit $\eta\colon \mathbb{1} \longrightarrow A$. The monoid multiplication μ of A is the following composite, with each $\mathbb{1}_?$ a copy of $\mathbb{1}$.

$$\begin{array}{ccc} A^{\otimes 2} & \xrightarrow{\quad\mu\quad} & A \\ {\scriptstyle \mathrm{id}_2}\big\downarrow & & \big\uparrow{\scriptstyle \phi_2^\#} \\ \big(\mathbb{1}_{\mathrm{id}_2} \amalg \mathbb{1}_{(1,2)}\big) \otimes A^{\otimes 2} & = & \mathsf{As}_2 \otimes A^{\otimes 2} \end{array} \tag{11.1.17}$$

The right vertical morphism $\phi_2^\#$ is the adjoint of the structure morphism

$$\mathsf{As}_2 = \mathbb{1}_{\mathrm{id}_2} \amalg \mathbb{1}_{(1,2)} \xrightarrow{\phi_2} \big[A^{\otimes 2}, A\big] = \mathsf{End}(A)_2.$$

The triple (A, μ, η) satisfies the monoid associativity and unity axioms in Definition I.1.2.8 by the corresponding properties (11.1.8) in As.

Conversely, given a monoid (A, μ, η), the assignment in the previous paragraph uniquely defines an operad morphism $\mathsf{As} \longrightarrow \mathsf{End}(A)$ by Theorem 11.1.7. By construction, these two assignments are inverses of each other. □

Remark 11.1.18. For symmetric monoidal categories that are not strict, we will use Proposition 11.1.15 along with Explanation 6.1.19. ◇

11.2. Detecting Ring Categories

In this section, we prove that the associative operad detects ring category structures on small permutative categories (Theorem 11.2.16). This result is due to Elmendorf-Mandell and is an application of the Coherence Theorem 11.1.7 for the associative operad.

Recall from Definition 11.1.1 and Lemma 11.1.4 that the associative operad As has $\mathsf{As}_n = \Sigma_n$, the nth symmetric group, and unit $\mathrm{id}_1 \in \mathsf{As}_1$. Its operad composition γ is given by the product

$$\gamma\big(\sigma, (\tau_1, \ldots, \tau_n)\big) = \overbrace{\sigma\langle k_1, \ldots, k_n\rangle}^{\text{block permutation}} \cdot \big(\overbrace{\tau_1 \times \cdots \times \tau_n}^{\text{block sum}}\big)$$

for $\sigma \in \Sigma_n$ and $\tau_j \in \Sigma_{k_j}$ for $1 \le j \le n$.

- By Theorem 11.1.7, As is operadically generated by $\mathrm{id}_0 \in \mathsf{As}_0$ and $\mathrm{id}_2 \in \mathsf{As}_2$ and is subject to unity and associativity relations.
- By Proposition 11.1.15, for a permutative closed category C with all small coproducts, As detects monoid structure. This means that a monoid structure on an object $A \in \mathsf{C}$ is uniquely determined by an operad morphism $\mathsf{As} \longrightarrow \mathsf{End}(A)$, with $\mathsf{End}(A)$ the C-enriched endomorphism operad in (11.1.14).

The unique object in the one-object multicategory As is denoted by $*$.

For each symmetric monoidal category $(\mathsf{C}, \otimes, \mathbb{1})$ with small coproducts, in which $\otimes$ commutes with small coproducts on each side separately, there is an adjunction

$$\mathsf{Set} \underset{\mathsf{C}(\mathbb{1},?)}{\overset{\amalg_? \mathbb{1}}{\rightleftarrows}} \mathsf{C}. \tag{11.2.1}$$

The right adjoint sends each object $A \in \mathsf{C}$ to the set $\mathsf{C}(\mathbb{1}, A)$ of morphisms $\mathbb{1} \longrightarrow A$ in C. The left adjoint is the functor in (11.1.13) that sends a set X to the coproduct $\amalg_X \mathbb{1}$. It is a strong symmetric monoidal functor. The associative operad As is regarded as a C-enriched operad via the left adjoint.

Example 11.2.2. For the symmetric monoidal category Cat of small categories and functors, the adjunction (11.2.1) becomes

$$\mathsf{Set} \underset{\mathsf{Ob}}{\overset{\mathsf{dis}}{\rightleftarrows}} \mathsf{Cat}. \tag{11.2.3}$$

The left adjoint sends each set X to the discrete category $\mathsf{dis}(X)$ with object set X. The right adjoint Ob is the forgetful functor that sends each small category to its set of objects. The associative operad As is also regarded as a Cat-enriched operad via the strong symmetric monoidal functor dis. So As_n is the discrete category with object set Σ_n. ◇

For the reader's convenience, here we recall from Chapter II.9 the definition of a ring category.

Definition 11.2.4. A *ring category* is a tuple

$$\big(\mathsf{C}, (\oplus, \mathbb{0}, \xi^{\oplus}), (\otimes, \mathbb{1}), (\partial^l, \partial^r)\big)$$

consisting of the following data.

The Additive Structure: $(\mathsf{C}, \oplus, \mathbb{0}, \xi^{\oplus})$ is a permutative category.

The Multiplicative Structure: $(\mathsf{C}, \otimes, \mathbb{1})$ is a strict monoidal category.

The Factorization Morphisms: ∂^l and ∂^r are natural transformations

$$\begin{aligned} (A \otimes C) \oplus (B \otimes C) &\xrightarrow{\partial^l_{A,B,C}} (A \oplus B) \otimes C \\ (A \otimes B) \oplus (A \otimes C) &\xrightarrow{\partial^r_{A,B,C}} A \otimes (B \oplus C) \end{aligned} \tag{11.2.5}$$

for objects $A, B, C \in \mathsf{C}$, which are called the *left factorization morphism* and the *right factorization morphism*, respectively.

To simplify the presentation, we often abbreviate $\otimes$ to concatenation, with $\otimes$ always taking precedence over $\oplus$ in the absence of clarifying parentheses. For example, the left factorization morphism is abbreviated to $AC \oplus BC \longrightarrow (A \oplus B)C$. The subscripts in $\xi^{\oplus}$, ∂^l, and ∂^r are sometimes omitted.

The above data are required to satisfy the following seven axioms for all objects A, A', A'', B, B', B'', C, and C' in C.

The Multiplicative Zero Axiom: The diagram of functors

$$\begin{array}{ccccc} * \times \mathsf{C} & \xrightarrow{\cong} & \mathsf{C} & \xleftarrow{\cong} & \mathsf{C} \times * \\ {\scriptstyle \mathbb{0} \times 1_{\mathsf{C}}} \downarrow & & \downarrow {\scriptstyle \mathbb{0}} & & \downarrow {\scriptstyle 1_{\mathsf{C}} \times \mathbb{0}} \\ \mathsf{C} \times \mathsf{C} & \xrightarrow{\otimes} & \mathsf{C} & \xleftarrow{\otimes} & \mathsf{C} \times \mathsf{C} \end{array} \tag{11.2.6}$$

is commutative. In this diagram, the top horizontal isomorphisms drop the $*$ argument. Each $\mathbb{0}$ denotes the constant functor at $\mathbb{0} \in \mathsf{C}$ and $1_{\mathbb{0}}$.

The Zero Factorization Axiom:

$$\begin{array}{ll} \partial^l_{\mathbb{0},B,C} = 1_{B \otimes C} & \partial^r_{\mathbb{0},B,C} = 1_{\mathbb{0}} \\ \partial^l_{A,\mathbb{0},C} = 1_{A \otimes C} & \partial^r_{A,\mathbb{0},C} = 1_{A \otimes C} \\ \partial^l_{A,B,\mathbb{0}} = 1_{\mathbb{0}} & \partial^r_{A,B,\mathbb{0}} = 1_{A \otimes B} \end{array} \tag{11.2.7}$$

The three equalities for ∂^l are called the *left zero factorization axioms*. The three equalities for ∂^r are called the *right zero factorization axioms*.

The Unit Factorization Axiom:

$$\begin{aligned} \partial^l_{A,B,\mathbb{1}} &= 1_{A\oplus B} \\ \partial^r_{\mathbb{1},B,C} &= 1_{B\oplus C} \end{aligned} \tag{11.2.8}$$

These are called, respectively, the *left* and the *right* unit factorization axioms.

The Symmetry Factorization Axiom: The following two diagrams in C are commutative.

$$\begin{array}{ccc} AC \oplus BC & \xrightarrow{\partial^l} & (A\oplus B)C \\ {\scriptstyle \xi^{\oplus}}\downarrow & & \downarrow{\scriptstyle \xi^{\oplus}1_C} \\ BC\oplus AC & \xrightarrow{\partial^l} & (B\oplus A)C \end{array} \qquad \begin{array}{ccc} AB \oplus AC & \xrightarrow{\partial^r} & A(B\oplus C) \\ {\scriptstyle \xi^{\oplus}}\downarrow & & \downarrow{\scriptstyle 1_A\xi^{\oplus}} \\ AC\oplus AB & \xrightarrow{\partial^r} & A(C\oplus B) \end{array} \tag{11.2.9}$$

These are called, respectively, the *left* and the *right* symmetry factorization axioms.

The Internal Factorization Axiom: The following two diagrams in C are commutative.

(11.2.10)

$$\begin{array}{ccc} AB\oplus A'B\oplus A''B & \xrightarrow{\partial^l\oplus 1} & (A\oplus A')B\oplus A''B \\ {\scriptstyle 1\oplus\partial^l}\downarrow & & \downarrow{\scriptstyle \partial^l} \\ AB\oplus(A'\oplus A'')B & \xrightarrow{\partial^l} & (A\oplus A'\oplus A'')B \end{array} \qquad \begin{array}{ccc} AB\oplus AB'\oplus AB'' & \xrightarrow{\partial^r\oplus 1} & A(B\oplus B')\oplus AB'' \\ {\scriptstyle 1\oplus\partial^r}\downarrow & & \downarrow{\scriptstyle \partial^r} \\ AB\oplus A(B'\oplus B'') & \xrightarrow{\partial^r} & A(B\oplus B'\oplus B'') \end{array}$$

These are called, respectively, the *left* and the *right* internal factorization axioms.

The External Factorization Axiom: The three diagrams in C below are commutative.

$$\begin{array}{ccc} ABC\oplus A'BC & \xrightarrow{\partial^l_{A,A',BC}} & (A\oplus A')BC \\ {\scriptstyle \partial^l_{AB,A'B,C}}\downarrow & & \Vert \\ (AB\oplus A'B)C & \xrightarrow{\partial^l_{A,A',B}1_C} & (A\oplus A')BC \end{array} \tag{11.2.11}$$

$$\begin{array}{ccc} ABC\oplus AB'C & \xrightarrow{\partial^l_{AB,AB',C}} & (AB\oplus AB')C \\ {\scriptstyle \partial^r_{A,BC,B'C}}\downarrow & & \downarrow{\scriptstyle \partial^r 1_C} \\ A(BC\oplus B'C) & \xrightarrow{1_A\partial^l_{B,B',C}} & A(B\oplus B')C \end{array} \tag{11.2.12}$$

$$\begin{array}{ccc} ABC\oplus ABC' & \xrightarrow{\partial^r_{AB,C,C'}} & AB(C\oplus C') \\ {\scriptstyle \partial^r_{A,BC,BC'}}\downarrow & & \Vert \\ A(BC\oplus BC') & \xrightarrow{1_A\partial^r_{B,C,C'}} & AB(C\oplus C') \end{array} \tag{11.2.13}$$

These are called, respectively, the *left*, the *middle*, and the *right* external factorization axioms.

The 2-By-2 Factorization Axiom: The following diagram in C is commutative.

$$\begin{array}{ccc} & A(B\oplus B')\oplus A'(B\oplus B') & \\ \nearrow^{\partial^r\oplus\partial^r} & & \searrow^{\partial^l} \\ AB\oplus AB'\oplus A'B\oplus A'B' & & (A\oplus A')(B\oplus B') \\ \downarrow{\scriptstyle 1\oplus\xi^{\oplus}\oplus 1} & & \uparrow{\scriptstyle \partial^r} \\ AB\oplus A'B\oplus AB'\oplus A'B' & & \\ \searrow_{\partial^l\oplus\partial^l} & & \nearrow \\ & (A\oplus A')B\oplus(A\oplus A')B' & \end{array} \tag{11.2.14}$$

This finishes the definition of a ring category.

Moreover, a ring category as above is said to be

- *small* if it has a set of objects and
- *tight* if ∂^l and ∂^r in (11.2.5) are natural isomorphisms. ◇

Definition 11.2.15. For a permutative category $(\mathsf{C}, \oplus, \mathbb{0}, \xi^{\oplus})$, a *ring category structure* on C is the additional data $(\otimes, \mathbb{1}, \partial^l, \partial^r)$ such that the tuple

$$\big(\mathsf{C}, (\oplus, \mathbb{0}, \xi^{\oplus}), (\otimes, \mathbb{1}), (\partial^l, \partial^r)\big)$$

is a ring category as in Definition 11.2.4. ◇

Recall from Section 6.6 that $\mathsf{PermCat^{su}}$ is the Cat-enriched multicategory with small permutative categories as objects. The category

$$\mathsf{PermCat^{su}}\big(\langle \mathsf{C}\rangle ; \mathsf{D}\big) = \mathsf{PermCat^{su}}\big(\langle \mathsf{C}_1, \ldots, \mathsf{C}_n\rangle ; \mathsf{D}\big)$$

has

- n-linear functors $\mathsf{C}_1 \times \cdots \times \mathsf{C}_n \longrightarrow \mathsf{D}$ (Definition 6.5.4) as objects and
- multilinear transformations (Definition 6.5.11) as morphisms.

Its operad composition is defined in (6.6.6) for multilinear functors and (6.6.9) for multilinear transformations. Also recall from Definition 5.1.12 the notion of an enriched multifunctor. The next result is [**EM06**, 3.4], which says that the Cat-enriched associative operad detects ring category structures on small permutative categories.

Theorem 11.2.16. *For each small permutative category* C*, there is a canonical bijective correspondence between*

- *ring category structures on* C *and*
- Cat*-enriched multifunctors*

$$F\colon \mathsf{As} \longrightarrow \mathsf{PermCat^{su}} \quad \textit{such that} \quad F(*) = \mathsf{C}.$$

Proof. By the adjunction (11.2.3), a Cat-enriched multifunctor $\mathsf{As} \longrightarrow \mathsf{PermCat^{su}}$ is equivalent to a multifunctor $\mathsf{As} \longrightarrow \mathsf{PermCat^{su}}$ with $\mathsf{PermCat^{su}}$ regarded as a multicategory without its Cat-enrichment, that is, multilinear transformations. Therefore, by Theorem 11.1.7, a Cat-enriched multifunctor

$$F\colon \mathsf{As} \longrightarrow \mathsf{PermCat^{su}} \quad \text{with} \quad F(*) = (\mathsf{C}, \oplus, \mathbb{0}, \xi^{\oplus}) \tag{11.2.17}$$

is equivalent to

- a 0-linear functor, that is, an object
$$F(\mathrm{id}_0) = \mathbb{1} \in \mathsf{C}$$
and
- a 2-linear functor
$$F(\mathrm{id}_2) = (\otimes, \partial^l, \partial^r) \colon \mathsf{C} \times \mathsf{C} \longrightarrow \mathsf{C}$$

such that the unity and the associativity relations (11.1.8) are preserved.

- The natural transformations
$$\begin{array}{ccc} (A \otimes C) \oplus (B \otimes C) & \xrightarrow{\partial^l_{A,B,C}} & (A \oplus B) \otimes C \\ (A \otimes B) \oplus (A \otimes C) & \xrightarrow{\partial^r_{A,B,C}} & A \otimes (B \oplus C) \end{array}$$
for $A, B, C \in \mathsf{C}$ are, respectively, the first and the second linearity constraints of the 2-linear functor $\otimes$ as in Definition 6.5.4.
- The preservation of the unity relation in (11.1.8) means the commutativity of the diagram
$$\begin{array}{ccccc} \mathsf{C} & \xrightarrow{\mathbb{1} \times 1_{\mathsf{C}}} & \mathsf{C} \times \mathsf{C} & \xleftarrow{1_{\mathsf{C}} \times \mathbb{1}} & \mathsf{C} \\ \| & & \downarrow{\scriptstyle (\otimes, \partial^l, \partial^r)} & & \| \\ \mathsf{C} & \xrightarrow{1_{\mathsf{C}}} & \mathsf{C} & \xleftarrow{1_{\mathsf{C}}} & \mathsf{C} \end{array} \tag{11.2.18}$$
of 1-linear functors. Here $1_{\mathsf{C}} \colon \mathsf{C} \longrightarrow \mathsf{C}$ is the identity 1-linear functor with the identity linearity constraint (6.6.1).
- The preservation of the associativity relation in (11.1.8) means the commutativity of the diagram
$$\begin{array}{ccc} \mathsf{C} \times \mathsf{C} \times \mathsf{C} & \xrightarrow{1_{\mathsf{C}} \times (\otimes, \partial^l, \partial^r)} & \mathsf{C} \times \mathsf{C} \\ {\scriptstyle (\otimes, \partial^l, \partial^r) \times 1_{\mathsf{C}}} \downarrow & & \downarrow {\scriptstyle (\otimes, \partial^l, \partial^r)} \\ \mathsf{C} \times \mathsf{C} & \xrightarrow{(\otimes, \partial^l, \partial^r)} & \mathsf{C} \end{array} \tag{11.2.19}$$
of 3-linear functors.

We now explain that the data $\{\mathbb{1}, \otimes, \partial^l, \partial^r\}$ and the commutative diagrams (11.2.18) and (11.2.19) are equivalent to a ring category structure on the small permutative category $(\mathsf{C}, \oplus, \mathbb{0}, \xi^{\oplus})$.

The commutativity of the diagrams (11.2.18) and (11.2.19) of *functors* means that $\otimes$ is strictly associative with $\mathbb{1}$ as the strict two-sided unit. In other words, $(\mathsf{C}, \otimes, \mathbb{1})$ is a strict monoidal category. At this point, the tuple
$$\big(\mathsf{C}, (\oplus, \mathbb{0}, \xi^{\oplus}), (\otimes, \mathbb{1}), (\partial^l, \partial^r)\big)$$
contains the data part of a ring category in Definition 11.2.4. It remains to check that

- the 2-linear functoriality of $(\otimes, \partial^l, \partial^r)$ and
- the linearity constraints of the commutative diagrams (11.2.18) and (11.2.19)

are equivalent to the ring category axioms (11.2.6)–(11.2.14).

First we consider the 2-linear functor axioms in Definition 6.5.4 for $(\otimes, \partial^l, \partial^r)$.

- Its unity axiom is the multiplicative zero axiom (11.2.6).
- Its constraint unity axiom is the zero factorization axiom (11.2.7).
- Its constraint associativity axiom (6.5.5) is the internal factorization axiom (11.2.10).
- Its constraint symmetry axiom (6.5.6) is the symmetry factorization axiom (11.2.9).
- Its constraint 2-by-2 axiom (6.5.7) with $(i,k) = (2,1)$ is the 2-by-2 factorization axiom (11.2.14). The case $(i,k) = (1,2)$ is equivalent to the case $(i,k) = (2,1)$ by the symmetry axiom (II.1.3.33) for the additive symmetry $\xi^{\oplus}$, which says $\xi^{\oplus}\xi^{\oplus} = 1$.

Next we consider the linearity constraints, in the sense of Definition 6.5.4, in the commutative diagrams (11.2.18) and (11.2.19) of multilinear functors.

- The linearity constraint of the identity 1-linear functor 1_{C} is the identity (6.6.1). Therefore, in terms of their linearity constraints, the commutativity of the diagram (11.2.18) of 1-linear functors is the unit factorization axiom (11.2.8).
- In the commutative diagram (11.2.19) of 3-linear functors, the equality of the first linearity constraints of the two composites is the left external factorization axiom (11.2.11). The equality of the second (respectively, third) linearity constraints is the middle external factorization axiom (11.2.12) (respectively, the right external factorization axiom (11.2.13)).

Therefore, a Cat-enriched multifunctor as in (11.2.17) is equivalent to a ring category structure on the small permutative category C. □

11.3. K-Theory Spectra of Ring Categories

In this section, we prove a theorem of Elmendorf-Mandell that says that the K-theory of each small ring category is a strict ring symmetric spectrum (Corollary 11.3.16). This result is a consequence of Theorem 11.2.16 and the Elmendorf-Mandell K-theory multifunctor. We begin by defining strict ring symmetric spectra and discussing some examples.

Strict Ring Symmetric Spectra. Recall from Definitions 7.6.1 and 7.6.8 and Theorem 7.6.15 the symmetric monoidal closed category

$$(\mathsf{SymSp}, \square_S, \mathsf{Hom}_S)$$

of symmetric spectra.

Definition 11.3.1. A *strict ring symmetric spectrum* is a monoid in SymSp. ◇

To explicitly explain the structure of a strict ring symmetric spectrum, first recall from Definitions 7.3.1, 7.3.3, and 7.4.1 the following concepts.

- sSet_*^{Σ} is the symmetric monoidal closed category of symmetric sequences.
- $\square$ is the Day convolution of symmetric sequences.
- $S \in \mathsf{sSet}_*^{\Sigma}$ is the symmetric sphere with monoid multiplication

$$S \square S \xrightarrow{\mu^S} S$$

determined by the canonical isomorphisms

$$S^p \wedge S^q \xrightarrow[\cong]{\mu^S_{p,q}} S^{p+q} \in \mathsf{sSet}_* \quad \text{for} \quad p, q \geq 0.$$

SymSp is the category of left S-modules as in Definition 7.4.2.

In the above context, we may unpack strict ring symmetric spectra as follows.

Proposition 11.3.2. *A strict ring symmetric spectrum is equivalent to a triple*

$$(X, \mu, \eta)$$

consisting of

- *a symmetric spectrum* (X, ρ) *and*
- *morphisms*

$$X \square X \xrightarrow{\mu} X \xleftarrow{\eta} S \in \mathsf{sSet}^{\Sigma}_*$$

such that the following four diagrams in sSet^{Σ}_* *are commutative.*

Compatibility of μ and S-action:

(11.3.3)
$$\begin{array}{ccccc}
(X \square S) \square X & \xrightarrow[\cong]{\xi \square 1} & (S \square X) \square X & \xrightarrow[\cong]{a} & S \square (X \square X) \\
{\scriptstyle a}\downarrow{\scriptstyle \cong} & & \downarrow{\scriptstyle \rho \square 1} & & \\
X \square (S \square X) & & X \square X & & \downarrow{\scriptstyle 1 \square \mu} \\
{\scriptstyle 1 \square \rho}\downarrow & & \downarrow{\scriptstyle \mu} & & \\
X \square X & \xrightarrow{\mu} & X & \xleftarrow{\rho} & S \square X
\end{array}$$

Compatibility of η and S-action:

(11.3.4)
$$\begin{array}{ccc}
S \square S & \xrightarrow{\mu^S} & S \\
{\scriptstyle 1 \square \eta}\downarrow & & \downarrow{\scriptstyle \eta} \\
S \square X & \xrightarrow{\rho} & X
\end{array}$$

Associativity:

(11.3.5)
$$\begin{array}{ccc}
(X \square X) \square X & \xrightarrow[\cong]{a} & X \square (X \square X) \\
 & & \downarrow{\scriptstyle 1 \square \mu} \\
{\scriptstyle \mu \square 1}\downarrow & & X \square X \\
 & & \downarrow{\scriptstyle \mu} \\
X \square X & \xrightarrow{\mu} & X
\end{array}$$

Unity:

(11.3.6)
$$\begin{array}{ccc}
S \square X & \xrightarrow{\rho} & X \\
{\scriptstyle \eta \square 1}\downarrow & & \| \\
X \square X & \xrightarrow{\mu} & X \\
{\scriptstyle 1 \square \eta}\uparrow & & \uparrow{\scriptstyle \rho} \\
X \square S & \xrightarrow{\xi} & S \square X
\end{array}$$

Proof. By Definition I.1.2.8, a monoid in $(\mathsf{SymSp}, \Box_S)$ is a triple

$$(X, \overline{\mu}, \overline{\eta})$$

consisting of

- a symmetric spectrum

$$\big(X, \rho\colon S \Box X \longrightarrow X\big)$$

 and
- morphisms

$$X \Box_S X \xrightarrow{\ \overline{\mu}\ } X \xleftarrow{\ \overline{\eta}\ } S \in \mathsf{SymSp}$$

such that the associativity and unity diagrams in SymSp below are commutative.

$$(11.3.7)\qquad \begin{array}{ccc} (X \Box_S X) \Box_S X & \xrightarrow[\cong]{a} & X \Box_S (X \Box_S X) \\ & & \downarrow{\scriptstyle 1 \Box_S \overline{\mu}} \\ \downarrow{\scriptstyle \overline{\mu} \Box_S 1} & & X \Box_S X \\ & & \downarrow{\scriptstyle \overline{\mu}} \\ X \Box_S X & \xrightarrow{\ \overline{\mu}\ } & X \end{array} \qquad \begin{array}{ccc} S \Box_S X & \xrightarrow{\ \rho\ } & X \\ {\scriptstyle \overline{\eta} \Box_S 1}\downarrow & & \| \\ X \Box_S X & \xrightarrow{\ \overline{\mu}\ } & X \\ {\scriptstyle 1 \Box_S \overline{\eta}}\uparrow & & \uparrow{\scriptstyle \rho} \\ X \Box_S S & \xrightarrow{\ \xi\ } & S \Box_S X \end{array}$$

By Proposition 7.6.2, the monoid multiplication $\overline{\mu}$ is equivalent to a morphism

$$(11.3.8)\qquad X \Box X \xrightarrow{\ \mu\ } X \in \mathsf{sSet}_*^{\Sigma}$$

such that the compatibility diagram (11.3.3) commutes. The monoid unit morphism $\overline{\eta}$ is a morphism

$$(11.3.9)\qquad S \xrightarrow{\ \eta\ } X \in \mathsf{sSet}_*^{\Sigma}$$

that is compatible with the left S-module action as in Definition 7.4.2. This compatibility condition is the commutative diagram (11.3.4). Since SymSp is the category of left S-modules, the diagrams in (11.3.7) commute in SymSp if and only if they commute in sSet_*^{Σ}. Using

- the structure morphisms μ in (11.3.8) and η in (11.3.9) and
- the universal properties of the coequalizer defining the smash product $\Box_S$ (Definition 7.6.1),

the diagrams in (11.3.7) commute in sSet_*^{Σ} if and only if the diagrams (11.3.5) and (11.3.6) commute. □

Explanation 11.3.10 (Strict Ring Symmetric Spectra)**.** As in Explanation 7.6.4, the multiplication morphism μ in (11.3.8) is equivalent to the family of $(\Sigma_p \times \Sigma_q)$-equivariant morphisms

$$(11.3.11)\qquad X_p \wedge X_q \xrightarrow{\ \mu_{p,q}\ } X_{p+q} \in \mathsf{sSet}_* \quad \text{for} \quad p, q \geq 0.$$

In terms of these morphisms, the compatibility diagram (11.3.3) is equivalent to the diagrams (7.6.5) and (7.6.7) with $f_{p,q} = \mu_{p,q}$. These diagrams are stated below, with $\Sigma_{p,n} = \Sigma_p \times \Sigma_n$ in the second diagram.

$$\begin{array}{ccc}
(S^n \wedge X_p) \wedge X_q & \xrightarrow[\cong]{a} & S^n \wedge (X_p \wedge X_q) \\
{\scriptstyle \rho_{n,p}\wedge 1}\downarrow & & \\
X_{n+p} \wedge X_q & & \downarrow{\scriptstyle 1\wedge \mu_{p,q}} \\
{\scriptstyle \mu_{n+p,q}}\downarrow & & \\
X_{n+p+q} & \xleftarrow{\rho_{n,p+q}} & S^n \wedge X_{p+q}
\end{array}$$

$$\begin{array}{ccc}
\Sigma_{(p+n+q)} \times_{\Sigma_{p+n,q}} \left(\Sigma_{(p+n)} \times_{\Sigma_{p,n}} X_p \wedge S^n\right) \wedge X_q & \xrightarrow{\xi_{p,n}\wedge 1} & \Sigma_{(p+n+q)} \times_{\Sigma_{n+p,q}} \left(\Sigma_{(n+p)} \times_{\Sigma_{n,p}} S^n \wedge X_p\right) \wedge X_q \\
{\scriptstyle a}\downarrow{\scriptstyle \cong} & & \downarrow{\scriptstyle \rho_{n,p}\wedge 1} \\
\Sigma_{(p+n+q)} \times_{\Sigma_{p,n+q}} X_p \wedge \left(\Sigma_{(n+q)} \times_{\Sigma_{n,q}} S^n \wedge X_q\right) & & \Sigma_{(p+n+q)} \times_{\Sigma_{n+p,q}} X_{n+p} \wedge X_q \\
{\scriptstyle 1\wedge \rho_{n,q}}\downarrow & & \downarrow{\scriptstyle \mu_{n+p,q}} \\
\Sigma_{(p+n+q)} \times_{\Sigma_{p,n+q}} X_p \wedge X_{n+q} & \xrightarrow{\mu_{p,n+q}} & X_{n+p+q}
\end{array}$$

The unit morphism η in (11.3.9) is equivalent to a sequence of Σ_p-equivariant morphisms

$$S^p \xrightarrow{\eta_p} X_p \in \mathsf{sSet}_* \quad \text{for} \quad p \geq 0. \tag{11.3.12}$$

The compatibility diagram (11.3.4) and the associativity diagram (11.3.5) in sSet_*^{Σ} are equivalent to the diagrams

$$\begin{array}{ccc}
S^p \wedge S^q & \xrightarrow[\cong]{\mu^S_{p,q}} & S^{p+q} \\
{\scriptstyle 1\wedge\eta_q}\downarrow & & \downarrow{\scriptstyle \eta_{p+q}} \\
S^p \wedge X_q & \xrightarrow{\rho_{p,q}} & X_{p+q}
\end{array}
\qquad
\begin{array}{ccc}
(X_n \wedge X_p) \wedge X_q & \xrightarrow[\cong]{a} & X_n \wedge (X_p \wedge X_q) \\
 & & \downarrow{\scriptstyle 1\wedge\mu_{p,q}} \\
{\scriptstyle \mu_{n,p}\wedge 1}\downarrow & & X_n \wedge X_{p+q} \\
 & & \downarrow{\scriptstyle \mu_{n,p+q}} \\
X_{n+p} \wedge X_q & \xrightarrow{\mu_{n+p,q}} & X_{n+p+q}
\end{array}$$

in sSet_* for $n, p, q \geq 0$. The top and bottom halves of the unity diagram (11.3.6) are equivalent to the diagrams

$$\begin{array}{ccc}
S^p \wedge X_q & \xrightarrow{\rho_{p,q}} & X_{p+q} \\
{\scriptstyle \eta_p\wedge 1}\downarrow & & \parallel \\
X_p \wedge X_q & \xrightarrow{\mu_{p,q}} & X_{p+q}
\end{array}
\qquad
\begin{array}{ccc}
\Sigma_{(p+q)} \times_{\Sigma_p\times\Sigma_q} X_p \wedge S^q & \xrightarrow[\cong]{\xi} & \Sigma_{(q+p)} \times_{\Sigma_q\times\Sigma_p} S^q \wedge X_p \\
{\scriptstyle 1\wedge\eta_q}\downarrow & & \downarrow{\scriptstyle \rho_{q,p}} \\
\Sigma_{(p+q)} \times_{\Sigma_p\times\Sigma_q} X_p \wedge X_q & \xrightarrow{\mu_{p,q}} & X_{p+q} = X_{q+p}
\end{array}$$

in sSet_* for $p, q \geq 0$. ◇

Using Explanation 11.3.10, we now discuss a few basic examples of strict ring symmetric spectra.

Example 11.3.13 (The Sphere Spectrum). Continuing Example 7.4.9, the symmetric sphere

$$S = \{S^p\}_{p\geq 0}$$

is a strict ring symmetric spectrum. The multiplication morphism $\mu_{p,q}$ in (11.3.11) is given by the canonical isomorphism

$$S^p \wedge S^q \xrightarrow[\cong]{\mu^S_{p,q}} S^{p+q} \quad \text{for} \quad p,q \geq 0.$$

The unit morphism η_p in (11.3.12) is the identity morphism of S^p. The diagrams in Explanation 11.3.10 for the symmetric sphere all follow from the bijection

$$(S^p)_n = \underline{n}^{\wedge p} \cong \underline{n}^p \quad \text{for} \quad n,p \geq 0$$

in Definitions 8.1.5 and 8.2.1 and the Symmetric Coherence Theorem 1.1.41 for sSet_*. ◇

Example 11.3.14 (Suspension Spectra of Simplicial Monoids). Suppose (K,μ,η) is a monoid in $(\mathsf{sSet}_*, \wedge, S^0)$. Continuing Example 7.4.10, the suspension spectrum

$$\Sigma^\infty K = \{S^p \wedge K\}_{p\geq 0}$$

is a strict ring symmetric spectrum. The multiplication morphism $\mu_{p,q}$ in (11.3.11) is given by the following composite.

$$\begin{array}{ccc} (S^p \wedge K) \wedge (S^q \wedge K) & \xrightarrow{\mu_{p,q}} & S^{p+q} \wedge K \\ & \searrow^{\xi_{\mathrm{mid}}} \qquad \nearrow_{\mu^S_{p,q}\wedge\mu} & \\ & (S^p \wedge S^q) \wedge (K \wedge K) & \end{array}$$

The morphism ξ_{mid} is the unique coherence isomorphism in the symmetric monoidal category sSet_* that swaps the first copy of K and S^q, as in Definition 1.3.2. The unit morphism η_p in (11.3.12) is the composite

$$\begin{array}{ccc} S^p & \xrightarrow{\eta_p} & S^p \wedge K \\ & \searrow^{\cong} \qquad \nearrow_{1\wedge\eta} & \\ & S^p \wedge S^0 & \end{array}$$

with $\cong$ the inverse of the right unit isomorphism in sSet_*. The first diagram in Explanation 11.3.10 for $\Sigma^\infty K$ is the outer diagram below in sSet_*, where $\wedge$ is omitted to save space.

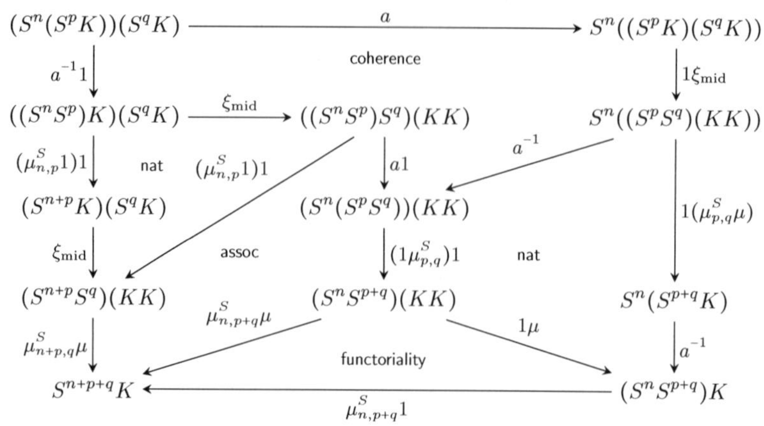

- The top region commutes by the Symmetric Coherence Theorem 1.1.41 for sSet_*.
- The left triangle commutes by the naturality of ξ_{mid}.
- The right trapezoid commutes by the naturality of the associativity isomorphism a.
- The lower left quadrilateral commutes by the associativity of μ^S and the functoriality of $\wedge$.
- The bottom triangle commutes by the functoriality of $\wedge$.

Other diagrams in Explanation 11.3.10 follow similarly from those for the symmetric sphere, the monoid axioms for (K, μ, η), naturality, functoriality, and Theorem 1.1.41 for sSet_*. ◇

Example 11.3.15 (Eilenberg-Mac Lane Spectra)**.** Suppose R is a ring with unit 1^R. Continuing Example 7.4.11, the Eilenberg-Mac Lane spectrum

$$HR = \left\{R \otimes S^p\right\}_{p\geq 0}$$

is a strict ring symmetric spectrum. The multiplication morphism $\mu_{p,q}$ in (11.3.11) is given on k-simplices by the following assignment on direct sum generators, with $r, r' \in R$ and $i_a, j_b \in \underline{k}^\flat$.

$$\begin{array}{ccc}
\Big(\bigoplus\limits_{(\underline{k^p})^\flat} R\Big) \wedge \Big(\bigoplus\limits_{(\underline{k^q})^\flat} R\Big) & \xrightarrow{(\mu_{p,q})_k} & \bigoplus\limits_{(\underline{k^{p+q}})^\flat} R \\
\big(r_{i_1,\ldots,i_p}\big) \wedge \big(r'_{j_1,\ldots,j_q}\big) & \longmapsto & (rr')_{i_1,\ldots,i_p,j_1,\ldots,j_q}
\end{array}$$

The unit morphism η_p in (11.3.12) is given on k-simplices by the following morphism.

$$\begin{array}{ccl}
(S^p)_k = \underline{k}^{\wedge p} & \xrightarrow{(\eta_p)_k} & \bigoplus\limits_{(\underline{k^p})^\flat} R \\
(i_1, \ldots, i_p) & \longmapsto & 1^R_{i_1,\ldots,i_p} \quad \text{if each } i_j \in \underline{k}^\flat. \\
(i_1, \ldots, i_p) & \longmapsto & 0 \quad \text{if some } i_j = 0.
\end{array}$$

The diagrams in Explanation 11.3.10 are commutative for HR because they are commutative on direct sum generators in $(HR)_p$ for $p \geq 0$. ◇

K-Theory Strict Ring Symmetric Spectra. Recall from Definition 10.3.32 the Elmendorf-Mandell K-theory multifunctor

$$\mathsf{K}^{\mathsf{EM}} = \mathsf{K}^{g} N_{*} \mathsf{J}^{\mathsf{EM}} \colon \mathsf{PermCat}^{\mathsf{su}} \longrightarrow \mathsf{SymSp}.$$

The following result is [**EM06**, 3.5] that describes the K-theory of ring categories.

Corollary 11.3.16. *For each small ring category* C, $\mathsf{K}^{\mathsf{EM}}\mathsf{C}$ *is a strict ring symmetric spectrum.*

Proof. Consider the multifunctors

$$\begin{array}{ccccc} \mathsf{As} & \xrightarrow{F} & \mathsf{PermCat}^{\mathsf{su}} & \xrightarrow{\mathsf{K}^{\mathsf{EM}}} & \mathsf{SymSp} \\ * & \longmapsto & (\mathsf{C}, \oplus, \mathbb{0}, \xi^{\oplus}) & \longmapsto & \mathsf{K}^{\mathsf{EM}}\mathsf{C} \end{array}$$

with F the multifunctor in Theorem 11.2.16 such that $F(*)$ is the underlying permutative category of C. Since As has a unique object $*$, the composite multifunctor $\mathsf{K}^{\mathsf{EM}} \circ F$ is equivalent to an operad morphism from As to the endomorphism operad of $\mathsf{K}^{\mathsf{EM}}\mathsf{C}$. By Proposition 11.1.15, such an operad morphism is equivalent to a monoid structure on $\mathsf{K}^{\mathsf{EM}}\mathsf{C}$. □

Example 11.3.17 (Endomorphism Ring Categories)**.** Suppose C is a small permutative category. Then

- the endomorphism ring category $\mathsf{Perm}^{\mathsf{su}}(\mathsf{C};\mathsf{C})$ in Theorem II.9.2.14 and
- the tight endomorphism ring category $\mathsf{Perm}^{\mathsf{sug}}(\mathsf{C};\mathsf{C})$ in Theorem II.9.2.20

are both small. Corollary 11.3.16 applies to each of them to yield a K-theory strict ring symmetric spectrum. ◇

Example 11.3.18 (Additive Distortion Category)**.** By Example II.9.1.18, the additive distortion category $\mathcal{D}^{\mathsf{ad}}$ in Section I.4.5 is a small ring category, which is also tight. Corollary 11.3.16 applies to $\mathcal{D}^{\mathsf{ad}}$ to yield a K-theory strict ring symmetric spectrum. ◇

Example 11.3.19 (Ring Categories via Strictification)**.** Suppose C is a small tight bimonoidal category (Definition I.2.1.2). By Corollary II.9.1.19, the equivalent

- right rigid bimonoidal category A in Theorem I.5.5.11 and
- left rigid bimonoidal category A_l in Theorem I.5.5.12

are small tight ring categories. Corollary 11.3.16 applies to each of A and A_l to yield a K-theory strict ring symmetric spectrum. ◇

11.4. The Barratt-Eccles Operad

In this section, we define the categorical Barratt-Eccles operad $E\mathsf{As}$ and describe it in terms of generators and relations (Theorem 11.4.14). As the first application of Theorem 11.4.14, in Proposition 11.4.26 we observe that the Barratt-Eccles operad detects permutative category structures. Moreover, Theorem 11.5.5 shows that the Barratt-Eccles operad detects bipermutative category structures on small permutative categories.

Translation Categories. In (11.2.3), we observed that the forgetful functor

$$\mathsf{Ob}\colon \mathsf{Cat} \longrightarrow \mathsf{Set}$$

has a left adjoint dis that sends a set X to the discrete category with object set X. The functor Ob also has a right adjoint given by the translation category functor in the next definition. The translation category functor will be used in the definition of the Barratt-Eccles operad.

Definition 11.4.1. The *translation category functor*

$$E\colon \mathsf{Set} \longrightarrow \mathsf{Cat}$$

is the functor defined as follows.

- For a set X, the *translation category* EX is the category with
  - object set X and
  - each morphism set a one-element set $*$.
- For a map $f\colon X \longrightarrow Y$ of sets, the functor

$$Ef\colon EX \longrightarrow EY$$

  is
  - the map f on objects and
  - the unique map to a one-element set on morphisms.

This finishes the definition of the translation category functor E ◇

Example 11.4.2 (Principle Bundles)**.** The classifying space (= the geometric realization of the nerve in Example 7.2.8) of the translation category EX is contractible for each set X, since the unique functor $EX \longrightarrow \mathbb{1}$ to the terminal category is an equivalence of categories. For a group G, the regular G-action, which is given by the product in G, induces a free G-action

$$EG \times G \longrightarrow EG. \tag{11.4.3}$$

Here the G-action is *free* in the sense that, for any $x, \sigma \in G$,

$$x\sigma = x \quad \text{implies} \quad \sigma = e$$

with $e \in G$ the identity element. For elements $x, y \in G$, we also denote the unique morphism by

$$yx^{-1}\colon x \longrightarrow y \in EG. \tag{11.4.4}$$

For each element $\sigma \in G$, the right σ-action in (11.4.3) is the functor

$$EG \xrightarrow[\cong]{\sigma} EG$$

that sends

- an object $x \in EG$ to $x\sigma \in EG$ and
- the unique morphism $yx^{-1}\colon x \longrightarrow y$ to the unique morphism

$$yx^{-1} = (y\sigma)(x\sigma)^{-1}\colon x\sigma \longrightarrow y\sigma. \tag{11.4.5}$$

Since the classifying space of EG has a free G-action and is contractible, it is the total space of a universal principal G-bundle. The reader is referred to [**MS74**] for more discussion of principal bundles. ◇

Example 11.4.6 (E as a Right Adjoint). There is an adjunction

$$\mathsf{Cat} \underset{E}{\overset{\mathsf{Ob}}{\rightleftarrows}} \mathsf{Set} \tag{11.4.7}$$

with

- left adjoint Ob the underlying object set functor in (11.2.3),
- E the translation category functor in Definition 11.4.1, and
- $\mathsf{Ob} \circ E = 1_{\mathsf{Set}}$.

Since Ob preserves products, if Q is a Cat-enriched operad, then $\mathsf{Ob}(\mathsf{Q})$ is an operad. By construction, or by the fact that it is a right adjoint, E preserves products. Therefore, if P is an operad in Set, then $E\mathsf{P}$ is a Cat-enriched operad such that

$$\mathsf{Ob}(E\mathsf{P}) = \mathsf{P}. \tag{11.4.8}$$

Moreover, the identity functor on Set induces a Cat-enriched operad morphism

$$\mathsf{P} \xrightarrow{\iota_{\mathsf{P}}} E\mathsf{P}. \tag{11.4.9}$$

On the left-hand side, P is regarded as a Cat-enriched operad via the discrete category functor dis in (11.2.3), which preserves products. ◇

Definition 11.4.10. The *Barratt-Eccles operad* is the Cat-enriched operad $E\mathsf{As}$, with

- As the associative operad in Definition 11.1.1 and
- E the translation category functor in Definition 11.4.1. ◇

For $i \neq j \in \{1,\ldots,n\}$, the transposition that swaps i and j is written as (i,j).

Explanation 11.4.11.

(1) For the Cat-enriched operad morphism

$$\mathsf{As} \xrightarrow{\iota_{\mathsf{As}}} E\mathsf{As} \tag{11.4.12}$$

in (11.4.9), each functor

$$\iota_{\mathsf{As}} \colon \mathsf{As}_n \longrightarrow E\mathsf{As}_n$$

is the identity function on objects. The object set operad of $E\mathsf{As}$ is the associative operad As. On morphisms, the operad structure of $E\mathsf{As}$ is uniquely determined by the fact that each morphism set in each translation category $E\mathsf{As}_n$ has only one element.

(2) Denote by

$$\mathrm{id}_2 \xrightarrow[\cong]{\tau} (1,2) \in E\mathsf{As}_2\big(\mathrm{id}_2; (1,2)\big) \tag{11.4.13}$$

the unique nonidentity isomorphism. Adjacent transpositions $(i,i+1)$ for $1 \leq i < n$ generate the symmetric group Σ_n. Moreover, there is an iterated block sum decomposition

$$(i,i+1) = \overbrace{\mathrm{id}_{i-1} \times}^{\varnothing \text{ if } i=1} (1,2) \overbrace{\times \mathrm{id}_{n-i-1}}^{\varnothing \text{ if } i=n-1} \in \Sigma_n.$$

By the definition (11.1.2) of γ in As, this iterated block sum is an operadic composite of $(1,2)$ and identities. So each morphism in $E\mathsf{As}_n$ decomposes into a categorical composite of isomorphisms of the form ϕv with

- each $v \in \Sigma_n$ and
- each ϕ an operadic composite of one τ and identity morphisms. ◇

Coherence of the Barratt-Eccles Operad. Theorem 11.1.7 characterizes the associative operad As in terms of generators, namely, $\mathrm{id}_0 \in \Sigma_0$ and $\mathrm{id}_2 \in \Sigma_2$, and the unity and associativity relations. The next result, which is [**Fre17**, 6.3.3], is the extension of that characterization to the Barratt-Eccles operad $E\mathsf{As}$. It is a coherence theorem in the sense that it characterizes the Barratt-Eccles operad in terms of a small number of generators and a few relations. The axioms (11.4.17)–(11.4.20) below are formally identical to those of a permutative category in Definition I.1.2.18. To simplify the notation, we write $y(x_1, \ldots, x_n)$ for the operad composition $\gamma\big(y, (x_1, \ldots, x_n)\big)$. An *enriched operad morphism* is an enriched multifunctor as in Definition 5.1.12 between two enriched multicategories with one object.

Theorem 11.4.14. *Suppose* $(\mathsf{P}, \gamma, 1)$ *is a* Cat*-enriched operad. Then a* Cat*-enriched operad morphism*

$$f \colon E\mathsf{As} \longrightarrow \mathsf{P}$$

is uniquely determined by

- *the objects*

(11.4.15)
$$\begin{aligned} \mathbb{1} &= f(\mathrm{id}_0) \in \mathsf{P}_0 \\ \mu &= f(\mathrm{id}_2) \in \mathsf{P}_2 \end{aligned}$$

  and
- *the isomorphism*

(11.4.16)
$$\mu = f(\mathrm{id}_2) \xrightarrow[\cong]{\xi = f(\tau)} f(1,2) = \mu(1,2) = \mu^{\mathsf{op}} \in \mathsf{P}_2(\mu, \mu^{\mathsf{op}})$$

  with

$$\tau \colon \mathrm{id}_2 \longrightarrow (1,2) \in E\mathsf{As}_2$$

  the isomorphism in (11.4.13).

The above data are subject to the following conditions.

Unity and Associativity on Objects: *The equalities of objects*

(11.4.17)
$$\begin{aligned} \mu(\mathbb{1}, 1) &= 1 = \mu(1, \mathbb{1}) \in \mathsf{P}_1 \\ \mu(\mu, 1) &= \mu(1, \mu) \in \mathsf{P}_3 \end{aligned}$$

hold.

The Symmetry Axiom: *The diagram*

(11.4.18)
$$\begin{array}{ccc} \mu & \xrightarrow{1_\mu} & \mu \\ & {\scriptstyle \xi}\searrow \quad \nearrow{\scriptstyle \xi^{\mathsf{op}}} & \\ & \mu^{\mathsf{op}} & \end{array}$$

in P_2 *is commutative, with*

$$\xi^{\mathsf{op}} = \xi(1,2) \colon \mu^{\mathsf{op}} \longrightarrow \mu$$

the image of ξ *under the right* $(1,2)$*-action.*

The Unit Axiom: *The diagram*

$$\begin{array}{ccc} 1 = \mu(1,\mathbb{1}) & \xrightarrow{\quad 1_1 \quad} & \mu(\mathbb{1},1) = 1 \\ & {\scriptstyle \xi(1,\mathbb{1})} \searrow \quad \quad /\!\!/ & \\ & \mu^{\mathsf{op}}(1,\mathbb{1}) & \end{array} \tag{11.4.19}$$

in P_1 *is commutative.*

The Hexagon Axiom: *The diagram*

$$\begin{array}{ccc} \mu(\mu,1) & \xrightarrow{\quad \xi \quad} & \mu^{\mathsf{op}}(\mu,1) \\ \| & & \| \\ \mu(1,\mu) & & \mu(\mu^{\mathsf{op}},1)(2,3) \\ {\scriptstyle \mu(1,\xi)} \searrow & & \nearrow {\scriptstyle \mu(\xi,1)(2,3)} \\ \mu(1,\mu^{\mathsf{op}}) & = & \mu(\mu,1)(2,3) \end{array} \tag{11.4.20}$$

in P_3 *is commutative.*

Proof. We need to prove the necessity and the sufficiency of the data (11.4.15)–(11.4.16) and the axioms (11.4.17)–(11.4.20).

Necessity. Suppose $f\colon E\mathsf{As} \longrightarrow \mathsf{P}$ is a Cat-enriched operad morphism. We define

$$\begin{cases} \mathbb{1} = f(\mathrm{id}_0) \in \mathsf{P}_0, \\ \mu = f(\mathrm{id}_2) \in \mathsf{P}_2, \quad \text{and} \\ \xi = f(\tau)\colon \mu \longrightarrow \mu^{\mathsf{op}} \in \mathsf{P}_2 \end{cases}$$

as in (11.4.15)–(11.4.16). Since f preserves the operad units, there is an equality

$$f(\mathrm{id}_1) = 1 \in \mathsf{P}_1.$$

- With ι_{As} as in (11.4.12), the composite

$$\mathsf{As} \xrightarrow{\ \iota_{\mathsf{As}}\ } E\mathsf{As} \xrightarrow{\ f\ } \mathsf{P}$$

  is a Cat-enriched operad morphism. So it preserves the operad relations (11.1.8) in As, which become the equalities (11.4.17) of objects in P.
- Before applying f, the analogues of the axioms (11.4.18)–(11.4.20) hold in $E\mathsf{As}$ because each morphism set $E\mathsf{As}_n(\cdot,\cdot)$ is a one-element set by Definition 11.4.1. Since f is a Cat-enriched operad morphism, after applying f, the axioms hold in P.

Therefore, the data (11.4.15)–(11.4.16) satisfy the axioms (11.4.17)–(11.4.20).

Sufficiency. Suppose given the data

- $\mathbb{1} = f(\mathrm{id}_0) \in \mathsf{P}_0$,
- $\mu = f(\mathrm{id}_2) \in \mathsf{P}_2$ as in (11.4.15), and
- $\xi = f(\tau)\colon \mu \longrightarrow \mu(1,2) = \mu^{\mathsf{op}} \in \mathsf{P}_2(\mu,\mu^{\mathsf{op}})$ as in (11.4.16)

such that the axioms (11.4.17)–(11.4.20) hold in P. We show that these data extend uniquely to a Cat-enriched operad morphism $f\colon E\mathsf{As} \longrightarrow \mathsf{P}$ in several stages.

Objects. By Theorem 11.1.7 and the unity and associativity relations (11.4.17), the objects

$$\mathbb{1} = f(\mathrm{id}_0) \quad \text{and} \quad \mu = f(\mathrm{id}_2)$$

uniquely determine a Cat-enriched operad morphism

$$\mathsf{As} \xrightarrow{\;f\;} \mathsf{P} \tag{11.4.21}$$

defined by

$$\begin{cases} f(\mathrm{id}_1) = 1 \in \mathsf{P}_1, \\ f(\mathrm{id}_2) = \mu = \mu_2 \in \mathsf{P}_2, \\ f(\mathrm{id}_n) = \mu(\mu_{n-1}, 1) = \mu_n \in \mathsf{P}_n \quad \text{for } n \geq 3, \text{ and} \\ \;f(\sigma) = \mu_n\sigma \in \mathsf{P}_n \quad \text{for } \sigma \in \Sigma_n. \end{cases}$$

By (11.4.8), this defines the object part of the Cat-enriched operad morphism $f\colon E\mathsf{As} \longrightarrow \mathsf{P}$.

Morphisms. The identity morphisms $1_{\mathrm{id}_0} \in E\mathsf{As}_0$ and $1_{\mathrm{id}_1} \in E\mathsf{As}_1$ must be sent by f to, respectively, the identity morphisms $1_{\mathbb{1}} \in \mathsf{P}_0$ and $1_1 \in \mathsf{P}_1$. To define f on the other morphisms in $E\mathsf{As}$, note that the coherence theorems for (symmetric) monoidal categories can be applied to

- the objects $\{\mathbb{1}, 1, \mu\}$, which are interpreted as, respectively, the monoidal unit, the identity functor, and the monoidal product, and
- the symmetry isomorphism ξ.

The reason is that the assumed axioms (11.4.17), and also (11.4.18)–(11.4.20) in the symmetric case, are formally identical to those of a (symmetric) strict monoidal category in Definitions I.1.2.1 and I.1.2.18. We apply these coherence theorems as follows.

- By (11.4.17) and Mac Lane's Coherence Theorem I.1.3.3, each iterated operadic composite in P_n involving only the objects

  $$\mu_0 = \mathbb{1} \in \mathsf{P}_0, \quad \mu_1 = 1 \in \mathsf{P}_1, \quad \text{and} \quad \mu_2 = \mu \in \mathsf{P}_2$$

  is equal to μ_n. For example, there are equalities of objects

  $$\begin{aligned} \mu_4 &= \mu\big(\mu(\mu, 1), 1\big) = \mu(\mu, \mu) = \mu\big(\mu(1, \mu), 1\big) \\ &= \mu\big(1, \mu(\mu, 1)\big) = \mu\big(1, \mu(1, \mu)\big) \in \mathsf{P}_4. \end{aligned}$$

- By (11.4.17)–(11.4.20) and the Symmetric Coherence Theorem 1.1.41, for each pair of permutations $\sigma, \theta \in \Sigma_n$, there is a *unique* isomorphism

  $$\mu_n\sigma \xrightarrow[\cong]{\;\theta\sigma^{-1}\;} \mu_n\theta \in \mathsf{P}_n \tag{11.4.22}$$

  that is a categorical composite of isomorphisms of the form ϕv, with $v \in \Sigma_n$ and ϕ an operadic composite of
  - one instance of $\xi = f(\tau)\colon \mu \longrightarrow \mu^{\mathrm{op}}$ in (11.4.16) and
  - identity morphisms.

Regarding $\xi = f(\tau)$ as corresponding to the permutation $(1,2) \in \Sigma_2$, the isomorphism in (11.4.22) corresponds to the permutation $\theta\sigma^{-1} \in \Sigma_n$. If

$$(\sigma, \theta) = (\mathrm{id}_2, (1,2)),$$

then the isomorphism in (11.4.22) is ξ.

By Explanation 11.4.11 (2), for each pair of permutations $\sigma, \theta \in \Sigma_n$, the unique isomorphism

$$\theta\sigma^{-1} : \sigma \longrightarrow \theta \in E\mathsf{As}_n$$

decomposes into a categorical composite of isomorphisms of the form $\phi\upsilon$, with $\upsilon \in \Sigma_n$ and ϕ an operadic composite of

- one instance of $\tau : \mathrm{id}_2 \longrightarrow (1,2)$ in (11.4.13) and
- identity morphisms.

The desired $f : E\mathsf{As} \longrightarrow \mathsf{P}$ must be levelwise a functor and preserve the operad structure, namely, the operad units, the equivariant structure, and the operad composition. So we must define the morphism

$$f(\sigma) = \mu_n \sigma \xrightarrow{f(\theta\sigma^{-1})} \mu_n \theta = f(\theta) \in \mathsf{P}_n \tag{11.4.23}$$

as the corresponding isomorphism in (11.4.22). The uniqueness of the isomorphism in (11.4.22) implies the following two statements.

- $f(\theta\sigma^{-1})$ is independent of the choice of a decomposition of $\theta\sigma^{-1} \in E\mathsf{As}_n$ in Explanation 11.4.11 (2).
- $f : E\mathsf{As}_n \longrightarrow \mathsf{P}_n$ is a functor for each n.

The uniqueness of each functor $f : E\mathsf{As}_n \longrightarrow \mathsf{P}_n$ is part of the definitions (11.4.21) and (11.4.23).

It remains to check that the functors $f : E\mathsf{As}_n \longrightarrow \mathsf{P}_n$ preserve the operad structure. Since $f : \mathsf{As} \longrightarrow \mathsf{P}$ in (11.4.21) is a Cat-enriched operad morphism, by (11.4.8) we only need to consider morphisms in $E\mathsf{As}_n$.

Equivariant structure. The assertion that f preserves the symmetric group action means that, for each permutation $\pi \in \Sigma_n$, the diagram

$$\begin{array}{ccc} E\mathsf{As}_n & \xrightarrow{\pi} & E\mathsf{As}_n \\ {\scriptstyle f}\big\downarrow & & \big\downarrow{\scriptstyle f} \\ \mathsf{P}_n & \xrightarrow{\pi} & \mathsf{P}_n \end{array} \tag{11.4.24}$$

of functors, with each horizontal arrow the right π-action, is commutative. Since f and π are functors, by the decomposition of morphisms in $E\mathsf{As}_n$ in Explanation 11.4.11 (2), it suffices to consider morphisms of the form $\phi\upsilon \in E\mathsf{As}_n$ with $\upsilon \in \Sigma_n$ and ϕ an operadic composite of one $\tau : \mathrm{id}_2 \longrightarrow (1,2)$ and identity morphisms. On such a morphism $\phi\upsilon$, the commutativity of (11.4.24) follows from

- the definition (11.4.5) of the equivariant structure on $E\mathsf{As}$ and
- the uniqueness of (11.4.22).

Operad composition. The assertion that f preserves the operad composition means the commutativity of the following diagram of functors.

$$\begin{array}{ccc} E\mathsf{As}_n \times \prod_{j=1}^{n} E\mathsf{As}_{k_j} & \xrightarrow{\gamma} & E\mathsf{As}_{k_1+\cdots+k_n} \\ {\scriptstyle f \times \prod_j f}\big\downarrow & & \big\downarrow{\scriptstyle f} \\ \mathsf{P}_n \times \prod_{j=1}^{n} \mathsf{P}_{k_j} & \xrightarrow{\gamma} & \mathsf{P}_{k_1+\cdots+k_n} \end{array} \tag{11.4.25}$$

To prove the commutativity of (11.4.25), we use the following facts:

- $E\mathsf{As}$ and P both satisfy the equivariance axioms (6.1.6) and (6.1.7).
- f and γ are functors.
- f preserves the equivariant structure.
- Morphisms in $E\mathsf{As}$ decompose as in Explanation 11.4.11 (2).

Using these facts, it suffices to consider (11.4.25) with

- an identity morphism in n of the $n+1$ factors in $E\mathsf{As}_n \times \prod_{j=1}^n E\mathsf{As}_{k_j}$ and
- in the remaining factor, an operadic composite of one $\tau\colon \mathrm{id}_2 \longrightarrow (1,2)$ and identity morphisms.

In this case, the commutativity of (11.4.25) follows from the uniqueness of (11.4.22). Therefore, $f\colon E\mathsf{As} \longrightarrow \mathsf{P}$ is a Cat-enriched operad morphism. □

Permutative Categories as Algebras. Proposition 11.1.15 provided a conceptual explanation of the associative operad As as the operad for monoids. The next observation is the analogue of that result for the Barratt-Eccles operad $E\mathsf{As}$ in Definition 11.4.10. We consider the category Cat of small categories and functors as a symmetric monoidal closed category with

- the Cartesian product $\times$ as the monoidal product,
- the terminal category $\mathbb{1}$ as the monoidal unit, and
- internal hom given by diagram categories, which have functors as objects and natural transformations as morphisms.

Similar to Convention 6.4.6, we use the same notation Cat for the associated Cat-enriched multicategory in Proposition 6.3.6. The unique object in $E\mathsf{As}$, as a one-object Cat-enriched multicategory, is denoted by $*$. For an object A in a permutative closed category C, recall the C-enriched endomorphism operad $\mathsf{End}(A)$ in (11.1.14), with nth object the internal hom object $[A^{\otimes n}, A] \in \mathsf{C}$.

Proposition 11.4.26. *For a small category C, a permutative category structure on C is uniquely determined by a Cat-enriched multifunctor*

$$f\colon E\mathsf{As} \longrightarrow \mathsf{Cat} \quad \textit{such that} \quad f(*) = \mathsf{C}.$$

Proof. A Cat-enriched multifunctor $f\colon E\mathsf{As} \longrightarrow \mathsf{Cat}$ such that $f(*) = \mathsf{C}$ is equivalent to a Cat-enriched operad morphism

$$E\mathsf{As} \xrightarrow{\;f\;} \mathsf{End}(\mathsf{C}) = \big\{\mathsf{Cat}(\mathsf{C}^{\times n}, \mathsf{C})\big\}_{n\geq 0}$$

to the Cat-enriched endomorphism operad of C. By Theorem 11.4.14, such a Cat-enriched operad morphism f is uniquely determined by

- the object
$$\mathbb{1} = f(\mathrm{id}_0) \in \mathsf{C},$$
- the functor
$$\otimes = f(\mathrm{id}_2)\colon \mathsf{C} \times \mathsf{C} \longrightarrow \mathsf{C},$$
and
- the natural isomorphism
$$\otimes \xrightarrow{\;\xi^{\otimes} = f(\tau)\;} \otimes(1,2) = \otimes^{\mathrm{op}}$$

such that the axioms (11.4.17)–(11.4.20) are satisfied.

- The unity and associativity axioms (11.4.17) state that $(\mathsf{C}, \otimes, \mathbb{1})$ is a strict monoidal category as in Definition I.1.2.1.
- The other axioms (11.4.18)–(11.4.20) state that $(\mathsf{C}, \otimes, \mathbb{1}, \xi^{\otimes})$ is a permutative category as in Definition I.1.2.18.

This finishes the proof. □

11.5. Detecting Bipermutative Categories

In this section, we prove that the Barratt-Eccles operad $E\mathsf{As}$ in Definition 11.4.10 detects bipermutative category structures on small permutative categories (Theorem 11.5.5). This result is due to Elmendorf-Mandell and is an application of the Coherence Theorem 11.4.14 for $E\mathsf{As}$. For the reader's convenience, here we recall from Chapter II.9 the definition of a bipermutative category.

Definition 11.5.1. A *bipermutative category* is a tuple

$$\big(\mathsf{C}, (\oplus, \mathbb{0}, \xi^{\oplus}), (\otimes, \mathbb{1}, \xi^{\otimes}), (\partial^l, \partial^r)\big)$$

consisting of

- a ring category

$$\big(\mathsf{C}, (\oplus, \mathbb{0}, \xi^{\oplus}), (\otimes, \mathbb{1}), (\partial^l, \partial^r)\big)$$

  as in Definition 11.2.4 and
- a permutative category structure $(\mathsf{C}, \otimes, \mathbb{1}, \xi^{\otimes})$, with $\xi^{\otimes}$ called the *multiplicative symmetry*.

These data are required to satisfy the following two axioms for objects $A, B, C \in \mathsf{C}$.

The Zero Symmetry Axiom: There is an equality of morphisms

$$\xi^{\otimes}_{A,\mathbb{0}} = 1_{\mathbb{0}} \colon A \otimes \mathbb{0} = \mathbb{0} \longrightarrow \mathbb{0} = \mathbb{0} \otimes A. \tag{11.5.2}$$

The Multiplicative Symmetry Factorization Axiom: The diagram

$$\begin{array}{ccc} (A \otimes C) \oplus (B \otimes C) & \xrightarrow{\partial^l_{A,B,C}} & (A \oplus B) \otimes C \\ {\scriptstyle \xi^{\otimes}_{A,C} \oplus \xi^{\otimes}_{B,C}} \downarrow & & \downarrow {\scriptstyle \xi^{\otimes}_{A \oplus B, C}} \\ (C \otimes A) \oplus (C \otimes B) & \xrightarrow{\partial^r_{C,A,B}} & C \otimes (A \oplus B) \end{array} \tag{11.5.3}$$

is commutative.

This finishes the definition of a bipermutative category. A bipermutative category is *small*, respectively, *tight*, if the underlying ring category is so. ◇

Definition 11.5.4. For a permutative category $(\mathsf{C}, \oplus, \mathbb{0}, \xi^{\oplus})$, a *bipermutative category structure* on C is the additional data $(\otimes, \mathbb{1}, \xi^{\otimes}, \partial^l, \partial^r)$ such that the tuple

$$\big(\mathsf{C}, (\oplus, \mathbb{0}, \xi^{\oplus}), (\otimes, \mathbb{1}, \xi^{\otimes}), (\partial^l, \partial^r)\big)$$

is a bipermutative category as in Definition 11.5.1. ◇

Recall from Section 6.6 that $\mathsf{PermCat}^{\mathsf{su}}$ is the Cat-enriched multicategory with small permutative categories as objects. The category

$$\mathsf{PermCat}^{\mathsf{su}}\big(\langle \mathsf{C}\rangle ; \mathsf{D}\big) = \mathsf{PermCat}^{\mathsf{su}}\big(\langle \mathsf{C}_1, \ldots, \mathsf{C}_n\rangle ; \mathsf{D}\big)$$

has

- n-linear functors
$$\mathsf{C}_1 \times \cdots \times \mathsf{C}_n \longrightarrow \mathsf{D}$$
in Definition 6.5.4 as objects and
- multilinear transformations (Definition 6.5.11) as morphisms.

Also recall from Definition 5.1.12 the notion of an enriched multifunctor. The next result is [**EM06**, 3.8], which says that the Barratt-Eccles operad detects bipermutative category structures on small permutative categories. It extends both

- Theorem 11.2.16 from As to the Barratt-Eccles operad and
- Proposition 11.4.26 from Cat to $\mathsf{PermCat}^{\mathsf{su}}$.

Theorem 11.5.5. *For each small permutative category* C*, there is a canonical bijective correspondence between*

- *bipermutative category structures on* C *and*
- Cat*-enriched multifunctors*
$$F \colon E\mathsf{As} \longrightarrow \mathsf{PermCat}^{\mathsf{su}} \quad \textit{such that} \quad F(*) = \mathsf{C}.$$

Proof. A Cat-enriched multifunctor

$$F \colon E\mathsf{As} \longrightarrow \mathsf{PermCat}^{\mathsf{su}} \quad \text{such that} \quad F(*) = (\mathsf{C}, \oplus, \mathbb{0}, \xi^{\oplus})$$

is equivalent to a Cat-enriched operad morphism

$$F \colon E\mathsf{As} \longrightarrow \mathsf{End}(\mathsf{C}) = \Big\{ \mathsf{PermCat}^{\mathsf{su}}\big(\langle \overbrace{\mathsf{C}, \ldots, \mathsf{C}}^{n}\rangle ; \mathsf{C}\big)\Big\}_{n \geq 0}$$

to the Cat-enriched endomorphism operad of C. By Theorem 11.4.14, such a Cat-enriched operad morphism is uniquely determined by

- the 0-linear functor, that is, object
$$F(\mathrm{id}_0) = \mathbb{1} \in \mathsf{C},$$
- the 2-linear functor
$$F(\mathrm{id}_2) = (\otimes, \partial^l, \partial^r) \colon \mathsf{C} \times \mathsf{C} \longrightarrow \mathsf{C},$$
and
- the invertible multilinear transformation
$$(\otimes, \partial^l, \partial^r) \xrightarrow{\ \xi^{\otimes} = F(\tau)\ } (\otimes^{\mathsf{op}}, \partial^r, \partial^l)$$

such that the conditions (11.4.17)–(11.4.20) are satisfied, with

$$(\mu, 1, \xi) \quad \text{interpreted as} \quad \big((\otimes, \partial^l, \partial^r), 1_{\mathsf{C}}, \xi^{\otimes}\big).$$

By Theorem 11.2.16 and the first paragraph of its proof, the data $(\otimes, \mathbb{1}, \partial^l, \partial^r)$ and the condition (11.4.17) are equivalent to a ring category structure on the small permutative category $(\mathsf{C}, \oplus, \mathbb{0}, \xi^{\oplus})$ in the sense of Definition 11.2.15. In particular, $(\mathsf{C}, \otimes, \mathbb{1})$ is a strict monoidal category. Next we consider the remaining data $\xi^{\otimes}$ and the axioms (11.4.18)–(11.4.20).

The invertible multilinear transformation

$$\xi^{\otimes}: \otimes \xrightarrow{\cong} \otimes^{\mathrm{op}}$$

is, by definition, a natural isomorphism that satisfies the two conditions in Definition 6.5.11.

- The three remaining axioms (11.4.18)–(11.4.20) state that $(\mathsf{C}, \otimes, \mathbb{1}, \xi^{\otimes})$ is a permutative category as in Definition I.1.2.18.
- The commutativity of the diagram (6.5.12) for $\xi^{\otimes}$ and $i = 1$ is the multiplicative symmetry factorization axiom (11.5.3) in a bipermutative category. In the presence of the symmetry axiom (11.4.18), the case $i = 2$ of the diagram (6.5.12) is equivalent to the case $i = 1$. So it does not impose any additional restriction.
- The second condition in Definition 6.5.11 states the equalities

$$\xi^{\otimes}_{-,\mathbb{0}} = 1_{\mathbb{0}} = \xi^{\otimes}_{\mathbb{0},-}. \tag{11.5.6}$$

The first equality in (11.5.6) is the zero symmetry axiom (11.5.2) in a bipermutative category. In the presence of the symmetry axiom (11.4.18), the second equality in (11.5.6) is equivalent to the first one. So it does not impose any additional restriction.

Therefore, a Cat-enriched operad morphism

$$F: E\mathsf{As} \longrightarrow \mathsf{End}(\mathsf{C})$$

is equivalent to a bipermutative category structure on C. □

11.6. K-Theory Spectra of Bipermutative Categories

In this section, we prove that the K-theory of a small bipermutative category is an E_∞-symmetric spectrum (Corollary 11.6.12). This result is due to Elmendorf-Mandell. It is a consequence of Theorem 11.5.5, the Elmendorf-Mandell K-theory multifunctor, and the fact that $E\mathsf{As}$ is an E_∞-operad (Proposition 11.6.3). Along the way, we record the relationship between strict ring symmetric spectra, E_∞-symmetric spectra, and commutative monoids in SymSp (Proposition 11.6.6).

E_∞-Operads. We first define E_∞-operads and discuss some examples. By Definition 7.2.3, the nerve functor

$$N: \mathsf{Cat} \longrightarrow \mathsf{sSet}$$

is a right adjoint. So it preserves products and takes a Cat-enriched operad to an sSet-enriched operad. A simplicial set is *contractible* if its geometric realization (Definition 7.1.14) is a contractible space.

Definition 11.6.1. An sSet-enriched operad P is an *E_∞-operad* if the following two conditions hold for each $n \geq 0$:

Contractibility: P_n is contractible.
Free Action: The Σ_n-action

$$\mathsf{P}_n \times \Sigma_n \longrightarrow \mathsf{P}_n$$

is free.

A Cat-enriched operad Q is an *E_∞-operad* if the sSet-enriched operad $N\mathsf{Q}$ is an E_∞-operad. ◇

Example 11.6.2. If P and Q are E_∞-operads, then so is the product operad $\mathsf{P} \times \mathsf{Q}$ because the Cartesian product preserves levelwise contractibility and free symmetric group action. ◇

Proposition 11.6.3. *The Barratt-Eccles operad $E\mathsf{As}$ is an E_∞-operad.*

Proof. The free symmetric group action on $E\mathsf{As}$ remains free after applying the nerve functor. It is levelwise contractible because each translation category EX, which has a unique element in each morphism set, is equivalent to the terminal category. □

Explanation 11.6.4 (Simplicial Barratt-Eccles Operad)**.** Using Explanation 7.2.4, we may describe the sSet-enriched Barratt-Eccles operad $N(E\mathsf{As})$ as follows. For $n \geq 0$, the simplicial set $N(E\mathsf{As}_n)$ has the set of k-simplices

$$N(E\mathsf{As}_n)_k = \Sigma_n^{\times(k+1)} \quad \text{for} \quad k \geq 0,$$

with Σ_n the symmetric group on n letters. Each $(k+1)$-tuple of permutations

$$(\sigma_0, \ldots, \sigma_k) \in \Sigma_n^{\times(k+1)}$$

records $k+1$ objects in the translation category $E\mathsf{As}_n$. Since each morphism set in $E\mathsf{As}_n$ is a one-element set, such a $(k+1)$-tuple determines a unique k-simplex in the nerve $N(E\mathsf{As}_n)$. The face and degeneracy maps are given by

$$\begin{aligned} d_i(\sigma_0, \ldots, \sigma_k) &= (\sigma_0, \ldots, \widehat{\sigma_i}, \ldots, \sigma_k) \\ s_i(\sigma_0, \ldots, \sigma_k) &= (\sigma_0, \ldots, \sigma_i, \sigma_i, \ldots, \sigma_k) \end{aligned}$$

for $0 \leq i \leq k$. In other words, the ith face map d_i removes σ_i, and the ith degeneracy map s_i repeats σ_i.

The simplicial operad unit is

$$\mathrm{id}_1 \in \Sigma_1 = N(E\mathsf{As}_1)_0.$$

The right Σ_n-action on $N(E\mathsf{As}_n)$ is given on k-simplices by the diagonal action

$$(\sigma_0, \ldots, \sigma_k) \cdot \pi = (\sigma_0\pi, \ldots, \sigma_k\pi) \quad \text{for} \quad \pi \in \Sigma_n.$$

The operad composition

$$N(E\mathsf{As}_n) \times \prod_{i=1}^{n} N(E\mathsf{As}_{k_i}) \xrightarrow{\ \gamma\ } N(E\mathsf{As}_{k_1+\cdots+k_n})$$

is given on p-simplices by

$$\begin{aligned} &\gamma\Big((\sigma_0, \ldots, \sigma_p); \big\{(\sigma_0^i, \ldots, \sigma_p^i)\big\}_{1\leq i\leq n}\Big) \\ &= \Big\{\gamma\big(\sigma_j, (\sigma_j^1, \ldots, \sigma_j^n)\big)\Big\}_{0\leq j\leq p} \in \prod_{j=0}^{p} \Sigma_{k_1+\cdots+k_n} \end{aligned}$$

for $p \geq 0$, $\sigma_0, \ldots, \sigma_p \in \Sigma_n$, and $\sigma_0^i, \ldots, \sigma_p^i \in \Sigma_{k_i}$. In the second line above, γ is the operad composition (11.1.2) in the associative operad. ◇

E_∞-Symmetric Spectra. Next we define E_∞-symmetric spectra and discuss a few basic examples. Recall from Definition 7.6.22 that each symmetric spectrum X has an *endomorphism simplicial operad* $\mathsf{End}(X)$, which is enriched over sSet_*. For $n \geq 0$, it has

$$\mathsf{End}(X)_n = \mathsf{SymSp}\big(X^{\wedge n} \wedge \Delta^?_+, X\big) \in \mathsf{sSet}_*$$

as its pointed simplicial set of n-ary operations.

Definition 11.6.5.

- An E_∞*-structure* on a symmetric spectrum X is an sSet-enriched operad morphism
$$\mathsf{P} \longrightarrow \mathsf{End}(X)$$
for some E_∞-operad P as in Definition 11.6.1.
- An E_∞*-symmetric spectrum* is a symmetric spectrum equipped with an E_∞-structure. ◇

Recall from Definition I.1.2.23 that a *commutative monoid* is a monoid with a strictly commutative multiplication.

Proposition 11.6.6.

(1) *Each commutative monoid in* SymSp *is an* E_∞*-symmetric spectrum.*
(2) *If a symmetric spectrum has an* E_∞*-structure via the Barratt-Eccles operad, then it is a strict ring symmetric spectrum.*

Proof. There are morphisms of sSet-enriched operads

$$\mathsf{As} \xrightarrow{\iota} N(E\mathsf{As}) \xrightarrow{\pi} \mathsf{Com},$$

with

$$\mathsf{As}_n = \Sigma_n \quad \text{and} \quad \mathsf{Com}_n = *$$

constant simplicial sets for $n \geq 0$. The morphism ι is the nerve of the Cat-enriched operad morphism ι_{As} in (11.4.12). The morphism π consists of the unique morphisms

$$N(E\mathsf{As}_n) \longrightarrow \mathsf{Com}_n = * \quad \text{for} \quad n \geq 0.$$

A commutative monoid structure on a symmetric spectrum X is equivalent to an sSet-enriched operad morphism

$$\mathsf{Com} \longrightarrow \mathsf{End}(X).$$

Precomposing with π yields an E_∞-structure, since $N(E\mathsf{As})$ is an E_∞-operad by Proposition 11.6.3. The second assertion follows by precomposing a given E_∞-structure

$$N(E\mathsf{As}) \longrightarrow \mathsf{End}(X)$$

with ι and using Proposition 11.1.15. □

Example 11.6.7 (The Sphere Spectrum)**.** The symmetric sphere

$$S = \{S^p\}_{p \geq 0}$$

is a commutative monoid in SymSp, so it is an E_∞-symmetric spectrum by Proposition 11.6.6. In more detail, by Examples 7.4.9 and 11.3.13, the symmetric sphere is a monoid in SymSp with the canonical isomorphism

$$S^p \wedge S^q \xrightarrow[\cong]{\mu^S_{p,q}} S^{p+q} \quad \text{for} \quad p, q \geq 0.$$

The commutativity of the multiplication

$$S \,\square_S\, S \xrightarrow{\mu^S} S \in \mathsf{SymSp}$$

follows from (i) the coequalizer in Definition 7.6.1 that defines $\square_S$ and (ii) the commutative diagram

(11.6.8)
$$\begin{array}{ccc} \Sigma_{(p+q)} \times_{\Sigma_p \times \Sigma_q} S^p \wedge S^q & \xrightarrow{\xi_{p,q}} & \Sigma_{(q+p)} \times_{\Sigma_q \times \Sigma_p} S^q \wedge S^p \\ & {\scriptstyle \mu^S_{p,q}} \searrow \quad \swarrow {\scriptstyle \mu^S_{q,p}} & \\ & S^{p+q} & \end{array}$$

in sSet_* for $p, q \geq 0$. ◇

Example 11.6.9 (Suspension Spectra of Commutative Monoids)**.** For a commutative monoid (K, μ^K, η^K) in $(\mathsf{sSet}_*, \wedge, S^0)$, its suspension spectrum

$$\Sigma^\infty K = \left\{S^p \wedge K\right\}_{p \geq 0}$$

is a commutative monoid in SymSp, so it is an E_∞-symmetric spectrum by Proposition 11.6.6. In more detail, by Examples 7.4.10 and 11.3.14, the suspension spectrum $\Sigma^\infty K$ is a monoid in SymSp with the multiplication morphism

$$\begin{array}{ccc} (S^p \wedge K) \wedge (S^q \wedge K) & \xrightarrow{\mu_{p,q}} & S^{p+q} \wedge K \\ & {\scriptstyle \xi_{\mathrm{mid}}} \searrow \quad \nearrow {\scriptstyle \mu^S_{p,q} \wedge \mu^K} & \\ & (S^p \wedge S^q) \wedge (K \wedge K) & \end{array}$$

for $p, q \geq 0$. By the coequalizer in Definition 7.6.1 that defines $\square_S$, the commutativity of the multiplication

$$\Sigma^\infty K \,\square_S\, \Sigma^\infty K \xrightarrow{\mu} \Sigma^\infty K \in \mathsf{SymSp}$$

follows from the commutative diagram

$$\begin{array}{ccc} \Sigma_{p+q} \times_{p,q} (S^p \wedge K) \wedge (S^q \wedge K) & \xrightarrow{\xi_{p,q}} & \Sigma_{q+p} \times_{q,p} (S^q \wedge K) \wedge (S^p \wedge K) \\ \downarrow {\scriptstyle \xi_{\mathrm{mid}}} & & \downarrow {\scriptstyle \xi_{\mathrm{mid}}} \\ \big(\Sigma_{p+q} \times_{p,q} (S^p \wedge S^q)\big) \wedge (K \wedge K) & \xrightarrow{\xi_{p,q} \wedge \xi} & \big(\Sigma_{q+p} \times_{q,p} (S^q \wedge S^p)\big) \wedge (K \wedge K) \\ & {\scriptstyle \mu^S_{p,q} \wedge \mu^K} \searrow \quad \swarrow {\scriptstyle \mu^S_{q,p} \wedge \mu^K} & \\ & S^{p+q} \wedge K & \end{array}$$

in sSet_* for $p, q \geq 0$, where $\times_{p,q} = \times_{\Sigma_p \times \Sigma_q}$.

- The top square commutes by the Symmetric Coherence Theorem 1.1.41 for sSet_*.
- The bottom triangle commutes by the commutativity of μ^K and (11.6.8).

◇

Example 11.6.10 (Eilenberg-Mac Lane Spectra)**.** Suppose R is a commutative ring. Then the Eilenberg-Mac Lane spectrum

$$HR = \left\{R \otimes S^p\right\}_{p \geq 0}$$

is a commutative monoid in SymSp, so it is an E_∞-symmetric spectrum by Proposition 11.6.6. In more detail, by Examples 7.4.11 and 11.3.15, the Eilenberg-Mac Lane spectrum HR is a monoid in SymSp with the multiplication morphism

$$\begin{array}{ccc} \Big(\bigoplus_{(\underline{k}^p)^\flat} R\Big) \wedge \Big(\bigoplus_{(\underline{k}^q)^\flat} R\Big) & \xrightarrow{(\mu_{p,q})_k} & \bigoplus_{(\underline{k}^{p+q})^\flat} R \\ \big(r_{i_1,\ldots,i_p}\big) \wedge \big(r'_{j_1,\ldots,j_q}\big) & \longmapsto & (rr')_{i_1,\ldots,i_p,j_1,\ldots,j_q} \end{array}$$

for $p, q, k \geq 0$, $r, r' \in R$, and $i_a, j_b \in \underline{k}^\flat$. By the coequalizer in Definition 7.6.1 that defines $\Box_S$, the commutativity of the multiplication

$$HR \,\Box_S\, HR \xrightarrow{\mu} HR \in \mathsf{SymSp}$$

follows from the commutative diagram

$$\begin{array}{ccc} \Sigma_{p+q} \times_{p,q} \Big(\bigoplus_{(\underline{k}^p)^\flat} R\Big) \wedge \Big(\bigoplus_{(\underline{k}^q)^\flat} R\Big) & \xrightarrow{(\xi_{p,q})_k} & \Sigma_{q+p} \times_{q,p} \Big(\bigoplus_{(\underline{k}^q)^\flat} R\Big) \wedge \Big(\bigoplus_{(\underline{k}^p)^\flat} R\Big) \\ {\scriptstyle (\mu_{p,q})_k} \searrow & & \swarrow {\scriptstyle (\mu_{q,p})_k} \\ & \bigoplus_{(\underline{k}^{p+q})^\flat} R & \end{array}$$

in which $\times_{p,q} = \times_{\Sigma_p \times \Sigma_q}$.

- On direct sum indices, commutativity follows from the commutative diagram (11.6.8).
- On direct sum generators in R, commutativity follows from the assumption that R is a commutative ring. $\diamond$

The E_∞-symmetric spectra in Examples 11.6.7, 11.6.9, and 11.6.10 are commutative monoids in SymSp. On the other hand, the K-theory E_∞-symmetric spectra in Corollary 11.6.12 below are, in general, not commutative monoids.

K-Theory E_∞-Symmetric Spectra. The Elmendorf-Mandell K-theory K^{EM} is the following composite of multifunctors.

$$\begin{array}{ccc} \mathsf{PermCat}^{\mathsf{su}} & \xrightarrow{\mathsf{K}^{\mathsf{EM}}} & \mathsf{SymSp} \\ {\scriptstyle \mathsf{J}^{\mathsf{EM}}} \searrow & & \nearrow {\scriptstyle \mathsf{K}^{\mathcal{G}}} \\ \mathcal{G}_*\text{-}\mathsf{Cat} & \xrightarrow{N_*} & \mathcal{G}_*\text{-}\mathsf{sSet} \end{array} \tag{11.6.11}$$

- J^{EM} is the Elmendorf-Mandell J-theory in Definition 10.3.25, which is a Cat-enriched multifunctor.
- $N\colon \mathsf{Cat} \longrightarrow \mathsf{sSet}$ is the nerve functor in Definition 7.2.3 from small categories to simplicial sets.
- N_* is the induced change-of-codomain symmetric monoidal sSet_*-functor in Theorem 9.2.19 (2).
- $\mathsf{K}^{\mathcal{G}}$ is the symmetric monoidal sSet_*-functor in Theorem 9.4.9.

The following result is the bipermutative analogue of Corollary 11.3.16 and is essentially [**EM06**, 3.9]. It describes the K-theory of bipermutative categories.

Corollary 11.6.12. *For each small bipermutative category* C, $\mathsf{K}^{\mathsf{EM}}\mathsf{C}$ *is an* E_∞-*symmetric spectrum.*

Proof. Consider the multifunctors

$$\begin{array}{ccccc} E\mathsf{As} & \xrightarrow{F} & \mathsf{PermCat}^{\mathsf{su}} & \xrightarrow{\mathsf{K}^{\mathsf{EM}}} & \mathsf{SymSp} \\ * & \longmapsto & (\mathsf{C}, \oplus, \mathbb{0}, \xi^{\oplus}) & \longmapsto & \mathsf{K}^{\mathsf{EM}}\mathsf{C} \end{array}$$

with

- K^{EM} the Elmendorf-Mandell K-theory multifunctor in (11.6.11) and
- F the Cat-enriched multifunctor in Theorem 11.5.5 such that $F(*)$ is the additive structure of C.

By a change of enrichment (Theorem 2.3.7) via the nerve functor $N\colon \mathsf{Cat} \longrightarrow \mathsf{sSet}$, $\mathsf{J}^{\mathsf{EM}}_N \circ F_N$ is an sSet-enriched multifunctor. Since $N(E\mathsf{As})$ is an E_∞-operad by Proposition 11.6.3, the composite sSet-enriched operad morphism

$$\begin{array}{ccccc} N(E\mathsf{As}) & & & & \mathrm{End}(\mathsf{K}^{\mathsf{EM}}\mathsf{C}) \\ \downarrow{\scriptstyle F_N} & & & & \uparrow{\scriptstyle \mathsf{K}^g} \\ N\mathrm{End}(\mathsf{C}) & \xrightarrow{\mathsf{J}^{\mathsf{EM}}_N} & N\mathrm{End}(\mathsf{J}^{\mathsf{EM}}\mathsf{C}) & \xrightarrow{N_*} & \mathrm{End}(N_*\mathsf{J}^{\mathsf{EM}}\mathsf{C}) \end{array}$$

gives $\mathsf{K}^{\mathsf{EM}}\mathsf{C}$ the structure of an E_∞-symmetric spectrum. □

The braided and E_n analogues of Corollary 11.6.12 are, respectively, Corollaries 12.5.3 and 13.5.2.

Example 11.6.13. By Theorem II.9.3.7, each of the following is a small bipermutative category:

- the finite ordinal category Σ in Proposition I.2.4.8,
- its variant Σ' in Proposition I.2.4.23,
- $\mathsf{Vect}^{\mathbb{C}}_{\mathsf{c}}$ of coordinatized finite dimensional complex vector spaces in Example I.2.5.9,
- each small right bipermutative category in Definition I.2.5.2, and
- each small left bipermutative category in Definition I.2.5.11, such as the distortion category $\mathcal{D}$ in Section I.4.2.

Corollary 11.6.12 applies to each of these small bipermutative categories to yield a K-theory E_∞-symmetric spectrum. See also Questions A.5.4 and A.5.6. ◇

Example 11.6.14 (Bipermutative Strictification). Suppose given a small tight symmetric bimonoidal category C (Definition I.2.1.2), such as

- a small distributive symmetric monoidal category in Proposition I.2.3.2,
- the symmetric bimonoidal groupoid Π in Theorem I.2.6.2, and
- the bimonoidal symmetric center of each small tight braided bimonoidal category in Theorem II.4.5.3.

By Corollary II.9.3.12, the equivalent

- right bipermutative category A in Sections I.5.2 through I.5.4 and
- left bipermutative category A_l in Theorem I.5.4.7

are small tight bipermutative categories. Corollary 11.6.12 applies to each of A and A_l to yield a K-theory E_∞-symmetric spectrum. ◇

Example 11.6.15 (Symmetric Center). Suppose C is a small braided ring category in which the left factorization morphism ∂^l is a natural epimorphism. By Theorem II.9.6.4, its symmetric center $\mathsf{C}^{\mathsf{sym}}$ is a small bipermutative category. Corollary 11.6.12 applies to $\mathsf{C}^{\mathsf{sym}}$ to yield a K-theory E_∞-symmetric spectrum. ◇

11.7. Notes

11.7.1 (The Associative Operad). In slightly different form, the material in Section 11.1 for the associative operad As can also be found in [**Yau16**, 14.2]. Further discussion and examples related to As, such as the operads for morphisms and general diagrams of monoids, can be found in [**Yau16**, Ch. 14] and [**Yau20**, Ch. 4.5 and 7]. Moreover, the Cat-enriched associative operad is called the *strict monoidal category operad* and denoted $\mathsf{MCat}_{\mathsf{st}}$ in [**Yau∞**, 18.5]. A more general analogue is the *monoidal category operad*, whose algebras are general, instead of strict, small monoidal categories [**Yau∞**, 18.2–18.3]. ◇

11.7.2 (The Barratt-Eccles Operad). The translation category EG of a group G is also known as the *action groupoid* of G acting on itself. The simplicial Barratt-Eccles operad $N(E\mathsf{As})$ in Explanation 11.6.4 is the simplicial E_∞-operad defined in [**BE74a**]. Due to its importance in infinite loop space theory, the Barratt-Eccles operad is discussed in many other papers, including [**BE74b, BE74c, Ber96, BF04, BF02, Smi89**]. In general symmetric monoidal categories, discussion of E_∞-operads in terms of the Boardman-Vogt W-construction and their algebras can be found in [**Yau20**, Ch. 7 and 11]. See also Question A.4.3.

Moreover, the Barratt-Eccles operad $E\mathsf{As}$ is called the *strict* S*-monoidal category operad* and denoted $\mathsf{MCat}^{\mathsf{S}}_{\mathsf{st}}$ in [**Yau∞**, 21.4]. A more general analogue is the S*-monoidal category operad*, whose algebras are general small symmetric monoidal categories, instead of small permutative categories [**Yau∞**, 21.4.7]. ◇

11.7.3 (Action Operads). The monoidal category operad and the S-monoidal category operad in Notes 11.7.1 and 11.7.2 are examples of the G*-monoidal category operad* $\mathsf{MCat}^{\mathsf{G}}$ for some *action operad* G [**Yau∞**, 19.1]. An $\mathsf{MCat}^{\mathsf{G}}$-algebra is a general small monoidal category with a compatible action by the action operad G on iterated monoidal products [**Yau∞**, 19.2]. Moreover, $\mathsf{MCat}_{\mathsf{st}}$ and $\mathsf{MCat}^{\mathsf{S}}_{\mathsf{st}}$ are examples of the *strict* G*-monoidal category operad* $\mathsf{MCat}^{\mathsf{G}}_{\mathsf{st}}$ in [**Yau∞**, 19.3]. ◇

11.7.4 (Ring and Bipermutative Structures). Theorems 11.2.16 and 11.5.5 are from [**EM06**, 3.4 and 3.8], but the proofs there are organized differently. More specifically, in [**EM06**], the proof of each of these two theorems has essentially three parts:

- Starting from a ring or bipermutative category structure, produce a multifunctor from As or EAs to $\mathsf{PermCat}^{\mathsf{su}}$.
- Starting from a multifunctor from As or EAs to $\mathsf{PermCat}^{\mathsf{su}}$, produce a ring or bipermutative category structure.
- Observe that the previous two constructions are inverses of each other.

On the other hand, our presentation in this chapter emphasizes that, in each case, the correspondence is a consequence of the coherence of the parameter operad that is *not* specific to $\mathsf{PermCat}^{\mathsf{su}}$.

- Both Proposition 11.1.15 and Theorem 11.2.16 use the Coherence Theorem 11.1.7 for the associative operad As.
- Both Proposition 11.4.26 and Theorem 11.5.5 use the Coherence Theorem 11.4.14 for the Barratt-Eccles operad EAs.

In particular, our proof of each of Theorems 11.2.16 and 11.5.5 produces the bijective correspondence in one step. The braided and general E_n cases (Theorems 12.4.5 and 13.4.12) will follow a conceptually similar pattern. ◇

11.7.5 (Ring Spectra and Homotopy). For discussion of the relationship between low-dimensional homotopy of ring symmetric spectra and their categorical algebra, see [**JP07, BJP08a, BJP08b, BM11, MT07**]. These papers develop categorical models and strict algebraic invariants for ring spectra whose nonvanishing homotopy groups are concentrated in dimensions 0 and 1. ◇

CHAPTER 12

K-Theory of Braided Ring Categories

In Corollaries 11.3.16 and 11.6.12, we observed that the Elmendorf-Mandell K-theory multifunctor (Definition 10.3.32)

$$\mathsf{PermCat}^{\mathsf{su}} \xrightarrow{\mathsf{K}^{\mathsf{EM}}} \mathsf{SymSp}$$

sends

- small ring categories to strict ring symmetric spectra and
- small bipermutative categories to E_∞-symmetric spectra.

This chapter extends the picture to the E_2 case by showing, in Corollary 12.5.3, that $\mathsf{K}^{\mathsf{EM}}\mathsf{C}$ is an E_2-symmetric spectrum for each small braided ring category C (Definition II.9.5.1). In Chapter 13, using E_n-monoidal categories and the n-fold monoidal category operad, the results in Chapter 11 and this chapter will be extended to the general E_n cases for $n \geq 1$.

Similar to the strict ring and E_∞ cases in Chapter 11, Corollary 12.5.3 is obtained by combining several key facts. First, the Elmendorf-Mandell K-theory multifunctor respects

- the categorical enrichment in the multicategory $\mathsf{PermCat}^{\mathsf{su}}$ of small permutative categories and
- the simplicial enrichment in the multicategory SymSp of symmetric spectra.

Therefore, a structure in $\mathsf{PermCat}^{\mathsf{su}}$ that is parametrized by a categorical operad passes along K^{EM} to symmetric spectra. For the E_2 case, the braid operad Br in Section 12.1 is an E_2-operad (Theorem 12.2.4) that parametrizes

- braided strict monoidal categories in Cat (Proposition 12.3.22) and
- braided ring categories in $\mathsf{PermCat}^{\mathsf{su}}$ (Theorem 12.4.5).

Both of these statements are consequences of the Coherence Theorem 12.3.10 for Br. Combining Theorem 12.4.5 with K^{EM} yields Corollary 12.5.3 about K-theory E_2-symmetric spectra. The following table summaries the main results in Chapter 11 and this chapter for the associative operad As, the braid operad Br, and the Barratt-Eccles operad $E\mathsf{As}$. The shorthand *smc* stands for *strict monoidal categories.*

operad	associative As (11.1.1)	braid Br (12.1.2)	Barratt-Eccles $E\mathsf{As}$ (11.4.10)
$E_?$-operad	E_1 (13.1.23, 13.2.1)	E_2 (12.2.4)	E_∞ (11.6.3)
coherence	11.1.7	12.3.10	11.4.14
in Cat	smc (11.1.15)	braided smc (12.3.22)	permutative (11.4.26)
in $\mathsf{PermCat}^{\mathsf{su}}$	ring (11.2.16)	braided ring (12.4.5)	bipermutative (11.5.5)
in SymSp	strict ring (11.3.16)	E_2 (12.5.3)	E_∞ (11.6.12)

Organization. Section 12.1 defines the braid operad Br. It generalizes the Barratt-Eccles operad EAs by using braids with specified underlying permutations as morphisms. In Definition 11.4.10, the associative operad As yields the Cat-enriched Barratt-Eccles operad EAs by applying the translation category functor E, which is a right adjoint (Example 11.4.6) and preserves products. On the other hand, the fact that Br is a Cat-enriched operad (Proposition 12.1.10) requires a longer proof. See also Note 12.6.4 for generalizations of the braid operad. Proposition 12.1.11 shows that the braid operad sits between the associative operad and the Barratt-Eccles operad.

Section 12.2 shows that the braid operad Br is an E_2-operad (Theorem 12.2.4). To define E_2-operads and E_n-operads in general, we first recall the little n-cube operad $\mathcal{C}_n$ of Boardman-Vogt and May. An E_n-operad is, by definition, a topological operad that is weakly equivalent to the little n-cube operad. This definition is extended to simplicial operads and categorical operads via, respectively, the geometric realization functor and the classifying space functor. The fact that Br is an E_2-operad is essentially due to Fiedorowicz [**Fie**∞] and is proved in [**Fre17**, 5.2.12]. See also Note 12.6.2.

Section 12.3 proves the Coherence Theorem 12.3.10 for the braid operad Br that describes Cat-enriched operad morphisms from Br. Similar to the case of the Barratt-Eccles operad, this coherence result depends on a decomposition of the morphisms in Br (Lemma 12.3.6). This decomposition is a consequence of the fact that braids are generated under braid products and sum braids by the generating braid $s_1 \in B_2$ and its inverse s_1^{-1}. As the first application of the Coherence Theorem 12.3.10, we observe in Proposition 12.3.22 that Br is the categorical operad for small braided strict monoidal categories.

Section 12.4 shows that the braid operad Br detects braided ring category structures on small permutative categories (Theorem 12.4.5). This is another application of the Coherence Theorem 12.3.10 for Br. Section 12.5 proves the main result about E_2-symmetric spectra using the braid operad Br. This result, Corollary 12.5.3, says that $\mathsf{K}^{\mathsf{EM}}\mathsf{C}$ is an E_2-symmetric spectrum for each small braided ring category C.

Reading Guide.

(1) Read Definition 12.1.2 and the statements of Propositions 12.1.10 and 12.1.11 for the braid operad Br.
(2) For the fact that Br is an E_2-operad, read Definition 12.2.3 and the statement of Theorem 12.2.4.
(3) For the coherence and K-theoretic properties of Br, read Definition 12.5.1 and the statements of Lemma 12.3.6, Theorems 12.3.10 and 12.4.5, Proposition 12.3.22, and Corollary 12.5.3.
(4) Go back and read the rest of this chapter.

12.1. The Braid Operad

In this section, we define the braid operad Br, which is the braided analogue of the Barratt-Eccles operad EAs in Definition 11.4.10. In Proposition 12.1.10, we prove in detail that Br is a Cat-enriched operad. In Proposition 12.1.11, we observe that the Cat-enriched operad morphism As $\longrightarrow$ EAs, from the associative operad to the Barratt-Eccles operad, factors through the braid operad. In subsequent sections, the braid operad will provide braided and E_2 analogues of Propositions 11.4.26 and 11.6.3, Theorem 11.5.5, and Corollary 11.6.12.

Recall the nth braid group B_n in Definition II.1.1.1. It is generated by $s_1, \ldots, s_{n-1}$, and is subject to the braid relations:

$$s_i s_j = s_j s_i \qquad \text{for } |i-j| \geq 2 \text{ and } 1 \leq i, j \leq n-1.$$

$$s_i s_{i+1} s_i = s_{i+1} s_i s_{i+1} \qquad \text{for } 1 \leq i \leq n-2.$$

With Σ_n denoting the nth symmetric group, the canonical group homomorphism

$$\begin{array}{ccc} B_n & \xrightarrow{\pi} & \Sigma_n \\ s_i & \longmapsto & (i, i+1) \end{array}$$

in (II.1.1.12) sends each braid b to its *underlying permutation* $\pi(b) = \overline{b}$. Also recall the notion of a Cat-enriched operad in Definition 6.1.1, with $(\mathsf{Cat}, \times, \mathbb{1})$ the symmetric monoidal category of small categories with the Cartesian product.

Motivation 12.1.1. In the category $E\mathsf{As}_n$, the objects are the permutations in Σ_n, and the same is true for Br_n. For two permutations $\sigma, \theta \in \Sigma_n$, there is a unique morphism

$$\theta\sigma^{-1} : \sigma \longrightarrow \theta \in E\mathsf{As}_n$$

as in (11.4.4). One way to understand this definition is that $\theta\sigma^{-1}$ is the unique permutation that sends σ to θ, in the sense that

$$(\theta\sigma^{-1})\sigma = \theta.$$

On the other hand, morphisms in the braid operad Br are braids. While each braid b has an underlying permutation $\overline{b}$, the latter does not determine the given braid. This fact accounts for the definition of morphisms in (12.1.3) below. ◇

Definition 12.1.2. The *braid operad* Br is defined as follows. Proposition 12.1.10 proves that it is a Cat-enriched operad.

Categories of Operations: For $n \geq 0$, Br_n is the groupoid defined as follows.

- Its object set is Σ_n.
- For permutations $\sigma, \theta \in \Sigma_n$, the morphism set is

$$\mathsf{Br}_n(\sigma; \theta) = \left\{ b \in B_n \mid \overline{b} = \theta\sigma^{-1} \in \Sigma_n \right\} \tag{12.1.3}$$

  with $\overline{b}$ the underlying permutation of b.
- For each $\sigma \in \Sigma_n$, the identity morphism

$$1_\sigma \in \mathsf{Br}_n(\sigma; \sigma)$$

  is the identity braid $\mathrm{id}_n \in B_n$.
- The categorical composition in Br_n is given by the multiplication in the braid group B_n, that is,

$$b' \circ b = b'b.$$

Unit: The unit is the identity braid

$$\mathrm{id}_1 \in \mathsf{Br}_1 \cong \mathbb{1}.$$

Symmetric Group Action: For each permutation $\upsilon \in \Sigma_n$, the right υ-action is the functor

$$\mathsf{Br}_n \xrightarrow[\cong]{\upsilon} \mathsf{Br}_n \tag{12.1.4}$$

defined by the assignments

$$\begin{cases} (\sigma \in \Sigma_n) \longmapsto (\sigma\upsilon \in \Sigma_n) \\ \big(b \in \mathsf{Br}_n(\sigma;\theta)\big) \longmapsto \big(b \in \mathsf{Br}_n(\sigma\upsilon;\theta\upsilon)\big). \end{cases}$$

Operad Composition: For $n \geq 1$, $k_1, \ldots, k_n \geq 0$, and $k = k_1 + \cdots + k_n$, the functor

$$\mathsf{Br}_n \times \prod_{j=1}^{n} \mathsf{Br}_{k_j} \xrightarrow{\gamma} \mathsf{Br}_k \tag{12.1.5}$$

is defined on objects by

$$\gamma\big(\sigma, (\sigma_j)_{1\leq j\leq n}\big) = \overbrace{\sigma\langle k_1, \ldots, k_n\rangle}^{\text{block permutation}} \cdot \overbrace{(\sigma_1 \times \cdots \times \sigma_n)}^{\text{block sum}} \in \Sigma_k \tag{12.1.6}$$

for $\sigma \in \Sigma_n$ and $\sigma_j \in \Sigma_{k_j}$. This is the composition in the associative operad As in (11.1.2). For morphisms

- $b \in \mathsf{Br}_n(\sigma;\sigma')$ and
- $b_j \in \mathsf{Br}_{k_j}(\sigma_j;\sigma'_j)$ for $1 \leq j \leq n$,

γ is defined by

$$\begin{aligned} \gamma\big(b, (b_j)_{1\leq j\leq n}\big) &= \overbrace{b\langle k_{\sigma^{-1}(1)}, \ldots, k_{\sigma^{-1}(n)}\rangle}^{\text{block braid}} \cdot \overbrace{\Big(\bigoplus_{j=1}^{n} b_{\sigma^{-1}(j)}\Big)}^{\text{sum braid}} \\ &\in \mathsf{Br}_k\Big(\gamma\big(\sigma, (\sigma_j)_{1\leq j\leq n}\big); \gamma\big(\sigma', (\sigma'_j)_{1\leq j\leq n}\big)\Big) \end{aligned} \tag{12.1.7}$$

with the sum braid and block braid in Definitions II.1.1.9 and II.1.1.20.

This finishes the definition of the braid operad Br. ◇

Explanation 12.1.8. In (12.1.7), $\gamma\big(b, (b_j)_{1\leq j\leq n}\big)$ is a sum braid followed by a block braid, as discussed in Explanations II.1.1.13 and II.1.1.25.

- The sum braid

$$\bigoplus_{j=1}^{n} b_{\sigma^{-1}(j)} \in B_k$$

  is geometrically obtained by placing the given braids $b_{\sigma^{-1}(1)}, \ldots, b_{\sigma^{-1}(n)}$ from left to right in the given order.
- The block braid

$$b\langle k_{\sigma^{-1}(1)}, \ldots, k_{\sigma^{-1}(n)}\rangle \in B_k$$

  is obtained from b by replacing its jth string by $k_{\sigma^{-1}(j)}$ parallel strings for $1 \leq j \leq n$.

As we will see shortly in the proof of Proposition 12.1.10, the permutation σ appears in (12.1.7) to account for the fact that b is a morphism $\sigma \longrightarrow \sigma'$ in Br_n. ◇

Example 12.1.9. Consider the groupoids Br_n.

- Br_0 and Br_1 are the terminal category $\mathbb{1}$ because Σ_0, Σ_1, B_0, and B_1 are the trivial groups.
- Br_2 has two objects, id_2 and $(1,2) \in \Sigma_2$. The braid group B_2 is the infinite cyclic group generated by s_1, with underlying permutation $(1,2)$. The morphism sets in Br_2 are as follows.

$$\mathsf{Br}_2(\mathrm{id}_2;\mathrm{id}_2) = \{s_1^{2k}\}_{k\in\mathbb{Z}} \qquad \mathsf{Br}_2((1,2);(1,2)) = \{s_1^{2k}\}_{k\in\mathbb{Z}}$$
$$\mathsf{Br}_2(\mathrm{id}_2;(1,2)) = \{s_1^{2k+1}\}_{k\in\mathbb{Z}} \qquad \mathsf{Br}_2((1,2);\mathrm{id}_2) = \{s_1^{2k+1}\}_{k\in\mathbb{Z}}$$

For $n \geq 2$, Br_n has the $n!$ permutations in Σ_n as objects. For any two permutations $\sigma, \theta \in \Sigma_n$, there are infinitely many braids $b \in B_n$ with underlying permutation $\theta\sigma^{-1}$. ◇

Proposition 12.1.10. Br *in Definition 12.1.2 is a* Cat*-enriched operad.*

Proof. We need to check that

- the various parts of Br are well defined, and
- the Cat-enriched operad axioms in Definition 6.1.1 are satisfied.

Br_n *is a category.* For morphisms

$$b_0\colon \sigma_0 \longrightarrow \sigma_1 \quad \text{and} \quad b_1\colon \sigma_1 \longrightarrow \sigma_2 \in \mathsf{Br}_n,$$

the morphism

$$b_1 b_0\colon \sigma_0 \longrightarrow \sigma_2$$

is well defined because its underlying permutation is

$$\overline{(b_1 b_0)} = \overline{b}_1 \overline{b}_0 = (\sigma_2\sigma_1^{-1})(\sigma_1\sigma_0^{-1}) = \sigma_2\sigma_0^{-1} \in \Sigma_n.$$

The unity and associativity of this categorical composition follow from the fact that B_n is a group.

γ *is well defined.* With the notation

$$\underline{k} = (k_1, \ldots, k_n)$$
$$\sigma\underline{k} = (k_{\sigma^{-1}(1)}, \ldots, k_{\sigma^{-1}(n)})$$

for lists of integers, the following equalities in Σ_k prove that the braid in (12.1.7) is a well-defined morphism

$$\gamma(\sigma, (\sigma_j)_{1\leq j\leq n}) = \sigma\langle\underline{k}\rangle \cdot \prod_{j=1}^{n} \sigma_j \longrightarrow \sigma'\langle\underline{k}\rangle \cdot \prod_{j=1}^{n} \sigma'_j = \gamma(\sigma', (\sigma'_j)_{1\leq j\leq n})$$

in Br_k as in (12.1.3).

$$\begin{aligned}
&\Big(\sigma'\langle\underline{k}\rangle\cdot\prod_{j=1}^{n}\sigma'_j\Big)\cdot\Big(\sigma\langle\underline{k}\rangle\cdot\prod_{j=1}^{n}\sigma_j\Big)^{-1}\\
&=\sigma'\langle\underline{k}\rangle\cdot\Big(\prod_{j=1}^{n}\sigma'_j\Big)\cdot\Big(\prod_{j=1}^{n}\sigma_j\Big)^{-1}\cdot\big(\sigma\langle\underline{k}\rangle\big)^{-1}\\
&=\sigma'\langle\underline{k}\rangle\cdot\Big(\prod_{j=1}^{n}\sigma'_j\Big)\cdot\Big(\prod_{j=1}^{n}\sigma_j^{-1}\Big)\cdot\sigma^{-1}\langle\sigma\underline{k}\rangle\\
&=\sigma'\langle\underline{k}\rangle\cdot\Big(\prod_{j=1}^{n}\sigma'_j\Big)\cdot\sigma^{-1}\langle\sigma\underline{k}\rangle\cdot\Big(\prod_{j=1}^{n}\sigma^{-1}_{\sigma^{-1}(j)}\Big)\\
&=\sigma'\langle\underline{k}\rangle\cdot\sigma^{-1}\langle\sigma\underline{k}\rangle\cdot\Big(\prod_{j=1}^{n}\sigma'_{\sigma^{-1}(j)}\Big)\cdot\Big(\prod_{j=1}^{n}\sigma^{-1}_{\sigma^{-1}(j)}\Big)\\
&=(\sigma'\sigma^{-1})\langle\sigma\underline{k}\rangle\cdot\Big(\prod_{j=1}^{n}\sigma'_{\sigma^{-1}(j)}\sigma^{-1}_{\sigma^{-1}(j)}\Big)\\
&=\overline{b}\langle\sigma\underline{k}\rangle\cdot\prod_{j=1}^{n}\overline{b_{\sigma^{-1}(j)}}=\overline{\gamma\big(b,(b_j)_{1\le j\le n}\big)}
\end{aligned}$$

γ is a functor. The fact that γ in (12.1.5) preserves identity morphisms follows from the equalities

$$\mathrm{id}_n\langle\sigma\underline{k}\rangle=\mathrm{id}_k=\bigoplus_{j=1}^{n}\mathrm{id}_{k_{\sigma^{-1}(j)}}\in B_k.$$

For morphisms

$$\sigma\xrightarrow{\;b\;}\theta\xrightarrow{\;b'\;}\varphi\in\mathsf{Br}_n$$
$$\sigma_j\xrightarrow{\;b_j\;}\theta_j\xrightarrow{\;b'_j\;}\varphi_j\in\mathsf{Br}_{k_j}$$

for $1\le j\le n$, the fact that γ preserves categorical composition is proved by the following equalities in B_k.

$$\begin{aligned}
&\gamma\big(b'b,(b'_jb_j)_{1\le j\le n}\big)\\
&=(b'b)\langle\sigma\underline{k}\rangle\cdot\Big(\bigoplus_{j=1}^{n}b'_{\sigma^{-1}(j)}b_{\sigma^{-1}(j)}\Big)\\
&=b'\langle\theta\underline{k}\rangle\cdot b\langle\sigma\underline{k}\rangle\cdot\Big(\bigoplus_{j=1}^{n}b'_{\sigma^{-1}(j)}\Big)\cdot\Big(\bigoplus_{j=1}^{n}b_{\sigma^{-1}(j)}\Big)\\
&=b'\langle\theta\underline{k}\rangle\cdot\Big(\bigoplus_{j=1}^{n}b'_{\theta^{-1}(j)}\Big)\cdot b\langle\sigma\underline{k}\rangle\cdot\Big(\bigoplus_{j=1}^{n}b_{\sigma^{-1}(j)}\Big)\\
&=\gamma\big(b',(b'_j)_{1\le j\le n}\big)\cdot\gamma\big(b,(b_j)_{1\le j\le n}\big).
\end{aligned}$$

The middle two equalities above use the equality $\overline{b}=\theta\sigma^{-1}\in\Sigma_n$. The first and last equalities are the definition (12.1.7).

For the Cat-enriched operad axioms in Definition 6.1.1 for Br, first observe that, on objects, Br is the associative operad As in Definition 11.1.1. So we only need to check the axioms on morphisms in Br, which are braids with specified underlying permutations.

Symmetric Group Action. To see that the symmetric group action (12.1.4) is an isomorphism of categories, first observe that its assignment on morphisms is well defined because

$$(\theta\upsilon)(\sigma\upsilon)^{-1} = (\theta\upsilon)(\upsilon^{-1}\sigma^{-1}) = \theta\sigma^{-1} = \overline{b}.$$

The functoriality of υ follows from the definition

$$\upsilon(b) = b\colon \sigma\upsilon \longrightarrow \theta\upsilon$$

for each morphism $b\colon \sigma \longrightarrow \theta$ in Br_n. This definition also implies that the symmetric group action on Br_n is compatible with the multiplication and identity in Σ_n.

Unity. The right unity axiom (6.1.4) holds in Br by the equalities

$$\begin{gathered}\overbrace{\mathrm{id}_1 \oplus \cdots \oplus \mathrm{id}_1}^{n} = \mathrm{id}_n\\ b\langle 1,\ldots,1\rangle = b\end{gathered}$$

in B_n. The left unity axiom (6.1.5) holds in Br because

$$\mathrm{id}_1\langle m\rangle = \mathrm{id}_m \in B_m.$$

Associativity. Suppose given morphisms

- $b \in \mathsf{Br}_n(\sigma;\sigma')$
- $b_j \in \mathsf{Br}_{k_j}(\sigma_j;\sigma'_j)$ for $1 \le j \le n$, and
- $b_{j,i} \in \mathsf{Br}_{l_{j,i}}(\sigma_{j,i};\sigma'_{j,i})$ for $1 \le j \le n$ and $1 \le i \le k_j$,

along with the following notation.

$$\begin{aligned} \underline{k} &= (k_1,\ldots,k_n) & k &= \sum_{j=1}^{n} k_j\\ \underline{l}_j &= \big(l_{j,1},\ldots,l_{j,k_j}\big) & L_j &= \sum_{i=1}^{k_j} l_{j,i}\\ \underline{l} &= \big(\underline{l}_1,\ldots,\underline{l}_n\big) & \underline{L} &= (L_1,\ldots,L_n) \end{aligned}$$

$$\theta = \gamma\big(\sigma,(\sigma_j)_{1\le j\le n}\big) = \sigma\langle\underline{k}\rangle\cdot\prod_{j=1}^{n}\sigma_j \in \Sigma_k$$

By definition (12.1.7), we have the following morphisms in Br.

$$\begin{aligned} \gamma\big(b,(b_j)_{1\le j\le n}\big) &= b\langle\sigma\underline{k}\rangle\cdot\bigoplus_{j=1}^{n} b_{\sigma^{-1}(j)}\\ &\in \mathsf{Br}_k\Big(\gamma\big(\sigma,(\sigma_j)_{1\le j\le n}\big);\gamma\big(\sigma',(\sigma'_j)_{1\le j\le n}\big)\Big)\\ \gamma\big(b_j,(b_{j,i})_{1\le i\le k_j}\big) &= b_j\langle\sigma_j\underline{l}_j\rangle\cdot\bigoplus_{i=1}^{k_j} b_{j,\sigma_j^{-1}(i)}\\ &\in \mathsf{Br}_{L_j}\Big(\gamma\big(\sigma_j,(\sigma_{j,i})_{1\le i\le k_j}\big);\gamma\big(\sigma'_j,(\sigma'_{j,i})_{1\le i\le k_j}\big)\Big) \end{aligned}$$

The following equalities in $B_{L_1+\cdots+L_n}$ prove the associativity axiom (5.1.4) for Br.

$$\begin{aligned}
&\gamma\Big(\gamma\big(b,(b_j)_{1\le j\le n}\big),\big((b_{j,i})_{1\le i\le k_j}\big)_{1\le j\le n}\Big)\\
&= \Big[\Big(b\langle\sigma\underline{k}\rangle\cdot\bigoplus_{j=1}^{n} b_{\sigma^{-1}(j)}\Big)\langle\theta\underline{l}\rangle\Big]\cdot\Big[\bigoplus_{j=1}^{n}\bigoplus_{i=1}^{k_{\sigma^{-1}(j)}} b_{\sigma^{-1}(j),\sigma^{-1}_{\sigma^{-1}(j)}(i)}\Big]\\
&= b\langle\sigma\underline{L}\rangle\cdot\Big[\bigoplus_{j=1}^{n} b_{\sigma^{-1}(j)}\langle\sigma_{\sigma^{-1}(j)}\underline{l}_{\sigma^{-1}(j)}\rangle\Big]\cdot\Big[\bigoplus_{j=1}^{n}\bigoplus_{i=1}^{k_{\sigma^{-1}(j)}} b_{\sigma^{-1}(j),\sigma^{-1}_{\sigma^{-1}(j)}(i)}\Big]\\
&= b\langle\sigma\underline{L}\rangle\cdot\bigoplus_{j=1}^{n}\Big[b_{\sigma^{-1}(j)}\langle\sigma_{\sigma^{-1}(j)}\underline{l}_{\sigma^{-1}(j)}\rangle\cdot\Big\{\bigoplus_{i=1}^{k_{\sigma^{-1}(j)}} b_{\sigma^{-1}(j),\sigma^{-1}_{\sigma^{-1}(j)}(i)}\Big\}\Big]\\
&= b\langle\sigma\underline{L}\rangle\cdot\bigoplus_{j=1}^{n}\gamma\Big(b_{\sigma^{-1}(j)},\big(b_{\sigma^{-1}(j),i}\big)_{1\le i\le k_{\sigma^{-1}(j)}}\Big)\\
&= \gamma\Big(b,\Big(\gamma\big(b_j,(b_{j,i})_{1\le i\le k_j}\big)\Big)_{1\le j\le n}\Big)
\end{aligned}$$

This proves that Br is a Cat-enriched operad. $\square$

The braid operad is closely related to the associative operad and the Barratt-Eccles operad. Recall

- the associative operad As with $\mathsf{As}_n = \Sigma_n$ in Definition 11.1.1,
- the Cat-enriched Barratt-Eccles operad $E\mathsf{As}$ with E the translation category functor in Definitions 11.4.1 and 11.4.10, and
- the Cat-enriched operad morphism

  $$\mathsf{As} \xrightarrow{\iota_{\mathsf{As}}} E\mathsf{As}$$

  in (11.4.12) that is levelwise the identity function on objects.

We will use Proposition 12.1.11 in Corollary 12.5.2 below.

Proposition 12.1.11. *There is a unique factorization of the* Cat*-enriched operad morphism* ι_{As} *as in the diagram*

$$\begin{array}{ccc}
\mathsf{As} & \xrightarrow{\iota_{\mathsf{As}}} & E\mathsf{As}\\
{}_{\iota^1}\searrow & & \nearrow_{\iota^2}\\
& \mathsf{Br} &
\end{array}$$

such that both ι^1 *and* ι^2 *are levelwise the identity functions on objects.*

Proof. By Definition 12.1.2, the object set operad of Br is the associative operad As. This defines the operad morphism ι^1. Since the object set operad of the Barratt-Eccles operad $E\mathsf{As}$ is also As, the operad morphism ι^2 is well defined at the object set level. For morphisms, ι^2 is uniquely defined by the fact that, in each translation category $E\mathsf{As}_n$, each morphism set has a single element. This fact also implies that ι^2 is levelwise a functor that preserves the operad unit, symmetric group action, and operad composition. $\square$

Explanation 11.6.4 describes the simplicial Barratt-Eccles operad $N(E\mathsf{As})$, with $N\colon \mathsf{Cat} \longrightarrow \mathsf{sSet}$ the nerve functor (Definition 7.2.3). The simplicial operad $N\mathsf{Br}$ will play an important role in Theorem 12.2.4 and Corollaries 12.5.2 and 12.5.3. We end this section with an explicit description of it.

Explanation 12.1.12 (Simplicial Braid Operad). For $n, k \geq 0$, a k-simplex in the nerve $N\mathsf{Br}_n$ is a diagram

$$\big((\sigma_i)_{i=0}^k, (b_j)_{j=1}^k\big) = \Big(\sigma_0 \xrightarrow{b_1} \sigma_1 \xrightarrow{b_2} \cdots \xrightarrow{b_k} \sigma_k \Big)$$

with

- each $\sigma_i \in \Sigma_n$ for $0 \leq i \leq k$ and
- each $b_j \in \mathsf{Br}_n(\sigma_{j-1}; \sigma_j)$ for $1 \leq j \leq k$.

In other words, $b_j \in B_n$ is a braid with underlying permutation

$$\overline{b_j} = \sigma_j \sigma_{j-1}^{-1} \in \Sigma_n.$$

In particular, the set of 0-simplices in $N\mathsf{Br}_n$ is

$$(N\mathsf{Br}_n)_0 = \Sigma_n.$$

The face and degeneracy maps are given as in Explanation 7.2.4.

The symmetric group Σ_n acts on $N\mathsf{Br}_n$ diagonally on objects, that is,

$$\big((\sigma_i)_{i=0}^k, (b_j)_{j=1}^k\big) \cdot \pi = \big((\sigma_i \pi)_{i=0}^k, (b_j)_{j=1}^k\big) \quad \text{for} \quad \pi \in \Sigma_n.$$

The unit in the simplicial operad $N\mathsf{Br}$ is

$$\mathrm{id}_1 \in (N\mathsf{Br}_1)_0 = \Sigma_1.$$

The operad composition

$$N\mathsf{Br}_n \times \prod_{\ell=1}^n N\mathsf{Br}_{k_\ell} \xrightarrow{\gamma} N\mathsf{Br}_{k_1+\cdots+k_n}$$

is given on p-simplices by

$$\begin{aligned}&\gamma\Big(\big((\sigma_i)_{i=0}^p, (b_j)_{j=1}^p\big); \big\{\big((\sigma_i^\ell)_{i=0}^p, (b_j^\ell)_{j=1}^p\big)\big\}_{\ell=1}^n\Big)\\ &= \Big(\big\{\gamma\big(\sigma_i, (\sigma_i^1, \ldots, \sigma_i^n)\big)\big\}_{i=0}^p, \big\{\gamma\big(b_j, (b_j^1, \ldots, b_j^n)\big)\big\}_{j=1}^p\Big).\end{aligned}$$

The last two γ's are those in, respectively, (12.1.6) and (12.1.7). ◇

12.2. The Braid Operad is an E_2-Operad

In this section, we observe in Theorem 12.2.4 that the braid operad Br in Proposition 12.1.10 is an E_2-operad. This is the E_2 analogue of Proposition 11.6.3, which says that the Barratt-Eccles operad is an E_∞-operad. This assertion will be used in Corollary 12.5.3, which says that the Elmendorf-Mandell K-theory of each small braided ring category is an E_2-symmetric spectrum. To make sense of the concept of an E_2-operad in Theorem 12.2.4, we first define the little n-cube operad $\mathcal{C}_n$ due to Boardman-Vogt [**BV73**, 2.49] and May [**May72**]. It is used to define E_n-operads for $n \geq 1$ in Definition 12.2.3.

Little n-Cubes. The ground symmetric monoidal category is Top, with compactly generated weak Hausdorff spaces as objects and continuous maps as morphisms. See Note 12.6.1 for related references. Suppose $\mathbb{R}$ is the topological space of real numbers. Denote by

- $[0,1] \subseteq \mathbb{R}$ the closed unit interval with interior $(0,1)$ and
- $[0,1]^{\times n} \subseteq \mathbb{R}^n$ the closed unit n-cube with interior $(0,1)^{\times n}$.

A *little n-cube* is a function

$$f = (f^1, \ldots, f^n): [0,1]^{\times n} \longrightarrow [0,1]^{\times n}$$

such that, for each $1 \le i \le n$,

$$f^i: [0,1] \longrightarrow [0,1]$$

is a linear function of the form

$$f^i(t) = a_i + t(b_i - a_i) \quad \text{for} \quad 0 \le t \le 1$$

and some $0 \le a_i < b_i \le 1$. We sometimes denote f^i by its image $[a_i, b_i]$ and a little n-cube by a product

$$[a_1, b_1] \times \cdots \times [a_n, b_n]. \tag{12.2.1}$$

For $k \geq 0$, a k-tuple $\underline{f} = (f_1, \ldots, f_k)$ of little n-cubes is said to have *pairwise disjoint interiors* if

$$f_i\big((0,1)^{\times n}\big) \cap f_j\big((0,1)^{\times n}\big) = \varnothing \quad \text{for} \quad 1 \le i < j \le k. \tag{12.2.2}$$

A k-tuple $\underline{f} = (f_1, \ldots, f_k)$ of little n-cubes with pairwise disjoint interiors is also regarded as a function

$$\coprod_{i=1}^{k} [0,1]^{\times n} \xrightarrow{\quad \amalg_i f_i \quad} [0,1]^{\times n}.$$

If $k = 0$, then this is regarded as the unique function $\varnothing \longrightarrow [0,1]^{\times n}$. The set of all continuous maps

$$\coprod_{i=1}^{k} [0,1]^{\times n} \longrightarrow [0,1]^{\times n}$$

is a topological space with the compact-open topology. The subset of k-tuples of little n-cubes with pairwise disjoint interiors is given the subspace topology and is denoted by

$$\mathcal{C}_n(k).$$

When $k = 0$, $\mathcal{C}_n(0)$ is the one-point space containing the function $\varnothing \longrightarrow [0,1]^{\times n}$. Below is a picture of an element with three pairwise disjoint little 2-cubes inside the unit square.

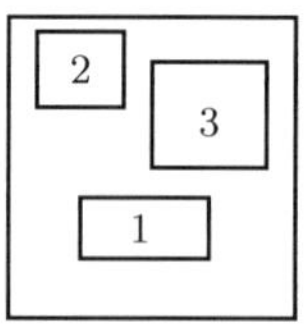

$\in \mathcal{C}_2(3)$

The Little n-Cube Operad. For each $n \geq 1$, the sequence of topological spaces

$$\mathcal{C}_n = \{\mathcal{C}_n(k)\}_{k \geq 0}$$

is a Top-enriched operad with the following data.

Unit: The unit is the identity map

$$\big(1: [0,1]^{\times n} \longrightarrow [0,1]^{\times n}\big) \in \mathcal{C}_n(1).$$

Equivariance: Given $\underline{f} = (f_1, \ldots, f_k) \in \mathcal{C}_n(k)$ and a permutation $\sigma \in \Sigma_k$, define

$$\underline{f}\sigma = \big(f_{\sigma(1)}, \ldots, f_{\sigma(k)}\big) \in \mathcal{C}_n(k).$$

Composition: For $m \geq 1$ and $k_1, \ldots, k_m \geq 0$ with $k = k_1 + \cdots + k_m$, the *composition*

$$\mathcal{C}_n(m) \times \prod_{i=1}^{m} \mathcal{C}_n(k_i) \xrightarrow{\ \gamma\ } \mathcal{C}_n(k)$$

is defined as the composite

$$\begin{array}{ccc} \coprod_{i=1}^{m} \coprod_{l=1}^{k_i} [0,1]^{\times n} & \xrightarrow{\ \gamma(\underline{f}, (\underline{g}_i)_{1\leq i\leq m})\ } & [0,1]^{\times n} \\ & {\scriptstyle \amalg_i \underline{g}_i} \searrow \quad \nearrow {\scriptstyle \underline{f}} & \\ & \coprod_{i=1}^{m} [0,1]^{\times n} & \end{array}$$

for $\underline{f} \in \mathcal{C}_n(m)$ and $\underline{g}_i \in \mathcal{C}_n(k_i)$ with $1 \leq i \leq m$.

This finishes the definition of $\mathcal{C}_n$.

- The unity axioms (6.1.4) and (6.1.5) hold because the unit in $\mathcal{C}_n(1)$ is the identity map.
- The symmetric group action and the equivariance axioms (6.1.6) and (6.1.7) hold because the symmetric group action on $\mathcal{C}_n(k)$ simply permutes the labels of the k little n-cubes in the k-tuples.
- The associativity axiom (6.1.3) holds because composition of continuous maps is strictly associative.

The Top-enriched operad $\mathcal{C}_n$ is called the *little n-cube operad.*

For example, with $\underline{f} \in \mathcal{C}_2(3)$ and $\underline{g} \in \mathcal{C}_2(2)$ on the left-hand side below, the composite $\gamma(\underline{f}, (1, \underline{g}, 1)) \in \mathcal{C}_2(4)$ is the picture on the right below.

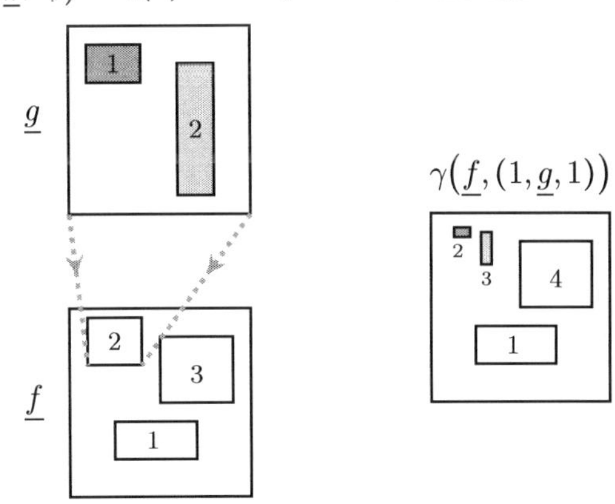

E_n-Operads. Recall from Example 7.2.8 that the *classifying space* $|N\mathsf{C}|$ of a small category C is the geometric realization of the nerve of that category (Definitions 7.1.14 and 7.2.3). The nerve N preserves limits, in particular, products, because it is a right adjoint. The geometric realization $|-|$ preserves *finite* limits, in particular, products. A detailed proof of this fact is in [**GZ67**, III.3.4]. Therefore, the classifying space functor $|N(-)|$ also preserves products and takes a categorical operad to a topological operad. In the next definition, a *zigzag* means a finite nonempty sequence of arrows, each pointing in either direction.

Definition 12.2.3. Two Top-enriched operads P and Q are *weakly equivalent* if there is a zigzag of Top-enriched operad morphisms

$$\mathsf{P} \xrightarrow{\sim} \mathsf{P}_1 \xleftarrow{\sim} \cdots \xrightarrow{\sim} \mathsf{Q}$$

connecting P and Q in which each operad morphism is levelwise a weak homotopy equivalence. Moreover, for $1 \leq n < \infty$, we define the following.

- A Top-enriched operad is an *E_n-operad* if it is weakly equivalent to the little n-cube operad $\mathcal{C}_n$.
- An sSet-enriched operad is an *E_n-operad* if its geometric realization is an E_n-operad.
- A Cat-enriched operad is an *E_n-operad* if its classifying space is an E_n-operad. ◇

The following result is the braid operad analogue of Proposition 11.6.3, which says that the Barratt-Eccles operad is an E_∞-operad. This result is essentially due to Fiedorowicz [**Fie**∞] and is proved in [**Fre17**, 5.2.12]. See also Note 12.6.2.

Theorem 12.2.4. *The braid operad* Br *in Proposition 12.1.10 is an* E_2*-operad.*

Proof. We refer the reader to [**Fre17**, Ch. 5] for the proof with nice illustrations. Here we provide an overview of the proof. The desired weak equivalence between the classifying space of the braid operad Br and the little 2-cube operad $\mathcal{C}_2$ is given by the following zigzag of Top-enriched operad isomorphisms $\cong$ and weak equivalences $\sim$.

$$\begin{array}{ccccc}
\mathcal{C}_2 & & & & |N\mathsf{Br}| \\
{\scriptstyle (6)}\big\uparrow{\scriptstyle \cong} & & & & {\scriptstyle \cong}\big\uparrow{\scriptstyle (1)} \\
\mathsf{Sy}(\widetilde{\mathcal{C}_2}) & & & & |N\mathsf{Sy}(E\mathsf{B})| \\
{\scriptstyle (5)}\big\uparrow{\scriptstyle \sim} & & & & {\scriptstyle \cong}\big\uparrow{\scriptstyle (2)} \\
\mathsf{Sy}(\mathsf{Q}) & \underset{\sim}{\xrightarrow{(4)}} & \mathsf{Sy}(|NE\mathsf{B}|) & \underset{\cong}{\xrightarrow{(3)}} & |\mathsf{Sy}(NE\mathsf{B})|
\end{array} \tag{12.2.5}$$

This diagram involves the following concepts and constructions.

Braided operads. Enriched *braided operads* have the same definition as enriched operads in Definition 6.1.1, except that each symmetric group Σ_n is replaced by the braid group B_n in Definition II.1.1.1. In the equivariance axioms (6.1.6) and (6.1.7), the block permutation and the block sum are replaced by, respectively, the block braid (II.1.1.21) and the sum braid (II.1.1.10).

B_∞-operads. Just as the symmetric groups form the associative operad As in Definition 11.1.1, the braid groups form a braided operad

$$\mathsf{B} = \{B_k\}_{k \geq 0}, \tag{12.2.6}$$

which is called the *braid group operad.*

- Applying the translation category functor E in Definition 11.4.1 yields a Cat-enriched braided operad $E\mathsf{B}$. It inherits from B the free braid group action in each level.
- Applying the classifying space functor, $|NE\mathsf{B}|$ is a Top-enriched braided operad with a levelwise free braid group action. Moreover, each space $|NEB_k|$ is contractible, since each translation category EB_k is equivalent to the terminal category.

A *B_∞-operad* is a Top-enriched braided operad R such that

- each space R_n is contractible, and
- each braid group action

$$\mathsf{R}_n \times B_n \longrightarrow \mathsf{R}_n$$

is free.

So $|NE\mathsf{B}|$ is a B_∞-operad.

Symmetrization. The kernel of the group homomorphism $\pi\colon B_k \longrightarrow \Sigma_k$ in (II.1.1.12) is called the kth *pure braid group* and is denoted by P_k. Elements in P_k are braids with identity underlying permutation. For an object with a braid group action, the quotient by P_k is called the *symmetrization* and is denoted by Sy. The symmetrization has the following properties.

- It inherits a symmetric group action, since

$$B_k/P_k \cong \Sigma_k.$$

- Sy turns a braided operad into an operad.
- When applied to a small category with a braid group action, Sy commutes with the nerve.
- When applied to a simplicial set with a braid group action, Sy commutes with the geometric realization, which is a coend.
- More conceptually, Sy is the left adjoint of the functor that sends an object with a Σ_k-action to the same object with the pullback B_k-action via the map $\pi\colon B_k \longrightarrow \Sigma_k$.

The commutativity of the symmetrization with the nerve and the geometric realization yields the isomorphisms of operads (2) and (3) in (12.2.5).

Br *as a symmetrization.* For each $k \geq 0$, there is a functor

$$EB_k \longrightarrow \mathsf{Br}_k$$

defined as follows.

- On objects, it is given by the underlying permutation $\pi\colon B_k \longrightarrow \Sigma_k$.
- Using the definitions (11.4.4) and (12.1.3), this functor sends a morphism

$$b'b^{-1}\colon b \longrightarrow b'$$

in the translation category EB_k to the morphism

$$b'b^{-1}\colon \pi(b) \longrightarrow \pi(b')$$

in the category Br_k.

These functors induce an isomorphism of Cat-enriched operads

$$\mathsf{Sy}(E\mathsf{B}) \xrightarrow{\cong} \mathsf{Br}.$$

Applying the classifying space functor yields the isomorphism of operads (1) in (12.2.5). This is [**Fre17**, 5.2.15].

Universal covering of $\mathcal{C}_2$. The little 2-cube operad $\mathcal{C}_2$ naturally extends to the B_∞-operad $\widetilde{\mathcal{C}}_2$, with each $\widetilde{\mathcal{C}}_2(k)$ the universal covering space of $\mathcal{C}_2(k)$. Passing to the symmetrization of $\widetilde{\mathcal{C}}_2$, the levelwise covering maps

$$\widetilde{\mathcal{C}}_2(k) \longrightarrow \mathcal{C}_2(k) \tag{12.2.7}$$

induce the isomorphism of operads (6) in (12.2.5). This is [**Fre17**, 5.1.6].

Connecting B_∞-operads. As noted above, $\widetilde{\mathcal{C}}_2$ and $|NE\mathsf{B}|$ are B_∞-operads. The sequence of spaces

$$\mathsf{Q} = \left\{\mathsf{Q}_k = \widetilde{\mathcal{C}}_2(k) \times |NEB_k|\right\}_{k\geq 0}$$

inherits the structure of a B_∞-operad, with the braid group acting diagonally in each level. Since $\widetilde{\mathcal{C}}_2(k)$ and $|NEB_k|$ are contractible, the projections

$$\widetilde{\mathcal{C}}_2(k) \xleftarrow{\sim} \mathsf{Q}_k \xrightarrow{\sim} |NEB_k|$$

are weak homotopy equivalences and respect the free braid group action. These projections form braided operad morphisms

$$\widetilde{\mathcal{C}}_2 \xleftarrow{\sim} \mathsf{Q} \xrightarrow{\sim} |NE\mathsf{B}|. \tag{12.2.8}$$

Passing to the symmetrization yields the weak equivalences of operads (4) and (5) in (12.2.5). □

12.3. Coherence of the Braid Operad

In this section, we prove a coherence theorem for the braid operad Br; see Theorem 12.3.10. This result is the braid operad extension of Theorem 11.4.14 for the Barratt-Eccles operad. Theorem 12.3.10 uses a decomposition of morphisms in Br in Lemma 12.3.6. Example 12.3.9 is a concrete illustration of that decomposition of morphisms in Br. As the first application, in Proposition 12.3.22, we observe that the braid operad detects braided strict monoidal category structure. It will be used again in Theorem 12.4.5 and Corollary 12.5.3 when we consider the K-theory of braided ring categories.

Decomposition of Morphisms.

Motivation 12.3.1. By Explanation 11.4.11 (2), morphisms in $E\mathsf{As}_n$ for $n \geq 2$ are Cat-enriched operadically generated by the isomorphism

$$\mathrm{id}_2 \xrightarrow[\cong]{\tau} (1,2) \in E\mathsf{As}_2\big(\mathrm{id}_2; (1,2)\big)$$

and identity morphisms. Lemma 12.3.6 below is the analogue for the braid operad Br. By Example 12.1.9, the braid operad Br contains the morphism set

$$\mathsf{Br}_2\big(\mathrm{id}_2; (1,2)\big) = \{s_1^{2k+1}\}_{k\in\mathbb{Z}},$$

where $s_1 \in B_2$ is the generating braid with underlying permutation $(1,2)$. In what follows, the isomorphism

$$\mathrm{id}_2 \xrightarrow[\cong]{s_1} (1,2) \in \mathsf{Br}_2 \tag{12.3.2}$$

will play the role of the isomorphism $\tau \in E\mathsf{As}_2\big(\mathrm{id}_2; (1,2)\big)$.

However, one should be careful that s_1^{-1} can mean either the inverse of the isomorphism in (12.3.2), which is denoted by

$$(s_1)^{-1} \colon (1,2) \longrightarrow \mathrm{id}_2,$$

or the different isomorphism

$$\mathrm{id}_2 \xrightarrow[\cong]{s_1^{-1}} (1,2) \in \mathsf{Br}_2. \tag{12.3.3}$$

They are related by the equality

$$(s_1)^{-1}(1,2) = s_1^{-1}\colon \mathrm{id}_2 \longrightarrow (1,2) \in \mathsf{Br}_2 \tag{12.3.4}$$

with $(s_1)^{-1}(1,2)$ the right $(1,2)$-action (12.1.4) on $(s_1)^{-1}$. The equality (12.3.4) holds because, by the symmetric group action (12.1.4) on the morphisms in Br, each side is a morphism $\mathrm{id}_2 \longrightarrow (1,2)$ represented by the braid $s_1^{-1} \in B_2$. There is also an isomorphism

$$(1,2) \xrightarrow[\cong]{s_1} \mathrm{id}_2 \in \mathsf{Br}_2, \tag{12.3.5}$$

which is equal to the right $(1,2)$-action on the s_1 in (12.3.2). ◇

Lemma 12.3.6. *Each morphism in Br_n for $n \geq 2$ decomposes into a categorical composite of isomorphisms of the form ϕv with*

- *each $v \in \Sigma_n$ and*
- *each ϕ an operadic composite of*
  - *one instance of*
  $$s_1^{\pm 1}\colon \mathrm{id}_2 \longrightarrow (1,2)$$
  in (12.3.2) and (12.3.3) and
  - *identity morphisms.*

Proof. First recall the following three facts:

- By definition (12.1.3), each morphism set $\mathsf{Br}_n(\sigma;\theta)$ consists of braids $b \in B_n$ with underlying permutation $\overline{b} = \theta\sigma^{-1}$.
- By Definition II.1.1.1, $s_i^{(n)} \in B_n$ for $1 \leq i < n$ are the group generators. The underlying permutation of $s_i^{(n)}$ is the adjacent transposition $(i,i+1) \in \Sigma_n$.
- By definition (II.1.1.10), there is a sum braid decomposition
$$\big(s_i^{(n)}\big)^{\pm 1} = \overbrace{\mathrm{id}_{i-1} \oplus}^{\varnothing \text{ if } i=1} s_1^{\pm 1} \overbrace{\oplus\, \mathrm{id}_{n-i-1}}^{\varnothing \text{ if } i=n-1} \in B_n$$
with $s_1 \in B_2$ the generating braid.

These facts imply that each morphism in Br_n decomposes into a categorical composite of isomorphisms of the form

$$\big(s_i^{(n)}\big)^{\pm 1} v\colon v \longrightarrow (i,i+1)v \in \mathsf{Br}_n \tag{12.3.7}$$

for some $1 \leq i < n$, permutation $v \in \Sigma_n$, and sum braid

$$\big(s_i^{(n)}\big)^{\pm 1} = \overbrace{\mathrm{id}_{i-1} \oplus}^{\varnothing \text{ if } i=1} s_1^{\pm 1} \overbrace{\oplus\, \mathrm{id}_{n-i-1}}^{\varnothing \text{ if } i=n-1} \colon \mathrm{id}_n \longrightarrow (i,i+1) \in \mathsf{Br}_n. \tag{12.3.8}$$

The block braid induced by an identity braid is an identity braid in the sense that

$$\mathrm{id}_n\langle \sigma \underline{k}\rangle = \mathrm{id}_k \in B_k.$$

By the definition (12.1.7) of γ in Br, the sum braid in (12.3.8) is an operadic composite of one

$$s_1^{\pm 1}\colon \mathrm{id}_2 \longrightarrow (1,2)$$

in (12.3.2) and (12.3.3) and identity morphisms. □

Example 12.3.9. To illustrate the decomposition in Lemma 12.3.6, consider

- the braid $b = s_2 s_1^{-1} \in B_3$ with s_1, s_2 the generating braids in B_3 and
- the cyclic permutations $\sigma = (1,3,2)$ and $\theta = (1,2,3) \in \Sigma_3$.

Since

$$\bar{b} = \sigma = \theta\sigma^{-1} \in \Sigma_3,$$

there is a morphism $b: \sigma \longrightarrow \theta$ in Br_3, as displayed below.

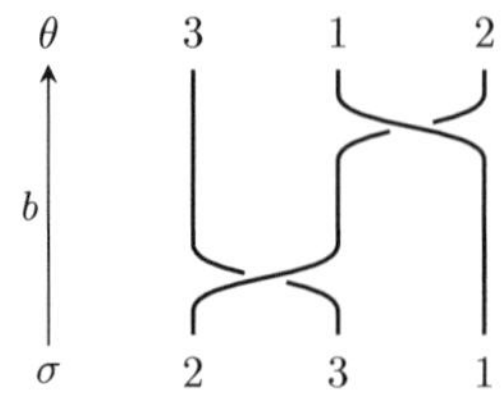

$$\begin{array}{c} \theta = (2,3)(1,3) \\ \Big\uparrow (\mathrm{id}_1 \oplus s_1)(1,3) \\ (1,2)\sigma = (1,3) \\ \Big\uparrow (s_1^{-1} \oplus \mathrm{id}_1)\sigma \\ \sigma \end{array}$$

The morphism $b \in \mathsf{Br}_3(\sigma;\theta)$ decomposes as the categorical composite

$$b = \big[(\mathrm{id}_1 \oplus s_1)(1,3)\big]\big[(s_1^{-1} \oplus \mathrm{id}_1)\sigma\big]$$

with $s_1^{\pm 1}: \mathrm{id}_2 \longrightarrow (1,2)$ in (12.3.2) and (12.3.3). Each of the two factors in the above decomposition has the form (12.3.7). This example illustrates that the permutation v and the isomorphism s_1^{-1} cannot be omitted from Lemma 12.3.6. $\diamond$

Coherence. The next result is a coherence theorem for the braid operad in the sense that it describes the braid operad in terms of a small number of generators and a few relations. It is the braid operad analogue of Theorem 11.1.7 for the associative operad As (Definition 11.1.1) and Theorem 11.4.14 for the Barratt-Eccles operad $E\mathsf{As}$ (Definition 11.4.10). The axioms (12.3.13) and (12.3.14) below are formally identical to those of a braided strict monoidal category in Definitions 1.1.1 and 1.1.14. An *enriched operad morphism* is an enriched multifunctor as in Definition 5.1.12 between two enriched multicategories with one object. This result is from [**Fre17**, 6.2.4 and 6.2.6]; see Note 12.6.3.

Theorem 12.3.10. *Suppose* $(\mathsf{P}, \gamma, 1)$ *is a* Cat*-enriched operad. Then a* Cat*-enriched operad morphism*

$$f: \mathsf{Br} \longrightarrow \mathsf{P}$$

is uniquely determined by

- *the objects*

(12.3.11)
$$\begin{aligned} \mathbb{1} &= f(\mathrm{id}_0) \in \mathsf{P}_0 \\ \mu &= f(\mathrm{id}_2) \in \mathsf{P}_2 \end{aligned}$$

  and

- *the isomorphism*

(12.3.12)
$$\mu = f(\mathrm{id}_2) \xrightarrow[\cong]{\xi = f(s_1)} f(1,2) = \mu(1,2) = \mu^{\mathsf{op}} \in \mathsf{P}_2(\mu, \mu^{\mathsf{op}})$$

  with

$$s_1: \mathrm{id}_2 \longrightarrow (1,2) \in \mathsf{Br}_2$$

  the isomorphism in (12.3.2).

The above data are subject to the following conditions.

Unity and Associativity: *The equalities of objects*

(12.3.13)
$$\begin{aligned} \mu(\mathbb{1}, 1) = 1 &= \mu(1, \mathbb{1}) \in \mathsf{P}_1 \\ \mu(\mu, 1) &= \mu(1, \mu) \in \mathsf{P}_3 \end{aligned}$$

hold.

The Hexagon Axiom: *The left and right hexagon diagrams*

$$\begin{array}{ccc} \mu(1,\mu) & \xrightarrow{\xi} & \mu^{\mathrm{op}}(1,\mu) \\ \| & & \| \\ \mu(\mu,1) & & \mu(1,\mu^{\mathrm{op}})(1,2) \\ {\scriptstyle \mu(\xi,1)}\downarrow & & \uparrow{\scriptstyle \mu(1,\xi)(1,2)} \\ \mu(\mu^{\mathrm{op}},1) & = & \mu(1,\mu)(1,2) \end{array} \qquad \begin{array}{ccc} \mu(\mu,1) & \xrightarrow{\xi} & \mu^{\mathrm{op}}(\mu,1) \\ \| & & \| \\ \mu(1,\mu) & & \mu(\mu^{\mathrm{op}},1)(2,3) \\ {\scriptstyle \mu(1,\xi)}\downarrow & & \uparrow{\scriptstyle \mu(\xi,1)(2,3)} \\ \mu(1,\mu^{\mathrm{op}}) & = & \mu(\mu,1)(2,3) \end{array} \tag{12.3.14}$$

in P_3 *are commutative.*

Proof. Similar to the proof of Theorem 11.4.14, there are two directions.

Necessity. Suppose $f\colon \mathsf{Br} \longrightarrow \mathsf{P}$ is a Cat-enriched operad morphism. We define

$$\begin{cases} \mathbb{1} = f(\mathrm{id}_0) \in \mathsf{P}_0, \\ \mu = f(\mathrm{id}_2) \in \mathsf{P}_2, \quad \text{and} \\ \xi = f(s_1)\colon \mu \longrightarrow \mu^{\mathrm{op}} \in \mathsf{P}_2 \end{cases}$$

as in (12.3.11)–(12.3.12). Since f preserves the operad units, there is an equality

$$f(\mathrm{id}_1) = 1 \in \mathsf{P}_1.$$

- The object set operad of Br is the associative operad As. So there is a Cat-enriched operad morphism

$$\mathsf{As} \xrightarrow{\iota} \mathsf{Br} \tag{12.3.15}$$

  that is levelwise the identity function on objects, that is, permutations in Σ_n. The composite

$$\mathsf{As} \xrightarrow{\iota} \mathsf{Br} \xrightarrow{f} \mathsf{P}$$

  is a Cat-enriched operad morphism. So it preserves the operad relations (11.1.8) in As, which become the equalities (12.3.13) of objects in P.
- Before applying f, the analogues of the left and right hexagons in (12.3.14) are commutative in Br. In each case, each of the two composites is the left, respectively, right, braid below, read bottom-to-top.

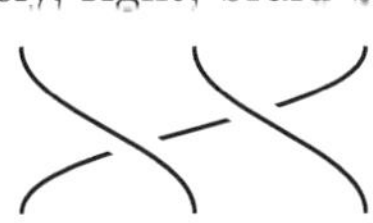

$s_2 s_1 \in B_3$

$s_1 s_2 \in B_3$

  Since f is a Cat-enriched operad morphism, after applying f, the axioms hold in P.

Therefore, the data (12.3.11)–(12.3.12) satisfy the axioms (12.3.13)–(12.3.14).

Sufficiency. Suppose given the data

- $\mathbb{1} = f(\mathrm{id}_0) \in \mathsf{P}_0$,
- $\mu = f(\mathrm{id}_2) \in \mathsf{P}_2$ as in (12.3.11), and
- $\xi = f(s_1)\colon \mu \longrightarrow \mu(1,2) = \mu^{\mathrm{op}} \in \mathsf{P}_2(\mu, \mu^{\mathrm{op}})$ as in (12.3.12)

such that the axioms (12.3.13)–(12.3.14) hold in P. We show that these data extend uniquely to a Cat-enriched operad morphism $f\colon \mathsf{Br} \longrightarrow \mathsf{P}$ in several stages.

Objects. By Theorem 11.1.7 and the axiom (12.3.13), the objects

$$\mathbb{1} = f(\mathrm{id}_0) \quad \text{and} \quad \mu = f(\mathrm{id}_2)$$

uniquely determine a Cat-enriched operad morphism

$$\mathsf{As} \xrightarrow{\ f\ } \mathsf{P} \tag{12.3.16}$$

defined by

$$\begin{cases} f(\mathrm{id}_1) = 1 \in \mathsf{P}_1, \\ f(\mathrm{id}_2) = \mu = \mu_2 \in \mathsf{P}_2, \\ f(\mathrm{id}_n) = \mu(\mu_{n-1}, 1) = \mu_n \in \mathsf{P}_n \quad \text{for } n \geq 3, \text{ and} \\ \quad f(\sigma) = \mu_n \sigma \in \mathsf{P}_n \quad \text{for } \sigma \in \Sigma_n. \end{cases}$$

By (12.3.15), this defines the object part of the Cat-enriched operad morphism $f\colon \mathsf{Br} \longrightarrow \mathsf{P}$.

Morphisms. The identity morphisms $1_{\mathrm{id}_0} \in \mathsf{Br}_0$ and $1_{\mathrm{id}_1} \in \mathsf{Br}_1$ must be sent by f to, respectively, the identity morphisms $1_{\mathbb{1}} \in \mathsf{P}_0$ and $1_1 \in \mathsf{P}_1$. To define f on the other morphisms in Br, note that the coherence theorems for (braided) monoidal categories can be applied to

- the objects $\{\mathbb{1}, 1, \mu\}$, which are interpreted as, respectively, the monoidal unit, the identity functor, and the monoidal product, and
- the braiding ξ.

The reason is that the assumed axioms (12.3.13), and also (12.3.14) in the braided case, are formally identical to those of a (braided) strict monoidal category in Definitions 1.1.1 and 1.1.14. We apply these coherence theorems as follows.

- By the axiom (12.3.13) and Mac Lane's Coherence Theorem 1.1.31, each iterated operadic composite in P_n involving only the objects

$$\mu_0 = \mathbb{1} \in \mathsf{P}_0, \quad \mu_1 = 1 \in \mathsf{P}_1, \quad \text{and} \quad \mu_2 = \mu \in \mathsf{P}_2$$

  is equal to μ_n.
- For each pair of permutations $\sigma, \theta \in \Sigma_n$, consider isomorphisms

$$\mu_n \sigma \xrightarrow{\ \cong\ } \mu_n \theta \in \mathsf{P}_n \tag{12.3.17}$$

  that are categorical composites of isomorphisms of the form ϕv, with $v \in \Sigma_n$ and each ϕ an operadic composite of
  - one

$$\xi = f(s_1)\colon \mu \longrightarrow \mu^{\mathsf{op}} \quad \text{or} \quad \xi^{-1}(1,2)\colon \mu \longrightarrow \mu^{\mathsf{op}}$$

    and
  - identity morphisms.

  The hexagon axiom (12.3.14) and the Braided Coherence Theorem 1.1.38 imply that two isomorphisms as in (12.3.17) are equal if their underlying braids are equal. In defining the underlying braids,
  - the permutations v are ignored, and
  - ξ and $\xi^{-1}(1,2)$ have underlying braids, respectively, s_1 and $s_1^{-1} \in B_2$.

  This property will be called the *braided uniqueness* of (12.3.17).

By Lemma 12.3.6, for each pair of permutations $\sigma, \theta \in \Sigma_n$, each isomorphism $b\colon \sigma \longrightarrow \theta$ in Br_n decomposes into a categorical composite of isomorphisms of the form ϕv with

- each $v \in \Sigma_n$ and
- each ϕ an operadic composite of one $s_1^{\pm 1}\colon \mathrm{id}_2 \longrightarrow (1,2)$ and identity morphisms.

The desired $f: \mathsf{Br} \longrightarrow \mathsf{P}$ must be levelwise a functor and preserve the operad structure, namely, the operad units, the equivariant structure, and the operad composition. So, by (12.3.4), we must define the isomorphism

$$f(\sigma) = \mu_n\sigma \xrightarrow{f(b)} \mu_n\theta = f(\theta) \in \mathsf{P}_n \tag{12.3.18}$$

as the corresponding isomorphism in (12.3.17) with

$$\begin{aligned} &f\big(s_1^{-1}: \mathrm{id}_2 \longrightarrow (1,2)\big) \\ &= \big(f(s_1): \mu \longrightarrow \mu^{\mathsf{op}}\big)^{-1}(1,2) \\ &= \big(\xi^{-1}(1,2): \mu \longrightarrow \mu^{\mathsf{op}}\big). \end{aligned} \tag{12.3.19}$$

The braided uniqueness of (12.3.17) implies the following two statements.

- $f(b)$ is independent of the choice of a decomposition of each morphism $b \in \mathsf{Br}_n$ in Lemma 12.3.6.
- $f: \mathsf{Br}_n \longrightarrow \mathsf{P}_n$ is a functor for each n.

The uniqueness of each functor $f: \mathsf{Br}_n \longrightarrow \mathsf{P}_n$ is part of the definitions (12.3.16) and (12.3.18).

It remains to check that the functors $f: \mathsf{Br}_n \longrightarrow \mathsf{P}_n$ preserve the operad structure. Since $f: \mathsf{As} \longrightarrow \mathsf{P}$ in (12.3.16) is a Cat-enriched operad morphism, by (12.3.15) we only need to consider morphisms in Br_n.

Equivariant structure. The assertion that f preserves the symmetric group action means that, for each permutation $\pi \in \Sigma_n$, the diagram

$$\begin{array}{ccc} \mathsf{Br}_n & \xrightarrow{\pi} & \mathsf{Br}_n \\ {\scriptstyle f}\downarrow & & \downarrow{\scriptstyle f} \\ \mathsf{P}_n & \xrightarrow{\pi} & \mathsf{P}_n \end{array} \tag{12.3.20}$$

of functors, with each horizontal arrow the right π-action, is commutative. Since f and π are functors, by the decomposition of morphisms in Br_n in Lemma 12.3.6, it suffices to consider morphisms of the form $\phi v \in \mathsf{Br}_n$ with $v \in \Sigma_n$ and ϕ an operadic composite of one $s_1^{\pm 1}: \mathrm{id}_2 \longrightarrow (1,2)$ and identity morphisms. On such a morphism ϕv, the commutativity of (12.3.20) follows from

- the definition (12.1.4) of the equivariant structure on Br_n,
- the fact (12.3.4) that
$$(s_1)^{-1}(1,2) = s_1^{-1}: \mathrm{id}_2 \longrightarrow (1,2),$$
- the definition (12.3.18) of f on morphisms, and
- the braided uniqueness of (12.3.17).

Operad composition. The assertion that f preserves the operad composition means the commutativity of the following diagram of functors.

$$\begin{array}{ccc} \mathsf{Br}_n \times \prod_{j-1}^{n} \mathsf{Br}_{k_j} & \xrightarrow{\gamma} & \mathsf{Br}_{k_1+\cdots+k_n} \\ {\scriptstyle f\times\prod_j f}\downarrow & & \downarrow{\scriptstyle f} \\ \mathsf{P}_n \times \prod_{j=1}^{n} \mathsf{P}_{k_j} & \xrightarrow{\gamma} & \mathsf{P}_{k_1+\cdots+k_n} \end{array} \tag{12.3.21}$$

By the equivariance axioms in Br and P and the functoriality of f and γ, it suffices to consider (12.3.21) with

- an identity morphism in n of the $n+1$ factors in $\mathsf{Br}_n \times \prod_{j=1}^n \mathsf{Br}_{k_j}$ and
- in the remaining factor, a morphism of the form ϕv with v a permutation and ϕ an operadic composite of one $s_1^{\pm 1}\colon \mathrm{id}_2 \longrightarrow (1,2)$ and identity morphisms.

In this case, the commutativity of (12.3.21) follows from the braided uniqueness of (12.3.17). Therefore, $f\colon \mathsf{Br} \longrightarrow \mathsf{P}$ is a Cat-enriched operad morphism. □

Braided Strict Monoidal Categories as Algebras. Proposition 11.1.15 shows that the associative operad As is the operad for monoids. Proposition 11.4.26 shows that the Barratt-Eccles operad EAs is the Cat-enriched operad for permutative categories. The following application of Theorem 12.3.10 is the analogue for the braid operad Br. See also Note 12.6.4.

Proposition 12.3.22. *For a small category* C*, a braided strict monoidal category structure on* C *is uniquely determined by a* Cat*-enriched multifunctor*

$$f\colon \mathsf{Br} \longrightarrow \mathsf{Cat} \quad \textit{such that} \quad f(*) = \mathsf{C}.$$

Proof. A Cat-enriched multifunctor $f\colon \mathsf{Br} \longrightarrow \mathsf{Cat}$ such that $f(*) = \mathsf{C}$ is equivalent to a Cat-enriched operad morphism

$$\mathsf{Br} \xrightarrow{\;\;f\;\;} \mathsf{End}(\mathsf{C}) = \big\{\mathsf{Cat}(\mathsf{C}^{\times n}, \mathsf{C})\big\}_{n \geq 0}$$

to the Cat-enriched endomorphism operad of C. By Theorem 12.3.10, such a Cat-enriched operad morphism f is uniquely determined by

- the object
$$\mathbb{1} = f(\mathrm{id}_0) \in \mathsf{C},$$
- the functor
$$\otimes = f(\mathrm{id}_2)\colon \mathsf{C} \times \mathsf{C} \longrightarrow \mathsf{C},$$
and
- the natural isomorphism
$$\otimes \xrightarrow{\;\xi^{\otimes} = f(s_1)\;} \otimes(1,2) = \otimes^{\mathsf{op}}$$

such that the axioms (12.3.13)–(12.3.14) are satisfied.

- The unity and associativity axioms (12.3.13) state that $(\mathsf{C}, \otimes, \mathbb{1})$ is a strict monoidal category as in Definition I.1.2.1.
- The hexagon axiom (12.3.14) state that $(\mathsf{C}, \otimes, \mathbb{1}, \xi^{\otimes})$ is a braided strict monoidal category as in Definition II.1.3.15.

This finishes the proof. □

12.4. Detecting Braided Ring Categories

In this section, we prove that the braid operad detects braided ring category structures on small permutative categories (Theorem 12.4.5). This result is an application of the Coherence Theorem 12.3.10 for the braid operad. In Section 12.5, we use this result to show that the K-theory of a small braided ring category is an E_2-symmetric spectrum (Corollary 12.5.3). For the reader's convenience, here we recall from Chapter II.9 the definition of a braided ring category.

Definition 12.4.1. A *braided ring category* is a tuple

$$\big(\mathsf{C}, (\oplus, \mathbb{0}, \xi^{\oplus}), (\otimes, \mathbb{1}, \xi^{\otimes}), (\partial^l, \partial^r)\big)$$

consisting of

- a ring category

$$\big(\mathsf{C}, (\oplus, \mathbb{0}, \xi^{\oplus}), (\otimes, \mathbb{1}), (\partial^l, \partial^r)\big)$$

  as in Definition 11.2.4 and
- a braided strict monoidal category structure $(\mathsf{C}, \otimes, \mathbb{1}, \xi^{\otimes})$, with $\xi^{\otimes}$ called the *braiding*, as in Definitions 1.1.1 and 1.1.14.

These data are required to satisfy the following axioms for objects $A, B, C \in \mathsf{C}$.

The Zero Braiding Axiom: There are equalities of morphisms as follows.

$$\begin{aligned} \xi^{\otimes}_{A,\mathbb{0}} &= 1_{\mathbb{0}} \colon A \otimes \mathbb{0} = \mathbb{0} \longrightarrow \mathbb{0} = \mathbb{0} \otimes A \\ \xi^{\otimes}_{\mathbb{0},A} &= 1_{\mathbb{0}} \colon \mathbb{0} \otimes A = \mathbb{0} \longrightarrow \mathbb{0} = A \otimes \mathbb{0} \end{aligned} \tag{12.4.2}$$

The Braiding Factorization Axiom: The diagram

$$\begin{array}{ccc} (A \otimes C) \oplus (B \otimes C) & \xrightarrow{\partial^l_{A,B,C}} & (A \oplus B) \otimes C \\ {\scriptstyle \xi^{\otimes}_{A,C} \oplus \xi^{\otimes}_{B,C}} \downarrow & & \downarrow {\scriptstyle \xi^{\otimes}_{A \oplus B, C}} \\ (C \otimes A) \oplus (C \otimes B) & \xrightarrow{\partial^r_{C,A,B}} & C \otimes (A \oplus B) \\ {\scriptstyle \xi^{\otimes}_{C,A} \oplus \xi^{\otimes}_{C,B}} \downarrow & & \downarrow {\scriptstyle \xi^{\otimes}_{C, A \oplus B}} \\ (A \otimes C) \oplus (B \otimes C) & \xrightarrow{\partial^l_{A,B,C}} & (A \oplus B) \otimes C \end{array} \tag{12.4.3}$$

is commutative.

This finishes the definition of a braided ring category. A braided ring category is *small*, respectively, *tight*, if the underlying ring category is so. ◇

Definition 12.4.4. For a permutative category $(\mathsf{C}, \oplus, \mathbb{0}, \xi^{\oplus})$, a *braided ring category structure* on C is the additional data $(\otimes, \mathbb{1}, \xi^{\otimes}, \partial^l, \partial^r)$ such that the tuple

$$\big(\mathsf{C}, (\oplus, \mathbb{0}, \xi^{\oplus}), (\otimes, \mathbb{1}, \xi^{\otimes}), (\partial^l, \partial^r)\big)$$

is a braided ring category as in Definition 12.4.1. ◇

Recall from Section 6.6 that $\mathsf{PermCat}^{\mathsf{su}}$ is the Cat-enriched multicategory with small permutative categories as objects. The category

$$\mathsf{PermCat}^{\mathsf{su}}\big(\langle \mathsf{C} \rangle ; \mathsf{D}\big) = \mathsf{PermCat}^{\mathsf{su}}\big(\langle \mathsf{C}_1, \ldots, \mathsf{C}_n \rangle ; \mathsf{D}\big)$$

has

- n-linear functors

$$\mathsf{C}_1 \times \cdots \times \mathsf{C}_n \longrightarrow \mathsf{D}$$

  as in Definition 6.5.4 as objects and
- multilinear transformations (Definition 6.5.11) as morphisms.

Also recall from Definition 5.1.12 the notion of an enriched multifunctor. The next result is the braid operad analogue of Theorems 11.2.16 and 11.5.5. It says that the braid operad detects braided ring category structures on small permutative categories. It extends Proposition 12.3.22 from Cat to $\mathsf{PermCat}^{\mathsf{su}}$.

Theorem 12.4.5. *For each small permutative category* C*, there is a canonical bijective correspondence between*

- *braided ring category structures on* C *and*
- Cat*-enriched multifunctors*

$$F\colon \mathsf{Br} \longrightarrow \mathsf{PermCat}^{\mathsf{su}} \quad \textit{such that} \quad F(*) = \mathsf{C}.$$

Proof. A Cat-enriched multifunctor

$$F\colon \mathsf{Br} \longrightarrow \mathsf{PermCat}^{\mathsf{su}} \quad \text{such that} \quad F(*) = (\mathsf{C}, \oplus, \mathbb{0}, \xi^{\oplus})$$

is equivalent to a Cat-enriched operad morphism

$$F\colon \mathsf{Br} \longrightarrow \mathsf{End}(\mathsf{C}) = \Big\{ \mathsf{PermCat}^{\mathsf{su}}\big(\langle \overbrace{\mathsf{C}, \ldots, \mathsf{C}}^{n} \rangle ; \mathsf{C}\big)\Big\}_{n \geq 0}$$

to the Cat-enriched endomorphism operad of C. By Theorem 12.3.10, such a Cat-enriched operad morphism is uniquely determined by

- the 0-linear functor, that is, object
$$F(\mathrm{id}_0) = \mathbb{1} \in \mathsf{C},$$
- the 2-linear functor
$$F(\mathrm{id}_2) = (\otimes, \partial^l, \partial^r)\colon \mathsf{C} \times \mathsf{C} \longrightarrow \mathsf{C},$$
and
- the invertible multilinear transformation
$$(\otimes, \partial^l, \partial^r) \xrightarrow{\xi^{\otimes} = F(s_1)} (\otimes^{\mathsf{op}}, \partial^r, \partial^l)$$

such that the conditions (12.3.13)–(12.3.14) are satisfied, with

$$(\mu, 1, \xi) \quad \text{interpreted as} \quad \big((\otimes, \partial^l, \partial^r), 1_{\mathsf{C}}, \xi^{\otimes}\big).$$

By Theorem 11.2.16 and the first paragraph of its proof, the data $(\otimes, \mathbb{1}, \partial^l, \partial^r)$ and the condition (12.3.13) are equivalent to a ring category structure on the small permutative category $(\mathsf{C}, \oplus, \mathbb{0}, \xi^{\oplus})$ in the sense of Definition 11.2.15. In particular, $(\mathsf{C}, \otimes, \mathbb{1})$ is a strict monoidal category. Next we consider the remaining data $\xi^{\otimes}$ and the hexagon axiom (12.3.14).

The invertible multilinear transformation

$$\xi^{\otimes}\colon \otimes \xrightarrow{\cong} \otimes^{\mathsf{op}}$$

is, by definition, a natural isomorphism that satisfies the two conditions in Definition 6.5.11.

- The hexagon axiom (12.3.14) states that $(\mathsf{C}, \otimes, \mathbb{1}, \xi^{\otimes})$ is a braided strict monoidal category as in Definition II.1.3.15.
- Consider the commutativity of the diagram (6.5.12) for $\alpha = \xi^{\otimes}$.
  - The case $i = 1$ is the top half of the braiding factorization axiom (12.4.3) in a braided ring category.
  - The case $i = 2$ is the bottom half of the braiding factorization axiom (12.4.3).
- The second condition in Definition 6.5.11 states the equalities
$$\xi^{\otimes}_{-,\mathbb{0}} = 1_{\mathbb{0}} = \xi^{\otimes}_{\mathbb{0},-}.$$
These equalities form the zero braiding axiom (12.4.2) in a braided ring category.

Therefore, a Cat-enriched operad morphism

$$F\colon \mathsf{Br} \longrightarrow \mathsf{End}(\mathsf{C})$$

is equivalent to a braided ring category structure on C. □

12.5. K-Theory Spectra of Braided Ring Categories

In this section, we prove that the K-theory of a small braided ring category is an E_2-symmetric spectrum (Corollary 12.5.3). This result is the E_2 analogue of Corollaries 11.3.16 and 11.6.12. It is a consequence of Theorem 12.4.5, the Elmendorf-Mandell K-theory multifunctor, and the fact that the braid operad is an E_2-operad (Theorem 12.2.4). Along the way, we record the relationship between strict ring, E_2-, and E_∞-structures on symmetric spectra (Corollary 12.5.2). To make sense of these results, we first define E_n-structure on symmetric spectra.

E_n-Symmetric Spectra. For each symmetric spectrum X, recall from Definition 7.6.22 that it has an *endomorphism simplicial operad* $\mathsf{End}(X)$, which is enriched over sSet_*. For $n \geq 0$, it has

$$\mathsf{End}(X)_n = \mathsf{SymSp}\big(X^{\wedge n} \wedge \Delta^?_+, X\big) \in \mathsf{sSet}_*$$

as its pointed simplicial set of n-ary operations.

Definition 12.5.1. Suppose $1 \leq n < \infty$.

- An *E_n-structure* on a symmetric spectrum X is an sSet-enriched operad morphism
$$\mathsf{P} \longrightarrow \mathsf{End}(X)$$
for some sSet-enriched E_n operad P as in Definition 12.2.3.
- An *E_n-symmetric spectrum* is a symmetric spectrum equipped with an E_n structure. ◇

Recall

- that a strict ring symmetric spectrum is a monoid in the category SymSp of symmetric spectra (Definition 11.3.1) and
- Definition 11.6.5, which is the E_∞ analogue of Definition 12.5.1.

Applying the factorization (Proposition 12.1.11)

$$\begin{array}{ccc} \mathsf{As} & \xrightarrow{\iota_{\mathsf{As}}} & E\mathsf{As} \\ & {\scriptstyle \iota^1}\searrow \quad \nearrow{\scriptstyle \iota^2} & \\ & \mathsf{Br} & \end{array}$$

and using Propositions 11.1.15 and 11.6.3 and Theorem 12.2.4, we obtain the following relationship between strict ring structure, E_2-structure, and E_∞-structure. This result is a refinement of Proposition 11.6.6 (2).

Corollary 12.5.2. *Suppose X is a symmetric spectrum.*

(1) *An E_∞-structure on X via the Barratt-Eccles operad $E\mathsf{As}$ induces an E_2-structure on X by restricting along ι^2.*
(2) *An E_2-structure on X via the braid operad Br induces a strict ring structure on X by restricting along ι^1.*

K-Theory E_2-Symmetric Spectra. Recall the Elmendorf-Mandell K-theory multifunctor in Definition 10.3.32,

$$\mathsf{K}^{\mathsf{EM}} = \mathsf{K}^{g} N_* \mathsf{J}^{\mathsf{EM}} \colon \mathsf{PermCat}^{\mathsf{su}} \longrightarrow \mathsf{SymSp}.$$

The following result is the E_2 analogue of Corollaries 11.3.16 and 11.6.12.

Corollary 12.5.3. *For each small braided ring category C, $\mathsf{K}^{\mathsf{EM}}\mathsf{C}$ is an E_2-symmetric spectrum.*

Proof. Consider the multifunctors

$$\begin{array}{ccccc} \mathsf{Br} & \xrightarrow{F} & \mathsf{PermCat}^{\mathsf{su}} & \xrightarrow{\mathsf{K}^{\mathsf{EM}}} & \mathsf{SymSp} \\ * & \longmapsto & (\mathsf{C}, \oplus, \mathbb{0}, \xi^{\oplus}) & \longmapsto & \mathsf{K}^{\mathsf{EM}}\mathsf{C} \end{array}$$

with F the Cat-enriched multifunctor in Theorem 12.4.5 such that $F(*)$ is the additive structure of C. By Theorem 12.2.4, $N\mathsf{Br}$ is an E_2-operad. As in the proof of Corollary 11.6.12, the composite sSet-enriched operad morphism

$$\begin{array}{ccccc} N\mathsf{Br} & & & & \mathrm{End}(\mathsf{K}^{\mathsf{EM}}\mathsf{C}) \\ {\scriptstyle F_N}\downarrow & & & & \uparrow{\scriptstyle \mathsf{K}^{g}} \\ N\mathrm{End}(\mathsf{C}) & \xrightarrow{\mathsf{J}^{\mathsf{EM}}_N} & N\mathrm{End}(\mathsf{J}^{\mathsf{EM}}\mathsf{C}) & \xrightarrow{N_*} & \mathrm{End}(N_*\mathsf{J}^{\mathsf{EM}}\mathsf{C}) \end{array}$$

gives $\mathsf{K}^{\mathsf{EM}}\mathsf{C}$ the structure of an E_2-symmetric spectrum. □

The E_n analogue of Corollary 12.5.3 is Corollary 13.5.2.

Example 12.5.4 (Braided Distortion)**.** The braided distortion category $\mathcal{D}^{\mathsf{br}}$ in Section II.5.2 is a small tight braided ring category by Example II.9.5.9. Corollary 12.5.3 applies to $\mathcal{D}^{\mathsf{br}}$ to yield a K-theory E_2-symmetric spectrum. See also Questions A.5.6 and A.5.7. ◇

Example 12.5.5 (Bimonoidal Drinfeld Center)**.** The bimonoidal Drinfeld center of each small tight ring category is a small tight braided ring category by Corollary II.9.6.1. Corollary 12.5.3 applies to the latter to yield a K-theory E_2-symmetric spectrum. ◇

Example 12.5.6 (Permbraided Strictification)**.** Suppose C is a small tight braided bimonoidal category (Definition II.2.1.29), such as

- a small abelian category with a compatible braided monoidal structure in Theorem II.2.4.22,
- $\mathcal{F}^{\mathsf{any}}$ of Fibonacci anyons in Theorem II.3.4.13,
- $\mathcal{I}^{\mathsf{any}}$ of Ising anyons in Theorem II.3.6.14, and
- the bimonoidal Drinfeld center of a small tight bimonoidal category in Theorem II.4.4.3.

By Corollary II.9.5.10, the equivalent

- right permbraided category A in Theorem II.6.3.6 and
- left permbraided category A_l in Theorem II.6.3.7

are small tight braided ring categories. Corollary 12.5.3 applies to each of A and A_l to yield a K-theory E_2-symmetric spectrum. ◇

12.6. Notes

12.6.1 (Compactly Generated Spaces). For more point-set level discussion of compactly generated weak Hausdorff spaces, which are used in Section 12.2 to define the little n-cube operads, the reader may consult [**Bro64, GZ67, Kel55, May72, May99, Ste67**]. In the literature, these spaces are sometimes called *k-spaces* and *Kelley spaces*. ◇

12.6.2 (Br is an E_2-Operad). Theorem 12.2.4 is [**Fre17**, 5.2.12] with one main difference. In [**Fre17**, Ch. 5], Fresse actually uses the *little 2-disc operad* $\mathcal{D}_2$ and its universal covering $\widetilde{\mathcal{D}}_2$, instead of the little 2-cube operad $\mathcal{C}_2$ and its universal covering $\widetilde{\mathcal{C}}_2$ as in Definition 12.2.3 and the diagram (12.2.5). The little 2-disc operad $\mathcal{D}_2$ has a definition similar to that of the little 2-cube operad, but it uses the closed unit 2-disc instead of the closed unit 2-cube. A little 2-disc is defined by radial contraction and translation. There is a weak equivalence, in the sense of Definition 12.2.3, between $\mathcal{C}_2$ and $\mathcal{D}_2$ via the Steiner operad [**Ber96, Ste79**]. Combined with [**Fre17**, 5.2.12], this yields Theorem 12.2.4.

Alternatively, without using the nontrivial weak equivalence between $\mathcal{C}_2$ and $\mathcal{D}_2$, one can directly adapt the argument in [**Fre17**, 5.1.6] for $\mathcal{C}_2$ and $\widetilde{\mathcal{C}}_2$ to obtain (4) and (5) in the diagram (12.2.5). A detailed description of the B_∞-operad $\widetilde{\mathcal{C}}_2$ is in [**Yau∞**, 5.4], which also contains a detailed description of $\widetilde{\mathcal{D}}_2$. Moreover, [**Yau∞**, Ch. 5] contains a general discussion of braided operads. Covering space theory, which is needed to define $\widetilde{\mathcal{C}}_2$ and $\widetilde{\mathcal{D}}_2$, is discussed in many books on elementary topology, such as [**Hat02, Mun00**]. ◇

12.6.3 (Braid Operad Coherence). Lemma 12.3.6 and Theorem 12.3.10 for Br are from [**Fre17**, 6.2.4 and 6.2.6] with the following differences.

(1) In [**Fre17**], those theorems have an extra unity axiom for the braiding

$$\xi_{\mathbb{1},1} = 1_1 = \xi_{1,\mathbb{1}} \in \mathsf{P}_1,$$

which is stated as

$$c(e, x_1) = id_{x_1} = c(x_1, e)$$

there. By Proposition II.1.3.21, this extra unity axiom is redundant because it is a formal consequence of the other axioms and data.

(2) In [**Fre17**, 6.2.4(a)], the role of the permutation v is not made explicit as in Lemma 12.3.6. The morphism

$$s_i^{(n)} \colon \mathrm{id}_n \longrightarrow (i, i+1) \in \mathsf{Br}_n$$

in (12.3.8) has domain the identity permutation id_n. In the categorical decomposition of a general morphism in Br_n, the domain of a typical factor is a general permutation and not id_n. Therefore, a general factor in the categorical decomposition has the form $\big(s_i^{(n)}\big)^{\pm 1} v$ in (12.3.7), with a permutation v.

(3) In [**Fre17**, 6.2.4(a)], the role of the isomorphism

$$s_1^{-1} \colon \mathrm{id}_2 \longrightarrow (1,2)$$

is not made explicit as in Lemma 12.3.6. While $s_1, s_1^{-1} \in B_2$ both have underlying permutation $(1,2)$, they are different as braids and represent two different morphisms $\mathrm{id}_2 \longrightarrow (1,2)$ in Br_2. So s_1^{-1} must be included in that lemma.

The necessity of the permutation υ and the isomorphism s_1^{-1} is illustrated in Example 12.3.9. ◇

12.6.4 (Operad for Braided Monoidal Categories)**.** The braid operad Br is called the *strict* B-*monoidal category operad* and denoted $\mathsf{MCat}_{\mathsf{st}}^{\mathsf{B}}$ in [**Yau**∞, 21.1.15]. A more general analogue is the B-*monoidal category operad* $\mathsf{MCat}^{\mathsf{B}}$, whose algebras are general, instead of strict, small braided monoidal categories [**Yau**∞, 21.1.7]. Further coherence properties of braided monoidal categories, in the more general context of action operads, are in [**Yau**∞, 21.2–21.3]. ◇

CHAPTER 13

K-Theory of E_n-Monoidal Categories

In Corollaries 11.3.16, 11.6.12, and 12.5.3, we saw that the Elmendorf-Mandell K-theory multifunctor (Definition 10.3.32)

$$\mathsf{PermCat}^{\mathsf{su}} \xrightarrow{\mathsf{K}^{\mathsf{EM}}} \mathsf{SymSp}$$

sends

- small ring categories to strict ring symmetric spectra,
- small bipermutative categories to E_∞-symmetric spectra, and
- small braided ring categories to E_2-symmetric spectra.

This chapter completes the picture with Corollary 13.5.2. It shows that $\mathsf{K}^{\mathsf{EM}}\mathsf{C}$ is an E_n-symmetric spectrum for each small E_n-monoidal category C (Definition II.10.7.2) for $n \geq 1$.

Similar to the strict ring, E_∞, and E_2 cases in Chapters 11 and 12, Corollary 13.5.2 is obtained by combining several facts. First, the Elmendorf-Mandell K-theory multifunctor respects

- the categorical enrichment in the multicategory $\mathsf{PermCat}^{\mathsf{su}}$ of small permutative categories and
- the simplicial enrichment in the multicategory SymSp of symmetric spectra.

Therefore, a structure in $\mathsf{PermCat}^{\mathsf{su}}$ that is parametrized by a categorical operad passes along K^{EM} to symmetric spectra. For the general E_n case, the n-fold monoidal category operad Mon^n in Section 13.1 is an E_n-operad (Theorem 13.2.1) that parametrizes

- n-fold monoidal categories in Cat (Proposition 13.3.18) and
- E_n-monoidal categories in $\mathsf{PermCat}^{\mathsf{su}}$ (Theorem 13.4.12).

Both of these statements are consequences of the Coherence Theorem 13.3.3 for Mon^n. Combining Theorem 13.4.12 with K^{EM} yields Corollary 13.5.2 about K-theory E_n-symmetric spectra.

Consider the first two cases of Corollary 13.5.2.

- If $n = 1$, then
  - Mon^1 is the associative operad As (Example 13.1.23), and
  - an E_1-monoidal category is a ring category (Example II.10.7.13).

  So Corollary 13.5.2 with $n = 1$ recovers Corollary 11.3.16.
- If $n = 2$, then Corollary 13.5.2 says that $\mathsf{K}^{\mathsf{EM}}\mathsf{C}$ is an E_2-symmetric spectrum for each small E_2-monoidal category C. So Corollary 13.5.2 with $n = 2$ is an alternative to Corollary 12.5.3 that also yields K-theory E_2-symmetric spectra. The difference is that Corollary 12.5.3 uses the E_2-operad Br and

braided ring categories, while Corollary 13.5.2 with $n = 2$ uses the E_2-operad Mon^2 and E_2-monoidal categories. By Theorem II.10.8.1, braided ring categories are special cases of E_2-monoidal categories.

The following table summaries the main results in Chapters 11 and 12 and this chapter for the associative operad As, the braid operad Br, the n-fold monoidal category operads Mon^n for $n \geq 1$, and the Barratt-Eccles operad $E\mathsf{As}$. The shorthand *smc* stands for *strict monoidal categories.*

operad	As (11.1.1)	Br (12.1.2)	Mon^n (13.1.12)	$E\mathsf{As}$ (11.4.10)
$E_?$-operad	E_1 (13.1.23, 13.2.1)	E_2 (12.2.4)	E_n (13.2.1)	E_∞ (11.6.3)
coherence	11.1.7	12.3.10	13.3.3	11.4.14
in Cat	smc (11.1.15)	braided smc (12.3.22)	n-fold monoidal (13.3.18)	permutative (11.4.26)
in $\mathsf{PermCat}^{\mathsf{su}}$	ring (11.2.16)	braided ring (12.4.5)	E_n-monoidal (13.4.12)	bipermutative (11.5.5)
in SymSp	strict ring (11.3.16)	E_2 (12.5.3)	E_n (13.5.2)	E_∞ (11.6.12)

For open questions related to E_n-operads, see Appendices A.2 and A.4.

Organization. Section 13.1 defines the n-fold monoidal category operad Mon^n and carefully proves in Proposition 13.1.20 that it is a Cat-enriched operad. The constituent categories $\mathsf{Mon}^n(k)$ are those in Definition II.10.5.13. In Theorem II.10.5.18, we saw that the categories $\mathsf{Mon}^n(k)$ provide a decomposition of the free n-fold monoidal category $\mathsf{FMon}^n(\mathsf{C})$ of a small category C. When $n = 1$, Mon^1 is equal to the Cat-enriched associative operad As.

Section 13.2 shows that the n-fold monoidal category operad Mon^n is an E_n-operad. This result is from [**BFSV03**, 3.14]. The assertion (Theorem 13.2.1) that the Cat-enriched operad Mon^n is an E_n-operad means that its classifying space is connected to the little n-cube operad $\mathcal{C}_n$ by a zigzag of topological operad weak equivalences. Explanation 13.2.12 compares (i) the zigzag in Theorem 12.2.4 that connects the classifying space of the braid operad Br with the little 2-cube operad $\mathcal{C}_2$ and (ii) the zigzag in Theorem 13.2.1.

Section 13.3 proves the Coherence Theorem 13.3.3 for the n-fold monoidal category operad Mon^n that describes Cat-enriched operad morphisms from Mon^n. While the details are different, the proof of Theorem 13.3.3 shares the same broad strategy as those of Theorems 11.4.14 and 12.3.10. In particular, the Coherence Theorem II.10.6.8 for $\mathsf{Mon}^n(k)$ is crucial in the sufficiency part of that proof. As the first application of the Coherence Theorem 13.3.3, we observe in Proposition 13.3.18 that Mon^n is the categorical operad for small n-fold monoidal categories.

Section 13.4 shows that Mon^n detects E_n-monoidal category structures on small permutative categories (Theorem 13.4.12). This is another application of the Coherence Theorem 13.3.3 for Mon^n. Section 13.5 proves the main result about E_n-symmetric spectra using the n-fold monoidal category operad Mon^n. This result, Corollary 13.5.2, says that $K^{\mathsf{EM}}\mathsf{C}$ is an E_n-symmetric spectrum for each small E_n-monoidal category C. Proposition 13.5.1 shows that an E_∞-structure induces an E_n-structure on a symmetric spectrum.

Reading Guide.

(1) For the n-fold monoidal category operad Mon^n, read Definition 13.1.12 and the statements of Proposition 13.1.20 and Theorem 13.2.1.

(2) For the coherence and K-theoretic properties of Mon^n, read the statements of Theorems 13.3.3 and 13.4.12, Proposition 13.3.18, and Corollary 13.5.2.
(3) Go back and read the rest of this chapter.

13.1. The Iterated Monoidal Category Operad

In this section, we define the n-fold monoidal category operad Mon^n in Definition 13.1.12. Proposition 13.3.18 and Theorem 13.4.12 will show that the Cat-enriched operad Mon^n is to n-fold monoidal categories and E_n-monoidal categories as

- the associative operad (Definition 11.1.1) is to monoids and ring categories (Proposition 11.1.15 and Theorem 11.2.16);
- the Barratt-Eccles operad (Definition 11.4.10) is to permutative categories and bipermutative categories (Proposition 11.4.26 and Theorem 11.5.5); and
- the braid operad (Definition 12.1.2) is to braided strict monoidal categories and braided ring categories (Proposition 12.3.22 and Theorem 12.4.5).

Moreover, Mon^n is an E_n-operad by Theorem 13.2.1. The fact that Mon^n is actually a Cat-enriched operad is proved in Proposition 13.1.20.

n-Fold Monoidal Categories. We first recall some concepts from Chapter II.10 about n-fold monoidal categories.

Definition 13.1.1. For $n \geq 1$, an *n-fold monoidal category* is a tuple

$$\left(\mathsf{C}, \{\otimes_i\}_{1\leq i\leq n}, \mathbb{1}, \{\eta^{i,j}\}_{1\leq i<j\leq n}\right)$$

consisting of the following data.

The Underlying Category: C is a category.
The Unit: $\mathbb{1} \in \mathsf{C}$ is an object, which is called the *unit.*
The Multiplicative Structures: For each $1 \leq i \leq n$,

$$(\mathsf{C}, \otimes_i, \mathbb{1})$$

is a strict monoidal category, which is called the *ith monoidal structure*, with $\otimes_i$ called the *ith product.*

The Exchanges: For each pair (i,j) with $1 \leq i < j \leq n$,

$$(A \otimes_j B) \otimes_i (C \otimes_j D) \xrightarrow{\eta^{i,j}_{A,B,C,D}} (A \otimes_i C) \otimes_j (B \otimes_i D) \tag{13.1.2}$$

is a natural transformation for objects $A, B, C, D \in \mathsf{C}$, which is called the *$(i,j)$-exchange.*

These data are required to satisfy the following equalities and commutative diagrams for objects A, A', A'', B, B', B'', C, C', D, and D' in C. The axioms (13.1.3)–(13.1.6) are defined for $1 \leq i < j \leq n$. The axiom (13.1.7) is defined for $1 \leq i < j < k \leq n$.

The Internal Unity Axiom:

$$\eta^{i,j}_{A,B,\mathbb{1},\mathbb{1}} = 1_{A\otimes_j B} = \eta^{i,j}_{\mathbb{1},\mathbb{1},A,B} \tag{13.1.3}$$

The External Unity Axiom:

$$\eta^{i,j}_{A,\mathbb{1},B,\mathbb{1}} = 1_{A\otimes_i B} = \eta^{i,j}_{\mathbb{1},A,\mathbb{1},B} \tag{13.1.4}$$

The Internal Associativity Axiom:

(13.1.5) 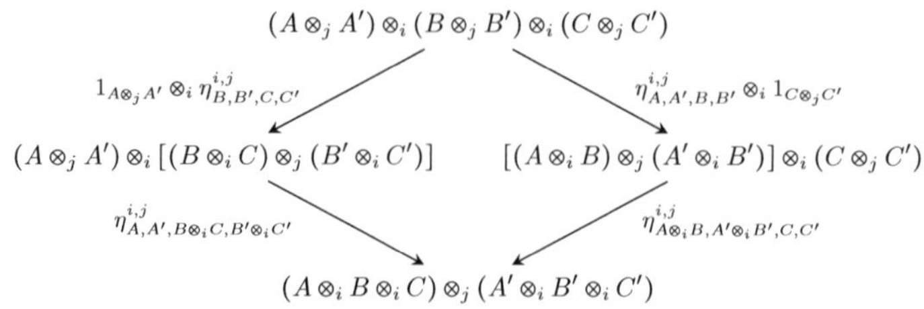

The External Associativity Axiom:

(13.1.6) 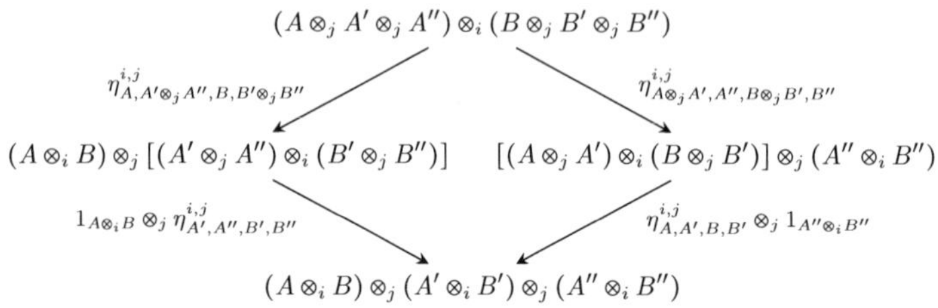

The Triple Exchange Axiom:

(13.1.7) 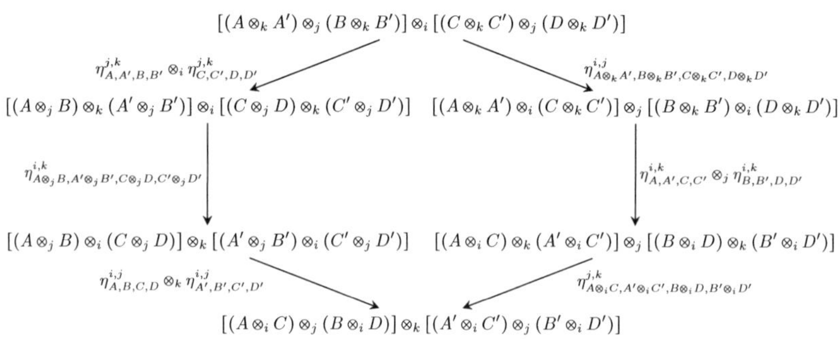

This finishes the definition of an n-fold monoidal category. It is *small* if it has a set of objects. ◇

By Propositions II.10.1.14 and II.10.1.21, braided strict monoidal categories are special cases of 2-fold monoidal categories, and permutative categories are special cases of n-fold monoidal categories for $n \geq 2$.

For n-fold monoidal categories C and D, an *n-fold monoidal functor* (Definition II.10.3.1)

$$\big(F, \{F_i^2\}_{i=1}^n\big) \colon \mathsf{C} \longrightarrow \mathsf{D}$$

consists of

- a functor $F \colon \mathsf{C} \longrightarrow \mathsf{D}$ and
- for each $1 \leq i \leq n$, a natural transformation

$$FA \otimes_i FB \xrightarrow{(F_i^2)_{A,B}} F(A \otimes_i B)$$

for objects $A, B \in \mathsf{C}$, which is called the *ith monoidal constraint*.

These data are required to satisfy the following two conditions.

Monoidality: For each $1 \le i \le n$,

$$(F, F_i^2) \colon (\mathsf{C}, \otimes_i, \mathbb{1}) \longrightarrow (\mathsf{D}, \otimes_i, \mathbb{1})$$

is a strictly unital monoidal functor.

The Exchange Constraint Axiom: The following diagram in D is commutative for all $A, B, C, D \in \mathsf{C}$ and $1 \le i < j \le n$.

(13.1.8)
$$\begin{array}{ccc}
 & (FA \otimes_j FB) \otimes_i (FC \otimes_j FD) & \\
\swarrow{\scriptstyle (F_j^2)_{A,B} \otimes_i (F_j^2)_{C,D}} & & \searrow{\scriptstyle \eta^{i,j}_{FA,FB,FC,FD}} \\
F(A \otimes_j B) \otimes_i F(C \otimes_j D) & & (FA \otimes_i FC) \otimes_j (FB \otimes_i FD) \\
\downarrow{\scriptstyle (F_i^2)_{A \otimes_j B, C \otimes_j D}} & & \downarrow{\scriptstyle (F_i^2)_{A,C} \otimes_j (F_i^2)_{B,D}} \\
F\big((A \otimes_j B) \otimes_i (C \otimes_j D)\big) & & F(A \otimes_i C) \otimes_j F(B \otimes_i D) \\
\searrow{\scriptstyle F\eta^{i,j}_{A,B,C,D}} & & \swarrow{\scriptstyle (F_j^2)_{A \otimes_i C, B \otimes_i D}} \\
 & F\big((A \otimes_i C) \otimes_j (B \otimes_i D)\big) &
\end{array}$$

There is a monoidal category $(\mathsf{MCat}^n, \times, \mathbb{1})$ of small n-fold monoidal categories and n-fold monoidal functors by Lemma II.10.4.2. Monoids in MCat^n are precisely small $(n+1)$-fold monoidal categories by Theorem II.10.4.5.

An n-fold monoidal functor $(F, \{F_i^2\}_{i=1}^n)$ is *strict* if each F_i^2 is the identity natural transformation. Equivalently, it is a functor that strictly preserves

- the unit $\mathbb{1}$,
- the monoidal products $\{\otimes_i\}_{i=1}^n$, and
- the exchanges $\{\eta^{i,j}\}_{i<j}$.

The wide subcategory of MCat^n with strict n-fold monoidal functors is denoted by $\mathsf{MCat}^n_{\mathsf{st}}$.

Free n-Fold Monoidal Categories. By Proposition II.10.5.9, there is an adjunction

(13.1.9)
$$\mathsf{Cat} \underset{U}{\overset{\mathsf{FMon}^n}{\rightleftarrows}} \mathsf{MCat}^n_{\mathsf{st}}$$

with

- U the forgetful functor that forgets about the n-fold monoidal structure and
- FMon^n the free n-fold monoidal category functor.

Suppose C is a small category.

- The objects in $\mathsf{FMon}^n(\mathsf{C})$ are generated by the objects in C under the strictly associative products $\{\otimes_i\}_{i=1}^n$, with a strict two-sided unit $\mathbb{1}$.
- The morphisms in $\mathsf{FMon}^n(\mathsf{C})$ are generated by
  - the morphisms in C,
  - the identity morphisms 1_A, and
  - the exchanges

$$(A \otimes_j B) \otimes_i (C \otimes_j D) \xrightarrow{\eta^{i,j}_{A,B,C,D}} (A \otimes_i C) \otimes_j (B \otimes_i D)$$

    for objects $A, B, C, D \in \mathsf{FMon}^n(\mathsf{C})$ and $1 \le i < j \le n$,

  under the strictly associative products $\{\otimes_i\}_{i=1}^n$ and composites.

These data are subject to the relations that make $\mathsf{FMon}^n(\mathsf{C})$ into an n-fold monoidal category.

Decomposition of Free n-Fold Monoidal Categories. For $k \geq 0$ and $n \geq 1$,

$$\mathsf{Mon}^n(k) \subset \mathsf{FMon}^n\{1,\ldots,k\}$$

is the full subcategory in which each object can be written as an iterated $\{\otimes_i\}_{i=1}^n$-product with each object in the discrete category $\{1,\ldots,k\}$ occurring precisely once (Definition II.10.5.13). Since $\{1,\ldots,k\}$ is a discrete category, the morphisms in $\mathsf{Mon}^n(k)$ are iterated composites of finite $\{\otimes_i\}_{i=1}^n$-products of identity morphisms and the exchanges $\{\eta^{i,j}\}_{i<j}$.

- For example, each of
$$\mathsf{Mon}^n(0) = \{\mathbb{1}\} \quad \text{and} \quad \mathsf{Mon}^n(1) = \{1\}$$
is the terminal category.
- The category $\mathsf{Mon}^n(2)$ has objects
$$1 \otimes_i 2 \quad \text{and} \quad 2 \otimes_i 1 \quad \text{for} \quad 1 \leq i \leq n.$$
Its morphisms are generated under composites by the exchanges
$$\tau(1) \otimes_i \tau(2) \xrightarrow{\eta^{i,j}_{\tau(1),\mathbb{1},\mathbb{1},\tau(2)}} \tau(1) \otimes_j \tau(2)$$
$$\tau(1) \otimes_i \tau(2) \xrightarrow{\eta^{i,j}_{\mathbb{1},\tau(1),\tau(2),\mathbb{1}}} \tau(2) \otimes_j \tau(1)$$
for $1 \leq i < j \leq n$ and $\tau \in \Sigma_2$.

A description of $\mathsf{Mon}^n(3)$ is in Example II.10.5.17.

There is a right Σ_k-action on the category $\mathsf{Mon}^n(k)$ that permutes the elements in $\{1,\ldots,k\}$ and changes the subscripts in the generating morphisms accordingly. By Theorem II.10.5.18, for each small category C, there is a natural isomorphism of categories

$$\coprod_{k\geq 0} \mathsf{Mon}^n(k) \times_{\Sigma_k} \mathsf{C}^{\times k} \xrightarrow[\cong]{\phi_{\mathsf{C}}} \mathsf{FMon}^n(\mathsf{C}) \tag{13.1.10}$$

that extends the isomorphism $\mathsf{Mon}^n(1) \times \mathsf{C} \cong \mathsf{C}$. For a small n-fold monoidal category C, the composite of the isomorphism (13.1.10) and the counit of the adjunction (13.1.9) yields the evaluation functors

$$\mathsf{Mon}^n(k) \times_{\Sigma_k} \mathsf{C}^{\times k} \xrightarrow{\theta_k} \mathsf{C} \tag{13.1.11}$$

for $k \geq 0$.

n-Fold Monoidal Category Operad. Recall from Definition 6.1.1 that a Cat-enriched operad is a Cat-enriched multicategory with one object, and $\mathbb{1}$ is the terminal category. We now assemble the categories $\mathsf{Mon}^n(k)$ into an operad.

Definition 13.1.12. For $n \geq 1$, define the data of a Cat-enriched operad

$$\mathsf{Mon}^n = \left(\{\mathsf{Mon}^n(k)\}_{k\geq 0}, \gamma, 1\right)$$

as follows.

Categories and Equivariance: For $k \geq 0$, $\mathsf{Mon}^n(k)$ is the small category with its right Σ_k-action in Definition II.10.5.13.

Unit: The *unit*

$$\mathbb{1} \longrightarrow \mathsf{Mon}^n(1) = \{1\}$$

is the unique functor.

Operad Composition on Objects: For $k \geq 1$, $j_1, \ldots, j_k \geq 0$, and $j = j_1 + \cdots + j_k$, the functor

$$\mathsf{Mon}^n(k) \times \prod_{i=1}^{k} \mathsf{Mon}^n(j_i) \xrightarrow{\gamma} \mathsf{Mon}^n(j) \tag{13.1.13}$$

is defined on objects by

$$\gamma\big(P, (A_i)_{i=1}^k\big) = P(A_1', \ldots, A_k'). \tag{13.1.14}$$

This object is obtained from

$$P = P(1, \ldots, k) \in \mathsf{Mon}^n(k)$$

by replacing each $i \in \{1, \ldots, k\}$ by

$$A_i' = A_i\big(\overline{j_i} + 1, \ldots, \overline{j_i} + j_i\big) \tag{13.1.15}$$

with

$$\overline{j_i} = \begin{cases} 0 & \text{if } i = 1 \\ j_1 + \cdots + j_{i-1} & \text{if } 1 < i \leq k. \end{cases}$$

So A_i' is obtained from

$$A_i = A_i(1, \ldots, j_i) \in \mathsf{Mon}^n(j_i)$$

by replacing each $r \in \{1, \ldots, j_i\}$ by $\overline{j_i} + r$.

Operad Composition on Morphisms: For morphisms

$$\begin{aligned} P &\xrightarrow{f} Q \in \mathsf{Mon}^n(k) \\ A_i &\xrightarrow{g_i} B_i \in \mathsf{Mon}^n(j_i) \end{aligned} \tag{13.1.16}$$

for $1 \leq i \leq k$, the morphism

$$\gamma\big(f, \{g_i\}_{i=1}^k\big) \colon P(A_1', \ldots, A_k') \longrightarrow Q(B_1', \ldots, B_k')$$

is defined as the composite

$$\begin{array}{ccc} P(A_1', \ldots, A_k') & \xrightarrow{f(A_1', \ldots, A_k')} & Q(A_1', \ldots, A_k') \\ & & \big\downarrow {\scriptstyle Q(g_1', \ldots, g_k')} \\ & & Q(B_1', \ldots, B_k') \end{array} \tag{13.1.17}$$

in $\mathsf{Mon}^n(j)$.

- The morphism $f(A_1', \ldots, A_k')$ is obtained from $f = f(1, \ldots, k)$ by replacing each $i \in \{1, \ldots, k\}$ by A_i' in (13.1.15).
- For each $i \in \{1, \ldots, k\}$,

$$g_i' = g_i\big(\overline{j_i} + 1, \ldots, \overline{j_i} + j_i\big) \tag{13.1.18}$$

  is obtained from $g_i = g_i(1, \ldots, j_i)$ by replacing each $r \in \{1, \ldots, j_i\}$ by $\overline{j_i} + r$.
- The morphism $Q(g_1', \ldots, g_k')$ is obtained from $Q = Q(1, \ldots, k)$ by replacing each $i \in \{1, \ldots, k\}$ by g_i'.

This finishes the definition of Mon^n. ◇

Example 13.1.19. An example of the composition

$$\mathsf{Mon}^n(2) \times \mathsf{Mon}^n(2) \times \mathsf{Mon}^n(3) \xrightarrow{\gamma} \mathsf{Mon}^n(5)$$

on objects (13.1.14) is

$$\begin{aligned}&\gamma\big(2 \otimes_i 1, \big(1 \otimes_j 2, (2 \otimes_k 3) \otimes_l 1\big)\big)\\ &= \big((4 \otimes_k 5) \otimes_l 3\big) \otimes_i (1 \otimes_j 2) \in \mathsf{Mon}^n(5)\end{aligned}$$

for $i, j, k, l \in \{1, \ldots, n\}$. Consider morphisms

$$\begin{aligned}P = 2 \otimes_i 1 &\xrightarrow{f = \eta^{i,i'}_{2,1,1,1}} 2 \otimes_{i'} 1 = Q \in \mathsf{Mon}^n(2)\\ A_1 = 1 \otimes_j 2 &\xrightarrow{g_1 = \eta^{j,j'}_{1,1,2,1}} 2 \otimes_{j'} 1 = B_1 \in \mathsf{Mon}^n(2)\\ A_2 = (2 \otimes_k 3) \otimes_l 1 &\xrightarrow{g_2 = \eta^{l,l'}_{(2\otimes_k 3),1,1,1}} (2 \otimes_k 3) \otimes_{l'} 1 = B_2 \in \mathsf{Mon}^n(3)\end{aligned}$$

with $i < i'$, $j < j'$, and $l < l'$ in $\{1, \ldots, n\}$. The morphism $\gamma\big(f, (g_1, g_2)\big)$ in (13.1.17) is the composite

$$\begin{array}{c} P(A_1', A_2') = \big((4 \otimes_k 5) \otimes_l 3\big) \otimes_i (1 \otimes_j 2)\\ \Big\downarrow f(A_1', A_2') = \eta^{i,i'}_{((4\otimes_k 5)\otimes_l 3),1,1,(1\otimes_j 2)}\\ Q(A_1', A_2') = \big((4 \otimes_k 5) \otimes_l 3\big) \otimes_{i'} (1 \otimes_j 2)\\ \Big\downarrow Q(g_1', g_2') = \eta^{l,l'}_{(4\otimes_k 5),1,1,3} \otimes_{i'} \eta^{j,j'}_{1,1,2,1}\\ Q(B_1', B_2') = \big((4 \otimes_k 5) \otimes_{l'} 3\big) \otimes_{i'} (2 \otimes_{j'} 1)\end{array}$$

in $\mathsf{Mon}^n(5)$. ◇

Proposition 13.1.20. Mon^n *in Definition 13.1.12 is a* Cat*-enriched operad.*

Proof. We must check that

- the composition γ in (13.1.13) is well defined, and
- the operad axioms in Definition 6.1.1 are satisfied.

Functoriality of γ. To see that γ is a functor, first observe that, if f and each g_i are identity morphisms, then $\gamma\big(f, \{g_i\}_{i=1}^k\big)$ in (13.1.17) is the composite of two identity morphisms.

To see that γ preserves composites, consider morphisms

$$\begin{aligned}P &\xrightarrow{f} Q \xrightarrow{\overline{f}} R \in \mathsf{Mon}^n(k)\\ A_i &\xrightarrow{g_i} B_i \xrightarrow{\overline{g}_i} C_i \in \mathsf{Mon}^n(j_i)\end{aligned}$$

for $1 \le i \le k$ and the diagram

$$
\begin{array}{ccccc}
P(A'_1,\ldots,A'_k) & \xrightarrow{f(A'_1,\ldots,A'_k)} & Q(A'_1,\ldots,A'_k) & \xrightarrow{\overline{f}(A'_1,\ldots,A'_k)} & R(A'_1,\ldots,A'_k) \\
 & & \big\downarrow{\scriptstyle Q(g'_1,\ldots,g'_k)} & & \big\downarrow{\scriptstyle R(g'_1,\ldots,g'_k)} \\
 & & Q(B'_1,\ldots,B'_k) & \xrightarrow{\overline{f}(B'_1,\ldots,B'_k)} & R(B'_1,\ldots,B'_k) \\
 & & & & \big\downarrow{\scriptstyle R(\overline{g}'_1,\ldots,\overline{g}'_k)} \\
 & & & & R(C'_1,\ldots,C'_k)
\end{array}
\tag{13.1.21}
$$

in $\mathsf{Mon}^n(j)$, with $j = j_1 + \cdots + j_k$.

- The square in (13.1.21) is commutative by repeated applications of the axiom (II.10.5.4), which is the naturality of the exchanges $\eta^{i,j}$.
- By the functoriality of each $\otimes_i$ in $\mathsf{Mon}^n(j)$, there is an equality
$$R(\overline{g}'_1 g'_1,\ldots,\overline{g}'_k g'_k) = R(\overline{g}'_1,\ldots,\overline{g}'_k) \circ R(g'_1,\ldots,g'_k).$$
So the top-right composite in (13.1.21) is the morphism
$$\gamma\big(\overline{f}f;\{\overline{g}_i g_i\}_{i=1}^k\big).$$
- The other composite in (13.1.21) is the morphism
$$\gamma\big(\overline{f};\{\overline{g}_i\}_{i=1}^k\big) \circ \gamma\big(f;\{g_i\}_{i=1}^k\big).$$

This shows that γ preserves composites and is a functor.

Equivariance. The equivariance axioms (6.1.6) and (6.1.7) hold because the right Σ_k-action on $\mathsf{Mon}^n(k)$ permutes the elements in $\{1,\ldots,k\}$ from the right.

Unity. The unity axioms (6.1.4) and (6.1.5) follow from the definitions of

- the unit
$$\mathbb{1} \xrightarrow{\cong} \mathsf{Mon}^n(1) = \{1\}$$
and
- γ on objects (13.1.14) and morphisms (13.1.17).

Associativity. For the associativity axiom (6.1.3), consider objects

- $P = P(1,\ldots,k) \in \mathsf{Mon}^n(k)$,
- $A_i = A_i(1,\ldots,j_i) \in \mathsf{Mon}^n(j_i)$ for $1 \le i \le k$, and
- $C_{i,l} = C_{i,l}(1,\ldots,m_{i,l}) \in \mathsf{Mon}^n(m_{i,l})$ for $1 \le i \le k$ and $1 \le l \le j_i$,

and the following sums, with the convention that an empty sum is 0.

$$\overline{m}_{i,l} = \sum_{s=1}^{l-1} m_{i,s} \qquad \overline{m}_i = \sum_{r=1}^{i-1}\sum_{l=1}^{j_r} m_{r,l} \qquad M = \sum_{i=1}^{k}\sum_{l=1}^{j_i} m_{i,l}$$

Using the prime notation in (13.1.15) to denote an expression whose labels are shifted up appropriately, the following equalities of objects hold in $\mathsf{Mon}^n(M)$.

$$
\begin{aligned}
&\gamma\Big(\gamma\big(P,(A_i)_{i=1}^k\big),\big((C_{i,l})_{l=1}^{j_i}\big)_{i=1}^k\Big)\\
&= \big(P(A'_i)_{i=1}^k\big)\big((C'_{i,l})_{l=1}^{j_i}\big)_{i=1}^k\\
&= P\Big(A_i\big(C_{i,l}(\overline{m}_i + \overline{m}_{i,l} + p)_{p=1}^{m_{i,l}}\big)_{l=1}^{j_i}\Big)_{i=1}^k\\
&= P\Big(\big(A_i(C'_{i,l})_{l=1}^{j_i}\big)'\Big)_{i=1}^k\\
&= \gamma\Big(P,\big(\gamma\big(A_i,(C_{i,l})_{l=1}^{j_i}\big)\big)_{i=1}^k\Big)
\end{aligned}
$$

This proves the associativity axiom (6.1.3) on objects.

For morphisms, in addition to the morphisms f and $\{g_i\}_{i=1}^k$ in (13.1.16), suppose given morphisms

$$C_{i,l} \xrightarrow{h_{i,l}} D_{i,l} \in \mathsf{Mon}^n(m_{i,l})$$

for $1 \le i \le k$ and $1 \le l \le j_i$. With the prime notation in (13.1.15) and (13.1.18), when applied to the morphisms

$$\Big(f, (g_i)_{i=1}^k, \big((h_{i,l})_{l=1}^{j_i}\big)_{i=1}^k\Big),$$

each composite in the associativity diagram (6.1.3) gives the following composite in $\mathsf{Mon}^n(M)$.

$$\begin{array}{c}
\big(P(A_i')_{i=1}^k\big)\big((C_{i,l}')_{l=1}^{j_i}\big)_{i=1}^k = P\Big(\big(A_i(C_{i,l}')_{l=1}^{j_i}\big)'\Big)_{i=1}^k \\
\Big\downarrow f\Big(\big(A_i(C_{i,l}')_{l=1}^{j_i}\big)'\Big)_{i=1}^k \\
\big(Q(A_i')_{i=1}^k\big)\big((C_{i,l}')_{l=1}^{j_i}\big)_{i=1}^k = Q\Big(\big(A_i(C_{i,l}')_{l=1}^{j_i}\big)'\Big)_{i=1}^k \\
\Big\downarrow Q\Big(\big(g_i(C_{i,l}')_{l=1}^{j_i}\big)'\Big)_{i=1}^k \\
\big(Q(B_i')_{i=1}^k\big)\big((C_{i,l}')_{l=1}^{j_i}\big)_{i=1}^k = Q\Big(\big(B_i(C_{i,l}')_{l=1}^{j_i}\big)'\Big)_{i=1}^k \\
\Big\downarrow Q\Big(\big(B_i(h_{i,l}')_{l=1}^{j_i}\big)'\Big)_{i=1}^k \\
\big(Q(B_i')_{i=1}^k\big)\big((D_{i,l}')_{l=1}^{j_i}\big)_{i=1}^k = Q\Big(\big(B_i(D_{i,l}')_{l=1}^{j_i}\big)'\Big)_{i=1}^k
\end{array}$$

This finishes the proof that Mon^n is a Cat-enriched operad. □

Definition 13.1.22. For $n \ge 1$, the Cat-enriched operad Mon^n in Proposition 13.1.20 is called the *n-fold monoidal category operad.* ◇

Example 13.1.23. Mon^1 is canonically isomorphic to the associative operad As in Definition 11.1.1, which is also regarded as a Cat-enriched operad as in Example 11.2.2. Indeed, by definition, each object in $\mathsf{Mon}^1(k)$ can be written as an iterated $\otimes_1$-product with each object in $\{1, \ldots, k\}$ occurring precisely once. Since $\otimes_1$ is strictly associative, the objects in $\mathsf{Mon}^1(k)$ canonically correspond to the permutations in Σ_k, with

$$\Sigma_k \ni \sigma \quad \text{corresponding to} \quad \sigma(1) \otimes_1 \cdots \otimes_1 \sigma(k) \in \mathsf{Mon}^1(k).$$

There are no generating exchanges $\eta^{i,j}$ when $n = 1$, so $\mathsf{Mon}^1(k)$ is the discrete category of Σ_k. Moreover, the right symmetric group action, the unit, and the operad composition (13.1.14) in Mon^1 correspond to those in As. ◇

Explanations 11.6.4 and 12.1.12 describe the simplicial Barratt-Eccles operad $N(E\mathsf{As})$ and the simplicial braid operad $N\mathsf{Br}$, with $N\colon \mathsf{Cat} \longrightarrow \mathsf{sSet}$ the nerve functor (Definition 7.2.3). The simplicial operad $N\mathsf{Mon}^n$ will play an important role in Theorem 13.2.1, Proposition 13.5.1, and Corollary 13.5.2. We end this section with an explicit description of it.

Explanation 13.1.24 (Simplicial n-Fold Monoidal Category Operad)**.** For $n \geq 1$ and $k, q \geq 0$, a q-simplex in the nerve $N\mathsf{Mon}^n(k)$ is a $(q+1)$-tuple of objects

$$(P_0, \ldots, P_q) \in \big(\mathsf{Mon}^n(k)\big)^{\times(q+1)}$$

such that

$$\mathsf{Mon}^n(k)(P_j, P_{j+1}) \neq \varnothing \quad \text{for} \quad 0 \leq j < q. \tag{13.1.25}$$

By the Coherence Theorem II.10.6.8 for n-fold monoidal categories, each morphism set in $\mathsf{Mon}^n(k)$ is either empty or has only one element. So the condition (13.1.25) specifies a unique q-simplex $(P_j)_{j=0}^q$ in $N\mathsf{Mon}^n(k)$. The face and degeneracy maps are given by

$$\begin{aligned} d_i(P_0, \ldots, P_q) &= (P_0, \ldots, \widehat{P_i}, \ldots, P_q) \\ s_i(P_0, \ldots, P_q) &= (P_0, \ldots, P_i, P_i, \ldots, P_q). \end{aligned}$$

In other words, the ith face map d_i removes P_i, and the ith degeneracy map s_i repeats P_i.

The symmetric group Σ_k acts diagonally on the q-simplices in $N\mathsf{Mon}^n(k)$, that is,

$$(P_0, \ldots, P_q) \cdot \pi = (P_0\pi, \ldots, P_q\pi) \quad \text{for} \quad \pi \in \Sigma_k.$$

Each $P_j\pi$ is obtained from $P_j \in \mathsf{Mon}^n(k)$ by permuting the elements in $\{1, \ldots, k\}$ that appear in P_j via π. The unit in $N\mathsf{Mon}^n$ is

$$1 \in \big(N\mathsf{Mon}^n(1)\big)_0 = \mathsf{Mon}^n(1) = \{1\}.$$

The operad composition

$$N\mathsf{Mon}^n(k) \times \prod_{\ell=1}^{k} N\mathsf{Mon}^n(m_\ell) \xrightarrow{\ \gamma\ } N\mathsf{Mon}^n(m_1 + \cdots + m_k)$$

is given on q-simplices by

$$\gamma\Big((P_j)_{j=0}^q; \big\{(P_j^\ell)_{j=0}^q\big\}_{\ell=1}^k\Big) = \Big\{\gamma\big(P_j, (P_j^1, \ldots, P_j^k)\big)\Big\}_{j=0}^q.$$

The γ on the right-hand side is the one in (13.1.14). ◇

13.2. The Iterated Monoidal Category Operad is an E_n-Operad

Recall from Definition 12.2.3 that a Top-enriched operad is an E_n*-operad* if it is weakly equivalent to the little n-cube operad $\mathcal{C}_n$. A Cat-enriched operad is an E_n-operad if its image under the classifying space functor $|N(-)|$ in Example 7.2.8 is an E_n-operad. Theorem 12.2.4 says that the braid operad Br is an E_2-operad. The following analogue for Mon^n in Definition 13.1.22 is the main theorem in [**BFSV03**, 3.14].

Theorem 13.2.1. *For $n \geq 1$, the n-fold monoidal category operad Mon^n is an E_n-operad.*

Proof. We refer the reader to [**BFSV03**, 3.14] for the full proof. Here we provide an overview of the proof. The desired weak equivalence between the classifying space of Mon^n and the little n-cube operad $\mathcal{C}_n$ is given by the following zigzag of Top-enriched operad isomorphisms $\cong$ and weak equivalences $\sim$.

$$\begin{array}{ccc} \mathcal{C}_n & & |N\mathsf{Mon}^n| \\ {\scriptstyle (1)}\uparrow{\scriptstyle \sim} & & {\scriptstyle \sim}\uparrow{\scriptstyle (5)} \\ \mathsf{D}_n & & \Big\{\operatorname*{hocolim}_{\mathsf{Mon}^n(k)} *\Big\}_{k\geq 0} \\ {\scriptstyle (2)}\uparrow{\scriptstyle \cong} & & {\scriptstyle \sim}\uparrow{\scriptstyle (4)} \\ \Big\{\operatorname*{colim}_{\mathsf{Mon}^n(k)} F_k\Big\}_{k\geq 0} & \overset{(3)}{\underset{\sim}{\longleftarrow}} & \Big\{\operatorname*{hocolim}_{\mathsf{Mon}^n(k)} F_k\Big\}_{k\geq 0} \end{array} \tag{13.2.2}$$

This diagram involves the following concepts and constructions.

The operad D_n. In the little n-cube operad $\mathcal{C}_n$, define *decomposable* elements inductively as follows.

- Each element in $\mathcal{C}_n(0)$ and $\mathcal{C}_n(1)$ is decomposable.
- Inductively, for $k \geq 2$, an element $f = (f_1, \ldots, f_k)$ in $\mathcal{C}_n(k)$ is decomposable if there exists a hyperplane $H \subset \mathbb{R}^n$ that satisfies the following conditions:
  - H is perpendicular to one of the n coordinate axes.
  - H does *not* intersect the interior of the little n-cubes f_i for $1 \leq i \leq k$.
  - H partitions f into two nonempty disjoint subsets

    $$S \coprod \big(\{f_1, \ldots, f_k\} \smallsetminus S\big)$$

    such that the following two conditions hold, where $|S|$ is the cardinality of S:
    * $S \in \mathcal{C}_n(|S|)$ lies in the negative side of H and is decomposable.
    * The subset

      $$\{f_1, \ldots, f_k\} \smallsetminus S \in \mathcal{C}_n(k - |S|)$$

      lies in the positive side of H and is decomposable.

For example, $\mathcal{C}_n(2)$ and $\mathcal{C}_2(3)$ contain only decomposable elements. For $k \geq 0$, define

$$\mathsf{D}_n(k) \subset \mathcal{C}_n(k) \tag{13.2.3}$$

as the subspace of decomposable elements. These subspaces form a topological suboperad

$$\mathsf{D}_n = \big\{\mathsf{D}_n(k)\big\}_{k\geq 0} \hookrightarrow \mathcal{C}_n \tag{13.2.4}$$

of the little n-cube operad. A general element $(f_1, \ldots, f_k)$ in $\mathcal{C}_n(k)$ can be shrunk toward the center of each constituent little n-cube f_i to yield a decomposable element. This gives the operad weak equivalence (1) in the diagram (13.2.2).

Separable little cubes. The functor F_k in (13.2.2) requires the following concepts from Definition II.10.6.1. For distinct elements $a \neq b \in \{1, \ldots, k\}$, the restriction functor

$$\mathsf{Mon}^n(k) \longrightarrow \mathsf{Mon}^n\big(\{a, b\}\big) \tag{13.2.5}$$

is the identity on $\{a,b\}$ and sends each $i \in \{1,\ldots,k\} \smallsetminus \{a,b\}$ to $\mathbb{1}$. For an object $A \in \mathsf{Mon}^n(k)$, we write

$$a \otimes_i b \in A$$

if the restriction functor in (13.2.5) sends A to $a \otimes_i b$.

Using the product notation in (12.2.1), suppose

$$\begin{aligned} f_- &= [a_1, b_1] \times \cdots \times [a_n, b_n] \\ f_+ &= [c_1, d_1] \times \cdots \times [c_n, d_n] \end{aligned}$$

are two little n-cubes. For $1 \le i \le n$, we write

$$f_- <_i f_+ \quad \text{if} \quad b_i \le c_i.$$

Geometrically, $f_- <_i f_+$ means that there exists a hyperplane H with the following three properties:

- H is perpendicular to the ith coordinate axis.
- The interior of f_- lies in the negative side of H.
- The interior of f_+ lies in the positive side of H.

For an object $A \in \mathsf{Mon}^n(k)$, an element $(f_1,\ldots,f_k) \in \mathcal{C}_n(k)$ is *A-separable* if

$$a \otimes_i b \in A \quad \text{implies} \quad f_a <_i f_b \quad \text{for} \quad a \neq b \in \{1,\ldots,k\}.$$

Define the contractible subspace

$$G(A) \subset \mathcal{C}_n(k)$$

of A-separable elements. There is an equality

$$\bigcup_{A \in \mathsf{Ob}(\mathsf{Mon}^n(k))} G(A) = \mathsf{D}_n(k) \tag{13.2.6}$$

of subspaces of $\mathcal{C}_n(k)$, where $\mathsf{D}_n(k)$ is the subspace of decomposable elements in (13.2.3).

The functor F_k. For $k \ge 0$, the functor

$$F_k \colon \mathsf{Mon}^n(k) \longrightarrow \mathsf{Top} \tag{13.2.7}$$

is defined by the subspace

$$F_k(A) = \bigcup_{\mathsf{Mon}^n(k)(Y,A) \neq \varnothing} G(Y) \subset \mathsf{D}_n(k) \tag{13.2.8}$$

for each object $A \in \mathsf{Mon}^n(k)$. This union is indexed by the set of objects $Y \in \mathsf{Mon}^n(k)$ such that the morphism set $\mathsf{Mon}^n(k)(Y,A)$ is nonempty. By the Coherence Theorem II.10.6.8 for $\mathsf{Mon}^n(k)$, each nonempty morphism set in $\mathsf{Mon}^n(k)$ has a unique element. Therefore, each morphism $B \longrightarrow A$ in $\mathsf{Mon}^n(k)$ has an associated subspace inclusion

$$F_k(B) \hookrightarrow F_k(A).$$

This defines F_k on morphisms in $\mathsf{Mon}^n(k)$.

For each object $A \in \mathsf{Mon}^n(k)$, the identity morphism of A gives the subspace inclusion

$$G(A) \hookrightarrow F_k(A).$$

This is actually a strong deformation retract. Therefore, since $G(A)$ is contractible, so is $F_k(A)$. Moreover, by (13.2.6), the subspace inclusions (13.2.8) induce an isomorphism

$$\operatorname*{colim}_{\mathsf{Mon}^n(k)} F_k \xrightarrow{\cong} \mathsf{D}_n(k)$$

for $k \geq 0$. These levelwise isomorphisms form the operad isomorphism (2) in the diagram (13.2.2).

Homotopy colimits. Each functor $F \colon \mathsf{C} \longrightarrow \mathsf{Top}$ with C a small category has a colimit, since Top has all coproducts and coequalizers. The *homotopy colimit* of F, denoted by $\mathsf{hocolim}_{\mathsf{C}} F$, is a modified construction that takes advantage of the notion of homotopy in Top by incorporating cylinders in the identification process. For example, the *homotopy pushout* of a pair of maps

$$X \xleftarrow{f} A \xrightarrow{g} Y$$

is the pushout

$$X \cup_f (A \wedge I_+) \cup_g Y$$

with I the unit interval. This pushout identifies, for each $a \in A$,

- $f(a) \in X$ with $(a, 0) \in A \wedge I_+$ and
- $g(a) \in Y$ with $(a, 1) \in A \wedge I_+$.

We refer the reader to [**Dug**∞, Part 1] and [**MP12**, 2.1] for an introduction to homotopy colimits. By collapsing the cylinders, there is a canonical map

$$\operatorname*{hocolim}_{\mathsf{C}} F \xrightarrow{\alpha_F} \operatorname*{colim}_{\mathsf{C}} F \tag{13.2.9}$$

from the homotopy colimit to the colimit. In general, α_F is not a weak homotopy equivalence. However, there are situations, including [**BFSV03**, 6.7–6.9], where α_F is a weak homotopy equivalence. The operad weak equivalence (3) in the diagram (13.2.2) is levelwise the canonical map α_{F_k} in (13.2.9) for the functor F_k in (13.2.7) for $k \geq 0$.

The constant functor. For $k \geq 0$,

$$* \colon \mathsf{Mon}^n(k) \longrightarrow \mathsf{Top}$$

is the constant functor at the one-point space. Since F_k is objectwise contractible, the natural transformation $F_k \longrightarrow *$ induces a weak homotopy equivalence

$$\operatorname*{hocolim}_{\mathsf{Mon}^n(k)} F_k \xrightarrow{\sim} \operatorname*{hocolim}_{\mathsf{Mon}^n(k)} * \tag{13.2.10}$$

by the homotopy invariance of homotopy colimits [**Dug**∞, 4.7]. These maps for $k \geq 0$ assemble to form the operad weak equivalence (4) in the diagram (13.2.2).

Homotopy colimits and classifying spaces. The operad weak equivalence (5) in the diagram (13.2.2) follows from the general fact [**Dug**∞, 4.5] that a homotopy colimit has the same weak homotopy type as the geometric realization of its simplicial replacement. For the constant functor $*$, the simplicial replacement is the nerve of the domain category $\mathsf{Mon}^n(k)$. □

Corollary 13.2.11. *The classifying spaces of*

- *the braid operad* Br *and*
- *the 2-fold monoidal category operad* Mon^2

are weakly equivalent Top*-enriched operads.*

Proof. By Theorems 12.2.4 and 13.2.1, both Br and Mon^2 are E_2-operads, so their classifying spaces are weakly equivalent to the little 2-cube operad $\mathcal{C}_2$. In more detail, by the diagrams (12.2.5) and (13.2.2) with $n = 2$, there is a zigzag of Top-enriched operad weak equivalences

$$|N\mathsf{Br}| \xleftarrow{\sim} \mathsf{Sy}(\mathsf{Q}) \xrightarrow{\sim} \mathcal{C}_2 \xleftarrow{\sim} \Big\{\operatorname*{hocolim}_{\mathsf{Mon}^2(k)} F_k\Big\}_{k\geq 0} \xrightarrow{\sim} |N\mathsf{Mon}^2|$$

that connects the classifying spaces of Br and Mon^2. □

Explanation 13.2.12. Let us compare

- the diagram (12.2.5) that connects $\mathcal{C}_2$ with $|N\mathsf{Br}|$ and
- the diagram (13.2.2) that connects $\mathcal{C}_n$ with $|N\mathsf{Mon}^n|$.

For the braid operad Br, the key concept in the diagram (12.2.5) is that of a B_∞-operad, which is a topological braided operad that is levelwise contractible and has a free braid group action. Both

- $|NE\mathsf{B}|$, which is the levelwise classifying space of the translation category of the braid group operad B in (12.2.6), and
- $\widetilde{\mathcal{C}_2}$, which is the levelwise universal covering of $\mathcal{C}_2$,

are B_∞-operads. Two B_∞ operads can always be connected by a zigzag of braided operad weak equivalences as in (12.2.8). This connection between B_∞-operads yields the key zigzag

$$\mathsf{Sy}(\widetilde{\mathcal{C}_2}) \xleftarrow{\sim} \mathsf{Sy}(\mathsf{Q}) \xrightarrow{\sim} \mathsf{Sy}(|NE\mathsf{B}|)$$

of weak equivalences (4) and (5) in the diagram (12.2.5).

For the n-fold monoidal category operad Mon^n, the key concepts in the diagram (13.2.2) are

- the functors
$$F_k\colon \mathsf{Mon}^n(k) \longrightarrow \mathsf{Top}$$
in (13.2.7) and
- the homotopy colimit of a functor
$$F\colon \mathsf{C} \longrightarrow \mathsf{Top}$$
with C a small category.

Homotopy colimit maps to colimit via the canonical map (13.2.9) and satisfies homotopy invariance, which yields the weak equivalence (13.2.10). These properties of homotopy colimits yield the key zigzag

$$\Big\{\operatorname*{colim}_{\mathsf{Mon}^n(k)} F_k\Big\}_{k\geq 0} \xleftarrow{\sim} \Big\{\operatorname*{hocolim}_{\mathsf{Mon}^n(k)} F_k\Big\}_{k\geq 0} \xrightarrow{\sim} \Big\{\operatorname*{hocolim}_{\mathsf{Mon}^n(k)} *\Big\}_{k\geq 0}$$

of weak equivalences (3) and (4) in the diagram (13.2.2). ◇

13.3. Coherence of the Iterated Monoidal Category Operad

Recall the coherence results for

- the associative operad As (Definition 11.1.1) in Theorem 11.1.7,
- the Barratt-Eccles operad $E\mathsf{As}$ (Definition 11.4.10) in Theorem 11.4.14, and
- the braid operad Br (Definition 12.1.2) in Theorem 12.3.10.

Each of these coherence results describes a small list of generators and relations for an operad morphism from the operad in question. The main result in this section, Theorem 13.3.3, is the analogue for the n-fold monoidal category operad Mon^n in Definition 13.1.22. It involves

- the objects

$$\begin{gathered} \mathbb{1} \in \mathsf{Mon}^n(0) = \{\mathbb{1}\} \\ 1 \otimes_i 2 \in \mathsf{Mon}^n(2) \end{gathered} \tag{13.3.1}$$

  for $1 \le i \le n$ and
- the generating exchange morphisms

$$(1 \otimes_j 2) \otimes_i (3 \otimes_j 4) \xrightarrow{\eta^{i,j}_{1,2,3,4}} (1 \otimes_i 3) \otimes_j (2 \otimes_i 4) \tag{13.3.2}$$

  for $1 \le i < j \le n$ in $\mathsf{Mon}^n(4)$.

We will use the juxtaposition notation

$$\gamma\big(y, (x_1, \ldots, x_n)\big) = y(x_1, \ldots, x_n)$$

in Definition 6.1.1 for operad composition. Recall that (i,j) denotes the transposition that swaps i and j. An *enriched operad morphism* is an enriched multifunctor as in Definition 5.1.12 between two enriched multicategories with one object.

Theorem 13.3.3. *Suppose $n \ge 1$ and $(\mathsf{P}, \gamma, \mathbb{1})$ is a Cat-enriched operad. Then a Cat-enriched operad morphism*

$$f \colon \mathsf{Mon}^n \longrightarrow \mathsf{P}$$

is uniquely determined by

- *the objects*

$$\begin{gathered} \mathbb{1} = f(\mathbb{1}) \in \mathsf{P}_0 \\ \otimes_i = f(1 \otimes_i 2) \in \mathsf{P}_2 \end{gathered} \tag{13.3.4}$$

  for $1 \le i \le n$ and
- *the morphisms*

$$\otimes_i(\otimes_j, \otimes_j) \xrightarrow{\eta^{i,j} = f(\eta^{i,j}_{1,2,3,4})} \otimes_j(\otimes_i, \otimes_i)(2,3) \in \mathsf{P}_4 \tag{13.3.5}$$

  for $1 \le i < j \le n$.

The above data are subject to the following conditions.

Unity and Associativity: *The equalities of objects*

$$\begin{gathered} \otimes_i(\mathbb{1}, 1) = 1 = \otimes_i(1, \mathbb{1}) \in \mathsf{P}_1 \\ \otimes_i(\otimes_i, 1) = \otimes_i(1, \otimes_i) \in \mathsf{P}_3 \end{gathered} \tag{13.3.6}$$

hold for $1 \le i \le n$.

n-Fold Monoidal Category Axioms: *The axioms* (13.1.3)–(13.1.7) *hold in* P *for the objects and morphisms in* (13.3.4)–(13.3.5).

Proof. As in the proofs of Theorems 11.4.14 and 12.3.10, there are two directions.

Necessity. Suppose $f\colon \mathsf{Mon}^n \longrightarrow \mathsf{P}$ is a Cat-enriched operad morphism. We define

$$\mathbb{1} \in \mathsf{P}_0, \quad \otimes_i \in \mathsf{P}_2, \quad \text{and} \quad \eta^{i,j} \in \mathsf{P}_4$$

as in (13.3.4)–(13.3.5). By definition, in Mon^n the following relations hold:

- Each $\otimes_i$ is strictly associative with $\mathbb{1}$ as the strict two-sided unit, in the sense of the following equalities.

$$\begin{aligned}\gamma\big(1 \otimes_i 2, (\mathbb{1}, 1)\big) = 1 = \gamma\big(1 \otimes_i 2, (1, \mathbb{1})\big) &\in \mathsf{Mon}^n(1)\\ \gamma\big(1 \otimes_i 2, (1 \otimes_i 2, 1)\big) = 1 \otimes_i 2 \otimes_i 3 &\\ = \gamma\big(1 \otimes_i 2, (1, 1 \otimes_i 2)\big) &\in \mathsf{Mon}^n(3)\end{aligned}$$

- The n-fold monoidal category axioms (13.1.3)–(13.1.7) hold.

Since f is a Cat-enriched operad morphism, the corresponding axioms, which are (13.3.6) and (13.1.3)–(13.1.7) for the objects and morphisms in (13.3.4)–(13.3.5), hold in P.

Sufficiency. Suppose given the data

- $\mathbb{1} = f(\mathbb{1}) \in \mathsf{P}_0$,
- $\otimes_i = f(1 \otimes_i 2) \in \mathsf{P}_2$ as in (13.3.4), and
- $\eta^{i,j} = f(\eta^{i,j}_{1,2,3,4})\colon \otimes_i(\otimes_j, \otimes_j) \longrightarrow \otimes_j(\otimes_i, \otimes_i)(2,3) \in \mathsf{P}_4$ as in (13.3.5)

such that (13.3.6) and the n-fold monoidal category axioms (13.1.3)–(13.1.7) hold in P. We show that these data extend uniquely to a Cat-enriched operad morphism $f\colon \mathsf{Mon}^n \longrightarrow \mathsf{P}$ in several stages.

Objects. Since f must preserve the operad unit and the symmetric group action, we must define

$$f(1) = 1 \in \mathsf{P}_1 \qquad f(2 \otimes_i 1) = \otimes_i(1,2) \in \mathsf{P}_2 \tag{13.3.7}$$

where, on the left-hand side, $1 \in \mathsf{Mon}^n(1) = \{1\}$ is the operad unit in Mon^n. We have already defined f on the objects in $\mathsf{Mon}^n(k)$ for $k \leq 2$.

To define f on the objects in $\mathsf{Mon}^n(k)$ for $k > 2$, first we make some preliminary observations.

- In each category $\mathsf{Mon}^n(k)$ with $k \geq 0$, an object X is said to be in *standard form* if the elements in $\{1, \ldots, k\}$ appear in X in increasing order from left to right. For example,

$$1 \otimes_i (2 \otimes_j 3) \in \mathsf{Mon}^n(3)$$

is in standard form, but $(2 \otimes_j 3) \otimes_i 1$ is not. Each object in $\mathsf{Mon}^n(k)$ can be written *uniquely* as

$$X\sigma \tag{13.3.8}$$

with $X \in \mathsf{Mon}^n(k)$ in standard form and $\sigma \in \Sigma_k$. For example,

$$(2 \otimes_j 3) \otimes_i 1 = \big((1 \otimes_j 2) \otimes_i 3\big)(1,2,3) \in \mathsf{Mon}^n(3),$$

with $(1 \otimes_j 2) \otimes_i 3$ in standard form.

- By Definition II.10.5.2 of $\mathsf{FMon}^n(\mathsf{C})$ and the definition (13.1.14) of γ on objects, each object $X \in \mathsf{Mon}^n(k)$ in standard form with $k \geq 3$ can be written as

$$X = \gamma\big(1 \otimes_i 2, (X_1, X_2)\big) \tag{13.3.9}$$

for some objects

$$X_1 \in \mathsf{Mon}^n(h) \quad \text{and} \quad X_2 \in \mathsf{Mon}^n(k-h)$$

in standard form, $1 \leq i \leq n$, and $1 \leq h < k$. For example,

$$(1 \otimes_i 2) \otimes_j (3 \otimes_l 4) = \gamma\big(1 \otimes_j 2, (1 \otimes_i 2, 1 \otimes_l 2)\big) \in \mathsf{Mon}^n(4),$$

with $1 \otimes_i 2$ and $1 \otimes_l 2 \in \mathsf{Mon}^n(2)$ in standard form.

Since f must preserve γ, if fX_1 and fX_2 are already defined, then, with X as in (13.3.9), we must define

$$fX = \gamma\big(\otimes_i, (fX_1, fX_2)\big) \in \mathsf{P}_k. \tag{13.3.10}$$

Well defined on objects. To see that (13.3.10) is well defined, suppose that X in (13.3.9) has another decomposition as

$$X = \gamma\big(1 \otimes_j 2, (X_1', X_2')\big) \tag{13.3.11}$$

for some $1 \leq j \leq n$ and objects X_1', X_2' in standard form. Then $i = j$, and, up to switching the two decompositions (13.3.9) and (13.3.11), X has the form

$$\begin{aligned} X &= (\cdots) \otimes_i (\cdots) \otimes_i (\cdots) \\ &= \gamma\big(1 \otimes_i 2 \otimes_i 3, (X_1, X'', X_2')\big) \end{aligned}$$

with

$$\begin{cases} X_2 = \gamma\big(1 \otimes_i 2, (X'', X_2')\big) \\ X_1' = \gamma\big(1 \otimes_i 2, (X_1, X'')\big). \end{cases} \tag{13.3.12}$$

For example, there are equalities in $\mathsf{Mon}^n(5)$ as follows.

$$\begin{aligned} (1 \otimes_j 2) \otimes_i 3 \otimes_i (4 \otimes_l 5) &= \gamma\Big(1 \otimes_i 2, \big(1 \otimes_j 2, 1 \otimes_i (2 \otimes_l 3)\big)\Big) \\ &= \gamma\Big(1 \otimes_i 2 \otimes_i 3, \big(1 \otimes_j 2, 1, 1 \otimes_l 2\big)\Big) \\ &= \gamma\Big(1 \otimes_i 2, \big((1 \otimes_j 2) \otimes_i 3, 1 \otimes_l 2\big)\Big) \end{aligned}$$

There are equalities in P as follows.

$$\begin{aligned} \gamma\big(\otimes_i, (fX_1', fX_2')\big) &= \gamma\Big(\otimes_i, \big(\gamma\big(\otimes_i, (fX_1, fX'')\big), fX_2'\big)\Big) \\ &= \gamma\Big(\otimes_i, \big(fX_1, \gamma\big(\otimes_i, (fX'', fX_2')\big)\big)\Big) \\ &= \gamma\big(\otimes_i, (fX_1, fX_2)\big) \end{aligned}$$

The first and the third equalities follow from (13.3.10) and (13.3.12). The second equality holds by the assumed associativity relation in (13.3.6). This shows that the definition (13.3.10) on objects in standard form is well defined.

Since f must preserve the symmetric group action, we must extend (13.3.10) by defining

$$f(X\sigma) = (fX)\sigma \in \mathsf{P}_k \tag{13.3.13}$$

for $X \in \mathsf{Mon}^n(k)$ in standard form and $\sigma \in \Sigma_k$. In particular, f is uniquely defined on the objects in Mon^n.

Preservation of operad axioms on objects. By (13.3.7) and (13.3.13), f preserves the operad unit and the symmetric group action.

To see that f preserves the operad composition on objects, we need to check that the diagram

$$\begin{array}{ccc} \mathsf{Mon}^n(k) \times \prod_{i=1}^{k} \mathsf{Mon}^n(j_i) & \xrightarrow{\gamma} & \mathsf{Mon}^n(j) \\ {\scriptstyle f\times\prod_i f}\downarrow & & \downarrow{\scriptstyle f} \\ \mathsf{P}_k \times \prod_{i=1}^{k} \mathsf{P}_{j_i} & \xrightarrow{\gamma} & \mathsf{P}_j \end{array} \tag{13.3.14}$$

is commutative on objects, where $j = j_1 + \cdots + j_k$. By

- the decomposition $X\sigma$ in (13.3.8) for objects in Mon^n,
- the fact that f preserves the symmetric group action on objects, and
- the equivariance axioms (6.1.6) and (6.1.7) for γ,

it suffices to consider objects in $\mathsf{Mon}^n(k) \times \prod_{i=1}^k \mathsf{Mon}^n(j_i)$ in standard form. Furthermore, by

- the decomposition (13.3.9) for objects in standard form,
- the associativity and unity axioms (6.1.3)–(6.1.5),
- the assumed unity relation (13.3.6) in P,
- the fact that $\mathbb{1}$ is a strict two-sided unit for $\otimes_i$ in Mon^n, and
- an induction on k,

it suffices to consider an object of the form

$$\big(1 \otimes_i 2, (X_1, X_2)\big) \in \mathsf{Mon}^n(2) \times \mathsf{Mon}^n(j_1) \times \mathsf{Mon}^n(j_2)$$

with $j_1, j_2 \geq 1$ and each of X_1 and X_2 in standard form. On this object, the diagram (13.3.14) is commutative by the definition (13.3.10) of fX.

Morphisms. The identity morphisms

$$1_{\mathbb{1}} \in \mathsf{Mon}^n(0) = \{\mathbb{1}\} \quad \text{and} \quad 1_1 \in \mathsf{Mon}^n(1) = \{1\}$$

must be sent by f to, respectively, the identity morphisms $1_{\mathbb{1}} \in \mathsf{P}_0$ and $1_1 \in \mathsf{P}_1$. By definition, each morphism $\varphi \in \mathsf{Mon}^n(k)$ with $k \geq 2$ decomposes into a categorical composite

$$\varphi = (\phi_r v_r) \circ \cdots \circ (\phi_1 v_1) \tag{13.3.15}$$

with

- each $v_l \in \Sigma_k$ for $1 \leq l \leq r$ and
- each ϕ_l an operadic composite of
  - one $\eta^{i,j}_{1,2,3,4}$ with $1 \leq i < j \leq n$ in (13.3.2) and
  - identity morphisms of objects in standard form.

Since f is to be a morphism of Cat-enriched operads, we must define

$$f\varphi = \big((f\phi_r)v_r\big) \circ \cdots \circ \big((f\phi_1)v_1\big) \in \mathsf{P}_k \tag{13.3.16}$$

as the corresponding categorical composite with each $f\phi_l$ the corresponding operadic composite of one $\eta^{i,j} = f(\eta^{i,j}_{1,2,3,4})$ in (13.3.5) and identity morphisms. In particular, if f is well defined on morphisms, which we will verify shortly, then it is uniquely defined on morphisms.

Well defined and functorial. To see that f is well defined on morphisms, note that the Coherence Theorem II.10.6.8 for $\mathsf{Mon}^n(k)$ can also be applied to the objects and morphisms

$$\big(\mathbb{1}, \{\otimes_i\}_{i=1}^n, \{\eta^{i,j}\}_{1\le i<j\le n}\big)$$

in P. The reason is that the assumed axioms (13.3.6) and (13.1.3)–(13.1.7) are formally identical to those in the categories $\mathsf{Mon}^n(k)$ for $k \ge 0$. We will refer to this property as the *n-fold monoidal coherence in* P. Suppose

$$\varphi = (\phi'_s v'_s) \circ \cdots \circ (\phi'_1 v'_1) \in \mathsf{Mon}^n(k)$$

is another decomposition of φ in (13.3.15) as in the previous paragraph. In the image

$$\big((f\phi'_s)v'_s\big) \circ \cdots \circ \big((f\phi'_1)v'_1\big) \in \mathsf{P}_k, \tag{13.3.17}$$

each $f\phi'_h$ is an operadic composite of one $\eta^{i,j}$ and identity morphisms. By the n-fold monoidal coherence in P, the two morphisms (13.3.16) and (13.3.17), which have the same (co)domain, are equal. This shows that f is well defined on morphisms. The n-fold monoidal coherence in P also implies that

$$f\colon \mathsf{Mon}^n(k) \longrightarrow \mathsf{P}_k$$

is a functor for each $k \ge 0$.

Preservation of operad units and equivariant structure on morphisms. By definition, f preserves the operad unit on morphisms. To see that f preserves the symmetric group action on morphisms, suppose $\varphi \in \mathsf{Mon}^n(k)$ is as in (13.3.15) and $\pi \in \Sigma_k$. The right π-action in each of Mon^n and P is a functor. Using this fact and the definition (13.3.16) of $f\varphi$, the following computation shows that f preserves the right π-action on morphisms.

$$\begin{aligned}
f(\varphi\pi) &= f\Big(\big[(\phi_r v_r) \circ \cdots \circ (\phi_1 v_1)\big]\pi\Big)\\
&= f\Big((\phi_r v_r)\pi \circ \cdots \circ (\phi_1 v_1)\pi\Big)\\
&= f\Big(\phi_r(v_r\pi) \circ \cdots \circ \phi_1(v_1\pi)\Big)\\
&= (f\phi_r)(v_r\pi) \circ \cdots \circ (f\phi_1)(v_1\pi)\\
&= \big((f\phi_r)v_r\big)\pi \circ \cdots \circ \big((f\phi_1)v_1\big)\pi\\
&= \big[(f\phi_r)v_r \circ \cdots \circ (f\phi_1)v_1\big]\pi\\
&= (f\varphi)\pi
\end{aligned}$$

Preservation of operad composition on morphisms. For the preservation of the operad composition γ by f on morphisms, we use the following facts:

- f is levelwise a functor and preserves the symmetric group action.
- γ in each of Mon^n and P is a functor and satisfies the equivariance axioms (6.1.6) and (6.1.7).
- Morphisms in Mon^n decompose as in (13.3.15).

Using these facts, to prove the commutativity of the diagram (13.3.14) on morphisms, it suffices to consider a morphism

$$\big(\phi,(\phi_i)_{i=1}^k\big) \in \mathsf{Mon}^n(k) \times \prod_{i=1}^k \mathsf{Mon}^n(j_i)$$

such that the following two conditions hold:

- There is only one nonidentity morphism in $\{\phi,\phi_1,\ldots,\phi_k\}$.
- The nonidentity morphism in $\{\phi,\phi_1,\ldots,\phi_k\}$ is an operadic composite of
  - one $\eta^{i,j}_{1,2,3,4}$ with $1 \le i < j \le n$ in (13.3.2) and
  - identity morphisms of objects in standard form.

On such a morphism, the diagram (13.3.14) is commutative by the n-fold monoidal coherence in P. □

n-Fold Monoidal Categories as Algebras. Recall the following results:

- The associative operad As is the operad for monoids (Proposition 11.1.15).
- The Barratt-Eccles operad $E\mathsf{As}$ is the Cat-enriched operad for permutative categories (Proposition 11.4.26).
- The braid operad Br is the Cat-enriched operad for braided strict monoidal categories (Proposition 12.3.22).

Proposition 13.3.18 below is the analogue for n-fold monoidal categories and is an application of Theorem 13.3.3.

Proposition 13.3.18. *For $n \ge 1$ and a small category C, an n-fold monoidal category structure on C is uniquely determined by a Cat-enriched multifunctor*

$$f\colon \mathsf{Mon}^n \longrightarrow \mathsf{Cat} \quad \textit{such that} \quad f(*) = \mathsf{C}.$$

Proof. A Cat-enriched multifunctor $f\colon \mathsf{Mon}^n \longrightarrow \mathsf{Cat}$ such that $f(*) = \mathsf{C}$ is equivalent to a Cat-enriched operad morphism

$$\mathsf{Mon}^n \xrightarrow{\ f\ } \mathsf{End}(\mathsf{C}) = \big\{\mathsf{Cat}(\mathsf{C}^{\times k},\mathsf{C})\big\}_{k\ge 0}$$

to the Cat-enriched endomorphism operad of C. By Theorem 13.3.3, such a Cat-enriched operad morphism f is uniquely determined by

- an object
$$\mathbb{1} \in \mathsf{C},$$
- a functor
$$\otimes_i\colon \mathsf{C}\times\mathsf{C} \longrightarrow \mathsf{C}$$
for each $1 \le i \le n$, and
- a natural transformation
$$(A \otimes_j B) \otimes_i (C \otimes_j D) \xrightarrow{\eta^{i,j}_{A,B,C,D}} (A \otimes_i C) \otimes_j (B \otimes_i D)$$
for objects $A,B,C,D \in \mathsf{C}$ and $1 \le i < j \le n$

such that the following two statements hold.

- For $1 \le i \le n$, $\otimes_i$ is strictly associative with $\mathbb{1}$ as the strict two-sided unit. In other words, each $(\mathsf{C},\otimes_i,\mathbb{1})$ is a strict monoidal category.
- The n-fold monoidal category axioms (13.1.3)–(13.1.7) hold in C.

This is the same thing as an n-fold monoidal category structure on C. □

13.4. Detecting E_n-Monoidal Categories

In this section, we prove that the n-fold monoidal category operad detects E_n-monoidal category structures on small permutative categories (Theorem 13.4.12). This result is an application of the Coherence Theorem 13.3.3 for the n-fold monoidal category operad. In Section 13.5, we use this result to show that the K-theory of a small E_n-monoidal category is an E_n-symmetric spectrum (Corollary 13.5.2). For the reader's convenience, here we recall from Chapter II.10 the definition of an E_n-monoidal category.

Definition 13.4.1. For $n \geq 1$, an *E_n-monoidal category* is a tuple

$$\Big(\mathsf{C}, (\oplus, \mathbb{0}, \xi^{\oplus}), \{\otimes_i, \partial^{l,i}, \partial^{r,i}\}_{1\leq i\leq n}, \mathbb{1}, \{\eta^{i,j}\}_{1\leq i<j\leq n}\Big)$$

consisting of the following data.

The Ring Category Structures: For each $1 \leq i \leq n$, the tuple

$$\big(\mathsf{C}, (\oplus, \mathbb{0}, \xi^{\oplus}), (\otimes_i, \mathbb{1}), (\partial^{l,i}, \partial^{r,i})\big) \tag{13.4.2}$$

is a ring category (Definition 11.2.4). The natural transformations

$$\begin{aligned} (A \otimes_i C) \oplus (B \otimes_i C) &\xrightarrow{\partial^{l,i}_{A,B,C}} (A \oplus B) \otimes_i C \\ (A \otimes_i B) \oplus (A \otimes_i C) &\xrightarrow{\partial^{r,i}_{A,B,C}} A \otimes_i (B \oplus C) \end{aligned} \tag{13.4.3}$$

for objects $A, B, C \in \mathsf{C}$, are called, respectively, the *ith left factorization morphism* and the *ith right factorization morphism*.

The n-Fold Monoidal Structure: The tuple

$$\big(\mathsf{C}, \{\otimes_i\}_{1\leq i\leq n}, \mathbb{1}, \{\eta^{i,j}\}_{1\leq i<j\leq n}\big) \tag{13.4.4}$$

is an n-fold monoidal category (Definition 13.1.1), with *(i,j)-exchange* the natural transformation

$$(A \otimes_j B) \otimes_i (C \otimes_j D) \xrightarrow{\eta^{i,j}_{A,B,C,D}} (A \otimes_i C) \otimes_j (B \otimes_i D) \tag{13.4.5}$$

for objects $A, B, C, D \in \mathsf{C}$ and $1 \leq i < j \leq n$.

These data are required to satisfy the following axioms for $1 \leq i < j \leq n$ and objects A, A', B, B', C, C', D, and D' in C.

The Zero Exchange Axiom:

$$\eta^{i,j}_{A,B,C,D} = 1_{\mathbb{0}} \quad \text{if} \quad A,\ B,\ C, \text{ or } D \text{ is } \mathbb{0}. \tag{13.4.6}$$

The Exchange Factorization Axiom: The following four diagrams are commutative. They are called, respectively, EF1, EF2, EF3, and EF4.

$$\begin{array}{ccc} & [(A \otimes_j B) \otimes_i (C \otimes_j D)] \oplus [(A' \otimes_j B) \otimes_i (C \otimes_j D)] & \\ \swarrow{\scriptstyle \eta^{i,j}_{A,B,C,D} \oplus \eta^{i,j}_{A',B,C,D}} & & {\scriptstyle \partial^{l,i}_{A\otimes_j B, A'\otimes_j B, C\otimes_j D}}\searrow \\ [(A \otimes_i C) \otimes_j (B \otimes_i D)] \oplus [(A' \otimes_i C) \otimes_j (B \otimes_i D)] & & [(A \otimes_j B) \oplus (A' \otimes_j B)] \otimes_i (C \otimes_j D) \\ \Big\downarrow {\scriptstyle \partial^{l,j}_{A\otimes_i C, A'\otimes_i C, B\otimes_i D}} & & \Big\downarrow {\scriptstyle \partial^{l,j}_{A,A',B} \otimes_i 1} \\ [(A \otimes_i C) \oplus (A' \otimes_i C)] \otimes_j (B \otimes_i D) & & [(A \oplus A') \otimes_j B] \otimes_i (C \otimes_j D) \\ {\scriptstyle \partial^{l,i}_{A,A',C} \otimes_j 1}\searrow & & \swarrow{\scriptstyle \eta^{i,j}_{A\oplus A',B,C,D}} \\ & [(A \oplus A') \otimes_i C] \otimes_j (B \otimes_i D) & \end{array} \tag{13.4.7}$$

$$
(13.4.8)\qquad
\begin{array}{ccc}
 & [(A\otimes_j B)\otimes_i (C\otimes_j D)]\oplus[(A\otimes_j B')\otimes_i (C\otimes_j D)] & \\
 \swarrow{\scriptstyle \eta^{i,j}_{A,B,C,D}\oplus\eta^{i,j}_{A,B',C,D}} & & \searrow{\scriptstyle \partial^{l,i}_{A\otimes_j B, A\otimes_j B', C\otimes_j D}} \\
[(A\otimes_i C)\otimes_j (B\otimes_i D)]\oplus[(A\otimes_i C)\otimes_j (B'\otimes_i D)] & & [(A\otimes_j B)\oplus(A\otimes_j B')]\otimes_i (C\otimes_j D) \\
\Big\downarrow{\scriptstyle \partial^{r,j}_{A\otimes_i C, B\otimes_i D, B'\otimes_i D}} & & \Big\downarrow{\scriptstyle \partial^{r,j}_{A,B,B'}\otimes_i 1} \\
(A\otimes_i C)\otimes_j [(B\otimes_i D)\oplus(B'\otimes_i D)] & & [A\otimes_j (B\oplus B')]\otimes_i (C\otimes_j D) \\
 \searrow{\scriptstyle 1\otimes_j \partial^{l,i}_{B,B',D}} & & \swarrow{\scriptstyle \eta^{i,j}_{A,B\oplus B',C,D}} \\
 & (A\otimes_i C)\otimes_j [(B\oplus B')\otimes_i D] &
\end{array}
$$

$$
(13.4.9)\qquad
\begin{array}{ccc}
 & [(A\otimes_j B)\otimes_i (C\otimes_j D)]\oplus[(A\otimes_j B)\otimes_i (C'\otimes_j D)] & \\
 \swarrow{\scriptstyle \eta^{i,j}_{A,B,C,D}\oplus\eta^{i,j}_{A,B,C',D}} & & \searrow{\scriptstyle \partial^{r,i}_{A\otimes_j B, C\otimes_j D, C'\otimes_j D}} \\
[(A\otimes_i C)\otimes_j (B\otimes_i D)]\oplus[(A\otimes_i C')\otimes_j (B\otimes_i D)] & & (A\otimes_j B)\otimes_i [(C\otimes_j D)\oplus(C'\otimes_j D)] \\
\Big\downarrow{\scriptstyle \partial^{l,j}_{A\otimes_i C, A\otimes_i C', B\otimes_i D}} & & \Big\downarrow{\scriptstyle 1\otimes_i \partial^{l,j}_{C,C',D}} \\
[(A\otimes_i C)\oplus(A\otimes_i C')]\otimes_j (B\otimes_i D) & & (A\otimes_j B)\otimes_i [(C\oplus C')\otimes_j D] \\
 \searrow{\scriptstyle \partial^{r,i}_{A,C,C'}\otimes_j 1} & & \swarrow{\scriptstyle \eta^{i,j}_{A,B,C\oplus C',D}} \\
 & [A\otimes_i (C\oplus C')]\otimes_j (B\otimes_i D) &
\end{array}
$$

$$
(13.4.10)\qquad
\begin{array}{ccc}
 & [(A\otimes_j B)\otimes_i (C\otimes_j D)]\oplus[(A\otimes_j B)\otimes_i (C\otimes_j D')] & \\
 \swarrow{\scriptstyle \eta^{i,j}_{A,B,C,D}\oplus\eta^{i,j}_{A,B,C,D'}} & & \searrow{\scriptstyle \partial^{r,i}_{A\otimes_j B, C\otimes_j D, C\otimes_j D'}} \\
[(A\otimes_i C)\otimes_j (B\otimes_i D)]\oplus[(A\otimes_i C)\otimes_j (B\otimes_i D')] & & (A\otimes_j B)\otimes_i [(C\otimes_j D)\oplus(C\otimes_j D')] \\
\Big\downarrow{\scriptstyle \partial^{r,j}_{A\otimes_i C, B\otimes_i D, B\otimes_i D'}} & & \Big\downarrow{\scriptstyle 1\otimes_i \partial^{r,j}_{C,D,D'}} \\
(A\otimes_i C)\otimes_j [(B\otimes_i D)\oplus(B\otimes_i D')] & & (A\otimes_j B)\otimes_i [C\otimes_j (D\oplus D')] \\
 \searrow{\scriptstyle 1\otimes_j \partial^{r,i}_{B,D,D'}} & & \swarrow{\scriptstyle \eta^{i,j}_{A,B,C,D\oplus D'}} \\
 & (A\otimes_i C)\otimes_j [B\otimes_i (D\oplus D')] &
\end{array}
$$

This finishes the definition of an E_n-monoidal category. It is *small* if the category C is small. $\diamond$

Definition 13.4.11. For a permutative category $(\mathsf{C}, \oplus, \mathbb{0}, \xi^{\oplus})$, an *$E_n$-monoidal category structure* on C for $n \geq 1$ is the additional data

$$\left(\{\otimes_i, \partial^{l,i}, \partial^{r,i}\}_{1\leq i\leq n}, \mathbb{1}, \{\eta^{i,j}\}_{1\leq i<j\leq n}\right)$$

such that the tuple

$$\left(\mathsf{C}, (\oplus, \mathbb{0}, \xi^{\oplus}), \{\otimes_i, \partial^{l,i}, \partial^{r,i}\}_{1\leq i\leq n}, \mathbb{1}, \{\eta^{i,j}\}_{1\leq i<j\leq n}\right)$$

is an E_n-monoidal category as in Definition 13.4.1. $\diamond$

Recall from Section 6.6 that $\mathsf{PermCat}^{\mathsf{su}}$ is the Cat-enriched multicategory with small permutative categories as objects. The category

$$\mathsf{PermCat}^{\mathsf{su}}\big(\langle \mathsf{C}\rangle\,;\,\mathsf{D}\big) = \mathsf{PermCat}^{\mathsf{su}}\big(\langle \mathsf{C}_1,\ldots,\mathsf{C}_n\rangle\,;\,\mathsf{D}\big)$$

has

- n-linear functors

$$\mathsf{C}_1 \times \cdots \times \mathsf{C}_n \longrightarrow \mathsf{D}$$

  in Definition 6.5.4 as objects and
- multilinear transformations (Definition 6.5.11) as morphisms.

Also recall from Definition 5.1.12 the notion of an enriched multifunctor. The next result is the Mon^n analogue of Theorems 11.2.16, 11.5.5, and 12.4.5 for, respectively, the associative operad As, the Barratt-Eccles operad $E\mathsf{As}$, and the braid operad Br. It says that the n-fold monoidal category operad detects E_n-monoidal category structures on small permutative categories. It extends Proposition 13.3.18 from Cat to $\mathsf{PermCat}^{\mathsf{su}}$.

Theorem 13.4.12. *For $n \geq 1$ and each small permutative category C, there is a canonical bijective correspondence between*

- *E_n-monoidal category structures on C and*
- *Cat-enriched multifunctors*

$$F\colon \mathsf{Mon}^n \longrightarrow \mathsf{PermCat}^{\mathsf{su}} \quad \textit{such that} \quad F(*) = \mathsf{C}.$$

Proof. A Cat-enriched multifunctor

$$F\colon \mathsf{Mon}^n \longrightarrow \mathsf{PermCat}^{\mathsf{su}} \quad \text{such that} \quad F(*) = (\mathsf{C}, \oplus, \mathbb{0}, \xi^{\oplus})$$

is equivalent to a Cat-enriched operad morphism

$$F\colon \mathsf{Mon}^n \longrightarrow \mathsf{End}(\mathsf{C}) = \Big\{\mathsf{PermCat}^{\mathsf{su}}\big(\langle\overbrace{\mathsf{C},\ldots,\mathsf{C}}^{k}\rangle\,;\,\mathsf{C}\big)\Big\}_{k\geq 0}$$

to the Cat-enriched endomorphism operad of C. By Theorem 13.3.3, such a Cat-enriched operad morphism is uniquely determined by

- the 0-linear functor, that is, object

$$F(\mathbb{1}) = \mathbb{1} \in \mathsf{C},$$

- the 2-linear functors

$$F(1 \otimes_i 2) = (\otimes_i, \partial^{l,i}, \partial^{r,i})\colon \mathsf{C} \times \mathsf{C} \longrightarrow \mathsf{C}$$

  for $1 \leq i \leq n$, and
- the multilinear transformations

$$(A \otimes_l B) \otimes_k (C \otimes_l D) \xrightarrow{\eta^{k,l}_{A,B,C,D} = F(\eta^{k,l}_{1,2,3,4})} (A \otimes_k C) \otimes_l (B \otimes_k D)$$

  for $1 \leq k < l \leq n$ and objects $A, B, C, D \in \mathsf{C}$, such that

- the unity and associativity conditions (13.3.6) and
- the n-fold monoidal category axioms (13.1.3)–(13.1.7)

hold in C.

The ring category structures. By Theorem 11.2.16 and the first paragraph of its proof, for $i \in \{1, \ldots, n\}$, the data $(\otimes_i, \mathbb{1}, \partial^{l,i}, \partial^{r,i})$ and the conditions (13.3.6) are equivalent to a ring category structure on the permutative category $(\mathsf{C}, \oplus, \mathbb{0}, \xi^{\oplus})$ as in Definition 11.2.15. In particular, each $(\mathsf{C}, \otimes_i, \mathbb{1})$ is a strict monoidal category.

The n-fold monoidal category structure. The n-fold monoidal category axioms (13.1.3)–(13.1.7) guarantee that the tuple

$$\left(\mathsf{C}, \{\otimes_i\}_{1 \le i \le n}, \mathbb{1}, \{\eta^{k,l}\}_{1 \le k < l \le n}\right)$$

is an n-fold monoidal category.

The E_n-monoidal category axioms. For $1 \le k < l \le n$, the multilinear transformation

$$\otimes_k(\otimes_l, \otimes_l) \xrightarrow{\eta^{k,l}} \otimes_l(\otimes_k, \otimes_k)(2,3) \in \mathsf{End}(\mathsf{C})_4$$

is, by definition, a natural transformation that satisfies the two conditions in Definition 6.5.11.

- For $\alpha = \eta^{k,l}$, the commutative diagrams (6.5.12) for $i \in \{1, 2, 3, 4\}$ are, respectively, the axioms (13.4.7)–(13.4.10) in an E_n-monoidal category. Here we use the explicit formulas (6.6.3) and (6.6.8) to unravel the linearity constraints of the domain and codomain 4-linear functors of $\eta^{k,l}$.
- The second condition in Definition 6.5.11 states that

$$\eta^{k,l}_{A,B,C,D} = 1_{\mathbb{0}} \quad \text{if} \quad A, B, C, \text{ or } D \text{ is } \mathbb{0}.$$

This is the zero exchange axiom (13.4.6) in an E_n-monoidal category.

Therefore, a Cat-enriched operad morphism

$$F \colon \mathsf{Mon}^n \longrightarrow \mathsf{End}(\mathsf{C})$$

is equivalent to an E_n-monoidal category structure on C. □

13.5. K-Theory Spectra of E_n-Monoidal Categories

In this section, we prove that the K-theory of a small E_n-monoidal category is an E_n-symmetric spectrum (Corollary 13.5.2). This result is the E_n analogue of Corollaries 11.3.16, 11.6.12, and 12.5.3. It is a consequence of Theorem 13.4.12, the Elmendorf-Mandell K-theory multifunctor, and the fact that the n-fold monoidal category operad is an E_n-operad (Theorem 13.2.1). Along the way, we observe that an E_∞-structure induces an E_n-structure (Proposition 13.5.1).

E_n-Symmetric Spectra. An *E_n-symmetric spectrum* (Definition 12.5.1) is a symmetric spectrum X equipped with an sSet-enriched operad morphism

$$\mathsf{P} \longrightarrow \mathsf{End}(X)$$

with

- P an sSet-enriched E_n-operad (Definition 12.2.3) and
- $\mathsf{End}(X)$ the endomorphism simplicial operad (Definition 7.6.22).

The E_∞ analogue is Definition 11.6.5. Recall from Proposition 12.1.11 that the Cat-enriched operad morphism

$$\mathsf{As} \xrightarrow{\iota_{\mathsf{As}}} E\mathsf{As}$$

in (11.4.12), from the associative operad As to the Barratt-Eccles operad $E\mathsf{As}$, factors through the braid operad Br. Recall that $E\mathsf{As}$ is an E_∞-operad (Proposition 11.6.3) and Mon^n is an E_n-operad (Theorem 13.2.1). The following result provides a factorization of ι_{As} through the n-fold monoidal category operad Mon^n for $n \geq 1$. See also Corollary 12.5.2.

Proposition 13.5.1. *There is a commutative diagram of* Cat*-enriched operad morphisms*

$$\begin{array}{ccccccc} \mathsf{As} \cong \mathsf{Mon}^1 & \xrightarrow{i^1} & \mathsf{Mon}^2 & \xrightarrow{i^2} & \mathsf{Mon}^3 & \xrightarrow{i^3} & \cdots \\ & \iota_{\mathsf{As}} = j^1 \searrow & \downarrow j^2 & \swarrow j^3 & \cdots & & \\ & & E\mathsf{As} & & & & \end{array}$$

with each i^n for $n \geq 1$ the inclusion. As a result,

- *each E_∞-structure on a symmetric spectrum via the Barratt-Eccles operad* $E\mathsf{As}$ *induces an E_n-structure by restricting along j^n, and*
- *each E_{n+1}-structure on a symmetric spectrum via* Mon^{n+1} *induces an E_n-structure by restricting along i^n.*

Proof. The isomorphism $\mathsf{As} \cong \mathsf{Mon}^1$ is from Example 13.1.23. Each $\mathsf{Mon}^n(k)$ is the full subcategory of $\mathsf{Mon}^{n+1}(k)$ consisting of objects that can be written without $\otimes_{n+1}$. This inclusion preserves the operad structure in Mon^n and Mon^{n+1} in Definition 13.1.12, so

$$\mathsf{Mon}^n \xrightarrow{i^n} \mathsf{Mon}^{n+1}$$

is a Cat-enriched operad morphism.

To define j^n, recall from (13.3.8) that each object in $\mathsf{Mon}^n(k)$ can be written uniquely in the form $X\sigma$, with

- $X \in \mathsf{Mon}^n(k)$ in standard form, which means that the elements in $\{1,\ldots,k\}$ appear in X in increasing order from left to right, and
- $\sigma \in \Sigma_k$ a permutation.

For $k \geq 0$, the assignment

$$\mathsf{Mon}^n(k) \xrightarrow{j^n_k} E\mathsf{As}_k$$

on objects is defined as

$$j^n_k(X\sigma) = \sigma$$

for $X \in \mathsf{Mon}^n(k)$ in standard form and $\sigma \in \Sigma_k$. These assignments define an operad morphism between the object set operads of Mon^n and $E\mathsf{As}$. On morphisms, j^n_k is uniquely defined by the fact that, in the translation category $E\mathsf{As}_k$, each morphism set has only one element. This property also implies that

- j^n_k is a functor and
- $j^n = \{j^n_k\}_{k\geq 0}$ preserves operad units, symmetric group action, and operad composition.

Therefore,

$$\mathsf{Mon}^n \xrightarrow{j^n} E\mathsf{As}$$

is a Cat-enriched operad morphism.

The equality

$$j^n = j^{n+1} i^n$$

follows from the fact that i^n is the inclusion. This proves the existence of the commutative diagram in the statement. The remaining assertions follow from the existence of the Cat-enriched operad morphisms i^n and j^n. □

K-Theory E_n-Symmetric Spectra. Recall the Elmendorf-Mandell K-theory multifunctor in Definition 10.3.32,

$$\mathsf{K}^{\mathsf{EM}} = \mathsf{K}^{\mathcal{G}} N_* \mathsf{J}^{\mathsf{EM}} \colon \mathsf{PermCat}^{\mathsf{su}} \longrightarrow \mathsf{SymSp}.$$

The following result is the E_n analogue of Corollaries 11.3.16, 11.6.12, and 12.5.3.

Corollary 13.5.2. *For $n \geq 1$ and each small E_n-monoidal category C, $\mathsf{K}^{\mathsf{EM}}\mathsf{C}$ is an E_n-symmetric spectrum.*

Proof. Consider the multifunctors

$$\begin{array}{ccccc} \mathsf{Mon}^n & \xrightarrow{F} & \mathsf{PermCat}^{\mathsf{su}} & \xrightarrow{\mathsf{K}^{\mathsf{EM}}} & \mathsf{SymSp} \\ * & \longmapsto & (\mathsf{C}, \oplus, \mathbb{0}, \xi^{\oplus}) & \longmapsto & \mathsf{K}^{\mathsf{EM}}\mathsf{C} \end{array}$$

with F the Cat-enriched multifunctor in Theorem 13.4.12 such that $F(*)$ is the additive structure of C. By Theorem 13.2.1, $N\mathsf{Mon}^n$ is an E_n-operad. As in the proofs of Corollaries 11.6.12 and 12.5.3, the composite sSet-enriched operad morphism

$$\begin{array}{ccccc} N\mathsf{Mon}^n & & & & \mathsf{End}(\mathsf{K}^{\mathsf{EM}}\mathsf{C}) \\ \big\downarrow{\scriptstyle F_N} & & & & \big\uparrow{\scriptstyle \mathsf{K}^{\mathcal{G}}} \\ N\mathsf{End}(\mathsf{C}) & \xrightarrow{\mathsf{J}^{\mathsf{EM}}_N} & N\mathsf{End}(\mathsf{J}^{\mathsf{EM}}\mathsf{C}) & \xrightarrow{N_*} & \mathsf{End}(N_*\mathsf{J}^{\mathsf{EM}}\mathsf{C}) \end{array}$$

gives $\mathsf{K}^{\mathsf{EM}}\mathsf{C}$ the structure of an E_n-symmetric spectrum. □

Example 13.5.3. By Proposition II.10.10.2, each small category C freely generates a small E_n-monoidal category $\mathsf{FE}^n(\mathsf{C})$. Corollary 13.5.2 applies to $\mathsf{FE}^n(\mathsf{C})$ to yield a K-theory E_n-symmetric spectrum. See also Questions A.4.2 and A.5.7. ◇

13.6. Notes

13.6.1 (Homotopy Colimits). In addition to [**Dug∞, MP12**], other references for homotopy colimits, which are needed in the proof of Theorem 13.2.1, include [**BK72, CS02, Hir03, Rie14, Shu∞a, Vog73, Vog77**]. ◇

13.6.2 (Iterated Monoidal Category Operads). The nerve of the n-fold monoidal category operad Mon^n in Theorem 13.2.1 is weakly equivalent to

- the nth stage of Berger's filtration [**Ber96**] and
- the nth stage of Smith's filtration [**Smi89**]

of the Barratt-Eccles operad and is closely related to Milgram's model of $\Omega^n\Sigma^n X$ [**Mil66**]. See [**BFSV03**, Section 3]. Moreover, the group completion of the classifying space of an n-fold monoidal category is an n-fold loop space by [**BFSV03**, 2.2 and 3.14]. Proposition 13.3.18, which says that Mon^n-algebras are small n-fold monoidal categories, is a remark in [**BFSV03**, p. 292]. ◇

Bibliography and Indices

APPENDIX A

Open Questions

In this chapter, we discuss open questions related to the topics of this work. These open questions provide additional motivation for the main text.

A.1. Bimonoidal Categories

The following questions are about bimonoidal, symmetric bimonoidal, and braided bimonoidal categories in Definitions I.2.1.2 and II.2.1.29.

Question A.1.1 (Functoriality of the Matrix Construction)**.** In Theorem I.8.15.4, we showed that, for each tight symmetric bimonoidal category C, the matrix construction $\mathsf{Mat}^{\mathsf{C}}$ is a symmetric monoidal bicategory. Denote by $\mathsf{Bi}^{\mathsf{tsy}}$ the full sub-2-category of the 2-category $\mathsf{Bi}^{\mathsf{sy}}$ in Proposition I.7.1.7, with small *tight* symmetric bimonoidal categories as objects. Regard $\mathsf{Bi}^{\mathsf{tsy}}$ as a tricategory with only identity 3-cells. It is claimed in [**SP**∞] that small symmetric monoidal bicategories are the objects of a tricategory, denoted by SMB.

- Extend the assignment

$$\mathsf{C} \longmapsto \mathsf{Mat}^{\mathsf{C}}$$

  to a trifunctor

$$\mathsf{Bi}^{\mathsf{tsy}} \longrightarrow \mathsf{SMB}.$$

Among other things, one should carefully verify the tricategory axioms for SMB. For a discussion of tricategories and a detailed verification of the tricategory of small bicategories, the reader is referred to [**JY21**, Ch. 11]. ◇

Question A.1.2 (Bimonoidal Bicategories)**.** Taking the categorification from (commutative) rigs to (symmetric) bimonoidal categories one step further, we could ask about two different monoidal structures, $\boxplus$ and $\boxtimes$, on a bicategory, with $\boxplus$ symmetric.

- Define such a *(braided/sylleptic/symmetric) bimonoidal bicategory*, generalizing the (braided/sylleptic/symmetric) monoidal bicategories in Sections I.6.4 and I.6.5.
- For a tight symmetric bimonoidal category C, prove that the symmetric monoidal bicategory $\mathsf{Mat}^{\mathsf{C}}$ extends to a symmetric bimonoidal bicategory.
- For a tight braided bimonoidal category C, prove that the monoidal bicategory $\mathsf{Mat}^{\mathsf{C}}$ extends to a bimonoidal bicategory.

More precisely, in Theorem I.8.15.4, the monoidal composition $\boxtimes$ in the symmetric monoidal bicategory $\mathsf{Mat}^{\mathsf{C}}$ involves the matrix tensor product in (I.8.6.3). There should be another symmetric monoidal bicategory structure on the matrix bicategory $\mathsf{Mat}^{\mathsf{C}}$ in Theorem I.8.4.12, whose monoidal composition uses the matrix direct sum as in Example I.2.5.9. These two symmetric monoidal structures on $\mathsf{Mat}^{\mathsf{C}}$

should make it into a symmetric bimonoidal bicategory. A similar discussion applies in the braided case, with Theorem II.8.4.7 showing that $\mathsf{Mat}^{\mathsf{C}}$ is a monoidal bicategory.

- Extend the tricategory SMB (Question A.1.1) to a tricategory SBB with small symmetric bimonoidal bicategories as objects.
- Construct a tricategory BB with small bimonoidal bicategories as objects.
- Extend the assignment $\mathsf{C} \longmapsto \mathsf{Mat}^{\mathsf{C}}$ in
  - Theorem I.8.15.4 in the symmetric case to a trifunctor

$$\mathsf{Bi}^{\mathsf{tsy}} \longrightarrow \mathsf{SBB}$$

    and
  - Theorem II.8.4.7 in the braided case to an analogous trifunctor with codomain BB. ◇

Question A.1.3 (Bimonoidal Bicategorical Centers). Extend the bimonoidal centers in Theorems II.4.4.3 and II.4.5.3 to the bimonoidal bicategorical setting in Question A.1.2. In other words:

- Extend the bimonoidal Drinfeld center in Theorem II.4.4.3 to a bimonoidal bicategory and show that it is a braided bimonoidal bicategory.
- Show that the center of a braided bimonoidal bicategory is a sylleptic bimonoidal bicategory.
- Extend the bimonoidal symmetric center in Theorem II.4.5.3 to a sylleptic bimonoidal bicategory and show that it is a symmetric bimonoidal bicategory.

There are simpler centers of Gray monoids [**BN96**], braided monoidal 2-categories, and sylleptic monoidal 2-categories [**Cra98**]. As in Theorem II.4.4.3, a tightness assumption is likely necessary for some of these center constructions. ◇

Question A.1.4 (Gray Rings and Bipermutative Gray Monoids). Recall from Sections I.6.6 and I.6.7 that a Gray monoid is a 2-category equipped with a monoid structure under the Gray tensor product. A permutative Gray monoid is a Gray monoid equipped with a compatible Gray symmetry. Symmetric monoidal bicategories can be strictified to permutative Gray monoids; see [**GJO17b**] and the discussion near the end of Section I.6.7.

- Define a *Gray ring* and a *bipermutative Gray monoid* that are analogous to, respectively, a right rigid bimonoidal category (Definition I.5.5.8) and a right bipermutative category (Definition I.2.5.2).
- Along the lines of Theorem I.5.5.11, prove a strictification result from bimonoidal bicategories (Question A.1.2) to Gray rings.
- Along the lines of Theorem I.5.4.6, prove a strictification result from symmetric bimonoidal bicategories to bipermutative Gray monoids.

A bipermutative Gray monoid should be a 2-category equipped with two compatible permutative Gray monoid structures, $\boxplus$ and $\boxtimes$, that interact via distributivity. In a Gray ring, $\boxtimes$ is a Gray monoid structure that is not assumed to be permutative. The following table summaries these (conjectural) concepts.

lax structure	strict structure	strictification
bicategories	2-categories	[**JY21**, 8.4.1]
monoidal bicategories	Gray monoids	[**GPS95, Gur13**]
symmetric monoidal bicategories	permutative Gray monoids	[**GJO17b**]
bimonoidal bicategories	Gray rings	conjecture
symmetric bimonoidal bicategories	bipermutative Gray monoids	conjecture

In each row, the left column can be strictified to the middle column. ◇

Question A.1.5 (Horizontal Bicategories of Double Categories)**.** A number of bicategories, including those of spans and bimodules, are the horizontal bicategories of some double categories.

- For a tight bimonoidal category C, is the matrix bicategory $\mathsf{Mat}^{\mathsf{C}}$ in Theorem I.8.4.12 the horizontal bicategory of a double category?
- If so, does the symmetric monoidal bicategory in Theorem I.8.15.4 arise from a symmetric monoidal structure on the double category?

See [**HS**∞] and [**JY21**, 12.3 and 12.4] for a discussion of (monoidal) double categories and their horizontal bicategories. ◇

Question A.1.6 (Braided Sheet Diagrams)**.** String diagrams are graphical reasoning tools in monoidal categories [**JS91a, Sel11**]. Sheet diagrams [**CDH**∞], which we mentioned in Notes I.2.7.5 and I.7.9.2 and Example I.3.10.9, are their analogues for tight bimonoidal categories.

- Develop sheet diagrams for
  - symmetric bimonoidal categories (Definition I.2.1.2) and
  - braided bimonoidal categories (Definition II.2.1.29).

This is, in fact, a coherence question with several parts. More precisely, a *bimonoidal signature* S consists of (i) a set of generating objects and (ii) a set of generating morphisms, each with (co)domain in the free $\{\oplus, \otimes\}$ algebra S^{fr} (Definition I.3.1.2). Given a bimonoidal signature S, one first defines the appropriate braided bimonoidal sheet diagrams and topological deformations corresponding to the axioms of a braided bimonoidal category. Then one constructs a braided bimonoidal category S' with object set S^{fr} and, as morphisms, braided bimonoidal sheet diagrams modulo topological deformations. Finally, one proves that S' is braided bimonoidally equivalent to the free braided bimonoidal category on S. A similar discussion applies in the symmetric case.

In view of the results in Chapter II.3, braided bimonoidal sheet diagrams can be used as graphical reasoning tools in (i) quantum group theory and (ii) the Fibonacci and Ising anyons in topological quantum computation. Sheet diagrams for tight bimonoidal categories in [**CDH**∞] involve the symmetric monoidal string diagrams in [**JS91a**]. Braided bimonoidal sheet diagrams will likely involve both the symmetric monoidal (for the additive structure $\oplus$) and the braided monoidal (for the multiplicative structure $\otimes$) string diagrams in [**JS91a**, Ch. 2–3]. The Braided Bimonoidal Coherence Theorem II.5.4.4 will be needed to check the axioms in Definition II.2.1.29 for braided bimonoidal sheet diagrams. Similarly, Laplaza's Coherence Theorems I.3.9.1 and I.4.4.3 will be needed to check the axioms in Definition I.2.1.2 for symmetric bimonoidal sheet diagrams.

The distributivity morphisms δ^l and δ^r in Definitions I.2.1.2 and II.2.1.29 are not invertible in general.

- Is it possible to replace the tightness assumption—that is, the invertibility of δ^l and δ^r—in the sheet diagrams in [**CDH∞**] with flatness in Definitions I.3.9.9 and II.5.4.5?

Related to Section I.7.9, to replace tightness with the much weaker assumption of flatness, one would need to work directly with a flat (symmetric/braided) bimonoidal category and *avoid* using the Strictification Theorems I.5.4.6, I.5.4.7, I.5.5.11, I.5.5.12, II.6.3.6, and II.6.3.7. The reason is that each of these theorems requires the tightness assumption. See Question A.2.8 for further problems about sheet diagrams. ◇

A.2. E_n-Monoidal Categories

The following questions are about the E_n-monoidal categories in Part II.2.

Question A.2.1 (Coherence of E_n-Monoidal Categories).

- Prove coherence theorems for ring categories (Definition II.9.1.2) along the lines of Theorems I.3.10.7 and I.4.5.8. Each such coherence theorem should say that any reasonable formal diagram in a ring category involving
$$(\oplus, \mathbb{0}, \xi^{\oplus}, \otimes, \mathbb{1}, \partial^l, \partial^r)$$
is commutative, with an assumption on either the common domain or the two paths.
- Prove coherence theorems for bipermutative categories (Definition II.9.3.2) along the lines of Theorems I.3.9.1 and I.4.4.3.
- Prove a coherence theorem for braided ring categories (Definition II.9.5.1) along the lines of Theorem II.5.4.4.
- More generally, prove a coherence theorem for E_n-monoidal categories (Definition II.10.7.2) along the lines of Theorem II.5.4.4. The coherence theorems for n-fold monoidal categories (Theorem II.10.6.8) should be relevant.

As in Theorems I.3.9.1 and I.4.4.3, one may need to assume a monomorphism or an epimorphism condition on the factorization morphisms ∂^l and ∂^r ◇

Question A.2.2 (n-Monoidal Categories). In an n-fold monoidal category (Definition II.10.1.1), each monoidal structure $\otimes_i$ is strictly associative with a common strict unit $\mathbb{1}$. There is a more general concept called an *n-monoidal category* in [**AM10**, Def. 6.1, 7.1, and 7.24]. It allows each monoidal structure $\otimes_i$ to be nonstrict and distinct monoidal units.

- Describe the free n-monoidal category of a small category, along the lines of Proposition II.10.5.9 and Theorem II.10.5.18.
- Generalize the Coherence Theorem II.10.6.8 to n-monoidal categories.

In [**AM10**, Section 6.2], it is stated without detail that, in a 2-monoidal category, each formal diagram is commutative. It is stated there that this coherence result can be deduced from the work in [**Lew72**]. So one possible first step in answering these questions would be to prove in detail this coherence result for 2-monoidal categories. ◇

Questions A.2.3 through A.2.7 below are all related to each other.

Question A.2.3 (Lax n-Fold Monoidal Categories). Between an n-fold monoidal category and an n-monoidal category (Question A.2.2) is a *lax n-fold monoidal category*. The latter allows each monoidal structure $\otimes_i$ to be nonstrict, and it assumes a common nonstrict monoidal unit $\mathbb{1}$. Analogous to Propositions II.10.1.14 and II.10.1.21, general nonstrict braided monoidal categories should be examples of lax 2-fold monoidal categories, and general nonstrict symmetric monoidal categories should be examples of lax n-fold monoidal categories for $n \geq 2$.

- Describe the free lax n-fold monoidal category of a small category, along the lines of Proposition II.10.5.9 and Theorem II.10.5.18.
- Extend the Coherence Theorem II.10.6.8 to lax n-fold monoidal categories.
- Can n-monoidal categories be strictified to lax n-fold monoidal categories?
- Can lax n-fold monoidal categories be strictified to n-fold monoidal categories?

There are two other possible variants of n-fold monoidal categories. The variant in [**For04**] corresponds to a lax n-fold monoidal category with a common *strict* monoidal unit. The variant in [**FSS07**] corresponds to an n-monoidal category with generally distinct but strict monoidal units. The following table summaries the strictness assumptions of n-fold monoidal categories and its four variants.

monoidal units	strict $\{\otimes_i\}_{1 \leq i \leq n}$	nonstrict $\{\otimes_i\}_{1 \leq i \leq n}$
common strict	n-fold monoidal	[**For04**]
common nonstrict		lax n-fold monoidal
distinct strict		[**FSS07**]
distinct nonstrict		n-monoidal [**AM10**]

The questions below refer to lax n-fold monoidal categories as defined above, with $\otimes_i$ generally nonstrict and a common nonstrict monoidal unit. ◇

Question A.2.4 (Unstable Periodic Table of Weak n-Categories)**.** The *periodic table* in [**BD98**] of k-tuply monoidal n-categories is a guiding principle for defining some versions of weak n-categories. In the $n = 1$ column in the periodic table, the values $k = 0, 1, 2$, and ≥ 3 correspond to, respectively, categories, monoidal categories, braided monoidal categories, and symmetric monoidal categories. On the other hand, by Propositions II.10.1.14 and II.10.1.21, braided strict monoidal categories and permutative categories are special examples of, respectively, 2-fold monoidal categories and k-fold monoidal categories for $k \geq 2$. Proposition II.10.2.8 has examples of 2-fold monoidal categories that are not braided strict monoidal categories. The k-fold monoidal category operad Mon^k (Proposition 13.1.20) is an E_k-operad that parametrizes k-fold monoidal categories (Theorem 13.2.1 and Proposition 13.3.18).

- Construct an *unstable* periodic table in which the $n = 1$ column consists of lax k-fold monoidal categories (Question A.2.3) for $k \geq 1$.

The $n = 1$ and $n = 2$ columns of the unstable periodic table should look like this:

	$n = 1$	$n = 2$
$k = 0$	categories	bicategories
$k = 1$	monoidal categories	monoidal bicategories
$k \geq 2$	lax k-fold monoidal categories	k-fold monoidal bicategories

The unstable periodic table does *not* stabilize like the periodic table in [**BD98**], where the $(n = 1, k \geq 3)$ entries are all symmetric monoidal categories. Moreover, the $(n = 2, k \geq 2)$ column in the unstable periodic table contains yet-to-be-defined *k-fold monoidal bicategories.*

- Prove that braided, sylleptic, and symmetric monoidal bicategories in Section I.6.5 are examples of k-fold monoidal bicategories for, respectively, $k = 2, 3$, and ≥ 4. ◇

This question may be regarded as both (i) a litmus test for the correct definition of a k-fold monoidal bicategory and (ii) a conceptual unification of braided, sylleptic, and symmetric monoidal bicategories. Further examples of k-fold monoidal bicategories should arise from the matrix construction in Question A.2.7.

Question A.2.5 (Iterated Gray Monoids)**.** This is a variation of Question A.1.4.

- Analogous to k-fold monoidal categories (Definition II.10.1.1) for $k \geq 1$, define the concept of a *k-fold Gray monoid* that satisfies the following statements:
  - A Gray monoid is precisely a 1-fold Gray monoid.
  - A braided monoidal 2-category [**Cra98**] is an example of a 2-fold Gray monoid, analogous to Proposition II.10.1.14.
  - A sylleptic monoidal 2-category [**Cra98**] is an example of a 3-fold Gray monoid.
  - A permutative Gray monoid is an example of a k-fold Gray monoid for $k \geq 4$, analogous to Proposition II.10.1.21.
- Prove a strictification theorem from k-fold monoidal bicategories (Question A.2.4) to k-fold Gray monoids. This should fit into the following table of strictification theorems.

bicategories	2-categories	[**JY21**, 8.4.1]
monoidal bicategories	Gray monoids	[**GPS95, Gur13**]
braided monoidal bicategories	braided monoidal 2-categories	[**Gur11**]
k-fold monoidal bicategories ($k \geq 1$)	k-fold Gray monoids	conjecture
symmetric monoidal bicategories	permutative Gray monoids	[**GJO17b**]

In each row, the left column can be strictified to the middle column. For the symmetric case, see the discussion near the end of Section I.6.7. ◇

Question A.2.6 (Laplaza E_n-Monoidal Categories)**.** The factorization morphisms $\{\partial^{l,i}, \partial^{r,i}\}_{1 \leq i \leq n}$ in an E_n-monoidal category (Definition II.10.7.2) go in the opposite direction as the distributivity morphisms in a bimonoidal category (Definition I.2.1.2). Moreover, in an E_n-monoidal category, the monoidal structures $\oplus$ and $\{\otimes_i\}_{i=1}^n$ are all strict.

- Define a *Laplaza E_n-monoidal category* with
  - a generally nonstrict additive structure $(\oplus, \mathbb{0})$,
  - a lax n-fold monoidal structure $\big(\{\otimes_i\}_{i=1}^n, \{\eta^{i,j}\}\big)$ (Question A.2.3),
  - for each $1 \leq i \leq n$, a bimonoidal structure $(\oplus, \otimes_i, \lambda_i^{\bullet}, \rho_i^{\bullet}, \delta_i^l, \delta_i^r)$ (Definition I.2.1.2), and
  - appropriate axioms relating the lax n-fold monoidal structure and the n bimonoidal structures, analogous to those in Definition II.10.7.2.

Laplaza E_n-monoidal categories should contain the following examples:

- An E_n-monoidal category with invertible factorization morphisms should be an example of a Laplaza E_n-monoidal category, analogous to Theorems II.9.1.15, II.9.3.7, and II.9.5.6.
- A Laplaza E_1-monoidal category should be precisely a bimonoidal category (Definition I.2.1.2), analogous to Example II.10.7.13.

- A braided bimonoidal category (Definition II.2.1.29) should be an example of a Laplaza E_2-monoidal category with $\otimes_1 = \otimes_2$, analogous to Theorem II.10.8.1.
- A symmetric bimonoidal category should be an example of a Laplaza E_n-monoidal category for $n \geq 2$ with $\otimes_1 = \cdots = \otimes_n$, analogous to Theorem II.10.9.1.
- Similar to Theorem II.2.4.22, an abelian category with a compatible lax n-fold monoidal structure should be a Laplaza E_n-monoidal category.

Moreover:

- Prove a coherence theorem for Laplaza E_n-monoidal categories, along the lines of Theorem II.5.4.4. This will certainly involve the coherence theorem for lax n-fold monoidal categories in Question A.2.3.
- Prove a strictification theorem for *tight* Laplaza E_n-monoidal categories, along the lines of Theorems II.6.3.6 and II.6.3.7. Here *tight* means that all the distributivity morphisms, δ_i^l and δ_i^r, are natural isomorphisms.
- Is there an analogue of Baez's Conjecture (Theorems I.7.8.1, I.7.8.3, II.7.3.4, and II.7.3.6) for Laplaza E_n-monoidal categories? ◇

Question A.2.7 (Matrix Construction)**.** In Theorem I.8.4.12 we showed that the matrix construction $\mathsf{Mat}^{\mathsf{C}}$ is a bicategory for each tight bimonoidal category C. Moreover, $\mathsf{Mat}^{\mathsf{C}}$ is (i) a monoidal bicategory if C is a tight braided bimonoidal category (Theorem II.8.4.7) and (ii) a symmetric monoidal bicategory if C is a tight symmetric bimonoidal category (Theorem I.8.15.4).

- Show that the matrix construction $\mathsf{Mat}^{\mathsf{C}}$ of a tight Laplaza E_{k+1}-monoidal category (Question A.2.6) is a k-fold monoidal bicategory.

This question asks for a refinement of the table in the introduction of Chapter II.8 as follows.

tight — category C	bicategory $\mathsf{Mat}^{\mathsf{C}}$	
bimonoidal	plain	I.8.4.12
braided bimonoidal	monoidal	II.8.4.7
Laplaza E_{k+1}-monoidal ($k \geq 1$)	k-fold monoidal	conjecture
symmetric bimonoidal	symmetric monoidal	I.8.15.4

The k-fold monoidal bicategories in the conjectural row refer to the $n = 2$ column in the unstable periodic table in Question A.2.4. Proving that $\mathsf{Mat}^{\mathsf{C}}$ is a k-fold monoidal bicategory will certainly involve the coherence theorem for Laplaza E_{k+1}-monoidal categories in Question A.2.6. The general picture of the table above is that it takes a sum $\oplus$ and a product $\otimes$ to construct the matrix bicategory $\mathsf{Mat}^{\mathsf{C}}$. So any further monoidal structures on the bicategory $\mathsf{Mat}^{\mathsf{C}}$ would have to come from further monoidal structures on C. ◇

Question A.2.8 (Higher Sheet Diagrams)**.** As in Question A.1.6, develop higher dimensional sheet diagrams for

- n-fold monoidal categories (Definition II.10.1.1),
- n-monoidal categories (Question A.2.2),
- lax n-fold monoidal categories (Question A.2.3),
- E_n-monoidal categories (Definition II.10.7.2), and
- Laplaza E_n-monoidal categories (Question A.2.6).

As discussed in Question A.1.6, each of these items is a coherence question with several parts. ◇

A.3. Enriched Monoidal Categories

The following questions are about the concepts in Chapters 1, 2, and 3.

Question A.3.1 (Enriched Lax n-Fold Monoidal Categories). In Section 1.4, with V a braided monoidal category, we defined monoidal, braided monoidal, and symmetric monoidal V-categories, with the latter two assuming that V is symmetric. See Lemma 1.3.23 and Explanation 1.3.25.

- For $n \geq 2$ and V a symmetric monoidal category, extend the lax n-fold monoidal categories in Question A.2.3 to the V-enriched setting.
- Extend the results in Chapter 2 to lax n-fold monoidal V-categories.

Analogous to Propositions II.10.1.14 and II.10.1.21, braided monoidal V-categories should be examples of lax 2-fold monoidal V-categories, and symmetric monoidal V-categories should be examples of lax n-fold monoidal V-categories for $n \geq 2$. Theorem II.10.4.5 says that small $(n+1)$-fold monoidal categories are precisely the monoids in the monoidal category MCat^n of small n-fold monoidal categories and n-fold monoidal functors.

- Extend Theorem II.10.4.5 to the V-enriched setting. ◇

Question A.3.2 (Centers and Enriched Centers). The Drinfeld center of a monoidal category is a braided monoidal category (Theorem II.1.4.27), and the symmetric center of a braided monoidal category is a symmetric monoidal category (Proposition II.1.5.3). Moreover, the Drinfeld center and the symmetric center are generalized to (i) the bimonoidal setting in Theorems II.4.4.3 and II.4.5.3 and (ii) the ring categorical setting in Corollary II.9.6.1 and Theorem II.9.6.4.

- Define a center construction that sends an n-fold monoidal category (Definition II.10.1.1) to an $(n+1)$-fold monoidal category.

Theorem II.10.4.5 should be relevant.

- Repeat the previous question for
  - n-monoidal categories (Question A.2.2),
  - lax n-fold monoidal categories (Question A.2.3),
  - E_n-monoidal categories (Definition II.10.7.2),
  - Laplaza E_n-monoidal categories (Question A.2.6), and
  - lax n-fold monoidal V-categories (Question A.3.1).

As a special case of the last item, the Drinfeld center of a monoidal V-category with V *strict* is studied in [**KYZZ∞, KZ18**]. ◇

Question A.3.3 (Autonomous Enriched Monoidal Categories). The definition of a monoidal V-category K in [**MP19**, 2.1] assumes that V is braided strict monoidal and K is strict monoidal, so it is the special case of Definition 1.4.2 with both V and K strict. The main theorem in [**MP19**, 1.1] shows that, using their definition, there is a bijective correspondence between (i) some autonomous monoidal V-categories and (ii) some braided oplax monoidal functors from V to the Drinfeld center of an autonomous monoidal category. This bijective correspondence is extended to a 2-equivalence between 2-categories in [**Del∞**, 1.2]. In bicategorical language (Definition I.6.3.9), *autonomy* means that each object $X \in \mathsf{K}$ is equipped with both a left adjoint X_* and a right adjoint X^* that satisfy the triangle identities. A (braided) monoidal functor (F, F^2, F^0) is *oplax* if its monoidal constraint F^2 and unit constraint F^0 go in the opposite directions as those in Definitions 1.1.6 and 1.1.17, with appropriately adjusted axioms. Discussion of autonomous monoidal categories can be found in [**FY92, JS91b, JS93**].

- Extend the 2-equivalence in [**Del∞**, 1.2] to general monoidal V-categories K, with V and K not necessarily strict, as in Definition 1.4.2.

To extend this 2-equivalence to the general nonstrict case, the coherence and strictification results of enriched monoidal categories in Sections 2.5 and 2.6 will likely be necessary. ◇

A.4. Homotopy Theory

Question A.4.1 (Homotopy Theory of Matrix Bicategories)**.** In Example I.8.15.5, we listed some examples of tight symmetric bimonoidal categories C, to which the Bicategorification Theorem I.8.15.4 may be applied to yield a symmetric monoidal bicategory $\mathsf{Mat}^{\mathsf{C}}$.

- What can be said about the homotopy theoretic properties of any of these symmetric monoidal bicategories?
- Consider the previous question for the symmetric bimonoidal bicategories in Question A.1.2.

For instance, for the finite ordinal category Σ in Section I.2.4, Mat^{Σ} may be related to a remark in [**JO12**] about the multiplicative structure on the categorical model for the sphere spectrum. ◇

Question A.4.2 (Categorical Model of BP)**.** The Brown-Peterson spectrum BP has an E_4 structure [**BM13**], but not an E_∞ structure at any prime [**Law18, Sen∞**]. By Corollary 13.5.2, the Elmendorf-Mandell K-theory of each small E_n-monoidal category is an E_n-symmetric spectrum.

- Is there a small E_4-monoidal category (Definition II.10.7.2) whose K-theory is BP?

A positive answer to this question would provide a categorical model of BP. If Question A.5.7 has a positive answer for the 4-fold monoidal category operad Mon^4, then this question also has a positive answer, at least up to weak equivalences. ◇

Question A.4.3 (Boardman-Vogt E_n-Operads)**.** In a symmetric monoidal category C, the commutative operad Com is the operad with each entry the monoidal unit $\mathbb{1}$ and structure morphisms given by the coherence isomorphism $\mathbb{1} \otimes \mathbb{1} \cong \mathbb{1}$. The algebras of Com are precisely commutative monoids. One model of an E_∞-operad (Definition 11.6.1) is WCom, where W is the Boardman-Vogt W-construction. In the topological setting, it was introduced in [**BV73**]. For a conceptual presentation of the W-construction in terms of coends in a general symmetric monoidal category, see [**Yau20**, Ch. 6–7].

- Describe E_n-operads (Definition 12.2.3) for $n \geq 1$ as a filtration of suboperads of WCom.
- Compare these models of E_n-operads to
  - the n-fold monoidal category operad Mon^n (Definition 13.1.12),
  - Berger's filtration [**Ber96**] of the Barratt-Eccles operad $N(E\mathsf{As})$ (Definition 11.4.10),
  - Smith's filtration [**Smi89**] of the Barratt-Eccles operad,

– Batanin's E_n-operads [**Bat07, Bat08**],
– Fiedorowicz's E_n-operads [**Fie∞b**], and
– the Fulton-MacPherson E_n-operads [**Fre17, FM94, GJ∞, Sal01**].

In [**Yau20**, 3.2.11 and 6.3.1], the W-construction of an operad in a symmetric monoidal category is defined entrywise as a coend indexed by a substitution category whose objects are trees and whose morphisms correspond to tree substitution. This applies, in particular, to the commutative operad Com. So describing E_n-operads as sub-operads of WCom would provide a combinatorial description of E_n-operads in terms of trees. ◇

A.5. Algebraic K-Theory

Question A.5.1 (Multifunctorial K-Theory of Pointed Multicategories)**.** Contrary to the claim in [**EM09**, Theorem 1.3], Elmendorf-Mandell J-theory J^{EM} (Definition 10.3.25) does *not* extend to a multifunctor on all of $\mathsf{Multicat}_*$, the category of small pointed multicategories, but only to the full subcategory $\mathsf{Mod}^{\mathcal{M}\underline{1}}$ of left $\mathcal{M}\underline{1}$-modules, via the symmetric monoidal Cat_*-functor $\mathsf{J}^{\mathcal{T}}$ (Theorem 10.3.17). Examples 10.2.15 and 10.2.16 present some small pointed multicategories that are not left $\mathcal{M}\underline{1}$-modules.

- Is there a K-theory multifunctor that is objectwise equivalent to Segal K-theory K^{Se} and extends to $\mathsf{Multicat}_*$ via the endomorphism multicategory End in Corollary 5.3.9 and Definition 6.5.1?

The key issue is about the monoidal units. In $\mathsf{Multicat}_*$ the smash unit is $\mathsf{S} = \mathsf{I} \coprod \mathsf{T}$ (Definition 5.6.20), which is different from the monoidal unit $\mathcal{M}\underline{1}$ in $\mathsf{Mod}^{\mathcal{M}\underline{1}}$. Unlike Definition 10.3.16 with $\mathsf{J}^{\mathcal{T}}(\mathcal{M}\underline{1})$, the object $\mathsf{J}^{\mathcal{T}}(\mathsf{S})$ is the terminal $\mathcal{G}_*$-category $*$. If the monoidal unit constraint $(\mathsf{J}^{\mathcal{T}})^0$ for $\mathsf{J}^{\mathcal{T}}$ were to be defined as the unique morphism $J \longrightarrow \mathsf{J}^{\mathcal{T}}(\mathsf{S}) = *$ to the terminal object, as stated in the last paragraph in [**EM09**, Section 5], then $\mathsf{J}^{\mathcal{T}}$ cannot satisfy the unity axioms (1.1.10). The reason is that a general left or right unit isomorphism for $\mathsf{J}^{\mathcal{T}}(-)$ does not factor through the zero morphism in $\mathcal{G}_*$-Cat. So with $J \longrightarrow *$ as the unit constraint, $\mathsf{J}^{\mathcal{T}}$ would not be a monoidal functor. ◇

Question A.5.2 (Comparison of K^{Se} and K^{EM} for $\mathcal{M}\underline{1}$-Modules)**.** The Segal and Elmendorf-Mandell K-theory constructions are defined as the following composites, respectively:

$$\mathsf{K}^{\mathsf{Se}} = \mathsf{K}^{\mathcal{F}} N_* \mathsf{J}^{\mathcal{M}} \mathsf{End} \quad \text{and} \quad \mathsf{K}^{\mathsf{EM}} = \mathsf{K}^{\mathcal{G}} N_* \mathsf{J}^{\mathcal{T}} \mathsf{End}.$$

The domain of $\mathsf{J}^{\mathcal{M}}$ is the category of small pointed multicategories, $\mathsf{Multicat}_*$, and the domain of $\mathsf{J}^{\mathcal{T}}$ is the category of left $\mathcal{M}\underline{1}$-modules within $\mathsf{Multicat}_*$. Thus both K^{Se} and K^{EM} can be expanded to $\mathsf{Mod}^{\mathcal{M}\underline{1}}$. We will write

$$\mathsf{Mod}^{\mathcal{M}\underline{1}} \xrightarrow{\tilde{\mathsf{K}}^{\mathsf{Se}} = \mathsf{K}^{\mathcal{F}} N_* \mathsf{J}^{\mathcal{M}}} \mathsf{SymSp} \quad \text{and} \quad \mathsf{Mod}^{\mathcal{M}\underline{1}} \xrightarrow{\tilde{\mathsf{K}}^{\mathsf{EM}} = \mathsf{K}^{\mathcal{G}} N_* \mathsf{J}^{\mathcal{T}}} \mathsf{SymSp}.$$

- Is there a (natural) level equivalence $\tilde{\mathsf{K}}^{\mathsf{Se}}\mathsf{P} \longrightarrow \tilde{\mathsf{K}}^{\mathsf{EM}}\mathsf{P}$ for each left $\mathcal{M}\underline{1}$-module P?

The level equivalence $\mathsf{K}^{\mathsf{Se}}\mathsf{C} \longrightarrow \mathsf{K}^{\mathsf{EM}}\mathsf{C}$ given in Theorem 10.6.10 for each small permutative category C depends crucially on the adjunctions of Proposition 10.6.7 and these, in turn, depend on Proposition 8.5.4, which gives a strong symmetric monoidal adjunction

$$\mathsf{Cat}_*(a, \mathsf{C}) \underset{R}{\overset{L}{\rightleftarrows}} \mathsf{Multicat}_*(\mathcal{M}a, \mathsf{End}(\mathsf{C}))$$

for each small permutative category C and each pointed finite set a. Therefore the proof of Theorem 10.6.10 does not immediately generalize to $\tilde{\mathsf{K}}^{\mathsf{Se}}$ and $\tilde{\mathsf{K}}^{\mathsf{EM}}$. ◇

Question A.5.3 (Homotopy Theory of $\mathcal{G}_*$-Objects)**.** Notes 8.6.4 and 8.6.6 summarize equivalences of homotopy categories from the work of [**BF78, Tho95, Man10**]:

$$\mathit{Ho}(\mathsf{PermCat}^{\mathsf{su}}) \simeq \mathit{Ho}(\Gamma\text{-}\mathsf{Cat}) \simeq \mathit{Ho}(\Gamma\text{-}\mathsf{sSet}) \simeq \mathit{Ho}(\mathsf{Sp}^{\mathbb{N}}_{\geq 0}),$$

where $\mathsf{Sp}^{\mathbb{N}}_{\geq 0}$ denotes the category of connective sequential spectra. The composite from the homotopy category of small permutative categories to that of connective spectra is induced by Segal K-theory, K^{Se}. Therefore, by Theorem 10.6.10, the composite on homotopy categories is equal to that induced by Elmendorf-Mandell K-theory, K^{EM}. However, K^{EM} factors through $\mathcal{G}_*$-Cat and $\mathcal{G}_*$-sSet instead of Γ-Cat and Γ-sSet.

As diagram categories, $\mathcal{G}_*$-Cat and $\mathcal{G}_*$-sSet have Quillen model structures defined similarly to those of Γ-Cat and Γ-sSet. The functor

$$\wedge\colon \mathcal{G} \longrightarrow \mathcal{F}$$

is described in Definition 9.1.15 and is used in the comparison of $\mathsf{K}^{\mathcal{F}}$ and $\mathsf{K}^{\mathcal{G}}$ in Proposition 9.3.16. The induced functor from Γ-objects to $\mathcal{G}_*$-objects is symmetric monoidal by Theorem 9.4.18.

- Are the induced functors

$$\Gamma\text{-}\mathsf{Cat} \xrightarrow{\wedge^*} \mathcal{G}_*\text{-}\mathsf{Cat} \quad \text{and} \quad \Gamma\text{-}\mathsf{sSet} \xrightarrow{\wedge^*} \mathcal{G}_*\text{-}\mathsf{sSet}$$

  (monoidal) Quillen equivalences?

Question A.5.7 poses a related question regarding categories of algebras over operads. ◇

Question A.5.4 (K-Theory of Matrix Bicategories)**.** There is a K-theory construction via the following composite on objects only.

$$\mathsf{Bi}^{\mathsf{tsy}} \xrightarrow{\mathsf{Mat}} \mathsf{SMB} \xrightarrow{\mathsf{strictify}} \mathsf{PGray} \xrightarrow{K} \Gamma\mathsf{2Cat}$$

Mat is the matrix construction in Theorem I.8.15.4 that sends a tight symmetric bimonoidal category to a symmetric monoidal bicategory. The middle arrow is the strictification of symmetric monoidal bicategories to permutative Gray monoids. The right arrow is the K-theory of permutative Gray monoids in [**GJO17b**].

- How does this compare with the K-theory of tight symmetric bimonoidal categories in [**BDR04**], which is defined using the direct sum instead of the tensor product of matrices?
- What extra structures on Γ-2-categories are there when it is the K-theory of the strictification of $\mathsf{Mat}^{\mathsf{C}}$ for some tight symmetric bimonoidal category C, such as those in Example I.8.15.5 and $\mathsf{Vect}^{\mathbb{C}}_{\mathsf{c}}$ in Example I.2.5.9? ◇

Question A.5.5 (K-Theory of Matrix Symmetric Bimonoidal Bicategories)**.** Repeat Question A.5.4 for

- the matrix symmetric bimonoidal bicategories in Question A.1.2 and
- the bipermutative Gray monoids in Question A.1.4. ◇

Question A.5.6 (K-Theory of Distortion Categories)**.** Recall the finite ordinal category Σ in Definition I.2.4.1. Here we consider Σ as a permutative category with respect to its additive structure $\oplus$. Quillen's +-construction $(B\Sigma)^+$ of the classifying space $B\Sigma$ is the sphere spectrum by the Barratt-Priddy-Quillen Theorem

[**BP72**]. A different way to say this is that the algebraic K-groups of Σ are the stable homotopy groups of the spheres.

- Can the algebraic K-groups of
  - the distortion category $\mathcal{D}$ in Section I.4.2,
  - the additive distortion category $\mathcal{D}^{\mathsf{ad}}$ in Section I.4.5, and
  - the braided distortion category $\mathcal{D}^{\mathsf{br}}$ in Section II.5.2

  be computed in similar terms?

By Examples 11.3.18, 11.6.13, and 12.5.4, respectively, $\mathsf{K}^{\mathsf{EM}}\mathcal{D}^{\mathsf{ad}}$, $\mathsf{K}^{\mathsf{EM}}\mathcal{D}$, and $\mathsf{K}^{\mathsf{EM}}\mathcal{D}^{\mathsf{br}}$ are strict ring, E_∞-, and E_2-symmetric spectra. Moreover, each of the distortion categories $\mathcal{D}$, $\mathcal{D}^{\mathsf{ad}}$, and $\mathcal{D}^{\mathsf{br}}$ is a Grothendieck construction over Σ by, respectively, Propositions I.4.6.5, I.4.6.7, and II.5.5.3.

- Does that yield a computation of their $(B?)^+$ and algebraic K-groups? $\diamond$

Question A.5.7 (Lifting K-Theory Equivalences to Algebras)**.** The Segal K-theory functor in Definition 8.5.1 induces an equivalence of homotopy categories, from permutative categories to connective symmetric spectra. See Question A.5.3 for further description and a related question about K^{EM}.

- Do the (Quillen) equivalences in Segal's K-theory lift to the categories of algebras over categorical operads, such as the E_2-operad Br and the E_n-operads Mon^n in Theorems 12.2.4 and 13.2.1?

For a categorical operad P and a Cat-enriched multicategory M, such as $\mathsf{PermCat}^{\mathsf{su}}$ in Section 6.6, a P-*algebra in* M is defined as a Cat-enriched multifunctor

$$F\colon \mathsf{P} \longrightarrow \mathsf{M}.$$

This is equivalent to a Cat-enriched operad morphism

$$\mathsf{P} \longrightarrow \mathsf{End}(A)$$

to the Cat-enriched endomorphism operad of the object $A = F(*)$, with $*$ the unique object in the multicategory P. If the answer to this question is yes for a categorical operad P, then Segal's K-theory induces an equivalence between the homotopy categories of (i) P-algebras in permutative categories and (ii) P-algebras in connective symmetric spectra.

To answer this question, it is tempting to use [**WY19**, Th. 4.4 and 4.6], which give sufficient conditions under which a Quillen equivalence between monoidal model categories lifts to a Quillen equivalence between the categories of algebras over some colored operads. This will *not* work because the domain of Segal's K-theory is the multicategory $\mathsf{PermCat}^{\mathsf{su}}$ (Section 6.6), which is not a symmetric monoidal category, hence also not a monoidal model category. $\diamond$

Bibliography

[Abe80] E. Abe, *Hopf algebras*, Cambridge Tracts in Mathematics, vol. 74, Cambridge University Press, Cambridge-New York, 1980, Translated from Japanese by Hisae Kinoshita and Hiroko Tanaka. (cit. on p. II.106).

[Ada95] J. F. Adams, *Stable homotopy and generalised homology*, Chicago Lectures in Mathematics, University of Chicago Press, Chicago, IL, 1995, Reprint of the 1974 original. (cit. on p. xvii).

[AM10] M. Aguiar and S. Mahajan, *Monoidal functors, species and Hopf algebras*, CRM Monograph Series, vol. 29, American Mathematical Society, Providence, RI, 2010. doi:`10.1090/crmm/029` (cit. on pp. I.430, I.431, II.307, II.314, II.315, III.508, III.509).

[Art47] E. Artin, *Theory of braids*, Ann. of Math. **48** (1947), 101–126. doi:`10.2307/1969218` (cit. on p. II.34).

[Awo10] S. Awodey, *Category theory*, second ed., Oxford Logic Guides, vol. 52, Oxford University Press, Oxford, 2010. (cit. on pp. xxx, I.18).

[BDR04] N. A. Baas, B. I. Dundas, and J. Rognes, *Two-vector bundles and forms of elliptic cohomology*, Topology, geometry and quantum field theory, London Math. Soc. Lecture Note Ser., vol. 308, Cambridge Univ. Press, Cambridge, 2004, pp. 18–45. doi:`10.1017/CBO9780511526398.005` (cit. on pp. I.298, I.315, I.325, I.437, II.321, III.515).

[Bae18] J. C. Baez, *rig category*, 2018, `https://ncatlab.org/nlab/show/rig+category`. (cit. on pp. xvi, I.255).

[BD98] J. C. Baez and J. Dolan, *Categorification*, Higher category theory (Evanston, IL, 1997), Contemp. Math., vol. 230, Amer. Math. Soc., Providence, RI, 1998, pp. 1–36. doi:`10.1090/conm/230/03336` (cit. on pp. I.431, II.315, III.509).

[BN96] J. C. Baez and M. Neuchl, *Higher-dimensional algebra I: Braided monoidal 2-categories*, Adv. Math. **121** (1996), no. 2, 196–244. doi:`10.1006/aima.1996.0052` (cit. on pp. I.428, II.312, III.506).

[BS11] J. C. Baez and M. Stay, *Physics, topology, logic and computation: a Rosetta Stone*, New structures for physics, Lecture Notes in Phys., vol. 813, Springer, Heidelberg, 2011, pp. 95–172. doi:`10.1007/978-3-642-12821-9_2` (cit. on p. I.45).

[BFSV03] C. Balteanu, Z. Fiedorowicz, R. Schwänzl, and R. Vogt, *Iterated monoidal categories*, Adv. Math. **176** (2003), no. 2, 277–349. doi:`10.1016/S0001-8708(03)00065-3` (cit. on pp. xvi, xx, xxvi, II.259, II.260, II.262, II.264, II.293, II.294, II.295, II.306, II.307, III.476, III.485, III.488, III.502).

[BR20] D. Barnes and C. Roitzheim, *Foundations of stable homotopy theory*, Cambridge Studies in Advanced Mathematics, vol. 185, Cambridge University Press, 2020. (cit. on p. xxxii).

[BW12] M. Barr and C. Wells, *Category theory for computing science*, Repr. Theory Appl. Categ. (2012), no. 22, xviii+538, Corrected reprint. (cit. on p. III.345).

[BE74a] M. G. Barratt and P. J. Eccles, *Γ^+-structures. I. A free group functor for stable homotopy theory*, Topology **13** (1974), 25–45. doi:`10.1016/0040-9383(74)90036-6` (cit. on p. III.447).

[BE74b] ———, *Γ^+-structures. II. A recognition principle for infinite loop spaces*, Topology **13** (1974), 113–126. doi:`10.1016/0040-9383(74)90002-0` (cit. on p. III.447).

[BE74c] ———, *Γ^+-structures. III. The stable structure of $\Omega^\infty\Sigma^\infty A$*, Topology **13** (1974), 199–207. doi:`10.1016/0040-9383(74)90011-1` (cit. on p. III.447).

[BP72] M. Barratt and S. Priddy, *On the homology of non-connected monoids and their associated groups*, Comment. Math. Helv. **47** (1972), 1–14. doi:10.1007/BF02566785 (cit. on pp. I.438, II.322, III.516).

[BM13] M. Basterra and M. A. Mandell, *The multiplication on BP*, J. Topol. **6** (2013), no. 2, 285–310. doi:10.1112/jtopol/jts032 (cit. on pp. I.435, II.319, III.513).

[Bat98] M. A. Batanin, *Monoidal globular categories as a natural environment for the theory of weak n-categories*, Adv. Math. **136** (1998), no. 1, 39–103. doi:10.1006/aima.1998.1724 (cit. on p. II.307).

[Bat07] ———, *Symmetrisation of n-operads and compactification of real configuration spaces*, Adv. Math. **211** (2007), no. 2, 684–725. doi:10.1016/j.aim.2006.07.022 (cit. on pp. I.436, II.307, II.320, III.514).

[Bat08] ———, *The Eckmann-Hilton argument and higher operads*, Adv. Math. **217** (2008), no. 1, 334–385. doi:10.1016/j.aim.2007.06.014 (cit. on pp. I.436, II.307, II.320, III.514).

[BM12] M. Batanin and M. Markl, *Centers and homotopy centers in enriched monoidal categories*, Adv. Math. **230** (2012), no. 4-6, 1811–1858. doi:10.1016/j.aim.2012.04.011 (cit. on p. II.307).

[BJP08a] H.-J. Baues, M. Jibladze, and T. Pirashvili, *Quadratic algebra of square groups*, Adv. Math. **217** (2008), no. 3, 1236–1300. doi:10.1016/j.aim.2007.08.007 (cit. on p. III.448).

[BJP08b] H.-J. Baues, M. Jibladze, and T. Pirashvili, *Third Mac Lane cohomology*, Math. Proc. Cambridge Philos. Soc. **144** (2008), no. 2, 337–367. doi:10.1017/S030500410700076X (cit. on p. III.448).

[BM11] H.-J. Baues and F. Muro, *The algebra of secondary homotopy operations in ring spectra*, Proc. Lond. Math. Soc. (3) **102** (2011), no. 4, 637–696. doi:10.1112/plms/pdq034 (cit. on p. III.448).

[Bec67] J. M. Beck, *Triples, algebras and cohomology*, Ph.D. thesis, Columbia University, 1967. (cit. on p. III.219).

[Bén65] J. Bénabou, *Catégories relatives*, C. R. Acad. Sci. Paris **260** (1965), 3824–3827. (cit. on p. I.207).

[Bén67] ———, *Introduction to bicategories*, Reports of the Midwest Category Seminar, Springer, Berlin, 1967, pp. 1–77. (cit. on p. I.207).

[Ber96] C. Berger, *Opérades cellulaires et espaces de lacets itérés*, Ann. Inst. Fourier (Grenoble) **46** (1996), no. 4, 1125–1157. (cit. on pp. I.435, II.307, II.319, III.447, III.473, III.501, III.513).

[BF02] C. Berger and B. Fresse, *Une décomposition prismatique de l'opérade de Barratt-Eccles*, C. R. Math. Acad. Sci. Paris **335** (2002), no. 4, 365–370. doi:10.1016/S1631-073X(02)02489-5 (cit. on p. III.447).

[BF04] ———, *Combinatorial operad actions on cochains*, Math. Proc. Cambridge Philos. Soc. **137** (2004), no. 1, 135–174. doi:10.1017/S0305004103007138 (cit. on p. III.447).

[BK00] A. J. Berrick and M. E. Keating, *Categories and modules with K-theory in view*, Cambridge Studies in Advanced Mathematics, vol. 67, Cambridge University Press, Cambridge, 2000. (cit. on pp. xxx, I.18, II.62).

[BKP89] R. Blackwell, G. M. Kelly, and A. J. Power, *Two-dimensional monad theory*, J. Pure Appl. Algebra **59** (1989), no. 1, 1–41. doi:10.1016/0022-4049(89)90160-6 (cit. on p. III.90).

[BG16] A. Blass and Y. Gurevich, *On quantum computation, anyons, and categories*, Martin Davis on computability, computational logic, and mathematical foundations, Outst. Contrib. Log., vol. 10, Springer, Cham, 2016, pp. 209–241. (cit. on p. II.106).

[BG20a] ———, *Braided distributivity*, Theoret. Comput. Sci. **807** (2020), 73–94. doi:10.1016/j.tcs.2019.11.027 (cit. on pp. xvi, xx, xxiii, xxiv, xxv, I.48, II.35, II.36, II.43, II.44, II.46, II.54, II.65, II.127, II.157).

[BG20b] ———, *Witness algebra and anyon braiding*, Math. Structures Comput. Sci. **30** (2020), no. 3, 234–270. doi:10.1017/s0960129520000055 (cit. on p. II.106).

[BM11] A. J. Blumberg and M. A. Mandell, *Derived Koszul duality and involutions in the algebraic K-theory of spaces*, J. Topol. **4** (2011), no. 2, 327–342. doi:10.1112/jtopol/jtr003 (cit. on p. III.411).

[Bly05] T. S. Blyth, *Lattices and ordered algebraic structures*, Universitext, Springer-Verlag London, Ltd., London, 2005. (cit. on p. II.307).

[BV73] J. M. Boardman and R. M. Vogt, *Homotopy invariant algebraic structures on topological spaces*, Lecture Notes in Mathematics, Vol. 347, Springer-Verlag, Berlin-New York, 1973. (cit. on pp. I.435, II.230, II.260, II.307, II.319, III.219, III.220, III.457, III.513).

[BO20] A. M. Bohmann and A. M. Osorno, *A multiplicative comparison of Segal and Waldhausen K-theory*, Math. Z. **295** (2020), no. 3-4, 1205–1243. doi:10.1007/s00209-019-02394-7 (cit. on pp. III.345, III.346, III.411).

[Bor94a] F. Borceux, *Handbook of categorical algebra 1: Basic category theory*, Encyclopedia of Mathematics and its Applications, vol. 50, Cambridge University Press, Cambridge, 1994. (cit. on pp. xxx, I.18).

[Bor94b] ———, *Handbook of categorical algebra 2: Categories and structures*, Encyclopedia of Mathematics and its Applications, vol. 51, Cambridge University Press, Cambridge, 1994. (cit. on pp. xxx, I.18, III.156, III.194, III.286, III.409).

[Bor94c] ———, *Handbook of categorical algebra 3: Categories of sheaves*, Encyclopedia of Mathematics and its Applications, vol. 52, Cambridge University Press, Cambridge, 1994. (cit. on pp. xxx, I.18).

[BF78] A. K. Bousfield and E. M. Friedlander, *Homotopy theory of Γ-spaces, spectra, and bisimplicial sets*, Geometric applications of homotopy theory (Proc. Conf., Evanston, Ill., 1977), II, Lecture Notes in Math., vol. 658, Springer, Berlin, 1978, pp. 80–130. (cit. on pp. I.437, II.321, III.287, III.315, III.515).

[BK72] A. K. Bousfield and D. M. Kan, *Homotopy limits, completions and localizations*, Lecture Notes in Mathematics, Vol. 304, Springer-Verlag, Berlin-New York, 1972. (cit. on p. III.501).

[BJS11] W. J. Bowman, R. P. James, and A. Sabry, *Dagger traced symmetric monoidal categories and reversible programming*, Proceedings of the 4th Workshop on Reversible Computation (RC 2011), 2011, available at https://www.williamjbowman.com/papers/index.html. (cit. on p. I.46).

[BP07] S. L. Braunstein and A. K. Pati, *Quantum information cannot be completely hidden in correlations: implications for the black-hole information paradox*, Phys. Rev. Lett. **98** (2007), no. 8, 080502, 4. doi:10.1103/PhysRevLett.98.080502 (cit. on p. I.46).

[Bro64] R. Brown, *Function spaces and product topologies*, Quart. J. Math. Oxford Ser. (2) **15** (1964), 238–250. doi:10.1093/qmath/15.1.238 (cit. on pp. III.286, III.473).

[Buc55] D. A. Buchsbaum, *Exact categories and duality*, Trans. Amer. Math. Soc. **80** (1955), 1–34. doi:10.2307/1993003 (cit. on p. II.62).

[CS16] J. Carette and A. Sabry, *Computing with semirings and weak rig groupoids*, Programming languages and systems, Lecture Notes in Comput. Sci., vol. 9632, Springer, Berlin, 2016, pp. 123–148. doi:10.1007/978-3-662-49498-1_6 (cit. on pp. I.20, I.45, I.46, I.47, I.48, I.49).

[CS02] W. Chachólski and J. Scherer, *Homotopy theory of diagrams*, Mem. Amer. Math. Soc. **155** (2002), no. 736, x+90. doi:10.1090/memo/0736 (cit. on p. III.501).

[CDH∞] C. Comfort, A. Delpeuch, and J. Hedges, *Sheet diagrams for bimonoidal categories*, 2020. arXiv:2010.13361v1 (cit. on pp. xvii, xxiii, I.49, I.131, I.293, I.429, I.430, II.313, II.314, III.507, III.508).

[Cor16] A. S. Corner, *Day convolution for monoidal bicategories*, Ph.D. thesis, University of Sheffield, 2016. (cit. on p. III.156).

[Cra98] S. E. Crans, *Generalized centers of braided and sylleptic monoidal 2-categories*, Adv. Math. **136** (1998), no. 2, 183–223. doi:10.1006/aima.1998.1720 (cit. on pp. I.253, I.428, I.432, II.312, II.316, III.506, III.510).

[Cru09] G. Cruttwell, *Normed Spaces and the Change of Base for Enriched Categories*, Ph.D. thesis, Dalhousie University, 2009. (cit. on pp. xxx, III.55, III.90).

[Cur71] E. B. Curtis, *Simplicial homotopy theory*, Advances in Math. **6** (1971), 107–209 (1971). doi:10.1016/0001-8708(71)90015-6 (cit. on pp. xxxii, III.286).

[Day70] B. Day, *On closed categories of functors*, Reports of the Midwest Category Seminar, IV, Lecture Notes in Mathematics, Vol. 137, Springer, Berlin, 1970, pp. 1–38. doi:10.1007/BFb0060437 (cit. on pp. xxx, I.18, III.139, III.156).

[DK69] B. Day and G. M. Kelly, *Enriched functor categories*, Reports of the Midwest Category Seminar, III, Lecture Notes in Mathematics, Vol. 106, Springer, Berlin, 1969, pp. 178–191. doi:10.1007/BFb0059139 (cit. on pp. xxx, I.18, III.156).

[DS97] B. Day and R. Street, *Monoidal bicategories and Hopf algebroids*, Adv. Math. **129** (1997), no. 1, 99–157. doi:10.1006/aima.1997.1649 (cit. on pp. III.55, III.156).

[Del∞] Z. Dell, *A characterization of braided enriched monoidal categories*, 2021. arXiv:2104.07747 (cit. on pp. I.434, I.435, II.318, II.319, III.512, III.513).

[Del20] A. Delpeuch, *A complete language for faceted dataflow programs*, Proceedings Applied Category Theory 2019 (ACT 2019), University of Oxford, UK, 15-19 July 2019 (J. Baez and B. Coecke, eds.), Electronic Proceedings in Theoretical Computer Science 323, 2020, pp. 1–14. doi:10.4204/EPTCS.323.1 (cit. on p. I.49).

[DP12] K. Došen and Z. Petrić, *Intermutation*, Appl. Categ. Structures **20** (2012), no. 1, 43–95. doi:10.1007/s10485-010-9228-x (cit. on p. II.306).

[Dri87] V. G. Drinfel'd, *Quantum groups*, Proceedings of the International Congress of Mathematicians, Vol. 1, 2 (Berkeley, Calif., 1986), Amer. Math. Soc., Providence, RI, 1987, pp. 798–820. (cit. on p. II.106).

[Dri89] ———, *Almost cocommutative Hopf algebras*, Algebra i Analiz **1** (1989), no. 2, 30–46. (cit. on p. II.106).

[Dug∞] D. Dugger, *A primer on homotopy colimits*, preprint available at https://pages.uoregon.edu/ddugger/hocolim.pdf. (cit. on pp. III.488, III.501).

[Dun97] G. Dunn, *E_n-ring categories*, J. Pure Appl. Algebra **119** (1997), no. 1, 27–45. doi:10.1016/S0022-4049(96)00015-1 (cit. on p. II.307).

[DS95] W. G. Dwyer and J. Spaliński, *Homotopy theories and model categories*, Handbook of algebraic topology, North-Holland, Amsterdam, 1995, pp. 73–126. doi:10.1016/B978-044481779-2/50003-1 (cit. on p. III.287).

[Ehr65] C. Ehresmann, *Catégories et Structures*, Dunod, Paris, 1965. (cit. on p. I.18).

[EK66] S. Eilenberg and G. M. Kelly, *Closed categories*, Proc. Conf. Categorical Algebra (La Jolla, Calif., 1965), Springer, New York, 1966, pp. 421–562. (cit. on pp. III.54, III.90).

[EML45] S. Eilenberg and S. Mac Lane, *General theory of natural equivalences*, Transactions of the American Mathematical Society **58** (1945), no. 2, 231–294. (cit. on pp. I.18, III.65).

[Elg21] J. Elgueta, *The groupoid of finite sets is biinitial in the 2-category of rig categories*, Journal of Pure and Applied Algebra **225** (2021), no. 11, 106738. doi:10.1016/j.jpaa.2021.106738 (cit. on pp. xxiii, I.48, I.203, I.259, I.292, I.293).

[EKMM97] A. D. Elmendorf, I. Kriz, M. A. Mandell, and J. P. May, *Rings, Modules, and Algebras in Stable Homotopy Theory*, Mathematical Surveys and Monographs, vol. 47, American Mathematical Society, Providence, RI, 1997, With an appendix by M. Cole. (cit. on pp. xvii, III.286).

[EM06] A. D. Elmendorf and M. A. Mandell, *Rings, modules, and algebras in infinite loop space theory*, Adv. Math. **205** (2006), no. 1, 163–228. doi:10.1016/j.aim.2005.07.007 (cit. on pp. xvi, xvii, xviii, xix, xx, xxvi, xxviii, xxix, I.49, I.173, I.204, II.229, II.230, II.231, II.232, II.244, II.247, II.257, II.258, II.261, II.307, III.250, III.253, III.314, III.317, III.345, III.346, III.347, III.389, III.408, III.409, III.410, III.413, III.423, III.431, III.440, III.445, III.447).

[EM09] ———, *Permutative categories, multicategories and algebraic K-theory*, Algebr. Geom. Topol. **9** (2009), no. 4, 2391–2441. doi:10.2140/agt.2009.9.2391 (cit. on pp. xvii, xviii, xix, xx, xxvi, xxviii, xxix, I.436, II.229, II.230, II.261, II.320, III.175, III.220, III.253, III.314, III.317, III.345, III.346, III.347, III.350, III.408, III.409, III.410, III.514).

[Eps66] D. B. A. Epstein, *Functors between tensored categories*, Invent. Math. **1** (1966), 221–228. doi:10.1007/BF01452242 (cit. on p. I.18).

[EGNO15] P. Etingof, S. Gelaki, D. Nikshych, and V. Ostrik, *Tensor categories*, Mathematical Surveys and Monographs, vol. 205, American Mathematical Society, Providence, RI, 2015. doi:10.1090/surv/205 (cit. on p. xxx).

[Fie∞b] Z. Fiedorowicz, *Constructions of E_n operads*, 1999. arXiv:math/9808089 (cit. on pp. I.436, II.307, II.320, III.514).

[Fie∞] ———, *The symmetric bar construction*, preprint available at `https://people.math.osu.edu/fiedorowicz.1/`. (cit. on pp. II.230, II.260, III.450, III.460).

[FSV13] Z. Fiedorowicz, M. Stelzer, and R. M. Vogt, *Homotopy colimits of algebras over Cat-operads and iterated loop spaces*, Adv. Math. **248** (2013), 1089–1155. doi:`10.1016/j.aim.2013.07.016` (cit. on pp. II.260, II.307).

[FV03] Z. Fiedorowicz and R. Vogt, *Simplicial n-fold monoidal categories model all loop spaces*, Cah. Topol. Géom. Différ. Catég. **44** (2003), no. 2, 105–148. (cit. on pp. II.260, II.307).

[For04] S. Forcey, *Enrichment over iterated monoidal categories*, Algebr. Geom. Topol. **4** (2004), 95–119. doi:`10.2140/agt.2004.4.95` (cit. on pp. xxx, I.431, II.306, II.315, III.54, III.509).

[FSS07] S. Forcey, J. Siehler, and E. S. Sowers, *Operads in iterated monoidal categories*, J. Homotopy Relat. Struct. **2** (2007), no. 1, 1–43, errata available at `http://www.math.uakron.edu/~sf34/`. (cit. on pp. I.431, II.271, II.306, II.307, II.315, III.509).

[FKLW03] M. H. Freedman, A. Kitaev, M. J. Larsen, and Z. Wang, *Topological quantum computation*, vol. 40, 2003, pp. 31–38. doi:`10.1090/S0273-0979-02-00964-3` (cit. on p. II.106).

[Fre17] B. Fresse, *Homotopy of operads and Grothendieck-Teichmüller groups. Part 1. The algebraic theory and its topological background*, Mathematical Surveys and Monographs, vol. 217, American Mathematical Society, Providence, RI, 2017. (cit. on pp. xxxi, I.436, II.307, II.320, III.434, III.450, III.460, III.461, III.464, III.473, III.514).

[FY92] P. Freyd and D. N. Yetter, *Coherence theorems via knot theory*, J. Pure Appl. Algebra **78** (1992), no. 1, 49–76. doi:`10.1016/0022-4049(92)90018-B` (cit. on pp. I.435, II.319, III.513).

[Fre03] P. J. Freyd, *Abelian categories*, Repr. Theory Appl. Categ. (2003), no. 3, 1–190. (cit. on pp. xxx, II.62, II.63).

[Fri12] G. Friedman, *Survey article: An elementary illustrated introduction to simplicial sets*, Rocky Mountain J. Math. **42** (2012), no. 2, 353–423. doi:`10.1216/RMJ-2012-42-2-353` (cit. on p. III.286).

[FP18] T. Fritz and P. Perrone, *Bimonoidal structure of probability monads*, Electronic Notes in Theoretical Computer Science **341** (2018), 121–149. (cit. on p. I.49).

[Ful97] W. Fulton, *Young Tableaux: With Applications to Representation Theory and Geometry*, London Mathematical Society Student Texts, vol. 35, Cambridge University Press, Cambridge, 1997. (cit. on p. II.307).

[FM94] W. Fulton and R. MacPherson, *A compactification of configuration spaces*, Ann. of Math. (2) **139** (1994), no. 1, 183–225. doi:`10.2307/2946631` (cit. on pp. I.436, II.307, II.320, III.514).

[GZ67] P. Gabriel and M. Zisman, *Calculus of fractions and homotopy theory*, Ergebnisse der Mathematik und ihrer Grenzgebiete, Band 35, Springer-Verlag New York, Inc., New York, 1967. (cit. on pp. xxxii, III.286, III.459, III.473).

[GH06] T. Geisser and L. Hesselholt, *On the K-theory of complete regular local $\mathbb{F}_p$-algebras*, Topology **45** (2006), no. 3, 475–493. doi:`10.1016/j.top.2005.09.002` (cit. on p. III.411).

[GJ∞] E. Getzler and J. D. Jones, *Operads, homotopy algebra and iterated integrals for double loop spaces*, 1994. arXiv:hep-th/9403055 (cit. on pp. I.436, II.307, II.320, III.514).

[GJ09] P. G. Goerss and J. F. Jardine, *Simplicial homotopy theory*, Modern Birkhäuser Classics, Birkhäuser Verlag, Basel, 2009, Reprint of the 1999 edition. doi:`10.1007/978-3-0346-0189-4` (cit. on pp. xxxii, III.260, III.286).

[GPS95] R. Gordon, A. J. Power, and R. Street, *Coherence for tricategories*, Mem. Amer. Math. Soc. **117** (1995), no. 558, vi+81. doi:`10.1090/memo/0558` (cit. on pp. I.429, I.432, II.313, II.316, III.55, III.507, III.510).

[Gra18] M. Grandis, *Category theory and applications: A textbook for beginners*, World Scientific Publishing Co. Pte. Ltd., Hackensack, NJ, 2018. doi:`10.1142/10737` (cit. on pp. xxx, I.18).

[Gra76] D. Grayson, *Higher algebraic K-theory: II (after Daniel Quillen)*, Algebraic K-theory (Proc. Conf., Northwestern Univ., Evanston, Ill., 1976), Springer, Berlin, 1976, pp. 217–240. Lecture Notes in Math., Vol. 551. doi:10.1007/BFb0080003 (cit. on pp. III.253, III.411).

[Gro57] A. Grothendieck, *Sur quelques points d'algèbre homologique*, Tohoku Math. J. (2) **9** (1957), 119–221. doi:10.2748/tmj/1178244839 (cit. on p. II.62).

[Gui10] B. J. Guillou, *Strictification of categories weakly enriched in symmetric monoidal categories*, Theory Appl. Categ. **24** (2010), No. 20, 564–579. (cit. on pp. I.204, I.293).

[GMMO∞] B. J. Guillou, J. P. May, M. Merling, and A. M. Osorno, *Multiplicative equivariant K-theory and the Barratt-Priddy-Quillen theorem.* arXiv:2102.13246 (cit. on p. III.411).

[Gur11] N. Gurski, *Loop spaces, and coherence for monoidal and braided monoidal bicategories*, Adv. Math. **226** (2011), no. 5, 4225–4265. doi:10.1016/j.aim.2010.12.007 (cit. on pp. I.432, II.316, III.510).

[Gur13] ———, *Coherence in three-dimensional category theory*, Cambridge Tracts in Mathematics, vol. 201, Cambridge University Press, Cambridge, 2013. doi:10.1017/CBO9781139542333 (cit. on pp. I.429, I.432, II.313, II.316, III.55, III.507, III.510).

[GJO17a] N. Gurski, N. Johnson, and A. M. Osorno, *Extending homotopy theories across adjunctions*, Homology Homotopy Appl. **19** (2017), no. 2, 89–110. doi:10.4310/HHA.2017.v19.n2.a6 (cit. on pp. III.315, III.350).

[GJO17b] ———, *K-theory for 2-categories*, Adv. Math. **322** (2017), 378–472. doi:10.1016/j.aim.2017.10.011 (cit. on pp. I.209, I.246, I.253, I.254, I.428, I.429, I.432, I.437, II.312, II.313, II.316, II.321, III.315, III.316, III.506, III.507, III.510, III.515).

[HRY15] P. Hackney, M. Robertson, and D. Yau, *Infinity properads and infinity wheeled properads*, Lecture Notes in Mathematics, vol. 2147, Springer, Cham, 2015. doi:10.1007/978-3-319-20547-2 (cit. on p. III.220).

[HS∞] L. Hansen and M. A. Shulman, *Constructing symmetric monoidal bicategories functorially*, 2019. arXiv:1910.09240 (cit. on pp. I.429, II.313, III.507).

[Hat02] A. Hatcher, *Algebraic topology*, Cambridge University Press, Cambridge, 2002. (cit. on p. III.473).

[Hin13] P. Hines, *Quantum speedup and categorical distributivity*, Computation, logic, games, and quantum foundations, Lecture Notes in Comput. Sci., vol. 7860, Springer, Heidelberg, 2013, pp. 122–138. doi:10.1007/978-3-642-38164-5_9 (cit. on pp. xvii, I.48, I.50).

[Hir03] P. S. Hirschhorn, *Model categories and their localizations*, Mathematical Surveys and Monographs, vol. 99, American Mathematical Society, Providence, RI, 2003. (cit. on pp. III.287, III.501).

[Hou07] R. Houston, *Linear logic without units*, Ph.D. thesis, University of Manchester, 2007. arXiv:1305.2231 (cit. on p. III.90).

[Hov99] M. Hovey, *Model categories*, Mathematical Surveys and Monographs, vol. 63, American Mathematical Society, Providence, RI, 1999. (cit. on pp. III.284, III.287).

[Hov01] ———, *Spectra and symmetric spectra in general model categories*, J. Pure Appl. Algebra **165** (2001), no. 1, 63–127. doi:10.1016/S0022-4049(00)00172-9 (cit. on pp. III.265, III.286).

[HSS00] M. Hovey, B. Shipley, and J. Smith, *Symmetric spectra*, J. Amer. Math. Soc. **13** (2000), no. 1, 149–208. doi:10.1090/S0894-0347-99-00320-3 (cit. on pp. xvii, III.265, III.286, III.287).

[JS12] R. P. James and A. Sabry, *Information effects*, ACM SIGPLAN Notices **47** (2012), no. 1, 73–84. (cit. on p. I.46).

[JP07] M. Jibladze and T. Pirashvili, *Third Mac Lane cohomology via categorical rings*, J. Homotopy Relat. Struct. **2** (2007), no. 2, 187–216. (cit. on p. III.448).

[JO12] N. Johnson and A. M. Osorno, *Modeling stable one-types*, Theory Appl. Categ. **26** (2012), No. 20, 520–537. (cit. on pp. I.435, II.319, III.513).

[JY21] N. Johnson and D. Yau, *2-Dimensional Categories*, Oxford University Press, New York, 2021. doi:10.1093/oso/9780198871378.001.0001 (cit. on pp. xxii, xxx, I.11, I.133, I.164, I.204, I.209, I.213, I.216, I.217, I.230, I.239, I.256, I.291, I.325, I.386, I.393, I.396, I.427, I.429, I.432, II.194, II.198, II.311, II.313, II.316, III.3, III.90, III.345, III.505, III.507, III.510).

[JS91a] A. Joyal and R. Street, *The geometry of tensor calculus. I*, Adv. Math. **88** (1991), no. 1, 55–112. doi:10.1016/0001-8708(91)90003-P (cit. on pp. I.49, I.131, I.429, II.313, III.507).

[JS91b] ———, *An introduction to Tannaka duality and quantum groups*, Category Theory (Como, 1990) (A. Carboni, M. C. Pedicchio, and G. Rosolini, eds.), Lecture Notes in Math., vol. 1488, Springer, Berlin, 1991, pp. 413–492. doi:10.1007/BFb0084235 (cit. on pp. I.435, II.319, III.513).

[JS91c] ———, *Tortile Yang-Baxter operators in tensor categories*, J. Pure Appl. Algebra **71** (1991), no. 1, 43–51. doi:10.1016/0022-4049(91)90039-5 (cit. on p. II.34).

[JS93] ———, *Braided tensor categories*, Adv. Math. **102** (1993), no. 1, 20–78. doi:10.1006/aima.1993.1055 (cit. on pp. I.18, I.435, II.17, II.19, II.34, II.156, II.319, III.513).

[JT91] A. Joyal and M. Tierney, *Strong stacks and classifying spaces*, Category Theory (Como, 1990) (A. Carboni, M. C. Pedicchio, and G. Rosolini, eds.), Lecture Notes in Math., vol. 1488, Springer, Berlin, 1991, pp. 213–236. doi:10.1007/BFb0084235 (cit. on p. III.284).

[KGB+19] A. R. Kalra, N. Gupta, B. K. Behera, S. Prakash, and P. K. Panigrahi, *Demonstration of the no-hiding theorem on the 5-qubit IBM quantum computer in a category-theoretic framework*, Quantum Inf. Process. **18** (2019), no. 6, Paper No. 170, 13. doi:10.1007/s11128-019-2288-4 (cit. on p. I.46).

[KV94] M. Kapranov and V. Voevodsky, *2-categories and Zamolodchikov tetrahedra equations*, Algebraic groups and their generalizations: quantum and infinite-dimensional methods (University Park, PA, 1991), Proc. Sympos. Pure Math., vol. 56, Amer. Math. Soc., Providence, RI, 1994, pp. 177–259. (cit. on pp. xvi, I.44, I.325).

[Kas95] C. Kassel, *Quantum groups*, Graduate Texts in Mathematics, vol. 155, Springer-Verlag, New York, 1995. doi:10.1007/978-1-4612-0783-2 (cit. on pp. xvii, II.34, II.106).

[KT08] C. Kassel and V. Turaev, *Braid groups*, Graduate Texts in Mathematics, vol. 247, Springer, New York, 2008, With the graphical assistance of Olivier Dodane. doi:10.1007/978-0-387-68548-9 (cit. on p. II.34).

[Kel55] J. L. Kelley, *General topology*, D. Van Nostrand Company, Inc., Toronto-New York-London, 1955. (cit. on pp. III.286, III.473).

[Kel64] G. M. Kelly, *On MacLane's conditions for coherence of natural associativities, commutativities, etc.*, J. Algebra **1** (1964), 397–402. (cit. on p. I.11).

[Kel69] ———, *Adjunction for enriched categories*, Reports of the Midwest Category Seminar, III, Lecture Notes in Mathematics, Vol. 106, Springer, Berlin, 1969, pp. 166–177. (cit. on p. III.156).

[Kel74] ———, *Coherence theorems for lax algebras and for distributive laws*, Category Seminar (Proc. Sem., Sydney, 1972/1973), Lecture Notes in Math., Vol. 420, 1974, pp. 281–375. (cit. on pp. I.133, I.169, II.34).

[Kel05] ———, *Basic concepts of enriched category theory*, Repr. Theory Appl. Categ. (2005), no. 10, vi+137, Reprint of the 1982 original (Cambridge Univ. Press, Cambridge). (cit. on pp. xxx, I.213, III.54, III.155, III.156, III.286).

[KML71a] G. M. Kelly and S. Mac Lane, *Coherence in closed categories*, J. Pure Appl. Algebra **1** (1971), no. 1, 97–140. doi:10.1016/0022-4049(71)90013-2 (cit. on p. III.155).

[KML71b] ———, *Erratum: "Coherence in closed categories"*, J. Pure Appl. Algebra **1** (1971), no. 2, 219. doi:10.1016/0022-4049(71)90019-3 (cit. on p. III.155).

[KML72] ———, *Closed coherence for a natural transformation*, Coherence in categories, Lecture Notes in Math., Vol. 281, 1972, pp. 1–28. doi:10.1007/BFb0059553 (cit. on p. III.155).

[KR68] C. G. Khatri and C. R. Rao, *Solutions to some functional equations and their applications to characterization of probability distributions*, Sankhya Ser. A **30** (1968), 167–180. (cit. on p. I.406).

[KYZZ∞] L. Kong, W. Yuan, Z.-H. Zhang, and H. Zheng, *Enriched monoidal categories I: centers*, 2021. arXiv:2104.03121 (cit. on pp. I.434, II.318, III.512).

[KZ18] L. Kong and H. Zheng, *Drinfeld center of enriched monoidal categories*, Adv. Math. **323** (2018), 411–426. doi:10.1016/j.aim.2017.10.038 (cit. on pp. I.434, II.318, III.55, III.90, III.512).

[Lac02] S. Lack, *Codescent objects and coherence*, J. Pure Appl. Algebra **175** (2002), no. 1-3, 223–241, Special volume celebrating the 70th birthday of Professor Max Kelly. doi:10.1016/S0022-4049(02)00136-6 (cit. on p. III.90).

[Lac10] ———, *A 2-categories companion*, Towards higher categories (J. Baez and J. May, eds.), IMA Vol. Math. Appl., vol. 152, Springer, New York, 2010, pp. 105–191. doi:10.1007/978-1-4419-1524-5_4 (cit. on p. I.209).

[LP17] V. T. Lahtinen and J. K. Pachos, *A short introduction to topological quantum computation*, SciPost Physics **3** (2017), 021. doi:10.21468/SciPostPhys.3.3.021 (cit. on p. II.106).

[Lam69] J. Lambek, *Deductive systems and categories II. Standard constructions and closed categories*, Category Theory, Homology Theory and their Applications, I (Battelle Institute Conference, Seattle, Wash., 1968, Vol. One), Springer, Berlin, 1969, pp. 76–122. (cit. on pp. III.219, III.220).

[Lap72a] M. L. Laplaza, *Coherence for distributivity*, Coherence in categories, Lectures Notes in Mathematics, Vol. 281, Springer-Verlag, Berlin-Heidelberg, 1972, pp. 29–65. (cit. on pp. xvi, xix, xxi, I.48, I.51, I.131, I.132, I.133, I.167).

[Lap72b] ———, *A new result of coherence for distributivity*, Coherence in categories, Lectures Notes in Mathematics, Vol. 281, Springer-Verlag, Berlin-Heidelberg, 1972, pp. 214–235. (cit. on pp. xvi, xix, xxii, I.48, I.135, I.167, I.168).

[Law18] T. Lawson, *Secondary power operations and the Brown-Peterson spectrum at the prime 2*, Ann. of Math. (2) **188** (2018), no. 2, 513–576. doi:10.4007/annals.2018.188.2.3 (cit. on pp. I.435, II.319, III.513).

[Lei∞] T. Leinster, *Basic bicategories*, 1998. arXiv:math/9810017 (cit. on p. I.209).

[Lei14] ———, *Basic category theory*, Cambridge Studies in Advanced Mathematics, vol. 143, Cambridge University Press, Cambridge, 2014. doi:10.1017/CBO9781107360068 (cit. on pp. xxx, I.18).

[Lew72] G. Lewis, *Coherence for a closed functor*, Coherence in categories, Lecture Notes in Math., Vol. 281, 1972, pp. 148–195. (cit. on pp. I.430, II.314, III.508).

[Lim59] E. L. Lima, *The Spanier-Whitehead duality in new homotopy categories*, Summa Brasil. Math. **4** (1959), 91–148 (1959). (cit. on p. III.287).

[Liu99] S. Liu, *Matrix results on the Khatri-Rao and Tracy-Singh products*, vol. 289, 1999, Linear algebra and statistics (Istanbul, 1997), pp. 267–277. doi:10.1016/S0024-3795(98)10209-4 (cit. on p. I.406).

[Lor21] F. Loregian, *(Co)end calculus*, London Mathematical Society Lecture Note Series, Cambridge University Press, 2021. (cit. on pp. xxx, I.18, III.156).

[ML63] S. Mac Lane, *Natural associativity and commutativity*, Rice Univ. Stud. **49** (1963), no. 4, 28–46. (cit. on pp. I.18, I.51).

[ML98] ———, *Categories for the working mathematician*, second ed., Graduate Texts in Mathematics, vol. 5, Springer-Verlag, New York, 1998. (cit. on pp. xxx, I.10, I.18, I.51, II.62, II.158, III.86, III.192, III.219).

[Maj91] S. Majid, *Representations, duals and quantum doubles of monoidal categories*, Proceedings of the Winter School on Geometry and Physics (Srní, 1990), no. 26, 1991, pp. 197–206. (cit. on pp. xvii, II.34).

[Maj93] ———, *Anyonic quantum groups*, Spinors, twistors, Clifford algebras and quantum deformations (Sobótka Castle, 1992), Fund. Theories Phys., vol. 52, Kluwer Acad. Publ., Dordrecht, 1993, pp. 327–336. (cit. on p. II.106).

[Maj95] ———, *Foundations of quantum group theory*, Cambridge University Press, Cambridge, 1995. doi:10.1017/CBO9780511613104 (cit. on p. II.106).

[MMSS01] M. A. Mandell, J. P. May, S. Schwede, and B. Shipley, *Model categories of diagram spectra*, Proc. London Math. Soc. (3) **82** (2001), no. 2, 441–512. doi:10.1112/S0024611501012692 (cit. on pp. xvii, III.286).

[Man10] M. A. Mandell, *An inverse K-theory functor*, Doc. Math. **15** (2010), 765–791. (cit. on pp. I.437, II.321, III.314, III.315, III.316, III.350, III.515).

[Man12] O. Manzyuk, *Closed categories vs. closed multicategories*, Theory Appl. Categ. **26** (2012), 132–175. (cit. on p. III.220).

[MSS02] M. Markl, S. Shnider, and J. Stasheff, *Operads in algebra, topology and physics*, Mathematical Surveys and Monographs, vol. 96, American Mathematical Society, Providence, RI, 2002. (cit. on p. xxxi).

[May72] J. P. May, *The geometry of iterated loop spaces*, Lectures Notes in Mathematics, Vol. 271, Springer-Verlag, Berlin-New York, 1972. (cit. on pp. xxxi, II.230, II.260, II.307, III.219, III.286, III.457, III.473).

[May74] ———, *E_∞ spaces, group completions, and permutative categories*, New developments in topology (Proc. Sympos. Algebraic Topology, Oxford, 1972), London Math. Soc. Lecture Note Ser., No. 11, 1974, pp. 61–93. (cit. on p. II.260).

[May77] ———, *E_∞ ring spaces and E_∞ ring spectra*, Lecture Notes in Mathematics, Vol. 577, Springer-Verlag, Berlin-New York, 1977, With contributions by Frank Quinn, Nigel Ray, and Jørgen Tornehave. (cit. on pp. xvi, I.48, I.49, I.171, I.173, I.203, I.204, III.411).

[May78] ———, *The spectra associated to permutative categories*, Topology **17** (1978), no. 3, 225–228. doi:10.1016/0040-9383(78)90027-7 (cit. on p. III.314).

[May82] ———, *Multiplicative infinite loop space theory*, J. Pure Appl. Algebra **26** (1982), no. 1, 1–69. doi:10.1016/0022-4049(82)90029-9 (cit. on p. III.411).

[May92] ———, *Simplicial objects in algebraic topology*, Chicago Lectures in Mathematics, University of Chicago Press, Chicago, IL, 1992, Reprint of the 1967 original. (cit. on pp. xxxii, III.286).

[May99] ———, *A concise course in algebraic topology*, Chicago Lectures in Mathematics, University of Chicago press, Chicago, IL, 1999. (cit. on pp. xxxii, III.286, III.287, III.473).

[May09a] ———, *The construction of E_∞ ring spaces from bipermutative categories*, New topological contexts for Galois theory and algebraic geometry (BIRS 2008), Geom. Topol. Monogr., vol. 16, Geom. Topol. Publ., Coventry, 2009, pp. 283–330. doi:10.2140/gtm.2009.16.283 (cit. on p. III.411).

[May09b] ———, *What are E_∞ ring spaces good for?*, New topological contexts for Galois theory and algebraic geometry (BIRS 2008), Geom. Topol. Monogr., vol. 16, Geom. Topol. Publ., Coventry, 2009, pp. 331–365. doi:10.2140/gtm.2009.16.331 (cit. on p. III.411).

[May09c] ———, *What precisely are E_∞ ring spaces and E_∞ ring spectra?*, New topological contexts for Galois theory and algebraic geometry (BIRS 2008), Geom. Topol. Monogr., vol. 16, Geom. Topol. Publ., Coventry, 2009, pp. 215–282. doi:10.2140/gtm.2009.16.215 (cit. on p. III.411).

[MP12] J. P. May and K. Ponto, *More concise algebraic topology: Localization, completion, and model categories*, Chicago Lectures in Mathematics, University of Chicago Press, Chicago, IL, 2012. (cit. on pp. xxxii, III.277, III.279, III.280, III.281, III.282, III.285, III.286, III.287, III.488, III.501).

[MS06] J. P. May and J. Sigurdsson, *Parametrized homotopy theory*, Mathematical Surveys and Monographs, vol. 132, American Mathematical Society, Providence, RI, 2006. doi:10.1090/surv/132 (cit. on p. III.286).

[MS76] D. McDuff and G. Segal, *Homology fibrations and the "group-completion" theorem*, Invent. Math. **31** (1975/76), no. 3, 279–284. doi:10.1007/BF01403148 (cit. on p. II.260).

[Mil66] R. J. Milgram, *Iterated loop spaces*, Ann. of Math. (2) **84** (1966), 386–403. doi:10.2307/1970453 (cit. on pp. II.307, III.502).

[Mil20] H. Miller, *Handbook of homotopy theory*, CRC Press/Chapman and Hall Handbooks in Mathematics Series, CRC Press, Boca Raton, FL, 2020. (cit. on p. xxxii).

[Mil71] J. Milnor, *Introduction to algebraic K-theory*, Princeton University Press, Princeton, N.J.; University of Tokyo Press, Tokyo, 1971, Annals of Mathematics Studies, No. 72. (cit. on p. xxxii).

[MS74] J. W. Milnor and J. D. Stasheff, *Characteristic classes*, Princeton University Press, Princeton, N. J.; University of Tokyo Press, Tokyo, 1974, Annals of Mathematics Studies, No. 76. (cit. on p. III.432).

[Mit64] B. Mitchell, *The full imbedding theorem*, Amer. J. Math. **86** (1964), 619–637. doi:10.2307/2373027 (cit. on p. II.63).

[Mit65] ———, *Theory of Categories*, Pure and Applied Mathematics, Vol. XVII, Academic Press, New York-London, 1965. (cit. on pp. xxx, I.18, II.62, II.63).

[MT10] I. Moerdijk and B. Toën, *Simplicial methods for operads and algebraic geometry*, Advanced Courses in Mathematics. CRM Barcelona, Birkhäuser/Springer Basel AG, Basel, 2010, Edited by Carles Casacuberta and Joachim Kock. doi:10.1007/978-3-0348-0052-5 (cit. on pp. III.219, III.220).

[Mon93] S. Montgomery, *Hopf algebras and their actions on rings*, CBMS Regional Conference Series in Mathematics, vol. 82, Published for the Conference Board of the Mathematical Sciences, Washington, DC; by the American Mathematical Society, Providence, RI, 1993. doi:10.1090/cbms/082 (cit. on p. II.106).

[MP19] S. Morrison and D. Penneys, *Monoidal categories enriched in braided monoidal categories*, Int. Math. Res. Not. IMRN (2019), no. 11, 3527–3579. doi:10.1093/imrn/rnx217 (cit. on pp. I.434, II.318, III.55, III.512).

[Mun00] J. R. Munkres, *Topology*, Prentice Hall, Inc., Upper Saddle River, NJ, 2000. (cit. on p. III.473).

[MT07] F. Muro and A. Tonks, *The 1-type of a Waldhausen K-theory spectrum*, Adv. Math. **216** (2007), no. 1, 178–211. doi:10.1016/j.aim.2007.05.008 (cit. on p. III.448).

[NSS+08] C. Nayak, S. H. Simon, A. Stern, M. Freedman, and S. Das Sarma, *Non-abelian anyons and topological quantum computation*, Rev. Modern Phys. **80** (2008), no. 3, 1083–1159. doi:10.1103/RevModPhys.80.1083 (cit. on p. II.106).

[Pac12] J. K. Pachos, *Introduction to topological quantum computation*, Cambridge University Press, Cambridge, 2012. doi:10.1017/CBO9780511792908 (cit. on pp. II.106, II.107).

[PP11] P. Panangaden and E. O. Paquette, *A categorical presentation of quantum computation with anyons*, New structures for physics, Lecture Notes in Phys., vol. 813, Springer, Heidelberg, 2011, pp. 983–1025. doi:10.1007/978-3-642-12821-9_15 (cit. on p. II.106).

[Pin14] C. C. Pinter, *A book of set theory*, Dover, New York NY, 2014. (cit. on p. III.282).

[Pro13] T. U. F. Program, *Homotopy type theory—univalent foundations of mathematics*, The Univalent Foundations Program, Princeton, NJ; Institute for Advanced Study (IAS), Princeton, NJ, 2013. (cit. on p. I.46).

[Qui67] D. G. Quillen, *Homotopical algebra*, Lecture Notes in Mathematics, No. 43, Springer-Verlag, Berlin-New York, 1967. (cit. on p. III.287).

[Qui73] ———, *Higher algebraic K-theory I: Higher K-theories*, Proc. Conf., Battelle Memorial Inst., Seattle, Wash., (1972), Lecture Notes in Math., Vol. 341, 1973, pp. 85–147. (cit. on pp. xxxii, III.253, III.411).

[Rad93] D. E. Radford, *Minimal quasitriangular Hopf algebras*, J. Algebra **157** (1993), no. 2, 285–315. doi:10.1006/jabr.1993.1102 (cit. on p. II.106).

[Rez∞] C. Rezk, *A model category for categories*, preprint available at http://www.math.uiuc.edu/~rezk. (cit. on p. III.284).

[Ric10] B. Richter, *An involution on the K-theory of bimonoidal categories with anti-involution*, Algebraic & Geometric Topology **10** (2010), no. 1, 315–342. (cit. on pp. xvi, xx, II.43, II.134, II.231, II.258).

[Ric20] ———, *From categories to homotopy theory*, Cambridge Studies in Advanced Mathematics, vol. 188, Cambridge University Press, Cambridge, 2020. (cit. on pp. xxxii, III.156, III.265, III.286).

[Rie11] E. Riehl, *Algebraic model structures*, New York J. Math. **17** (2011), 173–231. (cit. on p. III.287).

[Rie14] ———, *Categorical homotopy theory*, New Mathematical Monographs, vol. 24, Cambridge University Press, Cambridge, 2014. doi:10.1017/CBO9781107261457 (cit. on pp. xxxii, III.156, III.286, III.501).

[Rie16] ———, *Category theory in context*, Courier Dover Publications, 2016. (cit. on pp. xxx, I.18, III.194, III.219, III.220, III.286).

[Rom17] S. Roman, *An introduction to the language of category theory*, Compact Textbooks in Mathematics, Birkhäuser/Springer, Cham, 2017. doi:10.1007/978-3-319-41917-6 (cit. on pp. xxx, I.18).

[Ros95] J. Rosenberg, *Algebraic K-theory and its applications*, Graduate Texts in Mathematics, vol. 147, Springer Science & Business Media, 1995. (cit. on p. xxxii).

[Sal01] P. Salvatore, *Configuration spaces with summable labels*, Cohomological methods in homotopy theory (Bellaterra, 1998), Progr. Math., vol. 196, Birkhäuser, Basel, 2001, pp. 375–395. (cit. on pp. I.436, II.307, II.320, III.514).

[SPK11] J. R. Samal, A. K. Pati, and A. Kumar, *Experimental test of the quantum no-hiding theorem*, Phys. Rev. Lett. **106** (2011), 080401. doi:10.1103/PhysRevLett.106.080401 (cit. on p. I.46).

[SP∞] C. Schommer-Pries, *The classification of two-dimensional extended topological field theories*, 2011. arXiv:1112.1000 (cit. on pp. I.253, I.254, I.427, II.311, III.505).

[Sch03] B. S. W. Schröder, *Ordered Sets: An Introduction*, Birkhäuser Boston, Inc., Boston, MA, 2003. doi:10.1007/978-1-4612-0053-6 (cit. on p. II.307).

[Sch72] H. Schubert, *Categories*, Springer-Verlag, New York-Heidelberg, 1972, Translated from German by Eva Gray. (cit. on pp. xxx, I.18, II.62, II.63).

[Seg74] G. Segal, *Categories and cohomology theories*, Topology **13** (1974), 293–312. doi:10.1016/0040-9383(74)90022-6 (cit. on pp. xvii, xxviii, II.260, III.289, III.292, III.314, III.315).

[Sel11] P. Selinger, *A survey of graphical languages for monoidal categories*, New structures for physics, Lecture Notes in Phys., vol. 813, Springer, Heidelberg, 2011, pp. 289–355. doi:10.1007/978-3-642-12821-9_4 (cit. on pp. I.49, I.429, II.313, III.507).

[Sen∞] A. Senger, *The Brown-Peterson spectrum is not $E_{2(p^2+2)}$ at odd primes*, 2017. arXiv:1710.09822 (cit. on pp. I.435, II.319, III.513).

[SV02] A. Shen and N. K. Vereshchagin, *Basic set theory*, Student Mathematical Library, vol. 17, American Mathematical Society, Providence, RI, 2002, Translated from the 1999 Russian edition by Shen. doi:10.1090/stml/017 (cit. on p. III.282).

[Shu∞a] M. A. Shulman, *Homotopy limits and colimits and enriched homotopy theory*, 2009. arXiv:math/0610194 (cit. on pp. III.156, III.501).

[Shu∞b] ———, *Set theory for category theory*, 2008. arXiv:0810.1279 (cit. on pp. III.65, III.90).

[Sim11] H. Simmons, *An introduction to category theory*, Cambridge University Press, Cambridge, 2011. doi:10.1017/CBO9780511863226 (cit. on pp. xxx, I.18).

[Sly99] V. I. Slyusar, *A family of face products of matrices and its properties*, Kibernet. Sistem. Anal. (1999), no. 3, 43–49, 189. doi:10.1007/BF02733426 (cit. on p. I.406).

[Smi89] J. H. Smith, *Simplicial group models for $\Omega^n S^n X$*, Israel J. Math. **66** (1989), no. 1-3, 330–350. doi:10.1007/BF02765902 (cit. on pp. I.436, II.307, II.320, III.447, III.501, III.514).

[Sta15] S. Staton, *Algebraic effects, linearity, and quantum programming languages*, ACM SIGPLAN Notices **50** (2015), no. 1, 395–406. (cit. on p. I.49).

[Ste67] N. E. Steenrod, *A convenient category of topological spaces*, Michigan Math. J. **14** (1967), 133–152. (cit. on pp. III.286, III.473).

[Ste79] R. Steiner, *A canonical operad pair*, Math. Proc. Cambridge Philos. Soc. **86** (1979), no. 3, 443–449. doi:10.1017/S0305004100056292 (cit. on p. III.473).

[Swe69] M. E. Sweedler, *Hopf algebras*, Mathematics Lecture Note Series, W. A. Benjamin, Inc., New York, 1969. (cit. on p. II.106).

[Tho95] R. W. Thomason, *Symmetric monoidal categories model all connective spectra*, Theory Appl. Categ. **1** (1995), No. 5, 78–118. (cit. on pp. I.437, II.321, III.316, III.515).

[TS72] D. S. Tracy and R. P. Singh, *A new matrix product and its applications in partitioned matrix differentiation*, Statistica Neerlandica **26** (1972), 143–157. doi:10.1111/j.1467-9574.1972.tb00199.x (cit. on p. I.406).

[TTWL08] S. Trebst, M. Troyer, Z. Wang, and A. W. W. Ludwig, *A short introduction to fibonacci anyon models*, Progress of Theoretical Physics Supplement **176** (2008), 384–407. (cit. on p. II.106).

[Vog73] R. M. Vogt, *Homotopy limits and colimits*, Math. Z. **134** (1973), 11–52. doi:10.1007/BF01219090 (cit. on p. III.501).

[Vog77] ———, *Commuting homotopy limits*, Math. Z. **153** (1977), no. 1, 59–82. doi:10.1007/BF01214734 (cit. on p. III.501).

[Wal85] F. Waldhausen, *Algebraic K-theory of spaces*, Algebraic and geometric topology (New Brunswick, N.J., 1983), Lecture Notes in Math., vol. 1126, Springer, Berlin, 1985, pp. 318–419. doi:10.1007/BFb0074449 (cit. on pp. xxxii, III.411).

[Wan10] Z. Wang, *Topological quantum computation*, CBMS Regional Conference Series in Mathematics, vol. 112, Published for the Conference Board of the Mathematical Sciences, Washington, DC; by the American Mathematical Society, Providence, RI, 2010. doi:10.1090/cbms/112 (cit. on p. II.106).

[Wei13] C. A. Weibel, *The K-book: An introduction to algebraic K-theory*, Graduate Studies in Mathematics, vol. 145, American Mathematical Society Providence, RI, 2013. (cit. on pp. xxxii, III.411).

[WY18] D. White and D. Yau, *Bousfield localization and algebras over colored operads*, Appl. Categ. Structures **26** (2018), no. 1, 153–203. doi:10.1007/s10485-017-9489-8 (cit. on pp. III.219, III.287).

[WY19] ———, *Homotopical adjoint lifting theorem*, Appl. Categ. Structures **27** (2019), no. 4, 385–426. doi:10.1007/s10485-019-09560-2 (cit. on pp. I.438, II.322, III.351, III.516).

[WY20] ———, *Right Bousfield localization and operadic algebras*, Tbilisi Mathematical Journal, Special Issue (HomotopyTheorySpectra - 2020) (2020), 71–118. (cit. on p. III.287).

[Yau∞] D. Yau, *Infinity operads and monoidal categories with group equivariance*, 2019. arXiv:1903.03839 (cit. on pp. II.34, II.156, III.447, III.473, III.474).

[Yau16] ———, *Colored operads*, Graduate Studies in Mathematics, vol. 170, American Mathematical Society, Providence, RI, 2016. (cit. on pp. xxxi, III.197, III.219, III.447).

[Yau18] ———, *Operads of wiring diagrams*, Lecture Notes in Mathematics, vol. 2192, Springer, Cham, 2018. doi:10.1007/978-3-319-95001-3 (cit. on p. III.219).

[Yau20] ———, *Homotopical quantum field theory*, World Scientific, Singapore, 2020. (cit. on pp. I.435, I.436, II.319, II.320, III.219, III.250, III.447, III.513, III.514).

[YJ15] D. Yau and M. W. Johnson, *A foundation for PROPs, algebras, and modules*, Mathematical Surveys and Monographs, vol. 203, American Mathematical Society, Providence, RI, 2015. doi:10.1090/surv/203 (cit. on p. III.250).

[Yon60] N. Yoneda, *On Ext and exact sequences*, J. Fac. Sci. Univ. Tokyo Sect. I **8** (1960), 507–576 (1960). (cit. on p. I.9).

[Zak18] I. Zakharevich, *The category of Waldhausen categories is a closed multicategory*, New directions in homotopy theory, Contemp. Math., vol. 707, Amer. Math. Soc., Providence, RI, 2018, pp. 175–194. doi:10.1090/conm/707/14259 (cit. on pp. III.220, III.411).

List of Main Facts

Part I.1. Symmetric Bimonoidal Categories

Chapter I.1. Basic Category Theory

(1.1.11) An adjunction satisfies the triangle identities.
(p. I.7) A functor is an equivalence if and only if it is fully faithful and essentially surjective.
(1.1.14) Left adjoints preserve colimits. Right adjoints preserve limits.
(1.2.1) A monoidal category satisfies the unity axiom and the pentagon axiom.
(1.2.7) A monoidal category satisfies the left and the right unity properties.
(1.3.3) **Mac Lane's Coherence Theorem**. Any two words of the same length in a monoidal category are connected by a unique canonical map.
(1.3.5) **Mac Lane's Strictification Theorem**. Each monoidal category is adjoint equivalent to a strict monoidal category via strong monoidal functors.
(1.3.8) **Symmetric Coherence Theorem**. Any two permuted words of the same length in a symmetric monoidal category are connected by a unique permuted canonical map.
(1.3.10) **Symmetric Strictification Theorem**. Each symmetric monoidal category is adjoint equivalent to a permutative category via strong symmetric monoidal functors.
(1.3.12) **Epstein's Coherence Theorem**. For each (symmetric) monoidal functor $F\colon \mathsf{C} \longrightarrow \mathsf{D}$ and F-iterates $G, H\colon \mathsf{C}^n \longrightarrow \mathsf{D}$, there exists at most one F-coherent map $G \longrightarrow H$.

Chapter I.2. Symmetric Bimonoidal Categories

(2.1.2) A symmetric bimonoidal category has two symmetric monoidal structures, left/right multiplicative zero natural isomorphisms, and left/right distributivity natural monomorphisms, and satisfies 24 axioms.
(2.1.2) A bimonoidal category is defined in the same way as a symmetric bimonoidal category, but without the multiplicative symmetry $\xi^{\otimes}$ and the two axioms that involve $\xi^{\otimes}$. So a bimonoidal category is defined by the other 22 axioms.
(2.1.32) There is a tight symmetric bimonoidal category $\mathsf{Vect}^{\mathbb{C}}$ of finite dimensional complex vector spaces.
(2.2.13) Half of the 24 axioms in a symmetric bimonoidal category are formal consequences of the others.
(2.2.14) One axiom is redundant in a bimonoidal category, which is, therefore, determined by 21 axioms.
(2.3.2) Each distributive symmetric monoidal category yields a tight symmetric bimonoidal category, whose sum is the coproduct.

(2.3.3–2.3.5) Symmetric monoidal closed categories with finite coproducts, the category of modules over a commutative ring, and distributive categories are examples of tight symmetric bimonoidal categories.
(2.4.8) The category Σ of finite ordinals and permutations is a tight symmetric bimonoidal category.
(2.4.23) The variant Σ' of Σ is a tight symmetric bimonoidal category.
(2.5.7) Each right bipermutative category is a tight symmetric bimonoidal category.
(2.5.8–2.5.9) Σ' and $\mathsf{Vect}_c^{\mathbb{C}}$ are right bipermutative categories.
(2.5.16) Each left bipermutative category is a tight symmetric bimonoidal category.
(2.5.17) Σ is a left bipermutative category.
(2.6.2) There is a symmetric bimonoidal groupoid Π with syntax of finite types as objects and Π-terms and Π-combinators as morphisms.

Chapter I.3. Coherence of Symmetric Bimonoidal Categories

(3.1.6) In the elementary graph $\mathsf{Gr}^{\mathsf{el}}(X)$, δ^l and δ^r do not have formal inverses.
(3.1.8) Each prime edge involves at most one nonidentity elementary edge.
(3.1.9) The graph $\mathsf{Gr}(X)$ consists of the vertex set X^{fr} and prime edges.
(3.1.14) Each functor $\varphi: X \longrightarrow \mathsf{Ob}(\mathsf{C})$ extends additively and multiplicatively to a graph morphism $\varphi: \mathsf{Gr}(X) \longrightarrow \mathsf{C}$.
(3.1.18) The value in C of a path in $\mathsf{Gr}(X)$ is the composite of the images of its constituent prime edges under φ.
(3.1.25) An element in X^{fr} is regular if it has the same support as a formal polynomial whose monomials are distinct in the strict $\{\oplus, \otimes\}$-algebra X^{st}, and whose factors in each monomial are distinct elements in X.
(3.1.29) Any two elements in X^{fr} connected by a path in $\mathsf{Gr}(X)$ have the same support, and one of them is regular if and only if the other one is regular.
(3.2.15) For an element in X^{fr}, the size is equal to the rank if and only if it is a sum, with each summand either in X or a product of two elements in X.
(3.3.6) Each element in X^{fr} has a 0^X-reduction.
(3.3.11–3.3.12) Any two 0^X-reductions of an element in X^{fr} have the same codomain and the same value in a symmetric bimonoidal category.
(3.5.32) Each path in $\mathsf{Gr}(X)$ has a 0^X-reduction.
(3.5.33) Any two parallel paths in $\mathsf{Gr}(X)$ whose domain has the same support as 0^X have the same value in a symmetric bimonoidal category.
(3.6.5) An element in X^{fr} is a polynomial if and only if it is δ-reduced.
(3.6.9) Each element in X^{fr} has a δ-reduction.
(3.7.19) Each path in $\mathsf{Gr}(X)$ that does not contain 0^X has a $(0^X, \delta)$-reduction.
(3.8.5–3.8.7) Each element in X^{fr} has a 1^X-reduction. If the original element is δ-reduced, then all of its 1^X-reductions have the same codomain and the same value in a symmetric bimonoidal category.
(3.8.14) Each $(0^X, \delta)$-free path in $\mathsf{Gr}(X)$ whose (co)domain is δ-reduced has a 1^X-reduction.
(3.9.1) **Laplaza's First Coherence Theorem**. In each symmetric bimonoidal category C satisfying a monomorphism assumption, any two parallel paths in $\mathsf{Gr}(X)$ with a regular domain have the same value in C.
(3.9.9–3.9.10) Theorem I.3.9.1 applies to symmetric bimonoidal categories that are flat, in particular, tight.

(3.10.7) **Bimonoidal Coherence Theorem.** In each bimonoidal category C satisfying a monomorphism assumption, any two parallel paths in $\mathsf{Gr}^{\mathsf{ns}}(X)$ with a nonsymmetric regular domain have the same value in C.
(3.10.8) Theorem I.3.10.7 applies to flat, in particular, tight, bimonoidal categories.

Chapter I.4. Coherence of Symmetric Bimonoidal Categories II

(4.2.1) In the distortion category, each object is a finite sequence of nonnegative integers, and each morphism is a finite sequence of permutations.
(4.2.5) The distortion category is a groupoid.
(4.2.12) The additive structure of the distortion category is a permutative category.
(4.2.19) The multiplicative structure of the distortion category is a permutative category.
(4.2.29) The distortion category is a left bipermutative category.
(4.3.1) For a path in $\mathsf{Gr}(X)$, its distortion is defined as its value in the distortion category.
(4.4.3) **Laplaza's Second Coherence Theorem.** In each symmetric bimonoidal category C satisfying a monomorphism assumption, any two parallel paths in $\mathsf{Gr}(X)$ with the same distortion have the same value in C.
(4.4.5) Theorem I.4.4.3 applies to symmetric bimonoidal categories that are flat, in particular, tight.
(4.5.2) In the additive distortion category, each object is a finite sequence of nonnegative integers, and each morphism is a permutation.
(4.5.6) The additive distortion category is a groupoid and a tight bimonoidal category. It faithfully embeds into the distortion category.
(4.5.7) For a path in $\mathsf{Gr}^{\mathsf{ns}}(X)$, its additive distortion is defined as its value in the additive distortion category.
(4.5.8) **Bimonoidal Coherence Theorem II.** In each bimonoidal category C satisfying a monomorphism assumption, any two parallel paths in $\mathsf{Gr}^{\mathsf{ns}}(X)$ with the same additive distortion have the same value in C.
(4.5.9) Theorem I.4.5.8 applies to flat, in particular, tight, bimonoidal categories.
(4.6.5) The distortion category $\mathcal{D}$ is isomorphic to the Grothendieck construction $\int_{\Sigma} F$.
(4.6.7) The additive distortion category $\mathcal{D}^{\mathrm{ad}}$ is isomorphic to the Grothendieck construction $\int_{\Sigma} F^{\mathrm{ad}}$.

Chapter I.5. Strictification of Tight Symmetric Bimonoidal Categories

(5.1.1) A symmetric bimonoidal functor is a functor equipped with two symmetric monoidal functor structures, and satisfies two axioms.
(5.1.10) There is a 1-category $\mathsf{Bi}^{\mathsf{sy}}$ of small symmetric bimonoidal categories and symmetric bimonoidal functors.
(5.1.11) Each symmetric monoidal functor between distributive symmetric monoidal categories induces a symmetric bimonoidal functor.
(5.1.15–5.1.16) Σ and Σ' are isomorphic via symmetric bimonoidal functors.
(5.3.9) Each tight symmetric bimonoidal category C has an associated right bipermutative category A, whose objects are formal polynomials in the objects in C.
(5.4.6–5.4.7) **Bipermutative Strictification Theorems.** Each tight symmetric bimonoidal category is adjoint equivalent to a right/left bipermutative category via symmetric bimonoidal functors.

(5.5.1) A bimonoidal functor is a functor equipped with an additive symmetric monoidal functor structure and a multiplicative monoidal functor structure that satisfies four axioms.
(5.5.4) There is a 1-category Bi of small bimonoidal categories and bimonoidal functors.
(5.5.10) Each tight bimonoidal category C has an associated right rigid bimonoidal category A, whose objects are formal polynomials in the objects in C.
(5.5.11–5.5.12) **Rigid Strictification Theorems**. Each tight bimonoidal category is adjoint equivalent to a right/left rigid bimonoidal category via bimonoidal functors.

Part I.2. Bicategorical Aspects of Symmetric Bimonoidal Categories

Chapter I.6. Definitions from Bicategory Theory

(6.1.2) A bicategory has objects, (identity) 1-cells, (identity) 2-cells, vertical and horizontal compositions, an associator, and two unitors, and satisfies the unity axiom and the pentagon axiom.
(6.1.8) A 2-category is a bicategory whose associator and unitors are identities.
(6.1.10) A 2-category can be described by data and axioms.
(6.1.11) A locally small 2-category is precisely a Cat-category.
(6.1.16) A monoidal category is a one-object bicategory.
(6.2.1) A lax functor has an object assignment, local functors, and two laxity constraints, and satisfies the lax associativity axiom and the lax unity axioms.
(6.2.11) There is a 1-category Bicat with small bicategories as objects and lax functors as morphisms.
(6.2.14) A lax transformation has component 1-cells and natural component 2-cells, and satisfies the lax unity axiom and the lax naturality axiom.
(6.2.26) There is a 2-category $\mathsf{2Cat}$ of small 2-categories, 2-functors, and 2-natural transformations.
(6.3.1) A modification has component 2-cells and satisfies the modification axiom.
(6.3.7) For bicategories B and B' with $\mathsf{Ob}(\mathsf{B})$ a set, there is a bicategory $\mathsf{Bicat}(\mathsf{B},\mathsf{B}')$ with lax functors $\mathsf{B} \longrightarrow \mathsf{B}'$ as objects, lax transformations as 1-cells, and modifications as 2-cells. It is a 2-category if B' is a 2-category. It contains a full subbicategory $\mathsf{Bicat}^{\mathsf{ps}}(\mathsf{B},\mathsf{B}')$ with pseudofunctors as objects and strong transformations as 1-cells.
(6.3.9) An adjunction in a bicategory consists of two 1-cells and two 2-cells, and satisfies two triangle identities.
(6.4.1) A monoidal bicategory has a base bicategory, a monoidal composition, a monoidal identity, a monoidal associator, two monoidal unitors, a pentagonator, and three 2-unitors, and satisfies the non-abelian 4-cocycle condition and two normalization axioms.
(6.5.3) A braided monoidal bicategory is a monoidal bicategory equipped with a braiding and two hexagonators, and satisfies four axioms.
(6.5.7) A sylleptic monoidal bicategory is a braided monoidal bicategory equipped with a syllepsis that satisfies two axioms.
(6.5.9) A symmetric monoidal bicategory is a sylleptic monoidal bicategory that satisfies the triple braid axiom.
(6.6.12) The 1-category $\mathsf{2Cat}$ equipped with the Gray tensor product is a symmetric monoidal closed category Gray.

(6.6.13) A Gray monoid is a monoid in Gray.
(6.7.1) A permutative Gray monoid is a Gray monoid equipped with a Gray symmetry that satisfies three axioms.
(6.7.16) A permutative 2-category is a monoid in $(\mathsf{2Cat}, \times)$ that is equipped with a symmetry 2-natural isomorphism and that satisfies the same three axioms as for permutative Gray monoids.

Chapter I.7. Baez's Conjecture

(7.1.2) A bimonoidal natural transformation is a natural transformation that is a monoidal natural transformation for each of the additive structure and the multiplicative structure.
(7.1.7) There is a 2-category $\mathsf{Bi}^{\mathsf{sy}}$ of small symmetric bimonoidal categories, symmetric bimonoidal functors, and bimonoidal natural transformations.
(7.1.9) Each monoidal natural transformation between symmetric monoidal functors between distributive symmetric monoidal categories is a bimonoidal natural transformation.
(7.2.9) For each symmetric bimonoidal category C, there is a strong symmetric monoidal functor $F_{\oplus}: \Sigma \longrightarrow \mathsf{C}$ between the additive structures.
(7.3.28) For each flat symmetric bimonoidal category C, there is a symmetric monoidal functor $F_{\otimes}: \Sigma \longrightarrow \mathsf{C}$ between the multiplicative structures.
(7.4.4) For each flat symmetric bimonoidal category C, $F: \Sigma \longrightarrow \mathsf{C}$ is a robust symmetric bimonoidal functor.
(7.5.8) Epstein's Coherence Theorem I.1.3.12 has a bimonoidal analogue.
(7.6.2–7.6.3) For a symmetric bimonoidal category C and robust symmetric bimonoidal functors $G, H: \Sigma \longrightarrow \mathsf{C}$, there is at most one bimonoidal natural transformation $G \longrightarrow H$, which must be invertible if it exists.
(7.7.9) For each flat symmetric bimonoidal category C and robust symmetric bimonoidal functor $G: \Sigma \longrightarrow \mathsf{C}$, there exists a unique bimonoidal natural transformation $\theta: F \longrightarrow G$, which is, moreover, invertible.
(7.8.1) **Baez's Conjecture.** Σ is a lax bicolimit of the 2-functor $\varnothing \longrightarrow \mathsf{Bi}_{\mathsf{r}}^{\mathsf{fsy}}$.
(7.8.3) **Baez's Conjecture, Ver. 2.** Σ' is a lax bicolimit of the 2-functor $\varnothing \longrightarrow \mathsf{Bi}_{\mathsf{r}}^{\mathsf{fsy}}$.

Chapter I.8. Symmetric Monoidal Bicategorification

(8.1.1) For a category C, $\mathsf{Mat}^{\mathsf{C}}_{m,n}$ has $n \times m$ matrices of objects in C as objects and $n \times m$ matrices of morphisms in C as morphisms.
(8.1.8) The matrix product is a functor.
(8.2.2) There is a natural isomorphism $\ell_A: \mathbb{1}^n A \xrightarrow{\cong} A$ for each flat bimonoidal category C and $A \in \mathsf{Mat}^{\mathsf{C}}_{m,n}$.
(8.2.8) There is a natural isomorphism $r_A: A\mathbb{1}^m \xrightarrow{\cong} A$ for each flat bimonoidal category C and $A \in \mathsf{Mat}^{\mathsf{C}}_{m,n}$.
(8.3.1) There is a natural isomorphism $a: (CB)A \xrightarrow{\cong} C(BA)$ for each tight bimonoidal category C.
(8.4.12) For each tight bimonoidal category C, $\mathsf{Mat}^{\mathsf{C}}$ is a bicategory.
(8.4.14) $\mathsf{Mat}^{\mathsf{C}}_{n,n}$ is a monoidal category.
(8.6.7) The matrix tensor product is a functor.
(8.7.31) The triple $(\boxtimes, \boxtimes^2, \boxtimes^0)$ on $\mathsf{Mat}^{\mathsf{C}}$ is a pseudofunctor.
(8.8.49) The quadruple $(a^{\boxtimes}, a^{\boxtimes\bullet}, \eta^a, \varepsilon^a)$ is an adjoint equivalence.

(8.9.9) The quadruple $(\ell^{\boxtimes}, \ell^{\boxtimes\bullet}, \eta^{\ell}, \varepsilon^{\ell})$ is an adjoint equivalence.
(8.9.21) The quadruple $(r^{\boxtimes}, r^{\boxtimes\bullet}, \eta^{r}, \varepsilon^{r})$ is an adjoint equivalence.
(8.10.4) π is an invertible modification.
(8.11.4) μ is an invertible modification.
(8.11.9) $\lambda^{\boxtimes}$ is an invertible modification.
(8.11.14) $\rho^{\boxtimes}$ is an invertible modification.
(8.12.9) For each tight symmetric bimonoidal category C, $\mathsf{Mat}^{\mathsf{C}}$ is a monoidal bicategory.
(8.13.13) For $\sigma \in \Sigma_m$, there is a natural isomorphism $r^{\sigma}_{A}\colon A\mathbb{1}^{\sigma} \xrightarrow{\cong} A^{\sigma}$ for $A \in \mathsf{Mat}^{\mathsf{C}}_{m,n}$.
(8.13.16) For $\theta \in \Sigma_n$, there is a natural isomorphism $\ell^{\theta}_{A}\colon \mathbb{1}^{\theta}A \xrightarrow{\cong} {}_{\theta^{-1}}A$ for $A \in \mathsf{Mat}^{\mathsf{C}}_{m,n}$.
(8.13.20) The matrix tensor products $A \boxtimes B$ and $B \boxtimes A$ differ by a column permutation, a row permutation, and a multiplicative symmetry in each entry.
(8.13.44) The quadruple $(\beta, \beta^{\bullet}, \eta^{\beta}, \varepsilon^{\beta})$ is an adjoint equivalence.
(8.14.12) $R_{-|--}$ is an invertible modification.
(8.14.24) $R_{--|-}$ is an invertible modification.
(8.14.26) For each tight symmetric bimonoidal category C, $\mathsf{Mat}^{\mathsf{C}}$ is a braided monoidal bicategory.
(8.15.4) **Bicategorification Theorem**. For each tight symmetric bimonoidal category C, $\mathsf{Mat}^{\mathsf{C}}$ is a symmetric monoidal bicategory.
(8.15.5) Coordinatized 2-vector spaces $2\mathsf{Vect}_{\mathsf{c}} = \mathsf{Mat}^{\mathsf{Vect}^{\mathbb{C}}}$ form a symmetric monoidal bicategory.

Part II.1. Braided Bimonoidal Categories

Chapter II.1. Preliminaries on Braided Structures

(1.1.1) The braid group B_n on n strings is generated by $s_1, \ldots, s_{n-1}$ and subject to two braid relations.
(1.1.9) Sum braids generalize block sums of permutations.
(1.1.12) Each braid has an underlying permutation.
(1.1.20) Block braids generalize block permutations.
(1.2.4) Elementary block braids generalize interval-swapping permutations.
(1.2.14) Elementary block braids are compatible with sum braids.
(1.2.16) Elementary block braids satisfy the hexagon axioms.
(1.3.15) A braided monoidal category is a monoidal category equipped with a braiding that satisfies two hexagon axioms.
(1.3.18) A braided monoidal functor is defined in the same way as a symmetric monoidal functor.
(1.3.21) In each braided monoidal category, the left unit isomorphism uniquely determines the right unit isomorphism, and vice versa, via the braiding.
(1.3.28, 1.3.31) Each braided monoidal category satisfies the third Reidemeister move.
(1.3.36) A symmetric monoidal category is precisely a braided monoidal category whose braiding satisfies the symmetry axiom.
(1.4.27) The Drinfeld center of a monoidal category is a braided monoidal category.
(1.5.3) The symmetric center of a braided monoidal category is a symmetric monoidal category.

(1.6.3) **Braided Coherence Theorem.** Two braided canonical maps with the same (co)domain in a braided monoidal category are equal if their underlying braids are equal.
(1.6.5) **Braided Strictification Theorem.** Each braided monoidal category is adjoint equivalent to a braided strict monoidal category via strong braided monoidal functors.

Chapter II.2. Braided Bimonoidal Categories

(2.1.29) A braided bimonoidal category is a category equipped with an additive symmetric monoidal structure, a multiplicative braided monoidal structure, left/right multiplicative zero natural isomorphisms, and left/right distributivity natural monomorphisms, and satisfies twelve Laplaza's axioms and two additional axioms involving the braiding.
(2.1.37) Tight braided bimonoidal categories are equivalent to BD categories in the sense of Blass and Gurevich.
(2.2.1) Each braided bimonoidal category satisfies all 24 Laplaza axioms.
(2.2.3) A symmetric bimonoidal category is precisely a braided bimonoidal category whose braiding satisfies the symmetry axiom.
(2.3.2) In an Ab-category, composition with a zero morphism yields a zero morphism, and composition commutes with taking the additive inverse of a morphism.
(2.3.7 (1)) For any two objects in an Ab-category, a product, a coproduct, and a direct sum are equivalent.
(2.3.7 (2)) In an Ab-category, the direct sum morphism $f \oplus f'$ can be characterized in terms of the inclusions.
(2.3.7 (3)) In an Ab-category, the sum morphism $f + g$ factors as $\nabla_B(f \oplus g)\Delta_A$.
(2.3.7 (4)) A functor between Ab-categories whose domain has all direct sums is an additive functor if and only if it preserves direct sums.
(2.3.12) For any two objects in a preadditive category, the zero morphism is the unique morphism that factors through the zero object.
(2.3.15) An abelian category is an Ab-category with a zero object, a direct sum for any two objects, and a (co)kernel for each morphism, such that (i) each monomorphism is a kernel and (ii) each epimorphism is a cokernel.
(2.3.17) Each abelian category has all finite (co)limits, with (co)products given by direct sums.
(2.4.22) An abelian category with a compatible braided monoidal structure is a tight braided bimonoidal category.
(2.5.1–2.5.2) An abelian category with a compatible (symmetric) monoidal structure is a tight (symmetric) bimonoidal category.

Chapter II.3. Applications to Quantum Groups and Topological Quantum Computation

(3.1.19) A braided bialgebra is a bialgebra equipped with an R-matrix that satisfies two axioms. A symmetric bialgebra is a braided bialgebra in which the inverse of the R-matrix is its opposite.
(3.1.27–3.1.30) Each cocommutative bialgebra is a symmetric bialgebra with the R-matrix $1 \otimes 1$. Examples include group bialgebras, the universal enveloping bialgebra of a Lie algebra, and Sweedler's 4-dimensional non-(co)commutative bialgebra.
(3.1.33) Each anyonic quantum group is a braided bialgebra.

(3.2.6) The category of left modules over each bialgebra is a monoidal category under the tensor product.
(3.2.12) The category of left modules over each braided bialgebra is a braided monoidal category.
(3.2.13) The category of left modules over each symmetric bialgebra is a symmetric monoidal category.
(3.2.19) The category of left modules over each bialgebra is a tight bimonoidal category. The braided and the symmetric analogues are also true.
(3.3.27) The Fibonacci anyons form a monoidal category.
(3.4.5) The Fibonacci anyons form a braided monoidal category.
(3.4.13) The Fibonacci anyons form a tight braided bimonoidal category.
(3.5.27) The Ising anyons form a monoidal category.
(3.6.7) The Ising anyons form a braided monoidal category.
(3.6.14) The Ising anyons form a tight braided bimonoidal category.

Chapter II.4. Bimonoidal Centers

(4.2.6) The additive structure of the bimonoidal Drinfeld center is a symmetric monoidal category.
(4.3.3) The multiplicative structure of the bimonoidal Drinfeld center is a braided monoidal category.
(4.4.3) For each tight bimonoidal category, the bimonoidal Drinfeld center is a tight braided bimonoidal category.
(4.5.3) For each braided bimonoidal category, the bimonoidal symmetric center is a symmetric bimonoidal category.

Chapter II.5. Coherence of Braided Bimonoidal Categories

(5.1.2) A left permbraided category has an additive permutative structure, a multiplicative braided strict monoidal structure, and identities for $\lambda^{\bullet}$, $\rho^{\bullet}$, δ^l, $\xi^{\otimes}_{-,\mathbb{0}}$, and $\xi^{\otimes}_{\mathbb{0},-}$, and satisfies four braided bimonoidal category axioms.
(5.1.8) Each left bipermutative category is a left permbraided category.
(5.1.10) Each left permbraided category is a tight braided bimonoidal category.
(5.1.11) A right permbraided category has an additive permutative structure, a multiplicative braided strict monoidal structure, and identities for $\lambda^{\bullet}$, $\rho^{\bullet}$, δ^r, $\xi^{\otimes}_{-,\mathbb{0}}$, and $\xi^{\otimes}_{\mathbb{0},-}$, and satisfies four braided bimonoidal category axioms.
(5.1.17) Each right bipermutative category is a right permbraided category.
(5.1.19) Each right permbraided category is a tight braided bimonoidal category.
(5.2.7) The braided distortion category is a groupoid.
(5.2.13) The additive structure of the braided distortion category is a permutative category.
(5.2.21) The multiplicative structure of the braided distortion category is a braided strict monoidal category.
(5.2.28) In the braided distortion category, the right distributivity morphism δ^r has identity braid components.
(5.2.30) The braided distortion category is a left permbraided category.
(5.2.33–5.2.34) The braided distortion category is a tight braided bimonoidal category and satisfies all 24 Laplaza axioms.
(5.3.14) For a path in $\mathsf{Gr}(X)$, its value in a braided bimonoidal category is the composite of the images of its constituent prime edges.

(5.3.15) The braided distortion of a path in $\mathsf{Gr}(X)$ is its value in the braided distortion category.
(5.4.4) **Braided Bimonoidal Coherence Theorem.** In each braided bimonoidal category C satisfying a monomorphism assumption, any two parallel paths in $\mathsf{Gr}(X)$ with the same braided distortion have the same value in C.
(5.4.6) Theorem II.5.4.4 applies to flat, in particular, tight, braided bimonoidal categories.
(5.5.3) The braided distortion category $\mathcal{D}^{\mathsf{br}}$ is isomorphic to the Grothendieck construction $\int_{\Sigma} F^{\mathsf{br}}$.

Chapter II.6. Strictification of Tight Braided Bimonoidal Categories

(6.1.1) A braided bimonoidal functor is a functor equipped with an additive symmetric monoidal structure and a multiplicative braided monoidal structure, and satisfies two axioms.
(6.1.10) There is a category $\mathsf{Bi}^{\mathsf{br}}$ with small braided bimonoidal categories as objects and braided bimonoidal functors as morphisms.
(6.1.12) Each braided monoidal functor that is also an additive functor between abelian categories with a compatible braided monoidal structure canonically extends to a braided bimonoidal functor.
(6.1.15) Each symmetric monoidal functor that is also an additive functor between abelian categories with a compatible symmetric monoidal structure canonically extends to a symmetric bimonoidal functor.
(6.2.39) Each tight braided bimonoidal category has a canonically associated right permbraided category.
(6.3.6–6.3.7) **Permbraided Strictification.** Each tight braided bimonoidal category is adjoint equivalent to a right/left permbraided category via braided bimonoidal functors.

Chapter II.7. The Braided Baez Conjecture

(7.1.4) There is a 2-category $\mathsf{Bi}^{\mathsf{br}}$ of small braided bimonoidal categories, braided bimonoidal functors, and bimonoidal natural transformations.
(7.1.7) Each monoidal natural transformation between braided monoidal functors that are also additive functors between abelian categories with a compatible braided monoidal structure is a bimonoidal natural transformation.
(7.2.4) For each braided bimonoidal category C, there is a strong symmetric monoidal functor $F_{\oplus}\colon \Sigma \longrightarrow \mathsf{C}$ between the additive structures.
(7.2.9) For each flat braided bimonoidal category C, there is a braided monoidal functor $F_{\otimes}\colon \Sigma \longrightarrow \mathsf{C}$ between the multiplicative structures.
(7.2.11) For each flat braided bimonoidal category C, $F\colon \Sigma \longrightarrow \mathsf{C}$ is a robust braided bimonoidal functor.
(7.3.4) **Braided Baez Conjecture.** Σ is a lax bicolimit of the 2-functor $\varnothing \longrightarrow \mathsf{Bi}_{\mathsf{r}}^{\mathsf{fbr}}$.
(7.3.6) **Braided Baez Conjecture, Ver. 2.** Σ' is a lax bicolimit of the 2-functor $\varnothing \longrightarrow \mathsf{Bi}_{\mathsf{r}}^{\mathsf{fbr}}$.

Chapter II.8. Monoidal Bicategorification

(8.1.13) For each tight braided bimonoidal category C, $\mathsf{Mat}^{\mathsf{C}}$ is a bicategory.
(8.4.7) For each tight braided bimonoidal category C, $\mathsf{Mat}^{\mathsf{C}}$ is a monoidal bicategory.

Part II.2. E_n-Monoidal Categories

Chapter II.9. Ring, Bipermutative, and Braided Ring Categories

(9.1.15) Tight ring categories form a subclass of tight bimonoidal categories.
(9.1.19) Right and left rigid bimonoidal categories are tight ring categories.
(9.1.20) Each tight ring category is adjoint equivalent to a right, respectively, left, rigid bimonoidal category.
(9.2.14) Each small permutative category has an endomorphism ring category.
(9.2.20) Each small permutative category has an endomorphism tight ring category.
(9.3.7) Each tight bipermutative category yields a tight symmetric bimonoidal category.
(9.3.12) Right and left bipermutative categories are tight bipermutative categories.
(9.3.13) Each tight bipermutative category is adjoint equivalent to a right, respectively, left, bipermutative category.
(9.4.7) In a bipermutative category, about half of the ring category axioms are redundant.
(9.5.4) A bipermutative category is a braided ring category whose braiding satisfies the symmetry axiom.
(9.5.5) In a braided ring category, about half of the ring category axioms are redundant.
(9.5.6) Tight braided ring categories form a subclass of tight braided bimonoidal categories.
(9.5.10) Right and left permbraided categories are tight braided ring categories.
(9.5.11) Each tight braided ring category is adjoint equivalent to a right, respectively, left, permbraided category.
(9.6.1) The bimonoidal Drinfeld center of a tight ring category is a tight braided ring category.
(9.6.4) The symmetric center of a braided ring category with left factorization a natural epimorphism is a bipermutative category.

Chapter II.10. Iterated and E_n-Monoidal Categories

(10.1.9) A 1-fold monoidal category is a strict monoidal category.
(10.1.14) Braided strict monoidal categories form a subclass of 2-fold monoidal categories.
(10.1.21) Permutative categories form a subclass of n-fold monoidal categories for each $n \geq 2$.
(10.2.3) A totally ordered set with a least element forms a permutative category with identity symmetry.
(10.2.8) A totally ordered monoid whose unit is also the least element forms a 2-fold monoidal category.
(10.3.7) A 1-fold monoidal functor is a strictly unital monoidal functor.
(10.3.11) A braided strictly unital monoidal functor is a 2-fold monoidal functor.
(10.3.15) A symmetric strictly unital monoidal functor is an n-fold monoidal functor for each $n \geq 2$.
(10.3.20) The composite of two n-fold monoidal functors is an n-fold monoidal functor.
(10.4.2) MCat^n is a monoidal category.
(10.4.5) Monoids in MCat^n are small $(n+1)$-fold monoidal categories.

(10.4.13) A morphism of monoids in MCat^n is an $(n+1)$-fold monoidal functor with the last monoidal constraint the identity.
(10.5.9) $\mathsf{FMon}^n\colon \mathsf{Cat} \longrightarrow \mathsf{MCat}^n_{\mathsf{st}}$ is the left adjoint of the forgetful functor.
(10.5.18) $\mathsf{FMon}^n(\mathsf{C})$ decomposes into a coproduct $\coprod_{k\geq 0} \mathsf{Mon}^n(k) \times_{\Sigma_k} \mathsf{C}^{\times k}$.
(10.5.26) $\coprod_{k\geq 0} \mathsf{Mon}^n(k)/\Sigma_k$ is the free n-fold monoidal category on one object.
(10.5.28) There are evaluation functors $\theta_k\colon \mathsf{Mon}^n(k) \times_{\Sigma_k} \mathsf{C}^{\times k} \longrightarrow \mathsf{C}$ for each small n-fold monoidal category C.
(10.6.8 (1)) Each morphism set in $\mathsf{Mon}^n(k)$ has at most one morphism.
(10.6.8 (2)) There exists a morphism $A \longrightarrow B \in \mathsf{Mon}^n$ if and only if $a \otimes_i b \in A$ implies either $a \otimes_j b \in B$ for some $j \geq i$ or $b \otimes_j a \in B$ for some $j > i$.
(10.6.9) In each n-fold monoidal category, each formal diagram built from identity morphisms, the exchanges $\{\eta^{i,j}\}_{i<j}$, the monoidal products $\{\otimes_i\}_{i=1}^n$, and composites is commutative.
(10.7.13) An E_1-monoidal category is a ring category.
(10.8.1) Braided ring categories form a subclass of E_2-monoidal categories.
(10.9.1) Bipermutative categories form a subclass of E_n-monoidal categories for each $n \geq 2$.
(10.10.2) Each small category has a free E_n-monoidal category.

Part 1. Enriched Monoidal Categories and Multicategories

Chapter 1. Enriched Monoidal Categories

(1.1.31) **Mac Lane's Coherence Theorem.** Any two words of the same length in a monoidal category are connected by a unique canonical map.
(1.1.32) **Mac Lane's Strictification Theorem.** Each monoidal category is adjoint equivalent to a strict monoidal category via strong monoidal functors.
(1.1.38) **Braided Coherence Theorem.** Two braided canonical maps with the same (co)domain in a braided monoidal category are equal if their underlying braids are equal.
(1.1.39) **Braided Strictification Theorem.** Each braided monoidal category is adjoint equivalent to a braided strict monoidal category via strong braided monoidal functors.
(1.1.41) **Symmetric Coherence Theorem.** Any two permuted words of the same length in a symmetric monoidal category are connected by a unique permuted canonical map.
(1.1.42) **Symmetric Strictification Theorem.** Each symmetric monoidal category is adjoint equivalent to a permutative category via strong symmetric monoidal functors.
(1.1.44) **Epstein's Coherence Theorem.** For each (symmetric) monoidal functor $F\colon \mathsf{C} \longrightarrow \mathsf{D}$ and F-iterates $G, H\colon \mathsf{C}^n \longrightarrow \mathsf{D}$, there exists at most one F-coherent map $G \longrightarrow H$.
(1.2.1) For a monoidal category V, a V-category has hom objects in V satisfying associativity and unity axioms.
(1.2.4) A V-functor satisfies composition and identity axioms.
(1.2.7) A V-natural transformation satisfies a naturality axiom.
(1.2.13) There is a 2-category formed by small V-categories, V-functors, and V-natural transformations.
(1.2.16) The opposite of a V-category is defined if V is braided monoidal.
(1.3.3) The tensor product of V-categories is defined if V is braided monoidal.

(1.3.6) The tensor product of V-categories is 2-functorial.
(1.3.35) The underlying 1-category of V-Cat is monoidal if V is braided, and is symmetric monoidal if V is symmetric.
(1.4.2) A monoidal V-category has associator and unitor V-natural transformations that satisfy unity axioms and a pentagon axiom.
(1.4.7) Composition in a monoidal V-category has an enriched interchange.
(1.4.10) The definition of braided monoidal V-category requires that V be symmetric monoidal.
(1.4.10) A braided monoidal V-category has a V-natural braiding satisfying two hexagon axioms.
(1.4.13) A symmetric monoidal V-category is a braided monoidal V-category satisfying an additional symmetry axiom.
(1.4.17) A monoidal V-functor satisfies associativity and unity axioms.
(1.4.18) A braided monoidal V-functor satisfies a braid axiom. A symmetric monoidal V-functor is a braided monoidal V-functor whose domain and codomain are symmetric monoidal V-categories.
(1.4.22) A monoidal V-natural transformation satisfies monoidal naturality and monoidal unity axioms.
(1.4.25) There are 2-categories formed by each of: monoidal V-categories, braided monoidal V-categories, and symmetric monoidal V-categories with, in each case, the corresponding V-functors and V-natural transformations.
(1.5.1) We use the term Cat-monoidal 2-category to indicate monoidal V-categories when V = Cat.
(1.5.2) The underlying 1-category of a plain/braided/symmetric Cat-monoidal 2-category has the corresponding structure as a 1-category.
(1.5.3) A Cat-monoidal 2-category has a strict form of the data and axioms for a monoidal bicategory. Similar statements hold for the braided and symmetric cases.
(1.5.4) With the Cartesian product, Cat is a symmetric Cat-monoidal 2-category.
(1.5.5) For a braided monoidal category V, V-Cat is a Cat-monoidal 2-category. If V is symmetric, then so is V-Cat.

Chapter 2. Change of Enrichment

(2.1.2) Change of enrichment along a monoidal functor U is 2-functorial.
(2.1.7) The functor from V-Cat to Cat that takes underlying categories is injective on 2-cells.
(2.2.7) The assignment $E\colon \mathsf{V} \longmapsto \mathsf{V}\text{-}\mathsf{Cat}$ is 2-functorial with respect to monoidal functors and monoidal natural transformations.
(2.2.11) For small monoidal V, there is a 2-equivalence between V-Cat and $(\mathsf{V}_{\mathsf{st}})$-Cat.
(2.3.7) If U is braided moniodal, then the change of enrichment induced by U is a Cat-monoidal 2-functor. If U is symmetric, then so is the change of enrichment.
(2.3.9) The assignment $E\colon \mathsf{V} \longmapsto \mathsf{V}\text{-}\mathsf{Cat}$ of braided monoidal categories to Cat-monoidal 2-categories is 2-functorial. A similar result holds for symmetric monoidal categories with E producing symmetric Cat-monoidal categories.
(2.3.16) For small symmetric monoidal V, there is a symmetric Cat-monoidal 2-equivalence between V-Cat and $(\mathsf{V}_{\mathsf{st}})$-Cat.
(2.4.10, 2.4.15) For braided monoidal $U\colon \mathsf{V} \longrightarrow \mathsf{W}$, change of enrichment along U induces 2-functors between the 2-categories of monoidal V- and W-categories. If U is symmetric, then a similar result holds for braided and symmetric monoidal V- and W-categories.

(2.4.17) The underlying category of a monoidal V-category is monoidal. Similar statements hold for braided and symmetric cases, and for functors and natural transformations.
(2.5.1) Given the data of a monoidal V-category, the enriched monoidal category axioms are satisfied if and only if the underlying data satisfy the ordinary monoidal category axioms. Similar results hold for the braided and symmetric monoidal cases, and also for functors and natural transformations.
(2.5.6) **Enriched (Braided/Symmetric) Monoidal Coherence Theorem.** Any two V-words of the same length in a monoidal V-category are connected by a unique canonical V-map. Similar coherence results hold in the braided and symmetric cases.
(2.5.8) **Enriched Epstein's Coherence Theorem.** For each (symmetric) monoidal V-functor $F\colon \mathsf{K} \longrightarrow \mathsf{L}$, and F-iterates $G, H\colon \mathsf{K}^{\otimes n} \longrightarrow \mathsf{L}$, there exists at most one F-coherent map $G \longrightarrow H$.
(2.6.1) **Enriched Monoidal Strictification Theorem.** Each monoidal V-category is adjoint V-equivalent to a strict monoidal V-category via strong monoidal V-functors.
(2.6.3) **Enriched Braided Strictification Theorem.** Each braided monoidal V-category is adjoint V-equivalent to a braided strict monoidal V-category via strong braided monoidal V-functors.
(2.6.4) **Enriched Symmetric Strictification Theorem.** Each symmetric monoidal V-category is adjoint V-equivalent to a strict monoidal V-category via strong symmetric monoidal V-functors.

Chapter 3. Self-Enrichment and Enriched Yoneda

(3.1.11) Each symmetric monoidal closed V has a canonical enrichment over itself, $\underline{\mathsf{V}}$.
(3.2.1, 3.2.2) A category enriched over symmetric monoidal closed V has co/represented V-functors $\mathcal{Y}^X$ and $\mathcal{Y}_Y$ to $\underline{\mathsf{V}}$.
(3.3.2) The self-enriched category $\underline{\mathsf{V}}$ is symmetric monoidal as a V-category.
(3.3.4) The standard enrichment of a symmetric monoidal functor is symmetric monoidal in the enriched sense.
(3.4.12) **V-Yoneda Bijection Theorem.** For each V-functor $F\colon \mathsf{C} \longrightarrow \underline{\mathsf{V}}$ and each $X \in \mathsf{C}$, there is a bijection of sets $\mathsf{V}\text{-nat}(\mathcal{Y}^X, F) \cong \mathsf{V}(\mathbb{1}, FX)$. For each V-functor $G : \mathsf{C}^{\mathsf{op}} \longrightarrow \underline{\mathsf{V}}$ and each $Y \in \mathsf{C}$, there is a bijection of sets $\mathsf{V}\text{-nat}(\mathcal{Y}_Y, G) \cong \mathsf{V}(\mathbb{1}, GY)$.
(3.5.1) A V-coend is initial among V-cowedges. A V-end is terminal among V-wedges.
(3.5.5) If V is cocomplete, then V-coends are computed by a coequalizer in V. If V is complete, then V-ends are computed by an equalizer in V.
(3.5.12) For V complete symmetric monoidal closed, the mapping object for V-functors to $\underline{\mathsf{V}}$ is given by a V-end.
(3.6.9) **V-Yoneda Lemma.** For a V-functor $F\colon \mathsf{C} \longrightarrow \mathsf{V}$ with C small, there is a V-natural isomorphism

$$F \xrightarrow{\cong} \mathsf{Map}(\mathcal{Y}^{(-)}, F).$$

(3.7.3) For V complete and cocomplete symmetric monoidal closed, the Day convolution and hom diagram are given by a V-coend and a V-end, respectively.

(3.7.8) **V-Yoneda Density Theorem.** For a V-functor $X: \mathcal{D} \longrightarrow \underline{V}$ with $\mathcal{D}$ small, there is a V-natural isomorphism

$$\int^x \mathcal{D}(x,-) \otimes X_x \xrightarrow{\cong} X.$$

(3.7.13) There is an isomorphism $\mathsf{Map}(X,Y) \cong \mathsf{Hom}(X,Y)_e$, for V-diagrams X and Y.
(3.7.22) **Day Convolution Theorem.** For a small symmetric monoidal V-category $\mathcal{D}$, the category of $\mathcal{D}$-shaped diagrams in V is symmetric monoidal closed with the Day convolution product, internal hom, and monoidal unit $J = \mathcal{Y}^e$.
(3.7.28) Precomposition with a symmetric monoidal V-functor induces a symmetric monoidal functor between diagram V-categories.
(3.8.4) Change of enrichment along a symmetric monoidal functor $U: \mathsf{V} \longrightarrow \mathsf{W}$ induces a symmetric monoidal functor from $\mathcal{D}$-V to $\mathcal{D}_U$-W.
(3.9.3) If C is tensored and cotensored over V, then $X \otimes -$ and $Y^{(-)}$ extend uniquely to V-functors that are V-adjoint to the respective co/represented V-functors.
(3.9.8) If (F, U) is an adjunction of monoidal functors between symmetric monoidal closed categories and if F^2 is invertible, then F transfers tensor and cotensor structure over its codomain to corresponding structure over its domain.
(3.9.15) The symmetric monoidal closed diagram category $\mathcal{D}$-V is enriched, tensored, and cotensored over V.

Chapter 4. Pointed Objects, Smash Products, and Pointed Homs

(4.1.6) Smash product with respect to a terminal object T is given by a pushout from a monoidal product.
(4.2.1) Pointed hom with respect to a terminal object T is given by a pullback from an internal hom.
(4.1.5, 4.1.8, 4.2.3) Suppose C is complete and cocomplete symmetric monoidal closed. Then C_* is complete and cocomplete symmetric monoidal closed with respect to the smash product and pointed hom.
(4.3.11) For a small symmetric monoidal category $\mathcal{D}$ with a null object, its pointed unitary enrichment over $(\mathsf{V}_*, \wedge, E)$ is given by taking wedge sums of E over nonzero morphisms in $\mathcal{D}$.
(4.3.19) Assuming the basepoint of V is terminal and the basepoint of $\mathcal{D}$ is null, there is an equivalence of categories between pointed functors from $\mathcal{D}$ to V_* and V_*-enriched functors from the pointed unitary enrichment of $\mathcal{D}$ to the self enrichment of V_*.
(4.3.37) The category of pointed diagrams $\mathcal{D}_*$-V is complete and cocomplete symmetric monoidal closed. Moreover, it is enriched, tensored, and cotensored over V_*.

Chapter 5. Multicategories

(5.1.2) A multicategory has objects, n-ary operations, symmetric group actions, colored units, and composition that are subject to axioms for symmetry, associativity, unity, and equivariance.
(5.1.2) An operad is a multicategory with one object.
(5.1.11) Each small permutative category C has an endomorphism multicategory with the same objects and with n-ary operations given by morphisms out of n-fold sums in C.

(5.1.12) A multifunctor satisfies axioms for symmetric group action, units, and composition.
(5.1.17) A multinatural transformation satisfies a naturality condition.
(5.1.20) There is a 2-category consisting of small multicategories, multifunctors, and multinatural transformations.
(5.1.21) The initial operad I has a single object and only one operation, which is the unit on its one object.
(5.2.1, 5.2.2) The terminal multicategory T has a single object and a single n-ary operation for each $n \geq 0$. The terminal multicategory is also known as the commutative operad, Com.
(5.3.9) Taking endomorphism operads gives a 2-functor from $\mathsf{PermCat}^{\mathsf{su}}$ to $\mathsf{Multicat}_*$.
(5.4.1) A monad consists of an endofunctor together with multiplication and unit natural transformations such that the associativity and unit diagrams commute.
(5.4.2) A monad algebra consists of an object and a structure morphism such that associativity and unity diagrams commute.
(5.4.13) **Beck's Precise Tripleability Theorem**. An adjunction $L \dashv U$ is strictly monadic if and only if U strictly creates coequalizers for parallel pairs f, g for which (Uf, Ug) has a split coequalizer.
(5.4.18) If T is a monad on a complete and cocomplete category, and if T preserves filtered colimits, then the category of T-algebras is complete and cocomplete.
(5.5.1) A multigraph consists of vertices and multiedges.
(5.5.4, 5.5.9) The forgetful functor from small multicategories to small multigraphs has a left adjoint.
(5.5.11) The category of small multicategories is strictly monadic over the category of small multigraphs.
(5.5.14) The category of small multicategories is complete and cocomplete.
(5.6.9) The sharp product of multicategories is generated by operations $\phi \times d$ and $c \times \psi$ subject to symmetry and compatibility axioms.
(5.6.14) The tensor product of multicategories is generated by those of the sharp product, and subject to an additional interchange relation.
(5.6.14) A multifunctor out of a tensor product of multicategories consists of an assignment on objects that is multifunctorial in each variable separately and that preserves the interchange relation.
(5.7.2, 5.7.4) The internal hom for multicategories has operations given by transformations that satisfy a naturality condition.
(5.7.14) The category of small multicategories is complete and cocomplete symmetric monoidal closed with monoidal product given by the tensor product and closed structure given by the internal hom.
(5.7.22) The category of small pointed multicategories is complete and cocomplete symmetric monoidal closed with monoidal product given by the smash product and closed structure given by the pointed hom.
(5.7.23) The symmetric monoidal structure on $\mathsf{Multicat}_*$ does not restrict along End to a symmetric monoidal structure on $\mathsf{PermCat}^{\mathsf{su}}$.

Chapter 6. Enriched Multicategories

(6.1.1) A V-enriched multicategory has n-ary operation objects, symmetric group action, colored units, and composition morphisms in V. These satisfy axioms given by commutative diagrams in V for symmetric group action, associativity, unity, and equivariance.

(6.1.8) A V-enriched operad is a V-multicategory with one object.
(6.1.9) An object of an enriched multicategory has a V-enriched endomorphism operad.
(6.1.10) A V-enriched multifunctor satisfies axioms given by commutative diagrams in V for symmetric group action, units, and composition.
(6.1.14) An algebra c over a V-enriched operad P is given by a V-enriched operad morphism $\mathsf{P} \longrightarrow \mathsf{End}(c)$.
(6.1.15) A V-enriched multinatural transformation satisfies a V-naturality diagram in V.
(6.1.18) There is a 2-category consisting of small V-enriched multicategories together with V-enriched multifunctors and multinatural transformations.
(6.2.9) For a symmetric monoidal functor $U: \mathsf{V} \longrightarrow \mathsf{W}$, change of enrichment along U provides a 2-functor from small V-enriched multicategories to small W-enriched multicategories.
(6.3.3, 6.3.6) A symmetric monoidal V-category has a V-enriched endomorphism multicategory with V-objects of n-ary operations given by morphism objects out of n-fold left normalized products in K.
(6.3.10) A symmetric monoidal V-functor induces a V-enriched multifunctor between V-enriched endomorphism multicategories of its domain and codomain.
(6.4.3) The tensor product is a Cat-enriched symmetric monoidal product for the 2-category of small multicategories.
(6.4.4) The smash product is a Cat-enriched symmetric monoidal product for the 2-category of small pointed multicategories.
(6.4.5) Each of Multicat and $\mathsf{Multicat}_*$ has the structure of a Cat-enriched multicategory induced by the tensor and smash product, respectively.
(6.5.1) The Cat-enriched multicategory structure on $\mathsf{Multicat}_*$ induces a corresponding structure on $\mathsf{PermCat}^{\mathsf{su}}$.
(6.5.4) Multilinear functors of permutative categories consist of functors out of a Cartesian product together with linearity constraints. They are subject to axioms for unity, constraint unity, constraint associativity, constraint symmetry, and constraint 2-by-2.
(6.5.11) Multilinear transformations between multilinear functors satisfy multilinearity conditions with respect to linearity constraints and identities.
(6.5.10, 6.5.13) The categories of n-ary operations in $\mathsf{PermCat}^{\mathsf{su}}$ are canonically isomorphic to the corresponding categories of n-linear functors and transformations.
(6.6.13) The Cat-enriched multicategory structure of $\mathsf{PermCat}^{\mathsf{su}}$ is described explicitly in terms of multilinear functors and transformations.

Part 2. Algebraic K-Theory

Chapter 7. Homotopy Theory Background

(7.1.16) The geometric realization of the standard n-simplex is the topological n-simplex.
(7.1.19) The category of simplicial sets is symmetric monoidal closed with the monoidal product given by the levelwise Cartesian product.
(7.1.23) The category of pointed simplicial sets is symmetric monoidal closed with monoidal product given by the levelwise smash product.
(7.2.4) The nerve of a small category is a simplicial set with p-simplices given by strings of p composable arrows.

(7.2.5 (1)) A natural transformation between functors induces a simplicial homotopy on nerves.
(7.2.5 (2)) An adjunction of functors induces a simplicial homotopy equivalence on nerves.
(7.3.7) The category of symmetric sequences is symmetric monoidal closed with monoidal product given by Day convolution.
(7.4.5) The category of symmetric spectra is the category of left modules over the symmetric sphere.
(7.4.6) A symmetric spectrum consists of a symmetric sequence with structure morphisms satisfying unity, associativity, and equivariance axioms.
(7.5.5) The category of symmetric spectra is complete and cocomplete.
(7.6.1) The smash product of symmetric spectra is given by a coequalizer of actions by S.
(7.6.8) The internal hom for symmetric spectra is given by an equalizer of actions by S.
(7.6.15) The category of symmetric spectra is complete and cocomplete symmetric monoidal closed.
(7.8.8) Every level equivalence of symmetric spectra is a stable equivalence.

Chapter 8. Segal K-Theory of Permutative Categories

(8.1.8) A Γ-object in a pointed category C is a pointed functor from $\mathcal{F}$ to C.
(8.2.6) The construction $\mathsf{K}^{\mathcal{F}}$ is a functor from Γ-simplicial sets to symmetric spectra.
(8.3.13) For a small permutative category C, there are three variant constructions of Γ categories, $\mathsf{C}^{\mathcal{F}} = \mathsf{C}^{\mathcal{F}}_{\cong}$, $\mathsf{C}^{\mathcal{F}}_{\mathsf{lax}}$, and $\mathsf{C}^{\mathcal{F}}_{\mathsf{co}}$.
(8.3.21) For a small permutative category C, each of the Γ-simplicial sets $N_*\mathsf{C}^{\mathcal{F}}_{\cong}$, $N_*\mathsf{C}^{\mathcal{F}}_{\mathsf{lax}}$, and $N_*\mathsf{C}^{\mathcal{F}}_{\mathsf{co}}$ is special, and all three are levelwise weakly-equivalent.
(8.4.5) The partition multicategory $\mathcal{M}\underline{1}$ has two objects, $\varnothing$ and $\{1\}$, with operations given by partitions.
(8.4.7) The partition multicategory $\mathcal{M}$ defines a pointed functor from $\mathcal{F}^{\mathsf{op}}$ to $\mathsf{Multicat}_*$.
(8.4.8, 8.4.10) For a small permutative category C, there is an isomorphism of Γ-categories $\mathsf{J}^{\mathsf{Se}}\mathsf{C} \cong \mathsf{C}^{\mathcal{F}}_{\mathsf{lax}}$.

Chapter 9. Categories of $\mathcal{G}_*$-Objects

(9.1.7) The objects of $\mathcal{G}$ are tuples of objects of $\mathcal{F}$ subject to certain basepoint identifications.
(9.1.15) Smash product of pointed finite sets defines a strict symmetric monoidal functor from $\mathcal{G}$ to $\mathcal{F}$.
(9.2.1) A $\mathcal{G}_*$-object in a pointed category C is a pointed functor from $\mathcal{G}$ to C.
(9.2.15) If C is complete and cocomplete symmetric monoidal closed with terminal basepoint, then the category for $\mathcal{G}_*$-objects in C is complete and cocomplete symmetric monoidal closed with monoidal product given by Day convolution.
(9.2.19) The nerve induces a symmetric monoidal sSet_*-functor from small $\mathcal{G}_*$-categories to $\mathcal{G}_*$-simplicial sets.
(9.3.16) The construction $\mathsf{K}^{\mathcal{G}}$ is a functor from $\mathcal{G}_*$-simplicial sets to symmetric spectra whose restriction along $\wedge^*$ is equal to $\mathsf{K}^{\mathcal{F}}$.
(9.4.9) The functor $\mathsf{K}^{\mathcal{G}}$ is a unital symmetric monoidal sSet-functor.
(9.4.18) The functors $\wedge^*$ and $\mathsf{K}^{\mathcal{F}}$ are symmetric monoidal sSet_*-functors.

Chapter 10. Elmendorf-Mandell K-Theory of Permutative Categories

(10.1.6) The partition multicategory $\mathcal{M}$ is a symmetric monoidal functor, with monoidal constraint given by the partition product $\prod$.
(10.1.14) The category of left $\mathcal{M}\underline{1}$-modules is complete and cocomplete.
(10.1.15) The partition products $\prod_{\underline{1},b}$ and $\prod_{b,\underline{1}}$ are isomorphisms.
(10.1.28) The 2-category of left $\mathcal{M}\underline{1}$-modules in $\mathsf{Multicat}_*$ is a full sub-2-category and the smash product over $\mathcal{M}\underline{1}$ is isomorphic to that of $\mathsf{Multicat}_*$.
(10.1.36) The category of left $\mathcal{M}\underline{1}$-modules is symmetric monoidal in the Cat_*-enriched sense.
(10.2.14) If C is a small permutative category, $\mathsf{End}(\mathsf{C})$ has a canonical left $\mathcal{M}\underline{1}$-module structure. Taking this structure, End factors through $\mathsf{Mod}^{\mathcal{M}\underline{1}}$.
(10.2.15, 10.2.16) The category of left $\mathcal{M}\underline{1}$-modules is a proper subcategory of $\mathsf{Multicat}_*$. In particular, the monoidal unit S is not an $\mathcal{M}\underline{1}$-module.
(10.2.17) The symmetric monoidal structure on $\mathsf{Mod}^{\mathcal{M}\underline{1}}$ does not restrict to $\mathsf{PermCat}^{\mathsf{su}}$.
(10.2.22) The symmetric monoidal structure on $\mathsf{Mod}^{\mathcal{M}\underline{1}}$ is closed.
(10.3.3) The smash product of partition multicategories, $\mathcal{T}$, defines a pointed functor from $\mathcal{G}^{\mathsf{op}}$ to $\mathsf{Mod}^{\mathcal{M}\underline{1}}$.
(10.3.7) The functor $\mathcal{T}$ is strictly unital strong symmetric monoidal.
(10.3.13) The monoidal constraint for the partition J-theory $\mathsf{J}^{\mathcal{T}}$ uses the inverse monoidal constraint for $\mathcal{T}$.
(10.3.17) The partition J-theory $\mathsf{J}^{\mathcal{T}}$ is a symmetric monoidal Cat_*-functor.
(10.3.25) Elmendorf-Mandell J-theory $\mathsf{J}^{\mathsf{EM}} = \mathsf{J}^{\mathcal{T}} \circ \mathsf{End}$ is a Cat-enriched multifunctor.
(10.3.32) Elmendorf-Mandell K-theory $\mathsf{K}^{\mathsf{EM}} = \mathsf{K}^{\mathcal{G}} N_* \mathsf{J}^{\mathsf{EM}}$ is a sSet-enriched multifunctor.
(10.3.33) The multifunctor K^{EM} preserves enriched operad actions.
(10.4.18) For a small permutative category C there are three variant constructions of $\mathcal{G}_*$-categories, $\mathsf{C}^{\mathcal{G}} = \mathsf{C}^{\mathcal{G}}_{\mathrm{lax}}$, $\mathsf{C}^{\mathcal{G}}_{\cong}$, and $\mathsf{C}^{\mathcal{G}}_{\mathrm{co}}$.
(10.5.1) For a small permutative category C, there is an isomorphism of $\mathcal{G}_*$-categories $\mathsf{J}^{\mathsf{EM}}\mathsf{C} \cong \mathsf{C}^{\mathcal{G}}_{\mathrm{lax}}$.
(10.6.10) There is a level equivalence of symmetric spectra $\mathsf{K}^{\mathsf{Se}}\mathsf{C} \longrightarrow \mathsf{K}^{\mathsf{EM}}\mathsf{C}$ for each small permutative category C. It is natural with respect to strictly unital symmetric monoidal functors.
(10.7.16) There is a level equivalence $\mathsf{K}^{\mathsf{EM}}_{\cong}\mathsf{C} \longrightarrow \mathsf{K}^{\mathsf{EM}}_{\mathsf{lax}}$ for each small permutative category C.
(10.7.19 (4)) There is a level equivalence $\mathsf{K}^{\mathsf{EM}}_{\cong} \longrightarrow \mathsf{K}^{\mathsf{EM}}_{\mathsf{co}}\mathsf{C}$ for each small permutative category C.
(10.7.22, 10.7.27) The $\mathcal{G}_*$-category morphisms $\mathsf{C}^{\mathcal{G}}_{\cong} \longrightarrow \mathsf{C}^{\mathcal{G}}$ and $\mathsf{C}^{\mathcal{G}}_{\cong} \longrightarrow \mathsf{C}^{\mathcal{G}}_{\mathsf{co}}$ are components of Cat-enriched multinatural transformations.

Chapter 11. K-Theory of Ring and Bipermutative Categories

(11.1.4) As is an operad.
(11.1.7) As is generated by id_0 and id_2, which are subject to unity and associativity relations.
(11.1.15) As is the operad for monoids.
(11.2.16) As detects ring category structures on small permutative categories.

(11.3.2) A strict ring symmetric spectrum is a symmetric spectrum equipped with multiplication and unit morphisms in symmetric sequences that satisfy compatibility, associativity, and unity axioms.
(11.3.13–11.3.15) The sphere spectrum, the suspension spectrum of a monoid in pointed simplicial sets, and the Eilenberg-Mac Lane spectrum of a ring are strict ring symmetric spectra.
(11.3.16) $\mathsf{K}^{\mathsf{EM}}\mathsf{C}$ is a strict ring symmetric spectrum for each small ring category C.
(11.3.17) For each small permutative category, the K-theory of its (tight) endomorphism ring category is a strict ring symmetric spectrum.
(11.3.18) The K-theory of the additive distortion category is a strict ring symmetric spectrum.
(11.3.19) For each small tight bimonoidal category, the K-theory of its right/left rigid bimonoidal strictification is a strict ring symmetric spectrum.
(11.4.7) The translation category functor E is a right adjoint.
(11.4.11) Each morphism in the Barratt-Eccles operad decomposes into a categorical composite of ϕv with v a permutation and ϕ an operadic composite of one $\tau\colon \mathrm{id}_2 \longrightarrow (1,2)$ and identity morphisms.
(11.4.14) The Barratt-Eccles operad is generated by two objects and one isomorphism, which are subject to relations that are formally identical to those of a permutative category.
(11.4.26) The Barratt-Eccles operad is the Cat-enriched operad for permutative categories.
(11.5.5) The Barratt-Eccles operad detects bipermutative category structures on small permutative categories.
(11.6.3) The Barratt-Eccles operad is an E_∞-operad.
(11.6.6 (1)) Each commutative monoid in SymSp is an E_∞-symmetric spectrum.
(11.6.6 (2)) Each E_∞-symmetric spectrum via the Barratt-Eccles operad has a strict ring structure.
(11.6.7) The symmetric sphere is a commutative monoid in SymSp.
(11.6.9) The suspension spectrum of a commutative monoid in sSet_* is a commutative monoid in SymSp.
(11.6.10) The Eilenberg-Mac Lane spectrum of a commutative ring is a commutative monoid in SymSp.
(11.6.12) $\mathsf{K}^{\mathsf{EM}}\mathsf{C}$ is an E_∞-symmetric spectrum for each small bipermutative category C.
(11.6.13) The K-theory of each small right/left bipermutative category is an E_∞-symmetric spectrum. For example, this applies to the finite ordinal category Σ, its variant Σ', $\mathsf{Vect}^{\mathbb{C}}_{\mathsf{c}}$, and the distortion category.
(11.6.14) For each small tight symmetric bimonoidal category, the K-theory of its right/left bipermutative strictification is an E_∞-symmetric spectrum. For example, this applies to small distributive symmetric monoidal categories, the symmetric bimonoidal groupoid Π, and the bimonoidal symmetric center of a small tight braided bimonoidal category.
(11.6.15) For each small braided ring category whose left factorization morphism is a natural epimorphism, the K-theory of its symmetric center is an E_∞-symmetric spectrum.

Chapter 12. K-Theory of Braided Ring Categories

(12.1.10) The braid operad is a Cat-enriched operad.
(12.2.4) The braid operad is an E_2-operad.
(12.3.6) Each morphism in the braid operad admits a categorical decomposition into isomorphisms of the form ϕv with v a permutation and ϕ an operadic composite of one $s_1^{\pm 1}\colon \mathrm{id}_2 \longrightarrow (1,2)$ and identity morphisms.
(12.3.10) The braid operad is generated by two objects and one isomorphism, which are subject to relations that are formally identical to those of a braided strict monoidal category.
(12.3.22) The braid operad is the Cat-enriched operad for braided strict monoidal categories.
(12.4.5) The braid operad detects braided ring category structures on small permutative categories.
(12.5.2 (1)) An E_∞-symmetric spectrum via the Barratt-Eccles operad has an E_2-structure.
(12.5.2 (2)) An E_2-symmetric spectrum via the braid operad has a strict ring structure.
(12.5.3) $\mathsf{K}^{\mathsf{EM}}\mathsf{C}$ is an E_2-symmetric spectrum for each small braided ring category C.
(12.5.4) The K-theory of the braided distortion category is an E_2-symmetric spectrum.
(12.5.5) For each small tight ring category, the K-theory of its bimonoidal Drinfeld center is an E_2-symmetric spectrum.
(12.5.6) For each small tight braided bimonoidal category, the K-theory of its right/left permbraided strictification is an E_2-symmetric spectrum. For example, this applies to a small abelian category with a compatible braided monoidal structure, Fibonacci anyons, Ising anyons, and the bimonoidal Drinfeld center of a small tight bimonoidal category.

Chapter 13. K-Theory of E_n-Monoidal Categories

(13.1.20) Mon^n is a Cat-enriched operad.
(13.1.23) Mon^1 is the associative operad.
(13.2.1) Mon^n is an E_n-operad.
(13.3.3) Mon^n is generated by the objects $\mathbb{1}$ and $\{1 \otimes_i 2\}_{i=1}^n$ and the exchange morphisms $\{\eta_{1,2,3,4}^{i,j}\}_{1 \le i < j \le n}$, which are subject to relations that are formally identical to those of an n-fold monoidal category.
(13.3.18) Mon^n is the Cat-enriched operad for n-fold monoidal categories.
(13.4.12) Mon^n detects E_n-monoidal category structures on small permutative categories.
(13.5.1) The canonical Cat-enriched operad morphism $\mathsf{As} \longrightarrow E\mathsf{As}$ factors through Mon^n. As a result, an E_∞-symmetric spectrum via the Barratt-Eccles operad induces an E_n-structure. An E_{n+1}-structure via Mon^{n+1} induces an E_n-structure.
(13.5.2) $\mathsf{K}^{\mathsf{EM}}\mathsf{C}$ is an E_n-symmetric spectrum for each small E_n-monoidal category C.
(13.5.3) For each small category, the K-theory of its free E_n-monoidal category is an E_n-symmetric spectrum.

List of Notations

Part I.1

Chapter I.1	Page	Description
$\mathsf{Ob}(\mathsf{C})$, $\mathsf{Ob}\,\mathsf{C}$	I.3	objects in a category C
$\mathsf{C}(X,Y)$, $\mathsf{C}(X;Y)$	I.4	set of morphisms $X \longrightarrow Y$
1_X	I.4	identity morphism
$g \circ f$, gf	I.4	composition of morphisms
$\cong$, $\xrightarrow{\cong}$	I.4	an isomorphism
$F\colon \mathsf{C} \longrightarrow \mathsf{D}$	I.4	a functor
Id_{C}, 1_{C}	I.5	identity functor
$\mathbb{1}$	I.5	terminal category
θ_X	I.5	a component of a natural transformation θ
1_F	I.5	identity natural transformation
$\phi\theta$	I.5	vertical composition
$\theta' * \theta$	I.5	horizontal composition
(L,R,ϕ), $L \dashv R$	I.6	an adjunction
η, ε	I.6	unit and counit of an adjunction
$\mathsf{colim}\,F$	I.7	colimit
$\mathsf{lim}\,F$	I.7	limit
$\varnothing$, $\varnothing^{\mathsf{C}}$	I.8	an initial object
$\coprod$, $\sqcup$	I.8	a coproduct
$\otimes$	I.10	monoidal product
$\mathbb{1}$	I.10	monoidal unit
α	I.10	associativity isomorphism
λ, ρ	I.11	unit isomorphisms
(X,μ,η)	I.11	a monoid
(Y,Δ,ε)	I.12	a comonoid
(F,F^2,F^0)	I.12	a monoidal functor
ξ	I.14	symmetry isomorphism
$(\mathsf{Set},\times,*)$	I.15	category of sets
$(\mathsf{Cat},\times,\mathbb{1})$	I.15	category of small categories
$(\mathsf{Vect}^{\Bbbk},\otimes,\Bbbk)$	I.15	category of $\Bbbk$-vector spaces
$[-,-]$	I.15	internal hom
e, $-$, $u \,\square\, v$	I.15	words
$\sigma(-)$	I.17	a left permutation
$w\sigma$	I.17	a permuted word

Chapter I.2		
$(\oplus,\mathbb{0},\alpha^{\oplus},\lambda^{\oplus},\rho^{\oplus},\xi^{\oplus})$	I.21	additive structure
$(\otimes,\mathbb{1},\alpha^{\otimes},\lambda^{\otimes},\rho^{\otimes},\xi^{\otimes})$	I.21	multiplicative structure

$\lambda^{\bullet}$, $\rho^{\bullet}$	I.21	multiplicative zeros
δ^l, δ^r	I.21	distributivity morphisms
$\mathsf{Vect}^{\mathbb{C}}$	I.25	finite dimensional complex vector spaces
$\alpha^{-\oplus}$	I.26	inverse of $\alpha^{\oplus}$
$\mathsf{Mod}(R)$	I.33	category of R-modules
Σ	I.34	category of finite ordinals and permutations
Σ_n	I.34	symmetric group on n letters
$\sigma \oplus \tau$	I.34	block sum of permutations
M^T	I.38	transpose of M
Σ'	I.39	a variant of Σ
$\mathsf{Vect}_c^{\mathbb{C}}$	I.42	coordinatized version of $\mathsf{Vect}^{\mathbb{C}}$
$\mathbb{C}^m$	I.44	$\mathbb{C} \oplus \cdots \oplus \mathbb{C}$ with m copies of $\mathbb{C}$
Π	I.46	symmetric bimonoidal groupoid of syntax of finite types

Chapter I.3

S^{fr}	I.54	free $\{\oplus, \otimes\}$-algebra of S
$G = (V, E)$	I.54	a graph with vertex set V and edge set E
$u \longrightarrow v$	I.54	an edge with domain u and codomain v
$(e_n, \ldots, e_1)$	I.54	a path consisting of the edges $e_1, \ldots, e_n$
$v_0 \longrightarrow\!\!\!\!\rightarrow v_n$	I.54	a path with domain v_0 and codomain v_n
0^X, 1^X	I.55	additive zero and multiplicative unit in X
$\mathsf{Gr}^{\mathsf{el}}(X)$	I.55	elementary graph
$\mathsf{E}_{\mathsf{el}}(X)$	I.55	set of elementary edges
$\mathsf{E}_{\mathsf{el}}^{\mathsf{fr}}(X)$	I.55	free $\{\oplus, \otimes\}$-algebra of $\mathsf{E}_{\mathsf{el}}(X)$
$\mathsf{E}^{\mathsf{pr}}(X)$	I.56	set of prime edges
$\mathsf{Gr}(X)$	I.56	graph of X
$\varphi\colon \mathsf{Gr}(X) \longrightarrow \mathsf{C}$	I.57	graph morphism extending $\varphi\colon X \longrightarrow \mathsf{Ob}(\mathsf{C})$
φP	I.58	value in C of a path P
X^{st}	I.59	strict $\{\oplus, \otimes\}$-algebra of X
supp	I.59	support $X^{\mathsf{fr}} \longrightarrow X^{\mathsf{st}}$
norm	I.61	norm $X^{\mathsf{fr}} \longrightarrow \mathbb{Z}_+$
rank	I.62	rank $X^{\mathsf{fr}} \longrightarrow \mathbb{Z}_+$
size	I.62	size $X^{\mathsf{fr}} \longrightarrow \mathbb{Z}_+$
$P \oplus 1_c$, $1_c \oplus P$	I.69	sum of the paths P and 1_c
$P \otimes 1_c$, $1_c \otimes P$	I.69	product of the paths P and 1_c
(IH)	I.113	induction hypothesis
$\mathsf{Gr}^{\mathsf{ns}}(X)$	I.129	nonsymmetric graph of X
X^{ns}	I.129	nonsymmetric strict $\{\oplus, \otimes\}$-algebra of X
nsupp	I.129	nonsymmetric support

Chapter I.4

id_m	I.139	identity permutation in Σ_m
$\mathcal{D}$	I.139	distortion category
$\underline{r}$	I.139	a finite sequence $(r_1, \ldots, r_m)$
$\lvert\underline{r}\rvert$	I.139	length of $\underline{r}$
$\varnothing$	I.139	empty sequence
$\underline{\sigma}$	I.139	a morphism $(\sigma; \sigma_1, \ldots, \sigma_m)$ in $\mathcal{D}$
ϑ	I.152	$X \longrightarrow \mathsf{Ob}(\mathcal{D})$ and $\mathsf{Gr}(X) \longrightarrow \mathcal{D}$
$\mathcal{D}^{\mathsf{ad}}$	I.161	additive distortion category
ϑ	I.163	$X \longrightarrow \mathsf{Ob}(\mathcal{D}^{\mathsf{ad}})$ and $\mathsf{Gr}^{\mathsf{ns}}(X) \longrightarrow \mathcal{D}^{\mathsf{ad}}$
$\int_{\mathsf{C}} F$	I.164	Grothendieck construction of $F\colon \mathsf{C}^{\mathsf{op}} \longrightarrow \mathsf{Cat}$
$\Sigma^{\times n}$	I.165	n-fold Cartesian product of Σ

Chapter I.7

Chapter I.8

Part II.1

Chapter II.1

Chapter II.2

Chapter II.3

$\mathsf{Vect}^{\Bbbk}$	II.67	category of $\Bbbk$-vector spaces
$A^{\otimes n}$	II.67	$A \otimes \cdots \otimes A$ with n copies of A and $A^{\otimes 0} = \Bbbk$
$A^{\oplus n}$	II.67	$A \oplus \cdots \oplus A$ with n copies of A and $A^{\oplus 0} = \mathbb{0}$
$\mathsf{Vect}^{\Bbbk}_{\otimes}$	II.67	$\mathsf{Vect}^{\Bbbk}$ with the tensor product
$\mathsf{Vect}^{\Bbbk}_{\oplus}$	II.67	$\mathsf{Vect}^{\Bbbk}$ with the direct sum
$(A, \mu, \eta, \Delta, \varepsilon)$	II.68	a bialgebra
$\sum_i s_i' \otimes s_i''$	II.69	an element in $A^{\otimes 2}$
S^{op}	II.69	$\xi^{\otimes} S$
S_{12}, S_{13}, S_{23}	II.69	elements obtained from $S \in A^{\otimes 2}$ by inserting 1
$\sum_{(x)} x^{(1)} \otimes x^{(2)}$	II.69	Sweedler's notation for comultiplication
Δ^{op}	II.70	opposite comultiplication $\xi^{\otimes}\Delta$
$\Bbbk G$	II.71	group bialgebra of G
$U\mathfrak{g}$	II.72	universal enveloping bialgebra of $\mathfrak{g}$
H_4	II.72	Sweedler's 4-dimensional bialgebra
$\mathbb{C}\mathbb{Z}_n$	II.74	anyonic quantum groups
$\mathsf{Mod}(A)_{\otimes}$	II.76	$\mathsf{Mod}(A)$ with the tensor product
$\mathsf{Mod}(A)_{\oplus}$	II.79	$\mathsf{Mod}(A)$ with the direct sum
$\mathsf{Vect}^{\mathbb{C}}_{\mathsf{sk}}$	II.81	abelian category with linear maps $\mathbb{C}^m \longrightarrow \mathbb{C}^n$ as morphisms
$\mathcal{F}^{\mathsf{any}}$	II.82	Fibonacci anyons
$\mathbb{0} = (0;0)$	II.82	additive zero in $\mathcal{F}^{\mathsf{any}}$
$\mathbb{1} = (1;0)$	II.82	vacuum in $\mathcal{F}^{\mathsf{any}}$
$\tau = (0;1)$	II.82	non-abelian anyon in $\mathcal{F}^{\mathsf{any}}$
q	II.85	reciprocal of the golden ratio
z	II.87	$e^{3\pi i/5}$
$\mathcal{I}^{\mathsf{any}}$	II.91	Ising anyons
$\mathbb{0} = (0;0;0)$	II.91	additive zero in $\mathcal{I}^{\mathsf{any}}$
$\mathbb{1} = (1;0;0)$	II.91	vacuum in $\mathcal{I}^{\mathsf{any}}$
$\sigma = (0;1;0)$	II.91	non-abelian anyon in $\mathcal{I}^{\mathsf{any}}$
$\psi = (0;0;1)$	II.91	fermion in $\mathcal{I}^{\mathsf{any}}$
w	II.102	$e^{\pi i/8}$

Chapter II.4

$\overline{\mathsf{C}}^{\mathsf{bi}}$	II.112	bimonoidal Drinfeld center
$(\overline{\oplus}, \overline{\mathbb{0}}, \alpha^{\overline{\oplus}}, \lambda^{\overline{\oplus}}, \rho^{\overline{\oplus}}, \xi^{\overline{\oplus}})$	II.112	additive structure of $\overline{\mathsf{C}}^{\mathsf{bi}}$
$(\overline{\otimes}, \overline{\mathbb{1}}, \alpha^{\overline{\otimes}}, \lambda^{\overline{\otimes}}, \rho^{\overline{\otimes}}, \xi^{\overline{\otimes}})$	II.112	multiplicative structure of $\overline{\mathsf{C}}^{\mathsf{bi}}$
$\overline{\lambda}^{\bullet}, \overline{\rho}^{\bullet}$	II.112	multiplicative zeros of $\overline{\mathsf{C}}^{\mathsf{bi}}$
$\overline{\delta}^{l}, \overline{\delta}^{r}$	II.112	distributivity morphisms of $\overline{\mathsf{C}}^{\mathsf{bi}}$
$(A; \beta^A)$	II.112	an object in $\overline{\mathsf{C}}^{\mathsf{bi}}$
$\mathsf{C}^{\mathsf{sym}}$	II.124	bimonoidal symmetric center

Chapter II.5

$\mathcal{D}^{\mathsf{br}}$	II.135	braided distortion category
$\underline{r}$	II.135	an object in $\mathcal{D}^{\mathsf{br}}$
$\varnothing$	II.135	empty sequence in $\mathcal{D}^{\mathsf{br}}$
$\underline{\sigma}$	II.135	a morphism in $\mathcal{D}^{\mathsf{br}}$
$\int_{\mathsf{C}} F$	II.154	Grothendieck construction of $F\colon \mathsf{C}^{\mathsf{op}} \longrightarrow \mathsf{Cat}$
$\mathcal{B}$	II.154	braid category
F^{br}	II.155	functor $\Sigma^{\mathsf{op}} \longrightarrow \mathsf{Cat}$

Chapter II.6

$(F, F^2_\oplus, F^0_\oplus, F^2_\otimes, F^0_\otimes)$	II.160	a braided bimonoidal functor
$F_\oplus$	II.160	additive structure $(F, F^2_\oplus, F^0_\oplus)$
$F_\otimes$	II.160	multiplicative structure $(F, F^2_\otimes, F^0_\otimes)$
$\mathsf{Bi}^{\mathsf{br}}$	II.162	category of small braided bimonoidal categories
$(-)_{\mathsf{rt}}$	II.165	right normalized bracketing
$\underline{a}$	II.165	an object in A
π	II.165	realization function $\mathsf{Ob}(\mathsf{A}) \longrightarrow \mathsf{Ob}(\mathsf{C})$
$\cong^{\oplus}_{\mathsf{ML}}$	II.166	a Mac Lane coherence isomorphism
$\cong_{\mathsf{Lap}}$	II.168	a Laplaza coherence isomorphism
A	II.176	associated right permbraided category
π	II.177	functor $\mathsf{A} \longrightarrow \mathsf{C}$
ι	II.177	functor $\mathsf{C} \longrightarrow \mathsf{A}$

Chapter II.7

$\mathsf{Bi}^{\mathsf{br}}$	II.184	2-category of small braided bimonoidal categories
$\mathsf{Bi}^{\mathsf{fbr}}_{\mathsf{r}}$	II.185	full sub-2-category of $\mathsf{Bi}^{\mathsf{br}}$ with flat objects and robust 1-cells
Σ	II.187	category of $n \geq 0$ and permutations
$(-)_{\mathsf{lt}}$	II.188	left normalized bracketing
$\overline{n}$	II.188	left normalized sum of n copies of $\mathbb{1}$
θ^G	II.193	unique bimonoidal natural transformation $F \longrightarrow G$

Chapter II.8

$(\mathsf{B}, 1, c, a, \ell, r)$	II.199	a bicategory
$\mathsf{Mat}^{\mathsf{C}}_{m,n}$	II.200	category of $n \times m$ matrices in C
$A = (A_{ji})$	II.200	a matrix with (j, i)-entry A_{ji}
$\mathbb{0}_{m,n}$	II.201	0 matrix in $\mathsf{Mat}^{\mathsf{C}}_{m,n}$
BA	II.201	matrix product
$g \star f$	II.201	matrix product of morphisms
$\mathbb{1}^n$	II.201	$n \times n$ identity matrix
ζ^ℓ, ζ^r, ℓ, r, u	II.202	natural isomorphisms in $\mathsf{Mat}^{\mathsf{C}}$
$(1_\boxtimes, 1^2_\boxtimes, 1^0_\boxtimes)$	II.208	monoidal identity
$A \boxtimes B$	II.209	matrix tensor product
$(\boxtimes, \boxtimes^2, \boxtimes^0)$	II.211	monoidal composition
$(a^\boxtimes, a^{\boxtimes\bullet}, \eta^u, \varepsilon^u)$	II.217	monoidal associator
$(\ell^\boxtimes, \ell^{\boxtimes\bullet}, \eta^\ell, \varepsilon^\ell)$	II.219	left monoidal unitor
$(r^\boxtimes, r^{\boxtimes\bullet}, \eta^r, \varepsilon^r)$	II.221	right monoidal unitor
π	II.221	pentagonator
μ	II.222	middle 2-unitor
$\lambda^\boxtimes$	II.222	left 2-unitor
$\rho^\boxtimes$	II.223	right 2-unitor

Part II.2

Chapter II.9

$(\mathsf{C}, \oplus, \mathbb{0}, \xi^\oplus, \otimes, \mathbb{1}, \partial^l, \partial^r)$	II.232	a ring category
∂^l, ∂^r	II.233	left and right factorization morphisms
$\mathsf{End}(A)$	II.237	endomorphism rig of a monoid
$\mathsf{Perm}^{\mathsf{su}}(\mathsf{C}; \mathsf{D})$	II.238	category of strictly unital symmetric monoidal functors
$\mathsf{Perm}^{\mathsf{su}}(\mathsf{C}; \mathsf{C})$	II.239	endomorphism ring category
$(\boxplus, \mathbb{0}, \xi^\boxplus)$	II.239	additive structure of $\mathsf{Perm}^{\mathsf{su}}(\mathsf{C}; \mathsf{C})$
$\mathsf{Perm}^{\mathsf{sug}}(\mathsf{C}; \mathsf{C})$	II.243	a full subcategory of $\mathsf{Perm}^{\mathsf{su}}(\mathsf{C}; \mathsf{C})$

$(\mathsf{C},\oplus,\mathbb{0},\xi^{\oplus},\otimes,\mathbb{1},\partial^l,\partial^r)$	II.244	a bipermutative category
$(\mathsf{C},\oplus,\mathbb{0},\xi^{\oplus},\otimes,\mathbb{1},\partial^l,\partial^r)$	II.252	a braided ring category
$\overline{\mathsf{C}}^{\mathsf{bi}}$	II.256	bimonoidal Drinfeld center
$\mathsf{C}^{\mathsf{sym}}$	II.256	symmetric center

Chapter II.10

$(\mathsf{C},\{\otimes_i\},\mathbb{1},\{\eta^{i,j}\})$	II.264	an n-fold monoidal category
$\otimes_i$	II.264	ith product
$\eta^{i,j}$	II.264	(i,j)-exchange
$(\mathsf{C},\otimes_1,\otimes_2,\mathbb{1},\eta)$	II.266	a 2-fold monoidal category
$\leq$	II.269	a partial ordering
$x<y$	II.270	$x\leq y$ and $x\neq y$
$\max(x,y)$	II.270	maximum of x and y
$(M,\mu,\mathbb{1},\leq)$	II.270	a totally ordered monoid
$(M,\max,\mu,\mathbb{1},\eta)$	II.271	2-fold monoidal category of $(M,\mu,\mathbb{1},\leq)$ with $\mathbb{1}$ the least element
$(F,F^2_1,\ldots,F^2_n)$	II.273	an n-fold monoidal functor
F^2_i	II.273	ith monoidal constraint
MCat^n	II.278	category of small n-fold monoidal categories and functors
$\mathsf{MCat}^n_{\mathsf{sg}}$	II.278	wide subcategory of MCat^n with strong n-fold monoidal functors
$\mathsf{MCat}^n_{\mathsf{st}}$	II.278	wide subcategory of MCat^n with strict n-fold monoidal functors
$(\mathsf{MCat}^n,\times,\mathbb{1})$	II.279	monoidal category of small n-fold monoidal categories
(A,μ,η)	II.280	a monoid in a monoidal category
$(\varepsilon,\{\varepsilon^2_i\}_{1\leq i\leq n})$	II.280	unit n-fold monoidal functor
$(\otimes_{n+1},\{\eta^{i,n+1}\}_{1\leq i\leq n})$	II.280	multiplication n-fold monoidal functor
U	II.284	forgetful functor $\mathsf{MCat}^n_{\mathsf{st}}\longrightarrow\mathsf{Cat}$
$\mathsf{FMon}^n(\mathsf{C})$	II.284	free n-fold monoidal category of C
$\overline{F}$	II.286	$\mathsf{FMon}^n(F)$ for a functor F
$\mathsf{Mon}^n(k)$	II.288	a full subcategory of $\mathsf{FMon}^n\{1,\ldots,k\}$
ϕ_{C}	II.290	isomorphism $\coprod_{k\geq 0}\mathsf{Mon}^n(k)\times_{\Sigma_k}\mathsf{C}^{\times k}\longrightarrow\mathsf{FMon}^n(\mathsf{C})$
$\mathsf{FMon}^n(\mathbb{1})$	II.293	free n-fold monoidal category on one object
θ_k	II.293	evaluation functor $\mathsf{Mon}^n(k)\times_{\Sigma_k}\mathsf{C}^{\times k}\longrightarrow\mathsf{C}$
$R_{a,b}$	II.293	restriction functor $\mathsf{Mon}^n(k)\longrightarrow\mathsf{Mon}^n(\{a,b\})$
$a\otimes_i b\in A$	II.293	$R_{a,b}(A)=a\otimes_i b\in\mathsf{Mon}^n(\{a,b\})$
$\eta^{i,j}$	II.297	(i,j)-exchange in an E_n-monoidal category
$\otimes_i,\partial^{l,i},\partial^{r,i}$	II.297	ith product and left/right factorization morphisms
$\mathsf{FE}^n(\mathsf{C})$	II.303	free E_n-monoidal category of C

Part 1
Chapter 1

$(\mathsf{C},\otimes,\mathbb{1},\alpha,\lambda,\rho)$	III.4	monoidal category
(F,F^2,F^0)	III.5	monoidal functor
ξ	III.6	braiding
$[-,-]$	III.9	internal hom
C_{st}	III.10	strictification of C
$(\mathsf{V},\otimes,\mathbb{1},\alpha,\lambda,\rho,\xi)$	III.13	monoidal or braided monoidal category
C	III.13	V-category
m	III.13	composition in C
i_X	III.13	identity of X
F	III.14	V-functor
θ	III.14	V-natural transformation
$\phi\theta$	III.15	vertical composition
$\theta'*\theta$	III.15	horizontal composition

Chapter 2

Chapter 3

$\kappa\colon[\mathbb{1},W]\longrightarrow W$	III.108	composite of $\rho_{[\mathbb{1},W]}$ and eval
$\text{V-nat}(F,G)$	III.109	set of V-natural transformations $F\longrightarrow G$
$\overline{\eta}_Y\colon\mathsf{C}(X,Y)\longrightarrow FY$	III.109	composite induced by $\eta\in\mathsf{V}(\mathbb{1},FX)$
$\widetilde{\eta}\colon\mathcal{Y}^X\longrightarrow F$	III.109	V-natural transformation $\widetilde{\eta}_Y=(\overline{\eta}_Y)^{\perp}$
$\theta_0\colon\mathbb{1}\longrightarrow FX$	III.109	composite induced by V-natural $\theta\colon\mathcal{Y}^X\longrightarrow F$
$\theta\longmapsto\theta_0$, $\eta\longmapsto\widetilde{\eta}$	III.110	V-Yoneda bijection
(X,ζ)	III.113	V-cowedge of F
$(\int^{a\in\mathsf{A}}F(a,a),\omega)$	III.113	V-coend of F
(Y,δ)	III.113	V-wedge of F
$(\int_{a\in\mathsf{A}}F(a,a),\sigma)$	III.114	V-end of F
$\mathsf{Map}(F,G)$	III.116	mapping object between V-functors
$\theta^{\#}\colon Q\longrightarrow[P,R]$	III.116	transform of $\theta\colon P\longrightarrow[Q,R]$
C-V	III.118	mapping V-category of V-functors $\mathsf{C}\longrightarrow\underline{\mathsf{V}}$
m^{Map}	III.119	composition in mapping V-category
$\mathcal{Y}^{(-)}\colon\mathsf{C}\longrightarrow(\text{C-V})^{\mathsf{op}}$	III.121	Yoneda V-functor
$\mathsf{Map}(-,F)$	III.123	V-functor $(\text{C-V})^{\mathsf{op}}\longrightarrow\underline{\mathsf{V}}$ represented by F
$\phi_X\colon FX\longrightarrow\mathsf{Map}(\mathcal{Y}^X,F)$	III.125	morphisms induced by $F^{\#}_{X,Z}$
$\phi^{\perp}\colon F\longrightarrow\mathsf{Map}(\mathcal{Y}^{(-)},F)$	III.127	V-Yoneda isomorphism for F
$\mathcal{D}\text{-V}=\text{V-Cat}(\mathcal{D},\underline{\mathsf{V}})$	III.132	category of $\mathcal{D}$-shaped diagrams in V
$X\otimes Y$	III.132	Day convolution of $\mathcal{D}$-shaped diagrams X and Y
$\mathsf{Hom}_{\mathcal{D}}(X,Y)$	III.132	hom diagram from X to Y
ψ_s	III.133	morphisms induced by adjoints of $X_{x,s}$
$\psi^{\perp}$	III.133	V-Yoneda density isomorphism
ϕ_s, $\phi^{\perp}$	III.134	V-Yoneda isomorphism for X
$\gamma_{a,c;b}$, $\gamma_{b;a,c}$	III.134	density isomorphisms
$J=\mathcal{Y}^e$	III.135	unit diagram for Day convolution
α, λ, ρ, ξ	III.135	associativity, units, and symmetry for Day convolution
$F^*\colon\mathcal{E}\text{-V}\longrightarrow\mathcal{D}\text{-V}$	III.141	change-of-shape induced by $F\colon\mathcal{D}\longrightarrow\mathcal{E}$
$\mathsf{ev}_e\colon\mathcal{D}\text{-V}\longrightarrow\mathsf{V}$	III.142	evaluation at e
$U_*\colon\mathcal{D}\text{-V}\longrightarrow\mathcal{D}_U\text{-W}$	III.143	change of target induced by $U\colon\mathsf{V}\longrightarrow\mathsf{W}$
$\mathcal{D}_{\mathbb{1}}$	III.145	unitary enrichment of $\mathcal{D}$
$X\otimes A$, X^A	III.147	tensoring and cotensoring of C over V
$L_e\colon\mathsf{V}\longrightarrow\mathcal{D}\text{-V}$	III.153	symmetric monoidal functor $L_eA=A\otimes J$

Chapter 4

T, C_*	III.158	terminal object T and category of objects under T in C
$\iota_X\colon T\longrightarrow X$	III.158	basepoint of pointed object X
X_+	III.158	X with disjoint basepoint
$X\vee Y$	III.158	wedge product of X and Y
$\wedge$, $E=\mathbb{1}_+$	III.159	smash product and smash unit
$\mathsf{Hom}_*(X,Y)$	III.165	pointed hom
T, T^{V}	III.167	terminal object of V
VCat, $\mathsf{V}_*\mathsf{Cat}$	III.167	2-categories of small V-categories, respectively small pointed V-categories
$(\mathcal{D},\boxdot,e,\mathrm{T})$	III.168	symmetric monoidal category with null object T
$X^{\flat}$	III.170	punctured set
$\widehat{\mathcal{D}}$	III.170	pointed unitary enrichment of $\mathcal{D}$
$(\widehat{\mathcal{D}})\text{-}(\mathsf{V}_*)=\mathsf{V}_*\mathsf{Cat}(\widehat{\mathcal{D}},\underline{\mathsf{V}_*})$	III.174	V_*-enriched $\widehat{\mathcal{D}}$-shaped diagrams
$\mathcal{D}\text{-V}$	III.174	pointed diagrams from $\mathcal{D}$ to V
(L_e,ev_e)	III.174	strong symmetric monoidal adjunction

Chapter 5

$\mathsf{Prof}(C)$	III.177	C-profiles
$\langle c\rangle = (c_1, \ldots, c_n)$	III.177	C-profile of length n
$\oplus$	III.177	concatenation of profiles
$(\langle c\rangle ; c')$	III.177	C-profile $\langle c\rangle$ and element $c' \in C$
$(\mathsf{M}, \gamma, 1)$	III.178	multicategory, composition, and unit
$\mathsf{M}(\langle c\rangle ; c')$	III.178	n-ary operations with input profile $\langle c\rangle$ and output c'
$\langle c\rangle\sigma$	III.178	right permutation of $\langle c\rangle$ by σ
1_c	III.178	c-colored unit
$\mathsf{End}(X)$	III.181	endomorphism operad of X
$\mathsf{End}(\mathsf{C})$	III.181	endomorphism multicategory of C
$F\colon \mathsf{M} \longrightarrow \mathsf{N}$	III.181	multifunctor from M to N
$\alpha\colon F \longrightarrow G$	III.182	multinatural transformation from F to G
$\beta\alpha$	III.183	vertical composition
$\alpha' * \alpha$	III.183	horizontal composition
$\mathsf{Multicat}$	III.183	2-category of small multicategories
I	III.183	initial operad
$\mathsf{T} = \mathsf{Com}$	III.183	terminal multicategory; commutative operad
$(\mathsf{M}, *^{\mathsf{M}}, \iota^{\mathsf{M}})$	III.186	pointed multicategory
$\mathsf{Multicat}_*$	III.186	2-category of small pointed multicategories
$\mathsf{PermCat}^{\mathsf{su}}$	III.187	2-category with strictly unital symmetric monoidal functors
(T, μ, η)	III.190	monad T with multiplication μ and unit η
$\mathsf{Alg}(T)$	III.190	category of T-algebras
Δ_W	III.191	constant functor at W
$\mathsf{Vt}X$	III.196	vertices of a multigraph X
$X(\langle c\rangle ; c')$	III.196	multiedges with source $\langle c\rangle$ and target c'
MGraph, MGraph^C	III.196	category of small multigraphs; subcategory with vertex set C
$\mathsf{Multicat}^C$	III.197	subcategory of multicategories with object set C
(L^C, U^C)	III.197	left adjoint to forgetful $U^C\colon \mathsf{Multicat}^C \longrightarrow \mathsf{MGraph}^C$
f_0^*	III.197	change of object/vertex set by function $f_0\colon C \longrightarrow D$
(L, U)	III.198	left adjoint to forgetful $U\colon \mathsf{Multicat} \longrightarrow \mathsf{MGraph}$
$\langle c\rangle \otimes \langle d\rangle$, $\langle c\rangle \otimes^{\mathsf{t}} \langle d\rangle$	III.202	product and transposed product of tuples
$\xi^{\otimes}$	III.202	transpose permutation
$X \,\&\, Y$	III.202	product of multigraphs
$\mathsf{M} \# \mathsf{N}$	III.203	sharp product of multicategories
$\phi \otimes \psi$, $\phi \otimes^{\mathsf{t}} \psi$	III.204	product and transposed product of operations
$\xi^{\otimes}$	III.204	transposition bijection for operations
$\mathsf{M} \otimes \mathsf{N}$	III.204	Boardman-Vogt tensor product of multicategories
$\wedge$, $\mathsf{S} = \mathsf{I}_+$	III.207	smash product and smash unit for pointed multicategories
$\mathsf{M} \vee \mathsf{N}$	III.207	wedge product of pointed multicategories
$\langle Fc\rangle$, $\langle Fc\rangle^{\mathsf{t}}$, $\langle F\rangle\phi$	III.207	applying a tuple of operations $\langle F\rangle$ to a profile $\langle c\rangle$ and operation ϕ
$\mathsf{Hom}(\mathsf{M}, \mathsf{N})$	III.207	internal hom for multicategories
$\alpha\colon \langle F\rangle \longrightarrow G$	III.207	transformation in $\mathsf{Hom}(\mathsf{M}, \mathsf{N})$
$\mathsf{Hom}_*(\mathsf{M}, \mathsf{N})$	III.217	pointed hom multicategory

Chapter 6

$(\mathsf{V}, \otimes, \xi)$	III.222	permutative category, monoidal product, and symmetry
$(\mathsf{M}, \gamma, 1)$	III.222	V-enriched multicategory, composition, and unit
$\mathsf{M}(\langle c\rangle ; c')$	III.222	n-ary operation object with input profile $\langle c\rangle$ and output c'
$\langle c\rangle\sigma$	III.222	right permutation of $\langle c\rangle$ by σ
1_c	III.222	c-colored unit
$\mathsf{End}(c)$	III.224	endomorphism V-enriched operad of c

Part 2
Chapter 7

Chapter 8

Chapter 9

Chapter 10

Chapter 11

Chapter 12

Chapter 13

$\mathsf{hocolim}_{\mathsf{C}} F$	III.488	homotopy colimit of a functor $F: \mathsf{C} \longrightarrow \mathsf{Top}$
α_F	III.488	canonical map from the homotopy colimit to the colimit
$*$	III.488	constant functor $\mathsf{Mon}^n(k) \longrightarrow \mathsf{Top}$

Appendix A

$\mathsf{Bi}^{\mathsf{tsy}}$	III.505	full sub-2-category of $\mathsf{Bi}^{\mathsf{sy}}$ with tight objects
SMB	III.505	tricategory of symmetric monoidal bicategories
SBB	III.506	tricategory of symmetric bimonoidal bicategories
BB	III.506	tricategory of bimonoidal bicategories
$(B\Sigma)^+$	III.516	Quillen's +-construction of $B\Sigma$

Index

B

D

E

H

N

Q

S

Selected Published Titles in This Series

285 **Niles Johnson and Donald Yau,** Bimonoidal Categories, E_n-Monoidal Categories, and Algebraic K-Theory: Volume III: From Categories to Structured Ring Spectra, 2024

284 **Donald Yau,** Bimonoidal Categories, E_n-Monoidal Categories, and Algebraic K-Theory: Volume II: Braided Bimonoidal Categories with Applications, 2024

283 **Donald Yau,** Bimonoidal Categories, E_n-Monoidal Categories, and Algebraic K-Theory: Volume I: Symmetric Bimonoidal Categories and Monoidal Bicategories, 2024

281 **Horst R. Thieme,** Discrete-Time Dynamics of Structured Populations and Homogeneous Order-Preserving Operators, 2024

280 **Tadashi Ochiai,** Iwasawa Theory and Its Perspective, Volume 2, 2024

279 **Ellen Elizabeth Eischen, Wee Teck Gan, Aaron Pollack, and Zhiwei Yun,** Automorphic Forms Beyond GL_2, 2024

278 **Delaram Kahrobaei, Ramón Flores, Marialaura Noce, Maggie E. Habeeb, and Christopher Battarbee,** Applications of Group Theory in Cryptography, 2024

277 **George Metcalfe, Francesco Paoli, and Constantine Tsinakis,** Residuated Structures in Algebra and Logic, 2023

276 **Balázs Bárány, Károly Simon, and Boris Solomyak,** Self-similar and Self-affine Sets and Measures, 2023

275 **Alekos Vidras and Alain Yger,** Multidimensional Residue Theory and Applications, 2023

274 **Willi Freeden and M. Zuhair Nashed,** Recovery Methodologies: Regularization and Sampling, 2023

273 **Gennadiy Feldman,** Characterization of Probability Distributions on Locally Compact Abelian Groups, 2023

272 **Tadashi Ochiai,** Iwasawa Theory and Its Perspective, Volume 1, 2023

271 **Hideto Asashiba,** Categories and Representation Theory, 2022

270 **Leslie Hogben, Jephian C.-H. Lin, and Bryan L. Shader,** Inverse Problems and Zero Forcing for Graphs, 2022

269 **Ralph S. Freese, Ralph N. McKenzie, George F. McNulty, and Walter F. Taylor,** Algebras, Lattices, Varieties, 2022

268 **Ralph S. Freese, Ralph N. McKenzie, George F. McNulty, and Walter F. Taylor,** Algebras, Lattices, Varieties, 2022

267 **Dragana S. Cvetković Ilić,** Completion Problems on Operator Matrices, 2022

266 **Kate Juschenko,** Amenability of Discrete Groups by Examples, 2022

265 **Nabil H. Mustafa,** Sampling in Combinatorial and Geometric Set Systems, 2022

264 **J. Scott Carter and Seiichi Kamada,** Diagrammatic Algebra, 2021

263 **Vugar E. Ismailov,** Ridge Functions and Applications in Neural Networks, 2021

262 **Ragnar-Olaf Buchweitz,** Maximal Cohen–Macaulay Modules and Tate Cohomology, 2021

261 **Shiri Artstein-Avidan, Apostolos Giannopoulos, and Vitali D. Milman,** Asymptotic Geometric Analysis, Part II, 2021

260 **Lindsay N. Childs, Cornelius Greither, Kevin P. Keating, Alan Koch, Timothy Kohl, Paul J. Truman, and Robert G. Underwood,** Hopf Algebras and Galois Module Theory, 2021

259 **William Heinzer, Christel Rotthaus, and Sylvia Wiegand,** Integral Domains Inside Noetherian Power Series Rings, 2021

258 **Pramod N. Achar,** Perverse Sheaves and Applications to Representation Theory, 2021

257 **Juha Kinnunen, Juha Lehrbäck, and Antti Vähäkangas,** Maximal Function Methods for Sobolev Spaces, 2021

For a complete list of titles in this series, visit the
AMS Bookstore at **www.ams.org/bookstore/survseries/**.